The Great
Contemporary
Issues Series
Set 2 Vol. 9

ENERGY
AND
ENVIRONMENT

The Series, Set 2

The Great
Contemporary
Issues Series
Set 2 Vol. 9

ENERGY
AND
ENVIRONMENT

𝔗𝔥𝔢 𝔑𝔢𝔴 𝔜𝔬𝔯𝔨 𝔗𝔦𝔪𝔢𝔰

ARNO PRESS

NEW YORK / 1979

Stuart Bruchey

Advisory Editor

GENE BROWN

Editor

This softbound edition of Great Contemporary Issues is
distributed exclusively in the United States by
GROLIER EDUCATIONAL CORPORATION
Sherman Turnpike, Danbury, CT 06816

Library of Congress Cataloging in Publication Data

Main entry under title:

Energy and environment.

   (The Great contemporary issues)
   Articles from the New York times, 1907-1977.
   Bibliography.
   Includes index.
   1.   Environmental protection—United States.
2. Power resources—United States.   3.   Conservation
of natural resources—United States.   I. Bruchey,
Stuart Weems.   II.   Brown, Gene.   III.   New York times.
IV.   Series.
TD171.E53   333.7   78-7393
ISBN 0-405-12770-7

Manufactured in the United States of America

The editors express special thanks to The Associated Press, United Press
International, and Reuters for permission to include in this series of books
a number of dispatches originally distributed by those news services.

Book design by Stuart David

Contents

Publisher's Note About the Series

It would take even an accomplished speed-reader, moving at full throttle, some three and a half solid hours a day to work his way through all the news The New York Times prints. The sad irony, of course, is that even such indefatigable devotion to life's carnival would scarcely assure a decent understanding of what it was really all about. For even the most dutiful reader might easily overlook an occasional long-range trend of importance, or perhaps some of the fragile, elusive relationships between events that sometimes turn out to be more significant than the events themselves.

This is why "The Great Contemporary Issues" was created—to help make sense out of some of the major forces and counterforces at large in today's world. The philosophical conviction behind the series is a simple one: that the past not only can illuminate the present but must. ("Continuity with the past," declared Oliver Wendell Holmes, "is a necessity, not a duty.") Each book in the series, therefore has as its subject some central issue of our time that needs to be viewed in the context of its antecedents if it is to be fully understood. By showing, through a substantial selection of contemporary accounts from The New York Times, the evolution of a subject and its significance, each book in the series offers a perspective that is available in no other way. For while most books on contemporary affairs specialize, for excellent reasons, in predigested facts and neatly drawn conclusions, the books in this series allow the reader to draw his own conclusions on the basis of the facts as they appeared at virtually the moment of their occurrence. This is not to argue that there is no place for events recollected in tranquility; it is simply to say that when fresh, raw truths are allowed to speak for themselves, some quite distinct values often emerge.

For this reason, most of the articles in "The Great

Contemporary Issues" are reprinted in their entirety even in those cases where portions are not central to a given book's theme. Editing has been done only rarely and in all such cases it is clearly indicated. (Such an excision occasionally occurs, for example, in the case of a Presidential State of the Union Message, where only brief portions are germane to a particular volume and in the case of some names, where for legal reasons or reasons of taste it is preferable not to republish specific identifications.) Similarly, typographical errors, where they occur, have been allowed to stand as originally printed.

"The Great Contemporary Issues" inevitably encompasses a substantial amount of history. In order to explore their subjects fully, some of the books go back a century or more. Yet their fundamental theme is not the past but the present. In this series the past is of significance insofar as it suggests how we got where we are today. These books, therefore, do not always treat a subject in a purely chronological way. Rather, their material is arranged to point up trends and interrelationships that the editors believe are more illuminating than a chronological listing would be.

"The Great Contemporary Issues" series will ultimately constitute an encyclopedic library of today's major issues. Long before editorial work on the first volume had even begun, some fifty specific titles had already been either scheduled for definite publication or listed as candidates. Since then, events have prompted the inclusion of a number of additional titles, and the editors are, moreover, alert not only for new issues as they emerge but also for issues whose development may call for the publication of sequel volumes. We will, of course, also welcome readers' suggestions for future topics.

Introduction

This collection of articles on energy and the environment from the files of *The New York Times* covers a period of more than a century and reflects the periodic concern of various individuals and groups, both private and public, over the abuse of our natural resources. Appearing as it does at a moment of renewed controversy and debate it can only increase our sense of the gravity of the issues it addresses. One of the most fundamental facts underlying the economic development of the American people has been the lavish abundance of resources—in agricultural lands, forests, wildlife and fisheries, navigable and potable waters, minerals, oil, natural gas and other sources of energy—in relation to the size of the population at any given time. Indeed, the reality of abundance early gave rise to the Myth of Superabundance, to a widespread feeling that resources might be plucked like ripe fruit and without limit from the rich orchards of the environment. This feeling not only affected American attitudes and values, it also underlay our laws; furthermore, it inhibited the growth of technological knowledge. Even what may be called the First Conservation Movement— of the late nineteenth and early twentieth centuries—challenged only slightly if at all the optimism engendered by the Myth. What distinguishes the current renewal of that Movement is a deepening sense of danger and of limits. Even so, the issues are far from simple and there are those who hold strongly to the view that human knowledge and intelligence—and hence technology—remains our most important natural resource. There is much to be said in their favor.

Looking back, it is difficult to avoid the conclusion that generations of American farmers "wasted" their land. They practiced "extensive" rather than "intensive" agriculture, cleared the land of trees, planted and harvested crops till yields ran low, then abandoned the fields for new ones instead of attempting to restore the fertility of the old. Contemporary English observers of American agricultural practices had harsh words to say. The "first settlers, with the usual foresight of the Americans," charged the anonymous author of *American Husbandry,* written on the eve of the Revolution, "destroyed the timber, as if it was impossible they should ever want any." "[S]educed by the fertility of the soil on first settling,", he continued, "the farmers think only of exhausting it as soon as possible, without attending to their own interest in a future day: this is a degree of blindness which in sensible people one may fairly call astonishing." He thought American farmers generally to be "the greatest slovens in christendom." To the German traveller Peter Kalm it was clear that "their eyes are fixed upon the present gain, and they are blind to futurity."

Indeed they were. But could it have been otherwise, given the situation they confronted? A well-known farmer of Mt. Vernon, Virginia, named George Washington understood clearly the nature of this situation and described it in the following words to the English agricultural reformer, Arthur Young:

> An English farmer must entertain a contemptible opinion of our husbandry, or a horrid idea of our lands, when he shall be informed that not more than eight or ten bushels of wheat is the yield of an acre; but this low produce may be ascribed, and principally too, to [the fact] that the aim of the farmers in this country, if they can be called farmers, is, not to make the most they can from the land, which is, or has been cheap, but the most of the labour, which is dear; the consequence of which has been, much ground has been *scratched* over and none cultivated or improved as it ought to have been: whereas a farmer in England, where land is dear, and labour cheap, finds it his interest to improve and cultivate highly, that he may reap large crops from a small quantity of ground.

The "factor proportions" of American agriculture—cheap land and dear labor and capital—dictated an economic response that differed from that of English farmers, or American farmers at a later time, after immigration and capital accumulation had altered those proportions. In the closing decades before the Civil War marl, gypsum, and other rock fertilizers came into wide use, with the result that thousands of acres once in an exhausted condition were restored to productiveness.

But that is not the whole of the story. Resource renewal was not merely an automatic response to favorable change in economic conditions, nor was it always complete—in the sense that renewed resources equalled in quantity and quality those which had been spent. Changes in attitudes and values were an essential prelude to efforts to end waste. During most of the nineteenth century Americans believed it desirable to develop their economic resources as rapidly as possible. Abundant resources made it possible for men to "get rich quick" and particularly after the Civil War they competed vigorously with each other to obtain as large a share of the spoils as possible. Since the objects of competition were very often the forests, ranges and other resources of the public domain—that part of the national domain owned by the federal government—it is essential to trace briefly the origins of the public domain and note some of the important laws which provided for its transfer to private hands. As we shall see, those laws were laxly enforced and widely abused—a reflection of deep-seated individualism, small respect for government, and, not least, the fact of abundance.

The public domain originated in the surrender to the nation of claims to western lands which seven of the original thirteen states held on the basis of their colonial charters. Made between 1781 and 1802, the cessions amounted in all to more than 233 million acres. Title to Florida and to the remaining area west of the Mississippi (except for Texas) passed to the federal government at the time of their incorporation into the national domain. Jefferson's famed Louisiana Purchase from France (1803) enriched federal holdings by more than 523 million acres. Purchase of Florida from Spain in 1819 brought in an additional 43 million acres. Compromise of the Oregon boundary dispute with Great Britain in 1846 added some 180 million acres, and the treaty ending the war with Mexico in 1848 added 334 million more. Three purchases rounded out the accretions to the public domain: from Texas in 1850 (nearly 79 million acres), from Mexico in 1853—the Gadsden Purchase—(nearly 19 million acres), and from Russia in 1867—the Purchase of Alaska—(some 365 million acres). All told, the federal government acquired title to almost a billion and a half acres. Since the landed area of the United States—bodies of water excluded—consists of less than 2 billion acres, the tremendous size of the public domain can be seen at a glance. About three out of every four acres of land in the United States belonged to Uncle Sam. Policies adopted by Congress to dispose of these vast acres would deeply affect the country's economic development—and afford opportunities for substantial wealth to both private individuals and business groups.

To make a long and complex story short, the land laws passed during the years before the Civil War were increasingly liberal, requiring purchase of increasingly smaller size tracts at a general price of $1.25 per acre. A Preemption Act adopted in 1841 permitted prospective settlement on public domain that had been surveyed but not yet offered for sale ("unoffered land"), and the Homestead Act of 1862 granted 160 acres to settlers free of charge. Before Preemption the government had looked upon its lands primarily as a source of revenue. Preemption was a step in the direction of the West's ultimate objective of free lands. Thousands of settlers were to become owners of small farms through preemption of public land. The policy put the government on the side of the farm maker. It did so by holding out to the settler who had preempted the land, i.e., occupied it prior to its being offered for sale, a promise of preferential treatment at the land auction. After the Civil War, however, the quantity of land offered for sale was limited, and since unoffered land could only be obtained in small parcels (160 acres as a rule) under the preemption and homestead legislation, and later under the Timber Culture, Timber and Stone, and Desert Land Acts—the latter allowing 640 acres—all these laws became subject to major abuse by interests seeking to accumulate large holdings.

In the 1880s, a decade which witnessed the last great boom in purchases of land in large quantities, it was in particular the device of preemption that facilitated acquisition of rich timberlands. Individuals entered land under the Preemption Act, and also under the Homestead Act, not for the purpose of making farms for themselves but in order to resell the land to lumber interests. The longleaf pine and cypress lands of the South, the pine and redwood timberland of California, the Douglas-fir in Washington and Oregon, the remaining stands of white pine in Minnesota and Wisconsin—together with the best of the remaining lands suitable for grazing or wheat growing in Dakota Territory, Kansas, and Nebraska—attracted preemptioners. In the last four years of the 1880s unoffered or preemption sales exceeded sales of offered land by a margin of three to one. In 1887 and 1888, southern timberlands were swept into private ownership at a rapid rate by northern timber barons continuing to operate their northern mills while holding their southern lands for speculation or later development. At most of the pinery land offices public officials acted in collusion with speculators. "The fact is," writes the foremost authority on the development of public land laws, Paul W. Gates, "that western society held Federal ownership in little regard; few people questioned the right of the citizen or lumberman to take timber from public lands, to agree not to bid against others at the public sale, even to sign affidavits that lands being entered under the Preemption and Homestead Acts were intended for farms when they really intended to sell to lumbermen as soon as the patent was issued."

The land laws enacted by Congress were written by western members of that body in the interests of the West, and they were interpreted and administered by Interior Department and Land Office administrators in a way satisfactory to that region of the country. These laws enabled lumber companies, and also cattle companies, railroads, and speculative groups, to acquire great quantities of land in the 1880s. According to Gates, "most of the arable, grazing, and forest resources of significant value" were in private hands by 1891, frequently in large tracts.

One consequence was a high degree of concentration of ownership, the degree of land monopoly in California being "almost unexampled for the United States." Here the scramble for land of all kinds, arable, irrigable, or timberland, was intense, and most of the lands designated by law as intended for the support of public education and internal improvements fell into the hands of politicians and speculators. Because large operators controlled the necessary capital and labor to mine the land on a large scale it seems highly likely that concentration of ownership was more devastating than the damage to resources done by small farmers. The cumulative impact of the latter, however, cannot be ignored.

A second consequence was the destruction of many of the forest lands of the United States. In the words of Stewart L. Udall, lumbering, in its "raider phase," was a "strip-and-run business." The average life of a sawmill was the twenty years it took to strip a hinterland. As the brawling industry ran its course, such towns as Bangor, Albany, Williamsport, Saginaw, Muskegon, Eureka, and Portland in succession briefly boasted the title "lumber capital of the world." The waste of wood was enormous but the devastation was not caused by logging alone. Careless loggers started fires that burned as much as 25 million acres each year.

Other resources tell a similar tale of waste. Like lumber, oil also boasted its succession of towns across America claiming the title "oil capital of the world." When gushers came in they spewed oil at a fearful rate. Spindletop in Texas wasted 110,000 barrels a day before coming under control nine days later. Natural gas was also wasted. Oil producers did not realize that gas energy brought oil to the surface and they allowed it to dissipate. And as in the case of lumber, law fell on the side of waste. In early disputes over ownership, the courts held that oil belonged to anyone who could capture it. This law of Capture, Udall points out, "put a premium on speed, and most of the time the big rewards went to whoever struck the underground treasure first."

And so with still other resources. The hydraulic miners of gold in California sliced off the hills, moving the soil in massive quantities into the rivers that drained the Sierra Nevada. "For every ounce of gold collected, tons of topsoil and gravel washed into the river courses below. With the spring floods, clear streams became a chaos of debris, rocks, and silt; communities downstream were inundated with muck, and fertile bottomlands were blanketed with mud and gravel." Perhaps the most appalling waste of all was the erosion of topsoil due to overgrazing of the Western grasslands. "It was hard for a rancher to notice each spring that the grass on his range grew a little thinner, and that, as the years passed, the invasion of weeds expanded. But as the overgrazing continued, scrub growth took over, the slopes began to wash, small gullies developed into large ones, and ranchers found themselves the proprietors of homemade badlands." The farmers who tried to raise crops in the grassland country ran into similar difficulties. "Once the plains were plowed, the dry, upturned soil had no protection against the driving winds." Drought was a "recurrent fact of life on the Great Plains and when the dry years came, crops withered, the dust deepened, and whistling winds lifted thousands of tons of topsoil into the atmosphere." And thus the stage was set for the great Dust Bowl of the 1930s, in Udall's judgment "the most tragic land calamity ever to strike the North American continent."

But there were other tragedies—the massive slaughtering of beavers, the fur seal of the Northern Pacific, the whales of the Bering Sea, the buffalo of the Great Plains, and the passenger pigeon. At the beginning of the nineteenth century the number of these birds was estimated to be an incredible 5 billion. Succulent and easy to kill—some farmers used them for hogfeed—they were exterminated. In the early twentieth century the last survivor of the species died in the Cincinnati zoo. No one knows how many buffalo there were when the Louisiana Purchase brought their ranges into the public domain. Informed estimates vary between 10 and 100 million. "The big kill began when the Civil War ended. The army wanted them killed in order to starve out the Plains Indians; the cattlemen wanted them killed to save forage for their own livestock; the railroad men wanted them killed to supply profitable freight in the form of hides; the market hunters wanted them killed for their tongues and hides; and sportsmen came to kill them for trophies and for pleasure." By the mid-1880s only a few hundred scattered survivors could be found.

It is thus apparent that an appalling waste of America's natural resources took place in the nineteenth century. From the short-run point of view it undoubtedly often made "economic sense" for men and women to do what they did. Yet their actions must be judged not only within the context of their own present but also within our own. We can understand why people did what they did—after all the earth belongs to the living, as Jefferson said—but we can also deplore the impact of their decisions on generations not yet born. Some of the waste was retrievable—for example, a good deal of farm land—but not all. Some resources were badly depleted (buffalo, the grasslands of the plains) or dissipated altogether (passenger pigeons). In sum, however great the degree of our "understanding" that our predecessors were seeking to better their material and social positions in ways approved by the values and laws of their time; however full our realization that the legal system was constructed to facilitate a rapid transfer of resources from the public to the private sector (or was bent to that end by those who abused it), it is impossible not to regret many of the consequences of their actions.

The First Conservation Movement sought to use the legal system, and exhortation and education as well, to put an end to waste. "No man is a true lover of his country," President Theodore Roosevelt observes in one of *The New York Times* articles in this book, "whose confidence in its progress and greatness is limited to the period of his own life. . . ." In 1891 a General Revision Act repealed the Preemption and Timber Culture Acts, and, among other things, denied owners of more than 160 acres the right to make a homestead entry. Under President Roosevelt lands were withdrawn from the public domain and designated as forest wildlife refuges; National Parks were increased in number, and so too were so-called "National Monuments" (the Grand Canyon, for example). The best hydro-power sites on sixteen western rivers were reserved, as well as nearly 75 million acres of public coal and phosphate lands. In 1902 a vitally important Reclamation Act was passed, legislation which created an agency to cope with the water problems of the arid West. And fees were levied on stockmen whose herds grazed on the forest lands of the public domain.

Nevertheless the Movement did not turn its back on the goal of growth. Indeed, men like Roosevelt and Gifford Pinchot, the eminent forester, simply wished to put an end to waste, to use resources efficiently, to adopt, in the case of the forests, for example, the principle of sustained yields—of reforestation and the avoiding of cutting down more than was growing up. The resources of America would indeed prove endlessly abundant if managed with efficiency. Few men of the nineteenth century would have said what Albert Gallatin, Jefferson's former Secretary of Treasury, said in 1836: "I . . . prefer security to rapid growth."

Today economists have an answer to those critics of economic growth who warn, with differing dates for Armageddon, that a rising demand for such resources as petroleum, silver, gold, copper, lead, platinum, tin and zinc will outrun supply, that growth will peter out as its sinews wither. Not only the American economy but the world economy as well is capable of adjusting to shortages in materials and resources. The assumption that the world will end with a whimper instead of a bang largely reflects the omission of the variable of prices in projects of how resources will be used. When resources are scarce in relation to the demand for them, their price rises, and this induces users to substitute cheaper materials. Higher prices both stimulate efforts to conserve resource inputs and increase their supply through exploration.

The fact of the matter is that while there are exceptions in the case of some materials, the relative prices of natural resources as a class have not risen historically. Hence, the record of the past gives little evidence that the growth of the industrialized world is pressing against rigid raw material supply constraints. Mineral prices have roughly kept pace with industrial prices for the last one hundred years. The reasons for this are clear. In spite of the scarcity of high-grade ores, technical change has dramatically reduced costs of exploration and extraction. And technology has widened the opportunities to substitute plentiful materials for scarce ones.

The price rises of the recent past may well provide the needed incentive for efforts to speed the pace of substitution and, in addition, to discover and develop new sources of energy. Petroleum and natural gas will surely have to be supplemented by other energy sources in the near future. In a recent study on "The Allocation of Energy Resources" (February 1974) William D. Nordhaus predicts that domestic petroleum resources will be virtually exhausted by 1980, and that imported petroleum and natural gas will put a heavy drain on the United States balance of payments—besides being liable to interruption from political causes. After the year 2000, Nordhaus believes, the importance of imported energy sources will diminish, and that of liquefied coal and shale oil, augmented by light-water nuclear reactors, will grow. As the twenty-first century wears on, the breeder reactor will gradually take over; and solar, geothermal, gravitational, and perhaps other, as yet unforeseeable, new energy technologies may emerge to supplement it. After the exhaustion of the fossil fuels, perhaps in the twenty-second century, the economy will have to run on an electric hydrogen technology or some other exotic technology with a resource base that is virtually infinite, if high industrial civilizations are to survive. Nordhaus is optimistic. In his opinion, "we should not be haunted by the specter of the affluent society grinding to a halt for a lack of energy resources."

However, as the readings in this collection from *The New York Times* make clear, there are some who believe that technical progress will not keep pace with the pressures exerted by economic growth on agricultural and mineral resources and on the environment. These stresses, they argue, tend to multiply geometrically (or, in the language of mathematics, exponentially) while the capability of technical progress of accomodating to these rising demands is far more narrowly limited. The argument is basically similar to ideas expressed two centuries ago by the Reverend Thomas Malthus. Malthus's point was that since people tend to multiply exponentially, while, at best, the food supply increases at a constant rate, only continence, war, or starvation can redress the balance. Admittedly, the quality of life in the future will probably depend on willingness to slow rates of population growth. Yet one cannot avoid the conclusion that since the days of Malthus technological advance has, at the very least, helped postpone the confrontation between humanity and the resources of the world. And while, as Peter Passell and Leonard Ross acknowledge in *The Retreat From Riches,* no one can be certain about it, technical progress does not seem to be slowing down; indeed, the "best econometric estimates" indicate that it is growing exponentially.

Those critics of economic growth who deplore the effects of technological progress on the environment are on firmer ground. Among the "external diseconomies" of growth, particularly since the end of World War II, are air pollution, water pollution, solid waste pollution, congestion, noise, decline in security of persons and property, and general uglification of town and country. Highly productive factories use river water as a coolant, then dump it back "warmed and reeking with chemicals." Synthetic insecticides increase crop yields but are washed off millions of farms into the national water supply—threatening to make fish toxic to man. Electric power plants generate energy cheaply and efficiently, yet join with garbage incinerators and automobiles to foul the air of the nation's cities. Since the end of World War II the production of deadly mercury waste has increased dramatically, mainly because of the demand for chlorine. Tough plastic packaging, unlike frail cardboard, is impervious to chemical breakdown, so that litter remains intact until somebody picks it up. Phosphorous and nitrogen compounds from laundry detergents, processed sewage, and fertilizers are pumped into lakes and rivers, where they disrupt the balance of animal and vegetable life in large lakes. In consequence, some kinds of fish have disappeared altogether from Lake Erie, for example, and all catches are down sharply. The indictment, carefully drawn by Passell and Ross, could be expanded.

But is economic growth at fault? Or deficiencies in public policy? Economists answer that pollution comes not from growth but from "our perverse system of incentives to industry." And that is our answer too. Firms have been encouraged to poison the air and foul the water because society has regarded the environment as a free good. Now that pure air and water are becoming scarcer goods, society must insist that firms pay for their use. Taxation can compel firms to "internalize" social costs of production that were previously hidden or obscured. But since these costs will then be included among other costs of production, prices of goods and services must rise in consequence. Passell and Ross suggest that "halting growth would do far less to scrub the environment than a simple policy of making business put its money where its exhaust is." We should add to this that consumers must also be willing to pay the higher prices that must result. If they truly value environmental quality they will be willing to do so.

A final word. The external diseconomies of growth are not peculiar to the United States alone or to "capitalism." They are common to technologically advanced, industrialized and urbanized nations everywhere, to the Japanese and Russians as well as the Americans. Everywhere men and women must seek to strike the balance between growth and environmental protection that accords with their national history, resources, and needs. To poverty-stricken people in Africa, Asia, South America and the Middle East concern over environmental pollution is a luxury of the affluent. Americans must also keep in mind those who live in want in our own midst. The promise of American life has not been redeemed for blacks, Indians, and thousands of poor whites, all of whom would doubtless like to be able to afford a greater interest in the environment. Yet while the poor undoubtedly have a higher level of tolerance of environmental risk, the reality of that risk is not thereby lessened. And all alike must confront it. How to do so, how to decide on ways to minimize the impact on the poor of surrendering some portion of our material welfare for the sake of the environment which encloses us all in sickness or in health, in life or in death, is one of the great challenges of our times.

Stuart Bruchey

CHAPTER 1
Conservation

The Jamaica Bay Wildlife Refuge, part of the new Gateway National Park. To the west (background) is Manhattan; next door is Kennedy Airport.

Hosefros/NYT Pictures

OUR FORESTS.

The House Committee on Agriculture lately voted to report a bill for the collection of specimens of native woods in the United States, to be placed in the museum of the Agricultural Department in Washington. This step has not been taken too soon, for if the waste of wood continues at its present rate there may, a few years hence, be specimens of many kinds wanting. We have frequently called attention to the importance of forest preservation, and the short-sightedness of the present generation in respect to it, and are glad to see that the President is displaying a personal interest in the matter. The public is apt to lose sight of its importance so soon as the excitement and horror attending such terrible conflagrations as those which occurred in Wisconsin and this State two years ago die away.

In his report to the Land Office at Washington, in 1866, the Surveyor General wrote: "We have now reached a period when the demand for timber is rapidly on the increase and the supply diminishing. The demand is, indeed, undoubtedly increasing to an enormous extent." In 1867 the same officer wrote: "The subject of forest-tree culture has of late years attracted much attention in Europe on account of the inconvenient scarcity of all the more valuable kinds of timber, especially ship timber; and the subject is of no less importance in our own country, where regions exist completely destitute of trees, and where the supply of the more valuable kinds is limited, and be-

coming so scarce that it even now commands large prices. It is time that our best timber lands should be prized, not only in regard to present but future value." In a speech in Congress two years ago, Mr. HALDEMAN said it was computed that twenty million people are dwelling in wooden houses, their barns and out-buildings of wood, the fencing of wood. Then twenty thousand cords a day are daily consumed by locomotives, and the sixty thousand miles of road demand 2,500 ties to the mile. Some of the mills on Puget Sound have capacity to turn out daily 100,000 feet of lumber, and the present export of the Sound in prepared lumber, masts, and spars amount to over one and a half million dollars annually. Chicago, in the same time, sold nearly 1,000,000,000 feet of lumber, over 200,000,000 shingles, and 100,000,000 pieces of lath, and the enormous consumption of wood of that city in repairing the ravages of fire may well be imagined. Another tremendous devourer of wood is the mines. In the Comstock Mine, Nevada, there has been annually consumed of lumber and timber about 18,000,000 feet.

From Nebraska City, Mr. HAYDEN, United States Geologist, writes: "I would again speak of the importance of planting trees in this country. It is believed that the planting of ten or fifteen acres of forest-trees on each quarter of a section will have a most important effect on the climate, equalizing and increasing the moisture and adding greatly to the fertility of the soil.

The settlement of the country and the increase of the timber have already changed for the better the climate of that portion of Nebraska lying along the Missouri."

As matters stand, the consumption increases enormously every day with the increase of population, railroads, manufactures, and mining enterprise. Enormous quantities of trees in woods throughout the country are rendered valueless for timber by neglect. It is, indeed, melancholy when traveling to survey these wretched poles leaning one against the other, so cramped and choked that they can never develop. The various reports to the Land Office abound with valuable suggestions for the preservation and increase of timber. The subject, however, is not without its difficulties. Our Government does not command the centralized machinery which European countries employ to secure the preservation of their forests. The States in the far West are moving, but necessarily slowly, and not in unison. Several railway corporations, and notably the Pacific Railways, have taken practical steps to encourage the planting, and to determine the best means of carrying it on. There is a growing sentiment on the subject in the West, and Congress recently amended a law already in existence, extending special privileges on the public lands to those homestead settlers who would plant a certain proportion of their land with trees. These agencies work slowly, but much will yet be saved, and much done for the future. April 6, 1874

THE NEW FOREST RESERVES.

One of the most creditable acts of President HARRISON'S Administration, for which he, and especially Secretary NOBLE, should have hearty and unstinted commendation, has been the reservation of enormous forest tracts in various parts of the West. The withdrawal of these tracts will make them available in some cases for existing national parks, in others will found new pleasure grounds, and in all instances will preserve for public uses regions of great value, whether for their scenery, their remarkable tree growths, or their influence in preserving the sources of water supply. This good work was made possible through a wise provision in the act of 1891 for repealing the old timber-culture law, which provision gave the President the power to withdraw forest areas from acquisition for private uses. This power has been used with great energy, and sometimes against the influences brought to bear by settlers, sheep herders, miners, or other persons interested, from personal and pecuniary motives, in preventing its exercise.

We find, in looking over the more recent fruits of this judicious policy, that on the 20th of December last there was established in Southern California the San Gabriel timber land reservation of nearly 1,000,000 acres, near Los Angeles, including all the mountains at the north of Pasadena, from Salidad Cañon, where the Southern Pacific Railroad goes through the mountains, eastward to the Cajon Pass. This was followed, after a time, by another reservation of 800,000 acres, adjoining the San Gabriel, running eastward from Cajon Pass to San Gorgonia, and called the San

Bernardino Mountain Forest Reservation. On the 14th of February a third and far more extensive tract, called the Sierra reservation, and comprising over 4,000,000 acres, or more than 6,000 square miles, was set apart on the high Sierra extending southward from the line of the Yosemite National Park to the seventh standard parallel south. This great tract includes not only the existing Grant, Sequoia, Tule River, and Mount Whitney reservations, but the marvelous King's River Cañon, which Mr. JOHN MUIR has so well described. It is a region of splendid mountains and wonderful gorges, some of the mountains rising from 10,000 to 15,000 feet above the sea level. It contains most of the giant sequoias that had not already been reserved, with great sugar pines, cedars, and other valuable trees. Besides preserving natural scenery of exceptional grandeur, it secures the sources of water supply for the central California Valley from Mariposa to Bakersfield. Indeed, these three reservations just mentioned, taken with the new Yosemite National Park, established by Congress under Secretary NOBLE'S administration, and comprising, if we rightly remember, more than 1,000,000 acres, secure nearly the whole of the elevated forest chain which furnishes the water supply to the productive region of California south of San Francisco. Perhaps hereafter a mountain reservation may be formed north of the Yosemite to or toward Mount Shasta, thus conferring like benefits on the northern part of the State.

California is not the only part of the country that has profited by this intelligent

devotion to the preservation of the forests. In the State of Washington, around Mount Rainier, there is soon to be formed, we understand, a tract known as the Pacific Reserve, comprising about 1,000,000 acres. While, also, the recent work in California was going on, Mr. R. U. JOHNSON, who had been untiring in securing the Yosemite National Park legislation, called Secretary NOBLE'S attention to the importance of preserving the Grand Cañon of Colorado, and the President set apart there a great tract of 1,900,000 acres in a region that furnishes some of the most impressive and sublime scenery to be found on the continent. We may also recall that it was under this same act of March 3, 1891, that the President set apart the great forest tracts on the east and south boundaries of the Yellowstone National Park, aggregating 1,239,040 acres. Still another fine reservation was the White River tract in Colorado, comprising 1,198,080 acres. Two others in the same State, on Pike's Peak and Plumb Creek, aggregate 362,020 acres. The Pecos River Reserve in New-Mexico embraces 311,040 acres, and the timber land reserve in Oregon includes 142,080.

These facts will show the earnestness and success with which the policy of forest reservation has been carried out during the last two years. The value of the services thus performed can hardly be overestimated. The benefit is both local and national—in securing the sources of water supply and thus insuring irrigation and bountiful crops, and in preserving unimpaired the glories and beauties of natural scenery at the West.

February 27, 1893

FORESTRY AS A SCIENCE

GEORGE W. VANDERBILT'S BILTMORE AN OBJECT LESSON.

First Attempt on a Large Scale in This Country to Make the Science a Paying Investment—An Immense Arboretum, with a Library and School of Forestry, to be Established at Asheville in North Carolina.

If it were true, as an inspired tinker and Populist declared, that Mr. George W. Vanderbilt had established his vast estate of 10,000 acres, Biltmore, at Asheville, in North Carolina, for the purpose of self-aggrandizement only, or, at best, in the pride of ownership which marks the landed proprietor, he has vindicated himself from the accusation by connecting with his property one of the most important institutions of the country. Much has been said and written, first and last, about preserving the forests in this country after the methods which have prevailed for half a century and more in Germany, France, and other only valuable timber might be saved, but only valuable property might be saved, but safety from fire and flood should be had.

It is well understood that for these objects all dead wood, whether underbrush or high-reaching trees, must be cleared out of the way, and where they leave too wide spaces their places must be filled with other plantations. There is a scientific method and an unscientific way of doing this, and success depends wholly upon which one of these methods is adopted. For such reasons the science and profession of forestry has flourished in other countries than America, having extensive tracts of woodland for many years, in the same manner that gardening and botany have flourished.

Little or no attention, however, has been paid in America to forestry until recently, by either private or public persons. The States of New-York, Maine, and a few others with extensive forest lands, have paid some attention to the subject, but the Federal Government has only just awakened to its importance; and at the last session of Congress a commission of men who have studied the subject was authorized to investigate and report on the condition of the woodlands of the Northwest. The commission, consisting of several professors of arboriculture from the leading universities, like Harvard, Yale, Columbia, and others, with several Government officials, is now at its work in Montana.

Mr. Vanderbilt was an early exception to hhe apathy here on the subject of forest preservation, for shortly after the purchase of his estate at Asheville he began to form plans, not only for the preservation of the great domain of woodland which he found in existence, but for the rehabilitation of that portion of it which had been exhausted. The forest was broken and irregular in character, owing to the fact that the land had been divided among many small farmers, who had made frequent clearings, or had robbed the forest of its most vigorous and healthy trees. Scientific measures were required for the work of restoration, and Mr. Vanderbilt resolved to spare neither expense nor care in the scheme. It was a question whether at the end of a term of years he would have a noble forest of park-like character and a certain commercial value, or merely a barren and tangled woodland, gradually going to decay, and liable at any time to destruction by fire.

Mr. Vanderbilt accordingly sought for the best talent among those who had made dendrology a study, and was fortunate in obtaining the skillful services of Mr. Gifford Pinchot, a student of forest management in the best schools of Europe, and a man fully alive to the advantages and disadvantages of the different methods in their application in this country. Mr. Pinchot took hold of Biltmore Forest, of about 5,000 acres, which he found was composed mainly of oaks and other deciduous trees, mostly young, with scattered pines, which occasionally covered old and exhausted fields to the exclusion of other species. Nevertheless there was considerable present and prospective value in the timber and firewood of the forest, and in a report which he made of his findings, a year or so ago, he sketched a scheme in which he proposed three general objects, namely: A profitable production, which will give the forest direct utility; a nearly uniform annual yield, which will give steady employment to a trained force of foresters—woodchoppers and lumbermen; and a gradual improvement in the condition of the forest itself. These objects he proposed to obtain by dividing the estate into the high-forest system and the selection system. The rotation—that is, the length of time allowed for a second crop to become ripe on the same ground after the removal of the first crop—has been fixed at 150 years.

In a thoroughly equipped forest, managed under the high-forest system, there are as many subdivisions as there are years in the rotation, the trees of each subdivision being of an equal age, and only one year older or younger than those of the next subdivision. In this way it would be possible to cut every year one-one hundred and fiftieth of the whole area, thus securing a uniform annual crop during the whole period. In the selection system forest trees of all ages are mixed together, instead of being separated in groups according to their ages. The annual product is taken from all parts of the forest, the ripe trees being selected for cutting; but such a method necessitates in the case of a large forest area expensive transportation, and to avoid this Mr. Pinchot has adopted what he calls the location selection system, under which the annual yield is taken from a certain part of the forest during several years, then from another part, and so on.

Mr. Pinchot's balance sheet in his report above mentioned, covering the first year's operations of the Biltmore Forest, shows an expenditure of $9,911.76, with receipts amounting to $5,607.11, and material on hand worth at local market prices $3,911.25, or $9,519.36 in all, showing a deficit of only $392.40. In the year 1893 this deficit became a surplus of more than $1,200—a remarkable result, in view of the poverty of the forest he had to operate in and the difficulties which are always attendant upon the establishment of a new industry, especially in one like this, where all his assistants and workmen had to be formed from the very beginning.

But Mr. Vanderbilt has broader and more liberal views in his forest operations than an effort to make his investment pay by improving the property. He intends his forest, for one thing, as an object lesson in forest preservation to the country. Already preparations are on foot for a great arboretum at Biltmore, in which are to be gathered all the trees and shrubs of the temperate regions of the world, which will form a museum of the greatest interest. It will cover some 800 acres of land, distributed along both sides of a road twelve miles in length. Here the nurserymen and foresters of the entire country will be at liberty and have full opportunity to study and gain information as to the character and growth of important forest trees not to be obtained elsewhere.

In connection with this arboretum and the general scheme of forest management at Biltmore, it is said to be Mr. Vanderbilt's intention to establish and equip a school of forestry on or near his estate. Already a number of students are residing near the place, taking practical lessons in the science from Mr. Pinchot and his chief assistants, who are resident foresters. They also have free access to the notable collection of valuable books in the library connected with the arboretum, and which it is also said to be Mr. Vanderbilt's purpose to make a public one.

July 12, 1896

Irrigation in the West.

Written for THE NEW YORK TIMES by

FRANCIS G. NEWLANDS

Representative of Nevada.

The West has been for twenty years or more insisting that some legislation should be inaugurated by Congress looking to the reclamation of the arid west. The great difficulty has been that Senators and Representatives from the Eastern, Middle, and the humid States of the West have not understood the irrigation question.

They seem to regard it as something new, whereas, as a matter of fact, the question is as old as time. Almost all the extinct civilizations which history records were based upon irrigation, and irrigation is to-day practiced in more than one-half of the world. England has spent $300,000,000 in India on irrigation works, which have done much to render more certain the crops of that region. England is to-day expending millions of dollars on the Nile in extending the area of irrigation. Italy and Spain illustrate the practical results of scientific irrigation.

Now, what does irrigation involve? Irrigation is practiced only in arid and semi-arid countries, where the rainfall is either entirely lacking or is entirely insufficient to raise crops. In all such regions there is a heavy deposit of snow during the Winter on the mountains. These snows, melting during the Spring, form streams which are torrential in the Spring and early Summer and are dry, or nearly dry, in the late Summer and the Fall. Irrigation means that the water is taken out of such streams by ditches, which carry the water to the arid plains on a lower level. These waters are spread over the lands by sub-ditches, and thus a substitute is obtained for the waters which fall in the humid regions from the heavens.

The difficulty is that those waters are abundant when they are least needed, namely, in the Spring, and they are scarce when they are most needed, when it is necessary to ripen the crops. The storage of the torrential waters in the mountains is therefore resorted to. These waters are kept in artificial reservoirs, caused by the construction of dams at favorable places in the mountain valleys. These waters are let out into the stream in the very hot season when water is scarce. Storage thus enables the utilization of a greater amount of the torrential waters in irrigating the arid plains, as the stored waters supplement the torrential waters later on and ripen the crops which otherwise would be burned by the hot sun.

Storage involves the treatment of an entire watershed in a scientific way. Very large rivers have numerous tributaries, having their sources in the snows in the mountains, and the more water there is stored the greater the extent of the torrential waters that can be utilized in irrigation, and the greater the area of land that can be irrigated and cultivated.

The arid region extends from the hundredth parallel to the Pacific Coast. Draw a line through the middle of North Dakota, South Dakota, Nebraska, Kansas, and Oklahoma, and all to the west of it is either arid or semi-arid, the aridity increasing as you approach the Rocky Mountains. The western fringe of the area of the Pacific Coast States is humid, and the eastern portion of those States is semi-arid. There are thirteen States and three Territories included in the arid and semi-arid region.

To the question, What right has the West to demand Government aid in this matter? I reply: Simply because the Government is the owner of almost all of the arid land in that region. The Government owns 600,000,000 acres of arid and semi-arid land. We claim that it is the Government's duty to open this land to settlement, and to do what is necessary to promote settlement, namely, to conserve the snow and flood waters, and to construct such high-line ditches and canals as are necessary to bring the waters of that region within reach of the settlers; in other words, to make them available for settlement. The settlers will do the rest. They will actually reclaim the land, and the Government will simply render the waters of that region available.

The settlers cannot do this work, because it must be conducted on a very large scale. Large capital must be employed and scientific knowledge applied. A watershed must be treated as an entirety, regardless of State lines, for a river with its tributaries may pass through several States. It has not been the policy of the Government to grant this land in large tracts to corporations or individuals. In order, therefore, to promote settlement of small tracts for home builders it must itself make the waters of that region available so that they can be easily reached by settlers. The States could not do the work if they have no title to the land, and if the lands were granted to them by the Government they could not operate outside of their State lines, and oftentimes the lands to be reclaimed are in one State, while the waters, which ultimately reach them, must be stored at favorable places in an adjoining State.

The Government is a great land holder. Public policy demands that it should part with its lands to actual settlers in small tracts, who propose to make homes. The Government simply does the work and makes the expenditure that is necessary to promote settlement. It is not expected that it will actually reclaim the lands.

As to the amount of lands that can be reclaimed, it is estimated by the scientific men connected with the Geological Survey that enough snow falls in the mountain region to furnish water for the reclamation of about 10 per cent. of the arid area, or 60,000,000 acres. This would mean an area equal to the States of Iowa and Illinois. All the rest would remain forever arid, though it would produce certain grasses which are useful in fattening cattle.

The question as to how this work should be paid for has been a matter of great consideration with the Western men. They realize that there will be some indisposition to appropriating money from the Federal Treasury for the purpose, though there is no reason why, if Government money is to be expended on rivers and harbors to promote such navigation, such moneys should not also be spent on rivers to promote irrigation.

But the Western men concluded to urge a measure which would simply utilize the receipts from the sales of public lands in the arid region. So the Hansbrough-Newlands bill, which is now pending before Congress, provides that the receipts from the sales of such lands shall go into a special fund in the Treasury to be called the Arid Land Reclamation Fund, and that they shall be expended by the Secretary of the Interior in the construction

of storage and irrigation works, and that the lands for which the waters are made available shall be sold to actual settlers in tracts not less than forty, and not exceeding a hundred and sixty acres, the price to be so fixed by the Secretary of the Interior as to restore to the fund from the sales of lands granted to settlers the amount expended in each project. The price is to be paid by the settler in ten annual installments.

In this way a revolving fund is created out of the sale of the lands reclaimed, which is applied to new work, and thus, in the end, the West will reclaim itself. The Secretary of the Interior can make no contract for irrigation work unless the moneys therefor are in the fund. Thus the arid region will be reclaimed without taxation of the general public. Every guard is thrown around the bill so as to prevent land speculation or land monopoly. The purpose of the bill is to provide homes for actual settlers.

It is claimed by some that this work is premature, but it must be recollected that the public lands of the United States, which are watered from the heavens, have nearly all been taken up. The line of settlement has now advanced away beyond the Dakotas and Nebraska and Kansas. The quickness with which Oklahoma was settled up demonstrates this. Besides this, it will take at least fifty years to do the work, as irrigation work is exceedingly slow. We have been for many years disposing of from ten to fifteen million acres of public land annually. It is not probable that we can reclaim more than five hundred thousand or a million acres annually in the arid region.

Nor is there any danger of a disastrous competition with the existing farms of the East and West. It should be recollected that all the States outside of the original thirteen States were public land States. Since the beginning of this century the Government has disposed of over a billion acres of public lands, and it has thus developed the great States of the South and West. It might as well be contended that the people of the original thirteen States suffered from Western development, as that the eighty millions of people now occupying this country will suffer from the development of the arid region. The arid

region will simply furnish a market for Eastern manufactures and Eastern products. It will not compete with the Eastern or middle Western farms, because in the northern part of the arid region cultivation will be confined almost entirely to alfalfa, which is very useful in the fattening of cattle, and in the southern region cultivation will be confined largely to the citrus fruits and other products of a semi-tropical character.

It has only to be borne in mind that the entire area capable of reclamation does not exceed 60,000,000 acres, and that this area will only equal the area of the two States of Iowa and Illinois. If a nation of a few millions of people did not suffer from the development of Iowa and Illinois, how can it be contended for a moment that a country with 80,000,000 of people can suffer from the development of the arid region?

I am aware that some men who represent farming constituencies are opposed to the opening up of the arid land, but I cannot believe that they will adhere to their opposition when they reflect upon the history of this country, whose glory and strength have largely had their course in the development of the Western region. As there was no reason why settlement should terminate with the western boundary of the thirteen original States, and there was no reason why it should stop at the Ohio, and as there was no reason why it should stop at the Mississippi, or the Missouri, there is no reason why it should stop at the edge of the great American desert, which, by the hand of science, can be made a region of great fertility and wealth.

The bill which the Western members have instructed me to offer in the House, and Senator Hansbrough to offer in the Senate, is now under consideration in the Senate, and doubtless will pass in that body by an almost unanimous vote. I hope also for favorable consideration in the House, though I regret to say that many of the leaders of the House are at present opposed to the measure.

FRANCIS G. NEWLANDS.
Washington, March, 1902.

March 2, 1902

OUTLINE OF IRRIGATION BILL.

Representative Newlands of Nevada Discusses the Measure Signed by the President Yesterday.

Special to The New York Times.

WASHINGTON, June 18.—Representative Newlands of Nevada, who is one of the most influential supporters of the Irrigation bill, which was signed by the President to-day, and who has for so many years championed this subject in Congress, gave a brief outline of the effect of the bill to-day, in which he said that the best feature of the bill is its automatic action until the entire work is done without the necessity of further legislation by Congress.

"In this respect," said Mr. Newlands,

"it is perhaps the only bill ever passed which furnishes so complete, comprehensive and automatic a plan of action. Under its provisions at least $150,000,000 of the proceeds of the sales of public lands will be available in the next thirty years for irrigation works without further appropriation.

"The receipts from public lands for the last fiscal year, as well as the present, aggregating $6,000,000, are immediately available, and from this time on an amount of $3,000,000 per annum will be available, which sum will be constantly increased. The bill is carefully guarded. The Secretary of the Interior cannot let contracts unless the money is in the fund. Land monopoly is impossible."

Mr. Newlands believes the enactment of this legislation will be a greater benefit to the people of Nevada than anything since the discovery of silver in that State.

June 19, 1902

TIMBER FAMINE NEAR, SAYS MR. ROOSEVELT

Accident and Waste Rapidly Denuding the Country.

NATIONAL FOREST SERVICE

The President Repeats That One Is Needed—Civilization's Growing Need of Wood.

WASHINGTON, Jan. 5.—That this country is in peril of a timber famine, and that there should be a National forest service to assist in preserving forests was asserted by President Roosevelt this afternoon in an address before the American Forest Congress. In the course of his remarks the President said:

" The producers, the manufacturers, and the great common carriers of the Nation had long failed to realize their true and vital relation to the great forests of the United States, and forests and industries both suffered from that failure.

" But the time of indifference and misunderstanding has gone by. Your coming is a very great step toward the solution of the forest problem—a problem which cannot be settled until it is settled right.

" The great significance of this congress comes from the fact that henceforth the movement for the conservative use of the forest is to come mainly from within, not from without; from the men who are actively interested in the use of the forest in one way or another, even more than from those whose interest is philanthropic and general. The difference means to a large extent the difference between mere agitation and actual execution, between the hope of accomplishment and the thing done.

" The great industries of agriculture, transportation, mining, grazing, and, of course, lumbering, are each one of them vitally and immediately dependent upon wood, water, or grass from the forest. The manufacturing industries, whether or not wood enters directly into their finished product, are scarcely, if at all, less dependent upon the forest than those

whose connection with it is obvious and direct.

CIVILIZATION'S NEED OF WOOD.

" Wood is an indispensabe part of the material structure upon which civilization rests, and civilized life makes continually greater demands upon the forest. We use not less wood, but more.

" For example, although we consume relatively less wood and relatively more steel or brick or cement in certain industries than was once the case, yet in every instance which I recall, while the relative proportion is less the actual increase in the amount of wood used is very great.

" Thus the consumption of wood in ship building is far larger than it was before the discovery of the art of building iron ships, because vastly more ships are built. Larger supplies of building lumber are required, directly or indirectly, for use in the construction of the brick and steel and stone structures of great modern cities than were consumed by the comparatively few and comparatively small wooden buildings in the earlier stages of these same cities.

" Whatever materials may be substituted for wood in certain uses, we may confidently expect that the total demand for wood will not diminish, but steadily increase. It is a fair question, then, whether the vast demands of the future upon our forests are likely to be met.

PROSPERITY AND THE FORESTS.

" No man is a true lover of his country whose confidence in its progress and greatness is limited to the period of his own life, and we cannot afford for one instant to forget that our country is only at the beginning of its growth. Unless the forests of the United States can be made ready to meet the vast demands which this growth will inevitably bring commercial disaster is inevitable.

" The railroads must have ties, and the best opinion of the experts is that no substitute has yet been discovered which will satisfactorily replace the wooden tie. This is largely due to the great and continually increasing speeds at which our trains are run.

" The miner must have timber or he cannot operate his mine, and in very many cases the profit which mining yields is directly proportionate to the cost of the timber supply.

" The farmer, East and West, must have timber for numberless uses on his farm, and he must be protected by forest cover upon the headwaters of the streams he uses, against floods in the East and the lack of water for irrigation in the West. The stock man must have fence posts, and very often he must have Summer range for his stock in the National forest reserves.

" In a word, both the production of the

great staples upon which our prosperity depends and their movement in commerce throughout the United States are inseperably dependent upon the existence of permanent supplies from the forest at a reasonable cost.

FEARS A TIMBER FAMINE.

" If the present rate of forest destruction is allowed to continue, a timber famine is obviously inevitable. Fire, wasteful and destructive forms of lumbering, and legitimate use, are together destroying our forest resources far more rapidly than they are being replaced.

" What such a famine would mean to each of the industries of the United States it is scarcely possible to imagine. And the period of recovery from the injuries which a timber famine would entail would be measured by the slow growth of the trees themselves.

" Fortunately, the remedy is a simple one, and your presence here is proof that it is being applied.

" It is only as the producing and commercial interests of the country come to realize that they need to have trees growing up in the forest not less than they need the product of the trees cut down, that we may hope to see the permanent prosperity of both safely secured.

" I want to add a word as to the creation of a National forest service, which I have recommended repeatedly in messages to Congress, and especially in the last. I mean the concentration of all the forest work of the Government in the Department of Agriculture.

" As I have had occasion to say over and over again, the policy which this Administration is trying to carry out through the creation of such a service is that of making the National forests more actively and more permanently useful to the people of the West, and I am heartily glad to know that Western sentiment supports more and more vigorously the policy of setting aside National forests, the policy of creating a National forest service, and especially the policy of increasing the permanent usefulness of these forest lands to all those who come in contact with them.

" With what is rapidly getting to be the unbroken sentiment of the West behind this forest policy, and with what is rapidly getting to be the unbroken support of the great industries behind the general policy of the conservative use of the forest, we have a right to feel that we have entered on an era of great and lasting progress.

" I ask with all the intensity of which I am capable, that the men of the West will remember the sharp distinction I have drawn between the man who skins the land and the man who develops the country. [Applause.] I am going to work with, and only with, the man who develops the country. [Applause.] I am against the land skinner every time. [Applause.]

" Our policy is consistent to give to every portion of the public domain its highest possible amount of use, and of course that can be given only through the hearty co-operation of the Western people."

January 6, 1905

ROOSEVELT PLANS TO EMPLOY RIVERS

He Appoints a Commission to Investigate the Problem of Waterways.

WOULD CONTROL FRESHETS

Praises Railroads for Their Progress, but Says They Cannot Move the Nation's Commerce.

WASHINGTON, March 15.—President Roosevelt has decided to appoint an Inland Waterways Commission. Its duty will be a report a comprehensive plan for the improvement and control of the river systems of the United States. Eight public men have been asked to serve on it,

with Representative Burton of Ohio, Chairman of the Rivers and Harbors Committee in the last Congress, as Chairman. The President's letter is as follows:

The White House,
Washington, March 14. 1907.

My Dear Sir: Numerous commercial organizations of the Mississippi Valley have presented petitions asking that I appoint a commission to prepare and report a comprehensive plan for the improvement and control of the river systems of the United States. I have decided to comply with these requests by appointing an Inland Waterways Commission, and I have asked the following gentlemen to act upon it. I will be much gratified if you will consent to serve.

The Hon. Theodore T. Burton, Chairman; Senator Francis G. Newlands, Senator William Warner, the Hon. John H. Bankhead, Gen. Alexander Mackenzie, Dr. W. J. McGee, Mr. F. H. Newell, Mr. Gifford Pinchot, the Hon. Herbert Knox Smith.

Wants a Broad Plan.

In creating this commission I am influenced by broad considerations of National policy. The control of our navigable waterways lies with the Federal Government, and carries with it correspond-

ing responsibilities and obligations. The energy of our people has hitherto been largely directed toward industrial development connected with field and forest and with coal and iron, and some of these sources of material and power are already largely depleted; while our inland waterways as a whole have thus far received scant attention.

It is becoming clear that our streams should be considered and conserved as great natural resources. Works designed to control our waterways have thus far usually been undertaken for a single purpose, such as the improvement of navigation, the development of power, the irrigation of arid lands, the protection of lowlands from floods, or to supply water for domestic and manufacturing purposes. While the rights of the people to these and similar uses of water must be respected, the time has come for merging local projects and uses of the inland waters in a comprehensive plan designed for the benefit of the entire country.

Such a plan should consider and include all the uses to which streams may be put, and should bring together and co-ordinate the points of view of all users of water. The task involved in the full and orderly development and control of the river systems of the United States is a great one, yet it is certainly not too great for us to approach. The results which it seems to promise are even greater.

Railroads Cannot Give Relief.

It is common knowledge that the railroads of the United States are no longer able to move crops and manufactures rapidly enough to secure the prompt transaction of the business of the Nation, and there is small prospect of immediate relief. Representative railroad men point out that the products of the Northern interior States have doubled in ten years, while the railroad facilities have increased but one-eighth, and there is reason to doubt whether any development of the railroads possible in the near future will suffice to keep transportation abreast of production. There appears to be but one complete remedy—the development of a complementary system of transportation by water.

The present congestion affects chiefly the people of the Mississippi Valley, and they demand relief. When the congestion of which they complain is relieved, the whole Nation will share the good results.

While rivers are natural resources of the first rank, they are also liable to become destructive agencies, endangering life and property, and some of our most notable engineering enterprises have grown out of efforts to control them. It was computed by Gens. Humphreys and Abbott half a century ago that the Mississippi alone sweeps into its lower reaches and the Gulf 400,000,000 tons of floating sediment each year, (amount twice the amount of material to be excavated in opening the Panama Canal,) besides an enormous but unmeasured amount of earth salts and soil matter carried in solution.

Would Control Floods.

The vast load not only causes its channels to clog and flood the lowlands of the lower river, but renders the flow capricious and difficult to control. Furthermore, the greater part of the sediment and soil-matter is composed of the most fertile material of the fields and pastures drained by the smaller and larger tributaries.

Any plan for utilizing our inland waterways should consider floods and their control by forests and other means; the protection of bottom lands from injury by overflows, and uplands from loss by soil wash; the physics of sediment-charged waters and the physical or other ways of purifying them; the construction of dams and locks, not only to facilitate navigation, but to control the character and movement of the waters, and should look to the full use and control of our running waters and the complete artificialization of our waterways for the benefit of our people as a whole.

It is not possible properly to frame so large a plan as this for the control of our rivers without taking account of the orderly development of other natural resources. Therefore I ask that the Inland Waterways Commission shall consider the relations of the streams to the use of all the great permanent natural resources and their conservation for the making and maintenance of prosperous homes.

Praise for the Railroads.

Any plan for utilizing our inland water-ways, to be feasible, should recognize the means for executing it already in existence, both in the Federal Departments of War, Interior, Agriculture, and Commerce and Labor, and in the States and their sub-divisions; and it must not involve unduly burdensome expenditures from the National Treasury. The cost will necessarily be large in proportion to the magnitude of the benefits to be conferred, but it will be small in comparison with the $17,000,000,000 of capital now invested in steam railways in the United States—an amount that would have seemed enormous and incredible half a century ago. Yet the investment has been a constant source of profit to the people, and without it our industrial Progress would have been impossible.

The questions which will come before the Inland Waterways Commission must necessarily relate to every part of the United States, and affect every interest within its borders. Its plans should be considered in the light of the widest knowledge of the country and its people, and from the most diverse points of view. Accordingly, when its work is sufficiently advanced, I shall add to the commission certain consulting members, with whom I shall ask that its recommendations shall be fully discussed before they are submitted to me. The reports of the commission should include both a general statement of the problem and recommendations as to the manner and means of attacking it.

Sincerely yours,
THEODORE ROOSEVELT.

Mr. Bankhead has just finished his tenth term in Congress from Alabama; General MacKenzie is chief of engineers of the army; Mr. Newell is Director of the United States Reclamation Service; Mr. Pinchot is Chief Forester of the United States; Herbert Knox Smith is Commissioner of Corporations, and W. J. McGee is an anthropologist and geologist, formerly in charge of the Bureau of American Ethnology, and formerly President of the National Geographic Society.

March 17, 1907

GOVERNORS CHEER ROOSEVELT'S TALK

He Tells Them Conservation of All Natural Resources Needs One Coherent Plan.

PUTS JOHNSON IN CHAIR

Carnegie Pleads for More Careful Husbanding of Coal and Iron, Which He Says Are Being Wasted.

Special to The New York Times.

WASHINGTON, May 13.—The conference of Governors on the conservation of natural resources got under full swing at the White House this morning. They crowded the East Room and listened to a speech by the President with some of the old-time ring in it. The Governors had been sufficiently warmed up by the weather before they reached the White House to be ready for anything demanding enthusiasm, and they applauded the President with non-partisan liberality.

When he declared his intention to continue the Inland Waterways Commission and make it permanent whether Congress makes an appropriation for its expenses or not, there was vigorous cheering.

After delivering his speech, the President brought the conference right down to business by suggesting that there ought to be a committee on resolutions, and incidentally naming them. Then he paused to give the Governors opportunity to second the motion.

They looked around at one another, however, as if uncertain just what to do. Finally Gov. John Johnson came to the rescue with a motion to appoint the committee named by the President. At the afternoon session the President gave Johnson his reward by announcing that he would call on the Minnesota man to preside when he was not present himself.

At that there was a lively clapping of hands from all over the room, the Republican Governors giving Johnson the hand heartily. The President grinned as if he had been caught injecting politics into the game where it should not have been. Then he grinned some more and said: "And I am sure we all shall be glad to hear anything Mr. Bryan cares to say." Loud applause greeted that and Bryan walked to the front and talked to the President in a stage whisper.

Whereupon Mr. Roosevelt said:
"Mr. Bryan prefers to wait until nearer the close of the conference."

Two Important Ideas.

Two ideas resulted from the conference: The first is that a permanent organization between the States and the Nation is necessary, and will probably result from the present conference, to accomplish the end sought. The second, suggested by Secretary Root, is that there is no limitation by the Constitution to the agreements which may be made between the States, subject to the approval of Congress. The two ideas fully developed, it is predicted, would result in the conservation of the energies and resources of the nation through uniform and unconflicting laws, both National and State.

Forty-four Governors and 500 other persons taxed the capacity of the East Room. The others were Cabinet officers, Supreme Court Justices, Senators, Representatives, experts in all lines of industry. President Roosevelt's fifty-minute speech was many times interrupted by applause, and when he finally reached his point of praise of the Inland Waterways Commission, the Governors stood up and shouted.

The President's Speech.

President Roosevelt spoke in part as follows:

"The occasion for this meeting lies in the fact that the natural resources of our country are in danger of exhaustion if we permit the old wasteful methods of exploiting them longer to continue.

"The wares of the merchants of Boston of Charles, like the wares of the merchants of Ninevah and Sidon, if they went by water, were carried by boats propelled by sails or oars; if they went by land were carried in wagons drawn by beasts of draught or in packs on the backs of beasts of burden. The ships that crossed the high seas were better than the ships that had once crossed the Aegean, but they were of the same type, after all—they were wooden ships propelled by sails, and on land, the roads were not as good as the roads of the Roman Empire, while the service of the posts was probably inferior.

"The growth of this Nation by leaps and bounds makes one of the most striking and important chapters in the history of the world. Its growth has been due to the rapid development, and alas! that it should be said, to the rapid destruction, of our natural resources. Nature has supplied to us in the United States, and still supplies to us, more kinds of resources in a more lavish degree than has evtr been the case at any other time or with any other people.

"The wise use of all of our natural resources, which are our National resources as well, is the great material question of to-day.

"Disregarding for the moment the question of moral purpose, it is safe to say that the prosperity of our people depends directly on the energy and intelligence with which our natural resources are used. It is equally clear that these resources are the final basis of national power and perpetuity.

"Already the limit of unsettled land is in sight, and indeed but little land fitted for agriculture now remains unoccupied save what can be reclaimed by irrigation and drainage. More than half of the timber is gone. Many experts now declare that the end of both iron and coal is in sight.

"The enormous stores of mineral oil and gas are largely gone. Our natural waterways are not gone, but they have been so injured by neglect, and by the division of responsibility and utter lack of system in dealing with them, that there is less navigation on them now than there was fifty years ago. We have so impoverished our soils by injudicious use and by failing to check erosion that their crop-producing power is diminishing instead of increasing.

Questions Far Reaching.

"These questions do not relate only to the next century or to the next generation. It is time for us now as a Nation to exercise the same reasonable foresight in dealing with our great natural resources that would be shown by any prudent man in conserving and widely using the property which contains the assurance of well-being for himself and his children.

"In dealing with the coal, the oil, the gas, the iron, the metals generally, all that we can do is to try to see that they are wisely used. The exhaustion is certain to come in time.

"But in dealing with the soil and its products man can improve on nature by compelling the resources to renew, and even reconstruct, themselves, while the living waters can be so controlled as to multiply their benefits.

"On the average, the son of the farmer of to-day must make his living on his father's farm. There is no difficulty in doing this if the father will exercise wisdom. No wise use of a farm exhausts its fertility. So with the forests. We are over the verge of a timber famine in this country, and it is unpardonable for the Nation or the States to permit any further cutting of our timber save in accordance with a system which will provide that the next generation shall see the timber increased instead of diminished. Moreover, we can add enormous tracts of the most valuable possible agricultural land to the National domain by irrigation and by drainage.

One Coherent Plan Needed.

"We can enormously increase our transportation facilities by the canalization of our rivers so as to complete a great system of waterways on the Pacific, Atlantic, and Gulf Coasts and in the Mississippi Valley, from the Great Plains to the Alleghanies, and from the northern lakes to the mouth of the mighty Father of Waters. But all these various uses of our natural resources are so closely connected that they should be co-ordinated, and should be treated as part of one coherent plan, and not in haphazard and piecemeal fashion.

"We are coming to recognize as never before the right of the Nation to guard its own future in the essential matter of natural resources. In the past we have admitted the right of the individual to injure the future of the Republic for his own present profit. The time has come for a change.

Legislation for All.

"Any enactment that provides for the wise utilization of the forests, whether in public or private ownership, and for the conservation of the water resources of the country, must necessarily be legislation that will promote both private and public welfare.

"The opinion of the Maine Supreme bench sets forth unequivocally the principle that the property rights of the individual are subordinate to the rights of the community, and especially that the waste of wild timber land derived originally from the Sttae, involving as it would the impoverishment of the State

and its people, and thereby defeating one great purpose of government, may properly be prevented by State restrictions.

"Finally, let us remember that the conservation of our natural resources, though the gravest problem of to-day, is yet but part of another and greater problem to which this Nation is not yet awake, but to which it will awake in time, and with which it must hereafter grapple if it is to live—the problem of National efficiency, the patriotic duty of insuring the safety and continuance of the Nation."

Andrew Carnegie's Address.

Andrew Carnegie spoke on "The Conservation of Ores and Related Minerals" in part as follows:

"Iron and coal are the foundation of our industrial prosperity. The value of each depends upon the amount and nearness of the other.

"Unless there be careful husbanding, or revolutionizing inventions, or some industrial revolution comes which cannot now be foreseen, the greater part of that estimated 2,500,000,000,000 tons of coal forming our original heritage will be gone before the end of the next century, say two hundred years hence.

"Still more wasteful than our processes of mining are our methods of consuming coal. Of all the coal burned in the power plants of the country, not more than from 5 per cent. to 10 per cent. of the potential energy is actually used; the remaining 90 per cent. to 95 per cent. is absorbed in rendering the smaller fraction available in actual work. There is at present no known remedy for this. These wastes are not increasing; fortunately, through the development of gas-producers, internal-combustion engines, and steam turbines they are constantly decreasing; yet not so rapidly as to affect seriously the estimates of increase in coal consumption. We are not without hope, however, of discoveries that may yet enable man to convert potential into mechanical energy direct, avoiding this fearful waste. If that day ever come our coal supply might be considered unending.

"By 1938 about half of the original supply of iron ore will be gone, and only the lower grades of ore will remain, and all the ore now deemed workable will be used long before the end of the present century.

Iron Ore Going.

"I have for many years been impressed with the steady depletion of our iron ore supply. It is staggering to learn that our once supposed ample supply of rich ores can hardly outlast the generation

now appearing, leaving only the leaner ores for the later years of the century. It is my judgment, as a practical man accustomed to dealing with those material factors on which our National prosperity is based, that it is time to take thought for the morrow."

Mr. Carnegie spoke of the supplies of copper, zinc, lead, and silver, and of gold, saying, "Doubtless the duration of the supply will depend solely upon commercial conditions." Continuing, he said:

"It was not resources alone that gave this country its prosperity, but inventive skill and industrial enterprise applied to its resources. Individually we have been both forehanded and foreminded; Nationally we have been forehanded chiefly through the accident of discovery by John Smith and Walter Raleigh, but Nationally we are not yet foreminded. So far as our mineral wealth is concerned, the need of the day is prudent foresight, coupled with ceaseless research in order that new minerals may be discovered, new alloys produced, new compounds of common substances made available, new power-producing devices developed.

"I urge research into and mastery over nature.

"I urge on the Executives here assembled as our greatest need to-day the need for better and more practical knowledge. If our career of prosperity is to continue it must be on the basis of completer control of National sources of material and power than we have thus far exercised, a control to be gained only by research."

Speaking of the waste of natural gas, which he called the ideal fuel, I. C. White, State Geologist of West Virginia, said:

"At this very minute this unrivaled fuel is passing into the air within our domain from uncontrolled gas wells, from oil wells, from giant flambeaus, from leaking pipe lines, and the many other methods of waste at the rate of not less than one billion cubic feet daily, and probably much more."

He commended the Indiana statute for the conservation of natural gas. He also condemned the waste in the mining of coal.

John Mitchell, speaking of the waste of coal, said he thought that not more than 25 per cent. of the coal mined was lost.

"If American manufacturers and other great consumers were required to pay a higher rate for fuel, it would enable mining companies to produce and prepare for market countless millions of tons of coal which under present conditions are left in the ground, lost to the present and to future generations," he said.

He said that foreign consumers pay from one and one-half to two and one-half times as much for fuel as consumers here do."

May 14, 1908

ROOSEVELT DECRIES WASTE OF RESOURCES

Action Imperative, He Says in Forwarding Report of Conservation Commission.

WANTS BUREAU OF MINES

Work of His Administration, He Declares, Has Been Designed to Enthrone Justice.

Special to The New York Times.

WASHINGTON, Jan. 22.—President Roosevelt sent to Congress to-day a special message transmitting the report of the National Conservation Commission. In it he urged action by Congress for the preservation and development of the resources of the country. He also included a defense of his Administration.

Mr. Roosevelt says the report is " one of the most fundamentally important documents ever laid before the American people," and that " it contains the first inventory of its natural resources ever made by any nation." He goes on:

" The facts set forth in this report constitute an imperative call to action. The situation they disclose demands that we, neglecting for a time, if need be, smaller and less vital questions, shall concentrate an effective part of our attention upon the great material foundations of National existence, progress, and prosperity. * * * As it stands, it is an irrefutable proof that the conservation of our resources is the fundamental question before this Nation and that our first and greatest task is to set our house in order and begin to live within our means.

" The first of all considerations is the permanent welfare of our people; and true moral welfare, the highest form of welfare, cannot permanently exist save on a firm and lasting foundation of material well-being. In this respect our situation is far from satisfactory. After every possible allowance has been made, and when every hopeful indication has been given its full weight, the facts still give reason for grave concern.

" It would be unworthy of our history and our intelligence and disastrous to our future to shut our eyes to these facts or attempt to laugh them out of court. The people should and will rightly demand that the great fundamental questions shall be given attention by their representatives. I do not advise hasty or ill-considered action on disputed points, but I do urge, where the facts are known, where the public interest is clear, that neither indifference and inertia, nor adverse private interests, shall be allowed to stand in the way of the public good."

Policies of His Administration.

Mr. Roosevelt declares that " the policy of conservation is perhaps the most typical example of the general policies which this Government has made peculiarly its own during the opening years of the present century." He then enters into a discussion of various things done under his Administration.

" In whatever it has accomplished, or failed to accomplish," he says, " the Administration which is just drawing to a close has at least seen clearly the fundamental need of freedom of opportunity for every citizen. We have realized that the right of every man to live his own life, provide for his family, and endeavor, according to his abilities, to secure for himself and for them a fair share of the good things of existence, should be subject to one limitation and to no other. The freedom of the individual should be limited only by the present and future rights, interests, and needs of the other individuals who make up the community.

" The man who serves the community greatly should be greatly rewarded by the community; as there is great inequality of service so there must be great inequality of reward; but no man and no set of men should be allowed to play the game of competition with loaded dice. All this is simply good common sense.

" This Administration has achieved some things; it has sought, but has not been able, to achieve others; it has doubtless made mistakes; but all it has done or attempted has been in the single, consistent effort to secure and enlarge the rights and opportunities of the men and women of the United States. We are trying to conserve what is good in our social system, and we are striving toward this end when we endeavor to do away with what is bad. * * * All the acts taken by the Government during the last seven years, and all the policies now being pursued by the Government, fit in as parts of a consistent whole.

" We are building the Panama Canal; and this means that we are engaged in the giant engineering feat of all time. We are striving to add in all ways to the habitability and beauty of our country. We are striving to hold in the public hands the remaining supply of unappropriated coal, for the protection and benefit of all the people. We have taken the first steps toward the conservation of our natural resources, and the betterment of country life, and the improvement of our waterways. We stand for the right of every child to a childhood free from grinding toil, and to an education; for the civic responsibility and decency of every citizen; for prudent foresight in public matters, and for fair play in every relation of our National and economic life.

" In international matters we apply a system of diplomacy which puts the obligations of international morality on a level with those that govern the actions of an honest gentleman in dealing with his fellow-men. Within our own border we stand for truth and honesty in public and in private life; and we war sternly against wrongdoers of every grade. All these efforts are integral parts of the same attempt, the attempt to enthrone justice and righteousness, to secure freedom of opportunity to all of our citizens, now and hereafter, and to set the ultimate interest of all of us above the temporary interest of any individual, class, or group."

Urges Action by Congress.

In a brief review of the report of the commission, Mr. Roosevelt says:

" I urge that the broad plan for the development of our waterways, recommended by the Inland Waterways Commission, be put in effect without delay. It provides for a comprehensive system of waterway improvement extending to all the uses of the waters and benefits to be derived from their control, including navigation, the development of power, the extension of irrigation, the drainage of swamp and overflow lands, the prevention of soil wash, and the purification of streams for water supply. The cost of the whole work should be met by direct appropriation if possible, but if necessary by the issue of bonds in small denominations.

" I urge that provision be made for both protection and more rapid development of the National forests. Otherwise, either the increasing use of these forests by the people must be checked or their protection against fire must be dangerously weakened. The time has fully arrived for recognizing in the law the responsibility to the community, the State, and the Nation which rests upon the private owners of private lands. The ownership of forest land is a public trust. The man who would so handle his forest as to cause erosion and to injure stream flow must be not only educated, but he must be controlled.

" The remaining public lands should be classified and the arable lands disposed of to home makers. The use of the public grazing lands should be regulated in such ways as to improve and conserve their value. Rights to the surface of the public [land] should be separated from rights to [minerals] beneath [... se]parate phosphate rights still remaining with the Government should be withdrawn from entry and leased under conditions favorable for economic development.

" The accompanying reports show that the consumption of nearly all of our mineral products is increasing more rapidly than our population. Our mineral waste is about one-sixth of our product, or nearly $1,000,000 for each working day in the year. The loss of structural materials through fire is about another million a day. The loss of life in the mines is appalling. The larger part of these losses of life and property can be avoided.

" Our mineral resources are limited in quantity and cannot be increased or reproduced. With the rapidly increasing rate of consumption the supply will be exhausted while yet the Nation is in its infancy, unless better methods are devised or substitutes are found. Further investigation is urgently needed in order to improve methods and to develop and apply substitutes.

" It is of the utmost importance that a Bureau of Mines be established in accordance with the pending bill to reduce the loss of life in mines and the waste of mineral resources and to investigate the methods and substitutes for prolonging the duration of our mineral supplies."

In conclusion he asks for $50,000 for the expenses of the Conservation Commission.

Commission Seeks Co-operation.

The report of the commission dealt with the facts gathered by various sections of it, which were made public at the meeting of the commission in this city in December and summarized in THE NEW YORK TIMES. In conclusion the report says:

" The permanent welfare of the Nation demands that its natural resources be conserved by proper use. To this end the States and the Nation can do much by legislation and example. By far the greater part of these resources is in private hands. Private ownership of natural resources is a public trust; they should be administered in the interests of the people as a whole. The States and Nation should lead rather than follow in the conservative and efficient use of property under their immediate control. But their first duty is to gather and distribute a knowledge of our natural resources and of the means necessary to insure their use and conservation, to impress the body of the people with the great importance of the duty, and to promote the co-operation of all. No agency, State, Federal, corporate, or private, can do the work alone.

" The lack of co-operation between the States themselves, between the States and the Nation, and between the agencies of the National Government, is a potent cause of the neglect of conservation among the people. An organization through which all agencies, State, National, municipal, associate, and individual, may unite in a common effort to conserve the foundations of our prosperity is indispensable to the welfare and progress of the Nation. To that end the immediate creation of a National agency is essential. Many States and associations of citizens have taken action by the appointment of permanent conservation commissions. It remains for the Nation to do likewise, in order that the States and the Nation, associations and individuals, may join in the accomplishment of this great purpose."

August 12, 1909

PINCHOT IN DANGER OF LOSING HIS PLACE

Criticism of Reversal of Roosevelt's Policy on Water Power Surprises Taft's Friends.

HE HAS POWERFUL BACKING

Secretary Ballinger Compelled to Hear Himself Attacked at Spokane—Senator Chamberlain Defends Pinchot.

Special to The New York Times.

WASHINGTON, Aug. 11.—The belief is expressed here that Gifford Pinchot, United States Forester and Chairman of the National Conservation Commission, has imperiled his chances of long remaining in the Government service.

President Taft's friends were astonished this morning when they read the press dispatches quoting Pinchot as having made a speech yesterday at the Irrigation Congress at Spokane, attacking the Ballinger policy with regard to water power sites.

Pinchot takes the position and there is now in process of formation a gigantic water power trust in Montana and other Western States. Under Mr. Roosevelt Mr. Pinchot put many of these sites into the forest reserve, with the idea that they should be leased by the Government and not be permitted to pass to private ownership. Since he has been in office, Mr. Ballinger has taken many of these sites out of the reserve. This policy was apparently criticised by Mr. Pinchot in his speech before the Irrigation Congress, when he demanded a continuance of the Roosevelt policy. The speech is taken as a direct slap at Secretary Ballinger, who is also at the conference.

Before he was taken in as a member of the Cabinet Mr. Ballinger is understood to have looked upon the activities of the Forest Service with an unfriendly eye. When he was elevated to the control of the Interior Department, Mr. Ballinger found himself in a position to make public his dislike of the methods of the service.

It is not known exactly where President Taft stands with relation to the fight now being waged. No doubt is expressed that he will be forced into it before long, or the Forest Service and in fact the Land Office both will be demoralized. Pinchot has strong backing. His policy on water power was indorsed by the Governors of practically all the States and by prominent citizens at the White House conference called by President Roosevelt to discuss conservation of National resources.

SPOKANE, Wash., Aug. 11.—At the National Irrigation Congress this afternoon, Dr. George S. Pardee, formerly Governor of California, attacked Richard A. Ballinger, Secretary of the Interior, for his policy in dealing with public lands. Former Senator George Turner of Washington defended the Secretary. It was after Secretary Ballinger had read from a paper his ideas on land reclamation and the public domain, contending that what has been done by the Secretary of the Interior was under the law, that former Gov. Pardee took the platform to deliver a set address. He did not use his manuscript. He opened by saying that he was for Roosevelt and the Roosevelt policies. "Roosevelt was a President who did things first." he said, "and talked about them afterward. And that's the kind of men we should like to see in public office now."

Dr. Pardee told of the activities of ex-Secretary Garfield, who, under the instructions of President Roosevelt, withdrew from public entry many tracts of land under the belief that these lands should be held for the people. Now, he said, Secretary Ballinger had again put up for entry these lands, and each tract had in its boundaries a water power site.

The thing to do, said Dr. Pardee, is to withdraw the water power sites, as did Roosevelt, and hold them for the people. "When," pleaded the speaker, "are we ever going to have a chance for the common, hard-working citizen? Secretary Ballinger has said irrigation is not a proposition for a poor man. I take issue with him, and say it is particularly a poor man's proposition, and if there is any one trying to make it not so, let's find out about it."

Ex-United States Senator George Turner, replying to Pardee, said: "I think that the remarks of Gov. Pardee, following those of the Secretary of the Interior, are in bad taste. Mr. Ballinger has done in official capacity only what any man would do under the oath of office—he has obeyed the law. No man has the right to act first and read the law afterward."

Secretary Ballinger's Address.

Secretary Ballinger, in his address, said in part:

"While the government has invested more than $50,000,000 in irrigation works, many times that amount has been invested since the passage of the reclamation act by private enterprise, and it is safe to say that a large portion of these private investments have resulted from governmental example and encouragement; and let me say here that it has not been and is not the policy of the National Government in the administration of this act to hinder or interfere with the investment of private capital in the construction of irrigation works, but rather to lend it encouragement. This is particularly true in reference to irrigation under the Carey Act in the various States.

"The people of the West, who are familiar with the wonderful results of irrigation, are highly appreciative of the importance of the reclamation service, but the great difficulty which the service encounters is in finishing the projects now undertaken, as against the clamor for a diversion of the funds to new fields. In this respect the service has suffered.

"The danger which the Government is undertaking to overcome is the establishment of small irrigation projects in localities where, by such establishment, the larger opportunities are destroyed, thus preventing enormous areas of lands from over-acquiring the use of water. For lack of funds the Government is at present required to surrender possibilities in water appropriations which mean an enormous loss in future development of irrigation works."

January 23, 1909

TAFT TAKES STAND WITH BALLINGER

Dismisses Charges Against Interior Department Officials—Maker of Them to Go.

SECRETARY'S ACTS UPHELD

His Conservation Policy Not Out of Harmony with President's—Letter Given Out at Albany.

ALBANY, Sept. 15.—President Taft took his stand to-day on the controversy that has raged about Richard A. Ballinger, Secretary of the Interior, because of his policy on conservation of natural resources. The President stands with Mr. Ballinger. While here this evening he gave out a letter to Mr. Ballinger, in which he exonerates the Secretary and other officials in the matter of charges affecting Alaskan coal land, grants the

Secretary's request for the dismissal of L. R. Glavis, who made them, and in general approves Mr. Ballinger's policy.

Mr. Taft's letter, which is dated Beverly, Mass., Sept. 13, began:

"On the 18th day of August last L. R. Glavis, Chief of Field Division of the General Land Office, with headquarters at Seattle, Washington, called upon me here and submitted a statement or report relating to the conduct of the Interior Department and particularly to the action of yourself, Assistant Secretary Pierce, Commissioner of the General Land Office Dennett, and Chief of Field Service Schwartz in reference to the so-called Cunningham group of coal land claims in Alaska.

"Mr. Glavis's report does not formulate his charges against you and the others, but by insinuation and innuendo as well as by direct averment he does charge that each one of you while a public officer has taken steps to aid the Cunningham claimants to secure patents based on claims that you know or had reason to believe to be fraudulent and unlawful.

Glavis's Shreds of Suspicions.

"I have examined the whole record most carefully and have reached a very definite conclusion. It is impossible for me in announcing this conclusion to accompany it with a review of the charges and the evidence on both sides. It is sufficient to say that the case attempted to be made by Mr. Glavis embraces only

shreds of suspicions without any substantial evidence to sustain his attack.

"The whole record shows that Mr. Glavis was honestly convinced of the illegal character of the claims in the Cunningham group and that he was seeking evidence to defeat the claims. But it also shows that there was delay on his part in preparing the evidence with which to bring this with other claims to hearing, and that justice to the claimants required more speedy action than the department, through Mr. Glavis, seems to have taken.

"Mr. Glavis seeks by quoting from a single telegram in the department to show that at one time the department wished to delay him in his investigations of the Alaska claims, and at another time unduly to hurry him, and he attempts to prove these two circumstances as well by citing telegrams and correspondence without disclosing other circumstances and correspondence which he knew or had under his control, and which do show an entirely proper reason for the action which in each case was directed to be taken.

"In other words, the reading of the whole record leaves no doubt that in his zeal to convict yourself, Acting Secretary Pierce, Commissioner Dennett, and Mr. Schwartz, he did not give me the benefit of information which he had that would have thrown light on the transactions, showing them to be consistent with an impartial attitude on your part toward the claims in question.

Ballinger Got Only $250 Fee.

"Mr. Taft points out that Mr. Ballinger's employment in the Alaska cases involved a fee of only $250. He goes on:

"Your only action which could in any manner affect the Cunningham group of claims was an order made by you soon after assuming office, that the 30,000 claims pending and undisposed of in the Land Office should be pressed to final hearing and disposition as rapidly as possible consistent with justice, and these included the 931 Alaska coal claims, of which the Cunningham group numbered nineteen. As such expedition was essential both in the public interest and in that of the claimants, it could hardly be said to be action taken in the Cunningham claims.

"The record overwhelmingly establishes that, expressly because of your previous relation as counsel to one of the claimants, from this time you entered upon your duties of, the office of Secretary of the Interior until the present day, you have studiously declined to have any connection whatever with the Cunningham claims, or to exercise any control over the course of the Department in respect to those claims; that you have said so in written and verbal communications to your subordinates and to the claimants themselves. Moreover, in May last you came to me and made a similar statement to me of your course and intention in respect to those claims.

"The statement made by Mr. Glavis that while you did thus formally withdraw from any official connection with the Cunningham claims, you nevertheless continued to exercise your influence in regard to them, is not sustained by any evidence in the records produced.

Left Glavis in Charge of Cases.

"The truth is that had you or Commissioner Dennett or Chief of Field Service Schwartz, during the years of the pendency of these claims, been desirous, through dishonest motives and without regard to law and the interests of the public, of bringing them to patent, the opportunities for you to have done so were many, and the circumstance that speaks, not more conclusively than many others, but still most emphatically, against the accusatory statements of Mr. Glavis is the fact that though his conviction that the claims were fraudulent or illegal was well known in the department, he was allowed, during all the years of the pendency of these claims, to remain in charge of them as an agent of the department, when it would have been entirely easy for either you or Dennett or Schwartz to remove him to Portland or elsewhere and thus take the claims out of his jurisdiction.

"In your answer you request authority to discharge Mr. Glavis from the service of the United States for disloyalty to his superior officers in making a false charge against them. When a subordinate in a Government bureau or department has trustworthy evidence upon which to believe that his chief is dishonest and is defrauding the Government it is, of course, his duty to submit that evidence to higher authority than his chief.

"But when he makes a charge against his chief founded upon mere suspicions, and in his statement he fails to give his chief the benefit of circumstances within his knowledge that would explain his chief's action as on proper grounds, he makes it impossible for him to continue in the service of the Government, and his immediate separation therefrom becomes a necessity.

"You are therefore authorized to dismiss L. R. Glavis from the service of the Government for filing a disingenuous statement, unjustly impeaching the official integrity of his superior officers.

"I cannot close this letter without referring to certain other matters connected with your conduct of the Interior Department which have been unfairly used in the public press to support a general charge that you are out of sympathy with the declared policy of this Administration, following that of President Roosevelt, in favor of the conservation of National resources, especially in connection with coal lands, with waterpower sites, and with the system of reclamation of arid lands, which are all within the jurisdiction of the Interior Department.

Replies to Ex-Gov. Pardee.

"In the first place, it was charged on the floor of the Irrigation Convention at Spokane by former Gov. Pardee of California that you had restored to the public domain for settlement certain lands which had been withdrawn by the last Administration for the purpose of conserving waterpower sites, and that after complaint made thereof you had subsequently withdrawn some of the lands again from settlement; but that meantime, between the one act and the other, an opportunity had been given to the so-called 'Waterpower Trust' to file entries and obtain vested rights in valuable waterpower sites in the State of Montana.

A Cruel Injustice Done.

"When the facts are examined in this regard, it will be found that the persons responsible for the circulation of these charges have done you cruel injustice. The fact was that in January, 1909, in the last Administration, executive orders were made withdrawing from public settlement 1,500,000 acres at the instance of the Reclamation Service, for conservation of waterpower sites. Soon after you became Secretary of the Interior, you brought this order to my attention, and said that it included a great deal of land that had no waterpower sites on it, running back many miles from the rivers, and that it included much land which ought to be opened to public settlement; that you had applied to the Reclamation Bureau to know whether it was desired for reclamation purposes, and what their recommendation was in the premises, and that they recommended that it be returned to the public domain."

The order revoking the withdrawal of the 1,500,000 acres was made in April. Sufficient information was procured from the Geological Survey to permit an order withdrawing the land upon which were waterpower sites in May, and this withdrawal covered about 300,000 acres instead of 1,500,000. The form of the new order of withdrawal was such that it set aside all filings and entries of any kind which had been made prior to its going into effect; and, as a matter of fact, not one single filing has been attempted on any of the waterpower sites since the original order of withdrawal in January, 1909.

"It further appears from a report of the Director of the Geological Survey that the order of withdrawal of January, 1909, was hastily made by townships and by reference to inadequate maps, that it included large areas not within miles of any river or stream, and that it failed to include many valuable water power sites in the immediate vicinity.

"From the same reliable source it is learned that under the withdrawals made by your department from time to time, beginning in May last, there are now withheld from settlement awaiting the action of Congress, 50 per cent. more water power sites than under previous withdrawals, and that this has been effected by a withdrawal from settlement of only one-fifth of the amount of land.

Forestry Bureau Was Wrong.

"Another instance in your conduct of the department which has been mentioned as indicative of your purpose to block the general plan of conservation of natural resources is your refusal to carry out a contract made in the last Administration between the Secretary of the Interior and the Secretary of Agriculture, by which the Interior Department delegated to the Forestry Bureau of the Agricultural Department the power and duty to conserve the forests on the Indian reservations, and to expand under the control of the Forestry Bureau the money appropriated by Congress to be expended by the Indian Bureau for such conservation of Indian forests.

"Your declination to carry out the contract was made necessary by a ruling of the Controller, whose ruling is final and without appeal even to the President, that such an arrangement is a delegation of responsibility and authority for the expenditure of money which the appropriation by Congress for the Indian Bureau did not authorize.

"In my judgment he is the best friend of the policy of conservation of natural resources who insists that every step taken in that direction should be within the law and buttressed by legal authority. Insistence on this is not inconsistent with a whole-hearted and bona fide interest and enthusiasm in favor of the conservation policy. From my conferences with you and from everything I know in respect to the conduct of your department I am able to say that you are fully in sympathy with the attitude of this Administration in favor of the conservation of natural resources."

September 16, 1909

ELIOT IN CONSERVATION.

President of New Association Whose First Member Is Taft.

The National Conservation Association, with general offices here and in Washington, it is announced, will carry on a vigorous campaign under the leadership of Dr. Charles W. Eliot, President emeritus of Harvard. The general object of the association is to secure practical application through legislative and administrative measures by the States and by the Federal Government of the conservation principles adopted by the Governors of the United States at their conference with President Roosevelt in May, 1908.

President Taft has given the organization his hearty approval and is its first member. General offices were opened in the Fifth Avenue Building yesterday. Mr. Eliot is personally directing the work of the association, and the membership is open to every American citizen.

The officers of the association are: President—Dr. Charles W. Eliot of Cambridge, Mass.; Vice President—Walter L. Fisher of Chicago, and Secretary—Thomas R. Shipp of Indianapolis, Ind. Mr. Shipp was Secretary of the White House conference and Secretary of the National Conservation Commission. The Chairman of the Executive Committee is John F. Bass of Chicago, and the Executive Director is Royal L. Melendy, also of Chicago. A letter from President Taft was made public yesterday. It said:

I am glad to hear that the National Conservation Association has been formed under such a distinguished and capable leader and with a membership open to every American citizen.

Our people cannot do a more useful thing for themselves and for posterity than to give personal consideration to the great issues involved in what we have come to call the conservation movement. It is of the greatest importance that this movement should proceed both wisely and effectively, and the National Conservation Association should be a valuable instrumentality for accomplishing this result.

I shall be glad to have you enroll my name in its membership.

The association advocated legislation safeguarding the grant of water power rights.

The association declares that it desires to further all legislation designed to diminish sickness, prevent accidents and premature death, and increase the comfort and joy of American life.

October 28, 1909

ROOSEVELT AT ISSUE WITH TAFT IN SPEECH

Special to The New York Times.

ST. PAUL, Sept. 6.—Col. Roosevelt, in his address to the Conservation Congress to-day, took sharp issue with President Taft in the views he expressed on the question of Federal or State control. Both in manner and in language was the the contrast marked by all who had heard Mr. Taft yesterday. And it was the common judgment that there was a great difference in the reception of the two utterances by the same people.

The St. Paul and Minneapolis newspapers agree that the crowds yesterday were not overenthusiastic for the President, and that was the comment heard on all sides in the streets and hotels and at the auditorium this morning. But there was no question about the reception to Col. Roosevelt. It began with the first mention of his name by Gov. Stubbs of Kansas, and continued until the Colonel was back in his rooms at the hotel.

When he appeared in the vast hall the people leaped to their feet and cheered without cessation for two minutes and thirty-five seconds. They waved flags and banners, swung their hats, and even their bonnets, fluttered handkerchiefs, and stamped their feet—anything to help make a demonstration of satisfaction.

Crowd Hangs on His Words.

When he began to speak they hung on his words and cheered literally three-fourths of his sentences. By actual count there was applause at the end of seven of the first sentences he uttered, although he had not yet done more than open his address. As he went on it got so that frequently a sentence was interrupted by cheers. The great crowd was intent on showing that it liked the Colonel and was with him in what he was saying. And it was a red-hot conservation speech. There was no uncertainty about his position on the matters that have been in controversy among conservationists and that have been threatening to precipitate an angry fight in the meetings of the Conservation Congress.

Col. Roosevelt had prepared a speech which dealt with all the phases of conservation, and in which he took private ground along the lines supported by Gifford Pinchot and James R. Garfield. On the question of State or Federal control of water-power sites he was emphatically for National action, but, in view of the things that have been happening at the convention and the fight that was threatened on the Nationalists by the State righters, Col. Roosevelt made a number of digressions from the prepared address in which he went even further than he had originally intended to go in his declaration for Federal control.

Whether or not Col. Roosevelt intended this emphatic taking of advanced ground to be accepted by those who heard it as in contrast to the attitude of President Taft yesterday did not appear, but it is certain that the crowd immediately made that application of the Colonel's language and that it was tremendously popular.

The contrast was made all the more emphatic by the fact that in the very beginning of his speech Col. Roosevelt referred to the speech of the President yesterday and took occasion to praise some of the things that Mr. Taft had said. But it was faint praise at best, and the indorsement was limited to three things.

Praise of Taft Applauded.

"Much that I have to say on the subject," said the Colonel, after he had expressed his pleasure at being here, "will be but repetition of what was so admirably said by the President of the United States on this platform yesterday."

That was applauded, and then the Colonel went on with his more specific indorsement of the Administration, saying:

"In particular all true friends of conservation should be in hearty accord with the policy which the President laid down in connection with coal, oil, and phosphate lands."

That was the sum total of praise of the Administration in the speech. There was no word of criticism, but there was also no sparing of words in the setting forth of views and policies by Col. Roosevelt, which either are in open conflict with those expressed yesterday by President Taft, or concern points on which the President had remained silent. The fact that the President declared in terms several times that he was following the Roosevelt policies of conservation made all the more interesting the situation this morning, when Mr. Roosevelt himself was emphasizing the contrast between them.

Each side of the State and Federal control controversy has been accusing the other of attempting to pack the convention. If there has been a contest of that sort, the Federal control men certainly won it, for there can be no question of the tremendous popularity with the crowd of what the Colonel had to say this morning on that subject. He gave it to them straight from the shoulder, and did not hesitate to charge the big interests in unmistakable terms with seeking now to foster the State rights idea in the belief that they be able to control or prevent State action more easily than Federal action.

Col. Roosevelt stood for Federal control also in the matter of drainage of swamp lands, again disagreeing with President Taft.

"The only wise course is to have the Federal Government act," he said. "The land should be deeded back to the Federal Government by the States and then it should take whatever action is necessary. Much of this work should be done by the Federal Government, anyway."

Talks Right at Governors.

It was when he talked of the forests and water-power sites, however, that the Colonel came out most strongly for action by the Nation instead of by the States. Seated on the platform behind him were some of the Governors who had declared emphatically yesterday for State rights. The Colonel turned and talked directly to them, and laid down his proposition with all the vigor and emphasis he could muster, while the great crowd yelled and howled and cheered at every pause, often interrupting a sentence to voice its approval.

From his discussion of the forest question the Colonel turned to talk of the different commissions he had appointed while President, and which had been put out of business by the refusal of Congress to make appropriations for them. There he got in a blow at his old-time antagonist, Congressman Tawney of Minnesota, Chairman of the House Committee on Appropriations. It was Tawney who led the fight against these commissions, and the Colonel has never forgiven him.

President Taft has got along very well with Tawney, and it was to help the Congresman at home that the famous Winona speech was delivered last year. Col. Roosevelt did not name Tawney, but it was not necessary for him to do so in order to have the crowd know whom he meant. The application was instantaneous, and evoked considerable cheering from the progressive element in the audience.

Then the Colonel got back to the question of water-power sites, and began to hit out from the shoulder again on that subject.

"It isn't merely a question of State against Nation," he declared to an accompaniment of enthusiastic cheers. "It is really a question of special corporate interests against the interests of the people."

Crowd Cheers Frantically.

That fairly brought down the house. There was frantic cheering for half a minute—longer than any other cheering of the day except that upon the Colonel's entrance.

"If it were not for the special corporate interests," the Colonel went on, when quiet was restored, "you would never have heard of the question of State as against National control. Who can best regulate the special interests for the people's good?"

Instantly there was a reply from the gallery: "You can do it yourself," and then the crowd howled again.

"Most of these big corporations are financed and owned in the Atlantic States," went on the Colonel, "as it is a comic fact that the most zealous upholders of State rights in the present controversy live in other States."

Again he declared that he was for National control, because "the Federal Government is stronger; because it is better able to exact justice from the corporate interests, and also because it is less apt in some gust of popular passion to do injustice to the corporations."

September 7, 1910

FEDERAL CONTROL URGED IN PLATFORM

State Rights Supporters Beaten on Resolutions Adopted by Conservation Congress.

FIGHT OVER TAFT MENTION

Committee Refuses, 10 to 8, to Include His Name with Roosevelt's, and Both Are Finally Left Out.

ST. PAUL, Sept. 8.—The Conservation Congress adopted to-night a platform strongly favoring National control of natural resources. From this document the names of President Taft and Theodore Roosevelt were omitted by action of the Committee on Resolutions.

As first drafted the platform contained the name of Col. Roosevelt, and an attempt was made at the meeting of the committee this afternoon to include that of President Taft also. This was defeated after a bitter fight by a vote of 10 to 8. The attempt was then made to correct what some of the members declared was an injustice to Mr. Taft, and a motion to eliminate the name of Col. Roosevelt was also carried.

Through an inadvertence the copy of the resolutions given to the newspapers retained the name of Col. Roosevelt, and it was not until after the adjournment of the convention late to-night that the error was discovered. For a time it was supposed that Mr. Taft had thus been snubbed by the convention.

A stiff fight was also put up by the State rights advocates, but they were outnumbered. They threatened to carry the fight to the floor of the convention, but under vigorous use of the gavel this was prevented.

The principal planks in the platform are as follows:

Heartily accepting the spirit and intent of the Constitution and adhering to the principles laid down by Washington and Lincoln, we declare our conviction that we live under a Government of the people, by the people, for the people; and we repudiate any and all special or local interests or platforms or policies in conflict with the inherent rights and sovereign will of our people.

Recognizing the natural resources of the country as the prime basis of property and opportunity, we hold the rights of the people in these resources to be natural and inherent, and justly inalienable and indefeasible, and we insist that the resources should and shall be developed, used, and conserved in ways consistent both with current welfare and with the perpetuity of our people.

Recognizing the waters of the country as a great National resource, we approve and indorse the opinion that all waters belong to all the people and hold that they should be used in the interest of the people.

Realizing that all parts of each drainage basin are related and interdependent, we hold that each stream should be regarded and treated as a unit from its source to its mouth; and, since the waters are essentially mobile and transitory and are generally inter-State, we hold that in all cases of divided or doubtful jurisdiction the waters should be administered by co-operation between State and Federal agencies.

Recognizing the vast economic benefit to the people of water power derived largely from inter-State sources and streams no less than from navigable rivers, we favor Federal control of water power development; we deny the right of States or Federal Government to continue alienating or conveying water by granting franchises for the use thereof in perpetuity, and we demand that the use of water rights be permitted only for limited periods with just compensation in the interests of the people.

We hold that phosphate deposits underlying the public lands should be safeguarded for the American people by appropriate legislation, and we recommend the early opening of the Alaskan and other coalfields belonging to the people of the United States for commercial purposes on a system of leasing, National ownership to be retained.

We favor co-operative action on the part of States and the Federal Government looking to the preservation and better utilization of the soils by approved scientific methods.

We approve of the continuance of the control of the National forest by the Federal Government, and approve the policy of restoring to settlement such public lands as are more valuable for agriculture.

We indorse the proposition for the preservation by the Federal Government of the Southern Appalachian and White Mountain forests.

We realize that the fullest enjoyment of our natural resources depends upon life and development of the people physically, intellectually, and morally, and in order to promote this purpose we recommend that the training and protection of the people and whatever pertains to the health and general efficiency be encouraged by methods and legislation suitable to this end. Child labor should be prevented and child life protected and developed.

Realizing the waste of life in transportation and mining operations, we recommend legislation increasing the use of proper safeguards for the conservation of life. And we also recommend that, in order to make better provision for procuring the health of the Nation, a Department of Public Health be established by the National Government.

Gifford Pinchot and J. B. White, Chairman of the Executive Committee, declined to contest for the Presidency of the Congress. Henry W. Wallace of Des Moines was chosen President and D. Austin Latchaw Treasurer. Thomas Shipp was re-elected Secretary.

September 9, 1910

TARDY STEWARDSHIP.

"Sane, constructive stewardship" at Washington during the two Administrations of Mr. ROOSEVELT ought, certainly, to have assured the speedy development on public lands of electrical energy in the public interest. What did President ROOSEVELT do toward this with the efficient Federal instruments which, he says, are far superior to State instruments? He did nothing.

The transmission of electrical energy from waterfalls has long been a commercial possibility. Congress passed an act approved Feb. 15, 1901, permitting Federal sites to be used for this purpose under a license that might at any time be revoked by the Secretary of the Interior. No privilege was given, such as Mr. ROOSEVELT and Mr. PINCHOT now urge. The rights of companies on these sites were not protected. The developers of electrical energy did not know when the fee simple title to the land might pass to others. The Government exacted no revenue for its use, it furnished the holder no guarantee that his expensive plant should be built upon a site where some other applicant might not at any time acquire a title of property for an alien purpose.

That is the situation to-day, after two terms of the stewardship of ROOSEVELT and PINCHOT and a year under President TAFT. But the things that the Federal Government has failed to do, things which Forester PINCHOT and President ROOSEVELT tried tardily to persuade it to do, have been for years enforced under the laws of the State of Idaho. The Constitution and statutes of Idaho, while fostering their development, safeguard its water powers against monopoly. They require that a license be obtained from the State, and power sites gained under the license may be condemned if withheld from use by the private licensee, while the charge for the electrical energy he sells is fixed by the State. But the Federal Desert Land act of 1877 still permits the Water Power Trust to obtain entry upon Federal lands embracing water-power sites and hold them from use. The recent emergency measures withholding these lands from entry until Congress shall consider an amendment of the act still keep them from use. The public gains no advantage. Neither State nor Nation derives a revenue. Development is forbidden, awaiting tardy Congressional action.

Which is the better steward, the State or the Federal Government and the Federal Executive?

September 18, 1910

BALLINGER RESIGNS WITH TAFT'S PRAISE

The Object of an Unscrupulous Conspiracy, Declares the President, Aimed at Himself, Too.

W. L. FISHER HIS SUCCESSOR

He Is a Chicago Lawyer, a Friend of Pinchot, and a Leader In Conservation Movement.

Special to The New York Times.

WASHINGTON, March 7. — The long fight between the adherents of Former Chief Forester Gifford Pinchot and Secretary of the Interior Richard Achilles Ballinger ended to-day with the resignation of the Secretary and the acceptance of his resignation by the President.

With the news of Mr. Ballinger's resignation at the White House was given out the announcement of his successor, Walter Lowrie Fisher of Chicago, a lawyer and member of several Governmental commissions, who is considered acceptable to the conservationists.

While the Administration has realized for months that Mr. Ballinger's resignation was only a matter of time, his first formal written relinquishment of his office dates back six weeks. The publication of the news to-day is believed to come from a desire to distract attention from the Mexican situation and the mobilization of troops on the Texas border. On Jan. 19, basing his request 'on the condition of his health, he asked to be allowed to retire. The President at once replied, asking Mr. Ballinger to continue to give him his assistance in the Cabinet until the rush of the closing days of Congress was ended.

Under date of yesterday Mr. Ballinger formally renewed his request, and with every expression, of regret the President accepted it to-day. Mr. Taft, in the correspondence made public this afternoon, sweepingly assails Mr. Ballinger's enemies for entering into "one of the most unscrupulous conspiracies for the defamation of character that history can show," which, the President asserts, was also directed at him. He refers to the personal inconvenience and financial loss suffered by Mr. Ballinger in accepting, reluctantly, a Cabinet position, and declares that it is only because he does not feel justified in asking further sacrifices of his officer that he now consents to his resignation.

Avoids Democratic Attack.

As was suggested in these dispatches last night, Mr. Ballinger in getting out of the Department of the Interior now, does so just in time to avoid a serious move on the part of the Democratic House, if not on the part also of the Senate, where in the next session a majority might develop against him by a combination of insurgents and Democrats.

In the House beyond question, feeling is running and many of the Democrats have lately expressed themselves as favoring impeachment proceedings. It is doubtful that the House would have gone this length, but it is almost certain that they would have taken up the reports—adverse and favorable—of the joint committee which investigated the Ballinger-Pinchot controversy and pushed their consideration to a conclusion little to Mr. Ballinger's taste. Even with the

retiring Republican House the leaders never dared bring up the reports for final disposition.

The resignation of Mr. Ballinger to-day was preceded by the resignation of Oscar Lawler, Assistant Attorney General for Mr. Ballinger's department. Mr. Lawler's resignation was considered the first step toward a cleaning out of the department before the Democrats get to work on it, and it is expected that other resignations will be speedily made.

It has been talked of here for some time that Mr. Ballinger was put to heavy expense in defending himself at the recent inquiry. His fees to a single attorney—whom he did not select and who did not make the most favorable impression on the committee—went well into five figures, and incidental expenses were numerous.

Here is President Taft's letter accepting Mr. Ballinger's resignation:

The President's Letter.

Dear Mr. Secretary: I accept your resignation with great reluctance. I have had the fullest opportunity to know you, to know your standards of service to the Government and the public, to know your motives, to know how you have administered your office, and to know the motives of those who have assailed you. I do not hesitate to say that you have been the object of one of the most unscrupulous conspiracies for the defamation of character that history can show.

I have deemed it my duty not only to the Government, but to society in general, to fight out this battle to the end, confident that in the end your fellow-citizens would see that the impressions of you as a man and as the administrator of a high public office were false, and were the result of a malicious and unprincipled plan for the use of the press to misrepresent you and your actions, and to torture every circumstance, however free from detrimental significance, into proof of corrupt motive.

With the hypocritical pretense that they did not accuse you of corruption in order to avoid the necessity that even the worst criminal is entitled to, to wit, that of a definitely formulated charge of some misconduct, they showered you with suspicion, and by the most pettifogging methods, exploited to the public matters which had no relevancy to an issue of either corruption or efficiency in your office, but which paraded before an hysterical body of headline readers, served to blacken your character and to obscure the proper issue of your honesty and effectiveness as a public servant.

The result has been a cruel tragedy. You and yours have lost health and have been burdened financially. The conspirators, who have not hesitated in their pursuit of you to resort to the meanest of methods, including the corruption of your most confidential assistant, plume themselves like the Pharisees of old, as the only pure members of society, actuated by the spirit of self-sacrifice for their fellowmen.

Every fibre of my nature rebels against such hypocrisy and nerves me to fight such a combination and such methods to the bitter end, lest success in this instance may form a demoralizing precedent. But personal consideration for you and yours makes me feel that I have no right to ask you for further sacrifice. Of course, it has been made evident that I was and am the ultimate object of the attack; and to insist, against your will, on your remaining in office with the prospect of further efforts against you is selfishly to impose on you more of a burden than I ought to impose.

As I say farewell to you, let me renew my expressions of affection and sincerest respect for you, and of my profound gratitude for your hard work, your unvarying loyalty, and your effective public service. I hope and pray that success may attend you in your profession and that real happiness will come to you and yours when you return to that community where you live, and whose members know your worth as a man and a citizen, and who will receive you again with open arms. Sincerely yours,

WILLIAM H. TAFT.

The previous correspondence between Mr. Taft and the Secretary of the Interior begins with Mr. Ballinger's original resignation, which has been in the hands of the President since Jan. 19, and bears that date.

It follows:

Ballinger's First Resignation.

My Dear Mr. President: I have thought over the talk we had last Saturday. I am exceedingly grateful for your kind expression of appreciation of my work, and they certainly compensate me in the largest measure for what I have suffered.

Your attitude throughout has been a great source of comfort to me. In view of the fact that the condition of my health is such that I must ask to be relieved from office. I therefore again tender my resignation and ask you to accept it.

I should be untrue to all of my impulses if I did not seize the occasion to say that I have, at all times, striven conscientiously to meet the obligations imposed upon me, and to serve you, the Administration, and the country to the best of my ability. I am deeply sensible of the unfailing confidence which you have reposed in me under circumstances which have necessarily been trying to you, and the support and respect which you have never ceased to accord me I shall always remember.

I am anxious to retire as soon as I can properly do so, yet am unwilling to embarrass you, and therefore respectfully ask you to indicate, in regard to the matter, when it will best suit your convenience.

With renewed assurance of my highest regard, I am, faithfully yours,

R. A. BALLINGER.

To the President, Jan. 19, 1911.

To this, under date of Jan. 23, the President replied:

Dear Mr. Secretary: For reasons which have deeply impressed themselves in my heart and mind I would never consent to consider your resignation on any ground that was based on the good of the service or of helping me personally or politically, for no such ground is tenable by me. Only on the score of your health or personal convenience or to prevent further pecuniary sacrifice on your part will I consider the possibility of accepting your resignation.

But not even on the latter grounds can I consider it until after Congress adjourns, until after all unjust attacks are ended, until after I have had the benefit of your valuable and necessary aid during the remainder of a crowded session, and until we have reached the calm period which I hope will follow the present hurry and pressure and necessity for constant action and watching incident to the close of a short session. Then I'll take it up and answer you at length.

Sincerely yours,

WILLIAM H. TAFT.

Insists on Resigning.

It was not until yesterday that Secretary Ballinger renewed his request for the acceptance of his resignation. He did so in this letter, dated March 6:

My Dear Mr. President: As you fully appreciate, I entered the Government service under protest, and at great personal sacrifice.

While occupying the office of Secretary of the Interior my most earnest and conscientious attention has been given to the interests of the Government, and I feel that in constructive work and the advancement of the public service, under existing difficulties, I have nothing to regret in official administration except that my health and financial interests have greatly suffered to the extent that I can no longer sustain the burden.

Your constant support has always been a source of consolation during all the vicissitudes of my term of service, and I deeply appreciate the unfailing confidence you have reposed in me. I must, however, renew my appeal to be relieved as set forth in my letter of Jan. 19 last, and respectfully ask you to designate the time when my resignation shall take effect, which I hope may be immediately.

I have the honor to remain, faithfully yours,

R. A. BALLINGER.

It was in reply to this letter that President Taft wrote accepting Mr. Ballinger's resignation.

Mr. Ballinger bade good-bye to his fellow-Cabinet members at the White House to-day. As he was leaving he said: "I feel better than I have felt for two years. I shall leave for Seattle just as soon as I can possibly do so."

In a later statement he declared that it was his purpose "to prosecute the arch-conspirators who have been following me with the assassin's knife. The country shall know fully the injustice of the attacks upon me."

The appointment of Mr. Fisher gives to Chicago three Cabinet offices. The Secretary of War, Mr. Dickinson, though nominally a Tennesseean, is in realty from Chicago, as is Secretary of the Treasury MacVeagh. It is believed that Mr. Fisher's appointment will be entirely acceptable to Gifford Pinchot and the friends of conservation, who so bitterly have fought Mr. Ballinger. Mr. Fisher was the first President of the Conservation League, has played an important part in the traction fight in Chicago, and is a member of the Railroad Securities Commission.

March 8, 1911

FARMS WON FROM SWAMPS

THE irrigation projects in the United States for the reclamation of arid land have been so large in conception and so magnificient in engineering features that they have overshadowed the reclamation of swamp lands, which requires only the prosaic labor of digging ditches. But work on the wet lands has been going on steadily, not only by the diggers, but the scientific preparatory work of the Geological Survey has been very extensive, and, considering the cost involved, much larger in prospective returns than is possible from arid lands.

A total area of 130,000 square miles—about 83,000,000 acres—of wet lands in different sections of the country has already been surveyed and mapped by the Geological Survey, and it is estimated that of this area fully 80,000,000 acres of the most fertile of farming land will be available for cultivation.

The comprehensive plans of the Government engineers include making the main ditches large enough to afford navigation for small vessels, thus providing the cheapest possible means of transportation of the products of the reclaimed land to market points.

The cost of doing the work is estimated at from $2 to $10 per acre—as compared with the cost of reclaiming arid lands at from $12 to $60 per acre; and the natural fertility of the swamp lands, with their enormous accumulations of humus, is immeasurably greater than that of the arid lands.

The largest extent of wet lands in the United States in those States bordering on the Gulf of Mexico, Florida having the greatest area—about 19,000,000 acres—and Louisiana following with 10,000,000 acres, Mississippi with 5,750,000 acres, and Arkansas with 5,250,000 acres. Another large swamp area lies about the Great Lakes, Michigan having 3,000,000 acres, Wisconsin, 2,500,000 acres, Minnesota, 2,750,000 acres, and North Dakota, 3,000,000 acres.

A third large area lies along the Atlantic coast of North Carolina, South Carolina, and Georgia, each of these States having upward of 3,000,000 acres of wet land. The State of Maine has an equal area of reclaimable swamp land.

While the Government has not yet taken the active part in the reclamation of wet lands that it has in the arid lands of the Northwest, much work has been done by the States and by individual enterprise. In Minnesota alone, since 1900 more than 340 miles of drainage ditches have been dug by the State, at a cost of $585,000.

Large wet areas have been reclaimed in California, Florida, Louisiana, North Carolina, South Carolina, and Wisconsin. In the Gulf States a movement has been started to secure a general swamp reclamation law similar to the irrigation law which is working so successfully in the Northwest.

It is proposed that the Government shall dig the main ditches, or channels, to be navigable for small vessels, and that the branch channels and minor ditches be done by reclamation companies which would quickly take up the work in accordance with the recommendations of the Government engineers. The relatively small cost to the National Treasury at which a very large area of most valuable agricultural land can be made available seems to assure the acceptance of this plan by Congress.

June 5, 1910

Autos Win Way Into Yosemite.
WASHINGTON, April 30.—Secretary Lane to-day rescinded an order barring automobiles from the Yosemite National Park.

"This form of transportation has come to stay," said the Secretary, "and to close the park against automobiles would be as absurd as the fight for many years made by old naval men against the adoption of steam in the navy. Before we know it, they will be dropping into the Yosemite by airship."

May 1, 1913

THE NATION'S UNDEVELOPED RESOURCES

The report of Secretary of the Interior LANE is worth anybody's reading, and should be read in full by all business men. It is easy reading, for it has little detail, and is written in a spirit of constructive optimism. He attributes the deadlock in the development of the national resources to the fact that no substitute has been found for the system of exploitation of public resources for private benefit which the nation has resolved shall end. The Secretary thinks that a beginning should be made by ceasing to regard all acres as alike and disposing of Government land with respect to its suitable use. Farm lands should be sold to farmers, so that the nation should not be defrauded by not getting homesteaders when it sold homesteads. Forests should be sold to lumbermen who would develop them, rather than withhold them for a rise in prices due to scarcity. Grass land should be sold to stock raisers, and riparian lands with a view to utility for irrigation of power, and so on.

The Secretary's suggestions regarding Alaska are much more elaborate. He proposes a commission government for the Territory. That is to say he would have Alaska delivered to a Board of Directors who should administer all its assets primarily for the development of the Territory. Congress should indicate only the broad lines of policy, and the new Board of Directors should have nothing to do with Territorial politics. The duty of the board would be restricted to dealing with the property of the nation's territory as an entirety, with a separate budget, and on a co-ordinated scheme which should relate each undertaking to every other. Thus the States would be relieved of the burden of developing Alaska, and would profit by its self-development. The Secretary knows no better way of developing railways for Alaska than the time-honored method of granting alternate sections to the builders, so than the increase of value by the construction of the railway would be saved to the ungranted sections retained in Government ownership. The Secretary would allow private railways beside the public railways, subject to the Government's right of purchase. Rates he would fix rather with regard to the development of the country than regard to operation profits, and he would not apply the principle of the commodities clause so highly esteemed by many railway reformers among ourselves.

The Secretary would develop the coal lands in two ways. The mineral which is not suitable for transportation he would convert into power and transmit it by electricity, and he would grant to operators leases long enough to repay them for their costs of development, including building of railways in suitable cases. Time is of the essence of the contract in the case of Alaska, and what is done is doubled in value by expedition.

The Secretary is of another opinion regarding the development of the coal lands in the States. No less than 56,000,000 acres of coal lands are included in 66,000,000 acres of lands withdrawn from private development under the defective laws. The Secretary thinks that it is not for the public interest that these lands should be opened rapidly or ruthlessly. Neither should the lands be held idle from fears of exhaustion, since science now knows how to convert rain and snow into heat and light in competition with coal, or in substitution for it. A royalty basis, with precaution against waste in mining, is his suggestion, with a fair compromise between the needs of the present and the rights of posterity. In the development of oil lands the Secretary fears lest the present methods are defective in "not sufficiently rewarding the pioneer," and he proposes rewards for prospectors upon lands now withheld from development, with a royalty agreed upon beforehand for all discoveries.

The reclamation service has not given satisfaction, through excess of optimism on the part of both the Government and the farmers who bought irrigated lands. The ten years given for payment should be doubled, to allow them to earn themselves out of their troubles, by cumulating their earnings of the earlier years into investments for greater profits in the extended period. The Secretary thinks that $100,000,000 more can well be used in this manner, and he suggests that the work should be done in co-operation with the States benefited by the development of resources within their boundaries.

Mr. LANE has written what is almost a prospectus for the development of the national resources. He is an optimist as a speculator in futures, and he is cautious as a conservationist. Thus there is in his report a rare mingling of the methods of wealthy malefactors and of reformers more rabid than he. He balances his proposals as our recent President balanced his literary periods. In the pages of his report the oil and water are made to mix alluringly. Practical men will find some things to their liking, and those who think that conservation is more important than use will find something to their dislike.

December 24, 1913

WILSON FAVORS DAM BILLS.

Says He is as Much of a Conservationist as Any One.

WASHINGTON, July 20.—President Wilson today came out in support of the dam bills, agreed upon at a series of White House conferences and now pending in Congress.

In answer to inquiries, he said he believed them to be in accord with the best conservation ideas and in no sense party measures. The President said he regarded himself as much of a conservationist as any one else.

July 21, 1914

COAL AND WATER POWER.

The discovery of a thirty-foot vein of anthracite in Pennsylvania is good news for everybody except the immoderate conservationists. The Pennsylvania papers say that this is the second recent discovery in the Bear Ridge Mountain region, and the third of importance in the anthracite territory within a few years. If we do not know the resources of so familiar and thoroughly exploited a territory as Pennsylvania, the inference is strong that the world at large still holds unsuspected riches, even at the time that the alarm of exhaustion is raised anew. Forty years ago, when the annual production of anthracite ran about 25,000,000 tons, alarmists thought there would soon be no more. Now the output is about fourfold, and the discoveries continue.

The discovery of coal is nothing compared to the discovery of water power, about which a similar cry of waste is raised. Only a few days ago, when Mr. BURTON was delivering his valedictory in the Senate, he declared that there was thirty million horse power in the waterfalls and that the coal would be exhausted in a century. Here, he said, was a double waste, the failure to utilize the water power and the consumption of the coal, which never would be renewed. Nothing can justify waste, either of coal or water power, and conservation in reason is admirable. But the day of the exhaustion of coal is deferred with discoveries, and the waste of water power is economy of the capital necessary to make it available. Even according to Senator BURTON's figures it would cost six billions of dollars to harness the waterfalls, in addition to the cost of transmission lines and machinery. In short, the cost of the water power is not accurately known, and is enough to deter those who have the money and who find coal cheaper.

The water power is perpetual, it is true, but so is the cost of making use of it. There is waste in the utilization of both coal and water power, of capital in the case of water power and of mining and inefficient engineering in the use of coal. The question of reasonable conservation in our own times lies in the comparison of these costs. If we find it cheaper to use the coal than to develop water power, we may do it with a good conscience, for we will give to posterity more than we deprive it of. They will have the water power as the gift of nature. As a gift from the consumers of the coal they will have the developments of science. For example, the utility of water power is immeasurably increased by the development of electrical transmission. Water power remote from transportation or markets is only worth a tithe of water power annexed to markets and railways by the use of electricity for transmission to where economic use can be made of it and where the products can be carried to consumers economically.

That is the gift of our generation, and is worth incomparably more than the coal we have consumed. The world is richer by the use those who live upon it make of its natural wealth. We are enriching posterity rather than robbing it. Our children's children will bless us for our creations more than they would if we left the coal unmined. The wiser way is to deal with each case of either waste or development upon its individual merits and leave the aggregate of the decisions to make up the case for or against conservation. Those who take views too wide or too long for human foresight are liable to make blunders, as well as those who take short and individual views, and make their mistakes also. Infallibility is not for mortals. Those who aspire to advise everybody are apt to be carried away by their enthusiasm and their good intentions.

April 3, 1915

WATER POWER AID IN EVENT OF WAR

Hydroelectric Machinery Makers See in It Means of Making High Explosives.

NITROGEN FROM THE AIR

Manufacturers Organize In Effort to Obtain More Liberal Laws for Use of Streams.

Manufacturers of hydroelectric appliances yesterday formed the Water Power Development Association at a meeting held in the offices of I. P. Morris Company at 100 Broadway and announced the beginning of a campaign of publicity for obtaining more liberal laws governing the use of streams.

The corporations represented were:

Allis-Chalmers Manufacturing Company, Milwaukee, Wis.; American Rolling Mill Company, Mansfield, Ohio; American Smelting and Refining Company, New York City; Anaconda Copper Mining Company, New York City; Archbold-Brady Company, Syracuse, N. Y.; Buffalo Foundry and Machine Company, Buffalo, N. Y.; Goulds Manufacturing Company; James Leffel & Co., Springfield, Ohio; Locke Insulator Manufacturing Company, Victor, N. Y.; General Electric Company, Schenectady, N. Y.; Morris Company, Philadelphia, Penn.; Ohio Brass Company, Springfield, Ohio; Pelton Water Wheel Company, San Francisco, Cal.; Pittsburgh High Voltage Insulator Company, Derry, Penn.; Platt Iron Works, Dayton, Ohio; S. Morgan Smith Company, York, Penn.; Standard Underground Cable Company, Pittsburgh, Penn.; R. Thomas & Sons Company, East Liverpool, Ohio; Wellman-Seaver-Morgan Company, Cleveland, Ohio; Westinghouse Electric and Manufacturing Company, East Pittsburgh, Penn.

H. P. Hand, the Chairman of the newly created association, said that ever since 1913, when rigorous conservation laws were enforced, that the business of making apparatus for converting the power from rivers and falls into electrical energy had been waning. Government officials had practically stopped the giving of permits for the use of water on Government lands. Had it not been for the war orders all these industries would have suffered more than they have. The manufacturers are now taking account of the future of their industry. With the end of the demand for munitions they expect that many thousands of men will be idle.

It was given out by a representative of the new organization that there were 600 manufacturers interested, whose total capital was $200,000,000, and that in all they employed 400,000 men. The association says that the Government should encourage the hydroelectric machinery industry by giving more liberal use of the streams.

Cheap water power will, in its opinion, be the means of obtaining nitrogen from the air for making ammunition and high explosives should this country be shut off from the supply of the Chile nitrates. Germany has been extracting nitrogen in this way and the results have been very effective, according to the official statement of the new organization. The reproduction of nitrates from the fixation of the atmosphere depends upon the availability of cheap power.

It was pointed out that with the exception of the plant of the Southern Chemical Company, at Charlotte, N. C., of which James B. Duke is the head, that no plant is equipped for producing nitric acid in this way. The association also sees in the increase of water power a means of conserving the coal supply.

In order to bring before the public its views the organization has established a publicity office in the Marbridge Building in this city, and will shortly open another in the Munsey Building of Washington.

The manufacturers advocate the enactment into law of the Shields and Ferris bills.

The Shields bill, which has passed the Senate, permits the construction of dams in navigable rivers under licenses issued by the Secretary of War. It gives the right to operate power plants for fifty years, after which the Federal Government may take them over after giving two years' notice and paying a fair value.

The officers of the new organization are Mr. Hand, the Chairman; Calvert Townley of the Westinghouse Electric Manufacturing Company of East Pittsburgh, Vice Chairman; W. W. Nichols of the Allis-Chalmers Manufacturing Company of Milwaukee, Wis., Treasurer.

March 10, 1916

WAR AND POWER SITES.

Objections of Conservationists to Leases to Munition Plants.

To the Editor of The New York Times:

In your item " Du Ponts Offer Nitric Acid Plant " of March 30 a brief outline is given of a bill to be introduced in the Senate to permit fifty-year leases of power and dam sites in navigable streams or in public lands. These terms sound much like those of the Shields bill, which passed the Senate recently. This bill was opposed by the conservationists on the ground that it makes it impossible for the Government to recover the power rights except on payment of possibly enormous sums for plant and for increment in the value of land, or else by long and slow process of law.

Gifford Pinchot wrote an open letter to President Wilson pointing out these matters and also pointing out that only by means of water power could the Government produce cheaply the quantities of nitric acid needed for explosives.

The du Pont proposal is evidence that Mr. Pinchot's statement is correct. It is evidence that careful retention of Government power over our national resources is a necessary part of preparedness, so that the Government may draw on those resources when it needs them.

In the present case it may be better to lease power rights and then to buy back the product manufactured by them than to trust to the Department of the Interior or of the Army or of the Navy to handle the affair expeditiously and economically.

But when doing so let us not at the same time give away all other power sites under such terms that we cannot recover them even on conpensatory terms during fifty years, except by the slow process of condemnation, and that at the end of fifty years we shall still have some struggle to get them.

In case of war we may need water power sites for many purposes. Shall we find that what we now own we cannot get possession of except under two or three years' legal process? Is it not simpler to provide for repossession in the original bill?

The Shields bill has not yet passed the House. It can be amended or defeated in favor of a bill giving adequate protection to preparedness.

GEORGE S. JACKSON.

Boston, March 31, 1916.

April 3, 1916

FAVOR AGREEMENTS TO SAVE RESOURCES

Business Men of Nation Support Proposal for Co-operation by Timber and Ore Firms.

WANT NEW LEGISLATION

Support Plan of National Chamber of Commerce Committee by Overwhelming Majority.

Special to The New York Times.

WASHINGTON, Jan. 14. — American business interests affiliated with the Chamber of Commerce of the United States are in favor of combinations to conserve the natural resources of the country, according to an announcement made by that body tonight, based on votes cast in a referendum it has just conducted.

Elliot H. Goodwin, Secretary of the National Chamber, explained that 1,034 votes were cast in favor and 110 against the report of a special committee, which recommended that there be remedial legislation to permit co-operative agreements under Federal supervision in those industries which involve primary natural resources on condition that the agreements tend to conserve the resources, to lessen accidents, and to pro-

mote the public interest. All except four States were represented in the balloting, in which 413 commercial organizations participated. These organizations had from one to ten votes each, according to their membership.

The referendum will be considered at a meeting of the Board of Directors on Jan. 30. In announcing the result of the referendum the National Chamber made this statement tonight:

"The plan indorsed would make it possible for the Federal Trade Commission to go beyond its present powers of investigation and formulate constructive plans under which an industry might operate to the common benefit of consumers, workmen, and producers. The committee whose report was voted upon endeavored to point out a way by which the public interest might be safeguarded and promoted instead of being left to take care of itself, as at present. The recommendation is confined to timber, the ores, and deposits of useful metals, and deposits of minerals which are a source of heat, light, and power.

"The report of the committee, of which W. L. Saunders of New York, Chairman of the Board of Directors of the Ingersoll-Rand Company, and formerly President of the American Institute of Mining Engineers, is Chairman, advocated also increased safety for workmen. It was contended that in the last twenty-five years the deaths or injuries to life and limb had grown in breadth and intensity. The report went so far as to say that in the United States each year there were about 25,000 deaths from industrial accidents. In three industries—metal mining, coal mining, and lumber—with 1,400,000 employes, in 1913 the

number of fatalities was said to be almost exactly the same as among railway employes, although there were 300,000 more railway employes.

"In the proposed legislation, due attention would be given to property loss and the prevention of material waste. Statistics were presented in the National Chamber referendum which declared, for instance, that only about 35 per cent. of the total volume of lumber, as it stands in the forest, now reaches the ultimate consumer, most of the remainder being wasted. The Director of the Bureau of Mines was quoted as saying that something in the neighborhood of 40 per cent. of the coal in the seam is lost, so far as beneficial utilization was concerned. It was pointed out that millions of barrels of oil were lost, and a condition equally as bad existed in connection with natural gas and many other natural resources.

"In putting forth the question submitted in the referendum the committee developed the idea that any legislation enacted by Congress would in itself have a limitation, because Congress can do no more than deal with interstate trade. The committee had in mind remedial legislation which would in no sense encroach upon State jurisdiction, but which, when these enterprises through their entrance upon interstate trade became subject to Federal statute, would declare legal the situations that were in question in the event that certain conditions had been met.

"The form which remedial agreements among individual operators in the industries under discussion should take the committee did not undertake to suggest. It recognized that different industries may well require distinct forms of agreement. Thus by raising the question of a definite national policy for utilization and conservation of resources to which manufacturers generally must look for raw materials or fuel, the legislation advocated in the referendum would be of vital interest to consumers, producers, and operators."

January 15, 1917

URGES MORE FORESTS.

Federal Official Warns That Depletion Is Double New Growth.

WASHINGTON, Dec. 18.—A larger program of public acquisition of forests by the Federal Government, States, and municipalities and protection and perpetuation of forest growths on all privately owned lands which may not be used better for agriculture is recommended in the annual report of the Forester of the Department of Agriculture. This policy is made necessary, the report said, by the diminishing timber supply. The rate of depletion of the forests is more than twice what is being produced by growth in a form serviceable for purposes other than firewood.

"Already the supplies of all the great Eastern centres of production are approaching exhaustion, with the exception of the South," the report said, "and even there most of the mills have not over ten to fifteen years' supply of virgin timber. The Southern pine is being withdrawn from many points as a competitive factor and its place taken by Western timbers. This inevitably results in added freight charges, which the consumer must pay."

The report suggested that the Federal Government work primarily through State agencies.

December 19, 1919

UNREST LAID TO NERVES.

Social unrest is attributed to the requirements of modern efficiency in the annual report of the State Conservation Commission, which urges as an offset added attention to the conservation of wild places and wild life to the general physical and mental well being and happiness of the people.

With power driven machinery, steam and electric railways, telephones, the telegraph and wireless, and, finally, the airplane, all commonplaces of our daily existence, the race, it is pointed out, is prone to forget that our grandfathers lived under no such nervous strain as these inventions have brought with them, and that only three or four generations ago the high speed of modern life would have been actually inconceivable.

"It is a subject that is acquiring increasing importance, as the high tension of our social and industrial organization is becoming constantly more acute," continues the report. "Students of our social and industrial conditions are satisfied that the human organism has not yet adapted itself to the high nervous tension of modern life. Scarcely more than 125 years have sufficed to completely reorganize the mechanical details of production from the old domestic system under which our ancestors worked, to the highly developed factory system of today.

"Prior to the invention of the system of the steam engine in the latter part of the eighteenth century, industry was conducted largely in the home, in small towns or in the country. Practically every family lived in its own house, and cultivated land from which much of its food supply was obtained. The daily duties of each member of the family and of the servants or employes, who constituted practically a part of the family, were extremely varied and largely unhurried.

"The invention of the steam engine, and all of the other industrial advances that followed in its train, radically changed industrial and living conditions. Industry became concentrated in factories; factories required labor in large blocks, and readier access to lines of transportation; towns and cities grew in size, as the workers, driven for their livelihood from the country and small towns to the factories, were obliged to give up country life. In short, the domestic system of industry of the eighteenth century, with all the advantages that it held for healthful occupation and physical and nervous relaxation, gave place to the factory system of the nineteenth and twentieth centuries, as we know it today, with its high tension machine work, monotony of occupation, and decreased opportunity for healthful relaxation.

"In this short time wonderful strides have been made in the mechanical arts. The failure of the human organism, however, to adapt itself to the changed conditions constitutes one of the greatest physical and social menaces of today, and, in the opinion of many careful students of industrial and social conditions, threatens an actual breakdown of society.

"Relaxation from the physical and mental strain of modern industrial life is a vital necessity. This has never been more clearly apparent than since the war. Social and industrial unrest following the strain of the war has been world-wide, and relaxation of every sort has been apparent upon every hand. The entire situation is complicated to a high degree, and cannot readily be generalized in a short discussion. It is apparent, however, that relaxation is a fundamental need of all people, and that one of the most wholesome and popular methods of obtaining this relaxation is in out-of-door recreation."

May 2, 1920

SUBSTITUTING WHITE COAL FOR BLACK

New Water Power Act Opens Up Great Possibilities of Linking Isolated Sources of Energy.

AFTER ten years of effort, the door has been thrown open for "white coal" to serve industry and the home in this country. The signing of the water power bill by the President establishes, for the first time, a policy under which the vast water power resources of the country may be developed. Up to this time, by the lack of a law that would protect the interests of the people of the country and at the same time be liberal enough to attract the necessary investment of private capital, a heavy check has been placed on enterprise in harnessing water power. In a plea in behalf of the bill that now becomes a law, the Secretary of Agriculture said:

"The exigencies of war brought to light defects in our national utilization of power which had not been fully realized. Operating under statutes enacted when the electrical industry was in its infancy, we had permitted our vast water power resources to remain almost untouched, turning to coal and oil as the main source of power; for steam power could be developed more quickly and with fewer legal restrictions and with greater security to the investment. Probably not less than 85 per cent. of the power used for domestic, public and industrial purposes and for the operation of the railroads is produced from coal and fuel oil."

Fifty million horse power comprises the total now in use in this country. The low estimate is that this can be doubled by the utilization of the water power resources. High estimates place the potential water power as great as 200,000,000. In the previous uncertainty of the situation no satisfactory survey has ever been made, and one of the first needs under the new law will be to determine what these possibilities are. That they are extensive there is no doubt, and the supporters of the water power act are confident that the development of this resource is to be a factor of the first importance not only from a domestic standpoint but also in competing in the markets of the world where cost of production is to play a sharper and sharper part. Cost of power is one of these, and this must be supplied in one of three ways: By oil, by coal, by water.

To Supplement Coal and Oil.

In this country the production of oil has already passed its peak, and in the world as a whole it will in a comparatively few years become of diminishing importance as a power producer, according to experts. There are still large reserves of coal in this country, though the greater ones are in the distant West, and coal-producing countries are still far from exhaustion, but with the doubled price of coal, due in part at least to the increased cost of getting it out of the ground, coal-made power will be on the steady rise, it is predicted. In the creation of power, coal and oil are consumed, they cannot be replaced, and therefore, it is asserted, the country which in future competition depends wholly in great part on coal and oil power will be in a position of increasing disadvantage.

This fuel situation, to which attention has been increasingly called since the armistice, will result in a great stimulation of the development of water power the world over, wherever feasible, it is predicted. In this respect the United States is better off than any other leading industrial nation,

and at the same time, what it possesses has been the least exploited. Not more than 10,000,000 horse power derived from the flow of water is now in use in this country. It is not contended that water power can take the place of coal or that in some uses it can be a substitute, but that as a supplement it will add greatly to the total, pull down the average cost per unit, set electricity to work in the home far more extensively than at present, and bring about such public comforts as the substitution of electricity for coal on main railroad lines.

The aim of the engineers now planning for what they deem to be the beginning of a new era in power development is to connect in big trunk line systems—just as formerly disconnected railroads were linked together—steam and water power plants. An appropriation has been asked of Congress to make a survey for one from Boston to Washington via New York, to include all the systems now operating in this territory and to add to it the development of the unused water power. This is from an official statement at Washington of what may be brought about under this conception:

"Not only are water powers relatively unused, but we are far from accomplishing the economies which are practicable in the development of steam power. While individual steam stations have reached a high degree of mechanical efficiency, we have failed to realize the group efficiency of many stations operating in a system and we still pursue the crude method of transporting steam power by rail in its raw state as fuel instead of transmitting it over wires in its developed state as electrical energy. The power requirements of this country will not be met until we develop our water powers, tie them with the steam plants located at the mine itself and operate all in great interstate systems."

Development in the West.

Although the war saw an increase in horse power used in the United States, according to an expert at Washington, the demand is now 15 per cent. higher than it was then. California is the most highly developed State in water power and these producers are linked with the steam plants in one big system that extends from southern Oregon to the lower part of California. Engineers point to this as a model for the country. Electrical power is cheaper in California than anywhere else in an equal extent of territory, and the homes into which electrical wires do not run are reported to be comparatively few. In the thickly settled rural districts most of the farmers are provided with electricity, finding it cheaper than man power. Mills for grinding grain on the farm, washing machines, sewing machines are run by electricity. In the place of the oil lamp the farmer reads by an electric light.

O.C. Merrill, Chief Engineer of the Forestry Bureau at Washington, who receives the chief credit for the new law, having worked for it for ten years, said in Washington the other day:

"The law accomplishes two main things: First, it assures that the great water power reserves of the country will be held under public control. They can not be exploited at the expense of the public:

second, the development of the water power reserves, made possible under the law, means more and cheaper power, in the factory and in the home.

"This is the only way, because steam is going up all the time in cost. Until fuel began to advance, steam power was going down, with the increasing advance of coal, which has doubled in price in the last five years, the opposite result was brought about and every further increase in the price of coal will put up the price of steam. Since the bill passed the Senate there have been applications prepared for the development for a total of 500,000 horse power, and when the Water Power Commission is organized under the law, there will undoubtedly be many more applications for licenses. Under good business conditions I think 5,000,000 horse power from water will be added to our present output in five years. The way has been prepared; we cannot tell how fast development will follow. A new era is opening; I think we can safely say that. Much depends on what advantages capital sees under the law. That was what caused the long delay in getting the law, to frame a measure that would safeguard public interests and at the same time offer a fair return to capital. I believe this has been accomplished. There may be some extremists on both sides, the capitalist side and the conservationist side, but I think that the law is acceptable to most of these on each side and that it will be found a solution to the double problem that was faced. I am told that 90 per cent. of those concerns interested in the development of water power projects—that is the investment side—are satisfied and I know that some of the leading conservationists are.

"Up to the present time three departments— War, Agriculture and Interior—have had control over water powers, each distinct. A glance at the laws in force up to the passage of the new law will show how the development of water power was retarded. Water powers on navigable rivers have been administered by the War Department. The right to develop a power site on a navigable river was subject to the General Dam act of 1910. For each development a special act of Congress was necessary, and by the General Dam act the lessee was placed under rather heavy restrictions in insuring a supply of navigable water and the act of Congress which bestowed the right might be repealed at any time. Under such conditions capital was slow to invest in even the most promising propositions and the development of some of the best sites in the East has been held up on that account.

"Water powers on public lands have been under the control of the Department of the Interior and in the national forests under that of the Department of Agriculture. Here the laws in force were equally discouraging to those with capital to invest for development of such resources. The only way under the law covering this was that the license to develop a project be obtained under the revokable permit law of 1901. Under this law the permit could be revoked at any time at the discretion of the Secretary of the Interior, and naturally, the power industry was slow to build on any such uncertain foundation, so that capital was largely withheld. There was another way, how-

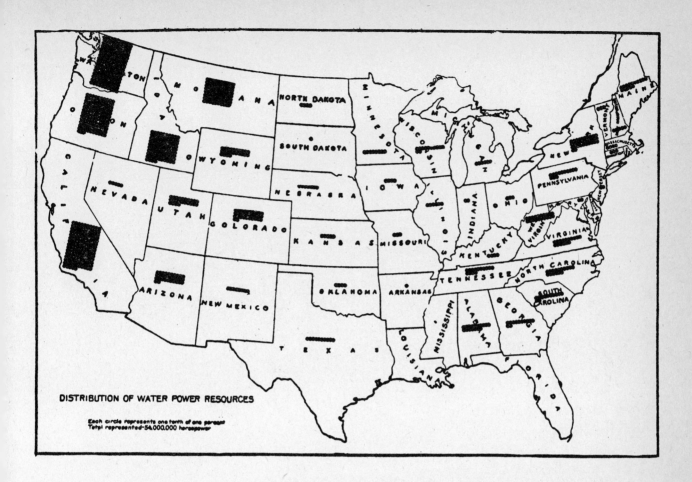

DISTRIBUTION OF WATER POWER RESOURCES

Each circle represents one tenth of one percent
Total represented—54,000,000 horsepower

ever, by which power sites could be developed on public lands. This was to acquire outright the land on which the site was situated through the homestead or mining laws. In order to prevent this being done large areas of lands have been withdrawn from entry for the last ten years, those public lands having on them valuable power sites.

The Water Power Commission.

"By the new law all water powers over which the United States has any jurisdiction will come under the control of the Water Power Commission, composed of the Secretaries of the Interior, War and Agriculture. Duplication will be avoided; a common policy will be pursued and the combined efforts of the three departments directed toward a constructive national program of intelligent, economical utilization of our water power resources.

"Licenses for power development run for fifty years. The commission fixes the amount the licensee must pay for the use of the water power and the price of the power to the consumer is fixed jointly by the commission and the State Commission, or, if a local agency does not exist, by the Federal Commission.

"At the expiration of a license the United States retains the right to take over and thereafter to maintain and operate the property upon payment of an amount which shall in no case exceed the original cost and which, on account of amortizing original cost out of earnings in excess of a specified rate, may in many cases be much less than the cost. The commission may permit the renewal of the license or the license may be transferred to another with the payment to the original lessee of what he is entitled to. Whatever licenses are issued for the development by private capital, full public control is retained over the construction, maintenance and operation of the project.

"Licenses issued to private or corporate applicants will require the payment of annual charges in an amount fixed by the commission for the purposes, first, of paying the cost of the administration of the act; second, for recompensing the United States for the use of its lands, and, third, of absorbing any excess profits that may be earned by the licensees. In fixing annual charges the commission shall avoid naming an amount which would result in increasing prices to consumers. One-half of the charges collected on account of public lands are to be paid into the reclamation fund, three-eighths to the States and one-eighth into the United States Treasury. One-half of the charges collected on account of licenses on navigable rivers are to be reserved as a special fund to be spent by the Secretary of War in the maintenance and operation of navigable structures or in the construction of headwater improvements on navigable rivers. The remaining half of charges from this source are to be paid into the Treasury. In the issuance of licenses preference is given to the States or municipalities

"The feature of outstanding public importance is the basing of property values upon cost. Under present conditions, wherever public agencies seek to regulate rates for services or to acquire private property for public use, the valuations are likely to include not only the original cost of the properties but appreciation in land values, franchises, 'going value' and every other element, tangible or intangible, that the ingenuity of the owners or their attorneys can devise."

Mr. Merrill said the water power resources of the country were divided between the East and the West in the proportion of about 25 per cent. for the East and 75 per cent. for the West. In the East, it was estimated that there were from 6,000,000 to 8,000,000 horse power in the St. Lawrence, 2,000,000 in New York State outside of St.

Lawrence, some in Maine and other New England States and large reserves in the Southeastern part of the United States.

"One of the first effects of the new law, I think," continued Mr. Merrill, "will be that it will open the way for the electrification of the railroads by the development of water power and the carrying of the power over long distances, as in the West. The railroads are overloaded with freight. A large proportion of their freight is coal, including coal for their own operation. Every little town has a steam power plant and the railroad has to haul coal for the towns along its line. These towns could also be served by long distance transmission, thus further lightening the burdens of the railroads."

June 27, 1920

FORESTRY AND LIVING COSTS.

According to the report of the Forest Service, timber in the United States is "being used and destroyed" four times as fast as new timber is growing. "Saw timber," which is the most valuable and the most necessary part of the stand, is being cut five and one-half times as fast as it grows. Lumber prices have risen far out of proportion to the general increase, heavy as that has been.

Already the fact has come home to every one. The manufacturer and the farmer feel it when they buy new vehicles and tool handles, when they repair old buildings or put up new ones. The universal shortage of housing is in a large measure due to the scarcity and the high prices of laths, shingles and lumber. The scarcity of wood pulp is felt in the cost not only of books and newspapers, but of every article that comes into the home in a paper wrapper or cardboard container. And unless drastic measures are taken to correct the ratio of cutting to the growth of timber the prices of all wood products are destined to advance even more sharply.

We cannot plead that this predicament is the result of any lack of resources. Outside of privately owned forests and of public reserves upon which cutting is forbidden there are no less than 80,000,000 acres of land, largely unfit for farming, but available for reforesting, which is at present an unproductive waste. A much larger area is only partly productive and, being widely scattered and unprotected, is yearly devastated by forest fires. At only a few scattered points is forestry practiced as it has been for generations in Europe, so that the timber is cut only when it reaches maturity and with reference to the steady productivity of the forest as a whole.

In no other field can conservation be practiced with such early and certain returns. The national resources in water power, coal and iron are limited; the best we can do is to prevent present waste. Forest lands are sufficiently abundant to supply the needs of the nation indefinitely. With adequate foresight and care, lumber should be among the cheapest of commodities and even a prime article of export to Europe. Private owners should be brought to a realization of the fact that a forest is not a bonanza to be exploited and then abandoned, but a property that can be made to yield large annual dividends in perpetuity. Waste lands which cannot be made productive in the present generation should be scientifically planted and cared for by the States and the nation.

Over a century and a quarter ago ALEXANDER HAMILTON said in a report to Congress: "Every nation ought to "endeavor to possess within itself all "the essentials of national supply. "These comprise the means of sub- "sistence, habitation, clothing and de- "fense." The Roosevelt Administration did much to advance the cause of conservation; but not until the late war did we realize the full meaning of HAMILTON's sentence. In the present crisis we have been obliged to appeal to Canada, and largely in vain.

September 8, 1920

BLOSSOMING DESERT

THE American desert blossomed last year to the extent of $68,-000,000 worth of crops, due solely to the efforts of the Reclamation Service, the Government's agent in making two blades of grass, an apple tree and half a dozen strawberry plants grow where one cactus grew before.

According to the Reclamation Record, since the Government works began delivering water for irrigation, the crops produced on the reclaimed land have exceeded $400,000,000 in value. These estimates do not include the large amount of land (1,100,000 acres in 1920) dependent upon private systems, which receive subsidiary supplies from the Government reservoirs in time of drought. These subsidies vary from a complete supply to a small percentage of the total. But such are the vicissitudes of dry farming that a small supply of water, delivered at the right time, may double the crop yield. It is often this vital modicum that the Reclamation Service gives.

The crop value per acre of the lands reclaimed by the Government is at present more than double the average for the whole country, doubtless because of the intensive cultivation of this virgin and hitherto arid soil. In 1920 the average yield for the United States was $23.44 an acre, while it was $38.80 for the reclamation "projects."

In addition to this direct increase in the value of the land, irrigation is also responsible for a large indirect investment. It is estimated that for every dollar the Government spends in irrigation, five to ten dollars is invested by farmers and the industries brought into being by the new communities.

Under these circumstances, land that sold for $10 an acre or less before irrigation, now sells for $200, $300 or $400. In rare instances the price has gone as high as $1,500 an acre on well improved lands. The average increase in value in the projects has been $200 an acre. Taking into account the increased value of lands in the settlements within the projects, the Reclamation Service has created land values of over half a billion dollars.

During 1920 the Government irrigated 1,225,000 acres through its own system and helped private companies supply 950,000 acres. The Government money with which irrigation is carried on is really a revolving fund, since the cost of constructing irrigation systems is paid back to the Government by the lands benefited and the cost of operation and maintenance is defrayed by an annual charge on the recipients of water. Construction charges are collected over a period of twenty years in percentage payments.

Loss accrues to the fund when a project is investigated and found unfeasible, when a project does not turn out as well as had been expected or when acreage is lost through waterlogging or some other cause. Each project must stand by itself, and the Service is not allowed to take a profit from the successful projects to cover losses. Yet in twenty years of service and the investment of more than $120,000,000 the Reclamation Service estimates that it has expended only $5,000,000, less than 5 per cent., for which there is no return in sight. Against this loss to the fund may be set down the tremendous increase in land values and large annual crops that result from the work of the Service.

July 17, 1921

FOOD FROM THE AIR

Projected Dams and Hydro-electric Stations Foundation for Great Future Industry

By DR. ROBERT CALVERT, Chemist, University of Southern California, Los Angeles.

"FOOD from air " was the optimistic prediction of a chemist twenty years ago. Food from air is a reality today, in Norway, in Germany and at Niagara Falls, Canada. Food from air is Henry Ford's idea for the Government plant at Muscle Shoals, Alabama. It should be one of our leading industries when great dams and hydro-electric stations preserve the water power now going to waste down the roaring canyons of our mountain slopes.

In addition to sunlight and water, plants require potash, phosphate and nitrogen. As these ingredients of our naturally fertile soil are exhausted, by intensive methods of cultivation, they must be replaced through fertilizers. We get our nitrogen from Chile. Why not from the air itself? Air is four-fifths nitrogen. Over every square foot of Manhattan Island there are 2,000 pounds of it which, if made soluble in water, is available for growing increased quantities of foods of all kinds, a change in solubility accomplished in Norway with cheap electric power. Our power from the rivers should be used by day for the usual commercial purposes, and by night, after the busy hours of peak load have passed, for giant electric furnaces for heating the air to temperatures at which the nitrogen will combine with oxygen or other materials for the production of fertilizer.

The water power possibilities of America are of fascinating extent. Niagara, the best known power site, has a drop of 162 feet; the Colorado River has a drop of 3,000 feet within a flowing distance of fifty miles, with narrow banks which make for economy in dam construction. On our side of Niagara there is developed about 800,000 horse power; one corporation's filings on Colorado River sites total 3,000,000 horse power.

Romances of Chemistry.

Development of cheap electric power on a large scale means more than food from air. Niagara with its relatively small surplus of power over that required for commercial purposes gives us more than fertilizer. Electricity is used there to decompose salt solution into caustic soda for our soaps and chlorine for purifying our drinking water and bleaching our paper and textiles, to decompose water into hydrogen for converting cottonseed oil to lard of unexcelled wholesomeness and into oxygen for acetylene welding, and finally to give aluminum for a thousand uses. There the processes of one chemist, Acheson, are used to make the world's softest lubricant, graphite, and the world's hardest abrasive, carborundum, which is exceeded in hardness only by the diamond.

These industries are each a romance of electrochemical achievement. And yet nitrogen, fertilizer from the air, is so overshadowing in interest and in importance that this discussion is limited strictly to the story of nitrogen.

Indispensable in agriculture at all times, nitrogen is the key industry of national defense in times of war. On the entry of the United States into the World War the politicians realized with a start that nitrogen, in the form of nitric acid or of nitrate of soda, is more necessary than armies themselves; that nitrogen enters into every pound of the important explosives, and that our supply comes almost entirely from a foreign shore. The best brains of the chemical fraternity were organized into an "Air Nitrates Division." And the War Department got busy. It could not function in war without nitrogen any more than a plant can grow without nitrogen or an automobile run without gasolene. A great nitrogen-from-air plant was built at Muscle Shoals.

The only misfortune is that the expenditure was made under emergency conditions, quickly, extravagantly, of necessity, and now the plant and location are about to be sold, probably to Henry Ford. The same money, spent under the sane conditions of peace, should have given us a great nitrogen industry, a bulwark of prosperity in time of peace and our first-line of defense in war.

As it is, we are still dependent for much of our nitrogen on the good-will of Chile and on keeping open against U-boat or airplane attacks the Panama Canal, through which move the never-ending cargoes of white treasure, nitrate of soda.

Chile's Nitrate Industry.

The nitrate of soda goes to fertilize our fields. Heated with sulphuric acid, on the other hand, it gives nitric acid, the pig iron of our dye, our celluloid and ivory pyralin, our motion-picture film and of our explosive industries.

This nitrate is found in the northern part of Chile, in a narrow strip of plateau, where the Andes Mountains flatten out before finally plunging into the sea. The nitrate lies on the top of the ground or near it; it is soluble in water, as sugar is in tea, and its preservation from geologic ages of long ago speaks eloquently for the dryness of that portion of the desert. One rides for days and sees hardly a living being save the men and beasts imported for the exploitation of the desert's resources. Vegetables are nonexistent and canned corn from the United States is a Sunday dinner delicacy at the nitrate works.

The ore contains about 20 per cent. of the nitrate. The ore is ground, then treated with boiling water to dissolve the soluble salts. The liquid, after draining from the solid material, is allowed to cool to deposit crystals of the nitrate of commercial purity.

From a few hundred thousand tons before the war, our imports of this Chile saltpetre rose in 1918 to 1,845,112 tons. Then, after the armistice, the demand for it in manufacture of munitions of war ceased; the imports fell to 407,457 tons in 1919. Wiseacres shook their heads: "We knew this nitrogen bubble would burst." But in 1920 the tremendous consumption in our peace-time industries drew down the stocks carried over from the war plants and required in addition fresh imports from Chile of 1,321,922 tons, valued at $63,129,196. Truly, a great industry is this nitrogen!

Known mineral resources are finite. As you remove from the earth, more does not grow, as do potatoes. Particularly is this true of nitrate of soda. With the growing appreciation of the value of nitrate fertilizer, the complete exhaustion of the Chilean beds is a matter of a few decades only. Then what? Will we turn to Germany and pay tribute in dollars and self-respect to her foresightedness and genius in drawing from the atmosphere an inexhaustible supply of nitrogen? Or will we harness the water power of our turbulent streams, add to it human ingenuity and keep at home the hundreds of millions of dollars yearly that would otherwise go abroad?

Nitrogen From the Air.

As long as a hundred years ago, with chemistry still in its infancy, an English chemist, Cavendish, found that the electric spark could be used to combine the nitrogen with the oxygen of the air. The observation of Cavendish lay unused for nearly a hundred years, until cheap electric power became available in Norway, and Dr. Samuel Eyde supplied the financial backing and the enthusiasm of youth. He built five factories from which, as early as 1912, he was shipping nitrates around the Strait of Magellan and selling on the West Coast in competition with nitrate from the West Coast of Chile.

Eyde's is called the arc process. Air is passed through a chimney containing a giant electric arc lamp which heats the passing air to 3,000 degrees Centigrade. One and a half per cent. of the nitrogen is combined with the oxygen. The resulting compound is allowed to add more oxygen and is then dissolved in water to give nitric acid. From the nitric acid can be made, if desired, a good nitrate fertilizer, such as lime nitrate.

Eyde at Nottoden is supposed to be using some 200,000 horse power. The possibilities at Muscle Shoals are thought to be ten times that amount of power.

Dr. Eyde was a conspicuous member of the last International Congress of Applied Chemistry in New York City. At a great testimonial meeting he ascribed his success to the "employment of young men, to the lack of experience, to disregarding the doubts and hesitations of the authorities." He gave thanks " to the young people, to their undaunted courage, energy, and love of action," which made the great new industry for the Norwegians.

For every pound of military explosive there is used, on the average, two pounds of nitrate of soda. Germany could never have faced a war, even of defense, with the ocean closed against her, except that she had working two good processes for utilizing atmospheric nitrogen. After the battle of the Marne and the destruction of her hopes for a quick, triumphal entry into Paris and world peace, the Kaiser's Government advanced 30,000,000 marks, then equivalent to $7,500,000, for the extension of the Haber process of the Badische Dye Company. What was the result? Today the German factories have a capacity twice as large as their total imports of nitrogen before the war. Compared with former imports of 750,000 tons nitrate yearly, she can now make as much fixed nitrogen as is contained in 1,500,000 tons. Her dependence has been changed toward domination and she is now in position to export nitrogen to whom she pleases.

The Haber process is the combination of nitrogen with hydrogen. The nitrogen comes from the air, hydrogen from water. Under the influence of heat and pressure and in the presence of an active form of iron, these two gases give ammonia. Ammonia is used as such in ice-making machinery; dissolved in water it is the drug store article of characteristic odor. But for the great uses of agriculture and manufacturing it is converted by oxygen of the air and water to nitric acid.

The American Cyanamid Company has done a great work in developing on this side of the Atlantic the process of combining nitrogen from air with the electric furnace product calcium carbide, which results from the action at very high temperatures of coke on lime. During the war our Government spent most of its millions at Muscle Shoals in building plants for this process. They were not fully built at the signing of the armistice. Under the terms of the contract with the cyanamid company there was to have been supplied from this plant at Muscle Shoals and from installations of the process elsewhere enough of the great high explosive, ammonium nitrate, to correspond to the fixed nitrogen in 500,000 tons of nitrate of soda.

March 5, 1922

NEW HOMES FOR 3,000,000 IN BIGGEST IRRIGATION PROJECT

Dotted Line Shows Area Affected by the Colorado River Project. Hundreds of Thousands of Acres of Arid Land Will Be Made Into Fertile Farms by the Development of the Colorado River, and Great Hydro-Electric Plants Will Distribute Power.

HARNESSING THE COLORADO

Will Open Vast Waste for Farming and New Industries

ANNUAL FLOODS TO END

Treaty Between Seven States and Nation Makes Possible Great Reclamation Plan.

CONGRESS soon will be asked to confirm a treaty between seven Western States, with the nation as an eighth partner, which is expected to clear the way for the greatest irrigation and power project ever undertaken in this country. This is the first interstate treaty ever reached between more than two States, and it was no small achievement to bring the seven together, each having vital interests in developing a vast stretch of arid land in the great Colorado River basin, where nothing much lives but mesquite, coyotes and rattlesnakes. The waste, according to the plans, is to be so converted that it will provide homes for at least 3,000,000 Americans in the next few years.

The Colorado River Commission was created by the Legislatures of Arizona, California, Colorado, Nevada, New Mexico, Utah and Wyoming, with Congress co-operating. Herbert Hoover was placed at the helm, and in record time the seven States reached a compact which will put an end to litigation that has held up development of the river basin for many years. That compact must yet be ratified by the seven Legislatures and officially approved by Congress. Then money will be needed to start the undertaking. A first appropriation of $70,000,000 has been asked. That is only a tithe of the ultimate sum required, but it is enough to insure flood control in the lower California valleys, the most pressing need.

The harnessing up of this river to do the work of man is destined to be one of the greatest engineering undertakings since the pyramids were reared in Egypt's sands. Just a few off-hand figures may help to visualize the picture. For one thing, 4,000,000 acres of desert are to be converted into a pleasant countryside. More than 6,000,000 horse power of electric energy is in sight from waste waters. The backlands in seven of our largest and most undeveloped States will be open to homebuilders, an empire greater than France. It has been estimated that 100,000,000 people might be sustained in this territory, which now has about 500,000. But for the present the figure is put at 3,000,000 by Government experts. That is to say, 3,000,000 as a beginning.

These are just a few of the things that are to be done in the subjugation of the Colorado River to the yoke of man, instead of letting it go on its wild, tumultuous course, as it has done for untold centuries, wasting power that could be made to turn the wheels of a considerable nation. While the Colorado has been plunging through its rocky gorge, half the length of the continent, land a few miles away has been scorched and seared for the lack of water. Just to complete the case against this great river it overflows every year, and does incalculable damage.

The Colorado River basin embraces 242,000 square miles in American territory, and laps over into Mexico for another 2,000. The basin takes in portions of Wyoming, Colorado, Utah, New Mexico, Nevada, Arizona and California. For a long time there was discussion about a comprehensive plan of development. Each State has done something on its own account, and there have been efforts at co-operation between one or more of them. But these plans usually wound up in controversy until the appointment of a commission by the Governors of the seven States, with Mr. Hoover at the helm, representing the Government. And in just eleven months this commission worked out a plan, got all of the members to accept it, and have had it signed, sealed and handed over to their respective Legislatures.

So much for the official part of the story. The real romance lies in the reclamation of this far-reaching territory which has been a waste place ever since the glaciers slipped on down into the Atlantic Ocean. These glaciers left behind a rugged country full of mountains, gorges, sand and general desolation. A long time afterward some primitive people came that way and managed to get a living out of the land. Probably they were Mayas. Cliff dwellings, bits of antiquities and other evidences of a remote people have been found all over the region, with traces of an irrigation system which antedated by some thousands of years that now projected.

Land of Romance.

The next comers were Spaniards, seeking those fabled cities of the Indians, where the streets were paved with gold. They came and suffered and departed. Another lapse, and we have the missionary, the hunter in his coonskin, and then the settler in his prairie schooner. But none of them lingered over long because it was a blighted land. They skirted around the edges for fairer places and the country has remained very much as it was in the time of the Mayas. But irrigation is going to introduce the era of the farm and power plants the age of industry.

The Colorado River is navigable in places and might be made so for a considerable distance. But navigation is to to be a minor issue in the big development plan. And agriculture for once is to have precedence over industry, which means that more attention will be devoted to irrigating farm land than to the creation of power. But there is such a flow of water between the river banks that there will be enough power for any conceivable purpose.

The Colorado River basin may be divided into three main sections—the upper, the middle and the lower. The first lies above the Arizona-Utah boundary. It is a rugged, inhospitable country, with narrow valleys and irrigable acreage considerably scattered. In 900 miles the river drops 3,500 feet. Some of the land in this zone already has been irrigated, supporting cattle and yielding sugar beets, potatoes, fodder, garden truck and fruit. There are 96,000 square miles in this section of the basin, 39 per cent of the drainage area, furnishing 87 per cent of all the water flowing through the Colorado's bed. In this territory, where most of the water comes from, there is less need of it at present than in the country to the south, but adequate provision is to be made for future development in both.

About 1,500,000 acres are under irrigation in the upper basin, and another 2,500,000 can be watered. In the four States of this upper zone, Colorado, New Mexico, Utah and Wyoming, there are eight reservoir sites suitable for the production of power, about 2,000,000 horse power, or one-third of the potential power of the river. And these will serve to stabilize the flow of the river when the water will be diverted for irrigation. And the third phase, flood control, will be closely linked up with the reservoirs and irrigation. Power and irrigation rank about equal in the plans for this section of the basin.

The middle reaches of the river, from the Arizona-Utah border to the Arizona-California boundary, some 500 miles, run through deep canyons for the most part. There is a drop of 3,000 feet in this stretch of the river, which cuts through a territory of some 77,000 square miles, 32 per cent of the whole basin. There is not much irrigable ground in this district, although some of the tributaries are promising, the land on either side being suitable for the same products as the upper basin. But 4,000,000 horse power of energy can be created, and this part of the river will be converted largely to generating power for the factories which are to rise and the electric railroads that will come.

Dam Sites Picked.

Three sites for big dams already have been picked out—at Lees Ferry, Diamond Creek and Boulder Canyon. At the first and the last the possibilities for dams larger than any in existence are excellent, and probably will be realized. It is believed possible to build a dam at Lees Ferry which would be 700 feet high, capable of storing 50,000,000 acre feet of water, developing 1,000,000 horse power. About 250 miles further down the river is Boulder Canyon. This dam, if built to its full estimated height, would be 600 feet high, storing 31,400,000 acre feet of water and developing 700,000 horse power. Smaller dams, including that at Diamond Creek, would be situated between the two monster reservoirs, helping to regulate the flow of water.

On reaching the third, lower and last reach of the river we find this part of the basin in Southern Arizona and Southeastern California, east and south of the Sierra Nevadas, where the Palo Verde, Coachella and Imperial Valleys lie. This area includes about 69,000 square miles, with a gentle climate, where continuous agriculture is possible. It is a rolling country much better suited for farming than either of the other two sections, and all activities in this zone are expected to be primarily agricultural. About 1,000,000 acres are under irrigation now and another 1,000,000 can be added.

It is the opinion of those who have studied the question that there will be sufficient water to meet the irrigation needs of all the States and at the same time develop 6,000,000 horse power, always including the construction of reservoirs as a necessary first step. These reservoirs would overcome the present variable flow and would be a benefit not only to the States in the lower basin, but to those in the upper in almost equal degree. During years of low water in the river bed the upper States could utilize it to the fullest degree, and the lower States could draw upon the water held in the reservoirs. In its natural state practically the whole flow of the river is absorbed.

The first and most pressing problem before the engineers and State economists is to check the annual floods which do so much damage in the Imperial, Coachella and Palo Verde Valleys. They suffer from Spring floods in the Colorado and Winter floods in the Gila River. To relieve this situation a bill has been introduced in Congress asking $70,000,000, $30,000,000 of which would be used to construct the dam at Boulder Canyon, putting a definite check to the Colorado floods and providing irrigation for 1,000,000 acres, all within a short time and almost out of hand. The other $20,000,000 would be devoted to a canal which is needed to insure water for the Imperial Valley by an all-American route instead of one partly through Mexican territory, as at present. It has been estimated that the cost of the whole operation would be repaid to the Government in forty years by the sale of power and the collection of fees for the delivery of water.

The natural flow of the Colorado River averages almost 20,000,000 acre feet a year. About one-third of this water is used, the lower basin utilizing some 3,700,000 acre feet. The compact between the States and the nation awards 8,500,000 acre feet, or more than double the present need, to the lower basin. This has been held enough to develop all feasible projects in that zone.

The upper basin will receive 7,500,000 feet, or 1,000,000 less than the lower, which also is double its requirement at this time and adequate for all contingencies. If this quantity of water is not consumed in irrigation the remainder will run down through the canyons for use in the lower basins. Some 4,000,000 acre feet of the total flow will be held in reserve for future apportionment or as unforeseen needs may appear. Thus the two basins are fully protected and have 4,000,000 acre feet on which to draw.

Biggest Engineering Feat.

Besides the Colorado River project the digging of the Panama Canal might be called a trifling matter and the Suez Canal merely a ditch between two oceans. Such engineering achievements as the Simplon Tunnel also fail to compare, and even the great Assouan Dam in Africa is of less importance. The water power and irrigation works on the Colorado will be in a series, extending for more than 1,000 miles, a greater undertaking in extent, difficulties and possibilities than anything since Cecil Rhodes projected the Cape-to-Cairo Railroad. As the work advances it will make agriculture possible in a large territory where it has been almost impossible. Instead of the country affected depending on cattle, mining and sheep, a dozen new avenues of activity will be opened up. There is little industry outside of mining in the territory, and none is probable on any broad scale, considering the cost of coal or oil and the difficulties and distance of transport. But with the erection of the new dams, storing millions of feet of horse power in "white coal," the whole region should awaken to industry, conducted with the lowest possible outlay for power.

The Colorado plan has so many possibilities that its realization will seem as though a fairy tale had come true. The 242,000 acres involved may contain untold deposits of minerals. In fact, they are known to have rich outcroppings of many veins, but the cost of development has been prohibitive.

The vision which made possible a co-operative agreement by the seven States may largely be credited to the League of the Southwest, an organization which has undertaken to support every movement looking to the betterment of that section. This organization brought pressure to bear on its Governors and other public men, ending in the appointment of the commission which made such an unusual record, transforming the matter from the theoretical to the brass-tack stage in eleven months. It is believed that all of the States will have passed the necessary ratification measures in three months, when the $70,000,000 appropriation will be urged in Congress. The bill is expected to pass, and soon afterward hammers should be ringing at Boulder Canyon in the first step toward conquering the Colorado.

URGENT NEED OF NATIONAL REFORESTATION

By PAT HARRISON,
U. S. Senator From Mississippi.

IT is hardly possible to conceive of any question more important than that of reforestation. No other civilized country has done so little toward the conservation of its forests and the adoption of a comprehensive policy of reforestation as has our own. Only a few decades ago we possessed what then seemed to be an inexhaustible supply of timber. Every section except the prairies and plains and arid lands of the great Middle West was fortunate in its rich supply of timber. But, with a ruthless hand and in practically all instances without a thought of the future —all of which was acquiesced in by the Government—forests have been denuded.

This movement first started in New England, then moved to the Great Lakes section, then to the South, and now has entered the forests of the great Northwest. At the present rate and under the present policy it will be only a few years until the supply of timber in the United States will be exhausted and our home-builders and industries will be compelled to seek their supplies from either Siberia or South America.

It is estimated that fully 5,000,000 acres of forests in the United States are being stripped annually, and that 10,000,000 acres are being burned annually. It is estimated that there are 460,000,000 acres of forest lands in the United States, and that already 70 per cent. has been logged of its virgin timber; that 24 per cent. contains today only a partial growth of culled timber of useful sizes; that 29 per cent. has been stripped clean of merchantable products, and that 17 per cent. of our forest area, or an aggregate of 81,000,000 acres, has been stripped clean of its merchantable timber, burned over, and is lying practically idle and known as unproductive land. With a situation such as these figures reveal confronting the American people, it is little less than a

crime for the Government to delay longer in solving the problem and adopting at the earliest possible date a fixed policy that will insure to the American people in the future an adequate supply of timber to meet the growing needs of our industries and home-builders.

When we are confronted with the facts revealing the rapidly decreasing supply of timber for all the purposes to which it is adapted, and then contemplate the constantly increasing price of timber and timber products, it should be quite enough to arouse us to what will happen unless the problem is met. The man of family of ordinary means will be compelled to go homeless, and the railroads and great industrial plants which are compelled to use timbers in laying their tracks or in the construction of their rolling stock or manufacturing plants, will increase the transportation rates or the price of their manufactured products because of the increased price they naturally will have to pay for the timber products they will be compelled to use.

In 1909 the United States produced its entire news print supply, but today we have become dependent upon foreign sources for at least two-thirds of our newsprint or its raw materials. Last year only about one-third of the American newspapers were printed upon the product of American forests. The needed supply of print paper for magazines, periodicals, books and newspapers is increasing daily. Our people are reading more than ever before. They are demanding more features in the newspapers and magazines than ever before. Business of every kind is employing the medium of the press or magazines in advertising more than ever before. All of these influences naturally increase the consumption in the United States of wood pulp. If this situation continues, and there is no reason to expect it to stop, the production of wood pulp upon our forest lands should proportionately

increase to meet the demands and needs of the future. We are confronted, however, with the exact antithesis of that situation. While the consumption of wood pulp is increasing, the production of wood pulp is decreasing, and year by year the growing importations of wood-pulp will become more necessary.

When we survey the situation with reference to the policy of the Federal Government and that of the States toward reforestation, we find that the Federal Government, excepting to procure some data and some information, much of which is valuable, with reference to the policy of other countries in the matter of reforestation, has restricted its administration, and I might say, study, to our national forests. Only a few of the States have given any thought to this important subject of reforestation or have adopted a policy of protection of the forests from fires. There is much that can be done by the large and small timber owners throughout the country if the State Governments and the Federal Government will co-operate in looking into this question in a broad, progressive and unselfish way, and adopt a policy whereby all the various interests as well as the State and Federal Governments can co-operate in conserving our forests, and through a policy of reforestation make possible the needs of tomorrow.

It has been shown that through a wise culling process in cutting timber, or through a policy of replanting and by protection from fire forests can be made to contribute splendid dividends upon the investment without destruction of the investment. These are some of the problems that the committee representing the Senate will undertake to study; and it is to be hoped that such a policy can be recommended to the Congress by this committee as will meet the approval of Congress and the American people.

February 25, 1923

PINCHOT SAYS FALL PLANNED HUGE GRAB

Declares There Was Scheme for Raid on Billions of Dollars' Worth of Resources.

ATLANTIC CITY, April 25.—Governor Gifford Pinchot of Pennsylvania told the American Society of Newspaper Editors tonight that after his protest to President Harding against the appointment of Albert B. Fall as Secretary of the Interior had failed he had enlisted the aid of the press of the country in a struggle for conservation of natural resources and was successful in his fight.

"Senator Fall's attack was on the conservation policy as a whole," Governor Pinchot said. "What he and his friends were after was not simply the navy's oil, but billions of dollars worth of other natural resources held by the Government for the benefit of all the people."

Once in office, Mr. Pinchot said, Fall had attacked the nation's forests.

"As a Western stockman the United States Forest Service had properly refused him privileges to which he had no right," continued the Governor. "As a means of getting even he undertook to secure the transfer of the national forests to his department. In the course of his campaign he succeeded in persuading President Harding to support him. The Brown commission for the reorganization of the Government fell in behind. The anti-conservation members of Congress were eagerly with him. But in spite of all of it he was defeated and defeated directly by the power of the press.

"Direct personal communication was established with the editors of five or six thousand of the most influential journals in America. A stream of editorials came pouring into Washington so definite, so forceful, so influential, that President Harding said to me in the White House, 'You are absolutely wrong in opposing the transfer of the national forests, but I pay you the compliment of saying that he cannot put it over against your opposition.'"

Governor Pinchot asserted that Fall arranged President Harding's trip to Alaska in order to turn the resources of that territory over to the "grabbers," but that President Harding defeated the scheme for "all time" in his last public utterance at Seattle.

April 26, 1924

AN AUTOMATIC SPRINKLER FOR LARGE GARDEN AREAS

A NOVEL system of irrigation has been devised which makes it possible to create showers of varying intensity over many acres. The shower may be varied from a slight drizzle to a heavy downpour, and be made to fall continuously or at any desired interval. The system is an elaboration of the primitive sprinkling pot or garden hose and works automatically. La Pluvoise, as the device is called, is the invention of a French engineer, Edmond Rolland, and is in successful operation in France.

An area of fifteen acres may be watered by one of these sprinklers in the care of a single workman. The system may be much further extended.

Every square foot of this area is covered by the artificial showers. The rainfall, which may amount to thousands of gallons of water every hour, is controlled by turning a single valve.

Simple in design, a long steel girder serves as a runway for the arm that distributes the water. To irrigate an area of fifteen acres a runway is constructed some 300 feet in length. The sprinkling device consists of two great metal arms, which roll up and down the runway, extending at right angles. It has been found that a length of 90 feet for each arm gives the best results. The water is sprayed from a series of jets placed at intervals along these arms, which move back and forth across the entire field.

The machine is operated wholly by the pressure of the water supplied. A slight pressure serves to keep the machine in constant motion. The water piped from an ordinary supply answers if it has sufficient pressure to rise 15 feet. When the water reaches this height it flows over a turbine wheel, which distributes it to the sprinklers and keeps the machine in operation. Once the water begins to flow, the long transverse arm begins to move down the long runway. The speed at which it moves is easily controlled.

As the arm moves, the water jets reach every inch of ground. When the arm has reached the end of the runway it is automatically reversed and returns, moving back and forth as long as the pressure keeps up. It can be checked at any point and made to move over a restricted area. The arm moves at the rate of 30 feet a minute. Once set in motion, it needs no watching, and the gentle rains from heaven fall continuously. The machine is guaranteed to operate for fifteen years.

September 2, 1923

DESERT ONCE, NOW YIELDS RICH CROPS

Transformation of Imperial Valley, California, Effected in Period of Twenty Years.

CLIMATE IS EXCEPTIONAL

Early Harvests Bring Fancy Prices for Out-of-Season Fruits and Vegetables.

California's Imperial Valley, which was nothing but a naked desert in 1900, is considered today the richest producing area in the world, with 534,674 acres under irrigation, a population in excess of 50,000, an assessed property valuation of $52,000,000, and a crop production worth $65,000,000 a year. This striking development, says George Law in a statement issued through the Los Angeles Chamber of Commerce, is the result of calculated investment accompanied by "such vision as has built railroads and manufacturing industries."

It was seen, Mr. Law explains, that this territory with a Summer as long as Alaska's Winter, could be capitalized to produce food all the time, in and out of season. Owing to the exceptional climate of the valley there can be produced in it vegetables and fruit to reach the market at periods when other sources are non-productive. The dividends on this kind of farming are correspondingly exceptional, ranging to 100 per cent. and more. People have gone to Imperial to make big money, says Mr. Law, and the fact that many have succeeded is responsible for the rapid development of the district.

Exceptional returns from plain crops which pay but a modicum elsewhere are due to early, or Winter, harvest. Cantaloupes ripen usually about May 15, when the Imperial growers can command fancy prices from consumers eager for early melons. Asparagus is harvested the middle of February, peas in March, onions, squash, tomatoes, potatoes in April. Lettuce is so planted as to ripen after the crop is out elsewhere, about Dec. 15, and to continues until the Coast lettuce comes in. While Imperial is the only source of supply for these popular vegetables, with the exception of small desert gardens elsewhere and hothouses, the receipts from sales, according to Mr. Law, are such as growers of vegetables never saw before. It is a limited period of cream without milk, provided the crops are not set back by pests or frosts, as sometimes happens.

A similar marketing advantage obtains in the case of certain fruits. The first table grapes can be shipped early in June; apricots about April 20, grapefruit Nov. 1, and others three to six weeks earlier than the regular California crops. Such an advantage, it is pointed out, is naturally fostering the planting of vineyards and orchards to the kinds of grapes and fruits that adapt themselves well to desert heat.

Happy Accident of Climate.

"This early ripening is a happy accident of climate," Mr. Law continues. "It would amount to nothing without accompanying essentials of successful production, such as farmers must bring to their task elsewhere. But Imperial Valley has its dramatic features. The fertilizing silt in the Colorado River water is deserving of as much stress as the climate. This adds nitrogen, potash and phosphorus and considerable organic matter to the soil, leaving little deliberate fertilizing for the farmer to do except turning under a green cover crop occasionally. With every acre-foot of water that is served to Imperial land, rich sediment from the Rocky Mountains finds a new resting place.

"In passing through the valley one cannot help but be impressed with the fact that not only the crops but even the ground is growing. The canals and ditches, in which dredges and crews of Mexican shovelers are continually at work, show the most conspicuous growth. The banks are gradually leaving the roadways in ravines. It is an accretion of land from the distant Rocky Mountains that is likely to prove embarrassing in time, notwithstanding Imperial's 180 feet below the sea. But the silt problem together with that of floods will be largely removed by the construction of dams to regulate the flow of the river.

"The first glance reveals Imperial Valley as a great investment garden—a garden to put money and energy into with the object of taking out quick and lordly returns. This inciting phase is a true and permanent characteristic; but the transient investor is giving way to the permanent rancher and home builder. Cultural conditions are in part compelling the change, but appreciation of the valley after twenty years of reclamation is the more potent factor.

"Now that time has established certain dependable facts about crops in Imperial Valley, the ranchers are establishing their gardens, fields, and orchards upon the sure foundation of those facts. In playing to the early markets the settler-rancher finds that a program of continued success follows better upon working a small tract of say forty acres thoroughly, by a plan of rotation, than in spreading out with less care over a greater acreage. The vegetables and the orchard fruits, which make the best returns, require the loamy or soft type of soil. The silt in the irrigating water, while helping to return fertilizer exhausted by such crops, tends to harden the soil. Hence the need, both for fertility and looseness of composition, of incorporating cover crops. The leasing practice ignores this necessity. Hence continued prosperity depends upon care and system, rarely bestowed except by the owner who is establishing both a permanent investment and a home.

Soft and Hard Land.

"The area of soft land, suitable for truck-gardening and orchards, is much less than that of hard clay land. The hard land, however, is excellent for alfalfa, grain and cotton. Fully a quarter of the present acreage under cultivation is in alfalfa, more than a quarter in barley, grain, sorghums and wheat. These two extensive industries, combined with dairying and the fattening of sheep and cattle from the desert ranges, form the great sure basis of the valley's prosperity. Ready markets for alfalfa, dairy products and livestock in Los Angeles, San Diego, and other centres of population along the coast, make the rewards to those engaged in these industries certain, though more moderate than returns from off-season produce. The hard land is cheaper, the whole investment less, and the hazard practically nil.

"The fattening of livestock is not at present as satisfactory as it should be, owing to insecure markets. The stock industry involves three parties, the range cattleman or sheepman, the farming finisher, and the packer. Circumstances at present tend to place the advantage in the hands of the packer. The farmers in Imperial Valley, who do the finishing, face a market that deprives them of a margin of profit. The future of this essential industry, which should be able to pay each party a satisfactory profit, depends, the farmers believe, upon the parties getting together to arrange a scale of prices which will take care of the first and second parties as well as the third.

"Cotton, which is being grown this year on about 50,000 acres, holds the promise of becoming a dependable industry for the vast area of cheap land. With the price around 30 cents a pound, the cotton growers are able to net profits comparable with those from deciduous fruits and walnuts. The steady development of cotton mills, textile industries and tire factories in Los Angeles guarantees the future of the Imperial Valley cotton industry. Woolen mills extend the increasing demand to the fatteners and shearers of sheep.

"Los Angeles and Imperial Valley stand in close reciprocal relations, although separated by 215 miles. Los Angeles money has made this desert garden; both are now growing in pace with each other's needs. The demand in Los Angeles grows as Imperial Valley increases the supply."

March 9, 1924

CONGRESS SEEKS LIGHT ON MUSCLE SHOALS

Puzzle of How to Utilize Effectively the Great Tennessee Water Power Project Comes Again Before Joint Committee—Remarkable Engineering Work, Inspired by War Fervor, Passes High Tests

By FRANK BOHN.

THE great dam at Muscle Shoals is probably the most efficient piece of construction work ever accomplished in this country. It will be worth much more than its cost, ultimately, as a producer of power and it is a splendid example of modern engineering skill, but so far it has been treated as a belated orphan produced by the economic orgy of war.

Probably no local issue in fifty years has perplexed Congress as much as the disposition of this property. The matter has been an issue for more than seven years. No executive and no Congress during that time has found a solution. No real buyer has appeared except Henry Ford, and Ford's offer did not satisfy Congress.

The next move is to be a public hearing on April 26 under a joint resolution of Congress. A committee of six, three from each house, will take testimony; prospective buyers will be there and representatives of the farm organizations, and agents of near-by cities and public utility corporations. Possibly somebody will suggest a solution of the Muscle Shoals puzzle.

The dam and power plant cost $55,000,000. There is, besides, one nitrogen plant (Haber method) which cost $12,000,000. This plant is probably worth its cost. But the cyanamid nitrogen plant, which cost $70,000,000, is now believed to be antiquated and will probably have to be junked.

Impressions of Muscle Shoals.

Is all this expenditure to be lost to the Federal Treasury and to the taxpayer? No. There is a value attaching to this great work that cannot be destroyed. It is concerned with the execution of the nation's larger policy of hydroelectrical development and should profoundly affect that extensive task through the generations to come.

An old field correspondent of an engineering publication recently made an extended visit to Muscle Shoals. He spent most of his nights on the dam.

"I'm enjoying an entirely new engineering experience," he said. "This job is so magnificent and wonderful that I have come to believe that it has a soul; so I've spent my nights cultivating an acquaintance with this strange new spiritual being. While I feel its presence in looking at it by day, I seem to sense the inner life of the job, in its peculiar power, only when I go and put my hands upon it at night."

Common laborers not a few have been heard to express themselves as moved with the same feeling that obsessed the engineering writer. This peculiar stimulation produced by the grandeur of a gigantic architectural masterpiece is, of course, a most ancient experience of the poetic mind. Those who wrought at designing and constructing the Gothic cathedrals of the Middle Ages were so thrilled by the vast conceptions of their work that their hearts as well as their hands moved in unison. When Cheops was done, probably the slaves who toiled upon it that day paused to look up and wonder.

Great Dam a Vast Work.

At Muscle Shoals, within a sweep of the eye, one can see what modern man can do with his machines toward the control of his environment. Since we did not make the mountains or the stars, our appreciation of them must ever be objective and limited. But in our great works of architecture and engineering man appears as creator. We have wrought ourselves into them—not ourselves as individuals, nor as mere groups of individuals—but as vast and complex civilizations.

At Muscle Shoals our contemporary America has constructed a modern and impressive work.

On the Colorado, we are told, we are to construct a dam 550 feet high, higher than the Washington Monument. On the St. Lawrence we are presently to throw dams across a most magnificent river. But the great dam at Muscle Shoals marks the present measure of our striving. Here, for reasons we shall come to presently, is a highly significant American contribution to contemporary civilization. If one starts out in the dawn—if one wanders over and about it, and looks and learns and thinks, until, beyond the clear waters of Wilson Lake above the dam, there comes the red flush of the next dawn—during even these twenty-four short hours one can begin to perceive both the statesmanship and the poetry inherent in modern hydroelectric engineering.

To understand this structure one must go back to its simplest form, the old mill dam on the creek. That little dam poured the waters of the stream over a wheel at the end of a race or funnel. Its wheel, in turning, turned the millstone by direct contact.

Mill Dam Grown Prodigious.

At Muscle Shoals the dam has grown to be nearly a mile long and ninety-five feet high. It is constructed of 1,400,000 cubic yards of concrete of the highest quality. At one end are two navigation locks, each 60 by 300 feet, made to lift vessels up over the dam. At the other end the electric power house has taken the place of the old mill wheel. This power house now contains eight units, which can develop 260,000 horsepower. Eventually there will be eighteen, capable of developing 612,000 horsepower.

All United States Government work of the Muscle Shoals sort is undertaken by the Corps of Engineers of the army. But when a peculiarly difficult task is under way the corps is permitted by law to retain expert civilian help. The corps and the civilian staff, when they work together as at Muscle Shoals, are designated the Department of Engineers.

The writer spent the better part of a day in company with the designing and supervising engineer, examining one of the 260,000 horsepower units. The great dam, considered as a simple instrument, functions through these colossal machines, the power of the falling water becoming the power of electricity. The machinery looks its part—a twentieth-century Frankenstein that has laid an unready Govern-

ment by the ears. Nobody at Washington appears to know what to do with it. The engineers on the job are prepared with the correct answer, but the politicians apparently are not seeking sound economic engineering advice regarding Muscle Shoals.

Conceived in Patriotism.

In 1917-18 certain members of Congress decided that it would be advisable for the Federal Government to build a power plant at Muscle Shoals. The ostensible purpose was to furnish power to run two nitrate plants, which were to turn out the product needed in the manufacture of high explosives for the war. Since the decision taken in 1917-18 to create Muscle Shoals for national defense purposes, synthetic chemistry has proved that water power is no longer necessary in the slightest degree for national defense or for peace-time fertilizer production.

The war ended Nov. 11, 1918; the first unit in the power house at Muscle Shoals was not assembled and ready for testing in December, 1925. Yet Congress wrought better than it knew. It furnished a skillful artist and teacher with his supreme opportunity to produce his masterpiece and so instruct the nation.

The designer, as he looked at the waters of the Tennessee rushing over the Shoals on a day early in 1918, perceived the basic facts of the situation. He said to himself:

"If I am indeed to work here instead of with the army in France, I will make my work count to the utmost in the service of my country." He would set an abiding example in engineering efficiency; with the support of the Government he would create a standard to which others, engaged upon similar tasks, must repair. Whereupon he bent himself to his task. And the details are as interesting, indeed, as the fundamental conception.

The first question an engineer must answer, in making his preliminary plans for a dam, concerns the structure of the river bed and the banks. Some might have been satisfied with the information obtained from a few score of borings in the rocky bed. Altogether over twelve hundred exploration holes were drilled. These varied from thirty to one hundred feet in depth. Where any cavity was found it was filled with concrete.

Water-bearing seams were filled and then retested by compressed air for the determination of their perfect tightness.

In the south bluff abutment section of the dam the explorations for the proving of tightness revealed unexpected porosity and made necessary the boring of over half a mile of tunnels. These tunnels revealed various unexpected and large cavities below the new pool levels, and altogether some 15,000 cubic yards of concrete were necessary to close up the cavities. Geological "dentistry" of this sort is always necessary in the construction of dams; but the stratified condition of the rock in the Muscle Shoals district required special attention in the

The Locks Above the Wilson Dam

The Wilson Dam, Seen From Down Stream

A Close-up of the Spillway From the Down Stream Side

exploration of the river bed. Tests after the dam was finished have proved, in so far as human observation of a fact of this sort can prove, that the geological dentistry over this entire mile stretch of river bottom was successful.

Everything considered, the damming up of a strong river is the most extraordinary task to which engineering skill can address itself. For, while any workman may sometimes deceive his employer or the most careful inspector, and a crafty lawyer may write a tricky phrase into a contract, there is at least one place on earth where tricks fail, errors are discovered and poor work cannot be disguised. Against the river no man can play the cheat. In the end some one must redress the balance.

Passes Efficiency Tests.

Here, upon the shoals of the Middle Tennessee, was to be a dam which could serve our children's children and not settle in the bed, slip a single inch down stream, or crumble at any point in all the years till then.

There are two "schools" of concrete making. One advocates concrete that is made soft in the mixer, and the other a product that is stiffer and harder. Soft concrete is made and placed more quickly than hard concrete. The latter, costing more time and effort, stands up better and lasts longer. The Wilson Dam is built of hard concrete.

The materials for making concrete consist of broken stone, or gravel, of about four and one-half parts, an acceptable grade of sand, two and a half parts, and an acceptable Portland cement, one part. The testing of these materials involves a broad experience of their use. After they have been assembled they must be mixed with just the right amount of water in order that the cement may be distributed throughout the mixture and thereafter complete its hydration, or set.

In the construction of the Wilson Dam every reasonable expense was allowed which might increase the efficiency of the work. Recent efficiency tests have fully justified the efforts put forth. The object in mind was never primarily to complete a certain part of the work on a certain date. Indeed, the largest ultimate pur-

pose throughout the work was, "Let us see first how well we can build." The dam should be a thing of beauty as well as of utility—a creation in which the whole nation might take pride. But its beauty should rise from the rock-bottom of its truth; in the foundations of its truth there should be no dark and hidden holes.

Ideals Wrought in Concrete.

This dam is comparable to a great cathedral, built when the builders wrought character and spirit into every carving and buttress. The likeness between the two structures is seen in all its clearness at the point to which we are now drawing attention. It is primarily because of the character wrought into this great work that it becomes so large a national asset. In constructing Notre Dame or Westminster Abbey each toiler conceived of his day's work in terms of the finished structure.

From beginning to end, in building Wilson Dam no contractor was permitted to interfere with this unity of purpose. All the work was done by the direct employment of all the help. Even the humblest was inspired with the vision without which no such work can attain perfection. So this colossal block of concrete is a kind of temple in which there is indwelling the pure ideals of the nation.

The top of the Wilson Dam is a bridge—a broad, smooth highway flung a mile across the Tennessee. It is a link in the Dixie Highway. At night this bridge is lighted up like Fifth Avenue. The lighting system was made as much a study by experts as boring the holes in the river bed. The lamps are placed in the walls which ensconce the roadway on either side. From above or below one sees no lights at all only an orange glow, like a halo, over the bridge.

From almost every aspect the purposes of the dam are always indicated in the seeing. Take the matter of the gates that control the flow of the river. These gates are fifty-eight

in number. They are raised and lowered by hydraulic pressure. A single man can open them all within two hours. When, in December, the writer looked down upon the flood below, the roaring waters poured through only three open gates. The purpose of the gates is to accomplish perfect flood control.

Purposes of the Dams.

In the power house the river is thrown against the collar and pulls the load. If there is more water than is needed at any time, the surplus flow automatically is directed through the gates and thrown away.

At the north end of the dam are the navigation locks. A small amount of power will serve to control their use. A generation ago the soldier and scholar General Robert of the Corps of Engineers of the Army dug the ship canal around the shoals. Time presses so fast in our generation that though the canal was the work of General Robert's later period of service he lived to see the beginnings of the dam which would hide his canal ninety-five feet deep under water.

Everywhere hydroelectric development is proving a benediction to the inland waterways. Instead of digging canals, dams are built to flood the shoals and shallow places in the rivers. And the hydroelectricity often pays for the improvement many times over. The completed dams, 1, 2 and 3, on the Tennessee River, will solve the problem of navigation at Muscle Shoals as it could not be solved by digging a canal. It may be noted here that dam 1, two miles downstream, is already completed. It has been constructed for navigation purposes only and is 10 feet high. Wilson Dam, 3, twenty miles above 2, will be 40 feet high and cover the shoals as far as Decatur. All three will account for over 140 feet of drop in the river.

These locks lend an added interest to the dam. With the improvement of dam 3, cargo and passenger ships and sporting craft of every description will ride the waves of the lake for fifty miles.

The Public Advantage.

That works built of concrete may have beauty as well as utility is demonstrated by the finished result at Muscle Shoals. "The designs must first be perfect from a utility standpoint, and then be made as beautiful and uplifting as possible, within reason"—this has been the daily motto from the time when one man visualized and began this construction, all the way through to the end, when eighty engineers and 4,000 workmen were employed. To succeed it was necessary to represent the simple

beauty of colossal strength.

How shall this power be used to the greatest public advantage? That happens to be at present a political problem; and toward efforts to solve political problems people as well as politicians are inclined to shirk responsibility.

The engineer went to Niagara and commanded the cataract to work for man. Now the engineer has gone to the Tennessee and built a new Niag-

ara. Such creative exploits have the touch of genius which inspired Shakespeare and Galileo. Most political leaders give the impression that they are one with Rip Van Winkle and Falstaff. The best answer for the question as to whether they are growing better or worse in this decade will soon be furnished by the disposition that Congress makes of the Muscle Shoals plant.

April 18, 1926

OUR NATIONAL PARKS BECOME UNIVERSITIES

In the First Decade of Their Organization Public Attendance Has Increased Fivefold—Nature Is the Teacher And Textbook in the Great Laboratories

OUR national parks may be considered among the few great unchanging things in America. The processes whereby nature shaped them in their magnificence go on, but at a rate of change attuned to the universe and hence infinitesimal in the experience of living men.

That human hands should make no material alteration is the very reason for their preservation as national parks. Their animals know no hunter's gun; their trees no woodman's axe. Roads are threaded in and out so painstakingly that scenery is conserved and buildings are designed and placed to harmonize or be inconspicuous against the background nature affords. Nevertheless, change has come—not in the revelations the parks have to offer, but in the manner in which they have been brought to offer them.

The parks are still playgrounds, the high Sierras still echo jazz strains and the careless laughter of city vacationists who come to keep cool and to entertain themselves. The Grand Canyon draws Summer campers who give the canyon itself but a casual look. Yet the national park system is evolving into something else. It has begun to be seen and accepted as a great out-of-doors university where nature is the "supreme teacher as well as the master textbook," in the words of Hubert Work, Secretary of the Department of the Interior.

The Parks and Education.

Appreciation of the educational value of the parks and their general use to that end have increased through the first decade of the

National Park Service, a milestone passed this month. Before the organization of the service in April, 1917, the national parks and monuments under the Department of the Interior were administered by a bureau charged with manifold activities. Their affairs were attended to if and when other things left time. Before Franklin K. Lane as Secretary of the Department of the Interior made a cause of national park work, the parks presented a conglomeration, it was said, rather than a system; and even Secretary Lane's improvements left many things to be improved.

A new era began with the creation of the service as a separate bureau of the Department of the Interior. The first decade has witnessed physical expansion, three parks— the Grand Canyon of the Colorado, Zion in Southern Utah and Lafayette on the coast of Maine—having been added to the system, bringing the total to nineteen. Four others have been authorized, though not yet established: Utah in the West, and Great Smoky Mountains, Shenandoah and Mammoth Cave in the East. The boundaries of several of the older parks have been extended, and the number of national monuments has been increased from nineteen to thirty-two. A national monument, generally speaking, is distinguished from a national park as a smaller area reserved by Presidential proclamation instead of Congressional act, to protect some particular object of historic and scientific value.

An even more striking development has been the concerted movement of people to the

parks in season, for which the automobile is largely credited. During the first year of the National Park Service's regime there were fewer than a half million visitors to the parks and monuments; last year were almost two and a half millions.

New Park Problems.

These throngs have given rise to questions of water supply, sanitation, housing, transportation, food and fuel that have assumed metropolitan proportions. In some of the parks it is necessary even to have United States marshals on duty to try cases of speeding and reckless driving. Yet the administration, with great and growing problems on its hands, has not been content merely to see the crowds safely and comfortably in, through and out, but has charged itself as well with teaching them something as they go.

In the conception of the national park system as a super-university of the natural sciences, each college presents a personality of its own, specializing on some particular natural endowment. For the study of glacier action the public may turn to Glacier National Park, in Montana, with its ruggedly carved mountains, its twisted, scooped-out valleys, its dozens of small glaciers and its mirror lakes. Or one may take the glacier course of equal advantage at the more accessible Rocky Mountain Park of Colorado, in the most magnificent section of the Great Continental Divide. Again, he may elect Mount Rainier, in Washington, capped by a frozen giant octopus stretching its ice tentacles down twenty-eight grooves at the rate of sixteen inches a day—the greatest single peak glacier system in the United States.

A vast geophysical laboratory lies in the Yellowstone. Here volcanic action of the past has left its trace on plains that steam from pools of rainbow tint and silvery sheen; geysers like "monstrous dancing ghosts" display their waterworks and hot springs ooze over terraces built up into mountains like tinted and frosted birthday cakes. Obsidian Cliff exhibits the black volcanic glass used for arrowheads by Indians who dared not brave the already hissing deities by venturing further into their vaporous domain, and Specimen Ridge shows fossil forests alternating with layers of lava to a height of 2,000 feet.

A college of stream erosion, of which the kindergarten form is the common roadside rain-washed gully, is the Grand Canyon, from the rim of which, a mile down and many miles away, may be seen the tiny silver string of the Colorado, fashioner of these vast amphitheatres of ever-changing hue, these myriad minarets and pinnacles and towers. Another course in the same study is offered in Yosemite Valley, cut to a depth of 3,000 feet by the Merced River before the ice age began and then quarried out by glaciers, leaving granite precipices rising five or six times the height of the Woolworth Building above the valley's wooded floor. Zion, a Yosemite painted in fiery vermilion and dazzling white, gives erosion's story yet another chapter.

Yosemite has its Mariposa Grove of big trees and Grand Canyon has its

STEPHEN T. MATHER

© Harris & Ewing, From Times Wide World.

Director of the National Park Service.

fossil plant and animal remains. But the college of dendrology is Sequoia, in California, where the 4,000-odd-year-old General Sherman tree, almost as high as the Flatiron Building and thick enough to accommodate a driveway and two street car tracks through its heart, is but one of an enormous grove. The college of paleontology is the Petrified Forests in

Arizona, where an area of forty square miles is strewn with prostrate tree stumps of stone. Mesa Verde, where prehistoric people hung cities on the sheer sides of sandstone cliffs, is an institution of anthropological studies.

Laboratories of Science.

Scientists have found the national parks, one after another, rich fields for endeavor, whether with whisk brooms among the habitations of cliff dwellers or with magnifying glasses among twisted strata. More recently schools and colleges have used them as laboratories.

The National Park Service has embarked on a more comprehensive educational program, however, than is involved in cooperation with these groups. As visitors have tended to extend their stay, often camping in the parks for weeks at a time, popular demand has arisen to know the whys of unusual formations; to learn the names of flowers and animals and birds. And in response the National Park Service has undertaken to set forth nature's textbook in terms the average man can understand.

Bulletin boards at the various stopping places speak to the tourists in a different manner than that of ten years ago. They invite him to hike with the nature guide and learn from him the full story of the trail. They give notice of a popular lecture for the evening on the habits of Rocky Mountain sheep or on glaciers and their origin. They advise a pageant setting forth the message of the giant trees or retelling some bit of ancient Indian lore. Rangers chosen for the parks are often naturalists, perhaps college professors or university students who have made specialties of the territory they cover; and scientists are engaged for lecture work and field courses.

Guides and Interpreters.

Interpretative guide service is now available in Yosemite, Yellowstone, Glacier, Mount Rainier, Sequoia, Grand Canyon, Zion, Crater Lake and Mesa Verde National Parks and at the Petrified Forest National Monument and Casa Grande, the ruins of great prehistoric dwellings built on the Arizona desert probably before the time of Columbus. In the Yellowstone, at Grand Canyon, Yosemite and Mount Rainier, natural trails, labeled with geological data and information on the birds, flowers and trees of the region, have been marked out for those who prefer to do their rambling by themselves.

Here and there, largely through the interest of the American Association of Museums and the donations of friends, museums have made their appearance at the parks and monuments. The museum may be a fine building with carefully displayed exhibits and classroom equipment, or it may be a few handmade boxes of specimens gathered under a tent. In any case it is a collection that acts as an index to the large museum outside; as a stimulus to interest and as an explanation of phenomena to be seen as they occur in nature on the spot. Yellowstone, Yosemite and Mesa Verde have full-fledged mu-

seums, and lesser exhibits housed in temporary structures are offered at Mount Rainier, Zion, Sequoia, Rocky Mountain, Glacier and Lafayette. In connection with some of these museums, libraries are being collected for the convenience of the research student.

Tourist Throngs Reached.

One hundred and fifty thousand of the 275,000 visitors to Yosemite were reached last year by the museum, nature guides, lectures and camp fire talks; and in the Yellowstone some 88,000 visitors came into contact with the ranger naturalists. When the other parks are taken in, hundreds of thousands of persons are seen to have availed themselves of the educational facilities offered where none existed before.

Tourist curiosity as to the underground workings that send geysers shooting into the air is not left to be diverted to other attractions or to satisfy itself with their imaginings. Explanation is to be had from the nature guide as to how water collects in the geyser crater at the bottom of a fissure in the form of a constricted tube connecting with the surface; how in the hot strata of earth it becomes sufficiently heated to expand, lifting the column of cooler water above it into the air; how the water later seeps back through the rocks to the crater and how the performance is repeated all over again.

Visitors to Glacier National Park do more than sigh ecstatically over the still depths of crystal lakes and the giddy heights of mountains clawed and scarred by time.

At the Petrified Forest the wayfarer learns how water carrying silica in solution seeped into living trees long ago, filling their cells with deposit and replacing their solid parts as they decayed, until a fabric of stone was wrought to duplicate the fabric of wood. At Mesa Verde the visitor learns the way of life, the building methods and the city planning ideas of America's ancient men.

Park Extension Courses.

The National Park Service in its lately found mission of ministering to the minds as well as to the bodies and the emotions of the traveling millions has pursued its educational policy with a thought also for those who may never see the parks. The University of Nature is gradually building up an extension department designed to reach far afield. Schools, clubs and organizations all over the country, says the director of the service, Stephen T. Mather, are seeking information on the parks in increasing numbers; are asking for lecturers, lantern slides, photographs and moving picture reels to bring the parks within their reach.

Whenever possible the service sends a representative to them, but on account of its limited personnel it has had more generally to adopt the practice of supplying the material and leaving the presentation to an outsider. Such assistance is extended so that as many as possible of those denied the privilege of residence in nature's university may nevertheless learn at her knee something of creation's story.

April 24, 1927

"WILDERNESS AREAS."

The "wilderness area" idea has received fresh impetus from the announcement of the Forest Service in Washington that forty-two tracts of lands in the National Forests, of 100,000 acres more or less apiece, have been set aside for the purpose of preserving the wilderness unspoiled by roads, railroads and camps. This is a distinct gain for a policy which has been slow in obtaining official support. To the credit of the Forest Service it must be said that the new interest in the wilderness area has been manifested despite the fact that the public demand is of only recent origin. The announcement indicates, however, that the Forest Service is not yet certain that the policy will meet with full support, as it points out that grazing and timber cutting will be permitted. Although there are foresters who contend that this can be so regulated as not to detract from the value of the wilderness areas, the public familiar with cut-over regions and overgrazed forest lands will reserve doubts as to the wisdom of this part of the program.

The Forest Service is handicapped by the obligation to make money so that the counties may have their share of the profits from timber rights and grazing to which they are entitled under the law and which are important items in their revenue. It is therefore difficult to decree that large tracts of lands shall be set aside by the Forest Service as non-productive. There are remote regions, it is true, where the nature of the land or the inaccessibility of the forests make lumbering operations unprofitable. These, however, are likely to be preserved in the natural course of events. What the public wants is that areas which might otherwise be cut over shall be preserved.

The Forest Service, like the National Park Service, has a further handicap of which the general public has until recently been unaware—large tracts within the boundaries of the areas designated on the maps as "National Forests" are privately owned. The Government does not control them, nor can it prevent their being logged. To include private lands in the wilderness areas is to run the risk of making the public believe that the lands are safe whereas their preservation depends only on the will of their private owners. The presumption is, of course, that the Forest Service will exclude such lands wherever possible.

Wilderness areas have been defined as regions more or less forested where primitive nature is modified to the least possible degree by human agencies. It is desirable that they have special natural attractions and plentiful game and fish. But most important is the condition that they be free from roads, railroads and settlements, so that they may be preserved in their original state. Such areas are already being set aside in the National Parks—in fact, the original purpose of the parks was to preserve entire tracts in their wild state. But in the parks, as in the forests, it has been necessary to provide for motor campers and tourists. In so doing the tendency has been to alter the wilderness nature of the regions. The policy of setting aside certain reserved areas within the parks and forests where motors may not penetrate and where nature will remain unspoiled would seem to be the most sensible way of adjusting this need of conservation to the necessity of development.

November 4, 1928

FOREST RESERVES TREBLED.

Review of 25-Year Period Shows Great Expansion in Use.

Special to The New York Times.

WASHINGTON, March 20.—Great expansion in the use of national forests and in the protection and development of the forest resources was reported in a review for the last twenty-five years issued today by the Forest Service of the Department of Agriculture.

Forest reserves in 1925 numbered sixty, with an area of approximately 59,000,000 acres. There are now 150 national forests, covering 160,000,000 acres.

In 1905 there were no fire towers or lookout stations and today 831 are maintained. A total of 1,186 public camp grounds have been improved in the twenty-five-year period. Since 1907 the mileage of national forest roads has been extended from 330 to 16,730. Forest trails have increased from 6,644 to 47,175 miles.

Receipts of the forest reserves in 1905 amounted to $85,000, all for timber sold. Revenue from all sources last year totaled $6,299,802.

In the latest year for which records are available 23,000,000 persons visited the forests.

March 21, 1930

HUGE DEVELOPMENT PLAN FOR SIX STATES IN SOUTH IS DRAFTED BY ROOSEVELT

By JAMES A. HAGERTY.

Special to THE NEW YORK TIMES.

WARM SPRINGS, Ga., Feb. 2.—Development of the entire Tennessee River watershed on a gigantic scale to link water power, flood control, reforestation, agriculture and industry in one vast experiment was proposed today by President-elect Roosevelt as a way to relieve unemployment and restore the balance between urban and rural populations.

Seated before the fireplace in the "Little White House," Mr. Roosevelt outlined to newspaper men what he characterized as "probably the widest experiment ever conducted by a government."

He declared that if the plan should be successful, it would be self-sustaining; he estimated that its adoption would put 200,000 men to work in the Tennessee River watershed alone.

The President-elect indicated that he hoped to extend the plan to other sections of the United States and reestablish American life on a basis that would end unemployment and decentralize industry.

Roosevelt's Outline of Plan.

As outlined by Mr. Roosevelt, the Tennessee River project, involving half a dozen States, will include:

1—Reforestation of the hillsides of the watershed, which alone would employ from 50,000 to 75,000 men.
2—Creation of flood control basins in the upper valleys of the Tennessee River watershed, of which the most important would be that at Cove Creek, not far from Knoxville.
3—Water-power development, beginning with full utilization of the plant already built at Muscle Shoals, to provide cheaper power for residents of cities, States and farms.
4—Reclamation for farm use of the fertile bottom lands of the river, in which farming is now prevented by frequent floods.
5—Elimination of the unprofitable agricultural lands by reforestation.
6—Improvement of navigation.
7—Stimulation of decentralized industry in the region by the supply of cheap power.

"Attacked from all angles, this project should give work eventually to about 200,000 men," he went on.

"We have been going at these projects piecemeal ever since the days of T.R. (Theodore Roosevelt) and Gifford Pinchot, who were pioneers of reforestation in this country. I believe it is now time to tie up all these various developments into one great comprehensive plan within a given area."

Mr. Roosevelt, who expressed satisfaction later in the day on being informed that the Washington and newspaper reaction to the project seemed to be favorable, said he did not believe that the reclamation of the bottom lands would increase the area of farm lands, since marginal lands now used for farming would be eliminated by reforestation.

Holds Foresting "Safe Investment."

Speaking of reforestation, in which he has a particular interest, Mr. Roosevelt explained that by it he meant more than planting seedling trees.

"When forest land is cut over and a second or third growth springs up, the growth should be thinned out by uprooting about four out of five of the young trees to give the remaining trees a chance to grow," he said. "Reforestation of this kind would supply jobs immediately.

"I regard it as a safe investment to buy and take care of land of this character. The money will come back through the sale of the tree crop. There always will be a market for timber and it is time we should act to get a tree crop, for we now are consuming three or four times as much timber as is produced annually."

There were three purposes in reforestation, with the growing of merchantable timber only one of them. The other purposes, he said, were prevention of soil erosion and the retarding of flood conditions.

"It is merely a case of applying the principle of city planning to a larger area," he said. "If it is successful, we can apply it to other watersheds, and there are watersheds in every part of the country."

Mr. Roosevelt's plan is the direct result of his visit to Muscle Shoals two weeks ago, when he inspected the government-owned plants there with Senator Norris, sponsor of a bill for using the power plant to its full capacity.

The evening after the visit the President-elect indicated in a speech at Montgomery, Ala., that he intended to propose to Congress a plan for development of the entire Tennessee valley.

Mr. Roosevelt announced that soon after he took office March 4 he would ask the various governmental agencies concerned to make surveys so that he might present the entire project to Congress as soon as possible. He asserted his belief that the project was "bankable" and that there was no doubt that bonds could be issued for the undertaking.

"If the project is successful, and I am confident it will be," he said, "I think the development will be the forerunner of similar projects in other parts of the country, such as in the watersheds of the Ohio, Missouri and Arkansas rivers and in the Columbia River in the Northwest.

Estimate 2,000,000 Horsepower.

"We now have about 12,000,000 or 13,000,000 wage earners unemployed, or about 30,000,000 of our population affected directly by unemployment. If we should return immediately to the high level of 1929, I think we still would have about 5,000,000 men out of work and on a dole. Our population is out of balance. If by government activity we can restore the balance we will have taken a great step forward.

"The normal trend now is a back-to-the-farm movement. For those who have had experience in agricultural work I think we would do well to provide a living."

Mr. Roosevelt said that the Tennessee River basin had an area of 64,000 square miles and included parts of Virginia, North Carolina, Tennessee, Alabama, Mississippi and Ohio. Flood control benefits would extend to parts of Illinois, Missouri, Arkansas, Mississippi and Louisiana.

He estimated that with a steady flow of the river brought about by reforestation and the construction of flood-control basins 2,000,000 to 3,000,000 horsepower could be produced at Muscle Shoals, where the government has already spent about $165,000,000 on power and nitrate plants.

February 3, 1933

SHOALS BILL HELD SECURITIES PERIL

W. L. Willkie Says Utility Holdings of $400,000,000 in South Would Be Worthless.

POWER LINES ARE OFFERED

Hearing Is Told That the Government Could Not Compete on Fertilizer.

Special to THE NEW YORK TIMES.

WASHINGTON, April 14.—Passage of a Muscle Shoals bill with Federal construction of transmission lines provided would make worthless $400,000,000 of securities of six Southern utilities companies, W. L. Willkie, president of the Commonwealth and Southern Corporation, declared today before the House Military Affairs Committee.

He said that enactment would virtually mean confiscation of property of the companies concerned, stripping them of constitutional rights.

"Every security house in New York," he testified, "has sent word to their customers that passage of the bill would ruin their securities. If the purpose of the measure is to distribute the greatest power to the greatest number at the lowest rate we offer you the means to do it.

"We will absorb the power as fast as we can. If it is necessary we will contract to carry your power and pass the savings on to the ultimate consumer."

Mr. Willkie reiterated the faith of his companies in the President's general development plan for the Tennessee Valley, but declared that it would be "worse than a crime" to destroy the existing companies by depriving them of their market.

"The officials of these companies are trustees and every single dollar of the $400,000,000 outstanding has been approved by a State agency in the State where they were sold," he said.

"Thousands of letters have come to us from disturbed investors and if this bill passes as now written, those securities will be destroyed."

Opposes Fertilizer Proposal.

Opposition from another source developed during the afternoon session, when Charles J. Brand, executive secretary of the National Fertilizer Association, insisted that fertilizer could not be manufactured at the Muscle Shoals plant as cheaply as by private industry.

Much of his testimony was a repetition of statements he has made before the committee during the last twelve years as to previous Shoals legislation, and at one point he expressed regret that Representative James, former chairman of the committee, had to listen again to this "tiresome" story.

Replying to Representative Goss of Connecticut, Mr. Brand said he knew that the government could not produce fertilizer as cheaply as private concerns, and that he had also been told that electricity could not be produced as cheaply by the government.

"The President wants to," Mr. Goss said.

"I wonder if he does," answered Mr. Brand.

He saw no occasion for putting the government into business, remarking:

"Don't knock out a window to kill a fly."

Replying to Mr. Goss again, he declared that he still held an opinion he expressed last year, that it would not be a bad arrangement to lease the shoals to private industry.

Representative Brown of Kentucky appeared this afternoon to ask that the legislation be amended to provide for development of the Cumberland River. He said that a dam near Jamestown would develop 800,000,000 kilowatt hours, which would care for the needs of the State. It would make the river navigable, he added, and benefit the mountainous section of Kentucky.

Trend Toward Steam Plants.

"During the last administration the Red Cross fed the hungry Republicans down in our State," he said, "and our Highway Department fed the Democrats. A lot of our people lost money in Insull Utilities investments and we want a chance for cheap power to help our people."

Mrs. Harris T. Baldwin, chairman of the Living Cost Department of the League of Women Voters, told the committee that her organization endorsed the general purposes of the bill, but objected to government manufacture of fertilizer on a large scale.

J. M. Barry, vice president and general manager of the Alabama Power Company, said gross revenue of the company in 1932 was $15,583,840, a decrease of $2,174,603 from 1931, and that his company paid $1,981,661 in taxes in 1932.

Outstanding in securities were $132,680,603, 51 per cent of which were owned by Alabama women. He said steam generating plants cost but $70 per kilowatt to build, whereas hydro-electric plants cost $160.

"Development of the past few years has been toward steam plants," he declared. "The trend has been toward lower construction costs and more efficient operation."

April 15, 1933

POWER MEN ACCUSED AT MUSCLE SHOALS; NORRIS BILL SIGNED

Alabama and Tennessee Companies Charged With Cheating and Damaging Plant.

CONGRESS INQUIRY LOOMS

Cummings Is Investigating Complaint — Companies' Heads Deny Charges.

Special to THE NEW YORK TIMES.

WASHINGTON, May 18.—An investigation by Congress of alleged misuse of government property at Muscle Shoals by two private power companies with the acquiescence of plant authorities was in prospect today, following disclosures made by Secretary Ickes as President Roosevelt signed the Norris bill providing direct government operation and control of the huge Federal development.

After a conference with Secretary Ickes this afternoon, the President called in Huston Thompson, former counsel for the Federal Trade Commission and special assistant to the Attorney General, who is understood to have been selected to investigate the charges.

Involved in the case, which took Washington by surprise, are charges that the Alabama Power Company and the Tennessee Power Company, which lease power from the government plant under a contract with the War Department, have been using its facilities to effect an interchange of surplus energy. In this way the two affiliates of the Commonwealth and Southern Corporation are alleged not only to have reduced the cost of their contract with the government, but also damaged government property and placed the policy of government operation in a bad light.

Charges Contained in Report.

These alleged irregularities are stated in a report made by Louis R. Glavis, special investigator for the Interior Department, which for a month has been in the hands of officials of the Department of Justice, who have extended the inquiry with a view to possible future prosecutions.

Secretary Ickes said the case was now in the hands of the Department of Justice, which was to decide whether any civil or criminal violations were involved.

"The practice complained of was first noticed," said Secretary Ickes, "by an electrical engineer who was in the party of President Roosevelt and Senator Norris when they went on an inspection trip of the Muscle Shoals plant last January. Nothing was done about it at the time, but the engineer subsequently wrote to the Senator about it, and his letter was turned over by Senator Norris to the President, who ordered me to make an investigation. I called in Mr. Glavis, who was formerly connected with the Federal Power Commission, to investigate.

The wartime power plant, costing $140,000,000 has been under the jurisdiction of the War Department, with army engineers directly in charge. Under an arrangement with this department, the Alabama Power Company was authorized to purchase government-produced power at the rate of 2 mills per kilowatt hour, but it was stipulated that there was to be no interchange of power between Alabama and Tennessee plants except in the event of a major breakdown.

The Alabama and Tennessee plants are connected with the Muscle Shoals power plant at Wilson Dam by power lines. The lines also connect with substation generators, and the government power, when used, is recorded on meters.

The practice complained of deals with the transfer of surplus energy from one of the privately owned plants to another by means of connecting equipment in the Muscle Shoals power house.

Through such transference, either of the two private plants would be able to dispense with the power produced at the government plant, although making use of government equipment and workers.

Secretary Ickes explained that several men were needed to place the interconnecting machinery in operation, with the result that some of them must have known of the alleged unauthorized use. The Secretary said he had been advised that the practice complained of was damaging to plant equipment, but he declined to make an estimate of the amount of the damage or the loss to the government.

Officials expect that Major Gen. Lytle Brown, chief of engineers, and possibly former Secretary Hurley might be called to testify if open hearings were held.

Secretary Ickes stated that nothing suspicious had appeared to the engineer who made the alleged discovery upon his inspection with the President's party, and gave it as his version that some one at the plant supposed Senator Norris had "left town" before he had actually gone. When the facilities used in the interchange are performing their proper function two lights are burning about the apparatus, but when used for transferring current from one privately owned plant to another only one light appears.

The engineer is alleged to have returned to the plant in the company of Senator Norris and to have detected an interchange by the appearance of only one light.

Senator Norris and Senator Black, who figured in a previous complaint to the War Department against the reported misuse of the plant facilities, were silent when asked about the disclosures, Senator Norris saying he was unaware of the engineer mentioned by Secretary Ickes.

General Brown made the following statement:

"The statement made that there has been injury to the government plant at Muscle Shoals or that there has been an obtaining of any power from Muscle Shoals without due payment is entirely unjustified and untrue so far as the knowledge of this office goes. The agreement made with the Alabama Power Company by the chief of engineers was to the effect that there should be no interchange of power between the Alabama Power Company and the Tennessee Power Company except in time of emergency. By emergency was meant a major breakdown."

The President signed, in the presence of legislators who for twelve years have fought for government operation of Muscle Shoals, the now famous legislation for governing its use in the production of electric power and fertilizer, and providing at the same time for the development of the entire Tennessee valley.

The President said it made him very happy to carry out the realization of a hope which he had long maintained. He took the opportunity, however, to warn innocent investors that they should not be caught in the clutches of conscienceless land speculators now promoting developments in the Tennessee valley.

Senator Norris, the father of the act and long an advocate of government operation on this project, headed a delegation of House and Senate members who witnessed the informal ceremony. A number of pens were used by the President and given to Senator Norris, Representatives Almon and McSwain and others.

"The Muscle Shoals legislation," said Senator Norris, "marks an epoch in the history of our national life. It is a monument to the victorious ending of a twelve years' struggle waged on behalf of the common people against the combined forces of monopoly and human greed.

"It is emblematic of the dawning of that day when every rippling stream that flows down the mountainside and winds its way through meadows to the sea shall be harnessed and made to work for the welfare and comfort of man. It establishes a new governmental policy which, when carried to its logical end, will bring blessings, peace and comfort to all our people.

"Long after all of us who have been participants in this controversy have passed away, generations now unborn, assembled around millions of happy firesides, will thank God for the vision of President Roosevelt and render praise to his memory for signing this bill.

"In behalf of a grateful people, who have watched this uneven struggle with intense interest, I desire to thank him and congratulate him."

As finally enacted the measure provides for hydroelectric power as well as the improvement of the Tennessee River for navigation purposes, and the creation of reservoirs to control flood waters of the Tennessee and Mississippi River basins.

The act creates a Tennessee Valley Authority similar to the New York Power Authority, and it is understood the President desires Frank P. Walsh, now head of the New York Power Authority, to carry out his ideas in the Tennessee Valley. Also mentioned for the position today was Dr. Arthur E. Magan, president of Antioch College, who had been active in the drafting of the Muscle Shoals legislation.

A board of three to be appointed by the President will direct the development, including the production of power and fertilizer, the building of a power dam on Cove Creek and the erection of transportation lines to dispose of the power to private utilities.

The chairman of the board is to receive a salary of $10,000 annually and his two associates $9,000 each.

The act stipulates that "it is the policy of the government, so far as practicable, to distribute the surplus power generated at Muscle Shoals equitably among the States, counties and municipalities within transportation distance of Muscle Shoals."

May 19, 1933

CONSERVATION OF OUR FORESTS GAINS RAPIDLY ON TWO FRONTS

Under the new lumber code, national lumbermen met with Secretary Wallace at Washington recently to recommend measures "for the conservation and sustained production of forest resources." The conference marked a further step in the conservation efforts of the Roosevelt administration. These efforts and their background are here discussed by Dr. Sherman, Associate Forester.

By E. A. SHERMAN,
Associate Forester, United States Forest Service.

AMERICAN forestry is at the beginning of a new epoch. Under the leadership of Franklin Roosevelt the forest conservation movement has taken on renewed vigor and already has made marked advances. Curiously, progress in forestry during the epoch that is just closed also owed much of its original impetus to a Roosevelt.

One of the distinctive policies of Theodore Roosevelt's administration was the conservation of natural resources, and particularly of forests. The underlying situation was quite different thirty years ago, however, from the situation today. The lumber industry, then at the peak of its development, was cutting timber faster than at any time before or since. The bulk of the virgin timber of the Northwest had been cut. The production of the Lake States pineries was rapidly declining, the Southern pine output was at its maximum, and large-scale cutting on the Pacific Coast was just beginning. Handling of forests on anything other than an

32

exploitation basis was a new idea to the great majority of the American people. Already, however, the evil results of forest destruction had become evident, and the conservation policy won wide popular support, particularly in the East.

Theodore Roosevelt's Policy.

In the field of forestry the chief accomplishment of the earlier Roosevelt's administration, under the able guidance of Gifford Pinchot, was the establishment on a firm basis of the national forests. The national forest system, which had been inaugurated in the early 1890s, was greatly enlarged by the addition of most of the remaining forests on the public domain, and a policy of administration and development of the national forests was adopted. Under this policy the forests were no longer to be withheld from use as "reserves," but their timber, forage and other resources were to be utilized to the full extent that might be compatible with the objective of the greatest good to the greatest number in the long run.

During the ensuing quarter century forest conservation made steady if not spectacular progress, although at times it appeared to be merely marking time. The organization and administration of the national forests were perfected. The Weeks Law was passed in 1911 to provide for Federal purchase and management of forest lands on the headwaters of navigable streams, particularly in the White Mountains and the Southern Appalachians. Its scope was later expanded to include lands primarily for timber production.

Recent Legislation.

The Clarke-McNary Law of 1924 expanded Federal cooperation with States and forest owners in protecting forests against fire. The McSweeney-McNary Act of 1928 set up a comprehensive long-time program of research in various aspects of forestry, including a thorough survey of the country's forest resources and requirements. Most of the States which have important forest areas organized forestry departments, and many of them embarked on programs of reserving or acquiring land for State forests.

By 1930, 20 per cent of the 495,-000,000 acres of commercial forest land in the continental United States was in public ownership or control. This included 75,000,000 acres in the national forests, nearly 11,000,000 acres owned by the States, counties and municipalities, and 13,000,000 acres in Indian reservations under government control or in the Federal public domain. The term commercial forest land, as used here, means land that is capable of producing timber of commercially valuable quantity and quality, and that is available for commercial timber growing. In addition, there were large areas of noncommercial forest, valuable chiefly for watershed protection or recreation, in the national and State forests and parks.

The total net area of land in the National Forests, not including Alaska and Puerto Rico, was about 139,000,000 acres.

Expansion of Sawmills.

During this quarter century the conservation of privately owned forests, which constituted four-fifths of the total forest area, made relatively little progress. Under the stimulus of unrestricted competition and the urge to liquidate timber investments, the sawmill industry expanded its capacity up to almost double the country's normal requirements for lumber. In the face of declining consumption of lumber the industry found itself obliged to meet carrying charges on this huge excess investment and at the same time, because of the prevailing cut-out-and-get-out method of operation, to retire not only its timber investment but also its plant investment as the timber was cut. Profits were thereby greatly reduced and in many instances were entirely eliminated.

Leaders in the industry realized that such conditions could not go on indefinitely, but they found it difficult to remedy the situation. Because of the anti-trust laws it was impossible for the industry to take concerted action to put an end to cut-throat competition and plan ahead for continuous operation. Much was said about forest conservation, but with a few notable exceptions cutting of privately owned forests continued under the same wasteful policy of liquidating the timber resource that had prevailed from the beginning. The future of the forest and the forest industries was not taken into consideration.

By 1930 at least 93 per cent of all the privately owned forest land east of the Rocky Mountains had been cut over. Of this 80 per cent was more or less restocked with timber growth, some of it large enough for cutting, but much of it inferior in quality and fit only for firewood or other low-grade products. Nearly 20 per cent, or 65,000,-000 acres, was almost completely denuded, or was restocking with weed trees of little value for present or future cutting. Large areas, abandoned by the owners, had become a sort of no man's land and were drifting into public ownership through the tax-delinquency route. The privately owned forests in the West were headed in the same direction, although, with half of the total area as yet uncut, they still contained 40 per cent of the national supply of saw timber.

New Roosevelt Measures.

The year 1933 marks the beginning of a new deal for American forests, both public and private. One of Franklin Roosevelt's first measures was the organization of the Civilian Conservation Corps, which at once put more than 300,000 men to work in improving and building up the forests in all parts of the United States. As a result of this project, not only are enormous public values being conserved and created, but the American people are gaining a better understanding than they ever had before of the objectives of forestry and of what forests mean in terms of national welfare.

The Forest Service, in response to a resolution of Senator Copeland, made an exhaustive study of the present status of forestry in the United States and early in the year submitted a comprehensive national plan for American forestry. Among other things this plan recommends the acquisition by the Federal and local governments of a very large area of forest land now in private ownership. Following this, and under the stimulation of the emergency conservation work project and the growing conviction that the public interest requires the maintenance of a forest cover on the millions of acres of idle land which is submarginal for agriculture and not likely to be utilized for any other purpose by private owners, the program of land acquisition has been greatly accelerated.

Purchases by the Government.

Within the last five months 942,-000 acres have been purchased or put under contract of purchase by the Federal Government, at a cost of $1,760,000. This is approximately one-fifth as much as was acquired during the preceding twenty-two years, since the enactment of the Weeks Law. A very much larger area is expected to be purchased during the next few months. Purchase areas have been approved in twenty-two States, all east of the plains.

Under the public works section of the National Industrial Recovery Act, $41,000,000 has been allocated for construction of roads, trails and other improvements and for other development work in the National Forests. This will not only give employment directly and indirectly to many thousands, but it will accomplish work that has long been sorely needed but for which funds have not been available.

Besides these important developments which affect the public forests mainly, an even more important development, revolutionary in its significance, is in sight for privately owned forests. Under the NIRA the lumber and timber products industries have been given a charter which will enable them to put their house in order and to establish themselves on a perpetual basis. The code of fair competition for these industries, approved by President Roosevelt on Aug. 19, makes it possible for them to take concerted action in ending the cut-throat competition which has brought them to the verge of ruin.

A Vital Code Article.

The most important and revolutionary provision of the code, is the statement of purpose "to conserve forest resources and bring about the sustained production thereof," and the definite commitment of the industry in Article X of the code, which reads as follows:

The applicant industries undertake, in cooperation with public and other agencies, to carry out such practicable measures as may be necessary for the declared purposes of this code in respect of conservation and sustained production of forest resources. The applicant industries shall forthwith request a conference with the Secretary of Agriculture and such State and other public and other agencies as he may designate. Said conference shall be requested to make to the Secretary of Agriculture recommendations of public measures, with the request that he transmit them, with his recommendations, to the President; and to make recommendations for industrial action to the authority, which shall promptly take such action, and shall submit to the President such supplements to this code as it determines to be necessary and feasible to give effect to said declared purposes.

Such supplements shall provide for the initiation and administration of said measures necessary for the conservation and sustained production of forest resources, by the industries within each division, in cooperation with the appropriate State and Federal authorities. To the extent that said conference may determine that said measures require the cooperation of Federal, State or other public agencies, said measures may to that extent be made contingent upon such cooperation of public agencies.

For the first time the industry as a whole has thus publicly committed itself to a policy of conservation and sustained yield of its basic resource, the forest. This new policy is striking evidence of a new and progressive attitude on the part of the leaders of the industry. It means that the traditional cut-out-and-get-out policy is to be abandoned, and that henceforth forest destruction is to be recognized by the industry as an unfair practice.

Cooperation Needed.

Of course, the declaration of policy is only a beginning. The translation of this policy into actual practice in tens of thousands of woods operations will be difficult and obviously cannot be accomplished overnight. It will require many adjustments within the industry and a patient process of educating not only the individual operators but also their many thousands of woods employes. It will require sympathetic cooperation on the part of the public.

In recognition of this new attitude the public may reasonably be expected to go further than it has gone hitherto in helping the industry to overcome economic and other handicaps. The conference with the Secretary of Agriculture, which was held on Oct. 24-26, considered not only the measures which the industry should adopt in order to carry out its commitments, but also the ways in which public agencies can promote the public interest by helping the industry.

Whether the forests and the industries dependent on them are to be perpetuated cannot be left solely to the self-interest of individual timber operators. It is decidedly a matter of public concern. The forest-products industries constitute one of the major industrial groups of the country. Being widely distributed in practically all sections, they form an exceedingly important element in the economic structure. In normal years their products are valued at more than $3,000,000,000. They employ more than a million persons, including part-time workers, and pay around a billion dollars in wages and salaries. Directly and indirectly they contribute largely to the support of local governments and institutions. Upon their continued existence depends the economic utilization of about one-fourth of our land area.

Forests' Wide Uses.

But forests are not only sources of raw material. They have other values, less easily measured in terms of money. They serve to ameliorate the climate. They hold the soil on hills and mountain sides and along streams. They modify extremes of run-off and protect the sources of water for domestic use, irrigation, power and navigation. They furnish a habitat for multitudinous species of birds, game animals, fur bearers and other wild life.

These services may far outweigh their material values, from the standpoint of society. For these services the public should be willing to do its part. As the forest-products industries demonstrate by their woods practices that they are going to cooperate in perpetuating the resource, the public can be counted on to meet them at least half way. December 24, 1933

LAND EROSION ATTACKED ON A SYSTEMATIC SCALE

Work Now Beginning in Selected Areas Will Serve to Demonstrate the Methods of Saving the Soil

By H. H. BENNETT,
Director, The Soil Erosion Service.

THE Soil Erosion Service, a branch recently established in the Department of the Interior at the direction of Secretary Harold L. Ickes, and operating with a $10,000,000 allotment from the Public Works Administration, is now proceeding with the actual field work of controlling erosion on a number of large representative areas scattered throughout the nation. The size of these demonstration areas, most of which will represent complete watersheds, ranges from about 100,000 acres to 15,000,000 acres in the instance of the project to be undertaken on the Navajo Indian Reservation.

The areas thus far selected—the starting points for a national program of soil conservation—and on which work has already begun, are located in the cotton-producing Piedmont Plateau of South Carolina and the Blackland region of Central Texas, the Palouse Wheat Belt of Washington and Idaho, the dairying section of Southwestern Wisconsin, the Corn Belt of North-Central Missouri and South-Central Iowa, the corn-small grain section of Central Illinois, the wheat-corn section of Northern Kansas, the cotton-small grain section of the Red Plains of Central Oklahoma, the bean-citrus region of Southern California and the Navajo Indian Reservation project covering a large area in Arizona, New Mexico, Utah and Colorado. Experimental areas are being set up on five of the principal soil types occurring within the watershed of the Tennessee Valley. Other watersheds are being selected in other States.

Methods of Stopping Soil Wash.

Every practical measure for controlling erosion will be used, according to the adaptability of the different kinds of land. Every acre will be treated in accordance with its particular needs. Steep, erosive slopes will be taken out of cultivation and planted to trees or grass or other thick-growing crops, such as are known to be highly resistant to soil washing.

Part of the cultivated land will be protected with the new system of strip-cropping, under which the clean-tilled crops, such as cotton, corn and tobacco, the real producers of erosion, will be grown between parallel bands of grass, lespedeza, sorghum and other dense crops planted across the slopes, on the level. These latter crops catch the rain water flowing down the slopes, spread it out and cause the suspended soil to be deposited, and much of the water to be absorbed by the ground, thus protecting the crops growing on the plowed strips below and conserving valuable supplies of moisture.

Field terraces will be employed where applicable, and in some localities the land will be scarified with a machine that scoops out 10,000 basin-like holes to the acre, each of which retains about five gallons of rain and causes it to sink into the ground where it falls, and rotations will be practiced, and cover crops and other control measures will be employed.

Briefly, the program calls for control of erosion, reduction of the flood hazard, protection of rich bottom lands from worthless sand and gravel washed out of the hills, prevention of silting of stream channels and reservoirs and readjustment of land-use practices. Efficient results will call for enthusiastic cooperation of farmers, ranchers, business men and every person having an interest in or love for the welfare of the land—the preservation of the most indispensable asset of the nation.

First Large-Scale Attempt.

Here is the first attempt in the history of the country to put through large-scale comprehensive erosion and flood-control projects, such as will apply to complete watersheds from the very crest of the ridges down across the slopes where floods originate, and on to the mouths of the streams. These will not be engineering projects or forestry projects or cropping projects, but a combination of all of these, operated conjointly with such reorganization of farm procedure as the character of the land indicates as being necessary.

In other words, well-rounded coordinated plans will be followed, the details of which will be based upon the best information in the possession of scientific agriculturists: the agronomist, the forester, the range specialist, the soil expert, the engineer, the economist and the wild-life specialist.

In the Wisconsin project, for example, some of the steep-timbered areas, now eroding because of excessive grazing, will be taken out of use and given complete protection in order to stop the excessive run-off of rain water, which has been speeding down across the cultivated slopes, ripping them to pieces or planing off the more fertile topsoil. Grass will be restored to these protected forest areas, and where the trees are too thin other trees will be planted. Small seedlings will be made of plants that furnish feed for quail and ruffed grouse.

Eventually, sportsmen will come from Milwaukee, St. Paul, Chicago and other places to pay the farmer for the privilege of hunting in his timbered lands that will be restocked with game. Below the forested land, those steep slopes now washing rapidly to a condition of low productivity will be taken out of the clean-tilled crops and put into permanent pasture to furnish the grazing that formerly was provided by the timbered areas. The grazing capacity of the farms will not be increased, but the crop area will be cut down to some extent. Better protection of the cultivated land from erosion will largely make up for the reduction of the cropping area, by way of higher average yields.

The work of the soil erosion experiment station, which was established two years ago by the Department of Agriculture, near La Crosse, has served to awaken the farmers of the region to the enormous cost of the evil.

Measurements made during the past year have shown that land of about average slope lost seventy-seven tons of soil from a single acre, where corn was grown. In addition, 22 per cent of all the rain that fell ran out of the fields in the direction of the Mississippi, carrying plant food, humus, rich soil, and sand to choke the channel-way of the river. Where bluegrass was grown the corresponding losses for the same period were only 6 per cent of the rainfall and the insignificant amount of .0063 of one ton of soil per acre. Thus, grass cut down the soil loss more than ten thousand times.

Scope of the Plan.

The national plan for conserving the soil will strike, primarily, not at reclaiming hopelessly worn-out, gullied land, but at saving the remaining areas of good land—those areas still retaining the topsoil or part of it. Much violently erosive land now in cultivation will be taken out of cultivation, where the farmers can be convinced of the logic of such procedure. These areas will go into trees and other thick-growing plants, representing the only possible means of controlling the washing on such areas.

With respect to methods of using the land, we have in the past been anything but discriminative. We have cut trees from steep, erosive slopes and planted the land to erosion-producing crops, such as cotton, corn and tobacco. We have farmed land which was adapted to trees and nothing but trees, to all sorts of crops, as if all kinds of land had the same natural adaptation.

Extent of the Reduction.

We have permitted our slopes to wash down to a mere geological skeleton of land, even to the point of destruction, thinking that erosion is a natural process not to be retarded by man, or without so much as devoting a serious thought to the matter. Of formerly cultivated land, 35,000,000 acres have been essentially destroyed by erosion, in so far as their having further value for crops concerned.

While we have an enormous area of land, it is by no means all good land. Most of our better soil is in cultivation and has been for some time. Much of this was only moderately productive the day it was cleared of its virgin stand of forest or broken out of its sward of prairie grasses.

Our estimates indicate that 3,000,000,000 tons of soil material are washed out of our fields, temporarily idle fields and pastures every year. More than 400,000,000 tons of suspended solid matter and many more millions of tons of dissolved matter pass out of the mouth of the Mississippi River annually. This comes largely from the farm lands of the Mississippi Basin. The greater part consists of super-soil (soil considerably richer than that of the Valley of the Nile); enough of it is lost every year to build 1,250 farms of 160 acres each, having a soil depth of 12 inches or twice the depth of the average upland soil of America.

March 11, 1934

ICKES MOVES TO END 'WASTE'

Special to THE NEW YORK TIMES.

WASHINGTON, Dec. 16.—A permanent public works program, intelligent planning of the use of land, water and mineral assets of the country and a rejuvenation of the agricultural technique of the nation were recommended in the report of the National Resources Board to President Roosevelt, made public today.

The report said that its recommendations, if adopted, would "end the untold waste of our national domain now and will measurably enrich and enlarge these national treasures as time goes on."

The board is composed of Harold L. Ickes, Secretary of the Interior, as chairman; Frederic A. Delano as vice chairman; George H. Dern, Secretary of War; Miss Frances Perkins, Secretary of Labor; Daniel C. Roper, Secretary of Commerce; Henry A. Wallace, Secretary of Agriculture; Harry L. Hopkins, relief administrator; Charles E. Merriam and Wesley C. Mitchell.

The report was described in the letter of transmittal as "the first attempt in our national history to make an inventory of our national assets and of the problems related thereto."

Almost half of the report was given over to a discussion of the advisability of presenting to Congress a public works program, drawn up on a basis of execution within six years, to be revised annually.

Program Looks to Future Need.

The idea behind the proposal was to have ready at all times a properly conceived and carefully studied outline of advisable public works to be put into execution at any time economic conditions might make it desirable.

"Based solely upon extension into the future of the average annual capital outlay for the entire United States (national, State and municipal) of about $2,400,000,000 for the ten years 1921 to 1930, the total of capital outlays for the ten-year period 1935 to 1944 may be estimated as approximately $24,000,-

000,000," the report said.

"Of course, such an extension of a past average makes no allowance for the factor of growth which has been so notable in past American expenditures upon public works and which presumably should and will continue in the future. Hence the ten-year total suggested may be regarded as a minimum estimate.

"Whether five billions a year for roads, parks, sewers, public libraries, forests, waterways, good housing and countless such other facilities is more than we should spend is a problem which we shall have to solve in the light of future developments touching the aggregate national income, public finances, modes of combating unemployment, popular demand for pleasures and protections afforded by public works, &c."

Human Values Considered.

The conservation and planned utilization of natural resources is not an end in itself, the report explained, saying that "it is too much to suppose that proper development of our drainage basin will of itself solve the problems of the perplexed body politic."

"Human resources and human values are more significant than the land, water and minerals on which men are dependent," it said. "The application of engineering and technological knowledge to the reorganization of the natural resources of the nation is not an end in itself, but it is to be conceived as a means of progressively decreasing the burdens imposed upon labor, raising the standard of living, and enhancing the well-being of the masses of the people.

"Unfortunately, this principle has not always been followed, even when declared; on the contrary, there has been tragic waste and loss of resources and human labor, and widespread spoliation and misuse of the natural wealth of the many by the few."

Looking at the agricultural development of the country in the next twenty-five years, the report

said that "if we assume the restoration of a volume of cityward migration such as prevailed during the decade ending with 1929, urban population will be nearly 20 per cent larger in 1960 than in 1930, while farm population will have slowly declined.

Sees Urban Decrease Possible.

"Assuming no net migration between city and country, urban population will increase less than 3 per cent by 1945 and will then slowly decrease, while farm population will increase nearly one-half by 1960 and rural non-farm by about one-fourth.

"The conclusions resulting from the two extreme assumptions as to the urban-rural migration suggest the great importance of the outlook for industrial recovery in determining our course in land utilization and policy. It is probable that during the next twenty-five years progress in the use of labor-saving machinery and in farming methods would make possible the production of our domestic supply and probable exports of farm products by a farm population little or no larger than at present."

The board found that American agricultural methods of production are inherently wasteful.

"The Japanese, for instance, employ little more than one-tenth as much crop land per capita in providing for domestic requirements as we do," it stated, "yet, with our larger acreage, millions of Americans have an inadequate and ill-balanced diet, especially lacking in milk, vegetables and fruit. In part this is due to poverty, but perhaps even more to ignorance and inertia."

No Increase in Crop Yields.

The conclusion was reached that "there has been no notable increase in crop yields for several decades. Progress in seed selection, fertilization and other improvements has been more or less offset by the influence of soil depletion, insect enemies and the crop diseases."

Taking into consideration all factors involved, the board's experts found that the acreage of land harvested, in order to meet domestic requirements, would develop as follows:

	Acres.
1930	359,000,000
1940	353,000,000
1945	368,000,000
1950	377,000,000
1955	383,000,000
1960	386,000,000

"Most of the serious maladjustments apply to land in private ownership," the report said, "and grow out of the virtually unlimited powers of use and abuse which we have permitted in our system of private property in land. Basically, however, absolute ownership still resides in the State (either individual commonwealth or nation) and under the police power the State may constrain the private use of land within bounds set by the public interest."

Water Developments Urged.

Specific projects for planning of water resources, as recommended by the board, included:
1. The Connecticut River power, flood control and stream pollution project in Vermont, New Hampshire, Massachusetts and Connecticut.
2. The Delaware River power, water supply and stream pollution project in New York, Pennsylvania and New Jersey.
3. A study looking to the coordination of the hydroelectric power to be developed in Northern New York, principally in the international section of the St. Lawrence, and the development of coal-generated, mine-mouth power in Pennsylvania, having in mind condensing water requirements, so as best to conserve the social, economic and industrial interests of the States of New Jersey, New York and Pennsylvania.

Surplus plant capacity and overproduction have led to serious wastage of mineral resources, the report found. Bituminous coal, petroleum and copper were listed as the chief commodities being produced under uneconomic conditions, for which future generations will suffer.

Permanent Board Advocated.

A national planning board was advocated, made up of five members appointed by the President, with a panel of consultants. Terms of the members should be indeterminate, like the Civil Service Commission, it was recommended.

"Standing apart from political and administrative power and responsibility," the board said, "but in close touch with the Chief Executive, and under the control of the political powers that be, such a group of men would have large opportunity for collecting the basic facts and for mature reflection upon national trends, emerging problems and possibilities, and might well contribute to those in responsible control facts, interpretations and suggestions of far-reaching significance."

December 17, 1934

ALL PUBLIC LAND BARRED FROM USE

Presidential Order Affects the Last 1,200,000 Acres of 165,695,000 Domain.

SURVEY WILL FIX FUTURE

WASHINGTON, Feb. 9 (AP).—In preparation for a nation-wide conservation program, President Roosevelt today withdrew all remaining public land from use.

His order, supplementing that of November, affects about 1,200,000 acres, and completes withdrawal from settlement, location, sale or entry of the entire 165,695,000 acres of public domain.

The November order was to make possible segregation of 80,000,000 acres as permanent livestock grazing areas under the Taylor act.

The President said that today's withdrawal of lands, applicable to twelve States, was ordered "pending determination of the most useful purposes to which they may be put in furtherance of the land program and conservation and development of natural resources."

He added that this land, not suited to profitable growing of crops, was destined for the conservation and development of forests, soil, and other natural resources, the creation of grazing districts, and the establishment of game preserves and bird refuges.

Although the Interior Department has not yet made final selection of the 80,000,000 acres of grazing land, Representative DeRoun of Louisiana, chairman of the House Public Lands Committee, has introduced a bill to extend the selection to the remaining areas suitable for livestock.

Little of the land withdrawn today was grazing acreage, and officials said that much of it would be used for forest and game preserves. A legislative program to end further homesteading and set up permanent uses for the acreage was authoritatively reported to have been drafted for submission to Congress soon.

A ban on further homesteading was one of the urgent recommendations of the National Resources Board, headed by Secretary Ickes. The board also urged withdrawal from cultivation of 75,000,000 acres of unproductive farm land.

The Federal Relief Administration land purchase division plans to complete the purchase of 7,000,000 acres by July. Officials explained that the prohibition of further settlement conformed to the board's remark that it was useless to buy up submarginal land while settlement of additional submarginal acreage was permitted.

In addition, the relief administration's rural rehabilitation division and the PWA subsistence homesteads agency have been purchasing more fertile land for their mingled industrial-agricultural communities.

Washington, with 692,751 acres, was the State chiefly affected by today's order. Land in eleven other States was affected as follows: Minnesota, 269,451 acres; Arkansas, 175,924; Florida, 32,303; Nebraska, 20,225; Alabama, Kansas, Louisiana, Michigan, Mississippi, Oklahoma and Wisconsin, small acreages.

The withdrawals were authorized under the land program section of the Recovery Act.

February 10, 1935

WILDLIFE CONSERVATION.

The Select Committee on Conservation of Wildlife Resources, created pursuant to a resolution of the Seventy-third Congress, has brought in a report that is disquieting. First, it is stated, as a result of its extensive investigations and hearings, that there is not only a steady decrease of game and game fish, but a corresponding increase in the number of hunters and fishermen, due partly to the increased number of the unemployed and partly to the increased interest in outdoor life. And as to land wildlife, there is an alarming decrease due to drainage, deforestation, erosion, fire, disease, water pollution and the increasing number of hunters. An ironic instance is furnished in the complete devastation (through drainage) of an area which, by proclamation of President THEODORE ROOSEVELT, was set apart as nurseries for wild fowl. " Ding," now " Mr. DAR-"LING, chief of the United States Bu-"reau of Biological Survey," gave like testimony about another area in the same State (Oregon), once the greatest breeding ground for wildlife and water fowl in the country. Water has been led off for irrigation purposes, but the extravagant enterprise has failed; meanwhile the wildlife has disappeared.

These are but instances of what is going on in many parts of the country. This means not only an economic loss by reason of the destruction of certain industries (" the fishing industry represents a billion dollar business "), but the devastation of a recreational estate that belongs to the whole nation.

A few years ago the Southern Newspaper Publishers Association, in gathering information concerning the relative news value of hunting and fishing material as compared with that in other outdoor sports, found that in fourteen States there were 4,420,876 people interested in hunting and fishing as compared with a combined total of 4,916,652 in baseball, football and golf. Statistics are also cited showing that 13,000,000 people in the United States go to the trouble of taking out hunting and fishing licenses. Besides these there are millions drawn to the parks and forests where " the major interest is in the wildlife there." The committee has a dozen or more specific recommendations to make, with the conclusion that

the time has come for the definite affirmation, not by words alone but by deeds and dollars, that all wildlife is an invaluable public resource, entitled not only to protective laws but also to effective aid.

Cannot " Ding " devote his extraordinary talent to this end?

March 22, 1935

New Yorker Is Awed and Shocked In the Fog of a Dust Bowl Storm

By GEORGE GREENFIELD
Special to THE NEW YORK TIMES.

IN KANSAS, ABOARD THE DENVER SPECIAL OF THE UNION PACIFIC, March 7.—I am in the midst of a dust storm. The conductor tells me it is the first bad dust storm of 1937, and one of the worst he has seen in two years.

Through the car window I can see telegraph poles, some twenty feet from the roadbed. Beyond that fences line a road. The fences are approximately 100 feet from the tracks. Beyond that is a thick fog—black, impenetrable, forbidding. We have passed through nearly 200 miles of man-made fog.

Looking out of the window, you might think it merely a cloudy, stormy day; that is, if you didn't know you were in a dust bowl. It is a cloudy day, but a different kind of clouds. Not the clouds New Yorkers know.

I had read of dust storms, but they were vague in my consciousness. Now I see one, and it is a terrible, an awesome thing.

They are clouds of dust—soil; soil blowing away from a ravaged and denuded land. A land raped by greed for $2 wheat.

I have had only one thought in my brain as I sit on the train and look out at this desert through which we have been passing for two hours. It is a saddening, almost a heart-tearing thought. It is the thought that right here, under my very eyes, I am seeing this country blowing away.

Can you, back in your secure and comfortable homes, picture it? No. You have to see it, feel it, sense it.

Here, in what was once the richest farm and stock land of the Middle West, I see the disintegration of that soil which has fed us.

Let me tell about it chronologically. I left Kansas City at 10:10 o'clock last night. When I awoke this morning in my berth I pulled up the curtain. It was around 7 o'clock and we were a little west of Ellis, Kan., which is 303 miles from Kansas City.

I looked out of the window and I saw a countryside bathed in bright sunshine. There was a tang of brisk air coming through the window. It was flat country, typical Kansas. But it looked pleasant in the sunshine. True, the ground seemed very bare and I could not see a human being for miles. But I thought that was because the day still was early. I didn't give it much thought, as I had breakfast in the diner and read a newspaper.

Darkening Over Barren Land

Then, about an hour later, I noticed the light was becoming dim. It seemed as if the sky was darkening, and I had the feeling that it was going to rain or snow. The whistle of wind sounded through the cracks of the window.

By now we were in the vicinity of Oakley, Kan. Suddenly I realized that we were in a dust storm. In the dust bowl of Western Kansas. What is a dust bowl? If it were not for a few gaunt, bare trees seen occasionally in the distance, I would think this country a flat Sahara Desert, except that the ground is hard and brown, and not rolling and sandy white.

It is not a desert in the sense that you see only vast reaches of sand, as in the Sahara. What makes you think of a desert is the lack of life. The land looks dead. I have not seen more than one or two automobiles driving on the road that parallels the railroad track for a hundred miles or more.

I have seen human beings only in the bleak, deserted-appearing villages, consisting of a dozen or so shacks, that we have passed or at which we halted to pick up water. Houses empty, yards empty. I have not seen a single child in these ghostlike, pathetic villages. The few persons I saw looked like a lost people living in a lost land.

Miles of Lifeless Terrain

I do not exaggerate when I say that this flat land, this country in which the soil is blowing away, and piling up in mounds, and filling your eyes, your mouth and nose with grit that irritates the throat and lining of your nostrils, I am not exaggerating when I say that it makes me feel I am looking at a dead land.

I see death, for there is no life; for miles upon miles I have seen no life, no human beings, no birds, no animals. Only dull brown land, with cracks showing; ground that looks like gray clay. Hills furrowed with eroded gullies—you have seen pictures like that in ruins of lost civilizations.

Trees, once in a while. But their branches, their naked limbs, are gray with dust. They look like ghosts of trees, shackled and strangled by this serpent, flinging their naked arms skyward as if crying for rescue from this encircling, choking thing.

I have just been talking to a fellow passenger. His name is Jacob Yakis. He lives in Denver and travels between his home and Kansas City for an oil concern. I told him of my shock at what I have been seeing. His face was grave, impassive. But he did not seem shocked. He had been through this for two years on some 100 trips between Kansas City and Denver.

"This isn't anything," said he. "I've been through this country in my automobile when the dust was so black that I couldn't drive. I stopped once and pulled up in a deserted stable for shelter and just sat there for three hours until it blew away enough to be able to see where I was going."

He shook his head and looked out of the window. A ray of sunlight was breaking through the murk, the first break in the thick cloud in an hour. The porter says pretty soon we will be passing out of the worst of it as we get closer to Denver.

Death Riding the Rails

Did I say a while back that all this reminded me of death? Let me tell you what a fellow passenger, who did not want his name published, told me.

"Two trainmen died not so long ago as a result of passing through these dust storms for two years," he said. "They got the dust in their lungs—dust pneumonia, they call it. Lungs couldn't function the way they should. Got real pneumonia and died."

These trainmen on lines going through dust bowl country risk their lives every time the wind blows hard. When the train stops at a village or town, they have to get out and load baggage, or flag, or do whatever trainmen have to do. Those few minutes of exposure in the dust-filled air do the.

damage. The dirt gets into their nostrils, their mouths and into their lungs.

I saw one of the trainmen get off at a stop a while back when we were in the midst of the thickest part of the storm. He had a silk handkerchief tied around his mouth as he walked beside the tracks. When he returned and the train got under way he cursed. He spat and made a wry face.

"Awful," he said, "awful."

When he goes out into the air at train stops and puts that handkerchief around his head, he keeps moistening it with saliva. That helps him to breathe. He also has a gadget like a gas mask. When the storms are real bad (they all say today's is not as bad as the 1935 one—and Lord knows what it is like when it's real bad) why, then he puts on his dust mask.

"I've been taking treatments that cost me $170," said the trainman. "Dust gets into the intestines, and that ain't so good."

His face was drawn and worried. He looked out of the window at a deserted village with an expression that I can't quite describe. He dreads the trip through Western Kansas as though it were the plague.

Death, did I say? I have been talking to people today who travel up and down this route, asking this man his opinion, another his experiences.

"I counted twelve telegraph poles on a trip through here last year," remarked one, solemnly, "and between those twelve poles saw not less than forty Kansas jackrabbits stretched out on the ground, deader than mackerel. Choked to death by dust. Why, these big jackrabbits—they were famous for their size—have been killed by thousands."

He paused to pound his right fist into his left palm, then continued:

"I mean by thousands! You don't see hardly any now. You don't see any wild life out here, 'cept maybe a few crows—and even those black bandits aren't here like they used to be. No, sir. No robins, no sage hens, no song birds and damned few animals of any kind. They're gone—gone. No place for 'em to live, nothin' to eat."

I haven't seen one bird today. A few crows and one rabbit I did glimpse.

Uprooting of Buffalo Grass

We passed a lonely farm a few miles back where seven cattle were grazing. They were the first cattle I had seen in several hours of riding.

"You see that buffalo grass?" said one of my companions in the parlor car. "There's still a little of that grass left there and those cattle look in pretty good shape. But that buffalo grass isn't here any more the way it used to be. The wind and dust have torn it up by the roots and left the soil as bare as your hand.

"Those few cattle used to be represented by vast herds so thick that you couldn't count them." He shook his head. "Yes, sir, right in this spot we're passing through they were as thick as flies. Now the stockmen are gone, too."

I looked out of the window again at the gray fog and bare ground.

"Why," my fellow-passenger continued, "a lot of this land was never tilled. It was stockmen's land, used for grazing. And yet look what has happened. The dust has blown on to it from other parts of the country and simply torn up the grass and vegetation and left it as dry as if it had been plowed up. And this was the cream of the country's soil.

"Why, just recently a fellow back in Ellis, Kan., put some of that blown soil in a bucket, watered it and planted radishes in it. He got the most beautiful radishes you ever saw. That shows what kind of richness this soil had. But the same fellow stuck a magnet into the bucket of dust and every time he did so some iron particles came up. That iron is what's cutting the land away. Don't know where that came from."

There was a puzzled, mystified look in his face. It is an expression I have seen on the faces of other people on the train. They can't understand this outburst of nature's wrath. I saw fear in their faces, as of impending disaster.

Coated by Pelting Blast

When we were going through the worst of the dust storm I wanted to get some pictures of the swirling, murky gray fog. I stepped out on the observation platform and my skin felt as if it were being pelted with tiny icicles. My throat clogged and my eyes stung, all in the space of a few minutes.

My snapshots taken, I stepped back into the car and one of the passengers remarked jocularly:

"That's going to cost you a suit of clothes."

I looked down at my shoes and trousers and they were covered with gray dust. The porter brushed me off, but it took twenty minutes to do it. He said I had better wash my hands and face with Sapolio, because ordinary soap won't get the stuff out of your pores. It grinds right into the pores.

"When I get to Denver," he said, "I'll have to take two baths—one with soap and water, and another with epsom salts. That gets the dirt out of your pores so your skin can breathe."

Right at this moment my throat is choking and I am coughing intermittently, even though we have by now passed out of the worst of the storm. Wonder if there's anything in liquid form on the train for a "dust bowl" throat? The man across the aisle is coughing. one eye is very red and the tears run down his cheek.

Some of the passengers on the train pooh-pooh this dust-storm talk. They say it's not so bad as the papers make out. They say it's bad for business, makes people afraid to invest in things out here and scares away people who have been living here. I suspect strongly this is native pride and shrewdness, coming to the defense of a threatened resource. Hope against hope.

They have no answer when I point out to the deserted-looking villages and the lifeless, utterly lifeless, land.

The windows in this train, all through the train, are sealed. It would take a crowbar to open them. The trains are air-conditioned throughout.

But the thin, gray dust covers the sill where I sit and write, seeping through sealed windows, and the porter bustles from chair to chair, with a towel in his hand, cleaning the stuff from the windows and the sills. He takes his broom and goes out on the platform and sweeps away a covering of sand, sand that looks like Jones Beach sand, only thinner, finer, lifeless-looking sand.

"I was in Oakley not long ago," said a thin, sallow-faced man, "when a cloud of dust hung over the town, and you could hardly see across the street. Then it started to rain, and, brother, I'm telling you it rained mud, just black mud.

And then it turned to sleet, and the ground was so slippery it was as much as your life to walk out in it."

"Yeah," said another passenger, "I know a town in this part of the country where they sent sixty-nine people to a hospital with dust inside 'em, and only nine came back. The rest are in the boneyard." His face wore a mask of puzzlement.

Blown Away by Erosion

The train is now within an hour of Denver. The sun is shining from a blue sky flecked with white clouds—real clouds. The air is free of dust again for the first time in several hours. But my throat still burns and my eyes and lips are irritated as though they had been cut by something steely fine and thin.

Soil erosion. Wind erosion. Those words are vague and nebulous to many of you in New York. I did not know about them, either, when I left New York. But I know now, for today I have seen. I have seen what waste, greed, exploitation have done to our land. Unwise use, short-sighted farming.

Literally, our land is blowing away, piling up in mounds of sand that make you think of the mounds of the Gobi Desert. Little holes in the land, about two or three feet in circumference, dot the bare countryside; places where the wind has struck and dug out the soil like so much feathery sand. No vegetation left to hold the land intact, to repulse the wind.

It is a frightening experience. A thing no one would believe or visualize unless he has gone through it. The last time I was out here, twenty years ago, there were cattle, and trees, and birds, and life.

I cannot help remember J. N. Darling's words at the conservation conference in St. Louis last week, where a great army was organized to fight these things that are wasting our grand country away. Ding, a man who had the vision of all this years ago, but to whom no one listened, said:

"We must stop soil erosion. We must stop it because no government, no matter whether it be Democratic, Republican, communistic or dictatorship, can withstand the demands of hungry men who search for food in vain."

Today I have seen the cold hand of death on what was one of the great breadbaskets of the nation.

March 8, 1937

DUSTBOWL TRANSFORMED

With the exception of a few counties in Texas, Oklahoma and Kansas the Dust Bowl is flourishing again, though wheat is selling around 60 cents instead of last year's dollar, and the prices for cotton and sorghum are low. Farmers who have seen little hard cash for years are actually beginning to think of paying their debts.

It is not alone the rain that has thus transformed an arid region which had become a national problem, but the new methods advocated to bind the topsoil and to check erosion. Farming methods which are designed to force available moisture into the soil and which involve the substitution of drought-resistant grain-sorghums for thirsty wheat and an increase in the growth of Russian thistles and other soil-binding weeds have played their part in bringing back something like

prosperity to what was once called the "Great American Desert." But the Soil Conservation Service of the Department of Agriculture finds it necessary to sound a note of caution. Because the rain has poured down like manna there is no reason to conclude that soil-blowing is ended.

Whether or not we believe in cycles—to many physicists they are merely the tools of a new astrology—Mr. J. B. Kincer of the Weather Bureau bases on them his prediction of the Great Plains' future. The present cycle of dryness began in 1930 and showed signs of tapering off into the present wet cycle in 1936. Dr. C. G. Abbot of the Smithsonian Institution reaches similar conclusions on the strength of his own set of cycles. Nature is full of cycles. All are as suspect as stock-market charts that purport to indicate the probable rise and fall of prices. Yet

cycles are all we have wherewith to predict climate. Assume that Mr. Kincer is right and that the Dust Bowl has seen the worst of the present cycle. More droughts must be expected, with the old whirling of dust across the United States to the Atlantic Ocean and a loss of values running into the hundreds of millions. There can be no end to this battle with the dry winds. Grass, trees, windbreaks must be planted to prevent the Great Plains from becoming an outstanding example of the devastation that follows subsoil farming. It is man who is responsible for what has occurred, and man must make good the havoc that he has wrought. Fortunately, a new science of soil conservation is evolving, a science which reckons with him and which demands from him as much restraint as it expects pliancy in Nature.

July 31, 1938

Concern for Natural Resources

PRESIDENT ASSAILS RESOURCES WASTE

Asks Congress to Conserve Energy Fuel Sources and Attack Water Pollution

Special to THE NEW YORK TIMES.

WASHINGTON, Feb. 16.—President Roosevelt called on Congress in one message today to consider methods of conserving and utilizing the nation's energy resources, and in another suggested a $2,000,000,000 Federal-State program for abatement of water pollution. Each subject was covered in bulky reports from the National Resources Committee.

Without adopting the Committee's recommendations for his own, the President pointed out that some legislation affecting coal, oil, natural gas and water power would expire at the end of this fiscal year and other similar measures would terminate after a few years. He offered the report as "a useful frame for legislative programs affecting these resources."

Energy resources are not inexhaustible, yet the nation permits waste in use and production, the President wrote, and apparent economies today mean in some instances that future generations will carry a burden of unnecessarily high costs, and be forced to substitute inferior fuels.

National policies, he said, must recognize the availability of all our vital resources, their location with respect to markets, and consider transport costs, technological developments to increase efficiency of production and use; the use of the lower grade coals; and "the relationship between the increased use of energy and the general economic development of the country."

Single Problem Presented

The time had come, the President said, to devise a policy treating the energy resources as a whole rather than individually. Since it could not be done overnight, it was necessary to recognize that each had a bearing on and affected the others.

The report noted that "the obvious fields of remedy with respect to conservation of energy resources seem to lie (1) in promoting greater efficiency in the production of the fuel resources from the standpoint of recovery; (2) in promoting greater economy in the use of fuels; and (3) in placing a larger share of

the energy burden on lower grade fuels and water power."

The president proposed a central technical agency to promote elimination of water pollution, and a system of Federal grants-in-aid and loans which would be integrated with other water resources and public works programs. "No quick and easy solution of these problems is in sight," the President said.

The committee, the message continued, estimated that an expenditure by public and private agencies of "approximately $2,000,000,000 over a period of ten to twenty years may be required to construct works necessary to abate the more objectionable pollution."

Responsibility for the development of the necessary works, the President said, rests primarily with municipal government and private industry, although many State agencies have forced remedial action where basic studies have shown it to be practicable.

Committee Recommendations

The energy resource recommendations of the Resource Committee were:

1. Establishment of a Federal Oil Conservation Board or Commission to draft "necessary rules and regulations" concerning production and distribution of oil and gas. The board would have power to require that gas and oil be so

produced as to avoid waste and protect the interests of producers.

2. Strict Federal regulation of the bituminous coal industry. The committe said the Bituminous Coal Act of 1937 has made an approach to this problem.

3. Maintenance of an active Federal policy of public development of water power to conserve petroleum, natural gas and high-grade coals; to make electrical energy more widely available; to bolster national defense by assuring an ample supply of electrical energy in time of war.

The committee's water pollution report recommended:

1. Designation of a Federal agency, probably the United States Public Health Service, to study water pollution.

2. Federal grants and loans to public bodies, and loans to industry.

3. Administration of appropriations or allotments under a Federal public works agency.

4. Pollution plans to be cleared through a Federal coordinating agency to insure conformity to existing regional plans for water use.

5. All pollution-abatement contracts to be approved by Congress before they are negotiated.

February 17, 1939

A GROWING SENSE OF URGENCY

MIDDLE WEST

Shelter-Belt Forest Plan Wins Support of Farmers

By HUGH A. FOGARTY
Special to The New York Times.

OMAHA, Feb. 7—The world's most ambitious "air-conditioning" program will make further great strides this year in a strip of land running from North Texas to the Canadian border.

The project is the shelter-belt plan, which has reached full maturity as it enters its fourteenth year. Some of the trees planted in the pioneer, drought-ridden year of 1935 are now taller than houses. The stately string of trees and shrubs stand as a growing monument to the late President Roosevelt, who firmly sponsored the project despite an initial barrage of doubts and coarse jokes.

The results of the work to date are monumental. Since the United States Forestry Service made the first planting some 300,000,000 trees have been started and approximately 40,000 individual windbreaks have been installed along 25,000 miles of Great Plains land.

Individual Effort

The average shelter-belt, planted at right angles to prevailing winds,

consists of seventeen to twenty-one rows of shrubs and trees covering an area about 150 feet wide and a quarter-mile to a mile long.

Gradually, the effort has tended to shift from one of governmental spoon-feeding to individual effort. At the outset, the Forestry Service and the Works Progress Administration did all the work and paid practically all the bills while the average farmer looked on in

skeptical tolerance of the supposed boondogglers' efforts. But the skepticism has disappeared, the plantings continue and the trees continue to grow.

There is still plenty of Government assistance. But the main effort now comes from farmer groups banded together in soil-conservation districts. The trees come cheap, largely a Federal-state gift. The Soil Conservation Service, which has introduced many refinements to the task, provides machinery to speed the planting.

Farm Groups Cooperate

Results of the program have been noticeable. Farmers and other soil-conservation devotees have been able to establish that the shelter-belts condition the air by slowing down wind and humidifying the air, that they retain winter snow, prevent wind erosion, lessen the danger of crop fires and slow down evaporation from the soil.

Most important, however, is the farmer's realization that the shelter-belts are not in themselves a cure-all. They are now regarded as one of many necessary steps in conserving the soil and maintaining its productivity.

February 8, 1948

38

SIGNS FLOOD CONTROL BILL

Truman Approves $641,575,666 for Army River, Harbor Work

WASHINGTON, June 25 (AP)—President Truman signed today an appropriation bill carrying $641,575,666 for Army civil functions for the year beginning July 1.

The bill provides a record $573,000,000 to the Army Engineers for the most extensive program of flood control and navigation projects ever undertaken in a single year.

Construction funds are provided for 200 flood control projects and seventy-eight rivers and harbors improvement jobs over the country. Additional sums are provided for planning future construction.

The bill also carries funds for maintenance and operation of the Panama Canal, the National Cemeteries, the Soldiers' Home and the Signal Corps communications system in Alaska.

In all the $641,575,666 appropriation is $96,228,634 under the budget estimate.

June 26, 1948

FARM CONSERVATION GAINS

Holders of Two-thirds of Land Aided Program in 1948

WASHINGTON, Jan. 2 (AP) — Conservation practices were carried on in 1948 by nearly half the nation's farmers operating some two-thirds of the country's cropland, the Department of Agriculture reported today.

The conservation efforts come under the agricultural conservation program, which is directed by farmer-elected committees in the more than 3,000 farm counties.

The department's preliminary report showed that more than fifty approved conservation practices were employed by some 2,500,000 farmers who cultivate 275,000,000 acres.

January 3, 1949

COURT EDICT HAILED ON TIMBER CONTROL

State's Right to Make Private Owner Reforest His Cutover Areas Is Affirmed

By WILLIAM M. BLAIR
Special to THE NEW YORK TIMES.

ST. LOUIS, Nov. 10—A decision made by the Supreme Court this week may prove another milestone in the conservation of the country's natural resources and strengthen the concept that man is the trustee of the land for the general welfare.

The ruling, reported today at the opening of the Soil Conservation Society of America's annual meeting, upheld the power of the State of Washington to compel persons engaged in commercial logging operations to reforest cut-over areas.

The decision was revealed by Fairfield Osborn, president of the New York Zoological Society and a leading conservationist, at a joint luncheon of the society, Friends of the Land and the St. Louis Farmers Club.

Calls Ruling "Startling"

Mr. Osborn, whose book "Our Plundered Planet" dealt with man's spendthrift policy on natural resources, said he had learned of the decision just before leaving New York. He said it was "somewhat startling."

In his address to the meeting he had pointed out that "the concept of trusteeship for the general good on the part of owners, whether large or small, of natural resources is a totally new concept for the great majority of Americans."

The Supreme Court's decision was contained in a two-line order handed down on Monday. It was on an appeal by the owner of 320 acres of timberland in Washington, Avery Dexter, from a ruling by the Washington State Supreme Court.

Mr. Dexter disputed the constitutionality of a state law that requires timberland owners to reseed or re-stock land to maintain sufficient forest reserves and cover. He contended the land was his to use as he saw fit.

The State Supreme Court wrote: "We do not think a state is required under the Constitution of the United States to stand idly by while its natural resources are depleted and high authority supports our view.

"Edmund Burke (eighteenth century English statesman) once said that a great unwritten compact exists between the dead, the living and the unborn.

"Colossal" Debt Cited

"We leave to the unborn a colossal financial debt, perhaps inescapable, but incurred, none the less, in our time and for our immediate benefit. Such an inviolate compact requires that we leave to the unborn something more than debts and depleted national resources."

"Surely," the Court stated, "where natural resources can be utilized and at the same time perpetuated for future generations, what has been called 'constitutional morality' requires that we do so."

It said there was sound authority for the stand that the state legislation was a proper exercise of police power to prevent land owners from wasting resources at will and noted that "one great purpose" of government would be defeated if the state and its people were helplessly impoverished.

It also noted that the trial court ably championed the rights of private property and private enterprise and pictured the effect of "bureaucratic controls on farms and other lawful business of the state."

"We are in accord with much that is said there (in the trial court decision) but it must be realized that private enterprise must utilize its property in ways that are not inconsistent with public welfare," it said.

Defines Obligation

Mr. Osborn told the meeting that conservation should be a primary obligation of the educational system, business, industry and labor, and within the interest of innumerable civic and social associations.

"As a nation," he said, "we have not lived within our renewable natural resources income since the days of President Polk. Certainly we cannot continue indefinitely by virtue of a series of technical shots in the arm which are merely palliatives for our mistakes."

The afternoon session was devoted to problems of water conservation and control.

November 11, 1949

Industry Is Threatened by Increasing Drain on Water Supply, Survey Shows

National Survey Shows High and Unplanned Rises, With a Lopsided Distribution

By KALMAN SEIGEL

The United States faces heavy economic losses in the foreseeable future because of unplanned, excessive increases in the use of its most precious mineral resource—water.

In recent years, overdevelopment of ground water resources by industry, more widespread use of air conditioning, overdevelopment for irrigation fostered by high farm prices and the growth and shift of population all have combined to create specific area shortages of a magnitude and variety that will proscribe economic development.

The chemical industry, which uses water as one of its principal raw materials, has been sharply affected in terms of plant explansion by the decreasing availability of adequate supplies of cool water.

Evidence gathered by geologists, Government officials and conservationists, coupled with a study of conditions throughout the country by THE NEW YORK TIMES, discloses that while there are abundant underground and surface supplies of fresh water, all areas of the country do not share equally in this asset.

It is this lopsided distribution, they agree, that nourishes the basic problem of an adequate supply where it is needed, when it is needed, and of the desired quality.

Likewise, it is overdevelopment of supplies in short areas, they add, that will force some communities, unable to augment their sources, to stand by idly while a part of their industry and population is liquidated.

Lack of Data a Factor

Of perhaps even greater significance in resolving the paradox of a famine in the midst of plenty are the serious gaps in the nation's basic data on water resources, in the opinion of members of the United States Geological Survey. This is an agency of the Department of the Interior charged with the investigation of water supplies.

The deficiencies, it is contended, have their roots in the absence of a comprehensive national plan on developing water resources. The development of a plan has been stymied in part by the number of Federal agencies involved in the total water problem, the multiple sources that finance investigations, widely divergent concepts of water development, and by the Government's passive attitude.

Staff members of the Interior Department admit freely that there are few, if any, problems

39

A Growing Sense of Urgency

that could not have been solved to facilitate total development of the nation's economic potential if the basic data on local and area conditions had been available prior to the present critical stage.

Greater Use Expected

The nation now uses about 700 gallons a day for each person, and even greater use can be expected in the next generation, with the ultimate extent of demand not yet in sight.

A vital and significant step toward meeting the demands and avoiding shortages, the nation's water scientists say, can be taken by adequate evaluations of supplies in areas of present and potential development.

Much already has been done in scattered areas, but more remains to be done. The nation's water scientists contend that with their present knowledge of hydrology, the development of superior engineering and pumping apparatus, there should be no excuse in the future for haphazard development, serious overdevelopment and waste of water resources.

A study of the known data concerning water shortage areas, under which this information will be assembled, analyzed and correlated, together with known methods for combating the shortages, is being conducted by the Conservation Foundation, 30 East Fortieth Street.

Seen as Valuable Guide

Begun last year in cooperation with various Federal and state agencies, under the direction of Dr. Harold E. Thomas of the Geological Survey, the study will be available to the public within a year.

According to Fairfield Osborn, president of the foundation, it already is possible to forecast that "the results will provide a useful guide to industry and public works organizations in their search for improved water supplies."

Basically, the United States has plenty of water. The average fall is sufficient to cover the country's 3,000,000 square miles to a depth of nearly 30 inches. Of this, the nation uses only about 30 per cent—but this fraction is unevenly distributed.

Reduced to its simplest terms, the problem is one of a humid East and an arid West, the line of demarkation being the line of 20-inch average annual rainfall that corresponds roughly to the 100th meridian; this dissects the western part of Texas and moves northward through the center of North and South Dakota.

East of this line the supplies normally exceed the requirements. West, however, there is a decreasing rate of precipitation, the region of greatest deficiency including Arizona, New Mexico, parts of Texas, Colorado and California. Utah and Nevada comprise an interior region of deficient rainfall known as the Great Basin.

More Demands on Wells

The country spends more than $1,000,000 a day for ground water. Since 1935 use of ground water from wells has risen from ten billion to twenty-five billion gallons daily.

Of this total irrigation consumes more than ten billion gallons; industry more than five billion, excluding water from municipal supplies; municipalities more than

three billion gallons, and rural use, excluding irrigation, more than two billion gallons a day. Ground water, however, accounts for only 20 per cent of all water used. The remaining 80 per cent is surface water, or lakes, streams and rivers.

How Problem Is Being Met

In some areas, the problems of over-consumption and under-supply are being met with a variety of ingenious remedies. In others, relief is being achieved through legislative controls, the expenditure of vast sums for remedial projects, and through recognized scientific techniques.

Basic to every technique, however, are knowledge of present supply and the geology of the surrounding terrain, and whether the technique has economic and engineering feasibility.

A study of conditions locally and throughout the country follows:

New York City

Ground water supplies became inadequate early in the city's history, and the water now used is drawn from the Catskill and Croton watersheds. The Department of Water Supply, Gas and Electricity warned only last week that further depletion of New York's water reserves might bring restrictive measures to reduce waste.

Appeals to the public to cut down on the use of water met with a disappointing response. The entire Catskill system, with a capacity of 150 billion gallons when full, was down to 68 billion gallons on Nov. 1; it now has fallen to 56 billion gallons.

Kensico, the Catskill "safety" reservoir, remains at a level of 13 billion gallons. Stephen Carney, commissioner of the department, explained that if public consumption could be reduced, it might be possible soon to draw more water into Kensico from storage than is taken out for consumption.

A new project is under way to utilize the waters of the Delaware River basin. Estimated to cost $375,000,000, the additional supply will be ready about 1956, and, on current estimates of population growth, will meet expanding needs only until about 1960. For a safe margin after that, the city will need supplies beyond those now planned.

Upstate New York

Special to THE NEW YORK TIMES.

ALBANY, Nov. 20—New York State's capital, which once drew its water supply from the more or less polluted Hudson River, now has a mountain source of such capacity that it has offered to share its surplus with the entire county of Albany.

A former state health commissioner, commenting on the heavily chlorinated water drawn from the Hudson, remarked that the Albany source was 18 per cent sewage. After an outbreak of typhoid fever in Albany, the city administration struck out into the Helderberg Mountains for a fresh supply.

That supply, in operation for seventeen years, provides 35,000,-000 gallons a day, 10,000,000 more than the city's needs. City officials announced recently that this could be stepped up to 65,000,000 gallons a day and that the city was willing to provide water for the whole county, some sections of

which suffered keenly during the drought last summer.

In view of the possibility of sharing its supply, Albany is considering regulations that would cut down waste, including standard rules for use of water in air conditioning.

Some of the near-by cities are not as fortunate as Albany. Schenectady depends for its supply upon an underground stream, whose full capacity is not known. On several occasions Schenectady has been forced to forbid use of water for sprinkling lawns.

The Catskill Creek in the Helderbergs, on which Albany has staked out rights, once was sought by New York City, but the state ruled that Albany had a prior claim.

New Jersey

Special to THE NEW YORK TIMES.

TRENTON, Nov. 20—With the water reserve in North Jersey at the lowest level on record, the State Water Policy and Supply Commission warned today that rationing lay ahead unless users conserved voluntarily.

Howard T. Critchlow, chief engineer for the commission, said the summer drought, which has persisted more or less since early June, had brought reservoir levels down to 25 per cent of capacity.

If sufficient rainfall does not come in the remainder of the fall before freezing weather arrives, he added, a serious situation may arise. Water consumption in the North Jersey metropolitan area has continued to increase each year. The withdrawal now totals 380,000,000 gallons a day.

Mr. Critchlow suggested three means of reducing water consumption. These include repair of all leaks in water systems, a large source of loss; care in closing taps when water is not needed, and reduction of such nonessential uses as car washings.

George R. Shanklin, assistant chief engineer, provided further data on the situation. He said the Wanaque Reservoir, serving Newark, Paterson, Passaic and Kearny, held a billion gallons less than normal at this time. The Pequannock Reservoir, another serving Newark, was a third of a billion gallons under normal.

On Oct. 31, Mr. Shanklin added, remaining reserves were nearly down to the lowest levels at the year-end of 1939 and 1941, two other serious drought years.

The situation is aggravated, he said, by the fact that consumption has risen 151 per cent since 1938. While industrial uses have expanded, he said, residential and individual uses also have shot up.

Philadelphia

Special to THE NEW YORK TIMES.

PHILADELPHIA, Nov. 20—The Delaware and Schuylkill Rivers carry more water through the Philadelphia metropolitan area, comprising five counties in Pennsylvania and three in New Jersey, than it will require for generations. Both streams, however, are polluted by industrial waste and sewage. As a result, the section's problem is not one of getting sufficient water, but one of making the present supply fit for human needs and in businesses requiring a high quality water.

James H. Allen, secretary of the

Interstate Commission on the Delaware River basin, reports that the area's demand for water from public supply agencies has increased about 100 per cent in the last quarter century, rising from 275,000,000 to 550,000,000 gallons a day. This, he pointed out, does not include water used by industrial plants, which probably use and then return to the rivers about 500,000,000 gallons daily.

In addition, Mr. Allen said, 100,000,000 gallons of ground water are used daily.

Philadelphia is spending $40,-000,000 to improve the taste and odor of its water, widely known as the "chlorine cocktail."

The Chester Municipal Water Authority is spending $15,000,000 for a new source of supply, and other communities in this area are spending another $5,000,000 for the same purpose.

Of special significance is the study now being made of possible dam sites in the Delaware River watershed. The survey is being made under the direction of Incodel, which received $200,000 from Pennsylvania, New York, and New Jersey, to find out whether the three states should join in the construction of a single project for their common use. That would include a future new source of supply for this city.

Also, the Federal Government, Pennsylvania, and New Jersey, are embarked on a three-year program, costing $60,000 annually, to find out the capacity of the ground water formations in this area. Then regulations are to be established so this supply will not be overdeveloped and depleted.

New England

Special to THE NEW YORK TIMES.

BOSTON, Nov. 20 — Although New England is suffering from its worst drought in twenty-five years, Federal, state and private engineers attribute it to a periodic rainfall deficiency and not to a permanent lowering of the water level.

They point out that, because of geographical and geological characteristics as part of the coastal watershed, the region depends on surface sources and streams for water. There was nothing wrong that a normal precipitation this winter could not improve, they said, recalling that last winter's snowfall was below normal.

Since New England had experienced similar conditions before, these sources held that with normal rainfall there would be more than enough water to supply the increased population and industrial activity of the last twenty-five years.

The Geological Survey Office of the Water Resources Branch, Department of the Interior, explained that a twenty-inch annual rainfall was considered the minimum requirement to sustain vegetation and that New England normally had twice that amount, or a twenty-inch "cushion."

Where suffering during the present drought had occurred, it was found in areas dependent on wells or inadequate storage reservoirs.

Little danger to shortages of water was seen in the Metropolitan Water District of Massachusetts, which serves twenty-two cities and towns within a ten-mile radius of Boston.

The Quabbin reservoir, seventy miles west, has 352,000,000 gallons in storage and the adjacent Wachusett reservoir has 49,500,000 gallons. This, state officials said, was enough to care for the metropolitan needs for more than 2,300 days of water consumption.

The South

Special to THE NEW YORK TIMES.

CHATTANOOGA, Tenn., Nov. 20 —The water resources of the South are good. Generally speaking, the yield of ground water (wells and springs) and surface water (streams and rivers) is plentiful. There are coastal areas with a threat of salt water intrusion due to overdrawing of ground water supplies, and in some highly industrialized sections there is a decline of water tables because of heavy pumpage and use of ground water.

However, the growing industrialization of the South has not yet seriously affected the water resources in a broad regional sense. It is important to note that the South has an excessive rainfall. The annual average rainfall is fifty-two inches in the seven states of the Tennessee Valley, and in this section an estimated 6,000 tons of water fall on each acre of land each year.

Furthermore, while the South is increasingly a site for industry, this expansion comes when the region has become fairly aware of conservation of natural resources under the watchful eye and advice of the Tennessee Valley Authority. The TVA system of twenty-eight dams has conserved great quantities of water.

The TVA has taught conservation of the soil and of the moisture content by contour and strip farming and especially by concentration on grass culture and pasturage.

In southwestern Louisiana there is a serious groundwater shortage. This results from the irrigation practices of rice growers. Years ago rice was irrigated by surface water, but now half of it is irrigated by ground water. The ground water levels have been drawn down by the heavy demand and there is salt water intrusion from the Gulf of Mexico. This situation has also developed in spots on the Florida and Georgia coasts.

The water tables in North Carolina, site of the textile industry, are at high levels. Only at New Bern, Kinston and Elizabeth City has increased pumpage lowered the levels. In Georgia water levels are high except for industrial and municipal pumpage lowering them in the Savannah and Brunswick areas. Water levels are normal in Alabama except for critical localities around Montgomery, Selma and Dothan with industrial pumpage. This general picture prevails throughout the South.

Progress, slow but definite, has been made in the region toward legislation for pollution control.

Central States

Special to THE NEW YORK TIMES.

CHICAGO, Nov. 20—Water resources of four central states, Illinois, Indiana, Michigan and Wisconsin, are in good condition except in "critical areas." These are invariably industrial areas, where pumpage exceeds natural recharge of ground water. Use of water has risen greatly in the last twenty-five years, calling for development of new sources, re-use by industry and reduction of pollution.

Although water levels have been dropping generally, new finds and new techniques in well-drilling give a better water supply than twenty-five years ago. In Indiana and Michigan the level has risen since the war. Lake Michigan is a good source for communities on its edge and some communities inland with depleted ground sources are pumping water in. The Illinois water survey division has a project at Peoria, its most critical area, for artificial recharge of ground sources from surface water, as yet in study stage.

Per capita water use has increased sixfold in critical areas in Illinois in the last twenty-five years. Artesian levels, which were sixty feet above ground ninety years ago (giving flowing wells) now are 600 feet below ground in some areas.

In Wisconsin the drop in water level has been great enough to make water appreciably more expensive to get. Surface water sources — lakes and streams — are good in all states except where pollution cuts availability, principally in Michigan and parts of Wisconsin.

In Michigan, where the Legislature has stiffened penalties on pollution, the $400,000,000 annual tourist industry is involved, since clear streams, lakes and beaches are a principal tourist drawing card.

Missouri Basin

Special to THE NEW YORK TIMES

ST. LOUIS, Mo., Nov. 20—In general, the water supply situation in this area and in the Missouri River basin is considered good. A watchful eye is being kept, however, on both ground and surface water levels, which are supplied from streams in good condition. There are some small communities, however, that are having difficulties because of a lack of stream flow, increasing pollution and dry weather.

Authorities do not consider the situation serious, but contend that in a growing number of communities the demand is greater than the recharge in ground water supplies because of industrial and population increases.

There is, however, a growing pollution problem and officials of the Missouri Resources and Development Department point out that this phase of the total water problem may outstrip that of lack of water.

In the western semi-arid reaches of the Missouri basin there are about 4,500,000 acres under irrigation, using 10,000,000 acre-feet of water annually. Most of this water is returned to streams, but with increases in irrigation in the next decade under the basin's development program, irrigation water is expected to pose a big problem.

Some irrigation reservoirs are depleted. It is too early to determine the mountain water situation, but many streams in western Nebraska, Montana, Wyoming, Colorado are low despite last winter's heavy snow cover.

The water table in the western area is reported high as a result of the "wet cycle" of the last seven or eight years, but fears have been expressed over a dry period during which the table will drop. There is range land in Montana and North and South Dakota that is very dry and water may be a serious problem there next year.

Federal and state officials say there is not enough water for present and future needs in the western area of the basin. This deficiency has caused substantial losses in agricultural production and has inhibited industrial development.

Los Angeles Area

Special to THE NEW YORK TIMES.

LOS ANGELES, Nov. 20—Southern California generally has been keeping ahead of its water requirements over the past twenty-five years despite natural aridity and unprecedented expansion of population and industry. In a few places the situation is tight and occasional rationing has been necessary.

Apprehension is being expressed about the future supply, although to some extent this is propaganda in California's long battle with Arizona over rights to Colorado River water.

Los Angeles and the other principal Southern California cities grouped in the metropolitan water district have, as their main source, the 238-mile Owens River aqueduct tapping the Sierra Nevada watershed. This has been supplemented in the last fifteen years by an aqueduct from the Colorado River, although at present this constitutes only a small fraction of Los Angeles' supply, and the district

as a whole has been drawing only about 16 per cent of its allotment from the river.

Los Angeles' population has increased in the last twenty-five years from 850,000 to 2,000,000, and its water consumption from 105 million gallons a day to 351 million. About 40 per cent goes to homes, 20 per cent to irrigation and less than 2 per cent to industrial uses.

Arizona's agriculture is concentrated mainly in a 1,200-square-mile irrigated desert area around Phoenix. The water comes about half from wells and half from the Salt, Gila and Verde Rivers. The supply is insufficient, and Arizona wants to divert 391 billion gallons a year from the Colorado to supplement it. California contends this would encroach on its river quota.

Pacific Northwest

Special to THE NEW YORK TIMES.

SAN FRANCISCO, Nov. 20 — The water situation in the Northern Pacific states is spotty. Vast semi-arid stretches of California, North Central Oregon and South Central Washington require irrigation and the water supply, especially in California's Central Valley, is inadequate to meet the needs without lowering of the underground water level.

For example, in a recent year when 1,550,000 acres were under irrigation in the upper San Joaquin Valley of California, the available water supply was adequate for irrigating an area of only 1,170,000 acres.

Neither Oregon nor Washington has any foreseeable threat of serious water shortages, except in a few coastal towns, because of ample rainfall in the western part of both states and heavy snow storage in the Cascade Mountains.

The situation has been made more critical, especially in northern and central California, by the great population movements of the 1930's, the war and postwar periods.

Industrial use of water has accentuated the shortages in such San Francisco Bay communities as Vallejo, home of the Mare Island Navy Yard, but industry's use of water is small compared with that for irrigation.

The present use of water for all purposes in northern and central California exceeds 10,000,000 acre feet annually. Ultimately an annual supply of more than 20,000,000 acre feet will be needed.

Oregon has been combating stream pollution by such industries as paper mills.

A dropping water level has become critical in some parts of the area.

November 21, 1949

Resources Report Warns of National Peril in Water Waste

By PAUL P. KENNEDY
Special to THE NEW YORK TIMES.

WASHINGTON, Dec. 17—This nation, lacking a uniform Federal policy for comprehensive development of water and land resources, appears to be slowly losing in its piecemeal battle to meet increasing water needs.

There are serious wastes in our precious water supply and if these wastes continue unchecked they will bring grave handicaps for us and for those who come after.

These warnings form the basis of a 445-page study made public today by the President's Water Resources Policy Commission. It is the first section of a report that, when completed in February, will comprise in three volumes the most comprehensive and authoritative water survey in the country's history.

Recognizing the interrelationships between land and water management, the commission calls attention to the increasingly apparent need to consider whole river basins in one broad and uniform policy rather than deal with the multiplicity of "one river, one plan" projects as at present.

"This commission," the study states, "is therefore recommending the achievement of the necessary coordination through unification of policy governing the actions of existing agencies. or of a single agency."

Policy Enactment Urged

It continues:

"This unification of policy should be assured through enactment of a single national water resources policy law controlling the activities of the government departments as presently organized or as they may be organized."

The commission, headed by Morris L. Cooke of Philadelphia, consulting engineer, was formed at the request of President Truman last year to consider the following in recommending a comprehensive policy of water resources development:

1. The extent and character of Federal Government participation in major water resources programs.
2. An appraisal of the priority of water resources programs from a standpoint of economic and social needs.
3. Criteria and standards for evaluating the feasibility of such projects.
4. Desirable legislation or changes in existing legislation.

President Truman, in a statement, voiced gratification for the report and endorsed its stress on the urgency of wise planning.

"Plans for water development," he said, "can no longer be made successfully by individual interests, whether they are private or public, whether they are local, state or Federal."

The commission centered its attentions on the problems of ten great river basins—the Columbia, Central Valley of California, Colorado, Rio Grande, Missouri, Ohio, Tennessee, Alabama-Coosa, Potomac and Connecticut. These were chosen over other basins because they presented the full range of all water and land problems known in this country.

Justification for the river-basin concept of planning is outlined in the report thus:

"Irrigation and drainage, navigation and flood control, the maintenance of underground water levels, the control of stream pollution resulting from human, animal and industrial wastes; the generation of electric power, the protection of salmon and other fish resources, the provision of ample domestic water supply —all of these purposes have legitimate claims within any one basin; but if one is developed without regard to its effect on the others, conflict and losses will result."

The commission recommended that all projects in progress be continued and those for which appropriations had been made go through as planned except where-

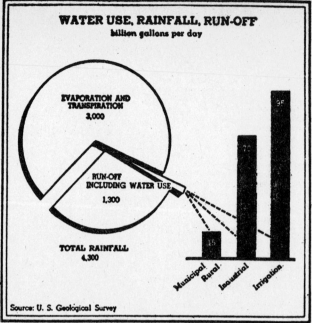

WATER USE, RAINFALL, RUN-OFF
billion gallons per day

EVAPORATION AND TRANSPIRATION 3,000

RUN-OFF INCLUDING WATER USE 1,300

TOTAL RAINFALL 4,300

Municipal Rural Industrial Irrigation

Source: U. S. Geological Survey

The New York Times   Dec. 18, 1950

Domestic and industrial use of water accounts for a comparatively small percentage of the total supply. The remaining 69.7 per cent is lost through evaporation and transpiration— exudation and absorption by vegetation—as noted on chart.

in they lent themselves readily to revision for integrating into other plans. Otherwise, it is generally recommended that no project be adopted until a full river-basin plan has been adopted.

Carrying over its basin-planning concept into the over-all economy of the nation, the report continues:

"Comprehensive, long-range plans must be worked out within each river basin. But a river basin is still not the final unit in an adequate use and conservation plan. The Federal Government has a substantial investment in existing water resources improvements; it will spend in 1951 another $1,200,000,000; it has authorization amounting to $12,000,000,000 in the years just ahead.

"These expenditures will be made for the good of the nation as a whole; and nothing less than the whole country can be the unit considered in the formulation of Federal policies."

The commission notes that in all our history there has never been a statement of national objectives in the whole field of water resources. It recommends, therefore, the following seven points as our long-range work sheet:

1. Preventing the ultimate decline of productivity through mismanagement and neglect by utilizing all phases of soil conservation, flood management, control of ground and surface waters and sound forestry.

Production Aim Stressed

2. Providing through increasing production of land and water resources a broader base for a steadily expanding national economy by transforming water resources from ineffective or destructive into beneficial agents, watering arid land, supplying municipal and industrial needs, improving channels for water transportation and generating hydroelectric power.

3. Providing opportunity for farms, urban homes and commercial establishments and industries to make full use of electric power, through a marketing policy for Federal power aimed at encouraging maximum use at the lowest possible rates.

4. Coordinating of water and land resources undertakings with specific plans to meet needs of national security.

5. Developing of balanced regional economies, offering maximum opportunity for farming coupled with non-agricultural rural employment.

6. Providing for expanding cultural opportunities, including all phases of recreational development from wilderness areas to wisely designed, artificial multiple-purpose reservoirs.

7. Protecting the public health, particularly through pollution abatement and control, mosquito control and all necessary provision for an abundance of high-grade municipal water supply.

One of the primary pressures on this attainment chart, the report indicates, lies in confusion on planning. It notes that separate agencies are increasingly concerned with the same river basin, particularly in the Western states.

"Authorizations, while remaining single-purpose in name, are becoming in fact multiple-purpose," it declares. "But the blunt truth is that instead of getting coordinated plans we stumble along amid a multiplicity of separate plans by the Federal, state and local agencies that finally are thrown together in an unsatisfactory planning. Again and again, projects are undertaken as if they were ends in themselves, instead of parts of a program designed to meet the needs of the land and of our people."

In this connection, the report comments:

"Another heavy drag on many comprehensive river - basin pro-

grams is the lack of effective state and local participation. This has been due in part to the fact that the state laws are no better adapted to comprehensive planning than are Federal laws."

To assure uniform planning in the use of water and related land resources the commission makes a series of recommendations, among which are:

1. Congress should provide that the unit for planning further water resources development should be the river basin and that a basin program should be prepared in relation to broader regional needs and assets.
2. A single appropriation should be made for each survey and planning operation. At each step in the planning, authorization and appropriation process the basin program should be treated as a single program for all purposes rather than as an aggregate of plans for separate purposes.
3. Congress should direct all Federal departments and agencies responsible for the development of water and land resources to promptly review as coordinated groups all existing plans and programs in cooperation with interested states. Plans already authorized by Congress should remain undisturbed unless this review results in specific recommendations for change.
4. Congress should authorize the creation of organizations to co-ordinate and prepare multiple-purpose plans for the several river basins.
5. Congress should authorize the creation of a board of review to analyze and review all projects and programs recommended by the basin commissions prior to their presentation to Congress.

To comply with these recommendations, it is indicated in the report, planning bodies should concern themselves chiefly with the following prime elements in water resources: watershed management, flood management, water supply, navigation, hydroelectric power, irrigation, pollution abatement, drainage, recreational use of water resources, fish and wildlife, sediment control, salinity control and insect control.

The report gives detailed attention to the highly complex problem of reimbursement for funds expended in water resources development. To the citizen's view the commission's basic thinking is expressed most simply, probably, in its recommendations on distribution of costs by specific functions. These are as follows:

1. **Hydroelectric Power, Municipal and Industrial Water Supply** —Prices charged should at least cover costs, including amortization of the investment with interest and payments in lieu of state and local taxes.

2. **Land Reclamation (Irrigation and Drainage)** —Primary beneficiaries of reclamation should repay without interest an amount assessed according to ability to pay based upon annual net income which the farmers derive from the project; under a formula adjusting repayment to production and marketing conditions, this amount to include the full costs of operation and maintenance facilities.

3. **Navigation Improvement**—On the condition that railroad rates cannot be set below costs in order to meet rates of water carriers, consideration should be given to establishing a system of waterway tolls designed to recover from the beneficiaries of waterways transportation a substantial portion of

the costs incurred for the development of waterways.

4. Flood Control—In case of local flood-control works, local communities directly affected to assume an appropriate share of cost, which could be apportioned on the basis of property values either by general taxation or by special assessment. In case of extensive flood-control works, involving large-scale upstreams storage and land management programs, states should establish special districts with powers to assume an appropriate share of costs.

5. Watershed Protection and Scientific Land-use Measures—Costs assumed by state and Federal Government, but with contribution by owners of farms and forests receiving private benefits.

6. Pollution Abatement—Principal costs to be borne by local communities and industrial firms, through provision of adequate sewage treatment facilities. In order to hasten construction of sewage and waste treatment works, the Federal Government should lend money at low rates to communities and industrial firms. Possibility of a combination Federal loan-grant arrangement should be considered.

7. Recreation, Fish, Wildlife—Preservation resources should be included in cost of providing major water resources development. Basic construction features should be the Federal responsibility, but where areas are not of national significance operation and maintenance should be paid for by state and local governments.

Admitting the impossibility of stating the cost of a resources investment program, the report declares:

"However, from studies and estimates by private and public sources it would appear that an adequate conservation and development program for renewable resources over the next generation might cost the nation in the neighborhood of $100,000,000,000."

Fresh water, the report notes, is our primary self-renewing source and our share of it in the United States is enormous. The total quantity in constant circulation, measured in precipitation, amounts to about 4,300,000,000,000 gallons daily or, roughly, ten times the average flow of the Mississippi River.

Some 3,000,000,000,000 gallons return daily to the atmosphere as vapor, leaving the remaining 1,300,-000,000,000 gallons for ground-water infiltration and surface runoff. Effective watershed management for the control of runoff and evaporation is recommended.

The report dwells on the extensive land holdings of the Federal Government and the miltiplicity of its management. It points out that within the forty-eight states the Government owns 458,-000,000 acres in important and significant watershed lands. Of this land, 407,-000,000 acres lie in eleven far western states. Eight Federal agencies administer 99 per cent of this land.

The report states:

"The problem these public lands raise is that they are becoming seriously eroded and are a continuing source of silt pollution in streams. No one Federal agency is responsible for their care, yet the Federal Government is the only source of management funds and programs."

Just over nine-tenths of the 1,900,-000,000 acres of land in the United States is devoted to crops, to pasture and grazing and to forest. There remains a total of 75,000,000 acres suitable for reclamation by flood protection, drainage and clearing. The report notes that a careful study of past and current rates of development and proposed programs of Federal agencies for the next few years indicates that probably 21,000,000 acres of average crop land equivalent may be brought into production through flood control, drainage and clearing by 1975.

Referring to water pollution, the commission says:

"Grave as it is now, the pollution problem is found to deepen as population and industry grow. This need not be. The losses and injuries we now sustain are by no means irretrievable if we act now."

In one of its six recommendations on pollution the commission suggests that control should be an integral part of comprehensive river basin programs. In another recommendation it is suggested that if Federal-state-local cooperative pollution control fails to provide clean rivers within a ten-year period the 1948 Water Pollution Control Act should be revised to provide Federal enforcement, without the requirement of state consent, where polluted streams are within the jurisdiction of Congress.

To achieve the objectives of hydroelectric power planning the commission recommends:

"Regional power resources should be viewed as a whole, regardless of ownership. It should be possible in every region to secure for all the benefits of carefully integrated power system development, either by agreement or by common ownership facilities."

Members of the commission besides Mr. Cooke are Paul S. Burgess, dean, College of Agriculture, University of Arizona; Lewis Webster Jones, president, University of Arkansas; Samuel B. Morris, Department of Water and Power, Los Angeles; Leland Olds, commissioner in charge of studies, New York; Ronald R. Renne, president, Montana State College, and Gilbert F. White, president, Haverford College.

December 18, 1950

2 GROUPS OBJECT TO WATER REPORT

Special to The New York Times.

WASHINGTON, Dec. 17 — The initial recommendations of the President's Water Resources Policy Commission prompted charges today that the study revealed a disregard for safeguarding local control and a hostility toward independent enterprise.

The American Watershed Council, attacking proposals for extension of federal controls, maintained that "safeguards to local control we had hoped to see in the report are conspicuously absent."

The American Watershed Council is an organization formed this year to provide a medium for conservationists, in and out of public life, to present their viewpoint on watershed problems. Interest of members has been concerned chiefly with the smaller American watersheds.

Speaking for the electric companies, James W. Parker, president of the Detroit Edison Company, contended that the report constituted "one more expression of hostile attitude toward independent enterprise." He said it was "mainly a restatement of an unfair Federal power and water resource policy favored by the 'public power' boosters in Government departments and bureaus."

Members of the executive committee of the American Watershed Council, according to the committee chairman, Clayton M. Hoff, "were appalled at the scope and tone of the report."

A member of the Interstate Commission on the Delaware River Basin, Mr. Hoff recalled that his organization, which testified before the President's advisory group maintained that a sound national land water policy could best be advanced by maintenance of local control.

Calls Report "Disappointing"

He contested the commission's conclusion that the nation was on the threshold of a tremendous increase in the volume of construction for Federal water projects. Such views in the report, coming in time of national crisis, were attacked as having a "business-as-usual tone."

"This report," the executive committee said, "will be most disappointing to the many local organizations which urged a greater local responsibility and less governmental paternalism."

Charging the commission with having "failed completely in its mission of establishing a workable pattern for an acceptable national power policy," Mr. Parker, representing the electric companies Public Information Program, declared:

"This group is headed by Morris L. Cooke and includes ex-Federal Power Commission Chairman Leland Olds. It clings, in its recommendation for development of hydro-electric power, to the false and oft-repeated thesis that the industrial economy of the United States has been retarded from the beginning by insufficient development of the country's power resources.

"The Cooke-Olds group takes little account of the fact that 55 per cent of the nation's hydro-electric power has been developed by private industry and that the industry is willing and competent to develop more—without expenditure of public money.

Cites Industry's Program

"Also the commission attaches far too little importance to the tremendous program of electric power development which has been and is being carried out by the electric industry—without help from the Government."

Mr. Parker maintained that the commission findings "amount to an extension of 'Government from Washington' despite overwhelming testimony urging water development and supervision by local administrators which the commission intself heard in a series of public meetings over the country."

The utilities' spokesman contended the commission disregarded much that it heard and "obviously has not considered that the establishment of new 'valley authorities' will produce tremendous dislocation of industry—which may some day result in a good many 'ghost towns.'"

Unless "hampered by excessive and ill-conceived interference," Mr. Parker said the industry, prepared to take on a 15,000,000,000 expansion program, "will be more than able to keep pace with the nation's industrial development both in war and peace."

December 18, 1950

Land in Irrigation Increases
WASHINGTON, Jan. 13 (UP)— Irrigated land in the United States increased nearly 25 per cent between 1940 and 1950, the Commerce Department reported today. A study of irrigation growth in connection with the 1950 census showed that 26,240,000 acres were under irrigation in seventeen Western states, plus Arkansas, Louisiana and Florida, in 1949. This was 5,100,000 more than in 1940.

January 14, 1952

RESOURCES WASTED, NEUBERGER CHARGES

The United States is using up its natural resources more recklessly than any nation ever has done, Senator Richard L. Neuberger declared here yesterday.

The Democratic Senator from Oregon, a journalist who has specialized in conservation, spoke at the twentieth anniversary luncheon of the Public Affairs Committee in the Biltmore Hotel. The committee is a nonprofit organization that publishes pamphlets on public questions.

About 400 supporters attended the luncheon and panel discussions on desegregation, family welfare, civil liberties and mental health during the day.

Senator Neuberger expressed alarm over a "new, wholesale attack" on conservation since 1953.

"We are sacrificing the full development of Hell's Canyon on the Snake River, the greatest power potential on this continent," he asserted.

"Development of energy reserves may decide the world armament race. What will our people in the future think of our shortsightedness, compared with the great Soviet dams on the Volga and Yenesei Rivers? The potential hydroelectric power of the Soviet Union is greater than that of the United States and Canada combined."

January 21, 1956

WOODLOTS GROW MONEY ON TREES

By JOHN J. ABELE

Owners of small woodlots are finding that money can grow on trees.

At the same time they are taking better care of their forest lands, thereby adding to the nation's timber resources.

These are some of the benefits of the national tree farm program, which completed its fifteenth year of operations in 1956. A tree farm is privately owned, tax-paying forest land devoted to production of repeated crops of commercial timber through good forest management.

Proper management of forest properties means healthier trees, better wood and—the pay-off—selective cutting that can bring profits to the owners.

Sponsored by American Forest Products Industries, Inc., representing the nation's wood-using industries, the tree farm program now embraces more than 41,000,000 acres of commercial timberland held by some 9,500 owners in forty-four states. About 4,000,000 acres were enrolled last year.

Farmers, ministers, teachers, mechanics and retired persons, as well as industrial concerns, are among the owners participating in the tree farm program. The farms come in a variety of sizes from a three-acre stand operated by a Georgia farmer to the 207,000-acre tract of an Oregon pulp company.

Industrial interests own about 32,000,000 acres of tree farm lands, with woodlot holdings and investment properties accounting for about 9,500,000 acres.

Although owners of small woodlots—those under 500 acres—are in the minority, potentially they are the more important element. About 61 per cent of the nation's commercial timberland is held by owners of small woodlots.

American Forest Products Industries estimates that the market value of forest products from these small holdings is about $700,000,000 a year. The products include sawlogs, pulpwood, posts, turpentine and Christmas trees.

INSPECTION DAY: John Rud, left, a tree farmer of Birchdale, Minn., checks aspen pulpwood on his land with Howard Post, forester of the Minnesota and Ontario Paper Company.

Some Examples Listed

Here are some examples of how woodlot farming pays off:

¶When his first daughter was born, a Georgia farmer planted 47,000 slash pine seedlings on 825 acres of wornout land. Two years later he bought an adjacent plot of 303 acres and planted an additional 150,000 seedlings. The major portion of his trees will be ready for harvest by the time his daughter is ready for college. Meanwhile, he is making about $1,500 a year from the sale of inferior trees for gum, turpentine and resin.

¶A Northeastern tree farmer has averaged $750 a year for the past fourteen years from the sale of sawlogs, pulpwood and fuel wood from his 150-acre woodlot.

¶A West Coast dairyman obtained 9,000 board feet of lumber for farm buildings with the first harvest from his thirty-acre tract of Douglas fir. A second cut three years later provided him with 42,000 board feet, enough to build a barn and two silos. A third harvest yielded 55,000 board feet.

Among the good forest management practices that tree farmers agree to follow are safeguards against insects, fire, disease and destructive grazing by animals. They also remove dead, diseased and deformed trees that could impair the growth and health of other timber.

The program also encourages "multiple use" of forest lands. This involves cooperation with water and soil conservation projects, permitting use of forest lands for recreational uses such as hunting, fishing and picnicking, and providing for the forest wildlife.

The tree farm idea reflects a desire to conserve forest resources, rather than preserve them in a permanently wild state. Absence of proper care and intelligent harvesting can result in the choking off of new growth and the weakening of older trees. Proper control measures insure the perpetual renewal of sources of forest raw materials.

Tree farm applicants have their lands checked by professional foresters before they are admitted to the program. The inspections are arranged by state committees of forest industry representatives or by forest industry associations. Reinspections are made periodically to make sure that proper management practices are being followed and to advise owners on special problems.

In some cases industrial concerns provide free professional advice on forestry problems to a group of tree farm owners. In return, the companies receive first call on the timber when it is ready to be harvested.

Aside from the tangible cash benefits they can derive from good forest management, tree farmers also gain a certain prestige by knowing they are contributing to better timber resources and by having their lands approved by experienced foresters.

January 20, 1957

Floods, and Their Control

The annual spring floods have already begun to devastate the inland valleys, with lost life and property and consequent suffering. March will bring more floods, and so will April. It has happened thus ever since man began deforesting the hills and draining the swamps. And soon the perennial demand for better flood control will be heard again.

When we talk of flood control we usually think of dams and deeper river channels, to impound the waters or hurry their run-off. Yet neither is the ultimate solution, simply because floods are caused by the flow of water downhill. If the hills are wooded, that flow is checked. If there is a swamp at the foot of the hills, the swamp sponges up much of the excess water, restores some of it to the underground supply and feeds the remainder slowly to the streams. Strip the hills, drain the bog-

lands, and you create flood conditions inevitably. Yet that is what we have been doing for years. And with our increasing urbanization, our "reclamation" of lowlands and our reckless population of the flood plains, we have created at least one new flood factor for every old one we bring under control.

There is no quick solution to the problems of flood. Dams help. Diversion channels help, sometimes; often they merely shift the flood problem to a neighbor downstream. Reforestation of watersheds helps a great deal, but it is a slow process. Conservation of swamplands, however, is both relatively easy and is fundamentally essential to any ultimate control. Every swamp that is drained, every bog that is filled, adds to the flood problem. Yet we go on draining and filling—and demanding better flood control. Why?

March 1, 1959

U. S. CROP LANDS FOUND SHRINKING

Million Acres Lost Yearly

By DONALD JANSON
Special to The New York Times.

WEST LAFAYETTE, Ind., Aug. 2—More than 1,000,000 acres are going out of crop land annually, largely to accommodate urban sprawl.

The figure is taken from a new national inventory of soil and water conservation needs.

The inventory indicates that the nation's crop land will be converted to city and other non-agricultural uses at an accelerating rate.

"This trend is likely to speed up as more and more people require more space in which to live, work and play," Donald A. Williams, administrator of the Department of Agriculture's Soil Conservation Service, told the Soil Conservation Society

of America this week.

The rush of city dwellers to the suburbs and beyond spreads the responsibilities for conservation to urban areas, he said.

"Soil and water conservation is not for farmers alone," he asserted.

Major Problem Seen

The inventory shows that 75,000,000 acres of highly erosive land should be removed from cultivation, Mr. Williams reported.

Cropping this land, much of it in the high plains of the West, depletes the soil and creates dust and wash-out hazards for neighboring farms and ranches.

Getting this marginal land permanently converted to pasture is "one of the major conservation problems of the future," Mr. Williams said.

The Great Plains Conservation Program, which offers farmers Federal technical and financial help in making the transition, has made some strides in recent years. But annual appropriations have been insufficient to enroll all the

marginal farmers who have sought aid, for example, in shifting from growing wheat to growing livestock.

Mr. Williams said the inventory showed that 444,000,000 acres were being cultivated, an equivalent amount is in pasture and approximately the same acreage is in woodland.

Two-thirds of the total, including two-thirds of the crop land, needs some kind of conservation treatment, he said.

Erosion is the major problem on 237,000,000 acres of crop land, the inventory shows. Walter C. Gumbel, president of the society, said agriculture was losing 400,000 acres of crop land a year through erosion.

Better Grass Needed

A third of the country's range land a year through erosion. grass, according to the inventory.

The inventory, begun in 1957 and completed except for state-by-state summaries, is the first of its kind. Thirty thousand people in 3,000 counties analyzed

the land and water resources in all of the nation's rural, privately owned property to produce it. The Government will publish a summary soon.

The inventory identifies 13,000 natural watersheds, the first time the nation has had a count. About half were found to need flood-control projects. Others need work to control erosion and drainage or to provide irrigation or recreation facilities, or to develop municipal and industrial water supplies. Many need multi-purpose projects.

Mr. Williams disclosed the findings at the closing session of the society's three-day annual meeting. Other speakers joined him in stressing the need for converting poor crop land to such soil-conserving uses as pasture and for more widespread use of such soil and water conservation practices as terracing and strip-cropping.

They agreed that removal of erosive land from wheat, corn and cotton production would help reduce Government-held surpluses in those crops.

August 6, 1961

UDALL URGES AID ON CONSERVATION

Asks White House Meeting to Bar Short-Term View

By JOSEPH A. LOFTUS
Special to The New York Times.

WASHINGTON, May 24—The White House Conference on Conservation, the first of its kind in fifty-four years, assembled today to hear the New Frontier's hopes and fears for America's natural resources.

It was a revival meeting of 500 conservationists, men and women, gathered to remind

themselves—and all others, they hope—that keeping the great outdoors great demands dedicated work and education.

Be not defensive, exhorted Interior Secretary Stewart L. Udall, because the forest and the clean stream pay no cash dividends tomorrow.

"It is a fallacy," said Mr. Udall, "to attempt a justification of conservation solely in short-run economic terms. Conservation of every resource cannot produce the same margin of profit as concentrated exploitation. Attempts to demonstrate the contrary do a disservice to our integrity and weaken our position.

"Conservation does not mean economic loss, but not all national forests would stand the test of a cost-benefit ratio and

a secluded glade will not produce income or taxes on a par with high-rise apartments or a filling station.

Some Sacrifice Needed

"It is time for the American people to assume the burdens of maturity. Social values must be equated with economic values; the overriding need of men for an environment that will renew the human spirit and sustain unborn generations requires some sacrifice of short-term profits."

Men have been leaving the land, and the great cities grow greater, but these men turn to the land for their recreation, for beauty and for escape from asphalt and steel. Unless the nation prepares to meet those needs, said Secretary Udall,

"we face an austere rationing of even those things which have been traditionally free in America—its waters, wilderness and space."

President Theodore Roosevelt, by calling the Congress of Governors in 1908, alerted the nation to the inroads that civilization was making upon its own natural estate. Today, as then, the watchword is education.

Freeman Tells Plans

Orville L. Freeman, Secretary of Agriculture, told the conference how his department, assuming assistance from Congress, proposed to develop multiple uses for private land.

Despite the far greater numbers of people to be fed and clothed in 1980, Mr. Freeman said, "we can meet all needs for crop products then with 50,000,000 fewer acres than we presently have available for cropping," if farm efficiency continues to advance as it has in the last decade.

Not one of these extra acres of cropland need be idle in 1980, he said.

"Every extra acre of cropland," he went on, "can be put to productive, economic use— for pasture and range, for timber, for fish and game and other wild creatures, for water conservation and supply, and for outdoor recreation.

"Multiple-use of privately owned land, as well as public land, can unlock the great outdoors to millions of Americans.

Outdoor recreation is one of the great unmet needs of the nation today. By the year 2,000 the demand for recreation should triple."

Secretary Freeman gave examples of farmers and ranchers who were doing well in the recreation business, providing vacation farms, picnicking and sports centers, fishing waters, camping and nature recreation areas, and hunting.

Hear Kennedy Today

President Kennedy, who will address the conference tomorrow, has proposed to Congress a program for the development of outdoor recreation resources, and conservation of public lands, soil, watershed, range

and timber resources.

The conference included traditional conservation groups such as the Izaak Walton League, the Sierra Club, National Audubon Society, North American Wildlife Foundation, the Boone and Crockett Club, and National Parks Association. It also attracted industry groups primarily interested in the impact that conservation measures have on their business growth, and some that have adopted conservation programs of their own.

Two examples are the Phillips Petroleum Company of Bartlesville, Okla., and Smith Research and Development Corporation of Lewes, Del., which will receive conservation

awards from Secretary Udall tomorrow.

Laurance S. Rockefeller, chairman of the Outdoor Recreation Resources Review Commission, was moderator of the meeting today. He will receive a plaque for his continuing contributions to conservation.

Others who will be honored for contributions in this field will be:

Former Senator Joseph C. O'Mahoney, of Wyoming; Percival P. Baxter, former Governor of Maine; Joseph W. Penfold, conservation director of the Izaak Walton League; Don G. Fredericksen of Gooding, Idaho, and M. D. Bryant of San Angelo, Tex.

May 25, 1962

RESOURCES OF U.S. FOUND ADEQUATE FOR THIS CENTURY

Some Sharp But Temporary Regional Shortages Are Seen in 5-Year Study

By JOHN W. FINNEY
Special to The New York Times

WASHINGTON, March 31 — The United States will have enough natural resources for at least the rest of the century to meet the demands of an increasing population and a rising national economy, it was reported today.

The nation, however, may face severe temporary and regional shortages of some resources, such as water and timber, and in the latter part of the century will probably be forced to turn to new sources for many raw materials.

These were the principal conclusions to emerge from a five-year study by Resources for the Future as reported today by the nonprofit research corporation. The study, entitled "Resources in America's Future," analyzes the requirements of the United States for land, water, energy and minerals over the rest of the 20th century and the supplies, domestic and foreign, available to meet the requirements.

Supplements 1952 Report

The 1,000-page report was described as "the first comprehensive

forward look at the nation's natural resources" since 1952 report of the President's Materials Policy Commission, which came to be known as the Paley Commission after its chairman, William S. Paley.

In some ways a descendant of the Paley report, the new study paints a generally more optimistic picture than its predecessor. The difference in outlook reflects a significant change in thinking that has been taking place in natural resources philosophy — a change brought about largely by the impact of modern technology.

Until the Paley report, natural resources thinking was dominated by the Malthusian doctrine of growing populations leading to physical scarcities. Out of the Paley report came the concept that the problem is not so much physical exhaustion of raw materials as the economic cost of recovering the materials.

In the new report is reflected a growing realization, implicit in the Paley report, that technology, in its ability to tap resources more efficiently or to find substitute products, is just as important as the physical quantities available in determining the future supplies of natural resources.

As the report puts it at one point: "The threat of scarcity has been held in check largely by technological progress" and technological advance, therefore, has become "the main escape hatch from scarcity."

The central question posed in the report is whether the United States can count on sufficient natural resources to meet its needs over the next 40 years. This is the period in which it projects that the population will nearly double to a level of 331,000,000 and the gross national product nearly quadruple

to an annual level of $2,000,000,-000,000.

The answer given is a qualified yes. The report gives this summary of its conclusions:

"Vastly greater quantities of natural resources will be required in the future. Neither a long view of the past, nor current trends, nor our most careful estimates of future possibilities suggest any general running out of resources in this country during the remainder of this century or for a long time thereafter.

"The possibilities of using lower grades of raw material, of substituting plentiful materials for scarce ones, of getting more use out of given amounts, of importing some things from other countries, and of making multiple use of land and water resources seem to be of sufficient guarantee against across-the-board shortage."

At the same time, the report warns that there is a "great likelihood" of "severe problems of shortage" developing from time to time in particular regions or segments of the economy for particular raw materials." For example, deficiencies, in either quantity or quality, are likely to occur in certain minerals, in land for recreation area and for lumber and water for certain western areas.

Suggestions for Action

The report concludes, however, that none of these potential shortages or difficulties is beyond the capacity of the economy or industry to overcome with intelligent, far-sighted policies. To prevent or reduce such resources deficiencies, the report suggests these three broad lines of action:

Maintain the flow of new and improved resource technology in discovery, production, transportation and use of natural resources.

Maintain and expand a world trading and investing system that can enlarge the opportunities of the United States and other coun-

tries for importing raw materials at low cost.

Conserve and use resources in accordance with soung ecological and economic principles.

For the major categories of resources the report presents the following analyses of the demand and supply over the next forty years:

LAND

Cropland acreage is adequate to satisfy food and fiber demand during the next four decades, with acreage surpluses likely for more than half of this period. However, the demand for other uses, such as recreation space, growing cities and reservoirs and watersheds, will mount so that by the year 2,000 there is expected to be a "land deficit" of 50,-000,000 acres. One solution for this deficit will be increased multiple use of land, as in devoting more forest and grazing land to recreation use.

This "severely tightening" land situation will come about even without taking into account the increasing demand for forest products. In fact, the estimated demand for forest products by the end of the century is expected to be so much larger than the domestic supply that something like 300,000,000 acres would have to be added to the existing 484,000,000 acres of commercial forest land to meet the requirements.

Since such an increase in forest land is unreasonable, the practical alternative is more intensive management of existing forests, intensified efforts to find substitutes and probably larger imports.

WATER

Some local and even regional stringencies of water supply have already appeared, and more will develop in the future. There is no prospect, however, of water shortage on a nationwide scale and little evidence to suggest that, given sufficient thought and effort, localized water shortages will be a serious impediment to continued national growth during the next four decades.

In the East, the main problem will be the quality rather than the quantity of water available. Statistically, the supply, is ample for the rest of the century; the question is how much of the supply will be fit to use. The two greatest needs will be increased storage capacity, especially to provide enough flow to dilute the concentration of pollution, and more thorough treatment of waste water.

In the West, in contrast, the big problem is using up of the available supplies of water. If the expected large growth of Western population is to be supported, therefore, the region will need not only new dams and reservoirs, long-distance transport of water, desalinization and water conservation but also better allocation of water resources among competing uses. For example, industrial and municipal use of additional supplies of water will probably have to be given higher priority over irrigation.

In the Pacific Northwest, the distinctive water problem is fuller realization, in cooperation with Canada, of the great hydroelectric potential of the Columbia River.

ENERGY

The total demand for energy is expected to triple by the year 2,000. No major difficulties are expected in meeting these energy requirements, but the pattern of supply is likely to change, particularly in the latter part of the decade.

At least through the 1970's it appears that the energy needs can be met with no significant increase in cost and with no sweeping changes in the present relative contributions of oil, natural gas and coal. Toward the end of the century, however, nuclear energy is expected to rise in importance in the generation of electricity and oil and gas to decline in relative importance.

The reserves of coal appear ample for a much longer period than the rest of this country. Coal requirements are expected to increase steadily throughout the next 40 years but to continue, although at a slower pace, the declining relative trend that in the last two decades has halved the share of coal in national energy use. Successful fasification or liquefaction of coal could slow or even reverse this trend.

Before the end of the century, it seems possible the century's petroleum and natural gas resources "will be sufficien-tly depleted flow of alternative energy sources." The incipient resources problem of domestic oil and gas, however, will be greatly mitigated, if not offset, by the increasing contribution of nuclear energy, exploitation of the oil shales of the Colorado Plateau and the tar sands of northern Canada, possibly gasification of coal into a high-energy fuel, long-range shipment of natural gas in liquid form, and the vast supplies of oil and gas known to exist in the Middle East and North Africa and probably in other parts of the world, thus far only sporadically explored.

NONFUEL MINERALS

In all but a few cases, domestic supplies of metals are inadequate to meet future requirements. Domestic requirements for most of the major metals are expected to grow much faster than population.

Steel requirements are expected to increase 200 per cent by the year 2,000; aluminum by 800 per cent and even lead, a slow-rising metal, by 100 per cent. The United States will be unable to meet these demands from domestic sources for more than a brief time, if at all, at present costs.

As a result, there will be a need for continued access to foreign mineral deposits and for advances in technology that will permit commercial exploitation of now costly or inaccessible deposits.

The study was written by Hans H. Landsberg, a member of the research staff of Resources for the Future; Leonard L. Fischman, economic consultant, and Joseph L. Fisher, president of the corporation. The study will be published in book form tomorrow by the Johns Hopkins Press.

April 1, 1963

1,000 DUNES ACRES CLEARED FOR MILL

Bethlehem Spurns Pleas by Indiana Conservationists

By DONALD JANSON
Special to The New York Times

GARY, Ind., April 11—While conservationists battle to save the Indiana Dunes through the time-consuming machinery of Congress, the Bethlehem Steel Company has taken the bulldozer by the horns.

Spurning pleas of supporters of pending legislation, which would make the last few miles of unspoiled Indiana dune land a national preserve, the company has stripped 1,000 acres of vegetation since the first of the year.

Now grading equipment is on hand to begin leveling the 10,-000-year-old dunes for a new steel mill.

Work on the $250,000,000 project is proceeding with "no holdup whatsoever," Arthur B. Homer, chairman of the corporation, said this week at Bethlehem's annual meeting in Wilmington, Del.

Conservationists have fought for years to save the shining dunes, which are unusual in their construction and "migrating" propensities. They want to preserve the dunes for the benefit of scientists and 7,500,000 city dwellers in the Chicago-Gary industrial complex, which is critically short of outdoor recreational assets.

Sand Swept by Winds

The sand hills, swept by the prevailing westerly winds off Lake Michigan, move as much as 60 feet a year.

Unlike the dunes of many other shores that are hills of earth covered with a veneer of sand, these are all sand. Over the centuries, they have formed ridges that now reach a mile and a half inland. Parts of them have snared a layer of soil and harbor a naturalist's paradise of trees, wildflowers and wildlife. All of them overlook the glistening beaches at the southern tip of the lake.

"Highest priority should be given to acquisition of areas located closest to major population centers," the Rockefeller Foundation's Outdoor Recreation Review Commission reported to Congress last year. "The need is critical—opportunity to place these areas in public ownership is fading each year as other uses encroach."

Bills to preserve 9,000 acres of the dunes, including the Bethlehem property, are before the Senate and House Interior Committees.

The elections last November resulted in a change in composition of the Senate Committee. Senator Paul H. Douglas, Democrat of Illinois, leader in the effort to save the dunes, expects favorable action this year where similar measures never got out of committee in the past.

"But the question now is whether it will come too late," said Thomas E. Dustin of Fort Wayne, a spokesman for a conservationist group called the Save the Dunes Council. He went on:

"Bethlehem has denuded 1,000 acres, much more than they need for the mill, in the best part of the proposed park in order to preclude any chance of passing the bill. They want to present Congress with a public-be-damned fait accompli."

In Bethlehem, Pa., a spokesman for the company said the tract was in land, zoned for heavy industry, that had been owned by industrial interests for more than 30 years. He pointed out that the site was outside an existing state park and other dune land.

Conservationists contend, however, that the state park is much too small for recreation needs and that its air and water would be polluted by full-scale steel operations.

Senator Douglas still hopes to stop Bethlehem. He said he believes that passage of the bill by July would precede a start in building the steel mill and that condemnation proceedings could be started at once.

Such a burst of speed on a conservation measure appeared unlikely, however, despite the support of the Department of the Interior for preservation of the dunes. Hearings are yet to be scheduled on the Douglas bill.

Meanwhile, a special session of the Indiana Legislature is wrestling with proposed tax measures that would include financing for a deepwater harbor at Burns Ditch, adjacent to the Bethlehem mill site.

Gov. Matthew E. Welsh and other Indiana politicians of both parties had hopes for Federal aid in financing a harbor. However, the Kennedy Administration refused to include in its budget for 1963-64 the $25,500,-000 requested, which would be in addition to $38,000,000 the Indiana Port Commission planned to raise in a bond issue.

Studies by the Army Corps of Engineers have determined that the port would primarily benefit only Bethlehem and another steel company, Midwest. Midwest has a mill on the other side of Burns Ditch, a channel that drains the Little Calumet River into the lake.

The corps is now studying another site, in an already industrialized area of Indiana nearer Chicago, that already has some of the required harbor facilities.

The prospects are considered slight that Indiana can finance a harbor by itself. The steel companies may have to build their own when they expand their dune land operations from rolling mills to fully integrated plants.

Proponents of an Indiana harbor say this would be a blow to the state's economy. It would also kill conservationists' hopes for a national preserve, because Bethlehem owns 4,000 acres of dune land on which to expand and build a harbor.

April 14, 1963

Park-Harbor Plan Approved for Indiana Dunes

U.S. Port Aid Is Conditional on Steel Output in Area

Compromise Provides 11,700 Acres for Recreation

WASHINGTON, Sept. 23 (AP)—The White House announced conditional approval today of a compromise proposal for a park-harbor development along the Indiana Dunes of Lake Michigan. The proposal has been the subject of a long and heated controversy.

Gov. Matthew E. Welsh of Indiana and Senators Vance Hartke and Birch Bayh, Indiana Democrats, were advised of the Kennedy Administration's conditional approval of the project at a White House meeting this morning.

Governor Welsh said the Budget Bureau would make public tomorrow a report on the recommended development of the Burns Ditch Harbor and the creation of an 11,700-acre recreational area to include the Indiana Dunes State Park.

The New York Times   Sept. 24, 1963
Cross indicates dunes area

Governor Welsh said approval of funds for port development would be conditioned on the construction of an integrated steel mill in the area and shipments totaling 10 million tons of coal annually.

An alternate plan would call for two integrated steel mills—mills that start with iron ore and turn it into a finished product—and 5 million tons of coal shipments yearly.

The secretary of the Army will determine when these conditions have been met, Governor Welsh said. He noted that the Bethlehem Steel Corporation had already announced plans for an integrated mill at the Burns Harbor site.

Mr. Welsh said the proposed recreation area would not include about 2,400 acres on which Bethlehem plans to expand its operations.

Under the plan, the Federal Government would build breakwaters and dredge the area at a cost estimated at $17 million to $25 million.

A deep-water port at Burns Ditch has been sought for decades by Indiana officials and industrial and civic leaders.

In recent years the proposal has been attacked by Senator Paul H. Douglas, Democrat of Illinois, and others who seek to preserve the area as a dunes park. Senator Douglas said today that the Budget Bureau report probably meant that "there will never be a Federal harbor in Porter County but that there is a good chance" for an 11,000-acre park there.

September 24, 1963

Critic at Large

Literature of Nature Is Changing as Fear of Man's Destructiveness Grows

By BROOKS ATKINSON

REPORT from an inveterate reader of books about the out-of-doors:

A form of literature that used to idealize nature is becoming a branch of natural philosophy. It is now concerned with the ominous future of the land. The littérateur is giving way to the zoologist and ecologist — to Rachel Carson, whose love of nature includes the responsibility of a scientist; to Lorus and Margery Milne, who portray their "The Valley" of today by beginning with the glacier. In current nature writing there is a lot of teaching in the selfless, affectionate manner of the late Aldo Leopold's "A Sand Country Almanac."

No doubt, the history of American civilization could be written in terms of our changing attitudes toward nature. In 300 years we have passed through three significant stages: (1) indifference or hostility to nature; (2) romantic delight in nature, and now (3) fear that man, the great predator, may destroy nature and civilization at the same time.

When the first colonists came here to found a nation in a wild country, they did not consciously love nature. Nature—specifically the forests — was an obstacle. Indians were constantly burning the forests, either to drive out the animals or enemies, or to clear the land for corn. Since the forests appeared to be inexhaustible, the settlers also burned them. In New England, white pine was so abundant that the settlers regarded it as a weed, and burned gigantic heaps of it to make clearings.

People do not consciously love nature until their civilization has been able to make a livable compromise with it — until they do not have to fight nature to stay alive. In the earliest stage of American civilization there was a sense of adventure about nature, as the novels of Cooper attest. But it was Emerson who opened the gate to the second stage by publishing "Nature" in 1836. In a fresh, beautifully written book he harmonized man with nature. "The misery of man appears like childish petulance when we explore the steady and prodigal provision that has been made for his support end delight on this green ball which floats him through the heavens," Emerson wrote in his most rhapsodic style. To Emerson, man and nature were in communion, though they remained apart, which is not the modern view.

In regarding nature as something, not to be fought, but to be embraced, Emerson cleared the way for his fellow townsman, Thoreau, who saw in nature a health that civilization lacked. Thoreau had the courage of Emerson's convictions. Out of nature Thoreau created an insurgent philosophy that the unhappy experience of the human race has been consistently confirming.

For a century the literature of nature has been charming and complacent. But the mood is changing. The third or present stage is full of dismal auguries. Note that the Audubon Magazine, once devoted to the casual pleasures of bird watching, is now primarily concerned with conservation, and maintains a representative in Washington to watch legislation affecting natural resources.

In 1836 Emerson thought that the world of nature was so impregnable that man's "chipping, baking, patching and washing" were too insignificant to have much effect on the whole. That is no longer a tenable point of view.

In "The Last Horizon" Raymond F. Dasmann, a zoologist and botanist, thinks that the pressures of overpopulation, if not controlled, will ultimately destroy our natural environment. Lois and Louis Darling conclude a study of the evolution and anatomy of birds ("Bird" is their title) with a chapter in the same somber key: "We squander in a few years the fossil fuel, coal and oil, which are the accumulation of untold ages. We poison the water and the air."

In "Face of North America" Peter Farb makes a similar conclusion: "The whole web of inter-relations developed in the wilderness over millions of years has been irretrievably lost."

If there is a fourth stage in American nature writing it will portray a world short of food, cramped for space and bereft of beauty.

July 23, 1963

To Save Wildlife and Aid Us, Too

By STEWART L. UDALL

"The squirrel has leaped to another tree, the hawk has circled farther off, and has now settled upon a new eyrie, but the woodman is preparing to lay his axe to the root of that, also."
—THOREAU'S JOURNAL (1851).

TWICE each year, at migration time, lovers of wildlife await the census count of North America's small band of whooping cranes, which fly between their summer nesting grounds in northwest Canada and their winter refuge on the Gulf Coast of Texas.

The whoopers have been poised on the edge of extinction for more than 30 years now. In 1938, at the low point, there were only 14. For a time the flock slowly increased, but after a survey last summer the Canadian wildlife experts reported with alarm that the nesting season was a failure and our whooping crane population had dropped in one year from 38 to 28.

The fight for existence of these rare birds symbolizes the plight of vanishing types of birds and animals everywhere. The degree of concern evoked in our minds by the threat to a species is one of the quiet tests of modern civilization. In the end, whether we provide conditions which will allow wildlife to coexist with us on this planet will be as significant a commentary on our progress as any of the feats of rockets and computers. Later this month President Kennedy will visit national parks and seashore and wilderness areas in 10 states across the country, drawing attention to the efforts of the Government to conserve and protect the nation's natural heritage.

It is estimated by the experts that more than 200 species of birds and mammals have already disappeared from the face of the earth and that nearly 250 species in various parts of the world are now on the danger list. The California condor, the polar bear, the woodland caribou, the manatee, the everglade kite, the Key deer, the sandhill crane are only a few of the imperiled species in our hemisphere.

Man-made threats to a species come from three sources. In too many parts of the world the destroyers are selfish or wanton individuals who ignore the elementary principles of conservation. A news story last year indicated that

STEWART L. UDALL, when he takes time off as Secretary of Interior, frequently goes camping or hiking in our wilderness areas.

hunters were using automobiles to corner and kill the few remaining Arabian oryx; in East Africa native poachers are making deep inroads into the finest big-game herds in the world.

"Progress" is the second and more subtle threat. Encroaching civilization daily destroys habitats that are essential for the survival of some species. Wildlife can thrive only when conditions favor reproduction; some creatures face eventual extinction the moment natural conditions are seriously unbalanced by man. Some animals and birds need space; others require solitude or special nesting conditions. But space and solitude are commodities increasingly in short supply in the twentieth century. Populations are exploding, and in our haste to exploit resources we have often ignored side effects that are detrimental to the welfare of wildlife.

A FINAL, and new, risk relates to those conquests and miscalculations of man which threaten the life process itself. Pollution of air and water is slowly reducing life expectancies of many forms of wildlife. This insidious process occurs quietly. The permanent loss of a stream or estuary by pollution is too often unnoticed until the damage is irreversible. In her provocative book, "Silent Spring," Rachel Carson called our time "an age of poisons" and urged a thorough evaluation of the use of commercial chemicals and pesticides. The entry of numerous poisons into the chain of life may be a fateful event for wildlife species—and perhaps for some members of the human species as well. Even the controlled killing of plants, insects and animals often has an unintended impact on nature's delicate balances.

The people of this country have passed through three stages in their treatment of wildlife. The first was the period of waste and slaughter that came to a culmination in the last half of the 19th century. The fate of the passenger pigeon and the buffalo symbolizes that age. Delectable and easy to kill, passenger pigeons were butchered by the millions. By 1915 a bird which once constituted nearly a third of the entire bird population of the United States had been relentlessly pursued to extinction. The vast buffalo herds that once roamed the great plains were the wildlife wonder of our continent; their near destruction also marks this savage and shortsighted hour in our history.

The crusade for wildlife protection in the eighteen-eighties began a protest against the slaughter and was spearheaded by such organizations as the Audubon Society and the Boone and Crockett Club, which was founded by Theodore Roosevelt and a group of his friends.

In the last four decades we have developed game management into a science. Hunters "crop" only the annual increase, and public opinion and public budgets support numerous wildlife protection programs. The main threat today arises from the side effects of advancing civilization. The draining of each swamp, the building of each new road, the indiscriminate broadcast of

pesticides, the widening circle of pollution, and the destruction of open space on the edges of our cities are now the clear and present danger to wildlife.

The Federal Government has many preservation programs under way, ranging from pesticide research to the purchase of wetlands to save habitat for the ducks and geese that travel the continental flyways. One exciting restoration project concerns the Aleutian Canada goose, a bird which was thought to be extinct. This small goose formerly nested throughout the Aleutians, but hunters made a fatal change in the bird's environment by introducing predatory foxes throughout the islands in an effort to increase the supply of fox pelts. Only 56 Aleutian geese were counted last year.

Last summer a few goslings were captured on Buldir Island and taken to the Monte Vista Wildlife Refuge in Colorado. While this temporarily transplanted flock increases its numbers, foxes will be removed from some of the islands so as to restore the habitat to its previous state. Later a covey of the captive geese will be taken back from Colorado to their ancestral islands and released, and the cycle of restoration will be completed. Through this process it is hoped that another species of wildlife will be rescued from extinction.

A SIMILAR project last year involved the nene (naynay) goose, Hawaii's state bird. This white-necked goose has largely forsaken swimming and flying for a life on the most improbable habitat imaginable—the barren lava beds of Hawaiian volcanoes. The nene's troubles began when it became prey to the dogs, cats, pigs, goats, rats and mongooses that had been brought to Hawaii by American settlers. By 1955, there were only 22 wild nene alive. By good fortune, however, some of these geese had been transplanted to the British Isles; a grant from the World Wildlife Fund, a worldwide voluntary organization devoted to saving threatened species, made it possible to ship 30 of these fine birds by air from England back to Maui last July.

Another example of bird restoration is that of the masked bobwhite quail, a bird which disappeared from its Arizona habitat a half century ago. Recently, some specimens were found in northern Mexico. A 640-acre tract was set aside in Arizona near Tucson by the Interior Department's bird experts and three pairs of the Mexican bobwhite quail were "planted" in the area where once their species flourished. If this rare quail re-establishes itself, it will be another lastminute victory for wildlife conservationists.

A Growing Sense of Urgency

THERE is hope that Congress will establish the Land and Water Conservation Fund bill recommended this year by President Kennedy. This historic conservation legislation will provide funds desperately needed for Federal and state governments to save choice lands for recreation and wildlife protection. It will take an all-out effort at both the state and Federal level, and by voluntary organizations also, to secure such lands for posterity.

However, there are signs of a new awakening on the conservation front. In the last three years a few states have established far-sighted buy-now, pay-later conservation programs. New Jersey's "Green Acres" plan, New York's $100-million-dollar acquisition program, and Wisconsin's effort to save its out of doors show what can be done at the state level if there is the right kind of leadership.

EVEN more encouraging are the efforts of local organizations to save marshes, seashores or forest refuges. Five years ago a group organized as "The Philadelphia Conservationists" purchased a 200-acre marsh and presented it to their city as a permanent wildlife sanctuary. In recent months an emergency committee of New Jersey citizens fought off the proposed intrusion of a jet airport by raising several hundred thousand dollars to save New Jersey's Great Swamp. This land, a natural museum piece, was later donated to the people of the United States, and stands as one of the rare instances on the Eastern Seaboard where wildlife won out over asphalt and concrete. Likewise, conservation commissions established in many Massachusetts communities are providing excellent examples of local initiative that is getting results.

Private philanthropy can also play a pivotal role. In May leaders of Boston's business community, joining together to encourage gifts of land and bequests of money for conservation projects, established a Fund for the Preservation of Wildlife and Natural Areas. Henry Thoreau once advised his New England neighbors that "a town is saved not more by the righteous men in it than by the woods and swamps that surround it." It is heartening that people of his neighborhood are, a century later, taking his advice at face value.

IT is inevitable, of course, that the worldly and cynical will propound their usual so-what questions. "What does wildlife contribute to our abundance, or add to the sum of human happiness?" they will ask in weary tones.

Such queries reveal minds that lack all reverence for the marvels of the natural world. These same individuals, who prate of the wonders of modern life and the new "conquests" of science, have, in incipient form, the attitudes that lead to what the ancient Greeks called hubris — the deadly arrogance of men who had lost their roots.

The truth is that there is a force at work in the world larger than ourselves, and the natural world is its outer garment. The creatures of nature have a claim to life as valid as our own. We need to accord them respect as much for our own mental and spiritual preservation as for theirs.

September 15, 1963

Critic at Large
Ecologists Add Soil and Water to the Cauldron of Mankind's Worries
By BROOKS ATKINSON

TO the layman, the ecologist's projection of the future makes a hopeless picture.

"Man is the only organism that lives by destroying the environment indispensable to his survival," William Vogt, a practising ecologist, declared in 1948. That sounds like a sentence of death.

But Dr. Vogt does not feel hopeless, despite a generally alarming report that he has just written on the destructive uses of land in Mexico and Central America. The report is published by the Conservation Foundation, 30 East 40th Street here. This 'Brief Reconnaisance," as Dr. Vogt calls it, constitutes a return visit to these places.

Since his last visit he has noted here and there more enlightened attitudes towards land that had been eroded and mined for years. In one corner of Mexico he found land that had been completely devoid of top soil now being restored to fertility by the building of terraces and dikes and the use of fertilizers. Although the Mexican department of conservation is ill-paid and inadequately staffed, it is making inroads on the ignorance of provincial farmers. In the United States as well as in Latin America, Dr. Vogt believes that attitudes towards the care of the land have improved a little since 1948 when he wrote a devastating book entitled "Road to Survival."

●

What is ecology, which has become an increasingly pertinent word in our civilization in the last quarter of a century? According to the dictionary it is "the branch of biology that treats of the relations between organisms and their environment." The word was first used about a century ago by Ernst Heinrich Haeckel, a German naturalist and philosopher who was an early advocate of the theory of evolution. Darwin was, at least in part, an ecologist. Ernest Seton Thompson wrote ecological studies, although no one called them that in his day. Dr. Vogt thinks that the first formal use of the word in this country was in 1913 when C.C. Adams, director of the New York State Museum, wrote a book entitled "Guide to the Study of Animal Ecology." In the ecologist's view, man is not a separate animal, as he seemed to be before evolution was understood, but one of the millions of organisms of nature, dependent on all the other organisms, as other animals are.

Ecology is moving into the area of medicine. Jacque May's "Ecology of Behavorial Diseases," now in preparation, will discuss the impact of a technological civilization on the human nervous system. For technological developments produce situations that create mental and nervous disorders—thus illustrating not only the facts of biology and the social sciences but the Greek theory of tragic irony. We lose something for everything we gain.

In Latin America, as well as North America, the basic problem is the population explosion.

●

"Since 1939, the year I began my field work in South America," Dr. Vogt observes, "the Latin American Population has grown more than 50 per cent; and despite considerable localized economic progress, there are probably more sick, hungry and desperate people now than there were then."

Populations increase not only because of a rising birthrate, but also because of a falling death rate; and increasing populations put increasing demands on natural resources.

Economists propose to aid underdeveloped nations by increasing capital assets and establishing industry. But ecologists have to begin with elementary soil fertility and water supply. As a Spanish philosopher, Adolph Munoz Alonso, has recently stated the mission of contemporary philosophy, ecologists are concerned with keeping man's "house humanly habitable." In both the Americas, North and South, land is being destroyed; water resources are dwindling; water is being polluted. At the same time, the populations are growing at a constantly accelerated rate. It has been estimated that by 1980 the standard of living in parts of our Middle West and South will begin to decline because of a shortage of water.

Taking all forms of organic life as one subject and making the whole world his laboratory, the ecologist has to give a judgment on the totality of life. The judgments he is now giving could hardly be more disturbing. The religious phrase, "reverence for life," has the greatest secular importance.

September 27, 1963

CONFLICTS BESET VALLEYS IN WEST

Recreation VS. Water Needs
Is a Troublesome Issue in
California Canyon Area

By WILLIAM M. BLAIR
Special to The New York Times

FRESNO, Calif., March 17—The road into Kings Canyon is not yet open. Snow still blocks a part of it but spring is in the chill night air. Soon the snow will be gone, the wildflowers will spring up and some visitors will ride into the deep gorge and perhaps hear about a controversy that points up a mounting conflict.

The conflict is how best to use land, of wild areas versus the economic needs of a booming agriculture and of cities and towns, of recreation versus dams for water and power and of political pressures for recreation on the Federal Government.

Kings Canyon and another wild valley, Tehipite Valley, are the focal point of the conflict. Both are immediately adjacent to Kings Canyon National Park, sticking like fingers into the park itself but not a part of the park.

Bill Before Congress

A bill before Congress would take them into the park. Farm and water interests in San Joaquin Valley, California's big, bustling agricultural heartland, want the Federal Government to keep its 20-year-old promise to retain the areas for water and power development.

The New York Times   March 20, 1964
disputed canyon area (cross).

This development would be through two dams thrown across forks of Kings River, one on the middle fork to use Tehipite Valley as a reservoir and one on the south fork across Kings Canyon in Cedar Grove.

A third small dam also would be constructed at the junction where the middle and south forks meet to form Kings River. A power plant would also be built at the junction. Another power plant would be constructed farther down stream. The cost is put at $140 million to $145 million.

The Unresolved Questions

The two great valleys were excluded when Congress created Kings Canyon National Park in the lofty Sierra Nevada on March 4, 1940. The exclusion was made as a condition by opponents of the park to withdraw their opposition. This was the promise that the sites be reserved for water development that

the water interests now want kept.

The major unresolved question is whether the dam projects are economically feasible and whether, in view of the ever-mounting demand for recreation, the inundation of the spectacular valleys is necessary.

The fight between the water users in San Joaquin Valley and the conservationists is a prime example of similar conflicts across the country. There is bitterness on both sides. Bothe sides seek to convince by hard figures that their side is right. However, in the end the fight boils down to preservation of a wild area with major esthetic values and the hard fact of the need for water and power to feed a growing use elsewhere.

A year ago, Representative B.F. Sisk, the Democrat who represents this area, introduced a bill in Congress to put Cedar Grove and Tehipite into the National Park System and thus preserve them. The two scenic valleys have been managed as a part of the system although they are in National Forestland.

The Park Service has put $1.5 million into Cedar Grove, the most readily accessible of the two areas. It has a small concession for supplies, some 500 campsites and the road on the floor of the eight-mile-long valley with great rock walls 1,200 feet apart. The Park Service concessioner, the Sequoia-Kings Canyon National Park Company, has been reluctant to go ahead with more public facilities until the status of the area is cleared.

2 Concessions Operated

The company operates the concessions in both the Sequoia and Kings Canyon National Park, home of giant Sequoia trees.

There is no public development in Tehipite Valley, a piece of glacier carved sculpture, more than 7,000

feet deep from the top of Tehipite Dome. It is now accessible only by foot or horseback, a two-day journey from a take-off point at the end of Cedar Grove, which also is the jumping off point for hiking and packing trips into other parts of the rugged back-country.

Both areas have been likened to Yosemite Valley, which encompasses a great array of natural wonders, including waterfalls, high wilderness country and other natural wonders. John M. Davis, superintendent of Sequoia and Kings Canyon National Parks, and his naturalist, James W. Corson, described the granite-lined gorges as "the equal of, if not superior in many ways to, Yosemite."

"The dam projects would destroy the valleys for all time," Mr. Davis said recently as he stood in Zumwalt Meadow near the end of Cedar Grove and listened to Kings River singing a clear song as it rushed toward San Joaquin Valley to irrigate farm land 70 miles away.

The Park Service's struggle is with the Kings River Water Association, an organization of 28 water users, 15 of them public irrigation districts formed by farmers. Both sides are marshalling arguments and lining up supporters.

Demonstrating the political implications of the fight is the recent directive of Secretary of the Interior Stewart L. Udall that the problem be restudied by the Bureau of Reclamation, although that dam-building agency had said earlier that the projects were not now needed.

The Los Angeles Department of Water and Power also recently decided that the project was not economically feasible and withdrew its reservations.

March 20, 1964

SAVING THE WILDS

New Preservation System Is Set Up
To Protect Nation's Wilderness

By WILLIAM M. BLAIR

BILLINGS, Mont.—A 60-year-old Montanan traced on a map the other day the boundaries of the Bob Marshall Wilderness Area. As he lifted his finger, a smile creased his face. He said:

"That's it. It has been pretty well explored for minerals, but you never can tell when some-

body might want to try something again. Let's hope nobody can touch it now. It's something that should have been done long ago."

His reference was to the creation of the National Wilderness Preservation System, as set down in Public Law 88-577, which went into effect Sept. 3. The act is the compro-

mise of a long and bitter fight to carve out of the national forests and other Federal lands a system that will, in the words of the legislation, "secure for the American people of present and future generations the benefits of an enduring resource of wilderness."

The compromise was between conservationists and those who would put forest resources to commercial and other-than-wilderness uses.

Area Defined

The Bob Marshall Wilderness Area lies in the Flathead and Lewis and Clark National Forests of Montana. It is one of several similar tracts, covering a total of 9.1 million acres,

to be placed in the wilderness system under the provisions of the new act.

These areas had been classified previously as "wilderness," "wild" and "canoe." Now they will be designated simply as "wilderness."

The act changes nothing so far as the forest lands are concerned. The Forest Service will continue its planned management and policing of the areas, and perhaps cut a few more trails for general public use, which has been increasing in the wilderness. However, instead of the wilderness being simply a part of a national forest or public land, subject to some controlled commercial activities, areas designated as wilderness will be "preserved."

A Growing Sense of Urgency

Problems Expected

Although the Forest Service will be carrying on what it has been doing in these areas for years, it is expected that the growing public use — by pack trips, on foot and by horse —will mean more tent-camping areas and problems such as sanitation around lakes and clean-ups to preserve the natural environment.

This may cause another problem, namely that of getting more money from Congress to do the work. This is because the act bars any appropriation for payment of expenses, or salaries for administration of the wilderness system as a separate Federal unit.

That there will be increasing public use is recognized in the act. Section 2 begins: "In order to assure that an increasing population, accompanied by expanding settlement and growing mechanization, does not occupy and modify all areas within the United States and its possessions * * *."

A Sometime Visitor

It pins this down by its definition that a wilderness, "in contrast with those areas where man and his own works dominate the landscape, is hereby recognized as an area where the earth and the community of life are untrammeled by man, where man himself is a visitor who does not remain."

Examples of public pressure against the wilderness can be found both in the act and elsewhere.

For example, the Gore Range-Eagle Nest Primitive Area of Colorado is made up of rugged, rock-climbing pinnacles and spectacular knife ridges. Forest Service officials had decided tentatively that an undetermined amount of land at the southern tip of this area, but not exceeding 7,000 acres, should be made available for Interstate 70.

Now the Secretary of Agriculture is authorized by the act to remove the acres, if he determines that the action "is in the public interest." This would be done when the Secretary reviews whether the primitive area should be included in the wilderness system. He also may recommend the addition of other lands, not now within the area, to replace the 7,000 acres that may be cut.

Tunnel Planned

And the Denver Water Board also has plans for a tunnel in the region. The tube would be used to transport water to meet the needs of the growing Denver metropolitan complex.

The act also provides that the wilderness areas shall remain open to prospecting for some 20 years, ending Dec. 31, 1983.

WHERE MAN IS A VISITOR

WASHINGTON—The following figures, supplied by the United States Forest Service, spell out the acreage and geographical distribution of the areas within the National Wilderness Preservation System as of Sept. 3, effective date of the new act. Asterisks indicate where adjacent states share a wilderness area, and totals reflect such overlapping.

State	Number of Areas	Net Acreage
Arizona	5	422,990
California	13	1,256,884
Colorado	5	274,859
Idaho	1*	987,910
Minnesota	1	886,673
Montana	5*	1,482,567
Nevada	1	64,667
New Hampshire	1	5,400
New Mexico	5	678,661
North Carolina	2	21,155
Oregon	9	662,847
Washington	3	583,196
Wyoming	4	1,812,012
	54	9,139,821

In this regard, conservationists, and other interested parties who had battled for more restrictive legislation on wilderness areas, already had begun voicing some fears—even before the ink on the act was dry. They accepted the compromise legislation, however, because they knew that some influential members of Congress would not accept closing the wilderness to new mining claims immediately.

While some officials, including the Forest Service workers on the spot, believe that the mining provision is not an open invitation for a new land rush, Forest Service officials in Washington believe otherwise.

Fears Expressed

Fears are being expressed that some attempts may be made, under the guise of mining claims, to obtain vacation retreats in the rugged wilderness, rather like private preserves with all the rights of hunting and fishing.

Some officials point to advertisements published in Western states to support these fears. The ads trumpet good fishing and hunting on small mining claims at prices of $2,000 to $5,000 each. Washington officials believe that some people may get the idea that they will not have to do any mining and will be safe.

Already, the Department of Agriculture's Forest Service has cautioned, in an interoffice memorandum, that it is "doubtful that the rate of claims locations will remain static." It expressed the belief that the act's limitation on future mining activity may cause an immediate "increase in prospecting and in the number of claims located."

Whether this will come about will not be known for some time, and will take an extensive survey to determine. Such a study would be expensive because mining claims are filed in county courthouses, rather than with a Federal agency.

The latest survey, made in July 1963, showed that 4,800 mining claims on some 100 million acres of wilderness were in force. At that time the rate of filing appeared to be about 15 a month. Patents of ownership have been awarded on about 140 additional claims covering 6,804 acres.

During hearings on the legislation, Representative Wayne Aspinall, Democrat of Colorado and chairman of the House Interior Committee, had elicited from some Western witnesses statements that all of the wilderness areas had been fairly well explored and that the likelihood of new mineral discoveries was as remote as some of the wilderness itself.

Author of Compromise

Mr. Aspinall, who long had fought to satisfy complaining mining interests, also worked out the compromise. This is because the act gave him a victory in his long fight to make Congress dominant over the Executive Branch in deciding what public land should be sealed off or opened up for development.

Mr. Aspinall had laid down this challenge to President Kennedy. The Executive Branch of the Government had held this power over public lands since

United States Forest Service

RUGGED—View from trail atop North Carolina's Linville Gorge.

IN MONTANA—Bob Marshall Wilderness Area will be protected under new act.

President Theodore Roosevelt seized it early in his conservation efforts.

Along with Mr. Aspinall, the chief architects of the wilderness system were Senator Clinton P. Anderson, Democrat of New Mexico, and Representative John P. Saylor of Pennsylvania, ranking Republican of the House Interior Committee.

Senator Anderson introduced the first wilderness bills. These passed the Senate twice, but were stalled in the House. Mr. Saylor succeeded in deleting from some of the earlier House bills several features, including one that would have permitted a commercial ski operation in a California wilderness area.

The act also is a tribute to the late Howard Zahniser, executive director of the Wilder-

ness Society, who fashioned the conservationists' long efforts to gain a wilderness system. He died in May.

Protective Feature

There is one deterrent to commercialism in the wilderness act. Service officials long have suspected that some mining claims are filed mainly for valuable timber stands. The act reserves to the Government title to the surface and products of the surface. This prevents the patent owner from lumbering or other commercial operations above the ground.

Prohibitions against commercial enterprises of any kind, plus those banning motorized equipment, road building, motor vehicles, aircraft and the construction of buildings, are written

into the law just as they were spelled out in the regulations set up by the Executive Branch previously. The Government may make exceptions in emergencies, including situations involving the health and safety of people within the areas and on valid mining claims.

Water Development

The President can authorize prospecting for water resources, the creation of reservoirs, power projects, transmission lines "and other facilities needed in the public interest, including road construction and maintenance essential to development and use thereof * * *."

Livestock grazing under Forest Service permits established prior to the effective date of the act will continue.

Exploration and prospecting for the purpose of "gathering information about mineral and other resources" can be continued if compatible with preservation of the wilderness environment. The act also directs that the area shall be surveyed by the Geological Survey and the Bureau of Mines, and a report made to Congress.

These agencies are Department of the Interior bodies. The Secretary of the Interior will consult with the Secretary of Agriculture on this work.

At the same time, the Senate-House conferees who whipped the legislation into final shape expected that the mining industry also would explore the existing primitive areas so as to give "Congress the benefit of professional technical advice as to the presence or absence of minerals in each area when Congress considers later whether the areas shall go into the wilderness system."

Other Acreage

In addition to the 9.1 million acres immediately put into the wilderness system, the act directs the Secretary of Agriculture and the chief of the Forest Service to determine whether any of 5.5 million more acres, presently classified as "primitive," can be included in the wilderness system. These acres were previously withdrawn from commercial use pending study, surveys and classification.

The Secretary of Agriculture is given 10 years to make recommendations. Not less than one-third of the "primitive" areas shall be reported on within three years, two-thirds within seven years and the remaining areas within the 10-year span.

A similar period of time is binding on the Secretary of the Interior. He is directed to determine whether additional public lands, including national parks, monuments and other units of the park system, plus wildlife refuges and game ranges, may be included in the wilderness system.

The act also limits increasing the size of the 34 existing primitive areas by more than 5,000 acres, unless Congress approves. Some wilderness advocates believe that this restriction might cause trouble in the future, in that it would be a bar to getting into the wilderness system

IN CALIFORNIA—A natural bridge in the John Muir Wilderness Area.

some choice land that should be preserved before commercial interests move in.

54 Wilderness Areas

The new 9.1 million acres of wilderness are within the 186 million-acre forest empire administered by the Forest Service. The wilderness is in 54 areas in 13 states, and ranges in size from 5,400 acres to 1.2 million.

All of the land is in Western states, except 5,400 acres in

New Hampshire, 21,155 acres in North Carolina and 886,673 acres in Minnesota.

The New Hampshire wilderness is on the Mount Washington slope in the White Mountains National Forest. The North Carolina wilderness areas are Linville Gorge, a deep, rough chasm in the Pisgah National Forest and Shining Rock in the same forest.

The Minnesota wilderness is the largest east of the Rocky Mountains. It is the Boundary

Waters and Canoe Area in the Superior National Forest, a region with hundreds of lakes and excellent fishing.

The Western states covered are Arizona, California, Colorado, Idaho, Montana, Nevada, New Mexico, Oregon, Washington and Wyoming. California, Montana and Wyoming have the largest acreages, but Idaho has the greatest potential for more wilderness.

October 18, 1964

PRESIDENT SIGNS SCENIC ROAD BILL

By JOHN D. POMFRET
Special to The New York Times

WASHINGTON, Oct. 22—President Johnson signed the Highway Beautification Act today and promised that "as long as I am President, what has been divinely given by nature will not be recklessly taken away by man."

The President, spending his first full day at the White House since surgeons removed his gall bladder and a kidney stone 15 days ago, looked trim and in good spirits.

He bounded into the ceremonial gold and white East Room of the White House, spoke about the glories of nature of nearly 20 minutes, stood for half an hour shaking hands with more than 100 guests, then chatted with them at a reception for a short time. He did all this without apparent fatigue.

Mr. Johnson is expected to go to

his ranch in Texas soon to recuperate from his operation. No definite time has been set for his departure.

Bill D. Moyers, the Presidential press secretary, said that Mr. Johnson was comfortable at the White House and could do there the work that had piled up while he was in the hospital.

Surgeon Returns Home

Mr. Moyers said that Dr. George A Hallenbeck of the Mayo Clinic in Rochester, Minn., the President's surgeon, felt that Mr. Johnson's condition was what he would expect of a patient of his age who had undergone the same operation. Mr. Johnson is 57 years old.

Dr. Hallenback returned to Rochester tonight and Mr. Moyers said that he felt comfortable about leaving.

The highway beautification measure, which cleared the House the night the President went to the Bethesda Naval Hospital for the

operation, is law as much because of Mrs. Johnson as because of her husband.

The fact that it passed at all was attributed by many to a Congressional reluctance to say no to a President's wife.

Mrs. Johnson suggested the measure to her husband, invited Congressmen to the White House, extolled highway beauty, and traveled around the country making speeches on the subject and planting trees.

Mrs. Johnson stood behind the President as he signed the measure. He bestowed upon her the first ceremonial pen and a kiss.

Calls Law 'A First Step'

Mr. Johnson noted in his remarks that the measure did not represent everything he had sought. He called it "a first step" and said there would be others.

Under the new law, the states must agree by Jan. 1, 1968, to control billboards and junkyards

along their 265,000 miles of interstate and primary highways or lose 20 per cent of their Federal highway grants.

Mr. Johnson told the audience of Administration officials, Congressmen and private citizens concerned with highway beauty, "There is the part of America which was here long before we arrived and will be here, if we preserve it, long after we depart."

"We have placed a wall of civilization between us and the beauty of our countryside," the President said. "In our eagerness to expand and improve, we have relegated nature to a week—end role, banishing it from our daily lives. I think we are a poorer nation as a result. I do not choose to preside over the destiny of this country and to hide from view what God has gladly given."

October 23, 1965

Nation's Shortages Bringing Nearer the Concept of Community Water Management

By GLADWIN HILL
Special to The New York Times

LOS ANGELES, Dec. 21—With many families, there comes a time of grim realization that hit-or-miss financing out of the sugar bowl is leading toward insolvency and that there has to be systematic coordination of income and outgo.

That juncture arrived for the United States this year in regard to water, as widening shortages and mounting pollution pose the specter of the world's most affluent nation possibly running short of the vital commodity.

The coming year, in the opinion of Federal water experts, therefore, will be an important one in the establishment of a new and inescapable feature of American life: the concept of community "water management."

"Water management" goes beyond just the procurement of water to deal with its supply, use and disposal as parts of an integrated problem.

Use Has Outstripped Nature

Throughout human history, man's practice has been to borrow small batches of water from nature's cycle of precipitation and evaporation, and return it to the cycle for automatic purification. But the rate of water use in the United States has outstripped nature's leisurely pace.

"The day is rapidly coming," says Secretary of the Interior Stewart L. Udall, "when the clouds simply will not be adequate as purveyors of water."

Arithmetically, the nation's mounting consumption of water could exceed the available supply within 20 years. Fortunately, that arithmetic is based on the assumption that water will be used only once; in reality, a gallon may circulate through several agricultural, domestic and industrial uses, thereby deferring the national day of reckoning.

Yet, across the country there are many places where demand already is overtaking the supply.

The solution is community water management.

One phase of water management is the tapping of natural water sources. When these are insufficient, there is a new alternative: to short-circuit nature's dilatory evaporation-precipitation purification cycle.

One example of such shortcircuiting is the desalting of sea water—a tactic that currently has stringent economic limitations and always will have geographic limitations.

Universal Possibilities

But another device, with universal possibilities, is the "renovation" or refining of used water so that it can be used over and over again.

"Up to now," says Dr. David Stephan, one of the United States Public Health Service's leading research scientists, in a recent interview, "we've been like the man in the story who turned in his cadillac when the ash trays got full. We think in terms of using water once and throwing it away. There's no more reason for throwing away water because it's dirty than to throw away a shirt because it's dirty. Both can be cleaned."

The idea of doing anything with sewage but throwing it away is traditionally repugnant.

"This feeling, however deep rooted, involves several misconceptions," Dr. Stephan, director of the Water Renovation Program, continued.

"Sewage actually is 99 per cent plain water. All the pollutants in it amount to less than 1 per cent. In sea water the contamination amounts to three times as much. We have processes for removing that 1 per cent—leaving water purer than when it came from nature."

He pointed to two bottles of water on his desk. One was crystaline. One had a faint cast of color, and some flecks of sediment. The latter was high quality tap water. The better water was processed sewage, or as the scientists prefer to call it, "waste water."

"Secondly," Dr. Stephan said, "the growth of population and the national volume of fluid wastes has made the notion of 'getting rid of waste water meaningless in many places. We discard it only to impose it on the other fellow.

"Processing waste water really is simply applying controlled scientific methods to what we now do haphazardly, when we dump waste water back into rivers or into the ground, and hope that nature will purify it—which it doesn't always do."

In the arid California town of Santee, where the Public Health Service has collaborated in an "advanced waste treatment" project, children frolic in a sparkling clean public swimming pool filled with refined waste water.

Sewage in Santee is given the primary and secondary treatment that is standard in up-to-date communities, is chlorinated, then held in aeration basins, and finally put through the natural filter of an ancient dry riverbed. The innovation was readily accepted by the citizenry.

Compact Indoor System

In the sewerage plant at Bijou, Calif., to reduce the drainage of extraneous compounds into adjacent Lake Tahoe, the same sort of purification is accomplished with a compact indoor system of chemical precipitators and carbon filters.

No one envisions the necessity of "renovated" water becoming the main drinking water supply anywhere, because there are so many other prospective uses for it that will release natural water for the drinking supply. Industry uses five times as much water as goes to domestic purposes, agriculture four times as much.

But countless thousands of Los Angeles residents already are getting refined waste water as part of their drinking supply.

On the outskirts of Los Angeles, in great open air "spreading beds," water piped 200 miles from the Colorado River is regularly poured out onto the ground to percolate down and replenish natural underground basins that feed city wells.

Along with this river water, there is spread up to 17 million gallons a day of processed waste water from the county's Whittier Narrows Reclamation Plant. This water is germ-free, and in terms of extraneous chemicals actually is purer than the Colorado River water.

The resultant saving in water procurement is rapidly paying off the $1.7 million cost of the reclamation plant.

This 17 million gallons is equal to the water needs of a community of 100,000 people.

A Plant in Nassau

A pilot plant of the same sort with a capacity of 43,000 gallons a day went into operation at the Bay Park Sewage Treatment Plant in Nassau County, L.I., in August. Studies are being made for another groundwater "recharging" operation at Riverhead in Suffolk County, L.I.

A community's sewage volume is about equal to its water intake. Accordingly, through renovation, a community theoretically could almost get by using the same water over and over again. In reality, there are several qualifications.

A fraction of any water supply is lost through actual consumption and through evaporation. Some water is lost during the removal of pollutants. And repeated re-use of water brings a gradual build-up of some particularly stubborn contaminants—notably salt—that eventually become unduly expensive to remove.

Nevertherless, the renovation of the bulk of a community's waste water seems quite possible.

With some 60 million of the nation's citizens now living in communities that provide the prerequisite of secondary sewage treatment, the potential volume of renovated waste water nationally is equivalent to the full supply of 10 cities the size of New York.

A Plant for New York

A study recently completed by the Department of Health, Education and Welfare for a waste water renovation plant suitable for New York City indicated that one capable of turning out 100 million gallons of drinkable water a day—one tenth of the city's requirement—could be built for $33 million.

This compares with the prospective cost of about $2.8 million for the Chelsea pumping station, designed to provide the same amount of water from the Hudson. However, the long term net cost of the renovated water was reckoned at 16 cents a thousand gallons, which is a nominal amount.

For many purposes, waste water does not have to be processed to anywhere near drinking water quality, reducing costs proportionally. Much of Baltimore's sewage, routinely treated and disinfected with chlorine, now goes to a steel plant. Throughout the arid Southwest disinfected sewerage plant effluent is used to water lawns, golf courses and some crops.

While the desalination of sea water and renovation of waste water at first glance seem quite disparate activities, actually they are converging sciences.

Desalination researchers are getting more and more into the removal of other mineral impurities, especially in connection with inland brackish water, which is perhaps desalination's most promising application in this country.

More than 1,000 communities are dependent on brackish ground water.

The Problem of Salt

Simultaneously, waste-water researchers, as they find ways of extracting other pollutants, are coming more and more up against the problem of salt, the distinctive contaminant in repeated use of water.

This convergence of research problems poses a political question. Desalination work is centered in the Department of the Interior. Water renovation work has been proceeding under the Public Health Service in the Department of Health, Education and Welfare.

The two endeavors may converge so closely soon that it will be logical to integrate them. That involves such a ticklish problem of interdepartmental rivalry that neither side wants to talk about it.

Waste-water renovation epitomizes "water management" in that at one and the same time it handles problems of supply and disposal.

"It's an inevitable fact of the future," says Dr. Stephan. "It will, I'm convinced, become one of man's chief environmental controls."

December 22, 1965

A Growing Sense of Urgency

U.S. URGES SAVING OPEN URBAN SPACE

Special to The New York Times

WASHINGTON, March 18— The Federal Government, local communities and private agencies must take swift action to save rapidly vanishing open space in urban areas, a Government report said today.

The report, published by the Department of Housing and Urban Development, is the first detailed study of the availability of open-space land in and around metropolitan areas.

In a scholarly and dispassionate tone, the 132-page document seeks to demonstrate that land for recreational and scenic purposes is rapidly diminishing under the fierce pressures of commercial and industrial development and it urges a number of steps to arrest the trend.

Three Reasons Given

The report was prepared for the department by Anna Louise Strong, acting director of the Institute of Legal Research at the University of Pennsylvania. It listed these major reasons for preserving open space:

¶To provide recreational space for a growing urban population whose demands for recreational opportunities "a half hour from home" are expected to increase tenfold by the end of the century.

¶To prevent the indiscriminate destruction of forests, wetlands, streams and agricultural areas.

¶To give the rapidly growing metropolitan areas a sense of pride and a touch of beauty. Unless remedial action is taken to incorporate open space in community planning, the report said, "human development and nature existing side by side will soon be no longer visible, except at the rapidly receding urban fringes."

The report also cited several studies that have pointed to a relationship between high-density living in crowded city areas and mental illness. Mrs. Strong noted widespread disagreement among psychologists on this point but suggested that the possibility of such a link was "sufficiently serious to require further research."

Land Shrinkage Illustrated

To illustrate the shrinkage of open-space land, the report noted that a million acres were lost each year to development for residential or commercial purposes; that in California alone, 375 acres a day, or 140,000 a year, of agricultural land is being turned to urban uses; that of the original 127 million acres of wetlands—marshlands —in the United States over 45 million have been destroyed by draining, filling, levees and dredging for residential or industrial uses.

The twin problems of diminishing space and growing demand are particularly critical, the report said, in the Northeast, which includes one-quarter of the nation's population but only 4 per cent of its recreation acreage.

The report defined urban open space as "that area within an urban region which is retained in or restored to a condition in which nature predominates." Such land can be found in the densely settled central city, the spreading suburbs, or on the edge of a metropolitan area "where the country still is dominant."

March 19, 1966

U.S. PANEL URGES $2.5-BILLION FUND TO CURB POLLUTION

By HAROLD M. SCHMECK Jr.
Special to The New York Times

WASHINGTON, June 12 — Americans must stop waging wholesale chemical warfare against themselves, a Government report on environmental pollution stated today.

The report called for $2.5-billion in Federal spending during the next five years to deal with all manner of problems such as air and water pollution, urban noise, odors and crowding and large-scale introduction of new products presenting unknown potential hazards to man.

"The task force [group making study] concludes that danger to environmental quality, particularly in the broad context that the task force has reviewed it, is among the most important domestic problems today," said the report. "It affects all Americans where they live, work, and play. It can materially damage their children and generations yet unborn."

Purpose of Study

The report was made to John W. Gardner, Secretary of Health, Education and Welfare, by a special group that he appointed six months ago to advise the department on how to cope with man's increasing power to foul the environment.

The study group listed 34 recommendations, of which 10 were described as "action goals" that are particularly urgent and in need of prompt action.

Longer range proposals included establishment of an environmental protection system to provide permanent means of dealing with hazards as they arise and a national council of advisers on man's relationships to his environment.

Need for Legislation

At a briefing today, Roy M. Linton, chairman of the study group of six persons, said he believed new legislation would be required for about half of the total recommendations and that spending beyond the first five years would probably be greater than the initial $2.5-billion.

The study group was outspokenly critical of the status of environmental pollution today, the attempts, so far to combat it, and the public attitude toward use of the nation's air, water and land.

Although it has been traditional to wait until pollution crises developed and then try to deal with them one by one, this system is no longer acceptable, the report said.

"The time has arrived," it declared, "when we must stop allowing human beings to serve involuntarily as guinea pigs in experiments with environmental change."

The 10 goals for immediate action included plans, to be in effect by 1970, to reduce by 90 per cent the emissions from industrial smokestacks and vehicle exhaust pipes.

Also among these goals was a plan to establish safety controls for general use of products having trace metals and chemicals that might be hazardous.

After 1970 the general use of any new synthetic material, trace metal or chemical would be prohibited until its safety was approved by the Department of Health, Education and Welfare.

Much the same type of controls would be set over consumer products of many kinds — including clothing and appliances.

An effort was recommended to test, by 1970, all public drinking water supply systems in the nation and to ensure health-approved drinking water for all of them.

Limited Authority

Present Federal authority to inspect and certify drinking water supplies is severely limited, the report said, noting that 50 million Americans today drink water that does not meet Public Health Service standards. Furthermore, the task force said it was not satisfied that these standards adequately reflected the needs of the people.

The immediate action goals also included research to establish facts on tolerable human levels of crowding, congestion, noise, odor and other specific stresses of urban life.

Another research effort was urged to determine, by next year, the effects of population trends on all the goals in environmental pollution abatement.

"The task force finds that today's urban environment is frequently placing severe physical and mental stresses on people," said the report.

In an earlier reference to problems caused by population growth, the report said the department might have to exert its influence here too.

"The task force feels it is entirely possible that the department will have to undertake, in addition to direct measures to facilitate family planning, programs to bring about a change in attitudes about the size of families," the report stated. "Such a shift in a basically personal attitude will not be achieved quickly or easily."

Unified Approach Urged

Because many state and local governments face difficulties in financing adequate pollution abatement measures, the study group recommended a White House conference on local government problems of this sort.

Also recommended were Congressional enactment of a unified environmental protection act and the establishment of a national council of ecological advisers comparable to the President's Council of Economic Advisors.

The report deplored past fragmentary efforts to deal with threats to the environment.

It said an individually acceptable amount of water pollution, added to a tolerable amount of air pollution, added to a bearable amount of noise and congestion — could add up to an environment unacceptable for health and well-being.

"It is entirely possible that the biological effects of these environmental hazards some of which reach man slowly and silently over decades or generations will first begin to reveal themselves only after their impact has become irreversible," the report said.

In addition to Mr. Linton, former staff director of the Senate Committee on Public Works, the study group included:

John J. Hanlon, director of public health of Detroit and Wayne County, Mich.

Samuel Lenher, a vice president of E. I. du Pont de Nemours & Co.

Harold L. Sheppard, a social scientist of W. E. Upjohn Institute for Employment Research, Washington.

Raymond R. Tucker, professor of urban affairs, Washington University, St. Louis.

Anne Draper, an economist of the A.F.L.-C.I.O. Research Department.

Speaking for Secretary Gardner at a briefing for reporters today, Wilbur J. Cohen, Under Secretary of Health, Education and Welfare, said the report would be studied with great care.

June 13, 1967

4 NEW LAWS ADD TO PUBLIC LANDS

By WILLIAM M. BLAIR
Special to The New York Times

WASHINGTON, Oct. 2—Two new national parks and a system of scenic rivers and trails were added to the nation's outdoor public preserve today.

President Johnson signed four bills to add, as he told witnesses to the ceremony in the East Room of the White House, "still more to the scenic wealth of our country."

The signing raised to an even dozen the number of national parks, national seashores and national lakeshores created in the Kennedy and Johnson Administrations since 1961.

It also marked the passage by Congress this year of more than a dozen major conservation bills. Some of them had long and bitter histories, such as the 58,000 acre Redwood National Park, which was established today.

In addition to the Redwood park, Mr. Johnson approved a North Cascades National Park in the state of Washington, two adjacent recreation areas and an adjacent wilderness area.

This area covers 1.1 million acres in what has been described as the American Alps, close to the populous areas of the Pacific Northwest.

The scenic trails measure provides for a system of urban and rural trails. The first components of the system are the Appalachian Trail in the East and the Pacific Coast Trail in the West.

The legislation also calls for a study of 14 other trails for possible inclusion in the system.

Wild Scenic Rivers System

The first Wild Scenic Rivers System, which also has had a stormy background, provides for preserving all or parts of eight rivers. The aim is to preserve the unspoiled sections.

The measure also names 27 rivers as potential additions to the system.

Mr. Johnson hailed the adoption of the Redwood park as one that "will stand for all time as a monument to the wisdom of our generation." He also said that it would be "remembered as one of the great conservation achievements of the 90th Congress."

However, in an offhand remark, he gave recognition to the hard fact that a long struggle still is ahead to round out the parks and other outdoor preserves. The Redwood park, for example, will not be completed until the process of acquiring private land within the park boundaries and obtaining title to state-owned land, including three state parks is completed.

Looking over a map marked with outdoor areas authorized since 1961, the President asked:

"Is there a member of the Appropriations Committee in the House?"

California to Trade

Political consideration also will play a part in the new park. Redwood timber companies will seek to make the best bargain they can under provisions involving the exchange of their holdings for Government-owned land near the park.

California will seek to make the best trades it can because the park bill provides that the Federal Government can acquire only by donation the three state redwood parks within the national park boundaries.

The three state redwood parks are the Jedidiah Smith and Del Norte, which make up the northern unit of the national preserve and Prairie State Park in a southern unit. The two sections are connected by a narrow strip of land along the Pacific Ocean. South of Prairie Creek State Park is still another section of land to be acquired. In this area, along Redwood Creek, are some of the world's tallest trees.

The measure authorizes the appropriation of $92-million from the Land and Water Conservation Fund, a congressionally approved law to help pay for conservation projects.

Congress tapped off-shore oil revenues this session to build up the fund, which has lagged in providing the funds needed for park acquisitions. The fund started with Park and other public recreation admission fees and the tax on motorboat fuels.

The $92-million is more than the cost of all the other national parks combined. The Redwood and North Cascades National Parks are the 34th and 35th national parks.

October 3, 1968

Visitors Are Swamping National Parks

By STEVEN V. ROBERTS
Special to The New York Times

YOSEMITE NATIONAL PARK, Calif., Aug. 31—They came in cars and trucks, in buses and campers and trailers, lumbering through the foothills of the Sierra Nevada, toward a Labor Day weekend away from the agonies of city life. But by Thursday evening they read this sign at the park entrance: "All campsites are full."

That warning has been sounded with increasing frequency this year, not only at the national parks and forests but also at thousands of other recreation areas across the country.

Facilities are staggering under a crush of humanity. Attendance at the national parks has been rising at least 7 per cent a year. One study indicates that even if population growth is discounted, four times as many people are visiting the parks today as did 20 years ago.

Last year, an estimated total of 40 million people visited the 32 operative national parks, and 157 million visitor days— one person staying 12 hours— were recorded at more than 12,000 national Forest Service units.

Park officials agree that the the figures add up to a major crisis. As Ernst W. Swanson, a professor emeritus of economics at North Carolina State College, said in a recent study:

"While we may grant that all Americans should have an opportunity to enjoy the many wonders our natural, historical and cultural assets hold for us, the most pertinent question ensues: Can we afford a burden of visits so immense as to threaten the existence of our parks?"

The outlook for the future is bleak.

"Yosemite will still be the same size 50 years from now," said Bryan Harry, the park's chief naturalist. "We can't make it any bigger or build another one. The population is not only growing, it is becoming more affluent and more mobile, and this land will become even more precious as other wild places continue to shrink. We have to find a way to cope with this problem—and we are open to suggestions."

The roots of the problem are fairly obvious. Affluence has spawned a whole new industry —the camping unit mounted on the back of a pickup truck— and the wilderness is now accessible to people who like running water and soft beds. (In lodges and cabins run by concessionaires, accommodations with baths and stoves are far more popular than rudimentary units.)

Vacations are longer, and new superhighways enable travelers to reach almost any park in the country within a few days. The growing congestion and lack of green space in urban areas is driving more people out in search of nature. But they wind up creating what they are trying to escape.

United Press International

Campers relaxing outside their trailers in Yosemite National Park. Only a day's drive from San Francisco and Los Angeles, it is considered the most overcrowded national park.

A Growing Sense of Urgency

Yosemite, only a day's drive from San Francisco and Los Angeles, is generally considered the most overcrowded park. Congestion reaches its peak on major holidays, and this Labor Day weekend was no exception.

A Constant Roar

The constant roar in the background was not a waterfall, but traffic. Transistor radios blared forth the latest rock tunes. Parking was at a premium. Dozens of children clambered over the rocks at the base of Yosemite Falls.

Campsites, pounded into dust by incessant use, were more crowded than a ghetto. Even in remote areas, campers were seldom out of sight of each other. The whole experience was something like visiting Disneyland on a Sunday.

"People who used to come to Yosemite for the beauty and the serenity stay away," Mr. Harry reported. "Those who come now don't mind the crowds; in fact, they like them. They are sightseers, and they come for the action. They don't come for what Yosemite really has to offer."

If Yosemite no longer appeals to the purist, it has become a "people's park," an enjoyable place for the average family to spend a few days.

"It's so quiet," said one young couple from Oakland as they stood in a crowd at Yosemite Falls. "At least it's quieter than the city."

It Beats Asphalt

Gary Yaeck of Bakersfield added: "It sure beats concrete and asphalt. My son caught three fish this morning—that's a pretty big thrill."

The problems of congestion are not limited to Yosemite, nor to holiday weekends. Overcrowding is a way of life for the national parks, giving rise to problems such as the following:

¶The fish stock at Yellowstone is so depleted that people over 12 are now prohibited from using live bait. Fish tend to swallow live bait and half of them die even when thrown back.

¶Lakes in Superior National Forest in Minnesota are becoming polluted from excessive use by campers. Algae fed by nutrients in human waste is beginning to form in even the remotest lakes, and forest officials are desperately trying to "toilet train" the flood of inexperienced campers.

¶Camping in the back country of the High Sierras is becoming so popular that officials of Sequoia National Park in California are thinking of installing sanitary facilities at the more popular campsites. In the back country of Yosemite there is no more room to bury garbage. Campers are now required to carry their garbage out with them.

¶More than 10,000 boatmen will ride the Colorado River through the Grand Canyon this year. Last year only 3,000 did it—as many as made the trip in all previous years combined. Trails leading from the rim to the canyon floor are in danger of eroding from overuse and must constantly be repaired.

¶Traffic jams are common on Trail Ridge Road, a highway rising more than 12,000 feet in the wilderness of Rocky Mountain National Park in Colorado. Cars are sometimes backed up for 20 miles in Great Smoky Mountains Park in Tennessee and North Carolina; commercial vehicles were recently banned from park roads and traffic lights were installed. Many parks have instituted one-way road systems.

¶Older parks, such as Yellowstone and Yosemite, are laboring under the handicap of antiquated sewage plants. Spring floods caused the system to break down here this year, and raw sewage poured into the Merced River. On big weekends, when the winds are wrong, traces of smog hang over the Yosemite Valley.

¶The normal destruction of plants and wildlife by huge crowds is compounded by vandalism. Signs are torn down, benches broken, doors ripped off their hinges. Pilfering became such a problem at the Petrified Forest in Arizona that officials bought some petrified wood and now give it out free to anyone tempted to smuggle home an exhibit from the park.

¶Serious crime rose 67.6 per cent in national parks last year, as opposed to 16 per cent for the country as a whole. Thievery and hooliganism are rising problems. After repeated nasty incidents, Sibina National Forest near Tucson, Ariz., had to be closed down after dark. In some areas, rangers travel only in pairs.

All this has park officials very concerned.

"Our challenge," said Fred Novak, Western regional director of the National Park Service, "is to strike a balance between protecting our resources and providing for their use and enjoyment."

In the past, the emphasis has been on use. Vast publicity campaigns urged people to visit the parks and officials took great pride in their burgeoning statistics. Now the stress is swinging toward protection.

"We used to say, 'Welcome, we're here to serve you,'" said one park official. "Now we say, 'Sorry, we have as many as we can take; you'll have to come back later.'"

Fixed Number of Sites

In the last year or two, national parks and forests have taken several steps toward regulating the use of their facilities. In the past, people were allowed to camp wherever they could find room. Now, most parks have a set number of campsites. When they are full, no one else is allowed to camp anywhere in the park.

Officials are also considering user fees for campsites (they are now free) and reservation systems (they are now on a first-come first-served basis). The theory is that fees would make some other places, such as private campgrounds, more attractive. And some people would stay home if they did not have reservations, rather than wander fruitlessly from place to place as they do now.

Many officials are considering plans whereby people would leave their cars outside a park and travel in by buses or other group carriers. An experiment will soon begin here in which tourists visiting the Wawona Grove of sequoia trees will park some distance away and be brought to the grove by bus.

Hard to Divert Tourists

Officials talk about efforts to divert people away from the more popular places, but that is extremely difficult.

"Everyone wants to see Yosemite Valley, the giant sequoias and Old Faithful," said one official. "These are the things they've always read about."

The State of Oregon is so concerned about the influx of tourists that it recently dropped its advertising campaign to attract more visitors. Gov. Tom McCall announced that gasoline tax revenues, formerly used for advertising, would be diverted to an antilittering campaign.

A national plan for outdoor recreation now being prepared for President Nixon is reported to suggest that industries and schools should stagger their vacation schedules. A second recommendation, according to park officials, is that more recreational facilities should be developed closer to urban areas. The true wilderness would then be left to those who are looking more for tranquility than recreation.

For Scattered Facilities

Another frequent proposal is that national parks should build more facilities in unused areas and not concentrate visitors in small sections. Officials point out, however, that funds for new construction are being devoured by rising maintenance costs, especially with so many inexperienced campers coming to the parks for the first time. The proposal also excites the wrath of conservationists.

"When you go outside Yosemite Valley," explained Mr. Harry, "the waterfalls you find are exciting because they are so wild and you're by yourself. If you build a road to them, they lose their charm. The beauty of the back country is in its solitude. It's an illusion

In increasing numbers, campers are being turned away from parks when there are no campsites left.

to say that all you have to do is open more territory. If you do, you'll ruin it."

The only real answer, park officials agree, is to buy more land. But as Professor Swanson points out, potential parkland is continually being consumed by logging companies and real estate developers. Large portions of the Florida coast have been turned into tract homes, he said, and stands of giant redwoods are being felled in California while negotiations creep along on establishing a new national park.

Funds for 265 Units

The budget for the National Park Service this year is almost $130-million, up from $125.5-million last year. But that is needed to maintain 265 different units, including the 32 national parks, covering more than 27 million acres. Professor Swanson estimates that the service could easily use double its current appropriation.

Even if new parkland is purchased, officials now believe that they will ultimately have to limit access to the parks. A study is being made to determine precisely when a park becomes saturated and starts to lose its essential character. The idea is abhorrent to many officials, but some of them are convinced that the only alternative is the gradual destruction of the national parks.

"I sometimes remember," said Mr. Harry, "a study an ecologist once did about muskrats. When they first moved into a marshland they lived in peace. But when the marsh filled to overflowing, the muskrats started pilfering and fighting with each other. Finally they killed the marsh and they all died. Maybe that's the point we are at."

September 1, 1969

58

PRESIDENT NAMES COUNCIL TO GUIDE POLLUTION FIGHT

Cabinet-Level Group Set Up to Combat Deterioration of Nation's Environment

NIXON VOICES OPTIMISM

Says the 'Energy and Skill' That Led to the Problems Can Help Conquer Them

By WALTER RUGABER
Special to The New York Times

MIAMI, May 29—President Nixon set up a Cabinet-level advisory group today to work against what he called increasing threats to "the availability of good air and good water, of open space and even quiet neighborhoods."

The President, at his waterfront home on nearby Key Biscayne, signed an Executive order establishing the Environmental Quality Council. He said it would be the "focal point" of an Administration effort to protect the nation's natural resources.

"The deterioration of the environment is in large measure the result of our inability to keep pace with progress," Mr. Nixon said in a statement. "We have become victims of our own technological genius."

Expresses Confidence

"But I am confident that the same energy and skill which gave rise to these problems can also be marshaled for the purpose of conquering them," he said. "Together we have damaged the environment and together we can improve it."

The President asked the new council to propose improved measures for the control of pollution, to coordinate efforts on different levels of government, and to anticipate problems that will arise.

Mr. Nixon will serve as chairman, Vice President Agnew as vice chairman and Dr. Lee A. DuBridge, the President's science adviser, as executive secretary.

The other members will be Secretaries Clifford M. Hardin of Agriculture, Maurice H. Stans of Commerce, Robert H. Finch of Health, Education and Welfare, George Romney of Housing and Urban Development, Walter J. Hickel of Interior and John A. Volpe of Transportation.

The President said that the structure of the new group "in some respects parallels" that of the National Security Council and the Urban Affairs Council, his top panels on foreign policy and social issues.

The environmental quality unit, like the urban affairs group that he also created, will give Mr. Nixon more administrative machinery in an area that, according to many authorities, badly needed it.

Dr. DuBridge, who met with the President for an hour this morning and then appeared at a briefing to explain the program, noted that the economic costs of improving the environment "can be very large and must be weighed against the benefits."

He said that in cities such as Los Angeles the "main source" of air pollution was the exhaust of automobiles and he added that the group would want to meet with manufacturers to discuss further controls.

"Can we still tolerate the combustion engine or is some other kind of propulsion mechanism going to be available?" Dr. DuBridge asked, suggesting a high-level interest in electric or steam cars.

He noted that several Government departments were engaged in studies of the pesticide DDT, whose sale was recently banned in Michigan when officials there decided it left dangerous residues on foodstuffs.

"We think we should bring these studies together to determine what action should be taken about the control and use of DDT and the development of substitute pesticides," Dr. DuBridge said.

He said there were "a couple of bills already being formulated in Congress which I think will be rather helpful." He described them as 'broad legislative statements of policy" rather than specific measures.

Senator Henry M. Jackson, Democrat of Washington, has proposed a statutory declaration of national environmental policy. He praised the President's move today.

Mr. Nixon's Executive order also established a 15-member Citizens Advisory Committee on Environmental Quality. The President named Laurance S. Rockefeller as chairman of the body.

Mr. Rockefeller served as chairman of former President Lyndon B. Johnson's Citizens Advisory Committee on Recreation and Natural Beauty. That group was abolished today by Mr. Nixon.

Its 12 members will serve on the President's new committee. The three additional members to be appointed by Mr. Nixon were not named today.

Mr. Nixon spent most of the day relaxing. His daughter and son-in-law, the David Eisenhowers, arrived this afternoon to spend the Memorial Day weekend with the President and Mrs. Nixon and their older daughter, Tricia.

May 30, 1969

NIXON PROMISES AN URGENT FIGHT TO END POLLUTION

Special to The New York Times

SAN CLEMENTE, Calif., Jan. 1—President Nixon pledged his Administration today to a "now or never" fight against pollution as he signed legislation creating a three-member Council on Environmental Quality.

"The nineteen-seventies absolutely must be the years," the President said in his first formal statement of the new decade, "when America pays its debt to the past by reclaiming the purity of its air, its waters and our living environment. It is literally now or never."

Speaking informally with reporters at the California White House this morning, Mr. Nixon expressed a particular anxiety about the toll that sprawling population and automobile traffic have already taken on his native Southern California.

Urges Protective Actions

He took a drive yesterday, he remarked, with his friend Charles G. Rebozo through the hills of Orange County, reflecting on the changes since his boyhood and speculating about the way it would look 10 years from now.

Unless protective measures are immediately implemented, Mr. Nixon said, "areas like this will be unfit to live in" by 1980.

He applied the same dire prediction to all of the nation's major metropolitan areas, specifically mentioning New York and Philadelphia. And he noted that leaders of other industrial nations shared his anxiety about the trend toward a "poisonous" world, a trend that is ironically a companion of growing affluence.

The President indicated he would have substantially more to say about environmental problems in his State of the Union Message on Jan. 22.

He gave no hints as to whom he would appoint to the new environmental council. But he said that the council would be assisted by a "compact staff," and would function with "the same close advisory relation to the President that the Council of Economic Advisers does in fiscal and monetary affairs."

Mr. Nixon noted that Congress was considering the creation of still another staff organization in the White House to deal with environmental problems.

"I believe this would be a mistake," he said. "The act I have signed gives us an adequate organization and a good statement of direction. We are determined that the decade of the seventies will be known as the time when this country regained a productive harmony between man and nature."

January 2, 1970

59

Pollution Fight Pressed Across Nation as One of Top Priorities

By WAYNE KING

Spurred by the emergence of conservation as a potent political issue, state and local government officials across the country are responding to growing public pressure with a flood of proposals for new environmental laws.

Governors and legislative leaders in more than a dozen states have put environmental protection at the top or near the top of the list of priorities for action this year.

Rhode Island, for example, last year considered only two air and water pollution bills. This year, 21 proposals have been introduced, plus four others dealing with pesticides.

Some 20 other states are considering a broad range of environmental bills, with new ones still to be introduced. Officials at the local level also are pressing for controls on pollution and appear more willing to enforce them.

Pressure for the new laws is coming from citizen groups like Chicago's SAVE (Society Against Violence to the Environment) and Pittsburgh's GASP (Group Against Smog and Pollution) that are proliferating across the nation.

In the San Francisco Bay area, an information switchboard listed 90 environmental groups in October and nearly 200 just a month later.

Politically, proponents of strong environmental laws range from members of conservative groups to radical adherents of the Students for a Democratic Society.

Confronted with such widespread grassroots feeling, Democrats, Republicans, liberals and conservatives are vying for a spot in the vanguard of the environmental crusade.

Typical of the political maneuvering is the situation in California, where Gov. Ronald Reagan, accused by critics of being a "Johnny-come-lately" to environmental problems, has proposed a program for "an all-out war against the debauching of the environment."

His chief political opponent, Jesse Unruh, the Democratic leader in the State Assembly who is a contender for the Governorship, has countered with a program of his own, including a plan for a state agency with "the power to reject any projects that do violence to the environment."

The Republican Speaker of the Assembly, Robert Monagan, has named a committee to formulate an Environmental Bill of Rights to guarantee Californians "the right to clean air, pure water, uncluttered beaches, scenic roads and the preservation of aesthetic values for himself anf future generations."

The political infighting in California has become so intense that a recent resolution urging the Federal Government to halt any new offshore oil drilling was delayed for a week while Democrats and Republicans squabbled over who would get credit for the bill. The impasse was resolved by amending the list of authors, after which the measure passed unanimously.

The popularity of the cause is beginning to bring warnings from longtime conservationists.

Philip Berry, president of the Sierra Club, a leader in conservation efforts, warned in California this month that "the popularity of our cause has attracted some whose motives must be questioned."

"Politicians paying lip service, industrialists laying down public relations smokescreens and anarchists voicing legitimate concerns about the environment for the ulterior purpose of attacking democratic institutions are all suspect," he said.

Other Favorable Action

But whatever the motives, legislators elsewhere are also responding with favor to environmental proposals.

A major conservation measure creating a Coastal Marshlands Protection Agency has passed both houses of the Georgia legislature before its adjournment Saturday. Intended primarily to protect Georgia's wetlands against developers, the measure was inspired by efforts two years ago by an Oklahoma company to mine phosphates in the areas, a move conservationists feared would poison the breeding ground for shrimp and other sea life.

A bill providing for a fine of $1,000 a day for polluting the air passed the New Mexico Legislature before its adjournment Thursday, with approval by Gov. David F. Cargo assured. Also awaiting the Governor's signature is a bill to ban the use of the pesticides DDT and DDD, except in an agricultural emergency.

The air pollution bill was weakened somewhat by reducing the fine from the $5,000 originally proposed, but the provision for a daily fine was retained. The bill also provides for a public hearing before variances from air pollution standards can be granted and establishes a petty misdemeanor penalty for minor violations.

A strengthened smoke control law providing for fines up to $300 or 30 days in jail for first offenses and a $1,000 fine or one year in prison for second offenses gained the unanimous approval of the Pennsylvania Legislature.

Washington Bills Signed

Gov. Daniel J. Evans of Washington yesterday signed several environmental measures, including one assigning unlimited liability to oil companies for spillage, even in "act-of-God" cases. Major oil company lobbyists had complained that insurance would be unobtainable under such terms.

Also approved by the Legislature before its adjournment Feb. 12 and signed by the Governor yesterday were a stripmining bill requiring miners t file plans for restoring the landscape, the creating of a Department of Ecology and establishment of an advisory agency to aid in selecting sites for nuclear power plants.

And in Maine, Gov. Kenneth M. Curtis recently signed legislation to control the sites of all major commercial and industrial developments and to levy a half a cent a barrel on petroleum products in transit to finance the cleaning up of spillages.

Some states, among them Michigan, Ohio, Pennsylvania and Wisconsin, have approved bond issues to fight pollution. The Wisconsin bonding program, signed last month by Gov. Warren E. Knowles, totals $200-million and is to be used primarily to help cities build facilities to abate water pollution.

In Illinois, where voters rejected a $1-billion bond issue proposal in 1968, Gov. Richard B. Ogilvie has urged that a $700-million proposal be put on the ballot in November.

A number of states have recently created or proposed new agencies or mergers of old ones to deal with environmental problems.

In one of them, Alaska, Gov. Keith H. Miller has created a new Division of Environmental Health and is also calling for development of a national institute of environmental science in the state to "develop new techniques for protection of the environment of all countries."

In New York last week, 41 legislators introduced a 175 page bill that they described as "the most comprehensive environmental bill in the history of New York State."

The bill is reportedly supported by 300 nature and conservation organizations and with bipartisan backing in both houses. It would strengthen controls on air, water and noise pollution and provide for citizens' lawsuits for damages caused by violators of state anti-pollution laws.

Included in 35 specific proposals is establishment of a state council on environmental quality.

A New Department

In Arizona, where smogplagued residents of Phoenix and Tucson are demanding stronger state controls on air pollutants, legislation has been introduced to define odor as a pollutant—a step in regulating the operation of cattle pens. Also proposed are measures to provide for research on automotive pollution, improved mass transit and tax credits for installation of pollution control equipment.

Gov. Russell W. Peterson of Delaware has created a new state department to deal with pollution and implemented a year-long study of the future of the Delaware River and Bay.

In Hawaii, Gov. John A. Burns—a Democrat accused by Republicans of being "soft on pollution"—has asked for an Office of Environmental Quality, pesticide controls, restrictions on building on beaches and a long-term program for sewage treatment.

Bills have been introduced to convert the state's fleet of automobiles to burn propane gas and to appropriate $500,000 to Honolulu to start work on a mass transit system. A related bill would ban the sale of vehicles with internal combustion engines after 1974.

Maryland's Gov. Marvin Mandel has proposed a broad nine-point environmental program to give the state more control over development, strengthen air and water pollution laws and provide for research and safeguards relating to atomic power plants on a Chesapeake Bay. Gov. Deane C. Davis of Vermont has proposed a priority 10-point program to back up his admonition to the Legislature that "the first and foremost goal of the 1970's is to make our lakes, streams and rivers clean again by 1980."

Gov. Francis W. Sargent of Massachusetts has outlined a six-point environmental program, led off by the introduction of a water pollution control measure last week. Another bill would allow individual citizens to bring suit to prevent damage to the environment, a proposal also being considered in several other states.

Gov. William G. Milliken of Michigan, has outlined a 20-point Charter for a Clean Environment, which ranges from compulsory litter containers for motor vehicles to the zoning of Great Lakes shorelands for erosion control.

A Top Priority

Colorado's Gov. John C. Love has put top priority on environmental problems, and more than 25 bills have been introduced in the Colorado Legislature dealing with air and water pollution, roadside advertising and the spread of urban blight.

Exhaust emissions pose an enforcement problem for many cities because of the difficulties of testing. A new law in Buffalo, also in use elsewhere, permits policemen to cite violators on the basis of "visual examination," A similar Erie County law resulted in citations against 90 violators in January, the first month it was in effect. Included were eight locomotives.

And in Detroit, Wayne County officials last month cited the Ford Motor Company for excessive smoke emissions from the company's River Rouge complex. The county also announced plans to bring suit against other industrial polluters.

Such efforts are apparently beginning to bring results. In Tacoma, Wash., for example, the Tacoma Kraft Mill of the St. Regis Paper Company recently announced a $17-million air-improvement program after being ordered by te Puget Sound Air Pollution Control Agency to come up with n acceptable plan to cut pollution ɔy this month.

February 24, 1970

REVISED POLICY FOR U.S. LANDS ASKED IN STUDY

FIVE-YEAR SURVEY

Major Shift Is Sought to Assist Mining and Timber Activities

By GLADWIN HILL
Special to The New York Times

WASHINGTON, June 23 — A Congressional commission, after a massive five-year study, recommended today that the one-third of the nation's land that is federally owned be largely retained by the Government, but that major changes be made in its management and use.

Foremost among the recommended changes were the following:

¶That Congress reassert its constitutional primacy in supervising the public lands and curb the President's power to shift public land from one use to another.

¶That public land laws be revised to help such commercial activities as mining, the timber industry and agriculture.

¶That land be made available to states for urban expansion.

¶That the United States Forest Service be shifted from the Department of Agriculture to the Department of the Interior.

A 342-page report containing these and 350 other recommendations was presented to President Nixon and Congressional leaders at noon by members of the Public Land Law Review Commission.

First in 2 Centuries

The study was the first comprehensive assessment of public land use in the two centuries of the nation's history, during which Congress and other agencies have passed thousands of laws and other enactments dealing piecemeal with the problem.

President Nixon, receiving the report in the White House Rose Garden, said it "will have without question a very great effect on the policy of this country."

"It is essential to plan now for the use of that land," he continued, "not do it simply on a case-by-case basis, but to have an over-all policy."

First reports from conservationists were unfavorable. Portions of the report appear to follow closely policies that have been advocated by the grazing, mining and timber industries and that have been criticized by conservation groups.

The 19-member commission was created by Congress in 1964 to chart a future for the 755 million acres of land—out of the nation's total of 2 billion acres—in the hands of Federal agencies.

Half of the Federal land is in Alaska, and nearly all of the rest is in the 11 contiguous Western states, although there are tracts in all the states.

The largest portion - some 450 million acres—is under the Interior Department's Bureau of Land Management, with 187 million more acres under the Forest Service and lesser amounts under Interior's National Park Service and Fish and Wildlife Service.

The commission's recommendations generally call for an array of new legislation to remedy what one official called "the chaotic jungle" of land laws going back to 1792.

The commission's chairman, Representative Wayne N. Aspinall, Colorado Democrat, said he hoped the 1971 Congress would start taking up the proposals and that the implementation process would be completed in "six or eight years."

This prospect was regarded by experienced Washington observers as uncertain. The recommendations of three previous Federal land study commissions in the last century were largely blocked by conflicting interest-groups. And many of the new proposals plainly contain the seeds of high controversy.

Judicious 'Multiple Use'

The report, while repeatedly stressing judicious "multiple use" of public lands with solicitude for environmental values, hewed closely to policies advocated by the timber, mining and grazing industries, which conservationists have denounced as overly exploitative.

The initial reaction of one conservation leader—Hamilton Pyles of the National Resources Council of America—to the report was that it was an "emasculation" of public land controls. Others suggested that the proposals were so disputable that they would provide a new rallying point for environment alist opposition.

The report, commission leaders acknowledged, deliberately by-passed the question of long-term conservation of such exhaustible resources as metals, coal and oil—leaving such considerations, a spokesman explained, "to the normal operations of the market place."

"The commission saw no reason for superimposing the views of Government executives on the decisions of business executives," the commission's director, Milton A. Pearl, a specialist in real estate law, told a news conference.

Representative Aspinall, chairman of the House Interior Committee, said the commission's constant concern had been to balance commercial uses of public land with recreational and environmental objectives, and that this accommodation was constructive because "nature is one of the worst offenders in regard to maintaining environment."

The commission comprised, in addition to Representative Aspinall, six Senators, six members of the House of Representatives and six non-Governmental appointees of President Johnson, among them Laurance S. Rockefeller, a leading conservationist.

President Nixon told Mr. Aspinall at the White House that he knew the report contained "a great deal of very helpful information and recommendations, many of which of course will be accepted and, we trust implemented."

Comparing it with the original Homestead Act of 1862, Mr. Nixon said he trusted it would "have the same vision and make the same contribution to the greater America that we all want for our children."

"I note," he added jocularly to commission members, "that you recommend selling some of these lands. You're not going to sell the White House, are you?"

Other major recommendations of the commission were the following:

¶Termination of grants of Federal land to states and institution of a system of "in lieu" payments to states that have large amounts of non-taxable Federal land.

¶Consolidation of scattered responsibility for land into one Senate committee and one House committee, and on the executive side into one Cabinet department.

¶"Phasing out" of the 20,000 existing vacation homes on public lands, and adoption of a policy of "occupancy uses" (settlement) "only where suitable private land is not abundantly available.

¶Revision, but not outright repeal, of the basic 1871 mining law, which some officials and conservationists have called a "scandalous" device for illegitimate acquisition of public lands.

61

Communities Supported

Another feature of the report was explicit support for communities depending on commercial activities on public lands, such as logging and grazing.

Although public lands provide only 3 per cent of cattle forage nationally, Mr. Pearl said there was no thought of phasing out this seemingly marginal activity because in some localities entire communities depended on it.

"Through its timber management and sales policies," the commission said, "the Federal Government over the years has in effect made a commitment to communities and firms that it will make timber available to assure their continued existence."

The commission's study was essentially a continuation of business that Congress left hanging in 1934, when it passed the Taylor Grazing Act prescribing management policies for public lands "pending their final disposition."

This has been construed as presupposing ultimate Federal relinquishment of public lands, and has generated perennial demands for transfer of land to states and private parties.

A Contrary Conclusion

The commission's conclusion, Mr. Pearl said, was that "the bulk" of public lands should be kept in Federal ownership—although that term did not appear in the report and the recommendations contained myriad ways in which tracts of public land could end up in other hands.

The report contained 18 broad policy "concepts," 137 numbered recommendations, and some 250 less formally stated ones.

It was described as representing the "consensus" of the 19 commission members, whose signatures were on it, although not corresponding with their views in every detail.

In one of the footnoted dissents to sections of the report, two members from Nevada, Senator Alan Bible and Representative Walter S. Baring, opposed the termination of Federal land grants to states. Federal land makes up 86.4 per cent of Nevada, the highest proportion after Alaska's 95.3 per cent.

The Federal Government "owes" nearly one million acres of land to 15 of the "lower 48" states under historic grant provisions still unconsummated. The commission recommended that these obligations be settled within 10 years.

For Alaska, the commission recommended "immediate establishment" of a joint Federal-state natural resources and regional planning commission to expedite the transfer to the state of the 104 million acres granted to it with statehood but delayed until Congress acts on the land claims of Alaskan natives — Eskimos, Indians and Aleuts.

The commission said "Con-gress has largely delegated to the executive branch its plenary Constitutional authority over the rental, management and disposition of public lands" and urged that Congress reassert this authority, "immediately review" existing executive decisions regarding land disposal, and delineate explicit ground-rules for Presidential latitude.

"Mineral exploration and development should have a preference over some or all other uses in much of our public lands," the commission said. "We recognize that [it] will in most cases have an impact on the environment or be incompatible with some other uses."

The panel recommended that successful prospectors on public lands be given title only to specified subsurface minerals rather than surface land (now obtainable for no more than $5 an acre), but should be able to buy surface acreage at "the market value."

The report advocated private development of oil shale lands in the West, a major reserve that has been frozen because of Federal uncertainty about viable policies.

Similarly, the commission proposed exploitation of the vast outer continental shelf oil reserves under "flexible methods of pricing" in place of the present bonus bid-fixed royalty system.

In regard to timber, much of which comes from national forests, the report was characterized by some observers as reflecting the philosophy if not the details of the proposed industry-sponsored National Timber Supply Act, which conservationists have denounced as sanctioning a "raid" on forest resources.

On up to one-fourth of national forest land, the commission said, commercial timber production should be formally classified as "the dominant use," while "those lands having a unique potential for other uses should not be included in timber production units."

The commission favored replacement of the century-old and virtually inoperative Homestead Act and related laws with "statutory authority for the sale of public lands . . . for agricultural purposes . . . at market value in response to normal market demand" with "no artificial and obsolete restraints such as acreage limitations on individual holdings, residency requirements and the exclusion of corporations as applicants."

Regarding the problem of expanding population, the commission recommended making some public land available for a prototype "new city" to explore policy.

The commission held 37 sessions, 10 of them regional public meetings, between July, 1965, and last April, and heard 900 witnesses. Its 40 special studies, made by staff members and contractors, make a stack of volumes a foot high and six-feet long. The commission spent $7,104,000—$286,000 less than its budget.

June 24, 1970

Citizens on U.S. Coasts Rally to Save Tidal Marshes

By BAYARD WEBSTER

An unorganized but extensive movement is under way in coastal areas to save the nation's salt marshes.

Citizens' groups displaying "Save the Wetland" banners are converging with increasing frequency on town councils, city officials, state legislatures and Congressmen in shore communities from Maine to California.

Moved by concern for beauty and ecological worth, they are seeking to prevent the further destruction of the eight to ten million acres of tidal marshes along the United States shoreline.

Their protests stem from the tremendous extent to which many towns, cities, villages and counties have committed themselves to filling in salt marshes for commercial or housing developments, using them for garbage dumps or generally allowing them to be despoiled in piecemeal fashion.

For the last few months in Douglaston, Queens, a small citizens group has mustered several hundred people to hold "marches on the marsh." These resulted in saving a 100-acre tidal marsh called Udalls Cove off Little Neck Bay from development as a parking lot and golf course.

In the San Francisco area a few weeks ago, a citizens group won a battle to save hundreds of acres of threatened bay marshes. Similar crusades and rallies for tideland causes have been held in hundreds of widely scattered coastal sections of the country.

Why this recent interest in saving salt marshes?

One reason is the obvious appeal of the wetlands—the feeling of wildness coupled with the peaceful aspect of the marsh with its constantly moving grasses and water that attract a variety of animal life as the seasons change.

But according to some marine biologists, it is not only the antijunkpile esthetic that moves people to seek to preserve the tidal marshes. Many people have recently become aware of the ecological worth of these areas.

"A few years ago 'salt marsh' was a phrase that hardly anyone really understood," said Derekson W. Bennett, conservation director of the American Littoral Society, a scientific organization devoted to the study of shore and tidal areas.

"But now," he said, "thousands of people are beginning to know what a salt marsh is and why it shouldn't be destroyed."

A salt marsh is an area of boggy, peat-like ground built up over a period of years along coastal estuaries that is periodically inundated by ocean or bay tides. A variety of grasses grow in it, ranging from the short cord and eel grasses to the tall, reedy types known popularly as cattails and foxtails.

Nutrients Added to Water

The grass stalks wither, die and deteriorate each winter, contributing nutrients to the water as they disintegrate and decay. These nutrients provide

The New York Times (by Ernest Sisto)

Citizens of Douglaston, Queens, walking in marshes of Udalls Cove, Little Neck Bay, during rally in February

food for microscopic animals, which in turn are consumed by tiny fish and shellfish such as oysters, clams, shrimp and scallops.

Some of these, in turn, satisfy the appetites of crabs, larger fish, waterfowl and aquatic mammals that forage in the shallows and among the reeds as the food chain, leading eventually to man, continues its cycle.

Several marshland studies have estimated that a tidal marsh can produce as much food an acre as does good beef cattle grazing land.

John R. Clark, a member of the staff of the United States Bureau of Sport Fisheries and Wildlife and founder of the American Littoral Society, states in his book, "Fish and Man," that two out of every three species of useful Atlantic fish — or two-thirds of the value (about $30,000,000) of the entire annual Atlantic Coast fish catch — depend in some way upon tidal marsh lands for their survival.

Mr. Clark notes that dozens of species of coastal fish depend upon the life-giving and sheltered marshes for "nursery areas" for their young.

He also points out that fish reared in one town's salt marsh may grow up to be caught offshore miles away. This means, he adds, that a salt marsh in one community can be an asset to many other communities.

Salt marshes, according to John and Mildred Teal, authors of "Life and Death of the Salt Marsh," also furnish homes for swamp rabbits, mink, muskrats, otters and raccoons, rail, gallinules, snipe and pheasant,

and provide nesting grounds and resting places for many waterfowl.

A Wintering Place

In the Chesapeake Bay marshes alone, it is estimated, more than 1,000,000 ducks and 500,000 Canada geese winter each year, according to James R. Goldsberry Jr., biologist of the Maryland Department of Game and Inland Fish.

Another relatively little-known virtue of the marshes is they are capable of controlling huge quantities of floodwaters that come from the sea. Lying in front of the shoreland, they take the brunt of blows delivered by storm waves, their finely intertwined root and reed growth dulling the ocean's force and preventing shore erosion.

Waves that can break apart solid breakwaters do only slight damage to the marshland, which soon repairs itself.

The awakened public interest in marsh preservation has arrived at a time when pressures on the United States coastline are at their greatest. Thirty-five per cent of the nation's population, for example, now lives within 150 miles of the East Coast.

Half of Lands Destroyed

In a report on marine resources of the Atlantic coastal zone, just completed for the American Geographical Society, George P. Spinner, the project director, says:

"Sites for new oil refineries, atomic power plants, pulp and paper mills and chemical plants, all of which prefer locations on or near the shoreline, must be found. It is evident that a

major planning effort will be required by all concerned if major damage to the coastal marine resources is to be prevented."

But in the 200 years since the Industrial Revolution began, extensive damage has already been done to these coastal areas.

Of the total of 12,841 square miles—or 8,218,240 acres—of estuarine salt marshes along the edge of the continental United States, it is estimated that 50 per cent has been destroyed in the last two centuries.

In addition to the competition among industry, housing developers and other interests for marsh land, pollution from industries, oil spills, urban sewage pollution, agricultural runoff of fertilizers and pesticides and thermal changes have contributed to the despoliation of thousands of acres of marshes and the animal life in them.

Although no conclusive evidence has yet been discovered, some marine scientists suspect that the recent decrease in some fish catches in several areas along the Atlantic, Gulf and Pacific coasts may be linked to a corresponding decrease in salt marsh acreage, coupled with pollution effects.

Some states, meanwhile, are belatedly preparing laws and taking other action to save their wetlands from development and pollution.

More Leisurely Pace

New Jersey, Maryland, Virginia and Delaware have recently purchased thousands of acres of salt marsh. And pres-

ent acquisition programs in New England and New York, where the coastal marshlands are relatively scarce, are intensive.

Conversely, in the four Southeastern states where almost 63 per cent of the coastal marshes are found, acquisition for preservation purposes has moved at a more leisurely pace and few areas have been set aside.

In California and Oregon, citizen efforts have resulted in the saving of thousands of acres of threatened marshes. In the Gulf Coast area, dredge-and-fill operations and high pollution indices are beginning to be reversed as the result of local conservationists' efforts.

On the Federal level, a start has been made on the classification of coastal marshes. Several surveys are being researched by the Fish and Wildlife Service.

The Department of the Interior is also preparing a report that will contain recommendations for protecting, conserving and restoring estuaries to maintain a balance between conservation and development.

And listed in a recently completed Federal study, "Shoreline Recreation Resources of the United States," is a virtue of the marshlands—beauty—that is not often cited.

In discussing how Southern California has developed some of its tidal marsh areas for recreational purposes, the report said: "These developments tempt one to suggest that the marsh shore can be managed easily to provide a recreation complex unmatched by almost any natural shoreline area."

July 13, 1970

Prairie Partisans Move to Save Grasslands

By JOHN NOBLE WILFORD
Special to The New York Times

PLAINFIELD, Ill., Oct 17 — There are people who love the prairie as others love the sea.

They love the feel of its black, pongy soil, the splendor of its many wild flowers and the sweep of its tall grass blowing in the wind. They feel, with Willa Cather, that "the grass was the country," and that its roots ran deep and shaped the heartland of a nation.

But those who love the prairie have to look hard nowadays to find any real prairie left to love. It has been plowed under, grazed over and covered by concrete and people.

Partisans on Offensive

Now, before it is too late, a small but fervent group of prairie partisans—scientists, conservationists, seed growers and businessmen—is making some progress in its efforts to hold on to whatever virgin prairie is left and to restore some lands to their former wild grandeur.

Prairie—the word is from the French for meadow—is a generally flat and fertile terrain covered with tall perennial grasses and an abundance of native flowering plants.

Partisans are buying up wild prairie ahead of the real-estate developers, fencing off relic prairies, restoring prairie-like fields for state parks and even attempting to establish a prairie national park.

"Man has probably misused the earth's grasslands more than he has misused any other plant environment vital to man," Dr. Roger C. Anderson, director of the University of Wisconsin Arboretum, told a recent conference of prairie experts in Madison, Wis.

The original prairie stretched from Indiana west to the Rockies, from Texas north through the Dakotas and into Canada. To the east the grass was tall; to the west, short. Only an occasional stand of cottonwood or grove of burr oak, usually by a stream, broke the sea of grass.

About all of the prairie left now are patches of grass and flowers on some rocky hillsides and sand ridges, along a few railroad rights-of-way and in old and forgotten cemeteries like the one in the cornfields out from this small farming community.

This rare relic of virgin prairie vegetation here was discovered not long ago by Dr. Robert F. Betz, a botanist and ecologist at Northeastern Illinois University

Steven C. Wilson from "Grass Land." © Wide Skies Press

The original prairie stretched from Indiana to the Rockies, from Texas to the Dakotas

in Chicago. He has found about 18 other prairie cemeteries in the Chicago area, all of which he hopes to fence off and preserve.

Reclamation Projects

In similar moves, several Midwestern universities are reclaiming patches of prairie for scientific, historical and esthetic reasons. The Nature Conservancy, a land conservation group, is buying prairie remnants in Illinois, Wisconsin, Minnesota, Nebraska and Kansas.

A group of Wisconsin citizens recently rescued 80 acres of prairie near Kenosha from the reach of real-estate developers. Illinois has acquired 2,000 acres of unplowed land 60 miles southwest of Chicago for the development of the Goose Lake Prairie state park, a restored prairie planned for research and recreation.

Jim Wilson, a mail-order seed grower in Polk, Neb., estimates he has supplied native grass and wildflower seed for some 3,000 acres of what he calls "soul plantings"—the mini-prairies established by school groups, landscape gardeners and homeowners.

One soul planter is Karl E. Wolter, who left the Bronx years ago to be a forester in Wisconsin. He is re-establishing prairie vegetation on his 12-acre farm near Madison simply because "a prairie in flower is a magnificent thing to see."

And on a grander scale, state leaders and scientists in Kansas are pressing the Federal Government to establish a prairie national park in the Flint Hills, a relatively unspoiled region south of Manhattan and near the eastern stretch of the Santa Fé Trail. In arguing their case, the Kansans point out that there are parks for the mountains, forests and seashore—but not for the prairie.

Attitude Changing

To Dr. Betz, the Northeastern Illinois professor, the new interest in saving the prairie reflects a gradual change in public attitudes toward nature.

"A few years ago we had a real crisis from our point of view." Dr. Betz said. "Almost none of the prairie had been saved. What was left was endangered, and the few of us who worried about it had almost no support from the public. We were just 'dicky-birds'—you know, nature lovers."

Now, Dr. Betz said he is in constant demand to speak to civic and school groups about the prairie. Through the Prairie Restoration Society, a branch of the Sierra Club organized by Chicagoans who are concerned with prairie preservation, he has raised money to try to save such remnants as the old cemeteries.

"Few of us living in Illinois, the so-called Prairie State, have ever seen a true native prairie," Dr. Betz said on the drive out to the cemetery near here. He pointed with mild contempt to the vegetation on the side of the road. "Ragweed, bluegrass, foxtail, Queen Anne's lace—immigrants all, just as we are, if you want to look at it that way."

The Indians managed the prairie very well, Dr. Betz explained. They were hunters and so never turned over the sod, exposing the soil to the erosion of wind and rain. They put up no fences and so the buffalo never overgrazed any one area. And, either by accident or to herd buffalo, they often set fire to the prairie grasses in the fall or spring, thereby killing off encroaching weeds and forest seedlings, but not the deeprooted perennials of the prairie.

"I remember growing up in Chicago and playing cowboys and Indians on vacant lots." Dr. Betz continued. "We called those lots prairies because of the weeds and grasses. I tried to learn all about those weeds because I thought I was learning the plants of the old prairie."

"It was not until 12 years ago, down on the Santa Fé tracks near Hodgkins, Ill., that I saw my first virgin prairie. When I saw it I fell in love with the prairie."

With enthusiasm and care, Dr. Betz parted the tall stalks and grass stems in the old cemetery. It was the resting place of settlers who came to Illinois from New England in the eighteen-thirties and first plowed the prairie under. In burying their dead, however, they preserved an acre of what they had destroyed all around.

Dr. Betz said that he had identified 80 to 90 species of prairie flowers in the cemetery—golden alexander, prairie lily, prairie clover, sunflower, silphium, goldenrod, blazing star, sky-blue asters, purple gentian and purple prairie phlox.

But as in the original prairie, most of the vegetation in the cemetery was grass—the big bluestem (sometimes called turkey-foot because of its three-pronged top), little bluestem and Indian grass.

Surviving Cemetery

The cemetery prairie survived because its land was never plowed like the surrounding cornfields and because it was burned off nearly every fall by a farmer down the road who wanted to discourage hunters. The burning kept the weeds out.

The edge of the cemetery showed the value of prairie vegetation as defense against erosion. At the edge there was a one-foot drop into the cornfield. In more than a century of plowing, water and wind have stripped a foot of soil off the cornfield—rich soil that was a product of millions of years of evolution through prairie growth and decay.

Having this "standard of comparison" between the unspoiled and the cultivated land. Dr. Betz said, is one of the reasons scientists are interested in saving relic prairies.

But since there are few relics like the old cemeteries, most scientists and conservationists attending the Second Midwest Prairie Conference, which was held last month at the University of Wisconsin, said that they were concentrating their efforts on prairie restorations—taking land that has never been plowed and only lightly grazed and planting it with

native grasses and flowers.

A model restoration is the 65-acre Curtis Prairie, developed on the outskirts of Madison by the University of Wisconsin. It was started in the nineteen-thirties by the old Civilian Conservation Corps. University students and professors combed hillsides, rural cemeteries and railroad right-of-ways to find the dozens of wild prairie plant species.

Only now, more than 30 years later, can the field be considered a reasonably authentic example of wild prairie, for the process of restoration is that slow and difficult.

Dr. Grant Cottam, chairman of the university's botany department, explained the process during a walk through the restored prairie. The field had been a pasture. If it had been cultivated for any length of time, Dr. Cottam said, it might have taken a century or more to restore it.

When the prairie seeds were sown, the first growth was mostly downward. For the first few years, prairie flowers and grasses concentrate on extending deep roots.

The Curtis Prairie is burned over nearly every fall. Besides killing off nonnative competition, Dr. Cottam said, the burning also somehow seems to increase the length of the growing season and "give us five times as many flowering plants the next spring."

Effects of the Railroads

Weeds—the annuals whose roots are shallow but whose seed are ubiquitous—are "by all odds, the toughest problem facing the prairie restorationist," said Jim Wilson, the Nebraska seedgrower.

Weeds compete with the native plants for moisture, nutrients and growing space. Despite the burnings, Dr. Cottam still found clumps of bluegrass in the Wisconsin restoration, vestiges of the field's days as a pasture.

Another problem is the dwindling sources of native seeds. Railroad right-of-ways used to be the best source. They were unplowed and were burned frequently to keep down the wild plants. But Southern Illinois University botanists reported at the prairie conference that railroads are increasingly using chemical herbicides to clear the right-of-ways.

The spraying destroys the natural vegetation and encourages the growth of weeds.

Most relic and restored prairies are too small to provide food and shelter for many of the original prairie wildlife, except for hawks, doves, sparrows, gophers and mice. The larger animals, especially buffalo and elk, have long ago vanished from the area.

At a 15,000-acre test site in northeast Colorado, scientists participating in the International Biological Program are attempting the first comprehensive study of grassland ecology. They are charting the living and eating habits of 60 species of mammals, 138 kinds of birds and more than 3,000 types of insects that depend on prairie grass.

In Illinois, a private citizens' group organized the Prairie Chicken Foundation to "preserve and perpetuate the prairie chicken." Though once an endangered species, the prairie chicken is now making a slow comeback, primarily because nesting sanctuaries were provided in areas of remnant prairie.

Leaders of the Kansas drive to establish a national prairie consider 30,000 acres "a minimum if the native fauna as well as the native flora is to be preserved."

The first proposed site for the park lay west of the Blue River, just north of Manhattan, Kan. But cattlemen objected, the Corps of Engineers went ahead with a dam and people started building cottages and marinas where the prairie park would have been.

On a visit to Kansas earlier this year, Walter J. Hickel, the Secretary of the Interior, had some encouraging words:

"I still think there is a need for the creation of a Prairie National Park and I think Kansas would be an ideal location for it."

Driving south of Manhattan, where the sky is big and the fields have never been plowed and bluestem grows by the side of the road, Dr. Lloyd Hulbert, a Kansas State University botanist, considered what a prairie has to offer.

"People go to see glaciers and mountains," Dr. Hulbert said. "Perhaps they would like to see what the prairie was like. It's part of our heritage, and it's a chance to get away from civilization. That's becoming a more important consideration all the time."

October 18, 1970

Nixon Reassures Industry, Bars Pollution 'Scapegoat'

Speaks to Businessmen

By JAMES M. NAUGHTON

Special to The New York Times

WASHINGTON, Feb. 10 — Officers of more than 200 major industries were assured by President Nixon today that they would not be made "scapegoats" of the drive for cleaner air and water.

"The Government—this Administration, I can assure you —is not here to beat industry over the head," the President told members of the National Industrial Pollution Control Council at a reception in the White House.

Mr. Nixon praised the corporation officers for voluntary efforts to curtail industrial pollution after receiving a series of reports in which the council pledged to continue its cleanup efforts.

"I do not see the problem of cleaning up the environment as being one of the people versus business, of Government versus business," the President said in impromptu remarks.

"I am not among those who believe that the United States would be just a wonderful place in which to live," Mr. Nixon went on, "if we could just get rid of all of this industrial progress that has made us the richest and strongest nation in the world."

He made the remarks two days after sending to Congress a message proposing broad initiatives by the Government to curtail pollution. The message detailed the need for new enforcement authority in such areas as strip and underground mining, ocean dumping and water pollution. It encouraged Congress to grant individual citizens the power to go to court to halt spoilage of water supplies.

Today, Mr. Nixon sent to Congress more than a dozen bills to carry out his proposals.

In Mr. Nixon's White House audience were the chief executives of several American corporations that had been the targets of recent Government lawsuits and enforcement proceedings to halt pollution.

They warmly applauded the President, in contrast to the more muted reception the businessmen gave several hours later to a stern speech by William D. Ruckelshaus, administrator of the Environmental Protection Agency.

"Industries and businessmen that must operate in the market place of free choice know that they must change, they must adapt, they must accommodate to changes in public attitudes or they will surely die," Mr. Ruckelshaus told the council members.

The National Industrial Pollution Control Council was established by the President last April to advise the Administration on environmental matters and to initiate voluntary cleanup programs in each industry. It has 63 members and about 200 panelists on subcouncils.

Environmental crusaders have been skeptical about the worth of the council, noting that its meetings until today had been closed to the public and news media, and that it was composed, as Senator Lee Metcalf, Democrat of Montana, had charged, of "the leaders of the industries which contribute most to environmental pollution."

Among the council members are officers of the Union Carbide Corporation, which has sparred with Mr. Ruckelshaus's agency about deadlines for reducing sulphur oxide emissions from its plant in Marietta, Ohio; the United States Steel Corporation, which accepted an Illinois court's timetable last month for a halt to waste dumping in Lake Michigan.

Also, General Motors Corporation, which, with other auto manufacturers, objected to Congressional legislation setting a 1975 deadline for 90 per cent reduction of auto exhaust pollutants, and Republic Steel Corporation, which told Federal investigators in 1969 that they had no legal right to question whether it was lagging behind schedule in reducing the dumping of waste into Cleveland's Cuyahoga River.

First Official Report

The council submitted to the President a list of 160 corporate pledges to help enhance the environment with specific actions. At the same time, in the council's first official report to Mr. Nixon, it warned of an "extreme position" appearing among environmentalists.

"The view that no material should be permitted to be released into the environment unless it can be shown to be harmless to man and his environment," the report said, would curtail industrial innovation and "could be likened to a requirement of proof of innocence by the accused rather than proof of guilt by the accuser."

Mr. Nixon said that he welcomed an opportunity, through the meeting today, to tell the American people that "American industry is also against pollution."

He praised the recent trend among corporations to advertise steps they were taking to combat pollution.

"This, of course, has an enormously good effect on the country," the President said, "to see that American industry is not the enemy of the good life, but, actually, because of what American industry does, because of what it is doing, it not only provides jobs and better incomes and better housing, better clothing for the American people, but also that American industry is concerned —just as concerned as the man in Government is—about a better life for our children, a cleaner America, an America in which we can have the clean air, the clean water and the quality of life that we believe is the American heritage."

February 11, 1971

The Environmental Conflict

Fresh air and clean water and a healthy environment are essential if human beings are to enjoy a life worth living on this crowded planet. But economic growth is necessary if the nation is to avoid worsening unemployment and a depressed standard of living. Last week President Nixon managed to position himself on both sides of this complex issue.

On Monday the President sent to Congress a message on the environment which was comprehensive and progressive. On Wednesday, speaking to the National Industrial Pollution Control Council, which includes the officers of many major polluters, he said reassuringly, "I do not see the problem of cleaning up the environment as being one of the people versus business, of government versus business."

These cheery informal remarks about harmony do not invalidate the specific recommendations of his message to Congress, but they do point up the problem inherent in the contemporary drive for improved environmental quality. Economics and ecology do not always go hand in hand. Installing expensive pollution-control equipment and processes may be corporate good citizenship but that is not necessarily good for profits. The costs cannot always be passed on to the consumer in the form of higher prices. A change-over may give a competitor a temporary but costly advantage. As Adlai Stevenson used to say, "There are no gains without pains."

There is the deeper problem of motivation. A corporate manager has his attention focused on profit targets and production schedules. He has a natural resistance to taking into account environmental costs, some of which may be invisible, incalculable or very long-term. A kind of reorientation or re-education is going to have to occur if ecological considerations are going to enter decisively into his thinking.

President Nixon perhaps did his business audience a disservice in suggesting to them that this transition can be smooth or effortless. Probably it can only come about, at least in part, by the clash—sometimes angry and noisy—of businessmen with government officials and citizen conservationists who instinctively put environmental values ahead of profits in every situation.

It is unhelpful to suggest, as the President did, that these battles involve a search for "scapegoats" or an effort "to beat business over the head." No sophisticated conservationist thinks of individual corporate managers as villains. Indeed, many dedicated conservationists are themselves business executives. What is involved is a conflict of values. It is the conflict between those who see man as masters of the earth, free to mine its topsoil or befoul the water or darken the sky if economic necessity requires such action, and those who see man as the earth's guests with the responsibilities and good manners of a guest.

In this philosophical conflict, the President refuses to commit himself. He proposes many excellent environmental measures which please conservationists, but he often talks the old-fashioned language of the profits-first businessman. This ambiguity extends to his actions. His relentless fight for the supersonic transport plane

goes counter to his proposals on air and noise pollution. Except for water sewage treatment plants, his budget requests are substantially below Congressional authorizations.

The harmony between business, conservationists and government in the environmental field which the President proclaims is theoretically attainable. But it can only come about through a fundamental reconciliation of the differing philosophies now held by the various participants. No such reconciliation is likely to evolve if the President and other leaders believe that the conflict between environmental values and growth-as-usual can be evaded or papered over.   February 14, 1971

Coast Desert a Vast, Littered Playground for Millions

By STEVEN V. ROBERTS
Special to The New York Times

BARSTOW, Calif., April 10 —All that glitters here in the California desert is not gold. Usually it's broken glass, or a discarded beer can, or a motorcycle leaping over a sand dune. Strange as it sounds, some experts fear the desert is dying.

The California desert covers more than 16 million acres of jagged mountains and sweeping plains, inhabited by small rodents that never drink water and with strands of drab cacti that burst forth with sudden stabs of color.

This arid wilderness is really two deserts, the Mojave and the Colorado, stretching from the Sierra Nevada and Death Valley on the north to Mexico on the south, from the Colorado River on the east to the mountains surrounding Los Angeles on the west. The total area is as big as West Virginia.

Once the desert was considered a barren, hostile place, the last barrier before reaching the Promised Land of Southern California. But today it is the great sandy backyard for 11 million people, a weekend refuge from Paradise-on-the-Pacific.

The Federal Bureau of Land Management, a part of the Department of the Interior, which supervises about three-quarters of the desert, put it this way: "Try to picture the California desert, not as a vast expanse of open space, but as a constantly shrinking landscape, surrounded by sprawling cities."

In 1970, according to the bureau, the desert absorbed 8 million visitor days—**one person spending one day. Only**

The New York Times   April 11, 1971
Shading denotes the 16-million-acre California desert

two years before the estimate was 5 million. This weekend alone, as on most holidays, more than 100,000 are expected. And these visitors, as they have so often done elsewhere, are threatening to ruin what they came to enjoy.

"Unless we act soon," the bureau said in a report last year, "the desert faces damage and destruction because of people pressure."

Few believe that the desert should be roped off, that people in the crowded coastal cities should not have an escape. But planning, officials feel, is essential; anarchy can only end in destruction. "We have to let people enjoy the public lands," said one, "and yet keep the natural environment intact. That is always the rub."

The damage takes many forms: erosion of the land, destruction of vegetation and wild-

life, the defacement of ancient Indian petroglyphs, the plunder of ghost towns, the loss of historic relics, and everywhere the mark of modern man—trash.

A Fragile Environment

"Everybody thinks of the desert as indestructible," said J. R. Penny, state director of the land management bureau, "but it's just the opposite. The desert is one of the most fragile environments we have."

It has been more than a year since the bureau warned of the imminent "destruction" of the desert. In that time, the spokesman conceded, it has "been able to do darn little" to stop the process. A ranger at the Joshua Tree National Monument south of Barstow said efforts to protect the desert "have been like throwing water up Niagara Falls."

The Bureau of Land Management has roughly one man in

the field for every one million acres, not nearly enough to patrol its vast domain. Moreover, the bureau has always been geared to supervise land, not people, and its agents have no training or authority to enforce the law. As a result, most use of the desert is completely unregulated.

The desert is hardly virgin territory. It has been violated for years by miners, squatters, ranchers, and people who just wanted a convenient dumping ground for an old automobile. But recently, a much more serious threat has developed, the off-road vehicle.

A novelty only five years ago, there are now 800,000 in the state—motorcycles, mini-bikes, jeeps, dune buggies, gyroplanes —most of them within striking distance of the desert.

What is happening to the land can be seen a few miles east of Barstow. Out here, ruts and gashes criss-cross the land like a child's aimless scribbles. Every arroyo has become a freeway, and on weekends the more popular paths remind one of the beach roads on a summer Sunday.

Plants Are Casualties

Scattered everywhere are the corpses of desert plants, like casualties on a battlefield, white and stick-like, deprived of life-giving moisture. Some of the remaining bushes are festooned with unnatural colors: orange ribbons marking racing trails, bright blue paper blown away from some abandoned campsite now caught in the sticky branches. In the dry desert climate, even paper lasts for years.

People often camp in large groups here, as if they were afraid of the desert or too used to cramped quarters to break the habit. And they leave their mark: an egg crate, a child's sneaker, the rusting hulk of an automobile, so many bottle caps and cans they almost look native to the region.

The refuse serves one purpose—as targets for amateur riflemen. When they run out of cans and cars they shoot at signs and rest stations and

The New York Times/D. Gorton

Off-road vehicles are a serious threat to the desert. Their tire marks crisscross the land; every path becomes a noisy highway.

rodents; nothing escapes their tell-tale pock marks.

"You get to the top of some hill and you think no one has been there in 200 years," said Bill Halligan, a young cyclist out for a weekday spin. "Then you see a beer can."

The most drastic impact of the vehicles is on vegetation. "Desert plants are not rugged, they're very delicate," said Ranger Donald M. Black of Joshua Tree, who is a naturalist. "They develop very shallow root systems and wax and cork coatings to retain moisture, and once they're broken, they don't usually heal before they dehydrate and die. The right conditions for growth might not come again for 10 or 20 years."

Moreover, the vehicles compact the soil, causing the rare desert rainfall to run off quickly, rather than to percolate into the soil. And their tracks are practically indelible. Scars left by General Patton when he trained his African armed unit here during World War II are still visible.

As people have invaded new areas of the desert, the wildlife has retreated. The big horn sheep, which are very sensitive to man, "are like an encircled band, restricted and confined more and more," according to Ranger A. B. Sansum of Joshua Tree. As he talked, Ranger Sansum got a call from a colleague. A young coyote was dying near a campground, apparently from eating human garbage and waste.

Another threat to wildlife are clubs of "varmint hunters" who track down coyotes, foxes and similar species. "There is a very delicate balance of nature between the varmints and other types—the varmints weed out the weaker animals," said

Ranger Black. "But for some reason these people feel they have a religious duty to kill varmints. I just don't understand them."

One of the major results of the off-road vehicle is noise. As Ranger Sansum put it: "People pay $2 a night to camp where it's peaceful and quiet. And then some dingaling comes in with his motorcycle and does nothing but drive around and around and around, often with the silencer off his muffler. He has no thought that people might be tired of all that noise they get all the time. On a heavy weekend, this place sounds like the Culver City speedway."

Off-road vehicle enthusiasts talk about the "freedom" they have and the chance to reach once inaccessible areas. But those areas often contain Indian petroglyphs — pictures scratched in the soft rock— valuable archaeological sites, and remnants of the white man's early days here. Many of these have been picked clean by rock hounds, bottle hunters, and plain vandals.

One example is a wooden plank road built across the desert by early pioneers. Most of it has now been burned for firewood.

What lies behind this destruction, and what can be done about it? Part of the problem is sheer numbers, and many officials feel that developing more facilities might make things worse, since they would only attract more people. Recently the Bureau of Land Management returned to the Federal Treasury $250,000 earmarked for a camp ground near Trona, because it felt the

ensuing visitors would destroy the area.

Another part of the problem is just ignorance, and the bureau hopes to get money for an educational program that would make people more aware of the limits of the desert.

But the issues go deeper. Many people who come here are not concerned with the esthetic value of the area but its recreational potential, according to Ranger Sansum. "They want to know how they can use it— collect rocks, or take plants, or drive over it," he said. "And this is where we bang heads with them."

There is also a small group of openly destructive vandals, the type who burn privys and shoot up signs. "I don't think they have a grudge against us," said ranger Black, "I think they're striking out at their situation in life. They can't get back at their boss, or a cop, so when they come out here, with no restraints, they strike out at anything."

Last Labor Day weekend, thousands of people camping out in the Imperial County sand dunes stopped a train and broke its windows. When a county sheriff went out to investigate they chased him away.

Officials agree that off-road vehicles must have some place to go; so do rockhounds and bottle hunters. The problem, as always, is balance and planning, reconciling conflicting demands, establishing some order, preserving the land even while it is being used.

Scouts Saving Painting

A few things are being done. A Boy Scout troop is fencing off a particularly valuable

Indian rock painting. Several large areas near here have been reserved for off-road vehicle use. But as Gordon W. Flint of the land management bureau office in Riverside said, "We haven't begun to scratch the surface of what's needed."

What is needed, according to conservationists, is a thorough survey of the desert, a plan to allocate different areas for different uses, and enough manpower and authority to enforce the regulations. The bureau asked Washington several months ago for $28-million to pay for drafting a master plan and setting up an "interim management" program.

But others add something else to the list of needs—a new attitude on the part of the public. As one writer put it:

"Up to now men have treated the deserts as if they made no difference. One of these days, when survival can no longer be taken for granted on a crowded, used-up earth, they may make all the difference."

Time is probably the rarest commodity in this ageless place. With 11 million people "poised like some gigantic tidal wave," as the bureau put it, eager to escape the smog and stress of daily life, crisis is a mild word.

"We can't afford to wait five years," said Mr. Flint. "Things like the petroglyphs will be totally destroyed, rare animal species will possibly become extinct, the danger to desert visitors who lack protection will increase. The beauty the California desert now holds will be diminished to the point where people won't want to use it. Would you want to stand up to your neck in beer cans?"

April 11, 1971

National Forests: Physical Abuse and Policy Conflicts

By GLADWIN HILL

The national forests, the country's biggest single reservoir of material and recreational resources, are in trouble.

Covering nearly one-tenth of the nation's area and worth untold billions of dollars, these preserves have long been considered limitless. But their integrity is now being threatened by both physical abuse and inponderables of national policy.

With increasingly bitter charges and countercharges being sounded over the management of the national forests, a six-month inquiry into the complexities of the controversy has shown that:

¶Bulldozers and tractors are boring into some of the nation's last remnants of pristine wilderness.

¶"Clear-cutting" of timber has left thousands of square miles of bald patches in the forests, some of which may never regrow.

¶Archaic laws and regulations are allowing choice expanses of the forests to be gouged and scarred by mining operations.

¶Nearly half the national forests' forage areas have been overgrazed or otherwise rendered substandard.

¶Preoccupation with the forests' commercial roles has severely retarded their development for public uses.

¶Pressures for increased timber production have led to extensive violations of the spirit, if not the letter, of laws intended to preserve the forests in perpetuity.

Propelled by the force of the "environmental revolution," the situation appears headed toward early showdowns within the Nixon Administration, in Congress and in litigation reaching all the way up to the Supreme Court.

The outcome as a whole could profoundly affect both the face of the nation and the quality of life of its citizens. At stake are 154 national forests, averaging 2,000 square miles each in area and spanning 40 states from Florida to Alaska.

They cover 187 million acres —nearly an acre for every man, woman and child in the country. The nation spends over $500-million a year to maintain and operate the forests—$1.5-

million a day. They yield about $300-million in revenues, mostly from the sale of timber.

The public's recreational use of the national forests is officially reckoned at double the use of the national parks, and is fast approaching the numerical equivalent of every citizen's spending a day in the forests each year.

Unlike the smaller national parks system, which is run by the Department of the Interior, which comprises only 28 million acres and which is dedicated entirely to recreation, the national forest system is run by the Department of Agriculture and is dedicated to "multiple use."

The forests by law are supposed to fulfill six functions: recreation, watershed maintenance, wildlife preservation, timber production, grazing and mining.

In one degree or another, the six-month investigation indicates these activities are being impared or compromised under the national forests' present administration.

"The Forest Service's management policies are wreaking havoc with the environment," Senator Gale McGee, a Wyoming Democrat, said recently. "Soil is eroding, reforestation is neglected if not ignored, streams are silting, and clear-cutting remains a basic practice."

Mismanagement Charged

He is in the forefront of a growing body of critics of the national forest administration. Their ranks include members of Congress, conservationists, citizen groups, scientists, economists and even, to an extent, the timber industry.

The gist of their complaints is that the national forests are being mismanaged — that overemphasis on the forests' commercial functions has led to excessive logging and long-term impairment of their noncommercial functions.

The Forest Service has acknowledged that many of the charges are true in some measure but it contends that such lapses are the exception rather than the rule.

Although generalizations about 154 separate national forests can be tenuous, evidence supporting the critics' principal allegations can be found from coast to coast.

In Montana's Bitteroot National Forest, there are whole mountainsides so skinned of centuries' growth of ponderosa pine and Douglas fir that they look more like man-made pyramids for a weird science-fiction film.

In West Virginia's Monongahela National Forest, clear-cuts up to a mile wide have defaced the landscape, disrupted drainage and routed wild game

from ancient haunts.

Wyoming's Bridger National Forest, like many others, is scarred by remains of badly built logging roads.

In the national forests of the Northwest are countless instances of erosion and siltation that even Federal water experts have been moved to deplore publicly.

Federal auditors over the last decade have found repeated instances of maladministration of the forests costing taxpayers millions in cash, recreational opportunities, watershed values and wildlife preservation.

Reconciling the national forests' "multiple uses" has always been inherently a problem. But during a half century of the nation's "economy of abundance" the 66-year-old United States Forest Service managed to juggle its conflicting responsibilities in relative obscurity and serenity.

Now, suddenly, the forests have been caught up in the "environmental crunch' '— the post-World War II population explosion, the recreation explosion, soaring demands for all the forests' varied resources and a sudden ascendance of esthetic values.

These converging currents have precipitated many collisions among the diverse functions of the forest, collisions that have been compounded into an unpublicized but palpable administrative crisis. Its major aspects include:

¶The Forest Service's traditional autonomy has run head-on into new laws making all Federal agencies publicly accountable for the environmental impact of everything they do.

¶Forest Service decisions, rulings and actions are being challenged in courts at all levels throughout the country.

¶Public criticism has engendered a cross-fire of conflicting legislative proposals affecting the forests.

¶From Congress and other quarters is coming a swelling stream of demands for a large-scale investigation of the whole national forest administration.

¶The Nixon Administration, in the executive reorganization plan it has placed before Congress, has acknowledged that the management of the national forests is less than optimal and needs revamping.

Imbalance Aggravated

A 10-year national forest development program outlined by President Kennedy to Congress in 1961 sought to right an admitted imbalance between commercial and noncommercial functions. Instead the imbalance has been aggravated.

Year after year, commercial functions of the forests have been receiving close to 100 per cent of projected financing, the noncommercial functions closer to 50 per cent or less.

The national forests are presumed to contain about 35 per cent of the nation's "inventory" of standing marketable timber, and have been providing about 18 per cent of the national timber supply.

Nominally, the Forest Service is cutting only 1.1 per cent of its timber annually, well within the rate calculated to maintain the forests indefinitely. But it admittedly achieves this ratio only by balancing excessive cutting in prime timber and prime locations against low levels elsewhere.

Under prolonged pressure to produce more timber, the Forest Service in 1964 adopted clear-cutting — the unsightly and ecologically disruptive practice of completely stripping forest tracts—as its primary "harvest" method in place of selective cutting.

'Subsidiary of Industry'

Sixty per cent of the national forest timber harvest recently has been performed by clear-cutting—more than double the proportion, according to the timber industry, on private and public lands generally.

"The Forest Service," Representative John Dingell, Democrat of Michigan, remarked recently, "is a wholly owned subsidiary of the timber industry."

A panel of specialists from the University of Montana investigated national forest management last year and reached a similar if less blunt conclusion.

"The de facto basis of Forest Service appropriations," the panel reported, "is the dollar income from national forests. . . . The direction of the Forest Service thus becomes primarily one of supplying building materials. . . ."

Another witness on the point is Dr. Edward C. Crafts, formerly second in command of the Forest Service.

"The Forest Service," he told a Congressional hearing in 1969, "is close to the brink with respect to timber management. Right now it is cutting about twice as much softwood sawtimber [the principal type of construction lumber] as it is growing. This situation cannot last."

Such criticism circulates also within the Forest Service. In a confidential intraservice memorandum, a high-ranking official of the agency last year lamented: "Our attention has largely been centered on industry problems related to contracts and valuation... Our measuring stick for the quality of timber sales is whether or not a forest has met its sell-and-cut goals."

Mining and Grazing

Other complaints, of long standing, have concerned mining and grazing.

Under public-land mining laws going back for a century, 140 million of the national forests' 187 million acres are open to anyone who wants to stake out a mining claim for specified minerals. Eventually, by complying with some simple regulations, a claimant can get a permanent title to the land involved.

The Forest Service shares jurisdiction with the Department of the Interior over such operations, and there is a lot of buck-passing. But conservation lawyers have been accusing the Forest Service of not exercising regulatory powers it plainly has.

The Monongahela National Forest is riddled with coal mining operations. A choice expanse of the Challis National Forest in Idaho currently is threatened by a molybdenum mining project. A phosphate mining project is threatening a portion of the Los Padres National Forest that is one of the last refuges of the rare California condor. Many other national forests are subject to similar disruptions and actual alienation of their land.

Over half of the 187 million acres of national forest land nominally is open to grazing, but only half of that—50 million acres—is considered suitable for use. Of this, 18 million acres is officially classified as in "poor" condition—the result, experienced Forest Service officials have said, of systematic over-grazing.

Another alleged victim of the Forest Service's solicitude for commercial activities has been the movement to preserve a small portion of the nation's wilderness in its pristine status in pursuance of a law enacted by Congress in 1964.

On many tracts proposed by conservationists for preservation, the Forest Service has moved conversely for the early introduction of logging and road building. The agency's seeming partiality to preserving timber-barren uplands has led to jokes about a policy of "wilderness-on-the-rocks."

Complaints of Industry

Amid all the allegations of excessive logging in the national forests, the timber industry itself—base of the $36-billion-a-year business that provides the nation with most of its wood and paper products—has also been bitterly unhappy.

Dependent in part on the national forests for its raw material, the industry contends the Federal preserves are not producing enough. It blames this on "inefficient" operation and on efforts of "preservationists" to restrict logging.

In 1969 the industry was instrumental in formulating the national timber supply bill. It called for national forest timber revenues, which now go into the United States Treasury, to be systematically plowed back into augmentation of national forest wood production.

A major argument advanced was that without more timber from the national forests, lumber prices would soar and crimp the Administration's housing program. The bill was energetically supported by both the White House and the Forest Service.

Conservationists called the measure a license to "raid" the national forests, and in February, 1970 persuaded the House of Representatives not to consider the bill.

Nevertheless, President Nixon thereupon passed on to the Forest Service a White House task force recommendation that the national forest timber harvest be increased 50 per cent by 1978—primarily to help meet projected needs for housing lumber.

Some experts question these purported needs and the national forests' responsibility for meeting them. In any case, conservationists say, they could largely be met simply by stopping timber exports, which now are running about 5 billion board-feet a year. But stopping the lucrative exports is a move that the timber industry, the White House and the State Department alike oppose.

Conservationists' resentment intensified when the Congressionally appointed Public Land Law Review Commission, after five years' study, recommended last year that up to half of the national forests' timber land be dedicated to the "dominant use" of timber production, with other uses subordinated.

Self-Criticism by Officials

For two years the Forest Service has been engaged in an extended exercise in self-reproach.

"Our programs are seriously out of balance," the agency's graying, 62-year-old chief, Edward P. Cliff, has acknowledged repeatedly, both in anguished Congressional testimony and in intra service communications exhorting his 20,000 personnel to change the service's direction and emphasis.

The essence of his defense is that the national forests like other governmental and nongovernmental entities, were caught short by the recent radical shift of public opinion about environmental values; that defects in the management of the national forests are the exception rather than the rule, and that there is nothing that cannot be rectified eventually either by the stream of administrative orders emanating from Forest Service headquarters in Washington or by bigger and better-balanced Congressional appropriations.

These explanations and reassurances Forest Service critics

reject out of hand. They contend the agency's troubles stem from a long-standing bureaucratic detachment from public sentiment that needs to be rectified by law.

Unquestionably, there are recurrent instances of at least a pronounced gap in rapport between the Forest Service and the public.

This spring some irate residents of the Shasta-Trinity National Forest area in California shot a Forest Service ranger who was simply carrying out his duties inspecting a mine.

Residents of Balsam Grove, N.C., were outraged when a ranger, in a land title dispute, bulldozed a large crater and buried a trailer that was a mountaineer's home.

Before the United States Supreme Court is a conservationists' suit to block a projected ski development in the Mineral King Valley in California's Sequoia National Forest.

Among other things, the plaintiffs complained that the project was initiated and inflated from a $3-million plan over an eight-year period without a single public hearing on its acceptability.

The first confrontation on national forest management came this spring at hearings on clear-cutting conducted by the Senate Interior Subcommittee on Public Lands.

Over a period of weeks, more than 100 witnesses from conservation and citizen groups, the Forest Service and the timber industry, along with forestry scientists, debated clear-cutting, without arriving at any consensus on whether it was good or bad.

However, bills were introduced in Congress calling for a moratorium on clear-cutting as well as for other rigorous regulation of both public and private forest lands.

And the indications are that this was only the opening skirmish of a long battle over the national forests.

November 14, 1971

National Forests: Timber Men vs. Conservationists

By GLADWIN HILL

The custodians of the national forests are coming under increasingly intense cross-pressures from rival blocs in the struggle over the future of the vast woodland preserves.

The timber industry, fortified by new support from the Nixon Administration, is demanding that the 187 million acres of national forests contribute more to the country's wood supply.

Conservationists, riding the crest of the "environmental revolution," are charging that the forests are already being overlogged and that recreational and wilderness needs are being shortchanged as well.

The debate has become so heated and the forces so formidable that the question of wilderness uses alone — how much should be maintained as playgrounds for city people, how much fenced off as "living museums" for hikers and campers—has become an exasperating dilemma for Federal administrators.

"The greatest conflict we face," says Edward Cliff, chief of the Forest Service, "is pressure for the preservation of wilderness as opposed to using it intensively for recreation, logging and purposes like that. It's an irreconcilable conflict."

Built-in Contradiction

Laws going back more than half a century make the national forests subject to overlapping commercial and noncommercial uses—logging, mining and grazing, on one hand, recreation, watershed and wildlife preservation, on the other —and striking a balance has always been a problem.

But now, with demand soaring for timber, recreational areas and land of all kinds for all uses, the management of the national forests is being challenged more fiercely than ever before. Even the Forest Service concedes its programs are misaligned.

Underlying the controversy, it was found in a six-month investigation, is a welter of questionable statistics.

Unlike any other renewable resource, the nation's wood supply comes from three diverse sources: forests owned by timber companies (25 per cent); other timber lands in the hands of 4 million private owners (50 per cent); and public lands (25 per cent).

The national forests are officially estimated to contain about 35 per cent of the country's standing commercial timber "inventory" and to be supplying about 18 per cent of the annual harvest.

'Timber Deficit' Foreseen

Currently the annual growth of all the country's commercial timber (officially defined as trees at least 5 inches in diameter 4½ feet above the ground) is about 17 billion cubic feet, and the annual harvest about 14 billion cubic feet.

This nominally leaves a surplus. But most species of trees take upward of 50 years to reach maturity. And some economists foresee, with increasing demands for wood, a possible national timber "deficit" less than 50 years hence—the start, unless policies are changed, of liquidation of the country's woodlands. Some critics contend this liquidation has started in the national forests already.

The Forest Service's annual "allowable cut" figure supposedly is based on a "sustained yield" rate that will maintain the national forests in perpetuity.

But the figure has risen steadily from 5.6 billion board feet in 1950 to 13.74 billion in 1971, despite a supposedly stable stock of standing timber.

The Forest Service has an explanation for this. Mr. Cliff, the agency's chief, told a Congressional hearing in 1969: "These increases reflect steady accomplishment over the years in construction of access roads, improved inventory data, changes in utilization standards, effective protection from fires, insects and disease and reforestation and timber stand improvement."

Reputable experts challenge this explanation. Gordon Robinson, forestry director of the Sierra Club, a leading conservation organization, told a recent Congressional hearing that the steady increases in the allowable cut "have not been earned through improved forest practices or enhanced growth, but through a long, dismal series of rationalizations invented to justify appeasement of the timber industry and obtain bigger appropriation from Congress."

'Accomplished by Computers'

He was supported by Charles H. Stoddard, former head of the Department of the Interior's Bureau of Land Management, the next largest Federal timber-producing agency.

"How the allowable cut was doubled with no increase in available standing timber or

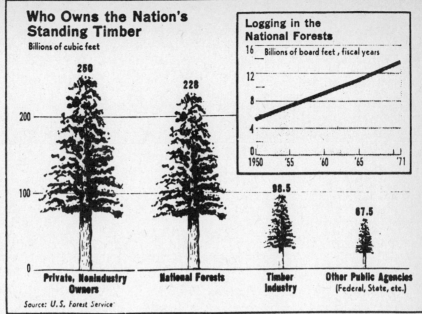

Who Owns the Nation's Standing Timber
Billions of cubic feet

250 — Private, Nonindustry Owners
228 — National Forests
98.5 — Timber Industry
67.5 — Other Public Agencies (Federal, State, etc.)

Source: U.S. Forest Service

Logging in the National Forests
Billions of board feet, fiscal years
16
12
8
4
0
1950 '55 '60 '65 '71

The New York Times/Nov. 15, 1971

growth rate is no mystery," he testified. "It is accomplished by computers: by reducing the rotation age of the next crop, by adding 'protection' forest land on steep slopes or poor soils which earlier surveys showed should not be logged."

Maintenance of "sustained yield" obviously means that logging in one place has to be balanced with growth in another place. The Forest Service used to do this balancing in "working circles" of a few thousand acres. This meant that logging had to be scattered.

This system has been quietly abolished. Each national forest is now treated as a "working circle," which means that contiguous miles can be denuded, against a "balance" of growth far away. But heavy concentrations of logging manifestly conflict with "multiple use" of forest tracts for recreation and wildlife and watershed protection.

To nominally conform with the Multiple Use-Sustained Yield Act of 1960, the Forest Service has had to stretch the "working circle" concept in one connection to include the whole national forest system.

In 1968, the last year for which comprehensive figures are available, 2.3 billion cubic feet of softwood sawtimber (construction lumber) was cut, chiefly in the West, against growth of only 1.9 billion cubic feet nationally. The books were balanced in terms of sustained yield only by counting in an excess of growth over cut in the less marketable hardwood forests of the southeast.

'Tree Farms' Deplored

The Forest Service's justification of such "deficit" cutting is that it has a large stock of old softwood (pine and fir) trees in the West that should be used

before they die and rot, to clear space for a new generation.

The timber industry contends the Forest Service is not doing enough of this, and that greatly increased yield could be achieved through "scientific forestry"—genetics, intensive fertilization, careful thinning, and elimination of low-value tree species.

But conservationists say the public is entitled to more than "manicured tree farms" in its national forests.

Thus, even when all the conventional factors in a national forest timber sale are worked out, there now is posed the question of a logging project's prospective impact on the nation's new scale of environmental values.

This tends further to impugn statistical or other generalizations about the availability of timber from national forests, and to leave unanswerable, pending more comprehensive studies than have ever been made, the question of how much these forests should contribute to the future timber supply.

Meanwhile, due presumably to chronic political and economic pressures on the White House, the Department of Agriculture and Congress, Federal appropriations continue to perpetuate what the Forest Service itself has called "a serious imbalance" of long standing in management of the public woodlands.

Nearly one-fifth of the newly voted $500-million Forest Service budget for the fiscal year 1971-72—over $90-million—was allocated for timber management, meaning primarily commercial timber production, which yields over $300-million a year in Federal revenue.

$40-Million For Recreation

For recreational facilities and services—for a clientele that is

double that of the national parks—the allocation was only $40,291,000.

This was in the face of at least a decade's backlog of unbuilt facilities. At last report, only 4,480 or 32,000 other recreational sites, projected in 1961, had materialized.

For wildlife management—the basis of both species protection and $500-million a year in private expenditures for hunting and fishing—the appropriation was $6,333,000—less than 2 per cent of the total budget.

For management of watersheds on which 1,800 cities and towns rely as water sources—and which the Forest Service says need much rehabilitation—the appropriation was $13,387,000: about 7 cents an acre.

An important element in the tug-of-war between commercial and noncommercial uses of the national forests is the battle over the preservation of pristine wilderness.

Of the nation's total area of 2.27 billion acres, the only Federal land permanently dedicated to public use is the 28 million acres in the national parks. Remaining Federal lands, including the 187 million acres in the national forests, is subject to commercial incursions and even alienation from public ownership.

Commercial Bias Seen

Congress in the 1964 Wilderness Act, projected the permanent preservation of up to 66 million acres of national park, national forest and Federal fish and wildlife tracts in their primitive states, roadless and devoid even of such amenities for campers as piped water.

The 10-year period for formally designating these lands is now two-thirds gone, and only about 10 million acres have

A Growing Sense of Urgency

been put into the National Wilderness Preservation System.

Conservationists attribute the slow pace of the program in part to commercial bias on the part of the Forest Service.

Although the agency initiated the wilderness preservation idea 50 years ago, it has developed mixed feelings on the subject. The service's staff handbook says:

"A majority of the people who go into the forest for recreation do not have the ability or desire to get away from the easy travel made possible by roads . . . Many feel that the wilderness classification is discriminatory because it permanently excludes them from areas which might otherwise be developed for their enjoyment."

Accordingly, conservationists say, more often than not when a pristine area is proposed for preservation, the Forest Service hastens to disqualify it by initiating logging and road-building programs on it. They have obtained Federal court orders to block some such development.

Of 13 "primitive" tracts in California, totaling over 1 million acres, suggested by the Sierra Club in 1968 for possible wilderness classification, the Forest Service's counter-recommendation in each case was that the area be wholly or partly "developed" with logging or roads.

The timber industry, on its part fighting what it calls "the withdrawal war," complains that national forest land open to logging has been progressively reduced to 73 million acres, or only 38 per cent of the total national forest area, and that the "preservationists" are going too far.

'Multiple Benefits'

"Dynamic forests where men and nature unite their efforts to intensify growth afford multiple benefits of wood products, pure air, clean water, fish and wildlife, recreation and forage, in abundance," says F. Lowry Wyatt, a recent president of the National Forest Products Association. "Forests consigned to the whimsical attentions of nature alone fall into neglect."

In rebuttal, the Sierra Club's executive director, Michael Mc-Closkey, says: "I think perhaps only 40 or 50 million acres of the national forests should be subject to full commercial logging.

"Possibly 30 million acres should be subject to limited cutting, 6 to 8 million acres preserved as wilderness, and 10 million acres protected for future study of their highest values."

That 6 to 8 million acres, he notes, would be less than one-half of 1 per cent of the nation's land.

Criticism descending on the Forest Service has caused a pro-

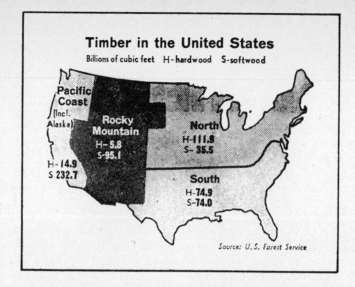

Timber in the United States
Billions of cubic feet H-hardwood S-softwood

Pacific Coast (Incl. Alaska)
H-14.9
S 232.7

Rocky Mountain
H- 5.8
S-95.1

North
H-111.9
S- 35.5

South
H-74.9
S-74.0

Source: U.S. Forest Service

longed flurry of corrective efforts.

In testimony to the Senate Subcommittee on Public Lands last spring, Mr. Cliff enumerated 30 "problem situations," running the gamut of Forest Service practices and policies, that were tacit admissions of mistakes.

A big shakeup at Forest Service headquarters in Washington in recent months brought in new men as Mr. Cliff's chief assistant and as heads of the service's principal divisions, including public relations.

The controversial practice of clear-cutting (stripping forest tracts bare), adopted as the principal logging method in 1964, has nominally been downgraded to an alternative to selective cutting only where circumstances particularly call for it—although that leaves the Forest Service wide leeway.

Along with numerous admonitions to establish better rapport with the critical public, service employs this year were given a 20-page "Guide to Public Involvement" detailing even sophisticated opinion-analysis techniques for assessing communications.

"We're snowed under with directives telling us to look out for environmental values," a Forest Service ranger in California remarked.

All this activity is intended to add up to a major reorientation of the Forest Service to meet the demands of the "Environmental Seventies."

However, there are still skeptics who ask whether the agency, one of the classic Federal bureaucracies, has really changed its outlook or is only paying lip-service to pressures of the moment; and whether the ills that have beset the national forests can be remedied by a flurry of administrative orders or call for new legislation.

"You can go into a tract of national forest and find that sheep have grazed it just flat," says Representative John Ding-

ell, Michigan Democrat, author of a major pending forestry bill. "Then you come back to Washington and the Forest Service says: 'It's being perfectly managed.' How can you be sure what's going on when you run into things like that?"

"The philosophies and policies of the Forest Service have not changed essentially in 20 years," says Brock Evans, northwestern representative of the Sierra Club. "In spite of all the environmental words, the emphasis is still primarily on timber cutting first."

An impending governmental showdown on national forest management involves action on three fronts: investigation, legislation and reorganization.

No one at present, perhaps, has an accurate, comprehensive picture of conditions in all the forests. Of the 154 national forests, only two—the Monongahela in West Virginia and the Bitterroot in Montana — have undergone intensive outside investigation in recent years.

Both inquiries—one by the West Virginia Legislature, the other by a panel of specialists from the University of Montana —found excessive logging, poor reforestation, bad road construction, erosion and other watershed damage, and impairment of wildlife preserves and recreational resources. The Forest Service tacitly acknowledged substance in these findings.

Similar conditions in a score of other national forests were attested and documented with photographs at Senate hearings on clear-cutting last spring.

The Senate Interior Subcommittee on Public Lands is continuing its inquiry. Meanwhile, Senator Gale McGee, Wyoming Democrat, supported by some leading conservation organizations and other members of Congress, is sponsoring a bill calling for a thorough investigation of the Forest Service by a "blue ribbon" commission of outside experts.

Among the numerous measures before Congress affecting the national forests are two bills that typify opposing viewpoints on how the forests should be managed.

One bill, sponsored by Senator Mark Hatfield, Oregon Republican, whose constituency includes big timber interests, is essentially a modification of the National Timber Supply Act that conservationists blocked in 1970 as a putative "raid" on the forests.

It would divert more than $300-million a year in timber revenues, which now go into the United States Treasury, into a fund to increase the forests' timber production and "enhance the forest environment." It also provides for Federal grants to states to promote private timber production.

Countermeasure Drawn Up

A countermeasure drafted by Representative Dingell in collaboration with conservation organizations requires the states to develop federally approved management plans for privately owned timber lands; provides for Government regulation of logging to protect environmental values; creates a timber-revenue fund for development of both commercial and noncommercial aspects of national forests, and bans timber exports unless national needs are assured five years ahead.

The Forest Service has grown, in 66 years, from a small Government bureau into a counterpart of today's big industrial conglomerates.

It has billions in assets, 20,-000 employes and a complex of differing activities at many locations, with 200 million "customers."

Some Federal management consultants feel there has not been commensurate growth in the agency's administrative machinery. Despite its contemporary complexity, for instance, the service traditionally is headed by a Civil Service employe whose principal credentials are in tree-culture and who —unlike heads of Government agencies of comparable importance—is virtually immune to changes in Presidents and Secretaries of Agriculture.

Various proposals have been advanced to improve national forest administration. Foremost among them is a recommendation by the President's Advisory Council on Executive Reorganization that was forwarded to Congress this year.

The council urged that the Forest Service—in recognition of its responsibility as a land management rather than crop-growing agency—be removed from the Department of Agriculture and put into a new Department of Natural Resources, one of four basic Cabinet divisions envisioned in the Executive Reorganization Plan now before Congress.

November 15, 1971

Study Finds Editorials Stress the Environment

A survey of 20,904 editorial in five leading American newspapers over a 12-month period shows that the environment was the major domestic issue of editorial concern.

The Public Issues Research Bureau, which analyzes editorial trends for government and industry clients, said that economic, urban, rights and transportation issues all received less sustained attention than the environment from October, 1970, to September, 1971, United Press International reported yesterday.

"Forty-four per cent of the editorials concerned social issues," the research bureau president, Burton Holmes, said. "Of these, 1,382, or 15.2 per cent, were on environmental subjects, with water quality receiving the most attention, followed by land-use policies, air quality and waste disposal."

January 16, 1972

An Earth-Exploring Satellite Is Orbited

By BOYCE RENSBERGER

A new era of earth exploration began yesterday with the successful lofting of an unmanned earth-orbiting satellite that will continuously scan the surface of the earth, radioing back many kinds of information on global environment and natural resources.

For many scientists in fields such as agriculture, ecology, forestry, oceanography and geology, the new Earth Resources Technology Satellite, known as ERTS-1, represents the long-awaited beginning of the application of sophisticated space technology to answering a wide range of questions about the earth.

The 1,965-pound satellite rode into space at 2:06 P.M., Eastern daylight time, aboard a two-stage Delta rocket from the National Aeronautics and Space Administration's facilities at Vandenberg Air Force Base at Lompoc, Calif.

The launching, which had been delayed two days because of difficulties with the rocket, placed ERTS-1 into an elliptical, nearly polar orbit as planned. Then, about an hour after the launching, the rocket's second-stage engine was restarted to change the orbit, pushing the satellite into a nearly circular path between 561 and 570 miles above the earth.

Although flight controllers were hoping for a perfectly circular orbit, they said that the path achieved was acceptable.

After the maneuvers, the nose cone of the 107-foot-tall rocket shed its covering and ejected the 10-foot-tall spacecraft. Then two huge paddles unfolded from the spacecraft like the wings of a butterfly to display solar cells that convert sunlight to electricity. With wings extended, the spacecraft was 11 feet wide.

Space agency officials said that the 103-minute orbit is designed so that, as the earth turns under it, the satellite's sensing devices will be able to scan every part of the planet's surface except small regions around the poles once every 18 days.

Thus, scientists will receive a complete picture of many environmental conditions with an update every 18 days. This will allow study of various changing processes such as the growth of crops, the advance of glaciers or the spread of water pollution over a period of time.

The first pictures and data from the spacecraft are expected sometime this morning.

ERTS-1 is designed to operate for one year, at a minimum. Space agency officials said, however, that it could last one to three years longer. It is to be replaced by a second ERTS satellite, containing more sophisticated equipment.

Although a detailed program has not been worked out, it is expected that the techniques tried in the experimental ERTS program will eventually lead to a generation of permanently orbiting satellites for environmental monitoring.

Additional earth-scanning experiments will be flown aboard Skylab, the manned mission in which astronauts will live aboard a roomy orbiting laboratory for up to 56 days at a time.

ERTS-1, now in orbit, will radio data gathered by a variety of sensors looking down on the earth from space and also relay signals picked up from about 150 automatic, ground-based scientific stations scattered around remote areas of North America.

Relaying the Data

The ground stations will monitor a number of things such as rainfall and snow levels, flow rates in rivers, soil moisture, temperature, wind velocity and levels of air pollutants. The ground stations, which are designed to operate for six months without attention, will transmit their measurements to ERTS-1 at least once every 12 hours.

The satellite, in turn, will relay the data to scientists on the ground.

On a global basis, the most important information from the ERTS mission is expected to come from the three television cameras and a multispectral scanner aboard the spacecraft.

The television cameras will simultaneously view a single 115-mile-by-115-mile square of the earth's surface, each camera sensitive to a select color. When the pictures are combined on earth, they are expected to yield a full-color view of a sharpness and color fidelity unattainable with conventional color cameras. The cameras will take a new still picture every 25 seconds.

In this way, the satellite can cover the United States with about 500 such pictures.

The multispectral scanner will view the same area covered by the television cameras but it will scan for reflected electromagnetic waves in four different bands—visible red, visible green and two near infrared bands.

Using this information, scientists said, it should be possible to identify surface features invisible to the naked eye. This is because objects reflect not just the visible light portion of sun but a wide range of invisible light as well.

This fact, for example, can make it possible to identify a certain type of water pollution by satellite. Visible light reflecting from the water may show nothing. But the polluted areas may reflect invisible wavelengths that only the multispectral scanner can detect.

While the satellite is in daylight, it will scan the earth's surface, recording information on tape. Then, when it passes into the night side of the earth, the tape recorder will play back the data, transmitting them to receivers on the ground.

Data from the receivers will be sent to the Goddard Space Flight Center in Greenbelt, Md. From there the data, in the form of still pictures and digitalized information, will be sent to the Interior Department's new Earth Resources Observation Systems Data Center at Sioux Falls, S.D.

There, all the information will be available to anyone at a nominal cost. A large number of research institutions and commercial concerns are expected to subscribe to regular mailings of the information.

Space agency officials explained that, unlike the practice in many earlier programs, scientific information will not be restricted to the use of select research groups. Instead, all the information will be available to anyone.

A Growing Sense of Urgency

So far, some 300 scientists from the United States and 31 foreign countries have signed up to use ERTS data for specific research projects.

Many of the initial projects will be concerned with learning how to identify various earth features by the characteristic combination of wavelengths they reflect. This combination, called a spectral signature, can then be used to determine the extent of the feature on the surface and to follow changes over a period of time.

Fields of Study

The information is eventually expected to yield a better understanding in many areas, including the following examples:

¶Agriculture. It should be possible to determine the type and size of various crops under cultivation and even to detect widespread plant diseases. Better mapping of climates and soil types could also help identify new areas appropriate for cultivation.

¶Forestry. ERTS-1 is expected to provide the first global inventory of timber, much of which grows in regions of the world difficult to reach regularly by conventional means.

¶Oceanography. Researchers at the University of Delaware, for example, are planning to use ERTS data to follow the changing distribution of oceanic algae, one of the chief sources of atmosphere oxygen.

¶Geology. "We are not going to detect minerals from space," said Dr. Paul Lowman of the Goddard Space Flight Center. "What you will find from ERTS imagery are areas which look promising. Then you can go in there and further investigate these areas from the ground."

¶Geography. Researchers plan, among other things, to monitor the changing boundaries of the urban areas around Boston and New Haven, hoping to use visible trends to guide policy-making on the ground.

¶Ecology. Virtually all ERTS data will be useful to ecologists but special studies of water pollution in the Great Lakes, New York Harbor, Southern California and Florida will be done. Scientists at Ohio State University also plan to monitor strip mining in their state.

¶Meteorology. One experiment scheduled for this winter is to study the freezing and thawing patterns of lakes in central North America for clues to the deep-water current patterns that can affect climate around the lakes.

Much of the same kind of research has been attempted from airplanes carrying similar scanning equipment. Short-term monitoring with airplanes is cheaper than by satellite, many scientists have found, but studies that require long-term repeated scannings are prohibitively expensive by this method. An orbiting satellite works out to be much cheaper in the long run, they say.

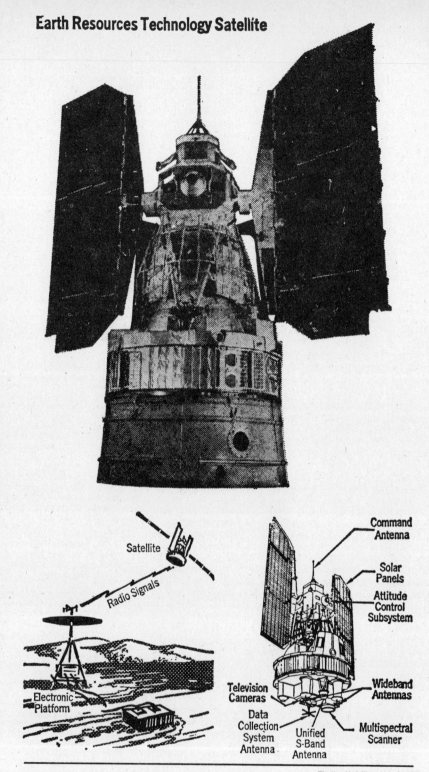

Earth Resources Technology Satellite

The New York Times/July 24, 1972

Earth Resources Technology Satellite, its solar panels spread like butterfly wings to draw energy from the sun, is to scan the earth's surface for new information on the global environment and natural resources. About 150 automatic sensing stations like the one pictured at lower left are to gather and send to the satellite data about such factors as stream flow and soil moisture. Satellite then relays data to earth. Multispectral scanner and television cameras, placed as shown in drawing at lower right, pick up infrared and other electromagnetic waves reflected by earth features. This information is beamed to the ground for conversion into detailed pictures.

74

July 24, 1972

A National Water Plan

A remarkable degree of boldness and imagination, as well as labor, has gone into the 1,122-page report of the National Water Commission. Few such agencies would have had the daring to take on, in effect, the Bureau of Reclamation, the Army Corps of Engineers and all those powerful private interests that benefit—out of all proportion to what they pay—from federally financed waterways, irrigation systems and flood control projects.

In various ways many of these developments have proved detrimental to the environment. To subsidize water transportation and aid some farmers, the Government has engaged in stream channelization and the diversion of rivers, at times to the ecological disruption of entire riverside areas and often at the cost of a stream's beauty and vitality. Irrigation has added to excess farmlands. And dams that are products of pork-barrel politics, though ostensibly for flood control, have crippled great rivers and the life they contain. Yet after eight billion dollars in such Federal expenditures—not to mention the tragic loss of life and property—flood damage continues to grow.

The chief conclusion of the commission, created by statute and headed by Charles F. Luce, is that these harmful and costly investments of public funds are stimulated by those who stand to gain at the general expense. These include barge companies that enjoy a right of way at no cost to themselves, farm corporations that obtain irrigation at far below cost and landowners whose property values skyrocket when a new dam upriver promises to protect inhabitants who ought not be occupying such lands to begin with.

If the report confined itself to this difficult question of flood plains alone, its contribution would be notable. It is something of a precedent for a national commission to conclude that the Government's cost-sharing policies have encouraged improper building on these vulnerable lands, encouraging "unwise economic developments in areas prone to periodic flooding and hurricane hazards" in addition to providing windfall gains to landowners while destroying valuable open space and wetlands.

The overwhelming emphasis of the commission's findings is that if the users and beneficiaries of project services were made to pay an appropriate share of the cost, there would be far less demand—in Congress and out—for needless ones. The unused backlog of authorized projects could be wiped out entirely.

It is doubtful how far this let-the-user-pay theory should be applied to the beneficiaries of water-treatment plants, and the commission itself seems willing to allow some leeway here to make up for the years of polluting that must be overcome. But this and many other aspects of the report will no doubt be argued at length in regional and national conferences to be held in January and February.

Meanwhile it is clear that the country owes itself a serious consideration of this ground-breaking report and a debt of gratitude to its authors.   December 24, 1972

Severe Shortage of Ores Feared if Waste Continues

Report by Geological Survey Warns Against Waste and Urges New Techniques

By HAROLD M. SCHMECK Jr.
Special to The New York Times

WASHINGTON, May 8—The ravenous American appetite for minerals will lead to severe shortages in the next few decades unless the nation stops wasting resources and starts employing better ways of finding and exploiting low-grade ores, according to a report from the United States Geological Survey.

The 722-page document, released today, is the first overall assessment of United States mineral use and resources in 21 years. The report covers everything from abrasives to zirconium, including coal, oil, natural gas and all the industrially important metals. The survey noted that these were the things on which civilization lived, not only in the United States, but throughout the world.

"The real extent of our dependence on mineral resources places in jeopardy not merely affluence, but world civiliza-tion," said the editors of the volume.

In a foreword, Rogers C. B. Morton, Secretary of the Interior, said the question of the magnitude of usable resources was of mounting concern. He noted that minerals and mineral fuels were "literally the cornerstones" of modern life.

A summary statement, released by the department with the report, said the nation's known deposits of mineral raw materials were seriously depleted and that future supplies must come, in many cases, from low-grade ores or from as yet undiscovered resources.

"For many minerals, our future production will depend on the mining of huge volumes of low-grade ores with adverse environmental impact unless we exert great care in their extraction and use," said Dr. V. E. McKelvey, director of the Geological Survey, which is a unit of the Department of Interior.

The report noted that increasing dependence on low-grade ores also meant increasing costs.

"The reader should bear in mind," the report said, "that as the mining industry turns to lower and lower grades of many ores, the cost and avail-ability of the required energy are probably the single most important factors that will ultimately determine whether or not a particular mineral deposit can be worked economically."

The report said that the United States now imported 29 per cent of its oil and gas requirements, about one-third of the iron ore needed and 87 per cent of the aluminum required.

Copper Supply Estimated

Imports account for only a small proportion of the two million tons of copper used every year in the United States, but the report said that the nation's known resources would be used up in 45 years at current rates of consumption. Known world supplies should last about 50 years.

Beyond that, the survey said, consumption, here and abroad, must depend on discovery of new deposits and on development of new methods for extracting very low-grade ore.

The United States produces about 9 per cent of the world's zinc, but uses three times that much. The metal is relatively plentiful, but the report estimates that between 1950 and 1970 the world used half of all the zinc ever produced up to that time.

"The phenomenal increase in the production and consumption of zinc in the 20th century cannot continue indefinitely and eventually production must decline as primary resources approach exhaustion," the report said.

In terms of quantities used, iron, aluminum, copper and zinc are, in that order, the most important industrial metals. They all have many uses without which modern industrial society would be hard pressed to survive.

Undersea Mining Urged

Of manganese, the fifth most widely used metal, the report said that the United States had no known reserves and little prospect of finding major domestic sources that could be used economically. The last manganese mine in the United States closed in 1970.

The metal is indispensable to steel production and has other uses as well.

"The element is essential to the whole industrial capacity of the world," said the report. "When we can do without steel, we can do without manganese."

The survey said world supplies were large in relation to consumption, but were irregularly distributed. In the case of this metal, and several others, the document said there was a need for more exploitation of undersea resources.

Many other minerals, used in relatively small quantities, are nevertheless essential to industrial civilization. In many of these, too, the survey shows the United States to be using much more than it produces. In mercury, for example, domestic production is less than one-third of need and the geologic outlook for discovering rich ores is rated poor.

The nation's use of silver for photography alone amounts to more than total domestic production.

The Geological Survey's document said that much valuable mineral raw material was being wasted. It said billions of cubic feet of the industrially important gas helium were being wasted because it was not being removed from natural gas.

"A major aspect of resources that appears in many of these chapters," said the introduction to the report, "is the extent to which many potential byproducts or coproducts are literally being wasted — lost forever — because there is no apparent economic incentive for recovering them."

The report disavowed both the prophets of doom, who see the end of civilization in sight, and the optimists who believe that greater need will continue to beget greater supplies forever.

Much remains to be learned about how much of the Earth's mineral resources can be made available for human use, the report said. Calculations from proven reserves are misleading on the side of pessimism, it said, because new discoveries are presumably still to be made. On the other hand, the document noted, "elements are available in the earth's crust in very finite amounts."

An example of industrial society's ability to change the natural resources equation was cited in the case of abrasives: About half of the nation's need for industrial diamonds is now met through man-made supplies, something that came into existence only about 15 years ago.

On the general outlook for mineral supplies, the report said that only a few commodities were available to the United States in quantities adequate to last for hundreds of years.

"By no means is it too early to become concerned about future mineral supplies—and to start planning," the report said.

In a chapter on mineral resource estimates and public policy, Dr. McKelvey expressed confidence that "for millenia to come we can continue to develop the mineral supplies needed to maintain a high level of living for those who now enjoy it and raise it for the impoverished people of our own country and the world."

But he conceded that there were others who did not share his views. He advocated a "deep review of resource adequacy," he said, because the world's population is already too large to survive without industrial civilization and the massive use of minerals it requires.

"If resource adequacy cannot be assured into the far-distant future," he said, "a major reorientation of our philosophy, goals and way of life will be necessary."

May 9, 1973

Mountain-Climbers Leave 'Mountains' Of Trash on Peaks

WASHINGTON (UPI) — Thirty years or so ago a party of amateur mountain-climbers arrived gasping at the top of Colorado's 14,256-foot Longs Peak.

They gasped some more when they saw, chained to a boulder on the peak's small summit, a large and heavy-looking steel trash can. Some ranger had lugged it there. Already on that day it was half full of litter.

So impressed were the climbers that some of them managed to stagger over to it and deposit their empty lunch bags. Others, too tired, just stuffed their refuse in adjacent crannies.

So, actually, it was not too surprising to learn from a recent article in Smithsonian Magazine that even the highest peak in North America, Alaska's 20,320-foot Mount McKinley, is soiled by trash.

Lofty snow-capped mountains look so undefiled when seen from below that it is hard to think they could be contaminated by man-made filth. Close inspection proves, however, that they are.

In the Smithsonian article, Doris Ewing, an "Oregon outdoorswoman," tells of a university of Oregon expedition by seven young men who set out to clean up litter left by previous climbers on Mount McKinley's slopes.

They Fall Short

The expedition, "unique in mountain climbing history," was kept by bad weather from getting to the top and was forced back to the 17,200-foot level. There the young men found a fantastic cache of trash

May 16, 1973

Vietnam Defoliation Scars Expected to Last a Century

By JOHN W. FINNEY
Special to The New York Times

WASHINGTON, Feb. 21—The American use of chemical herbicides in the Vietnam war, long a subject of controversy, caused wounds to the ecology of South Vietnam that may take at least a century to heal, the National Academy of Sciences has concluded.

The herbicides, according to an academy study, caused "serious and extensive damage" to the inland tropical forests and destroyed 36 per cent of the mangrove forests along the South Vietnamese coast.

In addition, the report said that there were indications that the herbicides when used for the destruction of crops, caused deaths among children of the montagnard tribes in the hills of western South Vietnam.

Controversy Spurred Study

These conclusions about the harmful, long-term effects of the use of herbicides in the war appeared in a report by National Academy of Sciences to the Defense Department that is to be submitted to Congress in the next few days.

The report is scheduled to be made public next week by the Senate Armed Services Committee, but meanwhile the contents were summarized for The New York Times by members of the academy, a prestigious scientific group that often serves as an arbiter for the Government in matters of scientific controversy.

The study was ordered by Congress in 1970 at a time of controversy over the impact of the extensive use of herbicides in the war. At that time, the military was defending the use of herbicides—the first time that such chemical agents had been used so extensively in warfare—against rising complaints from the scientific community of long-term destruction to the South Vietnamese ecology.

Between 1961 and 1971, the United States dropped more than 100 million pounds of herbicides—or about six pounds for every inhabitant—on South Vietnam. More than 5.7 million acres, or about one-seventh of South Vietnam's land mass, was sprayed with the chemical agents, which were generally far more potent than the herbicides commonly used for agricultural purposes.

The herbicides were largely used to strip away foliage in areas believed to be occupied by North Vietnamese or Vietcong troops, thus exposing them to attack. To a lesser extent, however, the herbicides were also used for destruction of crops.

Psychological Effect Cited

Aside from the ecological impact, the study concluded that the use of the herbicides had had an adverse psychological effect, turning Vietnamese opinion against the United States.

The principal Congressional mandate given to the academy, however, was to study the environmental impact of the herbicides. The study was made by 17 scientists from the United

States, Sweden, Britain and South Vietnam headed by Dr. Anton Lang, a plant pathologist at Michigan State University and director of the Atomic Energy Commission's plant research laboratory at the University.

None of the panel members had previously been particularly identified as conservationists or critics of the use of herbicides in Vietnam. Therefore, the general expectation in the Defense Department, which had been directed by Congress to commission the study, was that the panel would come up with a generally favorable report.

Pentagon Called Surprised

Defense officials were described by members of the scientific community as somewhat surprised by the generally critical approach taken in the report, which was carried out in secrecy and submitted nearly a month ago to the Pentagon.

The critical nature of the report may influence the debate in the Administration over including herbicides within the prohibitions of the 1925 Geneva treaty on chemical and biological warfare.

The Administration has taken the position that the treaty does not preclude the use of herbicides or tear gases. The Senate Foreign Relations Committee has refused to consider

the treaty, which has never been ratified by the United States, until the Administration interprets it to ban the use of tear gases and herbicides as well as poisonous gases.

The report did not deal at any length with whether the herbicides might have increased stillbirths or birth defects among Vietnamese children. Some scientific studies, including one conducted by the United States Army, have indicated that dioxin — a toxic, insoluble impurity found in defoliants — can cause liver damage, genetic changes and cancer.

One of the unexpected findings of the new study, however, was that herbicides apparently caused deaths among montagnard children. The military has consistently maintained that the herbicides had only a transitory effect on plant life and were used in such a way that they would not endanger human life.

This finding was based on interviews by Dr. Gerald C. Hickey, an anthropologist at Cornell University with wide experience in South Vietnam, with montagnard refugees from 12 villages in Pleiku and Kontum provinces.

Out of the interviews emerged this picture: After a plane had passed by, "spraying smoke," practically all the people in a village would show symptoms

such as abdominal pains, intense coughing and rashes like "massive insect bites."

The adults recovered, but after some of the attacks, which were directed at crops, some of the young children were said to have died. In one village, the refugees told Dr. Hickey, 38 children died. In another, the refugees said simply, "lots of children died."

In some cases, according to the refugee reports, the dead were babies carried on the back of their mothers into the fields after the herbicides had been deployed.

From the like symptoms reported by villagers, Dr. Hickey, in one chapter in the report, reached the preliminary, yet to him conclusive, finding that the illness and deaths were caused by the herbicides. One theory within the scientific community specifically implicates dioxin.

Within the panel some differences developed over the extent of the damage to the tropical forests in inland Vietnam that are the principal source of hard woods for the Vietnamese.

The majority concluded that the damage was "serious and extensive," but less than had previously been estimated in studies by the United States Agency for International Development. This conclusion provoked dissenting views from two panel members — Pham

Hoang Ho, a professor of botany at the University of Saigon, and Paul W. Richards, a biologist at the University College of North Wales, who is regarded as a pre-eminent authority on tropical rain forests.

The differences arose largely over the methods used to assess the damage. Because the inland forests are still largely under Vietcong control, it was necessary to rely on aerial photographs. One of the criticisms held that the photographs were inadequate for the measurement of the damage done beneath the dense canopy of a tropical forest. It was also said that the photographs had been improperly analyzed by graduate students at the University of Washington.

The report found that the damage in the tropical forests would persist over a "very long period." The normal period for a hard wood tree to mature in a tropical forest is 70 to 100 years.

When it came to the mangrove forests along the southeast coast of South Vietnam, the report was more precise in assessing the damage. It estimated that 36 per cent of the mangrove swamp land—one of the breeding grounds of fish for the Vietnamese diet—had been destroyed and that it would take well over 100 years to recover.

February 22, 1974

Many States Pass Environment Bills to Improve Quality of Life

By GLADWIN HILL

The nation's legislatures have resounded this year with debate on environmental problems that have led to many new laws.

A national sampling by The New York Times shows that virtually all the states, except for the few that had no legislative sessions, enacted measures aimed at bolstering the nation's physical "quality of life."

Although Congress rejected a program to give states grants for the planning of land-use as unwarranted interference with local prerogatives, land-use problems led the list of environmental interests in legislatures across the country. More than half the states debated scores of regulatory proposals.

Other subjects of legislation were air and water pollution, solid waste, noise, solar energy and smoking.

Proposals Weakened

In general, bills to protect environmentally choice land areas fared well, while proposals for comprehensive state regulation of land use encountered much resistance.

In Colorado, bills authorizing strong state regulation were watered down to affirm the authority of local governments to make decisions even about areas and activities of statewide interest.

In Utah, similarly, proposals to allow the state to impose land plans on local governments were weakened markedly after opponents picketed the State Capitol, denouncing proponents as "dictators who would tell us what color to paint our houses."

North Carolina enacted the most sweeping land-use legislation. The bills put coastal development under a permit system; required the 20 coastal counties to adopt zoning plans; provided for the classification of all the state's land and for state acquisition of environmentally choice areas, and appropriated $9.6-million for parkland acquisition and improvement.

A bill to regulate resort development in the mountains was blocked.

The most popular type of land-use legislation provided for state acquisition of park

lands. California legislators voted $63-million for this purpose, along with $3-million to develop trails and hostels in state parks and $1.5-million to acquire choice tracts around Lake Tahoe, where the pace of development has become highly controversial.

An intensive campaign by industry to weaken Delaware's three-year-old ban on heavy industry in a two-mile-deep strip along its hundred miles of coast did not make much headway. Two amending bills that were the subject of much legislative debate and statewide public hearings were blocked, in favor of a study to be conducted by the Delaware Society of Professional Engineers on what the exact effects of amendments might be.

Vermont's land-use program was stalled as it reached its third and final step. The state's 1970 "Act 250" provided for three steps — assessment of ecological constraints, formulation of a "capability and development plan" and application of that plan to the state's terrain.

The second phase was enacted last year over heavy op-

position from real estate interests. This year, they bottled up further action in committee, contending that the plan amounted to "statewide zoning" and would concentrate growth in "urban ghettos."

In Hawaii, which a decade ago was the first state to zone all its land, environmentalists were disappointed over the fate of a supplementary measure. A bill calling for environmental impact assessments of all shoreline development projects for 100 yards inland was weakened to apply only to projects in the belt 100 yards seaward from the shoreline.

New Laws

Laws governing surface mining were passed by New York, Indiana, Kansas, Kenucky, Montana, Ohio, South Carolina, Tennessee and Wisconsin.

The energy crisis engendered special legislation in only half a dozen states.

A California law establishing a five-member State Energy Conservation and Development Commission provided for a consumer tax of two-tenths of a mill per kilowatt hour on electricity—estimated to cost an' average householder 50 cents a year—to raise $16-million a year for research on energy conservation and development.

Connecticut established a state energy agency. Maryland gave the Governor special emergency powers. Ohio established an energy board with emergency powers. West Virginia and Minnesota also established energy commissions.

Several states passed bills that environmentalists considered retrogressive. Foremost among these was New Mexico's repeal of its Environmental Quality Act, which paralleled the Federal Environmental Policy Act in requiring advance

environmental impact assessments on major projects.

Arizona gave metal smelters a year's extension on compliance with regulations on air pollution. Illinois enacted a law providing for economic impact assessments along with environmental impact assessments, and also a law allowing industry to withhold technical data classifiable as "trade secrets."

Iowa granted a 10-year tax exemption on newly installed pollution control equipment.

Connecticut repealed its ban on phosphate detergents, but efforts to repeal Indiana's phosphate ban failed.

Of 60 bills considered by the Florida House of Representa-

tives' Environmental Protection Committee, 13—an exceptional proportion—were enacted into law. They applied to oil spill, water pollution control and solid waste measures.

Bills to facilitate solid waste disposal were enacted also in Connecticut, Georgia, Louisiana, Tennessee, West Virginia and Wisconsin, but were rejected in Arizona and Vermont.

A law like Oregon's and Vermont's to discourage nonreturnable beverage containers through mandatory deposits or other requirements was enacted by South Dakota, effective in 1976. But similar pro-

posals were blocked in 15 other states.

New York stiffened the maximum penalty for violation of its water pollution law. In civil cases, the fine was raised from $250 to $10,000 and in criminal cases, from $500 to $25,000, plus a year's imprisonment.

Other states that enacted water pollution abatement measures included New Jersey, Georgia, Florida, Virginia, West Virginia and Wisconsin.

Both Indiana and Arizona passed laws giving special tax exemptions on equipment installed to use solar energy for heating or cooling.

August 11, 1974

Chemists in Timber Revolution On Verge of Test-Tube Trees

By ANDREW H. MALCOLM
Special to The New York Times

FORT BRAGG, Calif.—Quietly like a tree growing, the nation's timberlands are today undergoing a major technological and scientific revolution that is radically changing the face of the country's vast woodlands.

Chemists in clean white coats are on the verge of creating test-tube trees, living growths with roots and needles created not naturally from seeds but chemically from the cells of another tree.

Other scientists in hard hats scour the woods for seeds and branch cuttings to crossbreed into "super trees," magnificent, fast-growing specimens bred especially for commercial attributes like the hybrid corns that revolutionized farming 20 years ago.

Sweating sawmill operators study computerized television monitors that automatically diagram the best cut for radically thinner sawblades to maximize the use of each log.

Elsewhere, helmeted helicopter pilots scatter tons of tree seeds artificially colored to fool hungry birds. Paul Bunyanesque machines roam the forests, snipping entire trees off at ground level and passing them through pulpers that consume 40-foot logs in nine seconds. Entire forests are fertilized and thinned like common carrot patches to produce harvestable growth decades sooner than nature can.

Significance for Consumers

These developments observed during several recent trips to timber camps, sawmills, plywood factories and forests in these heavily wooded areas of the Northwest, carry tremendous future significance for the

nation's consumers, their woodlands, their housing, the economy, balance of payments and land use at a time of growing concern over the world's ability to meet future food demands.

The United States with its vastness and varied climate zones has a unique global position. For this country is, as one expert put it, the Saudi Arabia of the timber world.

With 100 marketable tree species the United States has enough forest land to cover every inch of Belgium, Great Britain, Denmark, Portugal, the Netherlands, Italy, France, Spain, Japan, Jordan and both Germanys with enough trees left over to blanket all of Algeria, Austria, Israel and an extra Netherlands.

The Forest Service which presides over about one-quarter of the country's forest land, estimates that when Columbus landed, there were about one billion total acres of forests here. After almost 500 years of building, paving, burning, exporting and wasting by the nation's approximately two million timberland owners, about 75 per cent of that land is still tree-covered and about one-half is still available for commercial use. The rest has been set aside or cleared for homes or crops.

Of course, there are those who believe too much has been cleared already or cleared and replanted improperly. And there are those who see technology as only a more efficient threat to the pristine wilderness of the forests.

Such concerns are part of a fundamental split over what forest use provides the most social value—as a living wilderness museum, as a playground for city folk, or as a natural wood-producing factory turning out trees that make homes,

toys, paper, exports, jobs and profits.

"We have an emotional problem here," said Jere Melo, a forestry foreman. "People like trees. And we're cutting them down. But we're also planting them. We're not running a big park. But we are using—and replenishing—a natural resource. Which is more than the oil and coal people can say."

Renewal has not always been a major concern for the disparate timber interests whose 83,-000 companies are the descendants of businessmen and lumberjacks who gave the English language such expressions as "skidroad" and "cut and run."

But in recent years environmental threats and lawsuits have combined with rising lumber and land prices and improved long-range planning to focus greatly increased attention on renewal of the forests and the most efficient use of existing timber.

Until recent years this forest renewal was left largely to the trees themselves. "Nature," said W. D. Hagenstein of the Industrial Forestry Association, "does a first class job of growing trees. But she's awfully wasteful and time consuming. And with an earth full of people we have to do better."

And so Mr. Hagenstein, a former lumberjack, and other pioneers began vast tree farms and, more recently, the movement for "genetically improved" trees.

'Test-Tube' Trees Studied

The most dramatic step in that direction is the current research into "test-tube trees." Several projects are under way. Recently, the Oregon Graduate Center, a small independent research institute in Beaverton,

announced a $1.25-million grant from the Weyerhaeuser Company.

This research, well along already, will attempt to grow superior Douglas fir trees from single cells without the time-consuming and uncertain natural sequence involving seeds and pollination.

The procedure operates on the theory that every living cell has locked within itself the information to regenerate the larger organism of which it is part.

The research objective is to provide an environment — a combination of light, heat, nutrition, humidity and other factors—that will cause the cell's hormones and other messenger molecules to trigger regeneration not only of similar cells but of dissimilar cells to form wood, bark, leaves, roots and so forth.

"Each cell," said Dr. Doyle Daves, the center's chemistry department chairman, "has the potential to become whatever you want it to be by manipulating the environment. All the techniques we're using with conifers have been accomplished in other species, carrots, tobacco, soybeans, barley.

"In terms of domestication for timber," he continued, "trees are at the stage that corn, wheat or rice were 10,000 years ago. It doesn't take much imagination to picture the dramatic changes in production and physical appearance of trees in the years to come."

When a superior tree variety is found, scientists would normally have to wait years before the plant matured and produced seeds to get sufficient numbers of offspring for planting. The seeds, formed after pollination by another, possibly inferior, tree, would not all contain the same desirable traits.

By culturing individual cells, however, virtually any number of offspring genetically identical to the parent can be obtained from an immature plant, shortening a process of years to days.

The cone-bearing trees, primarily Douglas fir (named for David Douglas, a British botanist), were selected for research for their strength and workability as lumber.

"This single cell work is revolutionary research," said Ira Keller, the center's president, "they did it to develop the high yield strains of rice. They're doing it now with sugar cane cells."

Mass Tree Production

If successful, this technology will permit eventual mass production of strong, fast-growing trees highly resistant to certain diseases and insects. When grown under progressive forestry techniques including thinning and fertilizing, it is estimated that these new trees will produce more and better wood in perhaps 20 per cent less time. That means an extra tree harvest every five forest generations.

The same result is the goal of numerous new tree breeding programs. At Mosby Creek, Ore., for example, Philip Hahn, a research forester for the Georgia-Pacific Corporation, grafts the cone-producing tops of 50-year-old, well-formed Douglas firs onto the vibrant bottoms of two-year-old trees. The history, characteristics and growth rate of each tree is recorded in a kind of computerized yearbook for each nursery's graduating class.

The cones, carefully pollinated from trees with other desirable qualities, hopefully carry in their fertile seeds the best genetic traits of its parents' good girth, strong height, fast growth and adaptability to various altitudes, weather conditions and soils.

These seeds are meticulously planted by machines in sterile granite chips (to prevent fungi and rot) in holes drilled in styrofoam blocks.

The blocks sit on tables in vast greenhouses where fans suck in fresh air.

Within days millions of tiny green shoots, most with only a dozen needles on them, spring from the holes. In 40 years each fir can be harvested with enough wood fibers to make half a house.

The trees spend one to three years in the greenhouse, gradually being exposed to the elements to toughen them for their lives in distant woods.

There, they are planted by a special gun that digs a hole, drops in the treelet and tamps down the dirt. Here at Fort Bragg in recent days workers planted a quarter million redwoods, which will make handsome picnic tables or panel some den around the year 2026.

Some of the more potent and prolific "super trees," such as No. 62, nicknamed "Super Stud," have produced offspring that can adapt up to 1,000 miles from their original home and grow to be 20 inches tall in one year. This compares to five inches in some two-year-old "natural trees."

One latent danger is that these new hybrid trees might contain some unknown form of natural timebomb, a genetic weakness that could in, say, 20 years see vast forests fall victim to diseases or insects yet unknown with serious economic repercussions nationwide.

A similar catastrophe struck certain strains of hybrid corns in 1971. The exotic fungus cut harvests so much the bushel price of corn shot from $1.05 to $1.40 in weeks. A blight-resistant strain of corn was developed within months, but such steps for trees would take years.

Attention is also being focused on maximizing use of existing trees. Research, for instance, is under way on biological controls, helpful insects or bacteria used to control damaging ones, which according to some estimates destroy many more trees than fire.

Sawmills, spurred by rising prices, can now turn out usable lumber from logs up to 80 per cent rotten.

January 20, 1975

LAND SUBSIDENCE CALLED A THREAT

U.S. Agency Says Sinking Is Costly and Permanent

By HAROLD M. SCHMECK, Jr.
Special to The New York Times

WASHINGTON, Oct. 12 — Man's activities are causing land to sink in some parts of the United States in ways that are proving expensive and, at times, dangerous, according to experts of the Geological Survey.

This sinking — called land subsidence—is sometimes too subtle to be detected until damage has been done, they said. Furthermore, almost all of the effect on the land is permanent.

Most of the sinking results from the withdrawal of massive amounts of water, oil or gas from wells. The most dramatic case has been taking place for several decades in California's San Joaquin Valley. This sinkage has become one of the largest known man-made changes in the physical environment of the world, according to Joseph F. Poland of the survey.

The subsidence in the valley totals about 15.6 million acre feet—a volume equal to about half that of the Great Salt Lake, according to one recent report by Mr. Poland and several other experts of the survey.

Damage in the Valley

In some places in the valley the subsidence, which resulted from water-loss is approaching 30 feet, Ben E. Lofgren, one of the authors of the report, said in a recent telephone interview. He said that the sinking had caused damage to thousands of deepwater well, some of which had cost as much as $30,000 apiece to drill, had hampered the efficiency of aqueducts by changing the steepness of the grade and had caused some irrigation ditches to reverse their direction of flow.

Mr. Lofgren said that the The report said that an area 70 miles long and as much as 20 miles wide had been seriously affected.

"There are probably many areas of more subtle land subsidence in the United States that we have not detected yet," Mr. Poland said in a statement released yesterday by the survey, a unit of the Department of Interior.

"Certainly we can expect subsidence problems to multiply during the coming decades as we withdraw more and more of our underground water, oil, gas and other mineral resources to meet our growing needs," he said.

Another area of known major subsidence is the vicinity of Baytown, Tex., near Houston, where another survey report estimates that because of massive oil drilling, the land surface has subsided more than eight feet since 1920. The problem here is that the whole region is not much above sea level and areas of subsidence could be subject to flooding from the Gulf of Mexico when major hurricanes produce exceptionally high tides.

The statement from the survey said that several hundred square miles along Galveston Bay were less than 20 feet above sea level.

Studies in Three States

The Geological Survey is also studying subsidence problems in parts of Louisiana, Arizona and Nevada. Some experts suspect that land is also sinking in coastal areas of New Jersey, but this remains unproved to date.

Harold Misler, a geological survey officer in New Jersey, said that the water level in wells in some coastal areas of the state had dropped more than 100 feet in recent decades. He said that funds had not been available for installing equipment in some wells to measure subsidence if it is indeed taking place.

Mr. Lofgren said that the sinking of land in some coastal areas would be prevented by the invasion of sea water into underground areas from which large quantities of water or other fluids had been withdrawn. This, however, could sometimes contaminate water supplies.

Reports from the survey note that land sinking can be slowed or even halted by artificial replacement of the withdrawn fluid and that a shift from use of ground water to water brought in by aqueduct can also reduce the land-sinking problem.

Because of increased use of aqueduct-supplied water, subsidence has virtually ceased in some parts of the San Joaquin Valley, according to a release from the survey.

"With modern technology, such as artificial recharge to replace withdrawn water or oil, subsidence can be slowed or even stopped," said the release, quoting Mr. Poland. "But," it continued, "there is no known method yet for raising the land surface back to its former elevation once it has subsided As a result, the damage produced by subsidence is permanent, and especially in coastal areas, can require the construction of extensive systems of dikes, flood walls and pumping stations to protect developed industrial and population centers from flooding." October 13, 1975

Tucson, Dependent on Wells, Finds Itself in a Hole on Water

The New York Times/Sue Levy
While decorative fountains flow in plazas all over downtown Tucson, water tables are dropping as much as a foot a month and the Santa Cruz River, above, has been dry for years. Margot Garcia, right, cried after she and two other city council members were recalled for raising water charges.

By GRACE LICHTENSTEIN
Special to The New York Times

TUCSON, Ariz., Jan. 27—Decorative fountains of stone and marble gurgle festively all over downtown Tucson—on La Placita mall, in front of the new Federal building, along the Plaza De Las Armas.

Yet a well on Gov. Raul H. Castro's pony farm nearby ran dry two years ago. Municipal water tables are dropping as much as a foot a month. The city is ripping out green turf on street curbs to put in less thirsty palms and crushed brick. The City Manager is certain some eastern parts of Tucson will have low water pressure next summer.

Meanwhile, the city government is in an uproar over water rates. It reached a climax last week, when three City Council members were recalled by overwhelming margins in an election because they raised water charges last July.

Classic Southwest Controversy

It is all part of a classic Southwest water controversy, in which surrounding agricultural interests, rampant municipal growth and a lush life style are at war with the geologically implacable desert.

Tucsonans, whose number mushroomed by 250 percent in the past 20 years to about 400,000, live in one of only three major cities in the United States that depend entirely on well-pumped underground water. The others are San Antonio and Miami.

They have been told that they are taking water from their underground supply almost five times faster than it is replenished by nature. Yet they treat water the way Easterners treated oil and gas before the energy crisis—with profligate disregard for warnings that it will one day run out.

A 'Nonrenewable Resource'

Not immediately. "We are not going to wake up and find we're out of water," said H. Wesley Peirce, a geologist for the Arizona Bureau of Mines. "It's more accurate to say we are running out of cheap water. It's a nonrenewable resource the way we use it."

Experts cannot agree exactly how much water lies beneath the floors of southern Arizona's sparkling valleys. But there is no dispute that its cheap cost has led to wasteful practices, or that state groundwater laws favor irrigated cotton and pecan farms over the rapidly growing urban areas.

"Drive around and you'll see lots of grass, a lot of trees; people have brought their old environments with them here and they resent being told they're in the middle of a desert," Barbara L. Weymann, one ousted council member, said of the new residents who have moved to Tucson from wet Middle West and Eastern climes.

Joel D. Valdez, the City Manager, recalled that when he was a boy the Santa Cruz River, now a rubble-strewn sandy scar through the center of town, actually had water in it. In the late 19th century, he said, water sold for a penny a gallon. No one wondered about where it would come from in later centuries.

To fill the pipes in the growing town, the city dug wells, first in the Tucson Basin under the city, then in the Avra Valley, where it had to purchase water rights from farmers. In recent years it bought private water companies, too.

But the rates were still cheap compared with those in other parched places like Phoenix and Denver. Some wells and pipelines were in bad need of repair. The city subsidized home developers who needed to put in water lines to new subdivisions. The hook-up charge for a new house to a water line was $95 last spring, as against $1,500 in Denver.

Consultants told the city more than a year ago that a major capital improvement program to upgrade and maintain the system was needed. Bonds would have to be floated. The bonds, by law, required a certain level of revenue backing them up. The solution: a dramatic change in the water price structure.

The City Council was dominated for the past two years by slow-growth advocates who believed higher water rates were not only necessary but desirable to allow Tucson to deal with its flow of newcomers. It finally approved higher rates last June.

Bills Anger Residents

Residents got their first new bills in July, the heaviest water-use period of the year. Some water bills rose from $50 to $200, although others went up only a few dollars. Practically everyone protested vehemently.

"Fifteen people had called the clerk of the court the day I did, asking how to file recall petitions," said John Varga, a community college instructor and land-lord who became spokesman for the pro-testers.

With support from the Good Government League, a group of car dealers, merchants, real estate people and bankers who are Tucson's establishment, the pro-testers campaigned hard against the four City Council members who voted for higher water charges.

One member resigned. The protesters' slate for replacements won overwhelmingly last week.

Mr. Varga discussed the water issue in terms of a "socialistic" plot by university economists and planning advocates to force slow growth on Tucsonans who want a "free enterprise" system. "Government trying to take control of the natural resource as a weapon?" he said of the water rate increases. "I thought I lived in a free country."

Even supporters of the council felt it was politically naive to raise rates and to institute sudden rises in hook-up charges in the height of the summer, three months before state elections.

Mrs. Weymann said this week that the irony was that she and other so-called "slow growth" officials were really trying to help Tucson's growth by planning a more realistic, costlier water system for the future. Even more ironic, the new council members began backing away from pledges to roll back the water increases the very day they took office this week.

According to the City Manager, the city hydrologists and others, however, the problem is neither Tucson's alone nor Tucson's to solve.

Other cities, including Long Beach, Calif., and Baytown, Tex., have discovered that too much pumping of liquids from under the ground causes "subsidence," sinking land that can crack utility lines and undermine building foundations.

Nearby Land Dropping

Jerome J. Wright, the city hydrologist, said that subsidence created by over-pumping to irrigate farmland north of Tucson has dropped some land there more than seven feet in 35 years.

Moreover, he warned that "the deeper

Tucson goes" for its well water, "the less efficient the pumping is." And Mr. Peirce's studies indicate that after about 1,100 feet down, there may be no water to pump, merely rock.

Mr. Valdez, the City Manager, said that throughout Arizona, water was legally a private resource. Agricultural interests can pump as much as they want from the same underground sources Tucson draws from.

Recent court decisions as well as state law have consistently favored agriculture, which is a powerful state lobby. New state laws would have to be drawn before cities could plan for future water needs, he said.

As for the City Council fight, Mr. Valdez viewed it as "emotional." Tucsonans do not like to think of water as if it were nonrenewable coal or gas being "mined," he said, because "no one likes bad news."

When will Southwesterners listen to the water Cassandras?

"When they turn the tap on and there's a little trickle instead of a gush," he said

January 30, 1977

Outcry Greets Carter Plan to Curb Off-Road Vehicles on Public Lands
By PHILIP SHABECOFF
Special to The New York Times

WASHINGTON, March 28—A plan by President Carter to crack down on the use of snowmobiles, motorcycles, dune buggies and other off-the-road vehicles on environmentally sensitive public lands has stirred a hornet's nest of protest by users of such vehicles.

Since news of the proposed plan, which is expected to be included in the President's coming environmental message, was disclosed a few days ago, the White House, members of Congress, the Council on Environmental Quality and other agencies have been inundated with complaining letters, telephone calls and telegrams. A number of the communications have been vituperative and threatening, according to a Government aide who has had to reply to some of them.

President Carter is reported planning to amend an executive order issued by President Nixon in 1972, which authorized Federal agencies to prohibit the use of off-the-road vehicles in those parts of public lands where they threaten the environment.

Mr. Carter is reported planning to issue an order that would require, rather than authorize, the agencies to prohibit the use of motorcycles, motorbikes, snowmobiles, jeeps, dune buggies and other

off-the-road vehicles from areas of national parks, wildernesses, forests, seashores and other public lands where they damage the environment.

A spokesman for the Council on Environmental Quality said that the vehicles might cause severe damage to the natural habitat on public lands, destroying vegetation, wildlife and, often, Indian artifacts.

What the executive order would do, the official explained, would be to close down areas of public lands that are being jeopardized by the vehicles, not all public lands. He also cautioned that the proposal was still in draft form.

The motorbikers, snowmobilers and other riders believe that the President's contemplated move will keep them out of all but specially designated areas of public lands. One Government official said, "They are all screaming and yelling that their American rights are being violated."

A spokesman for the Council on Environmental Quality said that the protests had come from user groups, including a national snowmobile association, manufacturers of the vehicles, as well as from people who use them.

Senator James A. McClure, Republican

of Idaho, said: "The proposed ban on the use of off-road vehicles is absolutely ridiculous. Such a suggestion betrays a lack of understanding of the use of public lands in the Western United States. This smacks of Eastern establishment thinking."

But environmental groups and others who object to the use of the off-the-road machines in woods, at seashores and other natural surroundings are beginning to speak out in support of the proposed action.

Thomas L. Kimball, executive vice president of the National Wildlife Federation, said: "It is superb that [President Carter] is making that kind of commitment. The time has long since passed when we can let the off-road vehicles roam around at will. They have already destroyed a lot of the California desert by the overwhelming number of their motorbikes and jeeps."

Mr. Kimball said the vehicles should not be banned from public lands entirely but should be allowed in designated areas "where they do not have a deleterious effect on the ecology." The problem, he said, is that "none of these off-the-road people want to be restricted at all."

One Government official, a back-packer, recalled that he had been walking down a lovely, quiet trail in the woods with his family last summer when the silence was shattered by a group of motorbikers speeding by.

"They [drivers of motor vehicles] already have most of the world paved over," he said. "There ought to be some places where the rest of us can have peace and quiet."

March 29, 1977

CHAPTER 2

Pollution

A sample from the bottom of Mill Basin, part of Jamaica Bay in New York City.

NYT Pictures

SMOKELESS INDUSTRY.

That cities in which bituminous coal is used are getting rid of the smoke nuisance, and that black smoke issuing from a modern manufactory or power plant is a badge of bad engineering and management, is due to the economic saving effected in smoke abatement. Mr. A. S. ATKINSON declares in the current number of Moody's Magazine that "the modern mechanical stoker has saved millions of dollars' worth of coal."

The prohibitive laws against sending out dense columns of smoke above the cities of New York, Cincinnati, Baltimore, Washington, St. Louis, and Pittsburg, and their enforcement with the aid of civic bodies like the Manhattan Anti-Smoke League, have enabled engineers to discover a way to burn coal in the average furnace which gives at least 15 per cent. greater efficiency. In cities like Pittsburg and Cleveland, where 99 per cent. of the coal burned is bituminous, nearly all the great factories have installed, or are having installed, the modern stokers. These, provided with intelligent firemen, who, by scientific firing, save more than their wages in a month, will soon make grotesque the depictions of prosperous industrial towns belching forth from high chimneys black clouds of sulphurous smoke.

May 20, 1907

COMPLAIN OF SMELTER GAS.

WASHINGTON, Dec. 3.—A conference on the destruction in Montana of the forests and vegetation by the fumes from the copper smelting furnaces was held at the White House to-day, and it was decided to have representatives of the copper companies come to Washington to be heard on the matter before any action is taken.

Those taking part in the conference were the President, Attorney General Bonaparte, Special Counsel Ligon Johnson of the Department of Justice, who has been investigating the matter; Senator Dixon of Montana, and a number of agriculturists from Montana who came to Washington to make a protest in the matter.

The farmers desire the smelting companies to place gas consumers on their smelters, but the companies say this could not be done except at very heavy expense.

December 4, 1908

ANACONDA CO. GIVES IN.

Agrees to Shut Off Fumes That Have Injured Montana Forests.

WASHINGTON, May 1.—The Government suit against the Anaconda Copper Mining Company has been compromised. The company agrees that its smelters at Butte, Anaconda, and Great Falls shall be equipped to prevent the emission of poisonous gases, which the Government claims have done great damage to the National forests.

A committee composed of John Hays Hammond, mining engineer; Dr. John A. Holmes, Director of the Bureau of Mines, and Louis D. Ricketts, representing the copper company, has been named to see that the agreement is carried into effect.

Farmers of the vicinity have a separate suit against the company, which is not affected by the compromise.

May 2, 1911

DOCTORS PROTEST AGAINST AUTO SMOKE

Petition the Health Board to Abate the Nuisance Caused by Gasoline Vapors.

PERIL TO PEOPLE'S THROATS

A movement against the smoke nuisance created by automobiles was begun yesterday, when the National Highways Protective Society sent to the Board of Health a petition, signed by fifty well-known physicians of the city, asking the board to take up a crusade against offending chauffeurs.

Counsel for the Highways Society have discovered, after some investigation, that the city's Charter puts the smoke nuisance under control of the Board of Health, and in the petition the board is reminded of this. The petition reads:

To the Board of Health, City of New York:
We, the undersigned physicians of the City of New York, some of whom are members of the National Highways Protective Society, hereby petition your body to put a stop to the practice of operators of motor vehicles in ejecting smoke from the exhaust pipes of their cars, as it is injurious to the health of the citizens. As Section 1,171 of the Charter of the City gives you the power to do this, we trust that you will take strenuous measures to eliminate this nuisance.

Among the physicians signing the petition are Henry H. Curtis of 118 Madison Avenue; Nathaniel B. Potter, 48 West Fifty-first Street; Charles McBurney, 38 East Thirty-first Street; J. Ralph Jacoby, 54 West Eighty-eighth Street; Lawrence B. Bangs, George W. Jacoby, 44 West Seventy-second Street; John H. P. Hodgson, 29 Washington Square, and William M. Polk.

At the headquarters of the Highways Society it was said yesterday that protests against the nuisance have been received from others than physicians. Letters have been pouring in to the organization from men of prominence in the city urging the society to take some action that will stop the smoke nuisance. The Fifth Avenue Association and the Women's Municipal League have joined the fight, and both organizations have sent letters to the Highways Society pledging their full support.

For some time agents of the society have been watching the development of the smoke pest, and their reports show that it has been persistently increasing. The only real effort to stop the nuisance, of which the Highways Society has cognizance, was that made by Park Commissioner Smith, who a year ago prohibited the passing of smoking automobiles through Central Park.

It was a successful fight, as far as it went, for after some hundred arrests the belching of gasoline-laden smoke from machine running through the Park practically ceased. No effort was made to check the nuisance in other parts of the city, however, and gradually it has grown to great proportions.

According to the Highways Society's agents the taxicabs are general offenders, although other classes of vehicles run them close. The police have made no arrests in the streets, because there is no city ordinance covering the situation. In the case of the Park arrests, fines were imposed under a park ordinance dealing with nuisances in a general sense.

Until the Highways Society set its counsel at work on the problem there seemed no solution. A provision in the city charter making the Board of Health responsible for the extermination of nuisances was at length unearthed, and this, the society is convinced, will cover the situation provided the Health Board is willing to act.

Dr. Curtis, one of the signers of the petition, in explaining why physicians are interested in the effort to have the smoke nuisance abated, said yesterday that hundreds of New Yorkers are suffering from what he called vaso-motor rhinitis, which in plain terms is inflammation of the membranes of the nose and throat.

"It is absolutely dangerous for persons having any affliction of the nose, throat, or lungs to ride through the streets of New York behind one of the smoke-producing automobiles," said Dr. Curtis. "The air filled with gasoline smoke produces an inflammation of the membranes that makes any disturbance worse. Any one suffering with a cold needs fresh oxygen to breathe in, and gasoline vapor will not produce it.

"Invalids driving through the streets can derive no benefit from it if they are compelled to inhale these noxious fumes. I have treated hundreds of cases of vaso-motor rhinitis caused by this automobile nuisance. It is much the same as hay fever.

"New York is unfortunate in that the police have not been able to tackle this nuisance. In France the drivers of smoking automobiles are arrested and fined and the nuisance is reduced to a minimum. I hope the Board of Health takes up the matter, for it appears that it has absolute authority, and with a little effort it seems that the nuisance ought to be abated."

The solution of the smoke pest, according to those who have studied it, lies in compelling drivers of automobiles to regulate the amount of oil in the lubricating cylinders. By filling the cylinders to the brim clouds of smoke are produced, which the mufflers cannot stop. The agitators against the nuisance argue that if it could be checked in Central Park it can be abated everywhere in the city.

February 4, 1910

AUTOMOBILES MUSTN'T SMOKE

Health Board Amends Its Code to Forbid the Nuisance After July 1.

By an amendment to the Sanitary Code, adopted yesterday, the Board of Health believes it will be able to stop the use of smoking automobiles in city streets. That owners may have time to make any changes necessary to comply with the regulations it was made operative July 1.

The solution of the smoke problem was arrived at by adding the words "automobile" and "automobile enginemen" to

Section 181 of the Sanitary Code, which now reads:

No persons shall cause, suffer, or allow dense smoke to be discharged from any building, vessel, stationary or locomotive engine, or automobile, place, or premises within the City of New York, or upon waters adjacent thereto, within the jurisdiction of said city. All persons participating in any violation of this provision, either as proprietors, owners, tenants, managers, superintendents, captains, engineers, firemen, or automobile enginemen, or otherwise shall be liable therefore.

Health Commissioner Lederle said yesterday that an official statement of how the new regulation will be enforced will be sent to garages and auto owners.

April 28, 1910

SMOKE ABATEMENT PAYS.

Cleveland Expert Says It Gives a 4,000 Per Cent. Profit.

"Abatement of smoke in this city during 1913 has yielded the city a dividend of 4,000 per cent."

This estimate is a feature of the annual report of the division of smoke issued by City Smoke Commissioner E. P. Roberts of Cleveland, Ohio. He points out that the operation of his division cost $19,000.

The estimated saving of damage from smoke from stationary plants for 1912, 1913 and 1914 is $400,000. This, according to the commissioner's estimate, leaves a net profit of $381,000 and produced a dividend of 2,000 per cent.

The report adds that the plants have an actual life of ten years. Using $200,000 as an average annual saving, and $3,900 as the annual average cost, the annual dividend for ten years is put at 3,600 per cent. By adding the saving from elimination of railroad smoke the total of 4,000 per cent. is produced on a one year's basis, the commissioner asserts.

Another feature of the report is the statement that the damage resulting from the burning of one ton of bituminous coal in cities having a fair degree of smoke abatement is not less than $1.

The total amount of bituminous coal used in Cleveland according to the Chamber of Commerce report for 1913 was 3,500,000 tons. Using Chamber of Commerce data gathered in 1910 as a basis, the Smoke Commissioner says it is reasonable to state the damage is not less than $1 per ton even in cities such as Cleveland and Pittsburgh, where the conditions are much better per ton of coal than would be the case if smoke abatement had not received attention, and very much better than in numerous smaller cities where it has received no attention.

February 1, 1914

TOPICS OF THE TIMES.

Emergency Effects Persisting.

Whoever goes or is taken to a high place whence he can look the city over will be told by his eyes these days that the smoke ordinances are neither observed nor enforced with any approach to needful strictness. Apparently the burners of coal and the police have not recovered from the effects of the time when an almost desperate scarcity of anthracite justified a suspension of restrictions on the burning of soft coal. That time is past. Now anybody who has the price can get all the hard coal he wants and there is no excuse for the black clouds that daily pour out of many chimneys, to the pollution of our atmosphere and to the foul besmirching of many buildings that would remain beautiful if they had a chance.

Of course there is not, and never has been, a law against the burning of soft coal. It is only the making of much smoke that is forbidden, and bituminous coal does not do that if enough of intelligence and care are used in its combustion.

Related to the smoke nuisance in the city is the oil nuisance all along the coast. This comes in chief part, not from refineries ashore, but from ships that use the new fuel, and empty into the ocean the dregs from their tanks and the residuum from their furnaces. The stuff floats ashore, and not only is it a horror to bathers, but it kills both birds and fishes. It will take international action to stop this abuse, as the evil deeds are committed on the high seas.

March 26, 1924

CITY AIR LADEN WITH DIRT DESPITE ANTI-SMOKE LAWS

DIRT and acid still taint the air of our cities, says Dr. H. B. Meller of the Mellon Institute of Industrial Research, Pittsburgh, in a recent report. "Much anti-smoke agitation," he says, "has occurred in the United States in the last ten or twelve years. Surveys have been made and ordinances passed. The net result has been a decided decrease in the amount of dense smoke in city air, without a corresponding abatement of the evils of dirt and acids.

"Soot-fall studies have been completed in St. Louis, Cincinnati and Pittsburgh. In all these investigations all solid matter that was deposited was considered, while in legislative action, taken with a view toward abatement, dense smoke only has been prohibited. Excellent re-

sults have been attained in combating this nuisance, but dense smoke is responsible for only a small part of the precipitate.

"In London, where data have been gathered since 1914, little improvement is shown. In Pittsburgh the amount of tar has decreased materially, while the combustibles other than tar have increased. This increase in solid matter deposited is due partly to an increase in the number of stacks, but mainly, it is believed, to the fact that legislation has attempted to control dense smoke. The very means taken to eliminate dense smoke by increasing the draft have resulted in increasing the amount of solid material emitted from stacks, as greater draft means higher velocities and the carrying of more and larger solid par-

ticles into the atmosphere.

"One needs only to walk through some of the large plants to see and feel the amount of loose solid material that is being taken up into the air, carried a short distance and again deposited.

"While the regulations are enforced for locomotives, it is true that the locomotive of today, with its so-called 'self-cleaning front end,' which forces through the short stack practically all the solid material that reaches that front end, is responsible for much dirt and obnoxious gas in the vicinity of the railroads.

"Dense smoke represents a very small proportion of the nuisance. Surely, the time has come when remedial action should be taken."

November 23, 1924

FIND MORE SOOT HERE THAN IN PITTSBURGH

Sanitation Workers Say New York's Smoke Pall Costs City $96,000,000 a Year.

CHILD HEALTH MENACED

Mt. Vernon Gets Three to Nine More Hours of Light Each Month Than the Battery.

CITY ORDINANCE IS URGED

Engineer Declares Fuel Now Wasted Would Pay for Adequate Prevention.

"What Price Smoke?" a bulletin just published by the National Conference Board of Sanitation, in co-operation with the local Department of Health, contains the announcement that a smoke and soot survey of cities from the Atlantic Seaboard to St. Louis has shown New York City to be a smokier own than Pittsburgh. The bulletin declares also that this city's "minimum smoke tax" amounts to $96,000,000 a year.

Physicians are quoted as having said that smoke and soot injure health, retard the normal growth of children, increase skin diseases and destroy plant life. The bulletin points out that the city is spending $871,420 to rehabilitate Central Park, which is said to have been smothering under a blanket of black smoke that has been killing off its trees and shrubbery.

Engine - contributors to the bulletin say that New York City's smoke nuisance could be eliminated without industrial loss, for the fuel now being wasted in smoke and soot would, if saved, easily pay the cost of effective smoke prevention. O. P. Hood, chief mechanical engineer of the United States Bureau of Mines, says that the remedy is to provide such engineering supervision over all installations in the future as to insure non-smoking plants at the outset.

Drainage in Other Cities.

The bulletin sets forth, among other things, that Pittsburgh, until her recent "clean-up," had suffered an annual damage from smoke of $20 per capita; that more than a ton of soot per square mile falls every day in the central section of Rochester, N. Y.; that smoke keeps away 30 per cent. of Philadelphia's sunlight, and that smoke is injuring health very much in Chicago where it is obscuring germ-killing rays of sunlight.

The Missouri Botanical Garden at St. Louis and the Brooklyn Botanical Garden have reported, the bulletin declares, that smoke has shortened and, in some instances, destroyed plant life. It points out that the $300,000 estimated cost of enforcing the proposed New York City ordinance, which would require all fuel-burning appliances to be licensed and provide for the reorganization of the Sanitary Bureau of the Health Department, would be slight compared with the annual smoke losses here.

The bulletin asserts that St. Louis has brought about a two-thirds reduction of smoke in a year in a residence section of thirty blocks by inspectors of the Citizens' Smoke Abatement League who tried to aid every householder to adopt efficient and economical methods in fuel burning.

The injurious effects of smoke and chimney dust on health are referred to in statements by Health Commissioner Harris of New York; Dr. George W. Goler, Health Officer of Rochester; Dr. Frank H. Krusen of the Temple University School of Medicine of Philadelphia, United States Senator Royal S. Copeland of New York, Dr. Alfred S. Hess of New York City and Dr. Charles F. Pabst, also of this city.

Jersey Smoke Wafted Here.

Dr. Harris is quoted as having said that a study by the Weather Bureau showed that Mount Vernon, N. Y., has three to nine hours of sunlight more each month than has Battery Park. James W. Scarr of the Weather Bureau declares in the bulletin that New York City cannot rid itself of smoke unless cities in New Jersey cooperate, as 53 per cent. of the winds reaching here are westerly winds from the general direction of the great New Jersey "smoke zone" extending from Paterson to Amboy.

"It is not often," says Mr. Hood in the bulletin, "that smokelessness is one of the main objectives of installation. Capacity, convenience, efficiency and low cost come first, with a weak but laudable hope that smokelessness can be had also at no increase in cost. This order must be reversed in the public mind if we are to have clean air. Smokelessness must be the first requirement."

Miss Laura A. Cauble, Chairman of the National Conference Board of Sanitation, who has been leading the movement for cleaner air in cities, says in the bulletin that clean air will be obtained only when the public recognizes how great is the "smoke tax" on city dwellers, and then through city ordinances and their enforcement, and by the aid of architects and engineers.

April 1, 1928

Robot Samples Pittsburgh Air To Aid Carbon Monoxide Study

Special to The New York Times.

PITTSBURGH, Pa., June 28.—A chemical robot developed by the United States Bureau of Mines is now on duty twenty-four hours a day at Smithfield Street and Oliver Avenue, sampling the atmosphere for carbon monoxide, the invisible gas which poisons humans.

The robot is working as part of an investigation in which the Bureau of Mines, the Pittsburgh Department of Health, the Mellon Institute, the Department of Public Safety and the Better Traffic Committee will try to learn how much the downtown air is polluted by auto gases and how much the health of the city's residents is endangered.

June 29, 1931

BOYCOTT ST. LOUIS OVER SMOKE LAW

Southern Illinois Towns Demand End of New Ordinance

ST. LOUIS, Aug. 10 (P)—In a drastic effort to hold the soft coal market in this key Midwestern metropolis, a half-dozen Southern Illinois mining communities today began a long-threatened boycott of St. Louis merchandise in protest against the city's new smoke prevention program.

Warning that "we mean business," leaders of the movement declared that the boycott would spread over the entire Southern Illinois coal area unless the St. Louis smoke elimination ordinance was repealed.

The ordinance, based on recommendations for ridding the city of the smoke nuisance in three years, requires the universal use of smokeless fuel or mechanical equipment which can burn raw coal without smoke.

This means that unprocessed Illinois soft coal may be burned legally in the city only with stokers or other mechanical equipment.

The measure was passed several months ago over the protest of coal interests across the Mississippi River, who asserted that it would ruin scores of small operators not equipped for processing the Illinois product.

C. G. Stiehl of Belleville, Ill., president of a coal operators' association representing eighty mine owners employing 6,000 men, said that the boycott would center on wholesale foodstuffs, clothing and newspapers. He also said that the townspeople would also "cease shopping" in St. Louis for other merchandise.

It was pointed out that many Southern Illinois merchants buy almost exclusively from St. Louis wholesalers and jobbers.

Population of the six towns involved in the boycott move, all within a fifty-mile radius of St. Louis, is about 10,000. But Mr. Stiehl indicated that it might spread to the more heavily populated centers before the Fall heating season gets under way.

"These mining towns," Mr. Stiehl asserted, "have been organized 100 per cent by the coal operators' association in the boycott drive, with miners, merchants and civic organizations all cooperating."

The fight, however, has made Arkansas coal operators as happy as it has made those in Illinois unhappy.

A shift to Arkansas coal to furnish part of St. Louis's fuel supply under the new smoke requirements was made feasible by one railroad's reduction in freight rates to $2 a ton on trainload shipments.

Thousands of tons of Arkansas fuel have been shipped here in recent weeks. A fifty-four-car trainload arrived today.

August 11, 1940

CITIES ARE CURBING UNNECESSARY SMOKE

Pittsburgh Adopts Act to Stop Soft Coal Nuisance

An important step in contributing to the cleanliness and healthfulness of American cities has just been taken by Pittsburgh, points out Herbert U. Nelson, executive vice president of the National Association of Real Estate Boards, in the enactment of an ordinance requiring that soft coal be burned with mechanical combustion equipment or that smokeless fuel be used.

The regulations, he states, will go into effect Oct. 1 next for office buildings, apartments, mills and factories and on Oct. 1, 1942, for railroads. River craft and private homes must comply with the law by Oct. 1, 1943. Penalties for infringement involve fines ranging from $25 to $100 or imprisonment for thirty days. The act also provides that all new plants in Pittsburgh must have their plans for fuel-burning units approved by the bureau of smoke prevention and the sale and purchase of new equipment must also be reported.

"The Pittsburgh act," explains Mr. Nelson, "is very similar to the St. Louis ordinance enacted in 1940. Prior to enactment a city-wide campaign was conducted to ascertain the costs of unnecessary smoke upon health, comfort, realty values and also on taxes. In the first report issued regarding the results a 30 per cent saving in cleaning bills was noted.

"While smoke control is not new in city ordinances, effective enforcement is. In Birmingham, Ala., cooperative action of business and civic groups is creating wide public support for rigid enforcement of the smoke ordinance. Salt Lake City is another community acting to make itself a smokeless town. In that city an ordinance effective Oct. 1 provides that furnace grates of more than three square feet capacity must burn smokeless fuel or use mechanical firing equipment which improves combustion. Firemen and engineers must be examined and licensed by the smoke control division, and inspectors must approve mechanical equipment in all new buildings. Railroad switch engines must carry mechanical firing equipment.

September 21, 1941

PACIFIC STATES

'Smog' Perils the Sunshine, Glory of Los Angeles

By LAWRENCE E. DAVIES

SAN FRANCISCO, Nov. 2—It is no longer a secret along the West Coast that Los Angeles, the City of Sunshine, is worried about its "smog" problem. The war brought industrialization to Southern California and more than a year after the war's end thousands of hopeful persons from the Middle West and elsewhere pour daily into that land of promise. But the war also brought "smog."

The word was coined to provide a name for a combination of fog and smoke and chemical fumes which began in 1943 to cause concern to city officials and the Chamber of Commerce.

The Los Angeles Times, which this month has been campaigning vigorously for official action to rid the city and county of its "smog," stated blithely in its news columns that "for years now the sun has been something of a mystery here." The Bureau of Air Pollution Control has been busy and the District Attorney's office has filed suits against a long list of manufacturing companies under provisions of the State Code relating to public nuisances.

It would perhaps be unfair to report that residents of other West Coast cities are enjoying the plight of the "Sunshine Capital," where, they read, the "smog" on some days causes eyes to smart. But objective reporting requires that it be noted that some of the vehicles of information in other centers are not overlooking a bet.

The San Francisco News, published in a city which has taken a good deal of ribbing for its high and low fog, of the summer and winter varieties, sent a staff correspondent to the Southern California metropolis who found this "new type of super-colossal problem" upsetting the Chamber of Commerce.

November 3, 1946

20 DEAD IN SMOG; RAIN CLEARING AIR AS MANY QUIT AREA

Special to THE NEW YORK TIMES.

DONORA, Pa., Oct. 31—Several hundred asthma and cardiac sufferers remaining in this stricken town were evacuated to other areas tonight as a welcome rain helped to clear the air of a smog believed to have contributed to the deaths of twenty residents. The mysterious air-borne plague struck yesterday.

Low-hanging smog over an eight-mile area was considered a factor in the deaths, which were chiefly among elderly persons. A late check indicated that probably two more names would be added to the list of dead.

There also is the threat of pneumonia which might affect other residents of this community of 12,000 in the Monongahela River valley, about twenty-five miles southeast of Pittsburgh, according to Dr. I. Hope Alexander, city health director of Pittsburgh, who aided in treating stricken residents.

About 400 other persons were stricken by the mysterious smog but were treated in time to prevent fatal effects.

Norbert Hochman, a chemist attached to the Pittsburgh Smoke Prevention Bureau, advanced the theory that there was definitely enough sulphur trioxide to be toxic in the air in Donora, particularly close to the zinc works of the American Steel and Wire Company, a United States Steel Corporation subsidiary.

He explained that sulphur dioxide is formed in the process here. In contact with air this becomes sulphur trioxide, a deadly gas.

Donora had lived in a twilight world, under an unusual envelope of dense smog, since Wednesday, but no ill effects were noticed until early yesterday morning.

About 2 A. M. the town's eight doctors were swamped with telephone calls for help from asthma sufferers and anxious relatives. In a short time hospitals in the area were filled to capacity.

Doctors and emergency workers reported that patients showed similar symptoms, a gasping for air

87

and complaints of unbearable chest pains. At the height of the alarm the borough of Donora set up an emergency station in the community center, were victims were treated until they could be taken to the already overcrowded hospitals. Volunteer firemen from neighboring towns assisted Donora's fire department in getting oxygen to victims of the choking smog who were unable to reach aid otherwise.

Although earlier reports that chemical fumes were partially responsible for the death toll have not been confirmed, Dr. William Rongaus, a physician and member of the Donora Board of Health, was bitter in his denunciation of atmospheric conditions.

"It's murder," he said. "There's nothing else you can call it. There was smog in Monessen, too, but it didn't kill people there the way this did. There's something in the air here that isn't found anywhere else."

Monessen is about four miles from Donora on the other side of the Monongahela.

Young Persons Also Stricken

Dr. Rongaus said also that reports that only elderly persons were stricken were not entirely true. So far only persons aged about 60 have died.

"I treated many patients who were young and strong and never had any symptoms of asthma," Dr. Rongaus said. "All complained of severe pains in the lower chest. It seemed to me like a sort of partial paralysis of the diaphragm."

A Pittsburgh chest specialist said that there was a possibility that the dense smog alone might have caused the deaths of chronic asthma sufferers. He said that even normal amounts of smog or fog aggravated the condition and that when irritation was severe, bronchospasms resulted which in effect cut off the supply of oxygen in the lungs.

M. M. Neale, superintendent of the zinc works, told a meeting of the Board of Health that the plant was being shut down as a precautionary measure. He explained that the shutdown would be progressive, since it would require some time to reduce heat in the smelting ovens.

Spokesmen for the mill said also that they felt there was small chance that the mill was responsible. They pointed out that the process has been in use in the plant since 1917.

One doctor said the town's supply of medicine for such cases was virtually exhausted at the height of the emergency.

"We are using adrenalin and sedatives of all sorts, anything we thought would bring relief," he said.

Late this afternoon rain began to fall and the smog showed signs of lifting. Eighteen persons were still in Charleroi-Monessen Hospital, two of them in serious condition. Monongahela Hospital said fourteen were still under treatment.

The rain was viewed with some hope, although it was feared that it might produce another fog due to the humidity in the air. However, such a fog was expected to be a cleaner one.

Meanwhile, state, county and other officials joined with health authorities in an extensive investigation into the causes of the fatal smog. In Indiantown Gap, Pa., Gov. James H. Duff announced that a state investigation would be made to ascertain the cause.

November 1, 1948

LAG ON SMOG LAID TO INDUSTRIALISTS

McCabe of Mines Bureau Says They Fear Eradication Cost and Thwart Campaigns

2% EXPENSE HIS ESTIMATE

Official Mismanagement and Misguided Public Enthusiasm Also Blamed for Hazard

By GLADWIN HILL
Special to The New York Times.

PASADENA, Calif., Nov. 10—Blame for the persistence of noxious "smogs" in many American cities was placed by a leading air-pollution expert today on official mismanagement, misguided public enthusiasm and selfish industrialists.

Addressing the first national Air Pollution Symposium, Dr. Louis McCabe of the United States Bureau of Mines said that generally municipalities had underestimated the extent of the smog problem, citizens had been deluded by an extensive "folklore" about smog, and industrialists had thwarted campaigns against the nuisance.

Dr. McCabe, who organized the new Air Pollution Control Authority in Los Angeles, one of the most smog-ridden cities, is chief of the Bureau of Mines office of air and stream pollution prevention research.

The two-day symposium is being sponsored by the Stanford Research Institute, a unit of Leland Stanford University, in cooperation with the California Institute of Technology, the University of California, and the University of Southern California. Five hundred scientists, engineers and public officials from all parts of the country are attending.

"Diverting Papers on Minutiae"

While acknowledging "some commendable accomplishments" in the effort against air pollution, Dr. McCabe asserted that one main reason the effort was still a failure was "because industry believed that air-pollution control costs too much."

"There were 'cooperative' programs with the dual objectives of delay and defeat," he continued. "Engineers were assigned to write diverting papers on the minutiae of the problem, and trade journals editorialized on the unreasonableness of 'do-gooders'. These tactics haunt the sincere efforts of progressive industry today."

The Los Angeles situation has been officially ascribed in considerable measure to sulphurous fumes from oil refineries. The Stanford Research Institute recently made a study of the situation, sponsored by the Western Oil and Gas Association, which resulted in a report absolving the oil industry of any specific responsibility for smog. Asked what he thought of this report, Mr. McCabe said, "no comment."

The large-figure costs cited by industrial concerns for fume-suppressing equipment, Dr. McCabe said, usually turned out on inquiry to be a small fraction of their revenue. It had been authoritatively estimated, he said, that generally such installation would run only 2 per cent of plant investment and 2 per cent of operating costs.

Some Smog "Folklore" Cited

He cited as smog "folklore" the notions that occasional smogs were inevitable because of atmospheric freaks; that secondary factors like auto exhausts and even "particles from rubber heels" might be primarily responsible; and that because there was smog in coal-less cities, coal was not a smog factor elsewhere.

Control of atmospheric pollution, Dr. McCabe said, depended on sustained public opinion against it, and an effort as constant and comprehensive as was devoted to municipal sanitation like water purification and sewage disposal.

Asked how effective he considered New York's new municipal effort against air pollution, Dr. McCabe said: "No comment."

November 11, 1949

LOS ANGELES SMOG IS LAID TO PUBLIC

Industry-Sponsored Research Finds Population Produces Most of Impurities

HEATING, MOTORING CITED

By LAWRENCE E. DAVIES
Special to The New York Times.

PALO ALTO, Calif., Feb. 20—The "smog" problem, which has caused growing concern along with eye irritation and poor visibility in Los Angeles County in recent years, was labeled today one for which industry and the public must share responsibility.

After a survey lasting a year, the Stanford Research Institute concluded that the chief cause of the atmospheric condition known as smog was the "incomplete combustion of nearly 50,000 tons a day of fuels and rubbish." At least 2,280 tons of chemicals, it found, are released into the air every day by reaction and combination and these may result in the so-called smog effects such as eye irritation and reduced visibility.

Sixty per cent of these materials, according to the Institute's findings, result from activities of the public, including the driving of automobiles and buses, the burning of backyard trash, the heating of homes, stores and office buildings. Industries of Los Angeles County, it said, accounted for the remaining 40 per cent of the chemicals.

The Institute's study was made under the sponsorship of the Western Oil and Gas Association.

For several years, smog has caused increasing embarrassment and complaint, and a legislative committee has been studying the problem during the last year, taking testimony from industrialists, scientists and other witnesses.

In some quarters, the tendency had been to put much of the responsibility for smog conditions upon Los Angeles' growth as an industrial center.

The institute reported, however, that almost two-thirds of the 2,280 tons of chemicals entering the air daily over Los Angeles were organic in nature and that activities of the public accounted for 76 per cent of the organic part.

The biggest single source, it stated, was 550 tons of organic matter resulting from the burning of 4,000 tons of trash. Noting that 2,000,000 motor vehicles operated an average of 50,000,000 miles a day in Los Angeles County, it said that the exhaust fumes from them poured 350 tons of organic substances, thirty tons of aldehydes and forty tons of nitrogen oxides into the air. All of these, it added, contributed to smog conditions.

The institute concluded that public activities were responsible for 65 per cent of the aldehydes entering the atmosphere, 73 per cent of the ammonia, 40 per cent of the nitrogen oxides, 24 per cent of the sulphur oxides, 31 per cent of the acids and 46 per cent of the solids, with industry accounting for the rest.

February 21, 1951

POLLUTED AIR HELD NO HEALTH MENACE

3-Year Survey by Industrial Group, However, Proposes Government Control

FEAR TERMED UNJUSTIFIED

Regional Rule Is Proposed by Christy at Conference of Manufacturing Chemists

A three-year study by a chemical manufacturing industry committee produced yesterday a report holding that air pollution was "not a serious or critical menace to pub-lic health," but that Government control was necessary because some plants were making no effort to curb the nuisance.

The report was made public at the start of the two-day seventh air pollution abatement conference of the Manufacturing Chemists' Association at the Statler Hotel. The committee chairman, George E. Best, presented a seven-point program predicated on the establishment of state control bureaus, which could take no restrictive action in any community without the consent of a local control commission of representative citizens.

Some opposition to the state-local control proposal developed at the meeting, which was attended by 250 industry representatives and public officials.

Regional control under interstate agreement was held necessary by William G. Christy, chairman of the New York City Smoke Control Bureau, and Dr. Daniel S. Bergsma, New Jersey State Health Commissioner, who made the point that air pollution in a community sometimes originated beyond city or state boundaries.

"Right here we have a fine example of the need for at least re-gional control," Mr. Christy said. "New York suffers from New Jersey's air pollution. Directly across the river from us are six or seven heavily industrialized counties. The wind blows from the west 65 to 70 per cent of the time, and we get that pollution."

Remarking that he had been accused by the New Jersey Chamber of Commerce of "trying to make alibis for New York," Mr. Christy said he had aerial photographs of smoke and fumes blowing from New Jersey to Manhattan and Staten Island.

"The wind does blow in both directions," retorted Dr. R. H. Daines, secretary of the New Jersey Air Pollution Study Committee.

Tri-State Control Urged

Mr. Christy and Dr. Bergsma and Dr. Daines favored giving control over air pollution to the Interstate Sanitation Commission, whose fifteen commissioners, appointed in equal numbers by New York, New Jersey and Connecticut, have jurisdiction over water pollution. New York and New Jersey are each making available $30,000 for a study of the legal as-pects of such tri-state control. Mr. Christy told reporters it would take a year or two for such control to become effective because it would require approval by the legislatures of the three states and the Federal Government.

Opposition to any state or Federal control was voiced by Thomas C. Wurts, director of the Alleghany County Bureau of Smoke Control in Pennsylvania, who likened air pollution control to sewage disposal as an economic problem.

"The only economically efficient way in which to cope with the major sources of air pollution," Mr. Wurts said, "is to have industry itself inaugurate research which will result in finding practical means of controlling their own effluents. This is a far more economic approach than to have industry and citizens pay for an inefficient bureaucratic research organization through taxation."

Replying to Mr. Wurts, Mr. Best said research at the state level was more economical because data obtained from a study in one community might solve a similar problem in another community.

February 26, 1952

CHEMICAL CONCERN STARTS NEW PLANT

Consolidated Breaks Ground Near Houston for Sulphuric Acid Manufacturing Unit

Special to The New York Times.

BAYTOWN, Tex., Oct. 12—Ground has been broken here by Consolidated Chemical Industries for the construction of a sulphuric acid manufacturing plant that will use materials obtained from the refinery of the Humble Oil and Refining Company.

Called a major step in the reduction of air contamination at Baytown, where Humble operates one of the country's largest refineries, the new plant will use hydrogen sulphide recovered from crude oil at the refinery and also certain acid sludges formerly burned as fuels.

A spokesman for the two concerns said a part of the new plant, which will cost "several million dollars," would commence operating in about eight months and that the entire plant should be completed next October.

Second Plant in Area

The plant will be the second in the Houston area that is designed to combat air pollution at the same time that it disposes of waste materials. The Shell Chemical Corporation announced last week that production had begun on its sulphur recovery unit on the Houston Ship Channel. The unit will recover sulphur from waste refinery gases at a rate of about thirty-five tons daily.

In announcing the start of work on the Consolidated plant, the Humble Company said it had made capital expenditures of $3,000,000 over a five-year period in an effort to reduce air pollution at the Baytown refinery.

Humble said it was assisted in the pollution project by the Stanford Research Institute, a techni-cal group that contributed toward air pollution abatement in the Los Angeles area.

The Houston Chamber of Commerce, meanwhile, has retained the Kettering Institute to make an air pollution test of the growing Houston industrial area.

The decision to make the survey came after residents along the Houston Ship Channel complained of unpleasant odors emanating from the sulphuric acid plant operated by the Mathieson Chemical Corporation. The citizens took their complaint to the State Health Department and, following hearings, the Mathieson concern agreed to take measures to combat the problem.

October 13, 1952

4,000 LONDON FOG DEATHS

British Statistician Reports on Toll in December

Special to The New York Times.

LONDON, Feb. 12—About 4,000 persons died as a result of the four-day fog in London last December, according to Dr. William P. Dowie Logan, Britain's chief medical statistician, in a report published today in The Lancet, British medical journal.

By comparison, he said, the Meuse Valley fog episode in 1930 caused sixty-four deaths, and the Donora, Pa., episode in 1948 caused twenty deaths.

Dr. Logan said the air began to thicken at dawn on Friday, Dec. 5, and the maximum pollution rate reached 2.55 milligrams a cubic meter. The twenty-four-hour total of deaths reached 400, about twice the rate before the fog began. On Saturday 600 persons died and on Sunday and Monday, Dec. 7 and 8, more than 900 each day. By Tuesday the fog was lifting, but the death rate was still nearly 800.

February 13, 1953

LOS ANGELES SMOG PERSISTS 9TH DAY

LOS ANGELES, Oct. 15 (UP) —A choking, eye-searing smog blanketed the Los Angeles metropolitan area today for the ninth straight day and brought on a state of near-emergency.

The smog was blamed as a "contributing factor" in at least one death, that of a 10-year-old girl who choked to death.

Gov. Goodwin J. Knight interrupted his campaign for re-election to fly here and join in an investigation into the cause of the plainly visible pall of fumes, the worst in a year.

While the Governor began conferences with Los Angeles city and Council officials, Bernard Caldwell, chief of the State Highway Patrol, ordered his officers to set up roadblocks and cite all vehicles that put excessive fumes into the air.

The air pollution director Gordon P. Larson, summoned before the county grand jury's investigating committee, blamed inefficient combustion in automobiles for 90 per cent of the fumes in the downtown area. He said that when such fumes were combined with a temperature inversion it caused smog.

The Los Angeles area is surrounded by mountains that keep fumes bottled up in the region unless strong sea breezes disperse them. But in the fall of the year a layer of warm air settles over the region, preventing the cooler surface air from rising and dispersing.

Hospitals were deluged with calls for advice on the treatment of the eye-searing efect. In adjacent Pasadena, Mayor Clarence Winder called for public prayer "to deliver us from this scourage," while the Los Angeles City Council donned gas masks to demonstrate that the fumes were even entering the Council chambers.

October 16, 1954

SMOG IS TERMED A CANCER CAUSE

By BESS FURMAN
Special to The New York Times.

WASHINGTON, Nov. 18—Surgeon General Leroy T. Burney said today that a definite association between air pollution and diseases, notably cancer, had been established.

"Cancers can be produced with urban smog," the head of the Public Health Service told the First National Conference on Air Pollution at the Shoreham Hotel. He said he was speaking of animal experiments.

Dr. Burney also noted that human lung cancer deaths in the largest cities were twice as high as in nonurban areas.

After several years of close observation, Dr. Burney said, investigators are finding "a definite association between community air pollution and high mortality rates due to cancer of the respiratory tract, including the lung, cancer of the stomach and esophagus and arteriosclerotic head disease."

Warns of "Disaster"

Dr. Burney conceded that the case against air pollution as a cause of cancer had not been proved, but declared:

"In protection of human health such absolute proof often comes late. To wait for it is to invite disaster."

Dr. Chauncey D. Leake, Assistant Dean of Ohio State College of Medicine, representing the American Association for the Advancement of Sciences, termed nuclear weapon testing "a potentially very dangerous sort of air pollution involving all living things, now and to come."

"Unfortunately," he said, "the confusion over security matters has weakened popular confidence in the pronouncements, and in the judgment, of the Atomic Energy Commission."

He said that the smog, first identified with Los Angeles, can now be seen "hanging over cities everywhere" by all who travel by air.

"In Seattle," he said, "autos and trucks dump 100 tons of hydrocarbons, twenty to eighty tons of nitrogen dioxide, and four tons of sulfur dioxide into the city's air every day."

November 19, 1958

DELAYED EFFECTS OF SMOG STUDIED

By HOMER BIGART
Special to The New York Times.

DONORA, Pa., Nov. 28—Just over ten years ago a poisonous smog settled on this steel-mill town of 12,300, killing twenty persons and making nearly half the population ill.

A report on the long-term affects of that smog was made available today. It was a product of an investigation undertaken for the United States Public Health Service by the Department of Biostatistics of the University of Pittsburgh's Graduate School of Public Health.

The investigation revealed that the Donora residents who became ill during the 1948 smog had since shown a higher death rate and a higher frequency of disease than those unaffected by the smog.

It was found that the death rate for arteriosclerotic heart disease, the largest single cause of death, has been running more than twice as high among men in the smog affected group than for the non-affected males.

¶Within a year after the smog, it was already apparent that the affected group had a higher mortality rate. The difference in the mortality of the two groups increased steadily

The New York Times   Nov. 30, 1958
POISON SMOG: 20 persons died from polluted air in Donora, Pa. (cross), in '48.

until 1954, when it apparently stabilized.

¶The higher mortality rate for the affected group was generally, although not uniformly, consistent for all ages and both sexes.

¶Mortality among persons affected by the smog who gave no history of prior chronic illness was generally only slightly higher than mortality in the non-affected group.

¶Affected persons of both sexes in the 21-to-50 age group who had a history of heart disease, asthma, pneumonia or bronchitis prior to the smog showed a substantially higher subsequent mortality rate.

But persons who were 50 or older in 1948 and reported previous experience with those diseases showed no higher subsequent mortality rates than the non-affected of that age group.

Survivors among the affected persons showed a higher frequency of current illness and symptoms and a higher prevalence of chronic illness.

Exceptions Found to Be Few

The report noted that these differences were "reflected in each age and sex class with few exceptions" and "even the group under 20 years old in 1948 reflects this general pattern."

Survivors of the affected group were found to have a higher prevalence of cardiovascular and respiratory disease and more hospitalization. They were particularly prone to heart disease, asthma and bronchitis.

But the report left unanswered the question of whether the poison fog that blanketed Donora for five days starting Oct. 27, 1948, was linked to subsequent deaths and impairment of health.

The author of the study, Dr. Donovan J. Thompson, Professor of Biostatistics at the University of Pittsburgh, said his data were not sufficient to enable him to pin down the relationship between particular contaminants in the Donora air to specific symptoms.

Groups Distinguishable

Interviewed in Pittsburgh, Dr. Thompson said: "What impressed me most was the uniformity of data with respect to the affected and nonaffected groups. Today, after ten years, these groups are easily distinguishable."

He said that the role of the smog might have been one of aggravating existing heart and lung conditions in the victims and permitting the labeling of those cases that were the more severe or disabling.

But if that were true, one would expect to find a mortality differential in the elderly group, he said.

"That we didn't find the differential may be a chance occurrence, or may be that our time period is long enough to give competitive causes of death a chance to operate and thus obscure the picture," he added.

Dr. Antonio Ciocco, head of the biostatistics department, said that the data collected so far proved only that smog made sick people sicker.

The Pittsburgh investigators interviewer 4,092 persons. The study indicated that health conditions generally were no worse than elsewhere.

November 30, 1958

CALIFORNIA VOTES TO CONTROL SMOG

Legislature Approves a Bill Requiring Auto Devices to Limit Air Pollution

Special to The New York Times.

SACRAMENTO, Calif., April —The California Legislature has approved the first state smog-control bill in history. It is designed to control air pollution from automobile exhausts.

Smog control was one of the key measures in a special session of the Legislature called by Gov. Edmund G. Brown to run concurrently with the regular budget session. S. Smith Griswold, chief of the air pollution control district in Los Angeles, predicted that smog would disappear from Southern California within six years.

The legislation requires the installation of smog-control apparatus on all new automobiles sold in California a year after two such devices have been found to be acceptable.

Provisions of Measure

One year after approval, all used cars must be equipped with the devices except in counties that rule that they have no smog problem.

Two years after approval, all used commercial vehicles must have the smog reducers except in the exempt counties.

Three years after approval, all used cars must have the devices before they can be registered, again except in the exempt counties.

The counties that are excepted must make a smog survey every two years to prove that their situation has not changed.

The legislation provides for a thirteen-member motor vehicle board in the state Department of Public Health. It will include nine members appointed by the Governor, subject to confirmation by the State Senate.

The other members will be the directors of the state Departments of Public Health and Agriculture, the commissioner of the California Highway Patrol and the Director of Motor Vehicles. The members are to serve without compensation.

The board will set up standards for pollution-control devices and issue certificates of approval after tests.

An appropriation of $500,000 was provided for the Department of Public Health in carrying out the provisions of the bill, which was introduced in the Legislature at Governor Brown's request.

April 3, 1960

Air Pollution Is Eroding World's Stone Art

Ravages of Weather and Time Hastened —Concern Grows

By MILTON ESTEROW

Air pollution is hastening the work of weather and time in eroding outdoor stone art works in many parts of the world.

The deterioration of some of the finest monuments of antiquity and thousands of pieces of sculpture and carvings in the open and on public buildings and cathedrals is causing increasing concern.

The rate of decay, according to conservation specialists, has accelerated in recent years because of air pollution resulting from the rapid pace of 20th-century industrialization.

In Rome, the Coliseum, the Arch of Titus and many frescoes have been damaged recently.

In France, statues have been removed from the exterior of cathedrals and replaced with copies.

In Florence, the situation is described as disastrous.

In Athens, workmen check the Acropolis daily and fill in small cracks with cement.

In West Germany, the state of North Rhine Westphalia is spending $4 million annually to preserve monuments.

And at the Cloisters in Fort Tryon Park, nearly all works of great value are kept indoors.

"The world has so many problems of stone preservation that I can't begin to discuss them all," James J. Rorimer, director of the Metropolitan Museum of Art, which operates the Cloisters, said yesterday. "I have letters from almost every country in Europe that ask, 'What do you do to save stone?' The first answer is that you put it indoors."

Indoors, the problems of stone are comparatively minor, because the environment can be controlled. Outdoors, the remedy for decay caused by industrial pollutants is a simple one— wash the stone with water regularly.

However, it is expensive. To wash Trinity Church would cost $15,000 to $20,000. The bill for a 50-story skyscraper might be in six figures. Except for water scrubbings, there is no generally accepted method to prevent decay.

Historic and cultural monuments from ancient Greece to the Renaissance are not the only ones affected. Jean Carpeaux's statue of the dance at the entrance to the Paris Opéra, completed in 1869, is seriously threatened. A copy is being made. When it is completed, it

"The Conservation of Antiquities and Works of Art," by H. J. Plenderleith. Oxford, 1957.

Part of the frieze of the Parthenon in Athens reveals the deterioration of outdoor art in the twentieth century. Above: a plaster cast made in 1802, showing the marble after 2,240 years. Below: the marble as it actually looked in 1938, after 136 years of exposure.

will take the place of the original, which will be displayed in one of the foyers of the opera house.

Many Study Preservation

A growing number of Government officials, museum directors, scientists and those interested in preserving are studying methods to protect these works.

The most extensive research program was started several months ago at the Conservation Center of the New York University Institute of Fine Arts, 1 East 78th Street.

The institute is a graduate school devoted to research and teaching in art history and archeology and museum-personnel training.

"Because of these activities, we embarked on the study of conservation," Craig H. Smyth, director of the institute said. "The field of conservation as a whole has not been a university concern up to now."

The center was established in 1960 with a grant of $500,000 from the Rockefeller Foundation. Activities in stone are supported by the university and the foundation. Other foundations, including the Alfred P. Sloan Foundation, have indicated that they will support the program.

In France, experiments are under way in several university laboratories. Additional laboratories are being constructed at the Universities of Nancy and Strasbourg.

In Britain, the Building Research Center at Garston, Watford, is looking into the problem.

Another organization is the International Center for the Study of the Prevention and Restoration of Cultural Property in Rome. The center was created by the United Nations Educational, Scientific and Cultural Organization in 1958.

World Survey Under Way

The International Council of Museums, a branch of UNESCO, is making a worldwide survey of stone decay.

"Something must be done if we value these things," Mr. Smyth said. "The technical knowledge of the 20th century has not been brought to bear on conservation."

Freres Haine

MAINTENANCE program conducted by Belgian Government is intended to preserve stone landmarks, such as Hôtel de Ville, 15th-century City Hall of Brussels. Stages of its renovation are noticeable here.

EROSION is destroying figures on the facade of Exeter Cathedral in England.

Kayaert

DECAY over the centuries is evident on the 14th-century fresco by Giotto on the Scrovegni Chapel in Padua, Italy.

DANGER signals to experts are the cracks in the stone capitals on the Brussels City Hall. Unless the stone is replaced, it may fall to the street, to be shattered beyond recognition.

Seymour Lewin, Professor of Chemistry at New York University and a fine-arts professor at the institute who is conducting stone research, said:

"The rate of decay has increased greatly. The situation is getting more serious all the time. It's at its worst in highly industrialized cities. Many buildings have noticeably deteriorated in the last 20 years."

Dr. Lewin added: "The problem is not unlike that of finding a Salk vaccine for stone diseases. We want to find out how the process of decay is taking place and develop a technique to prevent it—to modify the surface of stone chemically so that it will be resistant to various elements.

"The technique should be inexpensive so that it can be used universally. It must be long lasting and easy for unskilled labor to apply.

"Our point of view is the same in principle as dentistry's addition of fluoride, which makes teeth more resistant to bacteria. Our approach is to make stone more resistant to corrosive agents."

Carelessness Blamed

In some cases, deterioration is due to lack of care in the handling of stone by builders. "Speedy construction doesn't lend itself to good workmanship," a specialist in stone preservation said. "You've got to keep the stone dry during the course of construction. If the stone is put on soaking wet, particularly limestone, it causes staining, which damages the stone."

However, the industrial age is considered the major villain. Industrial smoke produces corrosive elements that cause the stone to powder, crumble, flake, crack or chip.

Then there are automobile exhaust fumes and traffic vibration. "Weathering"—wind, rain, heat, snow and frost—hasten deterioration. Plant growths and pigeon droppings are also harmful.

Industrial fumes bring sulphur compounds into the atmosphere. These are converted into oxides of sulphur that, in combination with water, produce sulphuric acid, one of the strongest and most corrosive acids. Soot and dust particles deposited on the stone absorb the acid.

Frost also causes decay by freezing water in the pores of the stone, causing it to crack.

New York Times correspondents overseas found growing concern in many areas. Following are their reports:

ITALY

A severe winter, with days of below-freezing temperatures, did considerable damage to the Roman Forum.

In Florence, Ugo Procacci, superintendent of monuments, said that the situation was disastrous and that it would cost billions of lire to help. When Mr. Procacci was appointed to the post not long ago, he told.

a friend, "I've been asked to preside at a disaster."

He said that the more serious cases of deterioration were at the Ponte Vecchio, the Pitti Palace, the Palazzo Strozzi and the basilica of San Lorenzo.

Venice has many problems. Increased motorboat traffic on its canals is speeding deterioration by damaging foundations as a result of wave action. In addition, the salt air contributes to decay.

The government spends about $10 million annually in restoration. Throughout the country, there are many chapels with ancient frescoes that are decaying.

GREECE

The main concern is for the Acropolis in Athens with 2,400-year-old monuments like the Parthenon. "After a cold rainy night, you can see fallen pieces of stone around the base of columns," one observer said.

One corrosive agent is a moldy lichen carried by winds. It forms incrustations on marble and creeps into cracks, causing splintering and peeling. Until an effective cure is found, stopgap measures are taken—workmen visit the Acropolis daily and fill in cracks with cement or put back the largest of the fallen splinters.

FRANCE

There is a sense of urgency over the deterioration of "les vieilles pierres," as old stone works are called, because of the interest of André Malraux, Minister of Cultural Affairs.

Mr. Malraux is credited with having started a campaign several years ago for washing public buildings and monuments. A contest is being conducted over the French National Radio Network. Frenchmen are invited to submit a résumé of what they have done to save or improve monuments or distinctive historical buildings. Prizes include an Air France vacation trip.

The Church and Cloister of Moissac is in poor condition. Statues have been removed from the exterior of the Strasbourg Cathedral and replaced by plaster casts. The originals are on display in a building adjoining the cathedral.

ENGLAND

D. B. Honeyborne, an official at the Building Research Center in Garston, Watford, said that the main cause of decay was not only industrial fumes but also seaborne salt.

Britain's anti-pollution acts, which are aimed at banning coal fires, may cut down some pollution.

At Oxford University, a restoration program has been under way for six years and has already cost $4.2 million. The program is expected to continue another five years. Stone erosion was so bad in some cases that facades could be scraped through with a fingernail, disclosing the rubble behind. Hugh Arber, secretary of the Oxford Historic Buildings Fund, said that the quality of the stone may have been more significant than fumes.

The New York Times (by Henry Giniger)

THREATENED is Jean Carpeaux's statue, "La Danse," at entrance to the Paris Opéra. A copy of the work will take the place of the original, which will be moved inside.

In London, the statue of Nelson is given a scrubdown regularly, but the column on which it stands is not touched. It is believed that regular scrubbings might make the column less able to support Nelson.

BELGIUM

René Sneyers, a chemist and member of the International Museum Council, has been assigned by the Government to recommend techniques for treating Belgian buildings and monuments.

"The situation here is very serious," he said. The Government conducts a maintenance program for historic stoneworks. The Hôtel de Ville—Brussels's historic City Hall—in the Grand Place has been completely renovated over the years. The structure was built in the 15th century.

Work is going on at such landmarks as the Cathedral of St. Michel and St. Gudule. Both columns supporting the massive facade of the church are being replaced. Otherwise, Mr. Sneyers said, statues and other stone carvings over the entrance would crash to the street in a few years.

Mr. Sneyers is conducting an extensive study of stone art works for the museum council. He has sent questionnaires to 82 specialists in 33 countries. He began the project in September in Leningrad, where he examined the restoration being done at St. Isaac's Cathedral. The cathedral was built in the 18th century and was being overhauled when Mr. Sneyers was there.

WEST GERMANY

The postwar expansion of West German industry has gen-

The New York Times (by Clyde A. Farnsworth)

NETTING to keep out pigeons encloses a griffon in niche of the facade of St. Stephen's, in Vienna.

CORROSION is evident on the stone outcroppings of the South Tower of Notre Dame Cathedral, in Paris.

erated air pollution and made stone decay a major problem throughout the country.

In the state of North Rhine-Westphalia, which contains the industrial Ruhr, corroded arches and stone girders in churches and other buildings are constantly being replaced.

Stone sculptures that had been outside the Cathedral of Xanten on the lower Rhine have been placed inside. At the

Cathedral of Cologne, stone works are repaired by permanently assigned teams of masons and architects. Sculptures have been replaced by plaster casts at the Brühl Palace near Cologne.

Government anti-decay programs have been started in all West German states. Some observers say that many valuable old buildings are being ignored and doomed to ultimate destruction.

AUSTRIA

The most notable example of decay is St. Stephen's Cathedral in Vienna. The original building was burned in 1285. While elements of that structure remain, the cathedral was rebuilt in the following two centuries. No noteworthy single works of sculpture are decaying, but the edifice as a whole is suffering. The base of the main spire, which rises 450 feet above the heart of the city, has been in splints of metal scaffolding, and much of the church has been in a state of slow repair since World War II.

SPAIN

There are not many stone art works. However the Government is worried about castles in Spain. Two years ago it formed the Sociedad de Amigos de Castillos (Society of Friends of Castles), which is trying to persuade Spaniards to purchase castles on the condition that they restore the structures. The Government is restoring some castles itself, converting a few into hotels and encouraging village officials to help.

Here in New York, Mr. Rorimer said of the Cloisters: "Things are generally under special roofs, glass skylights or enclosed in winter to protect them from the elements. Where the capitals are unprotected during the winter, duplicates are available should replacements be required in the years to come. These are duplicates of originals and were made in the 14th and 15th centuries."

The New York Times

RESEARCH in preservation is conducted by Prof. Seymour Lewin, left, and Craig H. Smyth, who serves as the director of the New York University Institute of Fine Arts.

Most of the apse from the Church of San Martin at Fuentidueña, north of Madrid, which was put up at the Cloisters in 1961, is indoors. Cast-stone reproduction of two of the most important sculptures from the apse have been made and placed outdoors. The originals are indoors.

Egyptian sculptures in the Metropolitan Museum have been washed with water in large vats and treated with wax or other substances to harden the surface. Some have been steam cleaned or washed with soap to remove soluble salts.

"Every region has its own characteristic problems," Dr. Lewin said. "What might be true of New York would not be true of Cairo or New Delhi. For example, in Egypt, stone decay is not much of a problem. It is dry and warm, and because of an absence of a great deal of industry, there is no worry about air pollution.

"On the other hand, temples in the Orient are deteriorating not because of the industrial environment, but because of the natural environment — exposure to high humidity, fungi, lichens."

There are hundreds of types of stone. Among those commonly used are limestone, marble, granite and sandstone.

Within each kind of stone there are significant differences. Some of the stone in the Strasbourg Cathedral has been deteriorating rapidly. Yet other sections of the cathedral made apparently of similar stone are in much better condition.

Another example is the Washington Monument, which is having its first cleaning in 30 years. The lower third of the shaft is a lighter grayish white than the top two-thirds. The difference exists because work on the monument, which was begun in 1848, was interrupted between 1854 and 1880 for lack of funds. When work was resumed, the marble came from the same Maryland quarry, but it was from a different stratum.

The monument has not been subjected to much decay because of its flat surface. Rain washes it effectively, and layers of grime and acid do not have a chance to build on it.

Because of the differences in stone, the search for a remedy is difficult. The ancient Greeks used resins. Today, a major waterproofing technique is to spray or brush on silicic acid or synthetic resins. Another treatment is to impregnate the stone with wax. White beeswax is melted and brushed or sprayed onto the stone.

"Sometimes impregnation with wax or silica does more harm than good," Dr. Lewin said. "It forms a sort of skin on the surface of the stone to make it water repellent. But moisture sometimes gets trapped beneath this layer. When winter comes, the moisture freezes and causes the surface to crack. In a way, the silica or the wax prevents the stone from breathing.

Techniques Defective

"The trouble with most of the techniques is that the surface stays solid, but the deterioration process is not stopped. It's like tying a band of iron around the stone. The stone is still subject to decay but the pieces do not fall away."

A company here specializing in stone restoration is Nicholson and Galloway of 101 Park Avenue. It has restored many buildings, including Brooklyn's Borough Hall, Carlisle Cathedral in Britain and the Fifth Avenue Presbyterian Church.

"Stone is definitely deteriorating much quicker than in the past," said John Nicholson, head of the company. "There are a number of churches that have shown more advanced decay in the last 10 to 20 years than in the first 80."

The company does general maintenance work on Rockefeller Center's limestone exteriors. In the last decade all masonry joints of all buildings at the center have been repointed. Mortar between the stone blocks is cut out generally to a depth between ¾ of an inch and one inch, and new mortar is put in.

"Repointing is similar to dentistry," according to a Nicholson and Galloway promotional booklet. "If cavities in building stone are neglected, an operation may be required."

April 13, 1964

Index of Air Pollution To Be Printed in Times

Beginning today, The New York Times will publish an air-pollution index based on data supplied by the city's Air Pollution Control Laboratory. The laboratory will not provide data on weekends and therefore no index will be published in The Times on Sundays and Mondays.

From 1957 to 1963, the laboratory found that the air-pollution index could be calculated on the basis of average amounts of sulphur dioxide, carbon monoxide and smoke.

Using a mathematical formula expressing the usual relationships of the various elements in the air, the laboratory determined that the figure 12 could be considered as an average for the city's air quality. The number 50 has been tentatively set as the emergency level at which certain precautionary measures should be initiated.

Air Pollution Index

Period ended noon yesterday.	Average	Emergency level.
13.9	12	50

This index is an arbitrary computation based on the amount of sulphur dioxide, carbon monoxide, and smoke in the air recorded by the City Air Pollution Laboratory.

March 3, 1965

The Car and Smog: A Growing Controversy

Auto Industry Says Evidence Does Not Warrant Controls

By DAVID R. JONES
Special to The New York Times

DETROIT, April 4—A little-noticed controversy involving the family car, which seems destined to have significant influence on the welfare and pocketbooks of millions of Americans in the years ahead, is raging throughout the nation.

The controversy centers on the question of how much the automobile contributes to the nation's air-pollution problem, in which the major metropolitan areas, such as Los Angeles, face the prospect of eventually choking on their own exhaust.

What should be done to control the gases vehicles spew into the air? The question is pitting the automobile industry against Federal, state and local authorities.

This debate over automotive air pollution will be in the spotlight this week as the Senate Public Works Committee's subcommittee on air and water pollution begins three days of hearings Tuesday in Washington. Senator Edmund S. Muskie, Democrat of Maine, who is the subcommittee's chairman, has introduced a bill with broad support to impose nationwide exhaust controls on new cars.

Most government officials involved in the question believe the automobile is a growing pollution menace whose emissions must be strictly controlled. The official attitude was reflected in President Johnson's conservation message, in which he spoke of the need to substantially reduce or eliminate automotive air pollution.

The nation's automobile makers are equally convinced that there is insufficient evidence that cars are a national pollution problem to warrant controls. They urge further research, and say the taxpayer would get far more for his pollution dollar by spending it to expand controls on pollution from industrial plants and homes instead.

Devices Seen Likely

The most interesting aspect of this battle is that although the two sides squabble over the issue, both are coming to the conclusion that some kind of nationwide controls on auto exhaust on new cars within three years—merited or not.

The industry still believes this would be unjust and it continues to resist the prospect. However, most informed auto executives agree with one com-

Smog enveloping a section of downtown Los Angeles. Motor vehicle exhaust is said to be a major contributing factor to the condition, but industry sources question this theory.

pany's top pollution engineer, who laments:

"It's clear the politicians think this is for home and motherhood and it hasn't got a chance of losing."

This attitude, in fact, will prompt the industry in this week's hearings to take a more positive approach to the subject. Auto executives are expected to indicate a greater willingness to install control devices if Congress insists, although they still will argue that it would be premature to do so.

They may get some support from the Department of Health, Education and Welfare, which is expected to speak out in favor of the devices but to caution about the problems of inspection and the dangers of moving too quickly.

The relative merits of these divergent positions, and the outcome of the controversy, are of significance because almost every American will be touched by the results. A Federal law requiring exhaust-control devices could add $60 to $75 to the cost of most new cars and raise average maintenance bills by $10 a year.

Pollution experts in the United States contend that failure to install such devices would result in even more costly losses in pollution damage, not to mention health dangers.

Interest in controlling automotive pollution more closely has been mounting for several years, but it has gained new impetus in several states since the industry last August agreed to install exhaust-control devices on the 1966 models it will begin selling this fall in California.

"As soon as word came that the industry was doing something for California, there was a feeling elsewhere that if they could do this for California, they could do it for other states," says Donald A. Jensen, executive officer of the California Motor Vehicle Pollution Control Board in Los Angeles.

Alexander Rihm Jr., executive secretary of the New York State Air Pollution Control Board in Albany, discloses that the board last month voted to support legislation that would require exhaust-control devices on all new cars sold in the state. The board has not yet determined any specifics, but it could select elements from among a half-dozen exhaust-control bills already in the Legislature.

A subcommittee of the New Jersey Air Pollution Control Commission recently recommended exhaust controls for new cars in that state. This has led to a bill that has the backing of Gov. Richard J. Hughes.

The General Services Administration, which buys cars for the Federal Government plans to require exhaust-control devices on Government cars starting with 1967 models.

The direction and pace of any move to control exhaust on a nationwide basis should make itself clearer in the next few weeks.

The Muskie bill, which already has support within the Public Health Service, has 22 cosponsors. There is considerable Congressional sentiment for auto-pollution control. At present, however, there appears to be at least equal reluctance to impose too light controls on the auto industry.

Talks and Rebuttal

President Johnson is expected to make good before long on the promise in his conservation message to "institute discussions with industry officials and other interested groups" toward reducing or eliminating auto pollution. The industry is already preparing a rebuttal and bracing for what one executive expects will be "a little arm-twisting" from the President to get the companies to install control devices voluntarily.

There is a possibility that the industry would eventually install exhaust controls on a nationwide basis even if Federal legislation and Presidential persuasion failed. The industry will be installing devices on about 10 per cent of its market when it acts in California. If New York and a couple of other big states require devices, the industry could decide to install them voluntarily. Most auto men, in fact, would prefer Federal standards than a wide range of differing requirements in various states.

Until recently, most of the controversy over automotive air pollution has been confined to California, and particularly the sprawling Los Angeles region. There a combination of climate and geography have created repeated instances of photochemical smog, an irritating and damaging condition caused by the interaction of the sun's ultraviolet rays and chemical pollutants in the air.

Los Angeles has been fighting smog for 20 years, and first imposed strict controls on industry and homeowners to cut down on emissions into the atmosphere. California, which in 1953 pinpointed the reaction of sunlight and automotive emissions as the major air-pollution source, set up the state control board in 1960 to restrict auto emissions and bring air quality back to 1940 standards.

Authorities in other cities have begun in recent years to express greater concern about the problems of air pollution, and particularly auto pollution.

Anthony J. Celebrezze, Secretary of Health, Education and Welfare, in a report to Congress last December, called for immediate nationwide controls.

"Photochemical air pollution or smog is a problem of growing national importance and is attributable largely to the operation of the motor vehicle," the report declared. "Manifestations of this type of air pollution are appearing with increasing frequency and severity in metropolitan areas throughout the United States."

"Biological studies of animals show that the photochemical reaction products of automotive emissions produce adverse health effects," the report continued. "There is substantial evidence that these effects may appear in humans after extended exposure to air which is known to be polluted with these same products in many of the larger urban areas."

Nobody knows for sure how much automobiles contribute to the total problem of air pollution, which has been estimated to cost the United States as

much as $11 billion annually in agricultural damage, corrosion, soiling and similar effects. Some Government sources have held auto emissions responsible for about half of the national problem, but that is not certain. California studies indicate that about 10 per cent of the 67.3 million gallons of gasoline consumed by vehicles in the United States last year was emitted into the air.

Most cities have inadequate facilities for accurately determining how much pollution comes from autos. Arthur J. Benline, New York City's Commissioner of Air Pollution Control, estimates that cars cause one-third of the region's problem, but concedes this is only a guess.

Walter E. Jackson, Philadelphia's chief of air pollution control, says studies there indicate the auto causes 25 per cent of the city's problem.

James V. Fitzpatrick, director of Chicago's Department of Air Pollution, calculates from fuel statistics that half the organic gases in the atmosphere there come from vehicles.

Authorities concede that nationwide controls on auto emissions would primarily benefit dwellers in metropolitan areas and impose an unnecessary burden on the millions who reside in pollution-free rural areas. However, they point out that two-thirds of the population already lives in 212 standard metropolitan areas having only 9 per cent of the nation's land area. The Public Health Service says that any place with a population of 50,000 has enough vehicles to create a potential problem.

Impact Bound to Rise

The auto's contribution to pollution seems bound to spread. Between 1950 and 1959, the number of cars on the nation's highways rose at a rate double the population growth.

"This problem can easily become more widespread as the number of gasoline-burning vehicles increases — to 103 million by 1970 at the 1950-to-1959 rate of growth," according to the United States Surgeon General's office.

Gasoline-powered vehicles already discharge an estimated total of 92 million tons of deadly carbon monoxide into the air annually, not to mention millions of tons of the smog-forming hydrocarbons and nitrogen oxides that California is striving to control. The daily output

of carbon monoxide from vehicles, if confined over one area, would pollute the air to a concentration of 30 parts of carbon monoxide for each million parts of air to a height of 400 feet over 20,000 square miles — roughly the area of Massachusetts, Connecticut and New Jersey combined.

This concentration, for an eight-hour period, has been judged by California health authorities to cause discomfort, eye irritation, plant damage and reduced visibility. The level is often reached in Los Angeles, occasionally reached in Washington, and exceeded for at least an hour on some days in Chicago and Philadelphia, according to the Public Health Service's air-monitoring stations.

The monitoring network's studies also show that an average of 45 per cent of all hydrocarbons and 40 per cent of all nitrogen oxides discharged into the air in six selected cities come from automobiles. These chemical compounds, under the sun's ultraviolet rays, produce a photochemical reaction that creates oxidants, which are chemical substances such as ozone that have more of an oxidizing power than oxygen itself.

Measure of Pain

California health authorities say that when the oxidant level reaches 0.15 parts for each million of air for one hour, discomfort, eye irritation and plant damage can result. The air-monitoring network found that oxidants equalled or exceeded 0.18 parts for each million of air at least once in 1963 in Cincinnati, Chicago, New Orleans, Los Angeles and Washington. The level equalled or exceeded 0.25 parts for each million at least once in the first half of 1964 in Cincinnati, Philadelphia, St. Louis, Los Angeles, and Washington.

Although these levels were reached only a few days each year, most pollution experts say the frequency is increasing. The air-monitoring network already shows that oxidants in the first half of 1964 reached 0.10 parts for each million of air 34.3 per cent of the days in Los Angeles, 15.7 per cent in Chicago, 11.5 per cent in Washington, and more than 8 per cent in Philadelphia and St. Louis.

"Los Angeles no longer has, if it ever had, a monopoly on photochemical smog," according to a recent Department of Agriculture statement, which said the "characteristic symptoms

have been found in almost every metropolitan area in the country."

The entire coastal area from Washington to Boston "has come to rival Southern California for extent, severity, and economic loss to agriculture because of photochemical smog," it said.

Dr. John T. Middleton, director of the air-pollution research center at the University of California at Riverside, says ozone plant damage from photo chemical smog has been found in 22 states.

"The injury that was once confined to Los Angeles County now occurs in many states and causes economic loss estimated in excess of $25 million annually," he says.

Most stanch advocates of auto-pollution control believe vehicle emissions are a human health hazard, but they have been frustrated in their desire to prove it. The United States Surgeon General's office has said that years will be required before a definitive answer is found to that question.

However, a Public Health Service spokesman in Washington insists "we have bits and pieces of circumstantial evidence" that make controls desirable without proof.

"We think we must err on the side of caution when it comes to the protection of public health," he says. "If we have good reason to suspect the public is being exposed to a hazard that can be lessened or eliminated, it ought to be lessened or eliminated."

Dr. Dietrich Hoffman, a biochemist and air-pollution expert at the Sloan-Kettering Institute for Cancer Research in New York, has found that auto exhaust contains particles that, when applied to the skin of mice, have caused cancer. This does not prove that auto exhaust will cause cancer in humans, he explains, but he is working with the General Motors Corporation to reduce formation and emission of these particles — just in case.

"We have no evidence," he asserts, but the work with mice indicates this might be a worthwhile effort because "our goal is prevention."

"Nobody claims man will benefit from air pollutants," he says. "Everyone agrees we can only gain when we find ways to control it."

April 5, 1965

Atomic Age Has Altered the Habitat of Man

BY WALTER SULLIVAN

Twenty years after the first atomic bomb exploded over Hiroshima on Aug. 6, 1945, New York time, studies of fallout residues show that the physical, as well as the political, environment of the world has been permanently altered.

Even if there are no more bomb tests—which seems unlikely—the radioactivity of carbon in the air, the oceans and all living matter will remain twice normal for a prolonged period. It will not sink to its pre-bomb level for thousands of years.

Although large-scale testing of nuclear weapons in the atmosphere ceased at the end of 1962, the level of strontium 90 on the surface of the earth has continued to rise. According to a Weather Bureau study, it will probably reach its peak this year and then begin to sub-

side.

It has been rising because much of the bomb-produced fallout was injected into the stratosphere, where it has remained suspended, sinking to earth primarily during the spring. This reservoir is thought largely to have been drained by now, but a considerable amount of the bomb-produced strontium is "missing."

From the number of weapons exploded in the air it had been calculated that a certain amount of strontium 90 was added to the environment. Scouting of the upper air by U-2 aircraft, balloon samplers and other devices has enabled specialists to calculate the amount of bomb debris still up there.

Calculations Are Off

These observations, plus ground measurements, indicate that about 90 per cent of the observed strontium has fallen to the ground and only 10 per cent is still aloft. But when these two components are added, they fall far below the estimated total production. Apparently there is a gross error in the calculations or the observations.

The level of strontium is of special importance because it is picked up by the body and incorporated into bones and teeth. This arises from its chemical similarity to calcium.

While the ground burden of strontium 90 is still thought to be rising, the level in milk hit its peak in the spring of 1964.

Milk is the subject of special surveillance by the United States Public Health Service because it constitutes a rapid channel for delivery of fallout to the population. The bomb material falls on fields, is eaten by cows and can reach the consumer within days.

This is particularly critical where iodine 131 is concerned. This bomb product retains its radioactivity for so short a period that it is of no medical significance unless it reaches a person soon after the bomb blast. Its half life is eight days, which means that in such a period it loses half its radioactivity.

Half Life Is 28 Years

By contrast, the half life of strontium 90 is 28 years. That of carbon 14 is roughly 5,730 years. Such radioactive forms of various atoms are produced in the explosion and become at least a temporary part of the environment. Only in special cases have the levels become high enough to be of general medical concern.

Among examples of the special cases are the Alaskan Eskimos who live on caribou that eat reindeer moss. This moss has been comparatively rich in strontium from Soviet tests across the Arctic Ocean. The Eskimos are now under special observation.

Another source of concern has been the exposure of farmers living in valleys near the Nevada test site. On several occasions in recent years radioactive debris has been ejected from shallow underground explosions and been carried by wind beyond the test site.

On occasion, this has contaminated milk on nearby farms to the point where it was necessary to seize the milk and give the farmer a substitute supply.

Within a few days after the Chinese fired their first nu-

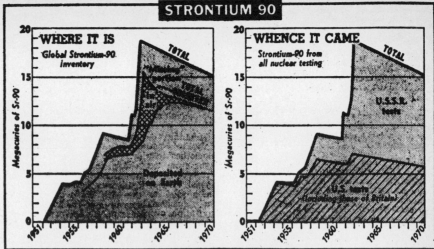

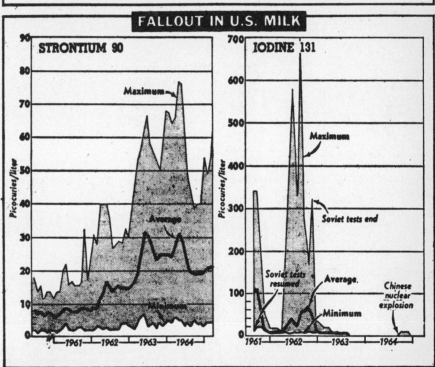

Aug. 7. 1965   From Weather Bureau and Public Health Service Reports

Upper diagrams depict year-to-year distribution of strontium 90. At right is relative contribution of nuclear tests. At left is strontium on ground and still airborne, plus estimated total from all tests. Air and ground samples fall short of estimate. Bottom diagrams show levels of radioactive strontium and iodine in milk. Since iodine decays rapidly, there has been almost none since 1963, except for small amount after Chinese tests.

clear explosion last Oct. 17 there was a sharp rise in the radioactivity of thyroid glands of cattle slaughtered in New York State. These glands are monitored by the Public Health Service in cooperation with the United States Department of Agriculture, since radioactive iodine entering the body moves quickly to the thyroid gland.

Rate Rose in November

The rise hit its peak in early November, then sank to normal in December. A similar jump was noted in the nationwide Pasteurized Milk Network of the Public Health Service. It operates 63 stations in the United States, with at least one in each state, plus Puerto Rico and the Canal Zone.

A similar rise was observed after the second Chinese blast, on May 15 of this year. However, the resulting levels were minor compared to those during the intense testing of 1962.

Among the various ways in which the Division of Radiological Health of the Public Health Service keeps an eye on fallout effects is the analysis of bones obtained from the newly deceased or where bone has been removed surgically. The program was begun late in 1961 and has shown strontium levels about half those expected.

This was stated by James G. Terrill, Jr. in his testimony a few weeks ago to a subcommittee of the Congressional Joint Committee on Atomic

Energy. Mr. Terrill is deputy chief of the Division of Radiological Health.

Strontium 90 levels in the bones of those less than five years old hit their peak in 1963. Diet studies, however, show that the maximum intake of fallout particles by the population was probably last year.

Students of the world's weather have continued to use fallout as a tool for tracing the movements of air masses.

Special tracers were included in certain of the high altitude explosions. For example, rhodium 102 was used to tag debris from a 1958 shot fired 27 miles above the Pacific. Cadmium 109 tagged fallout from a 1962 shot 250 miles up.

An unplanned tracer was added to the atmosphere when an orbiting reactor accidentally plunged into the atmosphere in April, 1964, dumping plutonium 238 into the air as it burned up. It has been found that it takes about two years for the stratospheric remains of such material to become uniformly distributed over the entire surface of the earth.

Prior to this there was little information on the extent to which there is an exchange of air between the Northern and Southern Hemispheres of the earth.

Likewise, the repeated observation that fallout descends from the stratosphere in spring has led to efforts to explain this phenomenon.

At Pennsylvania State University, Dr. Edwin F. Danielsen proposed that a fold of the stratosphere is sometimes pushed down into the troposphere, the lowest layer of the atmosphere, carrying fallout particles with it.

He thought that the Rocky Mountains might precipitate such folding of the lower boundary of the stratosphere. A project was organized by the Defense Atomic Support Agency the Atomic Energy Commission, the Weather Bureau and others to investigate the proposal.

Air sampling was done by B-50 and B-57 aircraft equipped to filter air en route and measure continuously the radio-activity of the collected material. The flights repeatedly traversed what appeared to be a fold of stratosperic air deep in the troposphere and every time they did so there was a sharp jump in radioactivity.

It is hoped that further studies of this sort will spell out in greater detail the manner in which this apparent folding takes place.

August 7, 1965

WARNING IS ISSUED ON LEAD POISONING

But Findings of Geochemist on Coast Are Disputed

By WALTER SULLIVAN

A California geochemist presented in detail yesterday his argument that lead was contaminating the environment to a dangerous degree.

A preliminary report by Dr. Clair C. Patterson, a research associate at the California Institute of Technology, was presented earlier in the week, and provoked the lead industry, the petroleum industry (which adds lead to gasoline) and at least one Public Health specialist to challenge his conclusions.

The chief issues are whether the lead content in human blood has sharply risen and whether the level is dangerous.

The current level is generally believed to be roughly half that at which obvious symptoms of lead poisoning begin to appear, although vulnerability to the substance varies radically among individuals. The lead level in a person's body is also generally believed to be affected by his environment.

Survey Taken

This was indicated in a survey recently published by the United States Public Health Service. Lead levels in the blood of 200 Cincinnati garage mechanics and parking lot attendants were found to be three times as high, on the average, as those of 25 Philadelphia suburbanites and California rural residents.

Lead is used in automobile fuel to promote uniform combustion and avoid "knocking."

Dr. Patterson's research has been supported by the Public Health Service and the Atomic Energy Commission. Details of his findings were made public in the Archives of Environmental Health, published by the American Medical Association.

His argument that the lead burden of human bodies has radically increased is a geochemical one. The chemistry of the body is closely related to the chemistry of the environment in which it has evolved. The most striking example of this is the resemblance of the mixture of salts in body fluids to those typical of sea water, indicating that man's ancestry stems from the sea.

Far Above 'Normal'

Dr. Patterson reasons that the proportions in the body of such metals as lead and barium should be related to their general abundance. From this he argues that Americans carry 100 times more lead than is "normal."

He believes a large part of this is caused by the use of lead in gasoline. He and his colleagues, who have studied ancient ocean sediments, calculate that the oceans are receiving lead at 50 times the rate of primitive times.

Likewise, they say the leads found in Los Angeles smog and California snow and gasolines all bear the same atomic fingerprint (in terms of isotopic abundance). This fingerprint differs basically from that of lead found in the old sediments.

The National Science Foundation reported earlier in the week that Dr. Patterson had found a sharp increase in the lead content of snow that had fallen on Greenland the last decade. This is being investigated further both there and in the Antarctic.

One critic of the findings was Dr. Leonard Goldwater, professor of occupational medicine at Columbia University. He has obtained blood and urine samples from 1,000 persons in the United States and 14 other countries in the Far East, Middle East, Europe and Latin America.

Other Reports Confirmed

His analysis of these specimens for lead, mercury and arsenic has shown no significant change in lead values from studies done some 30 years ago, he said. His project was carried out under a Public Health Service grant. His findings, however, confirmed other reports that lead levels in the body were affected by one's way of life.

This means that increased exposure to lead can produce an increase in the body burden of that metal, even though the body still sheds much of its lead intake.

Vehicle exhaust appeared to be the chief contributor of lead to the atmosphere, Dr. Patterson said, but he also noted other contributors. Among these was the use of lead arsenate insecticides on tobacco plants. Lead levels apparently are higher in those who smoke.

Other uses of lead that Dr. Patterson believed could be curtailed were in water pipes, kitchenware, paints and food-can solder.

In a statement on Dr. Patterson's report, the American Petroleum Institute said his findings had been based on geological studies "and it is impossible to interpret geological findings in biological terms." All "accepted medical evidence," it added, "proves conclusively" that lead in the environment presents no threat to public health.

Urban Situation Studied

The institute, as well as the Lead Industries Association, drew attention to the survey of lead in the atmosphere of three urban communities, published last January by the Public Health Service.

The survey was carried out by a team consisting of nine specialists from the petroleum, automotive and chemical industries and five Public Health officials. The highest lead levels found in those studied were "well within the presently accepted range of lead levels for humans," the report concluded.

Dr. Patterson, in his study, indicated that he did not consider this conclusion justified.

September 12, 1965

President Signs Bill to Control Automobile Exhaust Impurities

Special to The New York Times

WASHINGTON, Oct. 20—In his first ceremonial bill-signing since he entered the hospital for surgery, President Johnson today placed his signature on controversial legislation that will lead to exhaust controls on all new cars within two years.

Mr. Johnson signed the measure—whose formidable title is the Clear Air Act Amendments and Solid Waste Disposal Act—in his bedroom. The President was dressed in a dark blue, three-piece business suit, and sat behind a low writing table about 10 feet from his bed.

Behind him, in a semicircle, were Vice President Humphrey, Secretary of Health, Education and Welfare John W. Gardner, and 10 Senators and Representatives, including Senator Edmund S. Muskie, Maine Democrat, who was the principal architect of the bill.

The ceremony climaxed a long fight between the Congress and the automotive industry, a fight that occasionally found the Johnson Administration, which strongly supported the idea of controls—on both sides of the issue.

Control Standards Set Up

The legislation directs the Department of Health, Education and Welfare to establish standards for the control of automobile and diesel truck emissions in an effort to cut down on automobile air pollution.

As originally envisioned, however, the bill would have set specific limitations on allowable crankcase and exhaust emissions. In response to Administration requests, the specific standards were dropped in favor of discretionary power for the Health Secretary to set standards.

In addition, the Senate committee that wrote the bill deleted several smaller provisions at the Administration's request and also dropped the Nov. 1, 1966 deadline for emission standards.

Accordingly, the bill signed today does not specify any standards or set any date for their imposition. But Health Department officials have stated that they intend to set standards in a couple of months, to take effect Sept. 1, 1967.

This timetable, if carried out, means that all buyers of 1968 model cars will face added costs of between $18 and $45 a vehicle because of the controls. This would amount to more than $330 million annually on the basis of current new car sales of about 8.5 million units a year.

The timetable was something of a surprise to the industry, which until recently had hoped that the House would not act on the bill before next year. The house passed the measure late last month.

The industry said last April that it could install controls on 1968 models if Congress insisted, but it had been counting on extra time for research and hoped that no action would be necessary before the 1969 models.

Informed sources have said that there is a general understanding between Senator Muskie and Health Department officials that the standards would be similar to those going into effect now for new cars in California.

In signing the bill, Mr. Johnson said that "since the beginning of the Industrial Revolution we have systematically been polluting our air."

"We have now reached the point where our factories, our automobiles, our furnaces and our municipal dumps are spewing more than 150 million tons of pollutants annually into the air we breathe— almost onehalf million tons a day," the President said.

October 21, 1965

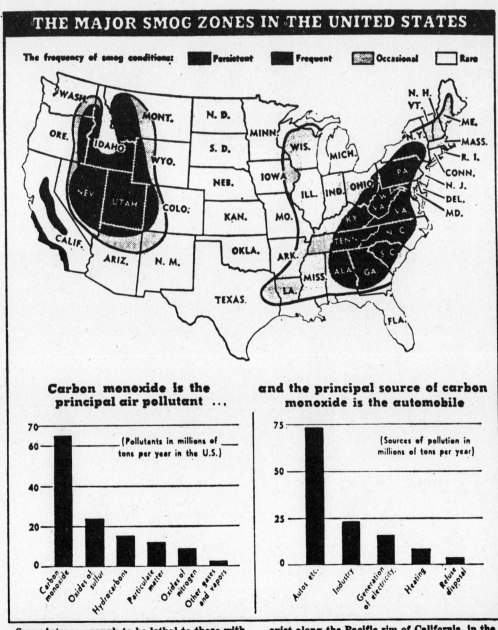

THE MAJOR SMOG ZONES IN THE UNITED STATES

The frequency of smog conditions: Persistent — Frequent — Occasional — Rare

Carbon monoxide is the principal air pollutant ...

(Pollutants in millions of tons per year in the U.S.)

and the principal source of carbon monoxide is the automobile

(Sources of pollution in millions of tons per year)

Smog intense enough to be lethal to those with lung ailments occurs when a combination of weather and terrain put a lid over a source of fumes, such as an industrial area or a heavily motorized city. Such areas of persistent danger exist along the Pacific rim of California, in the Utah area and in the mountains of Appalachia (map). However, not all these regions as yet generate enough smoke or vehicle exhaust to constitute a present-day health hazard.

April 3, 1966

Los Angeles: A Model for the Nation in Its Growing Struggle Against Air Pollution

CITY NEARLY FREE OF INDUSTRY SMOG

County Control Agency Cited by U.S. for Efficiency— Its Auto Rules Copied

Following is the third of four articles on air pollution, its effect on national life and what is being done to end it.

By GLADWIN HILL
Special to The New York Times

LOS ANGELES, Sept. 27— Next year, anyone in Los Angeles who wants to paint his house will have to use paint approved by the county. The county will tell retailers what kind of paint they can sell, which, in effect, is the same as telling manufacturers what kind of paint they can make if they are interested in the big Los Angeles market.

Some paints contain particularly noxious solvents that evaporate into the air. Applied to thousands of buildings simultaneously, they make a small, but appreciable, contribution to smog.

The ban on them, in favor of milder solvents, underscores how far Los Angeles has gone in its efforts to reduce air pollution. It also indicates the sort of restraints communities across the country may have to consider in the growing struggle with air pollution.

The general public will get its first taste of the restraints next fall, when new-car buyers will pay about $50 for special engine equipment to meet new Federal limitations on fume discharges.

Los Angeles Lesson

Perhaps few other places in the country will have to invoke the same measures as Los Angeles. But every community can profitably scrutinize Los Angeles and the state of California—first as the earliest, if not the worst, focus of pollution, and second, as prime examples of what can be done to solve the problems.

The Los Angeles County Air Pollution Control District, an enforcement agency with a 285-page manual of operations, has been cited by the United States Public Health Service as an administrative model.

The Motor Vehicle Pollution Control Board of California developed the automobile fume regulations that the Air Pollution Division of the Public Health Service is applying nationally next year. And Federal

officials plan to use California's automobile smog-control experience in the years ahead as a guide for national policies.

By dint of more than a decade of travail, Los Angeles County has eliminated nearly all its pollution from industry and other stationary sources—the principal producers of smog in many communities.

The county's continuing bouts of air pollution are attributed largely to its 4 million automobiles and the exaggerated chemical effect of California sunshine on their effluents.

Most of these cars do not have the fume-control equipment that California first demanded with the 1966 models As with the rest of the country, it will take 10 years for the older fume-spewing cars to be replaced.

Both Los Angeles and the state have done equally important pioneering in the critical nontechnical phase of pollution abatement: the political side.

By traditional standards, the paint-control regulation might have touched off a nationwide ruckus. Yet when Louis J. Fuller, director of the Los Angeles control agency, presented the proposal to the County Board of Supervisors in July it was ratified without a moment's debate. The groundwork had been laid in months of patient diplomacy with both the paint industry and public officials.

Things haven't always gone that smoothly. In its early days the district had to battle some industrial concerns all the way up to the United States Supreme Court to establish its regulatory authority.

The agency is before the State Supreme Court now in litigation with the oil industry over a regulation to limit the use of fuel oil in favor of cleaner natural gas.

But, over the agency's 19-year history, its combination of militancy and political adroitness has worked a revolution in matters of air cleanliness, which, at the end of World War II, precipitated a wave of civic outrage and alarm.

Coal burning, never extensive in Los Angeles, has been abolished entirely. Even oil burning —which produces sulphur dioxide fumes, as does coal—is banned seven months a year; natural gas must be used instead, and it must be used by industry when it is available.

Back-yard incinerators have been outlawed; municipal rubbish and garbage incineration abolished in favor of land-fill disposal, and building incineration ended except for a few expensive smokeless furnaces.

Fume emissions from stationary sources have been reduced from 6,375 tons a day to 1,375 tons—only about one-tenth as much as the discharge from cars.

In the 5,000 tons eliminated, the major items were 3,270 tons from oil refining, production and marketing, 600 tons from refuse burning, and 495 tons from fuel-oil burning.

Of the remaining 1,375 tons of emissions, it is estimated that 600 tons come from solvents, in uses ranging from dry cleaning to printing. The new solvents' regulation is designed to eliminate the 200 daily tons of solvent fumes that are considered controllable by any blanket measure.

1940 Levels Sought

"The program to abate air pollution from stationary sources was aimed at reducing pollution levels to those existing in 1940," says Mr. Fuller. "This goal has almost been reached," he says. "There is little more that can be done with stationary pollution sources. There are only two areas where significant improvements can be made—extension of the use of natural gas in large boilers and reduction of organic solvent emissions."

Mr. Fuller, an urbane but tough former police captain, succeeded S. Smith Griswold, the tall, stately and equally firm administrator who recently became chief of Federal abatement activities.

Mr. Fuller says that industry in Los Angeles County—including the nation's third-largest complex of oil refineries—has spent $125-million on pollution control equipment. Ancillary costs, such as installation, real estate and operating costs, may have cost another $125-million.

The grand total of all conceivable costs of Los Angeles's 19 years of pollution control, Mr. Fuller hazards, might run as high as $750-million.

This hypothetical cost amounts to only $7 per person per year. Mr. Fuller observes that the total is small when compared to the $11-billion in tangible damages which, Federal statistics indicate, uncontrolled air pollution would have inflicted on the county over the 19 years.

The control board, with jurisdiction over any installation that may produce fumes in the 4,000 square miles of the nation's second most populous metropolitan area, has a staff of 284. Its budget this year is $3,663,000.

Conducting inspections at the rate of 225,000 a year, it has issued 70,000 operating permits for equipment that could produce fumes. Violators of the more than 100 regulations are subject to misdemeanor penalties of up to six months in jail, with fines of as much as $500 for each day's violation.

In the last 10 years the agency has prosecuted 30,000 court cases, with a conviction rate of 96 per cent and fines totaling more than $700,000.

Initially, recurrent air pollution was so severe that the agency instituted a warning system, with three stages of "alerts."

The first stage signifies only that the air is approaching the "maximum allowable concentration" of 100 parts per million of carbon monoxide, three parts per million of sulphur oxides, three parts per million of nitrogen oxides, or .5 part per million of ozone.

Health Menace Graded

The second stage, based on approximately twice these concentrations, denotes a "preliminary health menace."

The third stage, based on three times the allowable concentrations signifies "a dangerous health menace."

At the second stage, the agency is empowered to shut down industry and stop automobile traffic. The third stage would call for state disaster action.

To date, conditions have never gone beyond a few annual first-stage alerts that are communicated to the public over radio and then canceled in a matter of hours or minutes. They arouse little notice.

There are eight other pollution control districts in California, covering 13 of the state's 58 counties.

It wasn't until the mid-1950's that scientists led by Dr. A. J. Haagen-Smit of the California Institute of Technology discovered that a major source of smog was automobile fumes, particularly the hydrocarbons, which are photosynthesized by sunshine into an array of particularly noxious gases.

It was found that one out of every 10 gallons of gasoline escaped unburned from a car. Every car in the nation, on an average, emits more than 100 cubic feet of fumes every day. Every thousand cars in a city emit more than 3½ tons of fumes.

To curb the discharges from its 9 million cars, California set up its Motor Vehicle Pollution Control Board in 1960. It is composed of 13 citizens appointed by the Governor, most of whom have special engineering, scientific or medical qualifications.

The board has an administrative and engineering staff of 16 and an annual budget of $485,000. Near its Los Angeles headquarters, it operates one of the world's most complete laboratories in the field of vehicular air pollution.

The newly created board promptly joined in a battle that the Los Angeles County agency had been waging for years with the automobile industry over fume controls.

Detroit Is Criticized

In a current Federal grand jury inquiry in Los Angeles, local officials have charged that.

the auto companies collusively dragged their feet for years on producing these controls. Detroit says it was engrossed in research.

When accessory manufacturers were on the verge of pre-empting the fume control business with exhaust attachments, Detroit announced in July, 1965, that it was prepared to turn out cars with built-in equipment.

This equipment is of two types. One is a pump that injects extra air into the exhaust manifold so that fumes are burned there. Chrysler accomplished the same fume reduction by special carburetor and ignition rigging.

The other is a "blow-by" tube. Since 1963, California has required that new cars have these tubes to carry troublesome crankcase gases back into the combustion chambers.

Starting last fall, with 1966 models, the exhaust controls were made mandatory on all new cars.

The two items, blow-by tubes and exhaust controls, are supposed to eliminate 70 per cent of the two worst automobile contaminants, hydrocarbon gases and carbon monoxide. Last March the Public Health Service felt confident enough of them to promulgate the California requirements nationally for next year.

The California board is considering tightening its restrictions by nearly 50 per cent before 1970. The new restrictions would reduce permissible hydrocarbon emissions from 275 parts per million to 180, and carbon monoxide from 1.5 per cent to 1 per cent.

The board is also moving toward control of another important pollutant family, oxides of nitrogen. Federal officials plan to follow close in applying such restrictions nationally.

California had hopes of developing exhaust controls for older cars, which make up 90 per cent of those on the road. This has been thwarted by both technical and legislative difficulties. The only requirement for pre-1963 models is that crankcase blow-by tubes must be put on cars when they change owners. There is no corresponding Federal requirement.

The achievement of control equipment is an oblique process. The state starts out by promulgating desired emission standards. The standards remain theoretical until industry comes up with the equipment to match the standards. This is the situation with oxides of nitrogen, for which controls have not yet been devised.

Another problem that will present itself in all 50 states is making sure fume control equipment operates effectively, something the Federal Government can't police.

Even California has only cursory random inspection by state highway patrolmen. Comprehensive checks on fume equipment can be incorporated in regular safety inspections— New York and New Jersey have laws calling for them -- but there is a problem of providing inspection stations with adequate testing devices.

California may also provide lessons for other states in the way it works out its administrative structure for pollution control.

There is now a three - way division of responsibility among the motor vehicle board, the county districts that deal with stationary pollution sources and the State Department of Health, which must work closely with the others.

Assuming that each contingent does its job perfectly, critics observe, there is no unified, statewide determination of priorities in expenditures and effort. However, Gov. Edmund G. Brown has just designated Eric Grant, the director of the motor vehicle board, as an inter-agency coordinator in a step toward unification.

Vernon MacKenzie, the recent director of the Air Pollution Division of the Public Health Service, is a soft-speaking man, inclined to understatement. But he recently remarked ominously:

"For the next decade or so, every urban area is going to have to control pollution in terms that might be considered radical even in Los Angeles."

September 28, 1966

SMOG EMERGENCY CALLED FOR CITY; RELIEF EXPECTED

Jersey and Connecticut Join Plea for Voluntary Action to Cut Down Pollution

CLEARING TODAY IS SEEN

By HOMER BIGART

The first stage of an air pollution emergency was declared yesterday in the New York metropolitan area, New Jersey and Connecticut as a stagnant air mass continued to envelop the Eastern Seaboard, trapping potentially lethal gases and smoke.

A sweep-out of the foul air after three days of smog was forecast by the Weather Bureau for today, when a cold air mass from the northwest is expected to start dispersing the gray, corrosive pall.

With the clean, cold air, due between 5 and 9 A.M., should come winds blowing in a southeasterly direction at 10 to 15 miles an hour.

Rain that began at 8:40 last night began clearing the atmosphere of fine dust particles.

Despite this relief, however, a spokesman for the Department of Air Pollution Control said that only the arrival of the cold air would spell an end to the smog.

Voluntary Controls Urged

Until the smog lifts, New Yorkers were asked through newspaper, radio and television announcements to comply voluntarily with these emergency antipollution measures:

¶Motorists were urged to use their automobiles only when absolutely necessary. Owners and operators of trucks were urged to curtail the use of their vehicles.

¶Owners of apartment houses were urged to shut down their incinerators.

¶Landlords using fuel oil or coal were urged to reduce indoor temperatures to 60 degrees. These fuels emit sulphur dioxide, the most lethal component in New York smogs. State law requires that between 6 A.M. and 10 P.M. an indoor temperature of at least 68 degrees be maintained if outside temperatures are below 50.

Temperature Sets Record

Heating was no problem, however, since the temperature here reached 64 degrees at 12:40 P.M. The mark broke the old record high of 63 for the date, set in 1946.

Austin N. Heller, the city's Commissioner of Air Pollution Control, made the recommendations at a City Hall news conference a few minutes after Governor Rockefeller proclaimed a "first alert advisory" for the metropolitan area.

The Governor acted on advice from State Health Commissioner Hollis S. Ingraham that air pollution levels in the metropolitan area had reached a danger point.

Similar voluntary first alerts were placed in effect for New Jersey and Connecticut. Residents of those states were asked to curtail the consumption of heating fuel, gasoline and electricity and to avoid burning trash or leaves.

At City Hall, Commissioner Heller said he saw slight chance that the temperature inversion—the lid of warm air that is preventing the pollution from escaping—would last long enough to warrant a second alert.

A second alert calls for more stringent "voluntary" restrictions on traffic and on industrial activity.

A third alert, declared only when the sulphur dioxide and carbon monoxide pollutants approach lethal levels, would empower the city to impose a brownout, throttling all but essential industrial activity and public transport.

New Yorkers went to work yesterday morning in acrid, sour-tasting air that was almost dead calm.

Many had headaches that were not the product, they thought, of holiday over-eating. Their eyes itched. Their throats scratched. But no deaths were attributed to the smog.

No Peril Seen Now

Commissioner Heller said he had checked 10 hospitals and was told that four of them had experienced an increase in asthmatic patients.

Deputy Mayor Robert Price, who stood by his side at the news conference, said that he had been in "constant touch with Mayor Lindsay, who is vacationing in Bermuda, and that the Mayor was "deeply concerned" but had decided not to rush home. The Mayor will return Monday as he had planned.

Dr. Aaron D. Chaves, the director of the City Health Department's Bureau of Tuberculosis, checked all the municipal hospitals and reported last night:

"In not one of them is a pattern emerging which would suggest we are dealing with an important health hazard as of this moment."

He said there had been an increase Thursday in the number of Bronchial asthmatic patients admitted to Sydenham and Metropolitan Hospitals. But other hospitals reported a decline in asthmatic admissions, and Dr. Chaves dismissed the Sydenham and Metropolitan figures as "random fluctuations."

Warning in 1963

Yesterday's air-pollution alert was not the city's first.

In October, 1963, a stagnant mass of warm moist air hung over the city for five days. On the fourth day, when pollutant counts had approached hazardous levels, the United States Public Health Service issued an "air-pollution warning" for the whole seaboard between Washington and New Bedford, Mass., extending inland to Altoona, Pa.

But the Public Health Service was powerless to enforce the alert and could only urge united action among the local governments in the metropolitan area.

The October, 1963, smog was swept away on the fifth day by a cold air mass from Canada, saving New York from a repetition of the disastrous smog

New York City, in a Shroud of Gray, Gasps for Air

incident of November, 1953, that accounted for more than 240 deaths.

The dirtiest hour yesterday was from 8 A.M. to 9 A.M., a period of dead calm, when the air-pollution index reached 48.9.

The Department of Air Pollution Control does not rec-basis of the index, which is based only on pollutant counts at the department's labratory at 170 East 121st Street. Other sections of the city might have filthier or cleaner air.

Interstate Group Acts

Yesterday's alert came after consultation between Commissioner Heller and members of the Interstate Sanitation Commission, which has been trying to clean up the air in the New York-New Jersey area.

Thomas R. Glenn Jr., chief engineer of the Interstate Sanitation Commission, said the first-stage alert had been recommended because of the high level of carbon monoxide and dust-carrying haze during the five-hour period prior to 11:25 A.M.

During this period, the carbon monoxide measurement exceeded 10 parts per million of air, and the haze exceeded 7.5.

The warning level was reached when the air sustained over 9 parts of carbon monoxide and 7.5 parts of haze for four hours. But the third dangerous component, sulphur dioxide, never passed the warning limit of 0.5 for any appreciable period, so action was delayed until after noon.

As a result of the rain last night, the carbon monoxide measurement fell from 20 parts per million at 8 P.M. to 5 parts per million at midnight. In the same four hours, the haze level dropped from 5.9 to 3.0, while the sulphur dioxide count tapered from .38 to .34.

It was bad news from the Weather Bureau—that despite some freshening of winds in the afternoon the stagnation would return in the evening—that impelled Commissioner Heller to urge the declaration of an alert.

Commissioner Heller pointed out that the city had already taken preparatory steps toward stemming the gush of pollutants into the atmosphere. The city's 11 incinerators were shut down, consuming just enough to keep the fires going.

Consolidated Edison said it had cut its emission of sulphur dioxide by 50 per cent on Thanksgiving Day and 40 per cent yesterday by using more natural gas at its power stations instead of fuel oil. The Long Island Lighting Company also voluntarily cut down on its use of fuel oil.

The Sanitation Department, however, faces an early crisis unless it can resume incineration. Commissioner Samuel J. Kearing Jr. said alternative means of disposal were expensive.

Each of the incinerators has a big storage pit, but these were filling rapidly. If the pits fill up and burning is still not permitted, the garbage would have to be hauled to distant land-fill operations in the

This is how New York appeared to Jerseyans looking across the Hudson from Hoboken

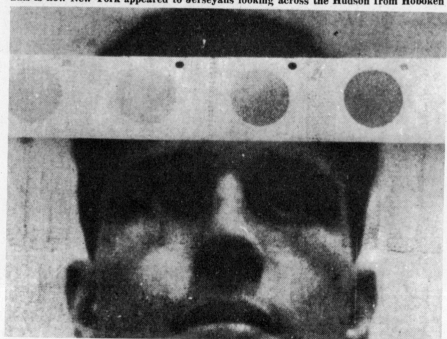

The New York Times (by Neal Boenzi and Ernest Sisto)

At the city's Air Pollution Control Center, 170 East 121st Street, a strip of test paper from an air-sampling device showed results of successive two-hour exposures to smog.

Bronx, Brooklyn and Staten Island. This would mean long trips for garbage trucks, often more than eight miles.

40 Scows Are Used

Refuse for the Staten Island fill at Fresh Kills is carried by barge, and there are only about 40 scows.

Commissioner Kearing said he might put some of them into night duty, for although none of the tugs are radar-equipped, they can navigate safely in the harbor, provided the smog is not too thick.

Some 600 extra sanitation men will be working tomorrow

to help get rid of the refuse. Commissioner Kearing said he was worried about the overtime bill.

The department's 11 incinerators are among the worst offenders of the city's air-pollution code.

It was revealed yesterday that something called "scrubbers" will soon be tried out at the incinerator at 74th Street and the East River. A "scrubber" is a device that throws a stream of air into gases as they proceed through a vent before reaching the stack. The air jet supposedly washes dirt particles out of the gas.

It will cost $200,000 to equip each of the furnaces with scrubbers, and the 11 incinerators have a total of 27 furnaces.

Commissioner Heller, who has promised a get-tough policy to clean up the air, has ordered public hearings starting Dec. 9 on proposals aimed at redesign-

ing some 17,000 private incinerators.

He has already applied to the United States Public Health Service for $545,000 in grants to help develop devices to reduce the outpouring of pollutants from these incinerators. The city would contribute $146,000 toward these projects.

A new law effective next May requires all incinerators—private and city-owned—to be modernized to comply with the air-pollution code.

Officials were thankful that the pollution crisis came on a holiday weekend, when fewer cars belch carbon monoxide and industrial activity is slowed. Although yesterday was a work day for most New Yorkers, traffic was abnormally light.

Frightened perhaps by stories of killer smogs, many New Yorkers apparently retired to the country for fresh air. No

revival of heavy traffic fumes is expected before Monday morning.

The effects of the smog were felt, to a lesser degree, in the suburbs and along the seaboard from Washington to Halifax, N. S.

Health officials in Nassau County reported no severe problems. "It is nowhere near the critical stage here as it appears to be in New York City," a spokesman said. "We are not alarmed by it."

In Suffolk County, the smog problem was called "almost nonexistent" by health officials, but the county asked the towns to halt all burning. Suffolk hospitals reported no increase in respiratory ailments.

A heavy haze hung over Westchester, but the county's Health Commissioner discounted any hazard.

"We have not received a sin-

gle report of anyone becoming ill, and I do not consider the smog to be serious at this time," said the Commissioner, Dr. William A. Brumfield Jr.

In Connecticut, State Health Commissioner Franklin M. Foote warned against the burning of leaves and rubbish. He said that some persons with bronchitis, emphysema, asthma or heart diseases were suffering.

The city of Greenwich reported that the smog had had no serious effects. The smog was light there compared with New York and other areas of Connecticut.

In New Jersey, the smog was said to be the heaviest in recent years, but there were no reports of an increase in serious illnesses. An 8 A.M. pollution reading taken in Elizabeth showed the level there was only half that of New York's.

November 26, 1966

Many Lands Wage War on Air Pollution

Smog Forcing Adoption of Strict Rules

White-collar workers in Germany's heavily industralized Ruhr District are getting used to carrying an extra shirt because they know the first one will be gray after half a day in the area's polluted air.

Motorists entering Paris from the long tunnel through the hills of St. Cloud find the city covered with a bowl of smog lumped from 4-million chimneys

The four ancient Greek bronze horses in Venice's St. Mark's Square are eaten away by polluted air, and Rome is often covered in the morning by a brown smog.

Japanese officials have discovered that cases of chronic bronchitis were up to four times more frequent in heavily polluted areas of Osaka and Yokkaichi than in other sections of those cities.

Families have begun to move out of Johannesburg because of a gray smog blotting out the blood-red South African sun.

Germany Takes Action

These are examples of international air pollution reported in a survey by New York Times correspondents in all parts of the world. The survey also revealed that new attempts to control pollution began in many areas during 1966 and more were planned this year.

The West German Government adopted extensive pollution-control regulations in 1959, but Ministry of Health officials said recently that new legislation was needed now.

D. P. A.

West Germany—Smoke pours from a plant in the state of North Rhine-Westphalia. Sign bars traffic on road between 6 and 10 A.M. and 4 and 8 P.M., because dangerous gases are being released then.

South Africa—Street scene in Johannesburg during smog attack. The Government labeled the city a smog-control area in 1966.

Paris-Match

Britain—Bobbies in London wore masks in 1962 smog that killed more than 100 persons.

"There still exists a whole range of plants and air-polluting substances still requiring special regulations," a ministry official said. He emphasized that the Government did not consider pollution as only a local, or even national, problem.

"Our regulations are the same for the most isolated factory as they are for a plant in the heart of the Ruhr District," he said.

Existing regulations were strict enough to reduce by more than 80 per cent the emissions from two of Germany's worst polluters — power plants and cement factories. But more restrictions were needed.

Workmen were waging a constand battle to save the historic Cologne Cathedral from the effects of air pollution, which was crumbling the building's sandstone. In southern Germany, the traditionally clear air of the Alps was filled with gases from new oil refineries in the region.

Pilot Project in Paris

Paris was the pilot city in a five-year-old French program to reduce air pollution. Most pollution there was caused by smoke and sulfur dioxide from home heating plants.

The city was divided into first and second-priority zones in 1964, with homes in the first zone required to use low sulfur coal and oil for heating. The same requirements went into effect this month in the second zone.

Sulfur dioxide emitted in home heating and power-plant operations forms sulfuric acid, which attacks the city's monuments and buildings. Paris is engaged in a long-range program to clean statuary and buildings.

Automobile exhausts also cause pollution problem in Paris and other French cities. Exhaust-recycling devices are required on cars this year.

There are problems outside the French cities, too. The countryside is dotted with smoldering town dumps. Fluoride fumes from metallurgical industries are blamed for the mass destruction of evergreens in the Alps.

Italy adopted her first air-pollution control law near the year's end. The legislation is designed to reduce pollution produced by home heating units. As in France, sulfur dioxide produced by burning coal and oil is the major cause of dirty air in Italy.

Rome, for example, has little industry for a major city, but on winter mornings the city is often covered with brown smog for two or three hours. The smog has the smell of coke and oil used to heat many buildings constructed in Rome since the end of the war.

Milan, where weather conditions often allow stagnant air to blanket the city for days, has the worst pollution problem in Italy. During such periods the city has recorded higher levels of sulfur dioxide than those normally found in London and New York.

The problems of Venice originate in the industrial area of Marghera on the mainland. Sulfur dioxide pumped from factory stacks combines with moisture in the air to form sulfuric acid, which goes to work on statues and buildings.

Japan has had dangerous pollution conditions since the end of the war and smog warnings were issued 64 times in Tokyo last year, compared with 51 in 1965.

The Welfare Ministry is preparing a new pollution law. Under it "nuisance" areas would be specified and industries within the areas would receive preferential treatment on financing and taxes for cooperating in eliminating pollution.

In Osaka and Yokkaichi, centers of the oil-chemical industry, the Welfare Ministry studied the incidence of chronic bronchitis in most polluted and relatively clean areas for eight months last year. It found that the disease occurred 2.8 to four times more in the polluted areas of Yokkaichi and 1.3 to 3.8 times in those of Osaka.

The ministry began enforcing a four-year plan last spring to reduce pollution. The first step was to establish 20 air-monitoring stations throughout the nation at a cost of $840,000.

South Africa, the most industralized nation in Africa, began its attack on pollution in 1966 with the passage of the Atmospheric Pollution Prevention Act.

A national commission has designated four cities as smog-control areas — Johannesburg, Durban Port, Elizabeth and Germiston. New control laws will go into effect in those cities this year.

Britain has been fighting smog for two decades. The battle assumed major proportions in 1953 after more than 4,000 London residents had died during a long pollution episode, in which sulfur dioxide and other pollutants piled up in the air for almost two weeks in stagnant weather.

A national investigating team concluded in 1954 that Britain should remove 80 per cent of the smoke in the air within 10 to 15 years. Officials concede that the goal has not yet been reached. But the Warren Spring Laboratory reported three months ago that the amount of solid matter in the air had been reduced 50 per cent since 1920.

However, the British, like the Americans, have had little success in removing invisible sulfur dioxide from the air. Warren Spring said the amount of sulfur dioxide in the air increased by 17 per cent between 1956 and 1965.

January 16, 1967

INTERSTATE GROUP BEING ORGANIZED TO SEEK CLEAN AIR

U.S., Pennsylvania to Join New York and New Jersey in Enforcing Standards

By RONALD SULLIVAN
Special to The New York Times

TRENTON, March 27 — The Governors of New York, New Jersey and Pennsylvania announced plans today to establish a new interstate agency to fight air pollution.

Called the Mid-Atlantic States Air Pollution Control Commission, the new agency will join New York, New Jersey, Pennsylvania, the Federal Government and probably Connecticut and Delaware in the country's first regional attack on air pollution.

As announced here at a news conference and in a statement issued by Governor Rockefeller in Albany, the new commission will have the power to establish and enforce air quality standards and investigate the causes and sources of air pollution.

Follows U. S. Proposal

The commission is a modified version of an interstate air pollution control agency that was proposed in January by the United States Department of Health, Education and Welfare.

Since the proposed commission will be an interstate compact, it needs the approval of the Legislatures of each member state as well as the Congress.

Specifically, the new commission would have the power to:

¶Investigate the causes and sources of air pollution, identify polluters and seek court orders that would mete out penalties of up to $1,000 a day in fines for violations.

¶Declare regional air pollution alerts and emergencies and carry out research on improving pollution abatement technology.

¶Establish air quality standards for emission from all pollution sources, including homes, industries, power plants and motor vehicles.

Won't Pre-empt Local Laws

State Health Commissioner Roscoe P. Kandle said the new commission would not pre-empt present state and local air pollution control laws in member states as long as they measured up to commission standards. Its major threat, he said, would be to apply uniform abatement controls over a regional area.

He said this would, for example, tend to create a large, uniform market for fuels with low amounts of sulphur and make an identical auto exhaust pollution device practical for the entire mid-Atlantic area.

Originally, the Federal Government recommended a strong agency composed only of New York, New Jersey and the Federal Government, with each having an equal vote.

However, Gov. Richard J. Hughes of New Jersey insisted that any commission would have to include Pennsylvania, which shares a serious air pollution problem with New Jersey in the Philadelphia-Camden area.

What the Governor had in mind was an agency patterned after the Delaware River Basin Commission, which was created in 1961 by a compact among New York, New Jersey, Pennsylvania, Delaware and the Federal Government to control the use of the Delaware watershed.

Governors in Agreement

Governor Hughes discussed the proposal on March 2, with Governor Rockefeller at a meeting of the commission in Dover, Del., and they agreed then that a similar commission offered the best means of fighting air pollution on a regional basis.

In letters today to Mr. Rockefeller and Mr. Hughes, Health Secretary John W. Gardiner said he was "pleased" by the announcement and commended them "for taking such a far-sighted step toward protecting the citizens of both states against the serious threat that air pollution poses."

As the plan was announced today at Gov. Raymond P. Shafer's office in Harrisburg, Pa., his spokesman, Hugh Flaherty, said that enabling legislation would be introduced there soon.

Similar announcements were made here by Senate President Sido L. Ridolfi, who is acting Governor while Mr. Hughes is in Florida on vacation, and in Albany by Jacob Underhill, assistant to Governor Rockefeller.

Gov. John N. Dempsey of Connecticut was said to be "strongly interested" in the commission, and Gov. Charles L. Terry Jr. of Delaware was reported "keenly interested" in having his state join too.

March 28, 1967

Oil Industry Pressing Search for Low-Sulphur Fuels to Meet Air-Pollution Rules

By GERD WILCKE

"It's like the stick held in front of a dog to train him to jump higher. Once he manages a certain level, the stick goes up a notch."

In this fashion, an oil official sought recently to describe the dilemma his industry is facing in trying to comply with more rigorous air-pollution standards.

Although there seems little illusion in the industry that it must comply, there also is widespread puzzlement as to what the eventual standards will be. Changing time targets, the oilmen insist, can produce costly miscalculations in investment planning.

New York's present plans call for a reduction in the permissable level of sulphur content in fuel oil from 2.2 per cent to 1 per cent by 1971.

What brought confusion to the suppliers of fuel oil is the recent recommendation by the United States Department of Health, Education and Welfare that the sulphur content of oil used here for space heating and other domestic, commercial and industrial purposes be limited to 0.3 per cent as of Oct. 1, 1969.

To be sure, it's still only a recommendation, but it helps explain the wariness of the oil people.

With practically every industrialized nation becoming rapidly pollution-conscious, a global search is under way for low-sulphur crude that can be refined to produce low-sulphur oil as well as improved technical processes that will make residual fuel oils "cleaner."

Areas where low-sulphur oil is available include Indonesia, Sumatra, North Africa, Nigeria, Colombia, Argentina, Western Canada and certain fields in the United States.

Unfortunately, these sources are not the exclusive domain for the United States consumer.

"There simply isn't enough low-sulphur fuel to meet the needs of all large cities," J. Cordell Moore, Assistant Secretary of the Interior, warned Congress recently.

"Air pollution is a serious problem," he continued, "but there is not enough natural gas or low-sulphur No. 2 oil to substitute for the high-sulphur heavy oil and coal used to generate heat and power."

As it is, northeastern cities, such as Boston, New York and Philadelphia, import between 80 and 90 per cent of their public utility and industry needs from abroad. This residual fuel oil has a sulphur content of 2.5 per cent on average.

Venezuela, the largest supplier before Canada, ships about 900,000 barrels a day to the United States. The high sulphur content of the oil (2.65 per cent) does nothing to abate the pollution problem here.

But shifting to a different source, even for a part of the supply, would have tremendous repercussions because the Latin-American nation derives about 90 cents in revenue money for each barrel of the oil sold here.

How can a solution be found that will bring cleaner air to the East Coast and at the same time prevent the fiscal collapse of a country?

D. N. Harris, chairman of the sulphur subcommittee of the American Petroleum Institute's committee for air and water

The New York Times

Heavy smoke pours from apartment building, adding to air pollution. Oil industry is seeking ways to reduce sulphur content of heating fuels. The map shows oil-shipping patterns throughout world. If priorities are given low-sulphur oil, patterns may be changed.

conservation, acknowledged in an interview that "expensive know-how is available to clean up the Venezuelan oil."

However, the A.P.I. official, who is associated with the Shell Oil Company, insists there is no technology as of now that could desulphurize oil to a level recommended by the Government.

As for Shell itself, the company has used segregation and selective blending to lower the sulphur content of Venezuelan oil from 2.7 per cent to 2.2 per cent at its refinery in Curaçao.

E. T. Layng, executive vice president of Hydrocarbon Research, Inc., a subsidiary of the Dynalectron Corporation, said his company, in cooperation with the Cities Service Company, had developed a process that could well serve refineries

to get the sulphur level down to 1 per cent.

End Result

The process, known as H-Oil hydrogenation, is employed at a Cities Service plant at Lake Charles, La. The plant can extract five pounds of sulphur for each barrel of oil and has a daily capacity of 2,300 barrels. What the process does is to

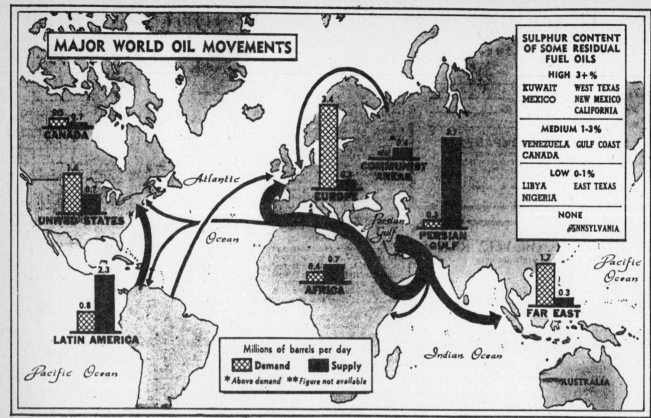

MAJOR WORLD OIL MOVEMENTS

SULPHUR CONTENT OF SOME RESIDUAL FUEL OILS

HIGH 3+%
KUWAIT WEST TEXAS
MEXICO NEW MEXICO
        CALIFORNIA

MEDIUM 1-3%
VENEZUELA GULF COAST
CANADA

LOW 0-1%
LIBYA   EAST TEXAS
NIGERIA

NONE
PENNSYLVANIA

Millions of barrels per day
Demand Supply
*Above demand **Figure not available

The New York Times — April 23, 1967

bring the residual oil into contact with hydrogen gas and a catalyst. The end result is to split the original fuel into low-sulphur oil and hydrogen sulphur gas, which in turn can be converted into marketable sulphur.

A similar plant, but with a capacity 10 times as big, has been designed for a refinery at Kuwait.

Hydrocarbon Research officials say that the process can easily remove 70 to 80 per cent of the sulphur in residual oil, bringing the content to about 1 per cent. But to get the content still lower would be prohibitively expensive.

Cost is increased about 10 to 15 per cent by the process currently used, after crediting the salable sulphur recovered, according to Hydrocarbon Research.

Although the Lake Charles plant has been in operation since 1963, a number of oil officials have questioned whether Hydrocarbon Research's cost estimates are applicable elsewhere.

How Much Is the Cost?

There is no doubt that someone has to foot the bill for the cleaner fuel oil. The question is how much the tab will be increased.

Consolidated Edison has estimated that it alone will have to invest more than $30-million to comply with the air-pollution program. To this come extra fuel costs of maybe $15-million a year, adding 14 cents a month to the electricity bill of small home owners.

Dr. José Antonio Mayobre, the Venezuelan Minister of Mines and Hydrocarbons, feels that investments totaling $100-million will have to be made by oil companies if Venezuelan fuel oil is to meet the standards on sulphur content recommended by the Inter-State Conference of

New York and New Jersey. He also feels that responsibility for finding a solution to the problem rests as much with the consuming countries as with the producers.

The Government official noted that about one-third of Venezuelan exports are affected by New York City's new regulations.

If the states of New York and New Jersey adopted similar rules, he said, refiners would have to embark on major structural changes.

A study made for the American Petroleum Institute by the Bechtel Corporation concludes that the additional outlay required to bring 2.6 per cent sulphur content in residual oils down to 0.5 per cent in a "typical" Caribbean refinery would amount to about double the original investment and an 80-cent-a-barrel increase in the cost of making the residual.

As cited by Platt's Oilgram, a trade news service, the study

found that "substantial additional capital investments are required to achieve even a moderate reduction in sulphur content. Thus, to reduce the sulphur content of the fuel oil from 2.6 per cent to 2 per cent requires an additional facilities investment of approximately $45-million."

To get the reduction down to a level of 0.5 per cent, the study found, would require an incremental investment of $100-million to $150-million.

"When compared to the present-day investment cost of the typical refinery (Bechtel assumed that a "typical" refinery in the Caribbean produced 281,-522 barrels a day) of $120-million, it can be seen that desulphurization to 0.5 per cent would approximately double the investment cost of the refinery," the report said.

April 23, 1967

Coal Group Fights for Delay in Clean Air Program

By ROBERT H. PHELPS
Special to The New York Times

WASHINGTON, May 7—The coal industry, backed by railroads and power companies, is waging a strong fight to delay the Administration's clean air program. Indications are that it has won a partial victory.

The campaign is designed to get the National Center for Air Pollution Control to put off recommendations that power

companies nationally reduce the emission of sulphur oxides.

Sulphur oxides, which are created by the burning of coal and oil, are the second most common air pollutants in the United States. Carbon dioxide from motor vehicles is the commonest.

Two months ago the pollution control center issued a 175-page report linking sulphur oxides in the air with respiratory, cir-

culatory and heart diseases, damage to vegetation, metal corrosion and reduced visibility. The criteria in the report are designed to guide state and local pollution agencies in setting standards for their localities.

The National Coal Policy Conference, Inc., headed by W. A. Boyle, who is president of the United Mine Workers union, would like to have the report

withdrawn on the ground that it is an invitation to power companies to stop using coal and switch to nuclear-generated energy or other fuels.

The conference is supported by coal companies, the United Mine Workers, coal-carrying railroads, mine equipment manufacturers and power companies.

Mr. Boyle, who will testify Tuesday before a Senate public

works subcommittee, agrees that something must be done to reduce air pollution.

"We have our house to clean up; we know it," he said in an interview. "But we need time; it can't be done overnight."

Mr. Boyle painted a dark picture of the consequences of what he called the "push-push-push" by the air pollution center, which is part of the Health, Education and Welfare Department, for stricter curbs on the emission of sulphur oxides.

"It will close down a great number of utility plants," he said, and throw thousands of miners out of work, hurt the railroads, damage industries dependent on a steady supply of power and reduce the amount of energy for home consumption.

"It's hard for me to believe that the President, Congress or even the people over at H.E.W. are out to put the coal industry out of business," he said.

Mr. Boyle will ask for assurances that regulations based on the criteria will be put off —five years is sometimes mentioned—until researchers find an economically feasible way of preventing the emission of sulphur oxides, either by removing them from coal and oil before burning or in chimney stacks. He said the Federal Government should help finance the research.

In response to the industry's demand, President Johnson has requested $2.7-million in a supplemental appropriation for the present fiscal year, ending June 30, for research on sulphur oxides. For next year he has asked $15-million.

Expects Higher Coal Use

Dr. John T. Middleton, director of the pollution control center, insisted in an interview that the coal industry "misunderstands" the situation.

"There is no threat to the coal industry at all," he said.

"Instead of using less coal we will have to use more coal; I fully expect it."

Nor, Dr. Middleton said, will the Federal Government try to set national standards regarding sulphur emission by power companies until researchers find economically feasible methods of preventing such pollution.

The pollution control center does not have the power now to set national standards, but under an Administration bill it would be allowed to do so.

Until a workable method of removing sulphur is discovered, Dr. Middleton said, intermediate steps must be taken. He termed sulphur pollution "dangerously high" in such metropolitan areas as New York, Philadelphia and Chicago.

One intermediate step should be the switch by power plants from high sulphur to low sulphur coal or oil, he said. That is what an air pollution conference for the New York metropolitan area recommended last January.

Dr. Middleton emphasized, however, that under present programs the nation would only hold its own in the battle against air pollution.

"In 1980 the air quality will be no different from today," he said. "We must have a significant breakthrough to make any improvement."

May 8, 1967

SURVEY FINDS PUBLIC UPSET ON POLLUTION

Air pollution is worrying the American people and they want action taken to clean the air they breathe, Louis Harris reported yesterday, in a survey in The New York Post.

However, by 46 to 44 per cent the public is unwilling to pay $15 more in taxes a year to alleviate air pollution, according to the survey.

Mr. Harris said 56 per cent of the American people believe there is considerable air pollution where they live. This figure rises to 72 per cent in the cities and 75 per cent in the suburbs.

Five of every 10 persons in the metropolitan areas feel air pollution is getting worse, compared with less than one in 10 who believes the situation is getting better.

By 2 to 1 margins, the public gives a negative rating to the job done in curbing air pollution by local industry and the automotive industry and by local, state and Federal Government, Mr. Harris said.

The public feels that air pollution is mostly caused by exhaust from automobiles and smoke from industrial plants. However, by 63 to 25 per cent, people also agree that some contamination of the air is necessary if industry is to provide jobs in their area, according to the survey.

A tax rise of $15 for air pollution control was most favored by the better educated younger people, those in the upper income groups, suburban dwellers and residents of the East.

July 25, 1967

CONGRESS PASSES A CLEAN-AIR BILL

President Gets Measure for 3-Year Pollution Fight

By WILLIAM M. BLAIR

Special to The New York Times

WASHINGTON, Nov. 14—Congress passed and sent to President Johnson today a bill that empowers the Federal Government virtually to immobilize a city if air pollution causes a serious health hazard.

The bill, which the President has included on his list of important legislation this year, provides for a three-year, $428-million program to combat air pollution. The Senate had approved a $700-million program but bowed to the House.

In both Houses, passage of the bill worked out by a conference of House and Senate conferees yesterday was by voice vote.

The bill also includes $125-million earmarked for research to determine the blame for dirty air that threatens urban dwellers.

Congress rejected the concept of national emission standards sought by Mr. Johnson but it authorized a two-year study.

The President had sought to convince Congress that a national standard should be set to cover all sources of pollution. This would include the amounts of pollutants that could be released into the air from factories, automobiles, apartments and other sources.

However, the amounts released would vary from city to city or region to region. A national standard, therefore, might be lower than needed for New York City but higher than necessary for Red Oak, Iowa.

The measure still leaves the initiative to the states to set up clean air standards and to plan cleanup campaigns. However, if the states do not act promptly, the Secretary of Health, Education and Welfare is authorized to step in and set standards.

The Secretary is also authorized within 18 months of passage of the bill to designate air quality regions for existing problem areas.

The bill grants the Government additional time to act as new areas emerge. The Secretary also could set up commissions to govern regions where air pollution problems overlapped states and the states did not act.

A significant enlargement of the Secretary's authority is contained in the provision for emergency power when he determines that a "substantial" health hazard exists in the polluted area, such as the heavy smog in New York at Thanksgiving, 1966.

If he decides that a serious health hazard exists, he would be empowered to shut down factories, halt automobiles and prevent pollution from any source.

To bring this about, he would seek Federal court injunctions. The suit could be brought against a specific polluter, such as a plant or plants owned by one or more companies, or an apartment house.

Senator Edmund S. Muskie, Democrat of Maine, who has spearheaded the drive for stronger pollution controls, has said that the injunctive power granted to the Secretary would be automatic unless the Secretary could be proved to be in error. The legislation, he said, carries practical, immediate relief for an area faced with dangerous health hazards from air pollution.

If the pollution stemmed from several sources, he could seek a blanket injunction to enable him to move against a city or region. With this power, he could halt traffic in New York, for example, if he was sustained in a finding that auto traffic was a serious health menace under given air conditions.

November 15, 1967

Car Exhaust Rules Are Tightened

By JOHN D. MORRIS
Special to The New York Times

WASHINGTON, June 3— The Health, Education and Welfare Department issued "new and stronger" standards today for controlling air pollution from motor vehicles.

The standards, based on proposals published by the department last January, will apply to 1970 models of American-made and imported passenger cars, trucks and buses.

They will reduce the exhaust tailpipe emissions of carbon monoxide and hydrocarbons by about 30 per cent below limits now in effect for the 1968 and 1969 model years, the department said.

Secretary Wilbur J. Cohen said he hoped application of the new limits would "stimulate a renewed effort to find improved ways of controlling emissions from motor vehicles." He added:

"Protection of the public health and welfare will no doubt require that more stringent standards be adopted in the future."

While the new standards on tailpipe emission are more stringent than the 1968-69 versions, they do not go so far in some other respects as the department's January proposals.

Responding to industry objections, the department dropped one major proposal that automobile manufacturers had vigorously opposed. This was for control of hydrocarbon evaporation from carburetors and fuel tanks.

The department said the evaporation control proposal "has been deferred from the 1970 to 1971 model year" because of a lack of sufficient data on the performance of control systems.

At the same time, however, despite industry objections, the department retained the authority to test samples of new cars for compliance with the antipollution regulations.

The manufacturers, while also opposing any raising of tailpipe emission standards, had conceded they would be able to comply.

Under the 1970 standards, tailpipe emissions from gasoline-fueled passenger cars and light trucks will be limited to 23 grams of carbon monoxide and 2.2 grams of hydrocarbons for each mile of driving.

The 1968-69 standards are applied in three classifications, depending on the engine's size. They range from 26 to 42 grams a mile for carbon monoxide and from 2.5 to 4.1 grams a mile for hydrocarbons.

The 1968-69 requirement for 100 per cent control of hydrocarbon emissions from the crankcase will remain in effect for all gasoline-fueled passenger cars, trucks and buses.

The 1970 standards include for the first time control of tailpipe emissions from large trucks and buses.

They will limit diesel smoke emissions "to a faint plume," the department said, and require a reduction of slightly more than one-third in present tailpipe emissions from gasoline-powered heavy trucks and buses.

June.4, 1968

DECREE SETTLES AUTO SMOG SUIT

Court Backs the Government on Antipollution Measure

By GLADWIN HILL
Special to The New York Times

LOS ANGELES, Oct. 28— United States District Judge Jesse W. Curtis approved late today a consent decree settling the Government's civil antitrust suit against the major automobile manufacturers in connection with the development of equipment to reduce smog.

The ruling came after a day-long hearing in which New York joined a score of other cities, states and individual plaintiffs in challenging the settlement as unjust.

Under the decree, the auto makers, without conceding that they conspired to delay the introduction of fume-suppression equipment, agree to follow new and more competitive ground rules in their developmental activity.

The plaintiffs told Judge Curtis that the proposed decree would allow the industry to escape without reproof what evidently has been 16 years of illicit activity, and would handicap the numerous parties who plan to press suits for triple damages.

Altogether, 20 lawyers argued against the decree's terms.

Judge Curtis gave his decision extemporaneously and said he would file a written opinion later.

He said he felt the decree met the basic criteria of being enforceable, being "consistent with" the original complaint, and not prejudicing the rights of any prospective damage claimants.

"It gives the Government all the relief it could have gained if it tried the case and won," he remarked. "It would be tragic if the court refused to sign the degree, the case went to trial, and the Government lost the case."

46 Members of Congress

Among those who have taken exception to the decree terms were 46 members of Congress, who were presented at the hearing by Thomas Sheridan, a former Assistant United States Attorney here. The group did not seek to intervene, but acted in response to the court's blanket invitation for comments.

Technically Judge Curtis's action is subject to appeal, and some of the plaintiff's counsel had talked of appealing. But as a practical matter, it was thought more likely that subsequent legal action would be centered on suits against the automobile companies.

Practices alleged in the Justice Department's original complaint, which the auto companies agreed not to engage in in the future, included any concerted action to delay installation of fume control equipment, to restrict individual company publicity on technical advances, to require joint appraisal of equipment, patents and to acquire patents on a pool basis, or to respond only jointly to governmental requests for information or proposals.

The defendants are required also not to engage in certain patent cross-licensing and information-exchanging arrangements considered restrictive of technological progress.

New York's corporation counsel, J. Lee Rankin, former United States Solicitor General, who presented the city's case, said outside the courtroom that a suit by the city for an undisclosed sum would be ready for filing in about two weeks.

The Justice Department's antitrust division lodged the charges against General Motors, Ford, Chrysler, American Motors, and the Automobile Manufacturers Association last Jan. 10 after a two-and-a-half year inquiry by a Federal grand jury here.

After what some members of Congress and public officials called "intense lobbying" by the industry, the Justice Department announced last Sept. 11 that it had reached agreement with the defendants on a consent decree under which, without conceding any guilt, they would agree to stop certain allegedly collusive practices in their efforts to develop fume - suppressing equipment for cars.

October 29, 1969

Chicago Has 6th Day Of Heavy Air Pollution

Special to The New York Times

CHICAGO, Nov. 11—Grimy, gray smog hung over this city today, as Chicagoans hacked, coughed and squinted through their sixth straight day of heavy air pollution.

The upper stories of buildings along Michigan Avenue were obscured from less than a block away. Incoming and outgoing flights at O'Hare International Airport were delayed.

The amount of sulphur dioxide in the air was measured this morning at 2½ times the danger level.

During the day, the Commonwealth Edison Company said that it had "markedly" reduced its burning of coal.

November 12, 1969

G.M. Redesigning Auto Engines For Operation on Unleaded Fuel

By JERRY M. FLINT
Special to The New York Times

DETROIT, Feb. 13—The General Motors Corporation is redesigning its car engines to operate on unleaded gasoline of lower octane ratings than those of most present fuels.

G. M. is expected to announce its timetable for the switchover soon, but it is believed the bulk of its 1971 models—next year's cars—will be redesigned for nonleaded fuel and the remainder switched by 1972.

The change to nonleaded fuel is one of the more dramatic indications of the power of the growing movement to save the environment.

The argument to remove lead from gasoline is based on the following:

¶Lead may be a dangerous pollutant.

¶Future Government proposals call for the removal of solid particles from auto exhausts, and these cannot be met unless the lead, much of the solid matter in the exhaust fumes, is removed.

¶Future pollution control devices to eliminate most other auto exhaust pollutants will not work well unless the lead

goes. This is probably the most important argument to the automobile industry.

General Motors' action probably will spur similar moves by the other car makers and send the oil companies scurrying to develop more powerful leadless gasoline.

G.M. would neither confirm nor deny formally its forthcoming move, but high-ranking company engineers made it clear to newsmen recently that the move was coming.

Also, Secretary of Transportation John A. Volpe, in Detroit today, said that G.M. officials had told him that the engine design change would be made by 1972.

This may cause some engine performance problems, but mainly with the larger, big-engine cars. The extent of the problems will depend on the ability of the gasoline companies to develop powerful fuels.

The lead ingredient is tetraethyl lead, developed by General Motors in the nineteen-twenties as an additive to gasoline. It serves to increase octane ratings — a measurement of gasoline-burning efficiency — making possible more powerful and more efficient engines.

The premium and regular grade gasolines used by most automobiles today contain lead,

but the premium fuels are approximately 100 octane and the regular grades are 94-96 octane.

Compression Ratios

Most American auto engines, even the smaller V-8's, have compression ratios of 8.5 to 1 or 9 to 1, and these operate on regular grade fuel.

The high-powered cars, the Cadillacs and the big-engine Mustangs, for example, have compression ratios higher than 9 to 1 and need the high octane premium fuels.

The higher the compression ratio—and that refers to the area into which the air-gasoline fuel mixture is squeezed in a cylinder—the more powerful the engine.

General Motors sources said the company planned to modify most of its standard engines to operate on an 8.2-to-1 compression ratio, which means the engines would need a 92 octane fuel.

Removing tetraethyl lead from premium, 100 octane gasoline would leave a 92 octane fuel; so it is possible that the changeover of fuel and the engine modifications for approximately 70 per cent of the new cars built would be relatively simple.

The effect on automobile performance, starting and acceleration, would be minor, according to General Motors.

"With the engines using regular fuel today, the customer will notice very little difference" in performance, Alex Mair, chief engineer of the Chevrolet Division of General Motors, said, adding, "You'd hardly notice it."

But the bigger engine cars —30 per cent of new-car production—need 100 octane gasoline, and what will happen to these is not yet clear.

The oil companies may be able to develop 100 octane fuel without lead.

The auto companies may even be able to modify their big engines so that they would operate on a less powerful fuel.

Or there may be combinations of both, meaning that big cars still would have big engines, but that the big engines would get perhaps 10 per cent less horsepower, which could result in less acceleration and lower gasoline mileage.

The switchover by General Motors and probably the other auto makers also may mean a two-fuel system, perhaps a three-fuel system at gasoline stations.

For some years there will have to be leaded gasoline available for the older model cars that need it today. Those engines cannot be modified.

Then there will be the unleaded fuel for cars that use regular grade fuel today, and perhaps a second unleaded fuel for cars that use premium today.

The General Motors move relieves the auto industry of the onus on the issue of delaying engine changes, and puts pressure on the oil companies to come up with less polluting fuels.

"We all know we're polluting the atmosphere," said Mr. Mair of Chevrolet. "But I think we've caught it in time, we can turn this around."

February 14, 1970

Eastern Europe Combats Air Pollution

By PAUL HOFMANN
Special to The New York Times

VIETNNA, Jan. 21 — A new state Commission for the Protection of Air Purity started its work in Budapest this week, exemplifying rising concern throughout Eastern Europe over the deterioration of the environment.

Palls of soft-coal smoke from millions of chimneys hangs over cities from Warsaw to Belgrade this winter, which has come early and with plenty of snow.

Clouds of exhaust gas from decrepit diesel-powered buses and autos without antipollution devices foul urban air in Eastern Europe all year around.

In the industrial districts of southern Poland, northern Bohemia and Moravia in Czechoslovakia, and western Hungary, pollution levels have been found comparable to that in New Jersey.

The coal and steel centers of Ostrava, Moravia, has sunsets as beautifully pastel — and ominous — as those in Los Angeles.

Air and water pollution problems are now also besetting regions in Rumania and other Balkan countries that a generation ago were rural stretches but now are dotted with chemical plants and machine shops.

While economic planners in the countries of Eastern Europe are earmarking more investment funds for industrial development, public opinion and the state authorities are catching up, belatedly, with the dangers of contaminated air and water.

According to the newspaper of the Hungarian trade unions, Nepszava, Hungary's economy loses an estimated $700-million every year because of air pollution. Diminished agricultural production in areas adjoining

chemical and drug plants is said to be a major factor.

From the standpoint of public health, Nepszava said, January and February were the worst months of 1969 with a marked increase in respiratory ailments in the cities. The sulphur content of the atmosphere in Miskolc, an industrial center in northern Hungary, was found to be double the permissible level, according to the newspaper.

The regional government of northern Bohemia warned recently that air pollution had reached "the limit where it can be endured." The regional authorities described the situation as serious and called for immediate measures.

According to the regional government, the air in the north Bohemian brown-coal basin, particularly the Chomutov and Teplice districts, was particularly foul because of smoke from three new thermal

electric power stations.

Research has shown that polluted air from the brown-coal belt was responsible for blight in agricultural areas. A report submitted to the regional authorities estimated at $20-million the population damage to the forests on the Erzebirge, or Ore Mountains, on the border between Czechoslovakia and East Germany.

From Poland to Rumania, efforts are under way to unify the several departments that so far have been in charge of environmental problems, and to give them more funds and greater powers to combat pollution.

Rumania, where industrial production has grown 1,000 per cent during the last 20 years, has just developed new dust collectors and filtering substances, to be used on a large scale. Two large water-purification plants have been commissioned.

January 3, 1970

Sato Plans Action to Meet Pollution Crisis in Japan

By TAKASHI OKA
Special to The New York Times

TOKYO, July 28—The pollution of soil, water and atmosphere has reached crisis proportions in this narrow land, where more than 100 million people are jammed together in an area the size of Montana.

Stung by criticisms, Premier Eisaku Sato's Government decided today to set up a central headquarters, with the Premier himself in charge, to coordinate measures to be taken to counter environmental disruption. The headquarters will be inaugurated on Friday.

During the last week, more than 8,000 people in Tokyo, many of them children, were treated in hospitals for smarting eyes, burning throats or other physical ailments because of the white smog that blanketed large areas of the capital for five days, municipal officials report.

Trees and shrubs are dying in the spacious, secluded gardens of the Imperial Palace, and worried court officials have urged Emperor Hirohito and Empress Nagako **to spend as much time as they can spare from their official duties away from Tokyo, in their mountain retreat at Nasu or their seaside villa at Hayama.**

Tagonoura on Suruga Bay, once a picturesque village with celebrated views of Mount Fuji, is today a highly polluted port, clogged by slime from waste discharged into the water by many papermaking factories.

Harbor Depth Reduced

The factories will not stop discharging the waste, while fishermen's groups refuse to let port authorities dump it into Suruga Bay. So the slime accumulates in the harbor, to a point where water that formerly was 28 feet deep now is only half that, and 10,000-ton ships can no longer berth at the pier there.

Kiyoshi Takizawa, a dairy farmer in Kurobe on the rugged Japan Sea coast, milks his seven cows daily, then pours the milk into a hole dug in the ground. He has had to do this since April, when inspectors declared a 325-acre tract of farm and pasture land polluted by cadmium poisoning.

The cadmium comes from the Nippon Mining Company's zinc refinery nearby. It has so impregnated the soil that no one will buy rice grown

Associated Press

Aerial view, Monday, of dense fog that had covered downtown Tokyo for nearly a week

on the 325 acres or drink milk from cows grazing there.

In Minamata on the southern island of Kyushu, fishermen have lost their eyesight or gone mad after eating fish caught in the bay. Mercury wastes discharged into the water by the Nippon Nitrogen Company are the cause. Forty-six victims have died in the last two decades, and some children in the area have been born mentally retarded and with defective physical coordination.

Contamination Kills Herons

At Sagiyama, north of Tokyo, 50 snowy herons, annual springtime visitors from Southeast Asia, died after eating fish that had been contaminated by insecticides, a high school teacher who has been observing them every spring reports.

Smog is everywhere — not only in major urban centers, but also in some rural areas victimized by the vagaries of wind and weather.

Last week, on an apparently clear, hot day, rural schoolchildren on the island of Shikoku suddenly complained of smarting eyes and sore throats. A freak wind, pollution experts surmise, must

have carried smog from the crowded Osaka - Kobe - Kyoto area across the Inland Sea to this distant region.

These are only a few examples of the kind of environmental disruption that is going on all through Japan, usually as a direct consequence of the nation's enormous expansion of industrial production in the last 10 years—an expansion that thrust Japan into a position of economic strength ranking immediately behind the United States and the Soviet Union.

Disregard for Public Charged

Up to now, environmental experts assert, the Government has followed economic growth policies that tended to favor industrial and business enterprises without considering tthe welfare of the public as a whole.

Each Government ministry has jealously guarded its prerogatives; for instance, in the case of the contaminated harbor at Tagonaura, the Ministry of International Trade and Industry set standards of waste disposal for the paper plants, while the Economic Planning Board

fixed standards for measuring pollution inside the harbor. Neither prevented the disaster that has befallen the port.

The Transport, Communications, Agriculture, Welfare and other Ministries all exercise a degree of control in matters pertaining to environmental disruption, but coordination has been lacking.

The Government, in deciding today to establish a special headquarters to deal with the situation, gave Sadanori Yamanaka, Minister of State in the Premier's Office, the additional assignment of serving as minister in charge of dealing with environmental disruption.

The Welfare Ministry has already announced a 10-year anti-pollution program in the three prefectures, or provinces, of Tokyo, Kanagawa, which adjoins the capital and includes Yokohama, and Osaka. But Tokyo and Osaka municipal officials say that the program is too slow and too vague, since these two cities have already set goals of achieving conditions relatively free from pollution throughout their metropolitan areas in three to five years.

THE IWAI BEACH in Tomiyama town, Chiba, southwest of Tokyo, where a long vinyl tube is used to protect bathing areas from floating refuse and oil contamination in Tokyo Bay.

More Funds Promised

Finance Minister Takeo Fukuda, who has hopes of succeeding Mr. Sato as Premier, promised that funds to control environmental disruption would be included in the budget for the 1971 fiscal year, which is now being compiled. This year Japan is spending about $126-million for various anti-pollution measures, and the amount is expected to go up steeply beginning next year.

Tokyo and Osaka officials estimate that at least $1.4-billion would be required to carry out a 10-year anti-pollution program for their cities.

The Ministry of International Trade and Industry plans to spend about $14-million over the next five years to develop electric automobiles as a means of reducing exhaust gas fumes. The ministry's goal is to develop a car capable of a speed of 48 miles an hour and of running at least 60 miles without recharging of its batteries.

Meanwhile, experts warn that so-called photochemical smog, the newest disruptive phenomenon, may continue in Tokyo throughout the hot season as automobile exhaust fumes are affected by ultraviolet rays to produce characteristic acrid white smog.

Traffic Controls Sought

Ryokichi Minobe, Tokyo's Governor, wants to have more control over traffic entering and leaving the downtown area, so that if worst comes to worst he can ban automobiles from the inner city. He has set up a smog warning system, under which a semi-alert will be sounded when the level of oxidants in the atmosphere rises to 0.15 parts per million, and a full alert when the level exceeds 0.3 parts per million. Under both types of alert, citizens will be asked to stay indoors.

The system went into effect yesterday. Today the smog that had hung over the city through the weekend lessened, and no alert was issued.

July 29, 1970

Electric-Car Research Surges in the Antipollution Age

By FARNSWORTH FOWLE

The electric car—the dowagers' delight of 60 years ago—is trying once again to challenge its vulgar rival, the complicated, noisy, dirty—but more efficient—internal combustion engine.

Persistent inventors are exploring various technological fronts in search of a battery to give the electric car the power, endurance and economy to match its ecological appeal in the age of antipollution.

They were spurred by a newly released survey backed by the electrical industry indicating that the youth market is even more eager than the over-60 set for a short-range limited-speed electric car costing $2,000 or less.

The study, commissioned by the Electric Vehicle Council, sought the response to such a vehicle with a top speed of 40 miles an hour and a range of 150 miles between pauses to recharge its batteries in a home charging unit.

American Motors hybrid Gremlin, one of 36 entries in electric-car race, is 2,000 pounds heavier than the standard Gremlin. The bulge in the hood houses the charging system.

Of those interviewed, 39 per cent of the men and 38 per cent of the women said they would be interested in buying such a car. And the percentage of interest was higher among younger than older people, the survey by the Opinion Research Corporation showed.

The council compared the figures with a 1967 study by the American Institute of Public Opinion and said the percentage of interested adults had risen from 32 per cent in that year. It noted further that those interested tended to be residents of cities of 100,000 or more population, and to be in the higher educational and income brackets.

The Edison Electric Institute, a leading trade organization, is a sponsor, along with Gulf General Atomic, of research into a rechargeable zinc-air battery system for vehicles. A prototype is scheduled for road testing late this year in an electric vehicle built by Joseph Lucas Industries of England.

Zinc-air batteries, which are said to last longer than conventional lead electric-storage batteries, are being used by the Army in specialized equipment such as back-pack radios, radar and night-vision devices.

The Sony Corporation of Japan has adapted the zinc-battery concept for use in vehicles. The zinc is powdered as a fuel and is used to generate new electricity as the power is applied. Sony says it can power a minicar for 5½ hours without refueling, using a booster battery for acceleration.

The American automobile industry, in the center of the pollution controversy, no longer scoffs at the electric-car enthusiasts, but encourages serious experimenters by lending them car bodies and chassis as test beds.

General Motors has its own experimental model, using zinc-air batteries with a lead-acid booster system in a modified Opel Kadette setting. It claims a 90-mile range at 55 miles an hour, but the weight and space for the battery system, as with the Sony car, leave little capacity for payload.

For several summers there has been a transcontinental electric-car race. This year it will have partial support of the Federal Government, a broader range of entrants and a new name—the Clean Air Car Race—under a contract between the National Air Pollution Control Administration and the Massachusetts Institute of Technology. The 36 cars due to start on Aug. 24 include some with all-electric power-plants and others with steam engines, low-pollutant combustion engines running on special fuels such as liquid natural gas, and several "electric hybrids," a term used for battery cars that also have combustion engines to build or supplement their electrical muscle as they go along.

One entrant, from the Worcester Polytechnic Institute, with help from American Motors, has a Gremlin body with a special bulge in the hood to house the charging system.

One inventor, Michel N. Yardney of New York, recently patented a control system for hybrid cars so that at low speeds they would automatically switch from engine to battery power. The purpose, he explained, is to keep exhaust pollution to a minimum.

Even a car-rental company will be involved in the futuristic Clean Air race. Avis will demonstrate the Electric Mark II tomorrow. It was made by Anderson Power Products, Inc., of Boston. It has a 200-mile range under city commuter traffic conditions. Its batteries are rigged for speedy replacement, but it takes nearly a ton of them to keep the Volkswagen-sized vehicle running.

August 16, 1970

AIR FOUND DIRTIER IN NORTH ATLANTIC

Contaminants Have Doubled Since Early in Century

WASHINGTON, Oct. 17 (UPI) —Scientists reported today that the air over the North Atlantic Ocean was twice as dirty as it was in the early nineteen-hundreds.

This is disturbing news for those weather experts who fear that air pollution, if it continues unchecked, will seriously affect the climate and perhaps bring on a new ice age.

Air pollution over the Indian Ocean off Southeast Asia also appears to have doubled in this century, but the South Pacific atmosphere remains as clean as it was.

The increase of air contamination over the North Atlantic, attributed to man-made pollutants, was reported by scientists of the National Oceanic and Atmospheric Agency.

They arrived at their conclusions after comparing data on atmospheric electrical conductivity obtained early in the century with data gathered on a recent global research cruise.

The data used for comparison purposes were collected between 1909 and 1929 by Carnegie Institution scientists aboard the sailing ships Carnegie and Galilee and in 1967 by scientists aboard the research ship Oceanographer.

"We know we are changing the atmosphere," a meteorologist, William E. Cobb, said. "We must know how and at what rate we are changing it."

October 18, 1970

Man's Best 'Friend', the Tree...

TREES help supply oxygen we need to breathe. Yearly each acre of young trees can produce enough oxygen to keep 18 people alive

TREES help keep our air supply fresh by using up carbon dioxide that we exhale and that factories and engines emit

TREES use their hairy leaf surfaces to trap and filter out ash, dust and pollen particles carried in the air

TREES dilute gaseous pollutants in the air as they release oxygen

TREES can be used to indicate air pollution levels of sulfur dioxide

TREES give us a constant supply of products—lumber for buildings and tools, cellulose for paper and fiber as well as nuts, oils and fruits

TREES lower air temperatures by enlisting the sun's energy to evaporate water in the leaves

TREES slow down forceful winds

TREES cut noise pollution by acting as barriers to sound. Each 100-foot width of trees can absorb about 6 to 8 decibels of sound intensity

TREES provide shelter for birds and wildlife and even for us when caught in a shower

TREE leaves, by decaying, replace minerals in the soil and enrich it

TREE roots hold the soil and keep silt from washing into streams

TREES salve the psyche with pleasing shapes and patterns, fragrant blossoms and seasonal splashes of color

Northeastern Forest Experiment Station

Gottscho-Schleisner

...But Air Pollution Threatens

By JOAN LEE FAUST

LAST spring, the unmistakable sounds of power saws were heard loud and clear in the San Bernardino National Forest, about 80 miles east of Los Angeles. They were made by lumbermen who contracted to cut down trees—dead trees—1,000 acres of ponderosa pines killed by air pollution.

This loss is significant because the pines are not growing in the thick of urban life, but in forests high in the mountains where the air usually is clear and invigorating.

Foresters first detected yellowing, browning, and dying of pine needles about 15 years ago. They wondered and delved in. Gradually, the evidence be-

came clear, the ponderosas were succumbing to toxic gases in the air.

*

Smog from the densely populated San Bernardino Valley, which is often trapped by an atmospheric lid at elevations of about 2,500 feet, is pushed by evening Pacific breezes up the slopes of the 5,000-foot-high mountains into the national forest. There among the firs and many pine species, grow the ponderosa pines, which are particularly vulnerable to ozone. This pollutant is formed by the action of sunlight on atmospheric nitrates and hydrocarbons formed by unburned gasoline from auto exhausts and by industrial plants. Mature ponderosas, which are native to

the Western states, grow to 150 and 180 feet and are important timber trees.

When the dead ponderosas were felled, the timber was sold at auction. The denuded areas were reseeded with sugar pines and giant sequoias, which have greater "survive-ability" in the polluted region. Although the dead trees left five very small holes in the vast 160,000 acres of ponderosas in the national forest, their loss is a tragic blow and another reason for urgency in solving the country's pollution problems. Foresters estimate that the ponderosas are succumbing to the smog at the rate of about 3 per cent a year.

Almost every state now turns in reports of plant damage from air pollution. One national estimate of injury to all types of vegetation is as high as a half billion dollars annually.

A survey made in New York last fall of woody plant damage caused by air pollution showed that many familiar trees and shrubs were showing signs of injury. Lilacs, beeches, lindens, dogwood, ginkgo and white pine foliage had flecking, scorch patterns and browning on the undersides of the leaves. The survey was conducted in Central and Prospect Parks and in the Brooklyn Botanic Garden.

Tree species differ in their susceptibility to damage from toxic gases and individuals within species show signs of injury in varying degrees. Many kinds of trees remain untouched. Air pollutants vary in every region and their cumulative effect on vegetation always reflects this variation.

What's so important about trees? They're nice to have around, to be sure, for their beauty, shelter, and greenery, especially in the big city canyons. Trees also happen to be an important part of this planet's life-support system.

Those little green bodies in the leaves, called chloroplasts, which give the foliage its green color are vital manufacturing plants. The leaves utilize carbon dioxide combined with water to manufacture the trees' food supply which is stored in a form of starch. The "exhaust" from this vital photosynthesis process is oxygen and the energy from the sun makes the whole system work.

When air pollutants contaminate this process, many trees are in trouble. A leaf's air intake is through tiny pores called stomates, most of which are located on the surface. When polluted air enters the leaves of susceptible trees, the plant chlorophyll is often destroyed, the photosynthesis process is disrupted and the plant is either stunted, chlorotic (yellowed) or in severe cases of long exposure as illustrated by the ponderosas, dead.

Two of the major pollutants that cause plant problems are ozone, primarily from auto exhausts and other unburned hydrocarbon sources, and sulfur dioxide emitted from coal and oil-burning power generating plants. Atmospheric fluorides, chlorine and other industrial effluents also cause varying degrees of plant damage.

Research is only beginning but programs are under way to develop geneticresistant species to survive air pollution. One of the Northeastern forests' prime conifers is the white pine which shows particular susceptibility to air contaminants in a malady described as chlorotic dwarfing. Scientists at the U.S.D.A. Forest Service Insect and Disease Laboratory in Delaware, Ohio, are developing superior strains of white pine that will have genetic potential to wither and toxic contaminants.

Researchers at Pennsylvania State University are working with Scotch pine, an important Christmas-tree crop in the state. The National Arboretum at Washington, D.C. is currently involved in a project to develop a hybrid London plane tree that will be suitable for urban environments.

Plant pathologists at Rutgers University are keeping close watch upon woody plants and their reaction to air pollutants. According to Miss Eileen Brennan, Associate Research Professor, preliminary evidence indicates that oaks, maples and London planes are good trees for polluted regions and so is sophora, the Japanese pagoda tree. For across the board use, however, a Rutgers favorite is ginkgo. Among the conifers, Miss Brennan suggests Douglas fir, junipers, hemlock and arborvitae.

*

Trees and all green plants are part of the earth's recycling system and because of this vital role, their potential as "air scrubbers" must not be overlooked.

A recent study of ozone uptake by plants at the Connecticut Agricultural Experiment Station in New Haven showed the possibilities of this air "scrubbing." Three staff researchers showed by computer analysis that plants can remove enough ozone from the air to benefit man. They studied what happens when a mass of polluted air containing 150 parts of ozone per billion parts of air passes over a forest of trees 15 feet tall. The computer analysis showed that if such an air mass stood over the forest for one hour, the air filtering down to the forest floor would have only 60 to 90 ppb of ozone remaining. The rest would have been taken up by the canopy of leaves. If the air mass remained over the forest for eight hours, the ozone level would have dropped to 30 ppb.

Efficient air "scrubbing" trees are going to be important parts of civic and commercial plantings for many years to come. As the data is accumulated by plant scientists, arboreta, botanic gardens, nurserymen and universities, trees and shrubs species with low pollution susceptibility will be in great demand.

Plant scientists who are close to pollution studies and plant susceptibility stress that amateur gardeners and botanists cannot correctly identify air pollution problems. There are striking similarities between visible injury from air pollution and plant damage from drought, frost, mineral deficiencies, insects and disease. Herbicide injury often causes the same type of mottling, stippling and stunting caused by exposure to toxic air gases. Investigators working with air pollution damage to plants should have extensive training and experience.

Despite air pollution and damage to trees by toxic gases, the earth is apparently not in danger of running out of oxygen. The amount of oxygen in the earth's atmosphere has remained constant for the past 60 years at 20.946 per cent by volume according to a report made this summer by the Environmental Science Services Administration. Oceanic plant life produces 60 per cent of the world's newly-released oxygen each year.

*

Trees have a big future cut out for them, to help with air "scrubbing" and to soothe the psyche of the urban man. But many more trees will be needed. Civic planners, tree commissions, landscape architects, garden clubs and park commissions can be influential in realizing this goal.

Dr. Henry M. Cathey, research horticulturist at the U.S.D.A. Crops Research Division in Beltsville, Md. puts it this way, "The decisions are in our hands. Green is the color of hope and in the green of plants is the hope of survival."

October 25, 1970

Rediscovering the Wheel

By EDWARD LINDEMANN

We all know that when the lights go green, every intersection in New York is barely more active than a parking lot. We are all aware that this mass of cars belches out exhaust containing, among other things, lead, carbon monoxide, and soot (which is approximately what "particulate matter" boils down to). Various ailments result from the exhausted air we breathe in New York.

Dean Myron Tribus of Dartmouth's School of Engineering warned three years ago that "We're on our way to a public catastrophe. . . . Carbon monoxide levels in New York City are approaching the lethal level." Aside from near-lethality, carbon monoxide adds accident danger, for it lessens alertness and warps drivers' judgment of time and distance.

A few rules are available for pedestrians, to lessen somewhat their inhalation of car fumes; walk near buildings, not near the street; stay back a good distance from the curb while waiting for a light; hold your nose when necessary. But they are clearly stopgaps.

Technological aid in reducing the outpourings of gasoline motors is still largely in the dither-and-argue stage, with little or no help from the most irresponsible of the major car manufacturers. The suggested substitutes— electric and steam cars — are still curiosities. And we continue to gasp.

Therefore we have to face a fact: New York City must — emphasizing must—switch its traffic as much as possible from cars to bicycles.

European cities have been bicycle-conscious for years. Already car-banning has begun in some outlying parts of the country. Yosemite Park has banned cars entirely from its Mariposa Grove, with buses being substituted and achieving a clear-cut reduction in exhaust. Last summer Yosemite said "no" to all cars in the eastern end of the valley, as a step toward a total car ban. A few cities, our own included, have experimented with brief carless periods in certain areas.

Is a total ban of combustion engines possible? Hardly that, since buses, supply trucks, and emergency vehicles are an absolute necessity. But we need to plan now for, say, an 80 per cent motorless city, through a car ban and a vigorous encouragement of bicycles. There are various crunches in such a development, and it is time to note them briefly so we can get to work on solving them:

1. Winter bike-riding is impractical at present, because of cold, rain, and snow. Therefore manufacturers need to develop a nonbulky weather enclosure for bikes—one that will weigh no more than two or three pounds,

and will be adaptable for the already existing lightweight, foldable bicycles as well as the standard ones.

2. The tired-out, aged, and invalids cannot be expected to bicycle. For them there must be more buses. Fumeless electric buses are now practical for crosstown use, and they may be quite usable also for the long north-south routes. Yet they are not part of our bus fleet. Why not?

Some older people could use the old-fashioned adult tricycle, manufactured primarily for invalids up till about 1900 or so, and in need of revival. This was, as indicated, a three-wheeler in which one sat back comfortably and pedaled with minimal effort—with of course no need to worry about balancing.

3. The exercise of cycling is fine for the health, yet exercise makes for deeper inhalations; doing this within the present car smog sounds ghastly. The answer would seem to be to go ahead with bicycles anyway, since the more they replace cars, the less smog there is to breathe.

4. What cyclists there are consistently ignore the fact that they are required by state law to observe the regulations governing cars. Almost without exception they go through red

lights and ride the wrong way on one-way streets. Injuries and even deaths have resulted. Enforcement would presumably mean licensing of bikes—a complicated headache. Yet it must be done—unless some workable new enforcement technique can be thought up.

5. Bicycle - stealing is increasing, often by ingenious new methods such as removal of the bike from its wheel that is locked to a fence or pole. Manufacturers themselves need to increase ingenuity of design to prevent this.

6. In a city geared to apartment-house living, bike storage can be a problem. Yet every apartment house has a cellar; and the foldable bicycle allows storage in at least some apartments. Considering that a bicycle uses about one-tenth the space of an average car, this is not a bad problem.

That covers most of the catches—none of them insoluble. Now, who will do something—before we all smother?

Edward Lindemann is science editor of a New York City publishing house —and a two-wheeler man.

December 5, 1970

BALTIMORE GETS STRICT AIR CODE

Antipollution Rules Termed Among Most Restrictive

Special to The New York Times

BALTIMORE, Dec. 26 — Dr. Neil Solomon, Maryland's Secretary of Health and Mental Hygiene, has signed air pollution regulations for the metropolitan Baltimore area that state officials consider among the most restrictive in the nation.

The regulations became effective Feb. 20. They are similar to Maryland regulations signed into effect Nov. 5 for Prince Georges and Montgomery Counties in suburban Washington.

Dr. Solomon announced the Baltimore-area regulations on sulphur oxides and particulate matter at a news conference this week. The announcement came almost two months after a public hearing in which city

officials had criticized the proposed curbs as too stringent.

The new regulations include the following antipollution curbs:

¶All visible emissions will be prohibited from any source, with the exception of a four-minute interval in any 16-minute period during certain operating conditions, including the starting of boilers.

¶Sulphur content for heavy fuel oil, currently limited to 1 per cent, will be limited to 0.5 per cent on July 1, 1975. Home heating oil, now also limited to 1 per cent, will be cut to 0.3 per cent by July, 1972.

¶Most small incinerators will be prohibited. Some will be banned in July, 1972; the rest, by July, 1973.

¶All open burning for land clearing wastes is prohibited immediately at any location closer than 200 yards from occupied buildings or public highways. The distance will be increased to 500 yards July 1, 1973. Leaf burning will be permitted only in areas where no public collection is provided.

¶All plants using heavy heating oil must have dust collec-

tors installed. For larger units, the deadline is Oct. 1, 1972; for smaller units, Oct. 1, 1973.

Dr. Solomon said that the regulations on visible emissions represented one of the most important aspects of the new code.

"For the first time in the world," he said, "there is a regulation on the books which says no visible emissions, except where unpreventable, will be allowed."

He said that the regulations, when implemented, would clean up 30 per cent of the air pollution now in the Baltimore area.

Dr. Solomon denied that the regulations would cause any industry to move out of the area. "These regulations are the same for surrounding counties," he said, "and soon we will end up blanketing the entire state with regulations like these."

Maryland's air pollution laws provide for a civil penalty of up to $10,000 a day for violators. Enforcement is a joint effort of state and local air control officials.

The Baltimore metropolitan area consists of the city and five surrounding counties.

December 27, 1970

President Signs Bill to Cut Car Fumes 90% by 1977

By JAMES M. NAUGHTON
Special to The New York Times

WASHINGTON, Dec. 31 — President Nixon, pledging vigorous enforcement, signed into law today a clean air bill that sets a six-year deadline for the automobile industry to develop an engine that is nearly free of pollution.

The President singled out the measure for ceremonial signing as he completed action on number of important bills. He hailed the pollution control legislation as the most important in the history of the search for cleaner air.

But the atmosphere at the White House ceremony appeared to be clouded with politics. Among those not invited to the ceremony was the legislation's chief architect, Senator Edmund S. Muskie, Democrat of Maine, a potential rival in the 1972 Presidential election.

"There were some 40 sponsors of the bill and we were physically unable to take care of all of them," Gerald L. Warren, deputy White House press secretary, told newsmen.

Mr. Warren said that Congress had been represented by the chairmen and ranking Republicans on the Senate Public Works Committee and the House Interstate and Foreign Commerce Committee, which dealt with the legislation. Until this morning, however, Mr. Nixon had not invited any members of Congress to participate in the ceremony.

One chairman, Senator Jennings Randolph, Democrat of West Virginia, reportedly telephoned the White House to ask, as a point of personal privilege, that he be invited.

The White House subsequently invited his Republican counterpart on the Public Works Committee, Senator John Sherman Cooper of Kentucky, and the senior members of the House committee, Representatives Harley O. Staggers, Democrat of West Virginia, and William L. Springer, Republican of Illinois.

Controversial Provision

All but Mr. Staggers were present, along with 18 Administration officials who crowded into the Roosevelt Room. The President, standing in front of a Frederic Remington painting of the charge of the Rough Riders at San Juan Hill, said that the clean air measure "came about by the President proposing" and the "cooperative efforts" of both political parties on Capitol Hill.

The most controversial section of the bill, setting strict 90 per cent reductions of hydrocarbons, carbon monoxide and nitrogen oxide in auto engine exhaust, had been vigorously opposed by the auto industry and by the Nixon Administration.

Senator Muskie appeared to be taking note of that fact in a statement he issued later today. "Although opposed in part by the Administration," he said, "[the act] was a nonpartisan Congressional effort and in final form was unanimously endorsed by both houses. If the Administration follows up on today's bill signing with a meaningful allocation of personnel and resources, the Congressional objectives can be achieved."

Extension of Deadline

The measure gives the auto industry until Jan. 1, 1975, to reduce emissions of hydrocarbons and carbon monoxide and until one year later to cut emissions of nitrogen oxide. It provides, however, for one-year extensions of both deadlines if the Environmental Protection Agency determines that they cannot be met by a "good faith" effort.

The industry is considered likely to seek and obtain the extensions, making the effective deadline five years for hydrocarbons and carbon monoxide and six years for nitrogen oxide.

Mr. Nixon said that the bill was "only a beginning." Turning to William D. Ruckelshaus, director of the Environmental Protection Agency, the President called him "the enforcer" and added, "You're going to be called a lot worse."

The President said that 1970 would go down in history as the "year of the beginning" and that 1971 would be "the year of action." He said he would submit "significant new recommendations" on the environment to the Congress early in the new year.

At a briefing for newsmen after the ceremony, Russell E. Train, chairman of the Council on Environmental Quality, said that he was "confident" the auto industry had the ability to meet the emission standard deadlines. Mr. Ruckelshaus said he would told the manufacturers to "a very strict burden of proof" before granting extensions of the deadlines.

Provides for Jail Terms

The clean air bill also provides for establishing strict national air quality standards for 10 major pollutants and permits suits by citizens against polluters. It also gives the Environmental Protection Agency authority to set limits as low as zero on emission of hazardous substances and to seek jail terms and fines in court cases against recalcitrant polluters.

Mr. Ruckelshaus said that he would enforce all the bill's provisions with vigor.

At the briefing, Mr. Warren was asked why Senator Muskie had not been invited to witness the signing of "The Muskie bill." Mr. Warren replied, "I don't believe it's been called that—in this room."

January 1, 1971

AIR QUALITY RULES FOR 6 POLLUTANTS GIVEN FOR NATION

By E. W. KENWORTHY
Special to The New York Times

WASHINGTON, April 30— William D. Ruckelshaus, administrator of the Environmental Protection Agency, announced today what he called "tough" national air quality standards for six principal pollutants.

Mr. Ruckelshaus conceded that the standards could be achieved only by "drastic" alterations in industrial practices and in the "commuting habits" of millions of persons living close to large urban centers.

The Clean Air Act of 1970 stipulated that the E.P.A. administrator must set standards for common pollutants. Under the law, the states have until Jan. 1, 1972, to submit plans for achieving the standards.

1975 Is Deadline

The environmental agency has until May 1, 1972, to approve or reject them. If it rejects a state plan, it has until July 1 to impose its own plan on that state. The states then have until July 1, 1975, to effect the plans.

The standards announced today were for sulfur oxides, particulates (soot and smoke), carbon monoxide, hydrocarbons, nitrogen oxides and photochemical oxidants.

As a prime example of the difficulties that lie ahead for some cities, Mr. Ruckelshaus cited the standards for particulates and sulfur oxides, which are spewed forth by the millions of tons annually, chiefly by electric power plants burning high-sulfur coal and oil, by coke ovens in steel mills, would have a hard time meeting the standards by 1975.

New York, he said, faces the greatest problem.

"We estimate," he said, "that to bring air pollution levels down to the standard for particulates [and sulphur oxides] in New York will require a 300 per cent increase in natural gas usage in the city." Natural gas is low in sulphur content and fly ash.

Need for Natural Gas

To meet these standards by the use of natural gas, the seven cities would have to increase the total national use of natural gas by about 15 per cent, and almost half that increase would go to New York City alone.

The difficulty with this solution is that, as the National Academy of Engineering has pointed out, the supply of natural gas is expected to decrease in 10 years unless large by smelters of nonferrous ores and by municipal incinerators.

Respiratory Illnesses

Sulphur oxides exacerbate respiratory illnesses; have increased death rates on several occasions and damage property. Smoke, soot and fly ash can injure the lungs.

The primary standard set today for sulphur oxides was 0.80 micrograms per cubic meter (1.03 parts per million of air) as an annual mean. The standard for particulates was 75 micrograms per cubic meter as an annual mean. (A microgram is a millionth part of a gram.)

Mr. Ruckelshaus said that most regions could meet the standards by switching to low-sulphur fuels and by requiring plants to install electrostatic precipitators to capture soot.

However, he said that seven cities — New York, Chicago, St. Louis, Baltimore, Hartford, Buffalo and Philadelphia —

new reserves are discovered.

Mr. Ruckelshaus also emphasized the difficulty that seven cities — New York, Chicago, Los Angeles, Denver, Philadelphia, Washington and Cincinnati — would have in meeting the carbon monoxide standard of 9 parts per million as a maximum eight-hour concentration not to be exceeded more than once a year.

Carbon monoxide, a by-product of the incomplete burning of carbon-containing fuels and of some industrial processes, is principally emitted by automobile exhausts in cities. The gas decreases the oxygen-carrying capacity of the blood, and, in accumulations found in many cities today, can impair mental processes.

Automobile Emissions

Mr. Ruckelshaus said that in the aforementioned cities, "where we have good enough data to make accurate predictions," only Cincinnati would "come close" to meeting the standard by 1975, and that it would not actually meet that

until 1977.

And this, he added, was assuming that auto manufacturers met this 1975 deadline under the Clean Air Act for producing engines whose emissions of carbon monoxide were 90 per cent below the allowable standards for 1970 models.

He said that if the legal deadline for carbon monoxide was to be met, some cities would have to make "drastic changes in their transportation systems," by developing rapid transit lines from the suburbs and limiting private cars in the inner cities.

Photochemical oxidants are produced in the atmosphere when hydrocarbons (chiefly from processing and using petroleum products) and nitrogen oxides (from high-temperature combustion processes, as in automobiles) are exposed to sunlight. The oxidants reduce resistance to respiratory infection, irritate muscous membranes and damage plant life.

The problem of the photo-

chemical oxidants—the smog that hangs over Los Angeles, Denver and other cities with much traffic and much sunlight—is really the problem of controlling hydrocarbon and nitrogen oxide emissions.

High Hydrocarbon Levels

An environmental agency aide illustrated the problem today by noting that in many cities it was common to find 2-3 parts of hydrocarbons to a million parts of air, whereas the primary standard set by Mr. Ruckelshaus called for a limit of 0.24 parts per million as a maximum three-hour average concentration not to be exceeded more than once a year.

The standard he set for nitrogen oxides was 0.05 parts per million as an annual mean.

Under the Clean Air Act, auto manufacturers also have a 1975 deadline for a 90 per cent reduction in 1970 standards for permissible hydrocarbon emissions, and a 1976 deadline for a 90 percent reduction in nitrogen oxides emitted by 1971 models.

Mr. Ruckelshaus said that the whole problem of photochemical oxidants was so complex "that it is difficult to predict whether or not the nation will meet the standards for these pollutants in the time allowed by the law."

The prospect for achieving control of nitrogen oxides from existing stationary sources by 1975 is "bleak," he said, but added that the Federal program for reducing emissions of nitrogen oxides and hydrocarbons from automobiles and controlling nitrogen oxides from new and modified electric power plants "should carry us a considerable distance down the road" to clean air.

Therefore, the administrator said, he has established standards for photochemical oxidants at levels approaching those "that occur fairly commonly in nature." The primary standard was 0.08 parts per million as a maximum one-hour concentration not to be exceeded more than once a year.   May 1, 1971

Jersey Cars Face Test Of Exhaust Emissions

By RONALD SULLIVAN
Special to The New York Times

TRENTON, Aug. 3—New Jersey environmental officials said today that they were going to start testing automobiles under the most comprehensive air pollution inspection system in the country.

At least a third of the state's 3.3 million cars are expected to fail the test.

The 33 state motor vehicle inspection stations will begin measuring the exhaust emissions on every registered car next year. Any car that fails will have a red sticker placed on the windshield and thus be banished from the state's roads unless the emission pollutants are eliminated within a two-week period. The driver of a car that is used after the grace period expires will be arrested.

Trucks and buses are covered by an antipollution code promulgated June 18.

Existing Federal automobile pollution responsibilities apply only to new cars and the antipollution devices that are now being built into them. In contrast, New Jersey's program will apply to all cars—from $50 jalopies to cars just off the showroom floor.

The new inspection system is expected to cost the motorist whose car exceeds the allowable pollution level $20 in repairs. Its political, social and ecological ramifications are likely to be significant in a state that is gaining a tough antipollution reputation.

California has a program of automobile emission inspections but it is a roadside, spot-check operation that applies only to 1966 models or newer vehicles.

The testing procedures here were developed by the State Department of Environmental Protection with the financial help of a Federal grant designed to test the effect of a full-scale auto emission inspection system.

New Jersey was selected by the Federal Government as a demonstration state because it is the most densely populated state with the highest urban concentrations in the nation and has the most densely traveled highways. Under the state-operated automobile inspection system, every registered car is passed or failed on an annual basis.

"Up to now we've mainly gone after the major polluters and the big smokestacks," remarked John Elston, the 31-year-old supervisor of the proj-

ect. "But now, for the first time anywhere, we are going to place the onus on the individual car owner, and not on the factory down the street from him."

As a result, state officials anticipate an angry reaction from car owners who face expensive repair costs.

Extent of Reduction

What it will mean for the average car owner whose car is not passed Mr. Elston said, is a $20 service station or garage repair bill for a partial motor tuneup. But he emphasized that better engine performance would offset the repair cost.

All told, the state expects to remove about 20 per cent of the 4.5 million tons of carbon monoxide gas and 32 per cent of the 750,000 tons of smog-producing hydrocarbons emitted by cars during the first year of the program's operation. Because inspections are spread throughout the year, with each vehicle assigned to a certain month, the effect of the program is not expected to be felt until late in the first year.

In the second year of operation, the state estimates that about 45 per cent of the cars now on the road will fail still higher antipollution standards that will be implemented.

The major impact will be felt by car owners with low incomes who generally operate the oldest and most rundown cars in the state.

Included in Chapter 15 of the New Jersey Air Pollution Control Code, the program will be the subject of a public hearing on Monday at the Teaneck campus of Fairleigh Dickinson University.

After the hearing, Commissioner Richard J. Sullivan of the Department of Environmental Protection plans to promulgate the new inspection code under the power granted him by the Legislature in 1966. The new regulations will take effect six months after Commissioner Sullivan signs the order.

According to Mr. Elston, the inspection will be carried out in 30 seconds by a machine—about the size of a gas pump—that will examine a car's exhaust as it goes through an inspection lane for other, usual tests, for such things as brakes and lights.

Visible smoke will mean automatic failure.

The department has formulated three standards. Cars built in 1967 or before will be rejected if their exhaust emission is 7.5 per cent carbon monozide or 1,200 parts for each million parts of hydrocarbons.

For cars made in 1968 and 1969—the years when federally mandated antipollution devices were first installed by car makers—the standards will be 5 per cent carbon monoxide and 600 hydrocarbon parts for each million parts.

For 1970 models and later, the standards will be 4 per cent carbon monoxide and 400 parts of hydrocarbon for each million. In 1973, the standards will be toughened.

The department says 38 per cent of the 1967 and older models and 36 per cent of the 1968 and 1969 models will fail. Eighteen per cent failure is expected for 1970 and 1971 cars.

August 4, 1971

Court Curbs 23 Companies In Alabama Pollution Peril

By JON NORDHEIMER
Special to The New York Times

BIRMINGHAM, Ala., Nov. 18—A Federal judge ordered 23 companies in this area to halt production today after air pollution climbed to about twice the level at which a danger alert would be given. The order was the first injunction against polluters made under the emergency provisions of the Clean Air Act of 1970.

It was sought by the Environmental Protection Agency after the companies allegedly ignored appeals from local health agencies to shut down voluntarily.

[A stagnant air mass that hung persistently over most of the Eastern Seaboard brought advisories of potential pollution dangers to half a dozen areas, including the New York metropolitan area.]

The United States Steel Corporation, which employs 11,000 workers at three plants, said that several steps had been taken to reduce the problem. But a local health official contended that the efforts were token "housekeeping" measures and that the concern had adamantly refused to curtail production.

After the order was issued early today by Federal District Judge Sam C. Pointer, U.S. Steel announced that its open hearth furnaces were cut back to 35 per cent of full capacity, a level a spokesman said was the minimum that could be obtained without damaging the equipment or causing a completely prolonged shutdown.

The emergency was eased today as a cold front pushed into the Southeast, unsettling the stagnant high pressure mass that had settled over the area for the last week.

The local air pollution count dropped to 410 this morning and rain was forecast for tomorrow, which would further cleanse the atmosphere.

The National Weather Service in Birmingham issued an air stagnation alert on Monday when calm atmospheric conditions limited the vertical mixing of pollutants over the city.

Dr. Douglas Ira Hammer of the E.P.A. testified that when suspended particulate levels of 1,000 micrograms per cubic meter exist for a 24-hour period an "imminent and substantial" danger to the public health

would occur. He said an equal danger existed when counts of 700 or more prevailed in an area for two consecutive days.

A hearing was scheduled in Judge Pointer's court tomorrow to determine if conditions would permit an easing of the order.

Health officials said that the pollution particulate count of soot and dust reached 771 micrograms per cubic meter of air on Tuesday and was recorded at 758 yesterday when the appeal was made to the E.P.A. to seek the temporary restraining order. No marked increase of respiratory ailments were reported in the Birmingham area, however.

An air pollution alert is issued when the particle count reaches the 375 level.

Dr. George Hardy, health officer for Jefferson County, said that by yesterday the 23 companies had been requested by telegram to curtail 60 per cent of their production.

He said that the response from five major concerns, including U.S. Steel, had been "inadequate" and that the firms had only been willing to make minor "housekeeping" cutbacks without interfering significantly with production.

Asked at a news conference if he believed that U.S. Steel had acted in the public interest during the crisis, Dr. Hardy replied, "I would say certainly not."

A U.S. Steel spokesman challenged the allegation and contended that the operation of the company's open hearth furnaces and coke ovens was rescheduled to spread emissions throughout the day in-

stead of in concentrated releases.

In addition, he said, incinerators were turned off, open fires banned and the use of gas was substituted for coal as fuel for boilers.

"We are certainly concerned with public health, and we are doing our part to eliminate this problem," he said.

It appeared during the day that little state or Federal legal apparatus existed to supervise strict compliance with the court order.

The State of Alabama had passed an air pollution law this year that got high marks from national experts, but Gov. George C. Wallace took no action to appoint a state commission that would enforce its provisions until last Monday as public alarm grew over the smog-filled air suspended over this industrial city.

The commission has yet to meet and appoint a director.

Judge Pointer issued the restraining order after midnight, and United States marshals notified the industries of its specifications as workers reported at plants for the morning shift.

By that time, westerly breezes were already breaking up the stagnant air mass, and the sun was cutting through the reddish-brown band of haze that has hung over the city for the last week.

Last April, the city experienced a similar pollution crisis. This led to state and Federal officials to call for tougher controls on local industry.

November 19, 1971

Utilities Reported to Lag in Air Pollution Controls

By DAVID BIRD

An independent, year-long study of private power companies across the country has concluded that the utilities have lagged far behind in installing available equipment that would keep them from polluting the environment.

The study, released yesterday by the Council on Economic Priorities, focused on 15 power companies that generate some 25 per cent of the nation's power. Included were the six largest utilities and nine others chosen for geographical distribution and representative size.

New York's Consolidated Edison was one of those criticized.

In Washington yesterday the Federal Power Commission published a national power survey predicting that the price of electric power would more than double by the year 1990 and

singling out environmentalism as one of the causes of actual or threatened power shortages.

The pollution study by the Council on Economic Priorities found that of the 124 major fossil-fuel plants — those that burn coal, oil or gas — operated by the 15 utilities in the study, only 45, or 36 per cent, adequately controlled particulate emissions, or soot.

Fifty-seven of the plants—46 per cent of the total—have continued to burn high-sulphur oil or coal, the study reported. The burning of high sulphur content fuels is a major cause of the presence in the air of sulphur dioxide, an irritating gas.

The study also found that 66 of the plants—81 per cent of the group—had no controls at all for emissions of nitrogen oxides, gases that are known to induce emphysema and are the main ingredients of smog.

Only one plant in the study group was absolved of causing any air pollution. It is the Geysers, a geothermal plant owned by the Pacific Gas and Electric Company that uses steam that comes naturally from the ground.

The Council on Economic Priorities, a nonprofit corporation founded in 1970, is financed partly by foundations such as the Rockefeller Family Fund and partly by the sale of its findings.

The council said the study was undertaken to develop information on how the corporations exercised their responsibility to preserve the environment. The information is gathered for consumers, prospective employes, investors and corporate managers.

Pollution and Advertising

In the face of lagging pollution control efforts, the study

said, the companies were spending only small amounts of money on research and development to find new ways to solve pollution problems as compared with expenditures on advertising and sales.

"In 1970, advertising and sales expenditures for the 15 companies," the study said, "totaled $126.9-million, 1.9 per cent of their combined gross operating revenues. The 15-company research spending, only a small part of which goes to develop pollution control techniques and more advanced generating methods, totaled $21.4-million in 1970 — one-sixth the amount spent for advertising and sales."

Some of the companies included in the study were sharply critical of the council's study in early replies. Others said they would have to study the 550-page document more carefully.

View Called Dated

The Edison Electric Institute, a New York-based trade association of privately owned power utilities, said in a statement, "The discussion of the research and development interest of the industry indicates that the council has not been following current developments in the industry."

The institute said that plans had been made to increase cooperative research ventures to a level of about $29-million a year.

The council study found that the companies' environmental accomplishments varied widely. It singled out three of the 15 companies as the worst air polluters — American Electric Power System, The Southern Company and Commonwealth Edison of Chicago.

American Electric, which sells more power than any other privately owned utility in the country, covers a seven-state area that stretches from Virginia to Michigan. According to the study American Electric's size may "put it in the best position to wreak maximum environmental havoc" with its plants that burn 6 per cent of all the coal burned in the country.

American Electric Power reacted sharply to the council findings. Donald C. Cook, the utility's chairman and president, called the study "a mis-leading, unscientific, distorted, partly false and highly prejudiced document."

He said that the study completely ignored "the relationship of a plant to its environment" and that it did not take into account "the fact that some companies' plants are located in urban areas, while others (like A.E.P.) are in sparsely settled rural areas."

Alice Tepper Marlin, co-director of the council, said that the study used a common yardstick for all utilities so as not to favor one company over another. The yardstick, she said, was whether the company had done as much as could be done with existing technology to clean up its emissions.

Errors Charged

A spokesman for the Southern Company in Atlanta said it had not completed its study of the council report but that preliminary examinations seemed to show there were "gross errors" in emission statistics. The council said it had double checked its figures with the Federal Power Commission and company officials.

Pollution from Commonwealth Edison, the study said, "may have the greatest impact on health as many of its plants are sited in the heart of downtown Chicago or in neighboring areas."

A spokesman for the Chicago utility said, "We believe the report destroys its own credibility with its many distortions, inaccuracies and misstatements." He said that only three of the utility's 13 plants were in Chicago "and none of them can be construed as being downtown."

New York's Consolidated Edison was singled out for special attention because it has been "living the utility industry's collective nightmare" of trying to produce power in one of the world's most densely populated areas.

But the report said, "In the face of its problems, Con Ed's commitment to R & D [research and development] has been abysmal. In 1970, the company's R&D outlay as a percent of gross revenue came to 1/7th of 1 per cent—tied for 11th place out of 15 companies in the study."

Con Edison, in a replying statement, said that the council report contained "errors of fact, contradictory statements, insupportable data and erroneous conclusions" and "completely omits any discussion of the economic, environmental social or health benefits of electricity."

The statement, by Robert O. Lehrman, Con Edison's vice president for public affairs, said that a "great deal" of information and data that the com-pany's environmental and technical experts had furnished the council had been "ignored or misunderstood."

In terms of "environmental responsibility" Southern California Edison was ranked at the top of the list of the 15 companies because, the study said, it had done the most to control its emissions.

"Almost on a par," the study said, is Pacific Gas and Electric of San Francisco.

As to the cost of achieving the best possible pollution control that technology could provide today, the study said the 15 companies would have to invest a total of between $1.3-billion and $2.2-billion, the heaviest expenditures being required by the study's three worst polluters.

The Edison Electric Institute's comment on the study said the costs of installing pollution control equipment "are seriously underestimated." The institute gave no figures of its own.

The council has its headquarters in New York at 456 Greenwich Street. The power company study was written by Charles Komanoff, Holly Miller and Sandy Noyes.

April 16, 1972

U.N. Parley Endorses Air Monitoring Net

By WALTER SULLIVAN
Special to The New York Times

STOCKHOLM, June 7—A proposal to set up a global network of 110 specially equipped stations to monitor changes in the world's climate and the level of air pollution was endorsed here today by delegates to the United Nations Conference on the Human Environment.

One hundred of the stations would monitor regional trends in the pollution of the atmosphere. The United States is establishing 10 of these—in Caribou, Me.; Atlantic City; Raleigh, N. C.; Salem, Ill.; Meridian, Miss.; Huron, S. D.; Alamosa, Colo.; Victoria, Tex.; Pendleton, Ore., and Bishop, Calif.

The remaining 10 of the 110 stations are to be called baseline stations. They will be in some of the most isolated clean parts of the earth to provide a standard for determining how bad pollution is elsewhere.

These stations are also to watch for changes in the atmosphere as a whole, such as in its burden of dust, its gradual increase in carbon dioxide content and the "health" of the ozone layer, 10 to 20 miles aloft, which acts as an umbrella against solar radiation.

The United States plans to operate four of the baseline stations, according to Dr. Robert M. White, head of the National Oceanographic and Atmospheric Administration in the Commerce Department. Two are already in operation—at the South Pole and on Mauna Loa, the giant volcano that dominates the island of Hawaii.

Funds have been provided for another station, Dr. White said, at Point Barrow, Alaska, and the fourth is to be in American Samoa.

The atmospheric monitoring program, which is to be coordinated by the World Meteorological Organization, a specialized agency of the United Nations, has been evolving for about two years. It was approved without dissent today in a conference committee, in which all 112 countries are represented, and no change in the vote is expected when the project comes up later at a plenary session.

While the Soviet Union has boycotted the conference here because of the exclusion of East Germany, which belongs neither to the United Nations nor to any of its specialized agencies, the Russians have taken part in planning the global monitoring network.

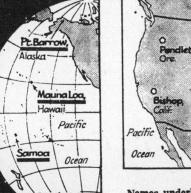
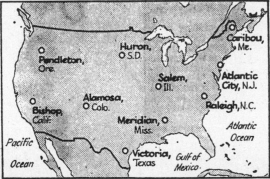

The New York Times/June 8, 1972

Names underlined on map at left indicate sites of the four "baseline" monitoring stations U.S. is to operate. The 10 stations that are projected for continental U.S. to monitor regional pollution are shown on map above.

With the United States taking care of 4 of the 10 projected baseline stations, the Soviet Union has reportedly offered to operate three and Canada the remaining three.

As for the regional monitoring stations, the Soviet Union has indicated its intention of setting up 10, the same number as the United States. Canada has pledged seven, and West European countries, 21. Kenya, with a plan for one station, is the only African country known to have made such a commitment.

Absorbs Heavy Radiation

The ozone layer, which is to be watched by the baseline stations, absorbs lethal wave lengths of ultraviolet radiation from the sun, and so life on earth is heavily dependent on its preservation.

Fears have been expressed that heavy supersonic air traffic in the stratosphere would set in motion chemical reactions depleting the ozone layer, which consists of molecules that have three oxygen atoms instead of two in normal oxygen gas.

Fears have also been expressed over the carbon dioxide that would be added to the atmosphere by heavy travel by supersonic transport. Carbon dioxide traps thermal radiation and holds it near the earth.

Would Record Variations

Sweden has expressed misgivings to the conference about possible effects of supersonic wave on the upper atmosphere. In one of the few critical remarks on the preparatory documents, Dr. Arne Engstrom of Sweden told the committee yesterday that it was unfortunate this problem was "not more extensively reflected" in the documents.

To keep track of what is happening in the ozone layer, the baseline stations will have to make observations up to 20 miles aloft. They will also record variations in dust content at various levels of the atmosphere, including trends in particle sizes, changes in the solar constant (the output of energy by the sun), and variations at different levels of carbon dioxide, water vapor and trace gases.

The regional stations will be in representative areas of the world, far enough from industrial centers to be free from heavy local pollution, yet close enough to record regional trends both from industrial activity and changes in land use

One of the most important outcomes of the conference here should be the establishment of a special environmental fund of $100-million that would help developing countries set up such stations. The United States has pledged $40-million to such a fund, if approved by the conference.

The most troublesome ele-element in today's debate arose from the apparent concern of some developing countries that monitoring stations would be set up on their territory for subversive purposes.

To satisfy such fears, a phrase was inserted in the recommendation specifying that stations were to be established only "with the consent of the countries concerned."

June 8, 1972

Steam Bus Cuts Smog

By ROBERT LINDSEY

The final report of the nation's most ambitious effort in more than 40 years to use steam power to propel motor vehicles has concluded that steam propulsion has enormous potential for reducing urban air pollution. But the report adds that enormous technical problems must be solved if the potential is ever to be exploited.

The report reviewed the design and experimental service last year of three different "steam bus" concepts in three California cities—Los Angeles, San Francisco and Oakland. The $7.9-million project was directed by the California Legislature and partially financed by the Federal Government.

The steam engines were built by Steam Power Systems of San Diego, the Lear Motors Corporation of Reno, and William M. Brobeck Associates of Berkeley, Calif.

None performed completely satisfactory, the report said, and none remained on regular passenger runs longer than 18 days before mechanical problems forced them out of service. Fuel consumption was inordinately high, and there were other problems.

Nevertheless, the California officials said the result were encouraging. Among other things, they said that, compared with a conventional diesel bus, the steam buses produced up to 30.5 per cent less carbon monoxide and up to 86 per cent less hydrocarbons and oxides of nitrogen. These three elements are the principal components of photo-

A steam-propelled bus in California

chemical smog.

"Some years—perhaps less than a decade if funding is adequate—of progressive engineering work will be required before the E.C.E. [external combustion engine] is ready for general application and acceptance," the report said.

A steam engine is called an "external combustion" engine because fuel is burned outside of the engine itself. The fuel (such as kerosene oil diesel oil) fires a burner, which heats water or other fluids and turns it into steam. The steam then drives a piston or turbine, and this energy is used to turn the wheels of the vehicle. The steam is later condensed, heated again, and the cycle is repeated. This is called a "Rankine" cycle engine.

In a conventional automobile engine, gasoline is burned "internally." A series of small, rapid-fire explosions of gasoline is harnessed to move pistons in cylinders rapidly back and forth. The reciprocating movement drives the wheels through a system of gears and a crank shaft.

Steam power proponents say external combustion of the fuel is much more efficient. This is because the fuel is more completely burned; thus, less drifts into the atmosphere. The result: much less emission of smog-creating pollutants.

Steam-powered cars and buses rivaled vehicles powered by internal combustion engines from about 1900 until the mid-nineteen-twenties. But "steamers" eventually lost out to the internal combustion engine because of lower cost and higher reliability.

The emergence of air pollution in many American cities —especially in California— led to a revival of interest in steam power in recent years.

In an effort to "prime the pump" of technology, the California Legislature's research staff urged companies to submit designs for a steam bus. The three companies responded. Conventional 40-passenger intracity buses were used; the conventional diesel engine was replaced with the steam engine.

In a 30-page report on the project, California officials said that, in addition to the data on low pollution, the design, construction and testing of the three buses had produced the following key findings:

¶Acceleration, speed and hill climbing are at least as good with steam power as they are with a conventional engine.

¶Exterior noise of the steam buses was less than a regular diesel bus, but interior sound levels were similar or higher.

¶Fuel consumption ran about three times that of a conventional bus.

Although the report gives comparative technical data for each of the three designs, it does not single out one as having the best performance. It does, however, give considerable praise to the design developed by William P. Lear, who made a fortune in electronics and aviation and has been trying to develop a feasible steam-powered car for more than five years.

The Lear engine was the only one of the three designs to use steam to power a turbine rather than a reciprocating piston. "It appears to be the first in history to be successfully propelled by a steam turbine," the report said of the Lear bus.

While the Lear bus had a few mechanical problems, the report said that it had relatively little trouble during 11 days of "revenue service" on a San Francisco bus route, one with steep hills. "During its brief exposure, the bus provided early indications of the potential for high system reliability," the report said.

Despite such encouraging results, the California Legislature research office stressed:

"We caution that many years of progressive and persistent engineering will be needed to make [steam powerplants such as those used in the trials] technically feasible."

Among other areas, it said that research and development were needed to improve the reliability of steam engines, and to improve fuel economy.

"Exploration of both turbine and reciprocating expanders" should proceed, the report said, "because it is not yet clear which form is superior for heavy duty, stop-and-go vehicles." At a minimum, the report said $20-million should be invested on further steam engine research over the next four years.

While the steam engines came off much "cleaner" than a conventional diesel engine regarding air pollution, the report said they could be refined to be even less polluting.

April 8, 1973

Auto Makers Win a Delay Of Year on Exhaust Curbs

Ruckelshaus Postpones Standards Set for 1975 but Lists Interim Rules— California Gets Stricter Controls

By E. W. KENWORTHY
Special to The New York Times

WASHINGTON, April 11 — William D. Ruckelshaus, administrator of the Environmental Protection Agency, gave automobile makers today an additional year to produce new cars meeting the 1975 standards for reducing emissions of hydrocarbons and carbon monoxide, the two major exhaust pollutants.

At the same time, as the law requires, Mr. Ruckelshaus set interim exhaust standards for new cars sold in 1975 in all states except California. These interim national standards will go "half the distance" toward those that were to have been met by 1975, he said.

For California, Mr. Ruckelshaus set interim standards that go two-thirds of the way toward the original 1975 Federal standards. The Clean Air Act of 1970, which requires the agency to set pollution standards, permits a difference for California provided such standards are stricter than the national ones. Last May Mr. Ruckelshaus rejected the manufacturers' plea to delay the 1975 standards, contending that the industry had not proved that the technology required to meet the standards was unavailable.

Today, acting under a court order to reconsider his decision, he explained he had decided that a suspension was in the public interest because manufacturers might not be able to produce enough properly equipped vehicles to meet the national demand. The delay's effect on the campaign against air pollution will be "minimal," he said.

Mr. Ruckelshaus said his decision would mean that meeting the interim national standards would increase the cost of a 1975 car about $100 over that of a 1973 model. In California, where a car would have to be equipped with a "catalytic converter," the cost would be increased about $160, he said. Industry sources have put the costs somewhat higher.

Converter-equipped cars require low-lead gasoline, which all major retailers must market by July 1, 1974. Mr. Ruckelshaus said that the converters would not increase gasoline consumption very much or reduce a car's performance.

Mr. Ruckelshaus told a crowded news conference:

"This is a terribly complex and important decision that involves the whole mix of our nation's struggle for a cleaner environment. Involved in this decision are billions of dollars, hundreds of thousands of jobs, probably the single most important segment of our economy, the largest aggregate manmade contributor to air pollution and the ambivalence of the American public's intense drive for healthy air and apparently insatiable appetite for fast, efficient and convenient automobiles.

2 Weeks of Hearings

"The ultimate effect of the decision will touch the lives of more than 200 million people."

The basis of his decision, Mr. Ruckelshaus said, was a determination, after more than two weeks of hearings last month, that the converter, which Detroit is banking on to reduce pollutants, is a workable device. The converter contains a catalyst, usually platinum or vanadium, that oxidizes carbon monoxide into harmless carbon dioxide, and hydrocarbons into water and carbon dioxide.

In hearings a year ago and again last month, representatives of the Ford Motor Company, the General Motors Corporation and the Chrysler Corporation testified to difficulties in building a system that would meet the 1975 standards over an automobile life of 50,000 miles, or even 25,000.

Mr. Ruckelshaus said today that most car models could meet the nationwide interim standards with devices other than the catalyst. He estimated that not more than 10 per cent of Detroit's production for sale outside California would require catalysts to meet the interim standards.

American-made cars sold in California, however, would require the catalyst to meet that state's stricter standards. California accounts for about 10 per cent of auto sales by domestic manufacturers.

Mr. Ruckelshaus said that the Japanese-made Honda with a dual carburetor or the Mazda with a rotary engine could meet the original 1975 standards.

He said that he had decided "to phase in the catalysts [in California] because of the potential societal disruption in attempting to apply this new technology across all car lines in one year."

"In weighing this potential against the minimal impact on air quality of this decision versus an outright denial of the suspension request, I believe it is the better part of wisdom to phase in the catalyst," he said.

In granting the auto companies' petitions today, the environmental administrator in effect met the proposals of both Ford and General Motors in concept but not in actual interim emission levels, either nationwide or for California.

The nationwide interim levels were tougher than the two companies had proposed, and the California interim levels were more stringent than Ford had advocated.

Senator Edmund S. Muskie said that the Senate Subcommittee on Air and Water Pollution, of which he is chairman, would hold hearings April 16, 17 and 18 on Mr. Ruckelshaus's decision.

"Mr. Ruckleshaus was faced with a very difficult decision, involving some very poor choices," the Maine Democrat said. "I am confident that he made the choice which, in his best judgment, would be most nearly consistent with the objectives of the Clean Air Act of 1970. His choices were limited by the inadequate response of the American automobile manufacturers to the challenge of the 1970 act.

Onus on the Industry

"The decision places the burden on the industry to produce what it has failed to produce thus far. The standard should be a fuel-efficient, clean car which is durable, drivable and economical."

Mr. Muskie's mentioning "poor choices" referred to the controversy that has arisen over the determination of United States companies to rely on development of the catalytic converter despite the fact that the Japanese Honda company has developed a dual-carburetor, stratified-charge engine that, on small cars, has met the 1975 standards, even after 50,000 miles.

In a report on Feb. 15, required by the law, the Committee on Motor Vehicle Emissions of the National Academy of Sciences said that the catalyst system was "the most disadvantageous with respect to first cost, fuel economy, maintainability and durability." The report termed the stratified-charge engine "the most promising system."

In the recent hearings, Mr. Ruckelshaus noted that if he insisted that the companies meet the 1975 standards on time, he might well be pushing them down the wrong path, because they have at present no system but the catalytic converter. On the other hand, he said, the companies gave him no assurance that, with a year's delay, they would develop an alternative technology.

When reminded today of the academy report and what he himself had said, Mr. Ruckelshaus replied that he had no authority to determine what was the best technology, and that this question must be left "to the market place."

The academy report stated that the incremental cost of a car equipped with a converter to meet the 1975 standards would be about $225 a year more that the cost of a 1970 car, including initial cost amortized over five years, maintenance, and extra fuel consump-

tion of about 5 to 15 per cent. By contrast, the report said, the annual cost of a stratified-charge, dual-carburetor engine, which imposes no fuel penalty, would be about $100, or less.

The law requires for 1975 model cars a 90 per cent reduction in emissions of hydrocarbons and carbon monoxide, as measured against emission levels of 1970 cars.

The law also requires a 50 per cent reduction in emissions of nitrogen oxides in 1973 as against emissions before 1968, and a 90 per cent reduction in 1976 from 1973 levels.

The act permits a year's postponement of both the 1975 and 1976 standards. However, emissions of nitorgen oxides were not at issue in the decision today.

California, which established standards of its own in 1966, before there were Federal standards, can get a waiver from the Federal standards as long as its own standards are more stringent.

The following table shows, in grams per mile, the precontrol levels, the emission levels

required to meet the 1975 requirement, the national interim levels and the California interim levels for 1975 established by Mr. Ruckelshaus:

	Hydro-Carbons	Carbon Monoxide
Pre-1968 cars (uncontrolled)	8.7	87
1975 standards	.41	3.4
1975 national interim	1.5	15
1975 California interim	.9	9

Mr. Ruckelshaus emphasized today that by 1976 all companies would have to meet the 1975 "clean air" standards.

General Motors had proposed national interim standards of 2 grams per mile of hydrocarbons and 25 grams per mile of carbon monoxide, considerably less stringent than those set by Mr. Ruckelshaus.

Ford had proposed national interim standards of 1.7 grams of hydrocarbons and 20 grams of carbon monoxide. Chrysler had proposed 1.7 grams for hydrocarbons and 24 grams for carbon monoxide.

For California interim standards, General Motors had proposed somewhat tougher, and

Ford somewhat less tough, standards than those proposed by Mr. Ruckelshaus. Chrysler opposed any separate interim standards for California, because it opposes any standards that require the use of catalytic converters.

To grant a suspension, the environmental administrator must make four determinations favorable to an applicant: (1) That suspension is in the public interest, (2) that the applicant has made a "good faith" effort to meet the standards, (3) that the applicant has proved technology is unavailable and (4) that a study by the National Academy of Sciences has indicated technology is not available.

Last May Mr. Ruckelshaus denied the first applications for suspension, largely on the ground that the companies had not demonstrated their principal contention that technology was unavailable. In February, the United States Court of Appeals for the District of Columbia Circuit ordered new hearings on two criteria — public interest and good faith.

April 12, 1973

Court Tells States They Cannot Permit Air Quality to Drop

By WARREN WEAVER Jr.
Special to The New York Times

WASHINGTON, June 11 — States cannot permit the quality of their air to deteriorate from present levels, under a decision that an equally divided Supreme Court affirmed today.

The Court's action will be effective even in those areas where national air quality standards are somewhat lower than existing conditions and would thus have permitted relaxation of air quality if the Court had not acted.

With Justice Lewis F. Powell Jr. not participating, the remaining Justices split 4 to 4 to hand a victory to environmen-

talists who had pressed the case and a defeat to the Nixon Administration's Environmental Protection Agency, which had opposed it.

As is customary when the Court divides evenly, there were no opinions, no announcement of which Justice was on what side of the issue and no legal precedent set. Such votes automatically affirm the ruling of the last court below.

The practical effect of the deadlocked vote, however, was immediate and national, prohibiting E.P.A. from approving any state air pollution plan that would allow a lower standard of purity in any area than presently exists there.

In the dispute, four environmental groups, seeking a stricter interpretation of the Clean Air Act of 1970, were opposed by industrial development forces, the United States Chamber of Commerce and mining

and utility interests. States divided over the issue, with New Mexico supporting the environmentalists and Arizona, California and Utah in opposition.

The Sierra Club and three other groups won a ruling in Federal District Court that the legislation had been "based in important part on a policy of nondegradation of existing clean air" and that Congress had not intended to allow states to let their air pollution levels rise.

When the United States Court of Appeals for the District of Columbia affirmed this decision without opinion, E.P.A. immediately disapproved all pending state air quality plans because none of them provided assurance that no deterioration would be allowed.

Economic Harm Feared

The Government asked the Supreme Court to overturn the lower courts on the ground that such strict antipollution controls would discourage economic expansion into areas with clean air, even making it impossible for population ex-

pansion into previously unoccupied land.

Laurence Moss, president of the Sierra Club, hailed the decision as meaning that "industry is going to have to solve its pollution problems at the source, rather than disperse them throughout the countryside" and affect the location of new towns.

Carl E. Bagge, president of the National Coal Association, called for Congress to rewrite the clean air act to modify the Court decision. Otherwise, he said, the impact of today's action will stop the construction of any new fossil fuel power plants in most of the United States" and "any prospect of producing synthetic natural gas or petroleum from the great coal and shale reserves of the nation."

Joining with the Sierra Club in bringing the action were the Metropolitan Washington Coalition for Clean Air, New Mexico Citizens for Clean Air and Water, and the Clean Air Council of San Diego County.

June 12, 1973

Pollution Curbs Still Hazy

By GLADWIN HILL

Many of the nation's motorists may have to resort for the time being to fortune tellers of tea-leaf reading if they want to know what their future driving patterns will be like.

To say that prospective automotive antipollution regulations in many big cities are in a state of flux would be an understatement.

As of early this year, there were 38 cities or urban areas, encompassing a large part of the nation's population, faced

with instituting "transportation control plans" (that is, auto traffic constraints) so their air would meet Federal air quality standards by the statutory deadline of May, 1975.

Engineering projections indicated that these places sim-

ply had so much auto traffic that they would still have too much air pollution even with all the required reductions from stationary sources and from the fume controls that auto manufacturers are required to put on cars.

Only 2 Solutions

This left only two solutions: make further reductions in the amount of fumes coming from cars (by such means as installing special controls on older, pre-regulation cars); or impose limitations or "disincentives" on auto traffic.

The problem was epitomized in the extreme case of Los Angeles, which was told last January that, to comply with the letter of the law, it would have to eliminate about 80 per cent of its auto traffic—a proposal that Enviromental Protection Agency officials admitted was inconceivable and which subsequently has been recast more realistically.

A score of possible ways to reduce aggregate emissions were suggested. They ranged from rationing gasoline to restricting central-city parking to discourage people from driving downtown.

Cities were invited to tailor their own solutions, and public hearings were held throughout the country to discuss alternatives.

Where do we stand now? Well, there's good news and bad news. The good news, for millions of people, is that nine cities have been taken off the list of 38, because subsequent recalculations of their prospective smog loads indicated that they would not be as heavy as originally estimated.

The Bad News

These cities are Birmingham and Mobile, Ala., Kansas City, Kan., and Kansas City, Mo., Dayton and Toledo, Ohio, Syracuse, Indianapolis, Austin-Waco, Tex. (considered a single air pollution area) and Corpus Christi, Tex.

The bad news is that 29 urban areas, containing 43 per cent of the nation's population, were, as of early october, still on the problem list. While most had formulated nominal strategies for solving their problems, the implementation of measures and their putative results remained somewhat conjectural.

29 Areas Listed

The 29 areas are Baltimore, Beaumont, Tex., Boston, Chicago, Cincinnati, Dallas, Denver, El Paso, Tex., Fairbanks, Alaska, Houston, Los Angeles, Minneapolis-St. Paul, New York City (central), New York-Connecticut - New Jersey (suburban).

Also, Philadelphia, Phoenix-Tucson, Ariz., Pittsburgh, Portland, Ore., Rochester, Sacramento, Calif., Salt Lake City, San Antonio, Tex., San Diego, San Francisco and San Joaquin Valley, Calif.; Seattle and Spokane, Wash.; Springfield, Mass., and Washington, D. C.

Each locality has tailored its own combination of remedial measures, and no two plans are exactly alike. However, the Environmental Protection Agency has compiled the following résumé:

¶Twenty areas are planning to reduce their smog totals by special controls on stationary-source activities— gasoline station fuel transfers, loading and unloading of fuel barges, commercial solvent use and dry cleaning.

¶Twenty-eight areas are planning to institute regular inspection of cars to prevent excess emissions.

¶Twenty-two plan to reduce fumes by improving traffic flow, improving public transit and organizing and promoting car pools.

¶Twenty-four plan to introduce "traffic disincentives," such as parking restrictions and surcharges, and no-vehicle ares.

¶Nineteen plan compulsory "retrofitting" of older cars with such equipment as exhaust catalysts, crankcase fume controls and ignition modifications.

¶Thirteen are considering limitations on gasoline sales.

Legally, the states, rather than the cities, are directly responsible to the environmental agency. States were supposed to file their transportation plans six months ago, in mid-April, but only a handful met that deadline.

Under the Clean Air Act of 1970 and a subsequent interpretive decision of the United States Court of Appeals, the environmental agency was supposed to impose plans by last Aug. 15 on any state that had not submitted satisfactory plans.

Deadline Extended

But the agency became so bogged down in this work that it had to ask the court for an extension of the deadline to Oct. 15, and then to Nov. 15, for completion of the work.

Meanwhile, because of demonstrated difficulties in instituting adequate transportation controls, the agency has granted extensions on the May, 1975, air quality requirement deadline of one year to Pittsburgh, Philadelphia and San Antonio, 18 months to New York City, and two years to Boston, Springfield, Houston, the Newark and Camden-Trenton portions of the New York metropolitan complex, and Los Angeles.

Two years is the maximum permitted by law, and in the aggravated case of Los Angeles, which sees compliance impossible before 1980, the E.P.A. has said it will ask Congress for a special dispensation.

What happens if a locality does not meet the air quality requirement by the deadline of May, 1975, or its extension?

Birmingham Case

It's a fairly sure bet that nobody will go to jail immediately. Among an array of legal and procedural possibilities, the main benchmark is the environmental agency's statutory responsibility—if it decides that a community's atmosphere is jeopardizing public health—to step in and shut off any source of pollution it deems advisable. It did this briefly in Birmingham two years ago.

A motorist who wants to know what local restrictions he may be subject to, or how far along they are toward going into effect, had best watch his local newspaper or ask local air pollution officials.

And if there is, at this juncture, some indefiniteness, he might keep in mind these recent words of Robert L. Sansom, assistant administrator of the environmental agency:

"The E.P.A. operates in two worlds—a legal world of court orders, mandatory deadlines and so forth, and the real world of reasonable and challenging goals."

October 21, 1973

Acid in Rain Found Up Sharply in East; Smoke Curb Cited

By BOYCE RENSBERGER

In the last two decades, rain falling on the eastern United States and Europe has increased in acidity to 100 to 1,000 times normal levels, two ecologists have found. They said that the change had come about despite the increased use of air pollution controls and, in large part, because of some methods now used to clean smokestack emissions.

The scientists said that the acid rain may be stunting the growth of forests and farm crops and accelerating corrosion damage to man-made structures.

Under normal circumstances, pure rainwater is only slightly acidic due to its reactions with carbon dioxide in the atmosphere. The acidity may be likened to that of a potato. In recent years, however, the average acidity of rainwater has increased to about that of a tomato. In occasional extreme cases, rains have been found to be as acidic as pure lemon juice.

The researchers said that much of the increased acidity could be traced to a rising use of antipollution devices that make many smokestacks appear to be no longer emitting smoke. The devices, which remove only visible particles of solid matter, and not gases, still permit the escape of sulphur dioxide and various oxides of nitrogen that are readily converted to sulphuric acid and nitric acid in the air.

Before the devices were used, the solid particles, which are capable of neutralizing acids, entered the atmosphere and largely balanced out acids derived from the gases. Now they can no longer do so.

The study was made by Dr. Gene E. Likens, an aquatic ecologist at Cornell University, and Dr. F. Herbert Bormann, a forest ecologist at Yale University. They reported their findings in the June 14 issue of Science magazine.

Problem 'Transformed'

The smokestack particle removers, and the increasing use of very tall smokestacks— some are nearly a quarter of a mile tall — to disperse pollutants over very large areas, the two scientists said, "have transformed local soot problems into a regional acid rain problem."

In a telephone interview Dr. Likens said that the acid rain problem illustrated the potential hazards in a piecemeal approach to solving air pollution problems. As yet, there is no widely accepted, reliable technology for removing sulphur dioxide from smoke although at least one pilot project testing a promising method is reported to be under way.

The most widely used method for lowering the output of sulphur dioxide, which is the chief contributor to acid in rain, has been to switch to fuels that contain less sulphur to begin with. This method led to a decline of about 50 per cent in sulphur dioxide emissions in major cities in the nineteen-sixties.

45% Increase Found

However, according to a report by Dr. John F. Finklea, director of the National Environmental Research Center, this improvement has been more than offset by rapidly growing industrialization of regions away from major cities that are burning sulphur-bearing fuels. The net change nationwide, Dr. Finklea found, has been a 45 per cent increase in sulphur dioxide emissions.

Dr. Likens said that while the ecological effects of acid rain are not well known, there are preliminary indications of a reduction in forest growth, which has been noted independently in northern New England and in Sweden.

Laboratory experiments in which acids equivalent to today's average rain were sprayed on growing trees found that pine needles grew to only half normal length. Birch leaves developed dead spots and grew in distorted shapes. Studies on tomatoes misted with the acid water found decreased pollen germination and lowered quality and production of tomatoes.

A number of lakes in Canada, Sweden and the United States have become increasingly acidic in recent years, and some have experienced serious fish kills associated with the acid levels, Dr. Likens said.

Although the ecologists did not try to estimate the corrosive effect of acid rain on bridges, buildings, outdoor statues and the like, they said that the nature of acids suggested that serious damage was being done.

Because of the chemical nature of acids, (all of which contain hydrogen ions) they tend to combine readily with atoms of other substances, forcing those atoms in effect, to switch their chemical bond from the original site to the hydrogen ion of the acid.

Thus, for example, the atoms of calcium, which form essential components of a limestone building, will lose their bonds to each other and attach themselves to the acid washing down the side of a building.

June 13, 1974

E.P.A. ASKS EASING OF STATES' CURBS ON AIR POLLUTION

By The Associated Press

WASHINGTON, Aug. 16 — The Environmental Protection Agency proposed today to let states pollute their air, if they decide industrial and economic growth is more important.

The Sierra Club, which had won a Supreme Court decision earlier forbidding "significant deterioration" of existing clean air, immediately promised a new court challenge to the E.P.A.'s proposal.

The agency's deputy administrator, John R. Quarles Jr., told a news conference that the proposal would let the states give industrial and economic growth priority over protection of pure air.

He said it would allow construction of huge 1,000-megawatt coal-burning power plants, petroleum refineries, oil shale processors, coal-gasification plants and other air-polluting installations, in such areas, particularly in the West where the air is so clear that a person can occasionally see for 100 miles in the distance.

Federal Courts' Stand

Mr. Quarles acknowledged that since 1972 the Federal courts have required that existing clean air must be given greater protection than is afforded by the nationwide general air quality standards protecting public health and the environment.

But Mr. Quarles also conceded that the proposed regulations would let the states pollute their air as much as permitted on the nationwide limits, for the sake of industrial and economic growth.

"I would expect there would be further litigation on this point," Mr. Quarles said. This view was backed by Bruce Terris, attorney for the Sierra Club, a nonprofit organization dedicated to protection of the environment. Mr. Terris, who also attended the news conference, said that E.P.A.'s justification of its proposed policy had already been rejected in the courts and would be rejected again.

However, Mr. Quarles said a court rejection of the proposal would probably prompt Congress to amend the Clean Air Act of 1970 and clarify its intentions on this thorny issue. He said that so far key members of the House and Senate had been reluctant to get involved in it.

But the whole issue, pitting the pressures for environmental protection against the pressures for economic growth, revolved around the interpretation of "congressional intent."

In the Clean Air Act, Congress clearly required establishment of federal standards forbidding all pollution harmful to human health, plus later, more stringent standards protecting animals, plants, property and environmental values. These standards have been established and are being carried out by E.P.A.

In 1972, however, the Sierra Club took the agency to court, arguing that the act's purpose, to "protect and enhance" air quality meant that air already cleaner than required by the national standards must not be polluted at all.

Federal courts agreed that Congress meant to forbid any "significant deterioration" of existing clean air — but they did not say how much pollution would be "significant."

The environmental agency provided its answer with the proposed regulations.

"Deterioration of air quality can be regarded as 'significant' only within the broader perspective of public expectations and desires concerning the manner in which a particular region should be developed," Mr. Quarles said. "Air quality alone should not dictate entire patterns of economic and social growth."

The agency proposed to establish the following categories of clean-air regions:

¶"Class I . . . where almost no change from current air quality patterns is desired," with tight limits on pollution increases.

¶"Class II . . . where moderate change is desirable but where stringent air quality constraints are nevertheless desired," with somewhat easier Federal pollution limits.

¶"Class III . . . where major industrial or other growth is desired and where increases in concentrations up to the national standards would be insignificant."

Like the national standards, the Class I and Class II standards would place specific limits on pollution levels of sulphur oxides and "particulates" (smoke and dust) in the region's air.

Mr. Quarles said his agency planned to put all clean-air regions into Class II at the outset, allowing, probably, for the construction of some major new industrial installations, provided they use the best available technology to control pollution.

The states then could reclassify any or all regions into the other classes — either imposing the tightest restrictions, through Class I designation, or removing restrictions through Class III designation.

He said it was "correct" that there would be nothing to prevent the states, through reclassification, from allowing existing clean air "to deteriorate all the way down to the national standard."

August 17, 1974

Court Backs E.P.A. on Air Quality Slash

WASHINGTON, Aug. 2 (UPI) —A three-judge appeals court today unanimously upheld Environmental Protection Agency regulations permitting some deterioration of air quality in growing communities.

The United States Court of Appeals for the District of Columbia rejected attacks on the regulations brought by the Sierra Club, utilities and the American Petroleum Institute. The court said that the regulations balance a need for continued community growth with the need for maintaining clean air.

Under the regulations, formally approved last year, states may divide communities into three categories when deciding whether to permit increased sulfur dioxide and particulate pollution in areas that already have cleaner air than the national standards required under the Clean Air Act of 1970.

The regulations were in response to a 1972 legal victory by the Sierra Club, affirmed by a divided Supreme Court. In that case, Federal courts ruled that the Clean Air Act gave E.P.A. responsibility to avoid "significant deterioration" of air that was already cleaner than minimum tests established by the act.

Appeals by Utilities

The regulations were approved at the district court level. Utilities, which contended that the regulations were improperly approved by E.P.A., appealed, asking that they be reconsidered under additional procedures of the agency. They also asked that the 1972 decision be reversed and that E.P.A.

authority be limited to areas that failed to meet the Clean Air Act's minimum standards.

The Sierra Club also appealed, contending that the classifications would permit air quality deterioration in violation of the law and that there

were insufficient tests for determining what additional pollution would be "insignificant."

Judge J. Skelly Wright said that Congress had intended E.P.A. to have authority over clean air and declined to over-

turn the 1972 decision. He also held that the new regulations were approved properly.

Now pending in Congress are various amendments to the Clean Air Act that would further strengthen E.P.A. authority over areas already meeting

the standards of the 1970 act. Other proposals in Congress would overrule the court decisions and limit E.P.A. auhority to those areas that fail to meet minimum air quality standards.

August 3, 1976

Air Pollution Drive Lags, But Some Gains Are Made

By GLADWIN HILL

The arrival of the first major target date in the nation's effort to eradicate air pollution finds the program embroiled in problems and running far behind schedule—yet solidly moving forward on the course set four years ago.

But it is now becoming evident that substantially clean skies, once a wistful goal for the nineteen-seventies, are unlikely to be achieved before the nineteen-eighties or possibly even the nineteen-nineties.

The target date, May 31, 1975, was set by the Clean Air Act of 1970 as the time when the states were supposed to have in effect federally approved pollution-abatement programs embodying compliance with Federal air quality standards.

While all states have comprehensive abatement programs under way, there is none today with full formal Federal approval, according to the Environmental Protection Agency.

Russell E. Train, the E.P.A. administrator, said at a news conference today that "significant progress" had been made, but he acknowledged that the nation still had "a long way to go" in trying to meet the Federal standards."

In the four years that the program has been under way, the Environmental Protection Agency estimates that $14-billion has been spent, by governmental entities, industry and individuals, mostly on equipment. The money has brought some gains toward the goal—and there have been some disappointments and reverses.

On the plus side there are the following:

¶There has been a 25 per cent decrease in sulphur dioxide, a major pollutant, in the nation's air since 1970—50 per cent in large metropolitan areas —according to an E.P.A. report covering the last half of 1974.

¶The E.P.A. doubled the pace of its enforcement efforts against stationary polluters (such as factories and power plants) in the second half of 1974, making 2,517 investigations that resulted in 234 enforcement actions.

¶Automobiles, the other major source of pollution, along with stationary sources, are theoretically producing 80 percent less pollution, under the 1975 emissions limits, than in 1970 (conclusive data are still lacking).

¶While there has been controversy over specific measures aimed at achieving the Federal standards, the standards themselves have not come under attack.

¶Public opinion, as indicated in several surveys, seems firmly behind the program.

On the debit side of the four-year-old program, in addition to the fact that no state yet has an abatement program federally approved, there are the following:

¶Although there has been general improvement in air quality from 1968 to 1973, according to a study made last year by a research organization, pollution levels have continued to rise in more than a dozen cities.

¶Statutory postponement of certain important compliance deadlines to as late as 1987 is being debated.

¶E.P.A. engineers have concluded that even with full compliance with stationary-source controls and with 90 per cent reduction in car emissions, there were such concentrations of cars in urban areas that 30 big cities could not meet the Federal standards unless they reduced traffic, or took special steps to further diminish auto emissions.

As the program has developed, the task of stifling atmospheric contaminants from millions of sources has proved far bigger and more complicated than anyone foresaw. Massive administrative, technical and legal problems have turned up.

Delay Means Toll

The consequent slow pace of the abatement program has meant prolongation of conditions, especially in cities, that the authorities say are exacting a heavy toll in health and money.

The National Academy of Sciences estimates that air pollution causes 4,000 deaths and four million days of illness every year. The Environmental Protection Agency estimates that, as of 1970, the monetary cost of air pollution in health and material damage probably amounted to over $12-billion annually.

But the four years since passage of the Clean Air Act seem to indicate that the clean-up program is both technically and economically feasible. On the economic side, the cost has not been prohibitive. For example, total national expenditures on the program in 1972 were $6.5-billion, according to the Department of Commerce— less than 1 per cent of the nation's total production of goods and services, and less than is being spent on water pollution.

And there have been few if any suggestions from the thousands of individuals officially and professionally involved in air pollution abatement that the program be abandoned or fundamentally altered.

The scope of the task is of course tremendous. The E.P.A. estimated that in 1970 200 million tons of pollutants were going into the nation's skies annually. This guess was based on projections and presumptions of auto, industrial and natural emissions, such as forest fires. A similar estimate has not been made in the years since then.

Six Pollutants

Conditions vary greatly from place to place even in the same metropolitan area, and from hour to hour. There are six basic pollutants—particulates, sulphur dioxide, hydrocarbons, nitrogen oxides, carbon monoxide and photochemical oxidants (sunlight-formed compounds of other pollutants). Many populated areas have problems with all of these; some with just some.

Monitoring — the measurement of pollutants by machines —indicates only the quantity at a particular spot at one moment. Converting cumulative measurements into even limited generalizations about

what people are actually breathing takes a year or more of arduous computer work. Hence the current picture has to be sketched in piecemeal.

Here are highlights of the E.P.A.'s semiannual report covering the last six months of 1974:

¶Sulphur dioxide. There has been a 25 per cent decrease nationally since 1970, 50 per cent in large metropolitan areas. Of the 247 Air Quality Control Regions into which the nation is divided, 134 were within Federal standards for sulphur dioxide, 42 had excessive emissions, and conditions in 71 were uncertain.

¶Particulates. "Significant progress, although few areas that had significant problems in 1970 have yet attained national standards."

¶Oxidants. Of 88 problem areas among the 247 regions, 74 had excessive amounts during the six months.

¶Carbon monoxide. Of 59 problem areas, 54 had excessive readings.

Data on hydrocarbons and nitrogen oxides were too fragmentary for generalization.

Roughly half of air pollution comes from stationary sources. Some 200,000 stationary sources are subject to Federal-state regulation. Of these, 20,000 "major emitters" (capable of releasing over 100 tons of contaminants a year) are estimated to produce 85 per cent of all stationary-source pollution.

Of 19,173 major emitters, the E.P.A. said in its report on the last half of 1974, 71 per cent were in compliance with the law—either observing emission limits or adhering to prescribed cleanup schedules. Eleven per cent (2,160 facilities) were not in compliance and were subject to enforcement proceedings. The status of 18 per cent (3,428 facilities) was indefinite.

While the Federal agency, moving against delinquents, was doubling its enforcement pace in the second half of 1974, states reported 81,160 investigations and initiated 7,205 enforcement actions in the period.

Under the law, nonwillful violations can be dealt with only through several injunction proceedings. Willful violations may bring criminal proceedings, with fines up to $25,000 a day and up to a year's imprisonment. This has rarely been resorted to.

Because of unforeseen problems, ranging from fume-control processes to equipment shortages and prospective fuel switching to save oil, the E.P.A. has asked Congress for authority to grant selective extensions on compliance by individual facilities to as late as 1987.

Car Status Indefinite

If stationary sources' status in the air pollution picture is somewhat indefinite, that of the automobile is even more so.

The Clean Air Act originally set 1975 as the basic target date for eliminating 90 per cent of exhaust fumes from cars. But, in case of problems, the law provided for as much as two years' delay, at administrative discretion, in the phased schedule of fume reductions.

The last of these possible administrative deferments came March 5, when the E.P.A. postponed until 1978 the final step in fume reduction. The agency is asking Congress to extend that deferment until 1982, to give time for improvement of fuel economy and resolve uncertainties about sulphur compound emissions caused by fume-control mufflers.

This leaves automobiles at the 1975 emissions limits, which means about an 80 per cent reduction theoretically under 1970 emissions.

But it has become evident that Congress greatly underestimated the automobile problem.

After the Clean Air Act was passed, E.P.A. engineers, working their slide rules, came up with their conclusion that 30 of the nation's largest cities, even with full compliance with controls, could not meet Federal air quality standards without reduction of the total volume of car emissions.

Urban Controls

This problem, under the terms of the Clean Air Act, gave rise to urban "transportation control plans," prescribed by the E.P.A. for all the affected cities. The plans comprised an array of special measures,

ranging from "retrofitting" cars with special fume controls to expanding rapid transit, promoting car pools and limiting parking spaces to discourage driving.

The transportation control plans have become the most controversial feature of the air pollution abatement program, since many of the measures, while mathematically necessary, are patently impractical The extreme case is Los Angeles, where, in addition to all other possible measures, the E.P.A. said air quality standards could be achieved only by eliminating 86 per cent of all motoring.

States are still only in the early stages of implementing E.P.A.-imposed transportation control measures, and the agency, recognizing that they could be unduly "costly and socially disruptive," has asked Congress to defer the implementation deadline to 1982, when cities' overall pollution problems will be less.

E.P.A. experts testified at a Senate hearing this month that, even with completion of the auto fume reduction program, in 1985 five out of 26 cities that have exceptional carbon monoxide problems would still have discharges in excess of Federal air quality standards; 23 of 30 cities that now have excessive hydrocarbons would still have them, and 9 of 10 cities would have excessive nitrogen oxides.

A study of air conditions in 43 cities, analyzing data for the decade 1964-73, was made last year by the Council on Municipal Performance, a nonprofit research organization in New York.

No Plan Approved

While the study indicated general improvement from 1968 to 1973, according to John Tepper Marlin, the council's executive director, "pollution levels have continued to rise in over a dozen cities, notably Boston,

Buffalo, Miami, New Orleans, New York, Portland, Ore., Phoenix, Syracuse and Washington."

Essentially what these plans consist of is the establishment of emission limits on stationary sources of air pollution; measures for limiting the combined atmospheric load of stationary and automobile emissions, and measures for staying within Federal air quality standards despite future growth of populations, automobiles and industry.

Even small details of a state's formal plans, because they have the force of law, have to be approved by the E.P.A. And, because of the massive problems involved, as well as unexpected legal technicalities, no state's plan has yet received full Federal approval, according to the agency. A multitude of the approvals are still in process.

Sixteen states, including New York and New Jersey, were granted delays extending to 1977 on parts of their plans because of special difficulties.

Many states were close to full approval when a succession of court decisions, such as one requiring protection of air already cleaner than Federal standards, forced the E.P.A. to call for inclusion of new elements in plans.

The issue of protection of already clean air, in which the Supreme Court upheld the requirement in a suit by environmentalists, is one of several subjects of litigation that the air cleanup program has evoked.

Another matter in litigation is whether pollution dispersion by tall smokestacks and "intermittent controls" (temporary shutdowns or fuel switching) can be used as permanent compliance measures or only as interim improvement measures, as contended by the E.P.A.

Matter of Authority

Also in question is whether the E.P.A. has the authority, as it asserts, to grant or deny permits on construction of

"complex sources" of pollution —facilities such as shopping centers, parking structures and auditoriums that might generate concentrations of pollution.

The National Association of Manufacturers is urging Congress to amend the Clean Air Act to eliminate any requirement for keeping air cleaner than Federal standards; to eliminate federally imposed urban transportation controls and complex-source restrictions; to authorize tall stacks and intermittent controls as permanent compliance measures, and to weaken the E.P.A.'s requirement that, where necessary for compliance, industrial facilities must employ flue gas desulphurization (smokestack "scrubber" equipment), whose feasibility the electric power industry has been challenging.

In assessing the outlook for the air pollution abatement effort, a fact that stands out is that the Federal air quality standards promulgated at the outset as the goal to protect public health have stood substantially unshaken.

Controversy has raged over what emission limitations are necessary, in particular circumstances, to achieve the standards, but not over the standards themselves.

A second salient fact is that public sentiment in favor of a sustained pollution abatement effort seems to remain high.

Daniel Yankelovich, who annually surveys opinion on policy questions affecting large corporations, reported early this year: "The desire for clean air continues strong and there appears to be little likelihood of major concessions to business."

Louis Harris, the pollster, reported in March that 46 per cent of respondents in a national survey ranked air pollution fourth among "very serious" national problem areas, after inflation, unemployment and water pollution. This represented a 12 per cent increase over a year before.

May 31, 1975

California Fines A.M.C. $4.2 Million Under Smog Code

By ROBERT LINDSEY
Special to The New York Times

LOS ANGELES, Jan. 5—California fined the American Motors Corporation $4.2 million today because the company allegedly submitted "totally false" test reports to the state and unlawfully sold cars that exceeded California's strict automotive pollution control standards.

The state also banned, effective at midnight tonight, the sale of three A.M.C. models powered by 304-cubic-inch V8 engines.

The fine is the largest ever levied on an automaker by a

state, although the Environmental Protection Agency fined the Ford Motor Company $7 million in 1973 for not complying with certain emission control regulations.

Action Is Criticized

American Motors called the action "unjustified" by the facts and said the amount of the fine was unreasonable.

The California Air Resources Board accused A.M.C. of violating two provisions of its air pollution code. It charged submission of false test data purporting to prove that its cars meet California smog-control

standards and sale of hundreds of cars with 304-cubic-inch V-8 engines that did not meet state limits on emissions of oxides of nitrogen and carbon monoxide.

Sales of new 1976 A.M.C. Matadors, Hornets and Gremlins with those engines were halted by the agency until modifications were made to make the vehicles conform to the regulations.

"We have never seen so many dirty cars," Thomas Quinn, chairman of the agency, asserted in announcing the fine. "But even more serious

is the fact that American Motors submitted false reports to the state which indicated that their cars were actually very clean.

"American Motors executives have denied any intentional wrongdoing and attributed the problem to lack of attention, poor maintenance of test facilities and neglect," he said.

"Whatever the case, we consider A.M.C.'s actions a major violation of California's anti-smog laws, and we intend to see that it never happens again."

Last year, the state agency.

fined the Chrysler Corporation $328,000 for alleged violations of the restrictions, and sales of certain Chrysler products were halted temporarily until modifications were made.

A spokesman for the agency said today that it was also monitoring the conformance of cars made by the General Motors Corporation and Ford Motor Company, but said that no action against them "is current."

The state is also rechecking, as a matter of routine, the test data submitted by G.M., Ford and Chrysler, the agency said.

In an unusual feature of the action, the state agreed to accept only $1,050,000 of the fine if A.M.C. signed an agreement to spend the remaining $3,150,000 to improve the emission control and fuel economy of its cars.

Reasons for Action

"We are considering this approach for two reasons," Mr. Quinn said. "First, we believe that payment of the full amount could damage A.M.C.'s financial base and lead to an eventual decline in competition among United States automotive manufacturers; and second, we feel that American Motors must substantially improve its fuel economy and emissions programs if it is to continue as a viable company."

The Air Resources Board said that, since the 1976 model year began late last summer, 1,239 of the A.M.C. cars with the V8 engine in question had been sold in the state, and it estimated that about 60 percent of these did not meet the state's pollution standards.

During the same period, it said, 11,244 A.M.C. cars were offered for sale in California that were covered by what the state claims were false data purporting to show compliance with the California smog laws.

Under state regulations, car manufacturers must test 2 percent of their cars and submit the findings to the state. The Air Resources Board also checks at least three cars of each model on its own.

Test Data Studied

Technicians in the agency said their suspicions were aroused because the test data submitted by A.M.C. indicated the cars were "among the cleanest produced anywhere in the world," while prototypes of the 1976 cars that had been tested in early 1975 only barely met restrictions.

This prompted the state to review the company's data, recheck additional A.M.C. cars, and come to the conclusion, Mr. Quinn said, that the A.M.C. test data were "totally false."

The state agency said rechecks of the company's 6-cylinder engine, used on many other cars sold in the state, indicated "a potential problem" also. However, at this point, they can continue to be sold pending additional tests now underway.

In a brief statement late today, American Motors said that it had not had an opportunity to study the agency's ruling, but called the decision to halt sales of cars with its 304-cubic-inch V8 engine "unjustified" and assailed the magnitude of the fine as "unreasonable."

"This engine was properly tested and certified prior to the start of the 1976 model year production," the company said. "The A.R.B. has raised questions regarding compliance with the applicable standards on the basis of limited testing."

State Prods Detroit

Since the mid-1960's, California has been something of a national pacesetter in prodding Detroit to clean up its engine emissions. In many cases, its regulations have preceded those later adopted nationally. But today's action was representative of a conflict between the industry and California that has been escalating during the last year under Mr. Quinn, a 33-year-old political activist who was appointed to his job after serving as a campaign manager for Gov. Edmund G. Brown Jr., a Democrat.

While the Federal E.P.A. has slackened its pressure on carmakers to further reduce emissions over the last year, partly because of doubts about the technical feasibility of some proposed avenues of doing so, Mr. Quinn has vowed to increase pressure on Detroit, in the belief it will force the industry to give smog prevention higher priority. Already, California's higher standards have prompted most manufacturers to produce some cars specially tailored for the California market, and other models are not available here at all.

California's stringent exhaust emission standards permit only 0.9 grams of hydrocarbon per mile, while Federal standards allow 1.5 grams.

A.M.C. said in Detroit that it had cooperated fully with the A.R.B. in trying to resolve questions connected with the emissions tests. "All of the board's inquiries have been responded to openly and forthrightly," the company said, adding that "A.M.C.'s test procedures had been only recently reviewed by the Air Resources Board technical representatives."

California conducted a series of 39 tests that showed 85 percent of A.M.C. cars tested failed California emission standards for either carbon monoxide or oxides of nitrogen.

January 6, 1976

POLLUTANTS IN AIR DRIFT 150 MILES

Balloon Crew Tracks Dirty Plume From St. Louis to Indiana Wheat Field

By EDWARD COWAN
Special to The New York Times

WASHINGTON, June 22—An experimental manned balloon flight earlier this month showed that polluted city air can drift 150 miles or more into rural areas with no thinning of pollutants, Government scientists reported last week.

The balloon, with a woman and three men riding in a gondola, took off from St. Louis on June 8, the third day of an atmospheric inversion that had trapped dirty air over the city. The balloon hovered in the stagnant air over St. Louis most of the day, then drifted with the plume of polluted air at night.

When it landed 24 hours later and 150 miles away in an Indiana farmer's wheat field, "we were seeing essentially the same concentrations of pollutants," Bernard Zak, scientific director of the experiment, reported.

Mr. Zak said the experiment, the second such flight of Project da Vinci, showed that air pollution was not simply a local problem for people who lived near coal-burning plants, factories that burned heavy fuel oil or heavy commuter traffic.

Over Long Distances

"High concentrations of pollutants can be transported long distances and can be visited upon people" not responsible for the pollution who live "one, two or three states away," he said.

Mr. Zak and the crew of the Gondola reported on the flight at a news conference at the National Geographic Society. The society has contributed about $40,000 toward the project's cost, said a spokesman for the Energy Research and Development Administration, which is spending almost $1 million on it. Other sponsors are the Environmental Protection Agency and the National Oceanic and Atmospheric Administration.

The first da Vinci flight, named for the Italian Renaissance painter and inventor, who was among many things a balloonist, took place in the very clean air of New Mexico in November 1974. It established scientific data against which later findings could be compared.

St. Louis was chosen as the takeoff place for the second flight and a third one next month because its air is usually dirty and because it is the site of continuing air pollution studies.

Ton of Instruments

The effort is intended to learn how pollutants behave over long distances and periods of time, Mr. Zak said. Crammed into the 10-foot-square, two-level gondola was more than a ton of instruments.

From thousands of readings, air samples and experiments the scientists at the energy research agency's Sandia laboratories in New Mexico expect to sketch a detailed picture of what happens to a plume of polluted air as it moves across several states. Eventually, air currents disperse pollutants and rain washes them to earth, but how long such dispersion takes is highly variable.

The energy research agency is interested in air pollution because of policy issues on emission control at power plants and oil refineries.

Rudolf J. Engelmann, the scientist in the gondola, told the news conference that the flight picked up a significant quantity of sulfur dioxide as it crossed the Shell Oil Company's Wood River refinery at Alton, Ill.

Earlier, on the Missouri side of the Mississippi River, Mr. Zak recounted, the crew observed that the smokestacks of the Union Electric Company's coal-burning Portage des Sioux power plant were shut down as the 70-foot-high, helium-filled balloon approached. Whether the plant was reacting to the approach of the scientific team was not clear.

The pilots, Vera Simons, who designs and builds balloons, and Jimmie Craig, reported that they, Otis Imboden, a National Geographic photographer, and Mr. Engelmann were constantly busy throughout the flight.

Miss Simons described the air plume as a "dirty mess" and "really murky," adding, "Things off in the distance had a fuzzy outline."

Mr. Zak said it was impossible to estimate the probability that a given rural area 150 miles from a city would get pollutants but it could happen routinely for some places.

June 23, 1976

E.P.A. Sources Say Agency Ended Efforts to Cut Auto Smog in Cities

By GLADWIN HILL

A major element of the Federal air-pollution abatement program, mandatory measures to reduce auto traffic in New York and most other big cities, is being largely abandoned according to authoritative sources in the Environmental Protection Agency.

The reason is that even though Federal law calls for the institution of urban "transportation control plans" as part of states' compliance with the Clean Air Act of 1970, there has been so much resistence from states, cities and citizens that it has become clear that the controls cannot be imposed nationally without an impossible mass of litigation.

Instead, the E.P.A. plans to leave adoption of the measures up to voluntary action by states and localities. Pollution experts consider the measures necessary to protect public health adequately.

The measures involve about a dozen basic "strategies" for reducing the volume of auto fumes, ranging from systematic development of car pools to the ultimate weapon of gasoline rationing. Three years of efforts to institute these measures across the country have been marked by controversy, delays, and legal jousting.

Voluntary Cooperation

"Essentially, what we've done is abandon all these plans," an agency spokesman acknowledged this week. "There's been so much static the conclusion has been that it's better to go the route of voluntary cooperation."

This unannounced policy change injects further uncertainty into the already cloudy question of when the nation will achieve the air quality standards called for in the 1970 act. The statutory target date, long since conceded to have been impracticable, was mid-1975.

The shift has special significance for New York City, which has had some of the worst automobile air pollution in the country, and the corresponding amount of litigation about implementing transportation controls.

The principal exception to the new "voluntary policy is a measure that Federal experts consider to have the largest practical promise for reducing fumes: mandatory periodic inspection and maintenance of individual automobiles' smog controls. The agency filed a test case suit in November to force the adoption of this practice in Cincinnati.

The agency's current head, Russell E. Train, an appointee of President Nixon, is expected to be replaced in the Carter administration. This could mean changes in any of the agency's policies. However, current agency officials suggested that the reasons for shifting to the "voluntary" approach on transportation plans were so compelling that alternatives would be hard to devise.

The agency's original calculations were that, in addition to all the regular stationary and mobile smog controls required by the law, 31 metropolitan areas could never meet air quality standards without special restrictions on auto traffic. Consequently, the list has been tentatively enlarged to involve 63 cities.

Early Experiences

Automobiles are responsible for a sizable portion of air pollution generally—70 percent of the carbon monoxide, 50 percent of the hydrocarbons and 30 percent of the nitrogen oxides.

The agency has estimated that transportation controls can reduce the total volume of car fumes in a city by as much as 30 percent.

Transportation controls first came to the fore in 1973 as a mandatory part of overall state plans for carrying out the terms of the Clean Air Act. If states did not produce satisfactory urban transportation control plans, the agency was empowered to formulate and impose them.

The agency at first demurred at doing this, because requisite fume reduction appeared to involve such impracticalities as eliminating nearly 90 percent of the auto traffic in Los Angeles. However, environmental organizations got court decisions compelling the agency to carry out the letter of the law.

This action in turn generated hundreds of lawsuits challenging the agency's authority to thus dictate state actions. Differing rulings by Federal lower courts led to several appeals currently before the United States Supreme Court.

Although the program has been under way three years, in terms of concrete accomplishments an agency official said ruefully this week, "The fact of the matter is that not a lot has occurred. If a state doesn't have a real commitment for transportation controls, there's not a lot the E.P.A. can do about it."

The agency's deputy administrator, John R. Quarles Jr., said legal challenges had dealt "a crippling blow" to the program, and that "unless some major changes are made, auto pollution may never be brought under control."

He said that initial court-imposed deadlines had necessitated some hasty and ill-advised directives and that the agency "has learned a great deal about how not to run a transportation control program."

'Understandable Hostility'

State and local officials had reacted to compulsory attempts with "understandable hostility," he said, adding, "Without local support, even a court-ordered plan has little chance of success—if real air pollution cleanup is to be made, the local community must agree that it's worth the effort."

A draft agency "strategy" document for attaining Federal air quality standards now being circulated cites only vehicle maintenance-and-inspection as a promising transportation-control measure. The paper says that "it is the E.P.A.'s policy that new and revised transportation strategies be developed through the normal urban transportation planning process."

Among the original 31 cities, the agency calculated that to meet Federal air quality standards, 28 would have to institute inspection-maintenance programs, 24 would have to restrict parking as a "disincentive" to city driving, and 19 would have to require "retrofitting" of older cars with special fume controls.

Other prescribed measures, with the number of cities needing them, were: preferential lanes for buses and car-pool vehicles—19; expanded public transit—10; systematic promotion of car pools—7; banning cars from some streets—7; restrictions on motor cycles—7; limited truck delivery hours—5; parking surcharges—3; gasoline rationing—13.

The agency dropped the parking-surcharge idea from the program because of the instant furor it aroused and unanticipated legal complications.

Car-Pooling Programs

What has actually materialized out of the program has been car-pooling incentive programs in several states, preferential laning in a number of cities, comprehensive vehicle inspection-maintenance programs only in New Jersey, Portland, Ore. and Phoenix, Ariz., a pilot program in Riverside, Calif., and voluntary programs in some other places.

An agency official said that because "we're in very much of a holding pattern" on the transportation controls program, the agency did not have a comprehensive list of what measures had actually been put into effect in what places.

An agency report last November on accomplishments in pollution abatement cited, in regard to transportation controls, only the handful of inspection-maintenance and car-pooling programs.

The agency has estimated that on post-1974 cars, typical annual inspection charges will range from $1.20 to $5.00 and average repair charges, if deficiencies are found, will run about $30.

Typical of the frustrations the program has encountered have been two major controversies in California. Southern Californians spent most of 1975 disputing over a state regional law requiring "retrofit" devices on older cars, which finally was repealed.

Last year a five-month experiment in restricting a Los Angeles freeway high-speed lane to buses and car pools during rush hours caused a public uproar. The innovation was halted by a Federal court ruling that environmental impacts had not been adequately studied.

New York City formulated a plan involving 37 variations and applications of basic traffic-reduction strategies, including special bridge and tunnel tolls and limitations on taxi cruising.

But the plan's main trial has been in marathon court proceedings, culminating in recent rulings that the plan should be put into effect, on an unspecified time schedule.

According to a recent report by the National Resources Defense Council, a leading plaintiff in the litigation, "New York City sits in clear violation of the law, and nothing is being done about it."

January 14, 1977

Expanding Plants Face Pollution Tradeoffs

Special to The New York Times

HOUSTON, July 31—Harry Birdwell, manager of the Oklahoma City Chamber of Commerce, was ecstatic last year when the General Motors Corporation announced plans to build a big automobile assembly plant in his city because it would create 5,000 jobs and bring a badly needed economic lift.

But it also thrust Mr. Birdwell into a ticklish new area of industrial development: He had to persuade air polluters already in the area to clean up enough of their emissions to make room for G.M.'s.

Debate Over Controls

Air pollution has become almost a commodity and men like Mr. Birdwell are its brokers, buying and selling so that a new plant can move into an area, as in G.M.'s move into Oklahoma City, or so that an existing plant can expand. Nearly 50 companies in 20 states, according to a spokesman for the Environmental Protection Agency, have had to clean up enough pollution caused by other plants to offset increases in their own output.

This budding corps of pollution traders had their roles created late last year in an E.P.A. ruling that sought to permit continued industrial growth without undermining whatever strides a region had made in cleaning up befouled air. The environmental agency said that it would permit a new pollution-causing industry, but only if enough existing emissions were cleaned up so that there would be no net increase in air pollution.

The ruling has heightened debate over whether the Federal Government should enforce strict air pollution standards at the possible expense of industrial growth. In the meantime, however, states with industries moving in and expanding have found various ways to comply with or circumvent the policy.

In California, where strict standards have already reduced air pollution from large factories, Standard Oil of Ohio has proposed buying the pollution of several dry cleaning plants and other small factories to offset the emissions of its planned station in Long Beach where oil from Alaska would be unloaded .

U.S. Steel's Example

Pennsylvania officials, eager to attract Volkswagen's first American plant to its state last year, switched chemicals in the substance used to fill potholes on Pennsylvania highways to cut down on hydrocarbon emissions near New Stanton, where the factory is.

The Ford Motor Company has agreed to add more pollution controls to its Louisville, Ky., assembly plant as a trade-off for adding a second work shift. But plant officials have yet to resolve the problem of additional auto emissions caused by more traffic to and from the plant each day.

In St. Paul, Minn., local government engineers have found a way to cut back on emissions from an expansion of the local sewage system. New control equipment and a reduction in the use of fuel oil will more than offset the emissions expected from the plant's sludge incinerator, local officials say.

Tough California Policy

The United States Steel Corporation's factory in Birmingham, Ala., has been asked by the local air pollution board to cut back on emissions before adding to its facilities. Plans for construction of a blast furnace there forced company officials to determine the overall environmental impact of the facility, and more modern controls will be required before the state board will allow the expansion.

In Oklahoma City, Mr. Birdwell shopped around for companies with crude oil storage tanks willing to install "internal floating covers" to reduce hydrocarbon vapors. Four companies agreed, and the E.P.A. approved the trade-off in May.

Had the companies refused to cooperate voluntarily, however, either General Motors or the Chamber of Commerce would have been forced to "buy" pollution from smaller sources—install equipment and clean the air at its own expense. "We would have bought it if we'd had to," Mr. Birdwell said.

Standard Oil of Ohio may make such a purchase in the case of its planned unloading dock in the Long Beach harbor. Sohio officials approached dry cleaning plants in the Long Beach region with proposals to clean up their pollution.

California, which exerts far tougher control than most states, imposes a two-to-one trade-off ratio—for every unit of pollution added, two must be removed. That naturally upsets Sohio officials.

"Trade-offs have almost become a commodity," said Samuel Baker, a spokesman for Sohio. "How responsible are we for what California permitted 10 or 20 years ago?"

The only other company to receive the E.P.A.'s approval for a trade-off was Volkswagen. The assembly plant in southwestern Pennsylvania will emit 720 tons of hydrocarbons annually, and the trade-off, worked out by the state's Department of Environmental Resources and Department of Transportation, followed Gov. Milton J. Shapp's order to state officials to find and clean up enough air pollution to equal Volkswagen's output.

While Pennsylvania is going along with the ruling, Texas is fighting it. John A. Hill, the Texas Attorney General, has sued the E.P.A. over the ruling, and the regional E.P.A. office in Dallas now reviews all permits issued by the state control board, the only one in the country that refuses to recognize the ruling.

An E.P.A. spokesman said four violation notices have been issued in Texas. Civil or even criminal proceedings could be brought against the violators if they fail to comply, the spokesman said.

August 1, 1977

CONFEREES SETTLE ON 2-YEAR EXTENSION OF CAR EXHAUST CURB

By ERNEST HOLSENDOLPH
Special to The New York Times

WASHINGTON, Aug. 3—A House-Senate conference committee reached a compromise agreement early today on new air pollution restrictions that ended the threat of an auto industry shutdown but produced mixed feelings among manufacturers and environmentalists.

The bill that emerged will extend for two more years—for the 1978 and 1979 model years—the existing tailpipe exhaust standards for three key pollutants —hydrocarbons, carbon monoxide and nitrogen oxide. House leaders cleared the way today for action on the agreement tomorrow, with likely concurrence by the Senate expected Friday.

The settlement, reached after a weary seven hours of maneuvering, punctuated by angry outbursts, appeared to be a considerable victory for Senator Edmund S. Muskie, Democrat of Maine, the Senate's leading environmentalist, who managed to head off an attempt by the industry supporters to win acceptance of a softer House bill.

Defeat for Dingell

It was also considered a defeat for Representative John D. Dingell, Democrat of Michigan, who had worked hard for easier pollutant limitations and stirred a mild uproar when he told the panel that health hazards were not a serious issue in the limitation of emissions.

In Detroit, auto industry executives issued cautiously worded statements of praise for the compromise. They said they would now make every effort to meet schedules in building 1978 model cars.

The production of 1978 car models can now proceed. Previously the manufacturers had asserted that they could not make cars this year to meet the 1978 emission standards and would shut down, idling hundreds of thousands of workers, rather than expose themselves to penalties.

Accommodation for Most

In a sense, most of the interests associated with air pollution problems received some accommodation in the legislation, which consisted of a series of amendments to the 1970 Clean Air Act. The manufacturers won more time while the environmentalists closed off what they felt were possible future gambits by industry to outflank the clean air laws.

"I think it's a reasonable compromise," Mr. Muskie told reporters this morning. "Finally we may have gotten a deadline [on auto emissions] that will be met."

Aside from auto pollution, the conferees also made provision for states to get a waiver on air pollution restrictions to permit the construction of power plants. Although the name of the Intermountain Power Project, a giant proposed electric utility plant, was never mentioned by name by the legislators, the waiver was put through mostly to help its developers get around Federal opposition.

The project, which will operate on low sulfur coal, is to be constructed next to Capitol Reef National Park in central Utah to provide power to 23 communities in the state, as well as to six cities in southern California, including Los Angeles.

The waiver provides that the plant can be built with the concurrent approval of the Interior Department and the Governor, but if they disagree, the President may allow them to construct it.

The legislation could force President Carter to make hard choices between environmental cleanliness and his goal of using coal to lessen dependence upon imported petroleum.

"We were disappointed in a number of things," said Chris Goddard, who spoke for the National Clean Air Coalition, a network of environmental groups. "For instance, there was some exemption for smelters and the authority to regulate indirect sources of pollution, such as shopping centers and the cars they attract."

She said, however, that the coalition supported the conference report, largely because it had confidence in the President and the new leadership of the Environmental Protection Agency to enforce it strictly.

The final seven-hour meeting of the conference was an exhaustive series of proposals, followed by caucuses, counter-proposals and more caucuses.

At one point Mr. Dingell asserted that there was no medical evidence to support allegations that auto emissions, particularly carbon monoxide, were injurious to health.

Blames Car Exhausts

Wendell R. Anderson, a Minnesota Democrat who is a new Senator and a physical fitness enthusiast, said that since moving here he has had trouble breathing and has suffered from nausea and burning eyes.

"Sometimes, driving in from Virginia, I have been unable to see the Washington Monument," he said, "and that's from cars, because there's no industry here."

Getting angrier as he talked, Mr. Anderson cut Mr. Dingell off when he tried to interrupt: "Don't point your finger at me. I've listened to you; now listen to me. We've given the auto industry 10 years to comply and they haven't come around."

Mr. Muskie, who was chairman of the meeting, then joined the criticism, saying: "Every time we give something to this industry, they ask for more. Now they ask us to yield on carbon monoxide. I won't do it."

From that point on the proponents of tougher restrictions, mostly from the Senate side, gained strength, and it became clearer as the night wore on that the conference would reach a decision.

August 4, 1977

WATER

THE FOULING OF WATER COURSES.

The pollution of streams by the sewage of towns has been the subject of much discussion in many States, some of which began some time ago to take measures for the prevention of this disagreeable and dangerous fouling of water courses. A recent decision of the Superior Court in Connecticut indicates that injured riparian owners there have a remedy at law. This decision, with repeated complaints about such pollution in several parts of Connecticut, brings the question clearly before the people of the State and may cause a radical change in the methods of sewage disposal now in use in several of the smaller cities.

Danbury, a city of about 20,000 inhabitants, discharges its sewage into the Still River, a small tributary of the Housatonic. This stream is commonly so low in Summer that a ten-inch pipe would carry the entire flow of water. The pollution has become a nuisance, not only on account of the repulsive odor and the possible effect of the contamination upon the health of adjoining residents, but also because the accumulating sediment clogs the stream and is filling up two or three mill ponds which are not far below the city.

Certain injured riparian owners applied to the court for an injunction, and this has been granted by Judge WHEELER, whose opinion in the case is very elaborate and instructive. It appears that the city has set up a plant for the deodorization or sterilization of the sewage, but the Court holds that, even if deodorization should be accomplished by it, the complainants would still be injured, because decomposition would be excited in the sewage so treated, owing to its association with other sewage or polluted matter which the stream receives from other sources.

There is now pending in the Connecticut courts a suit against the City of Waterbury for damages caused by the discharge of its sewage into the Naugatuck River, and it is quite probable that the plaintiffs will be successful. This stream, it appears, is very badly polluted. The Derby (Conn.) Transcript of the 11th ult., calling attention to the prevalence of typhoid fever in the group of manufacturing towns on the Naugatuck, near its junction with the Housatonic, said:

"The more our epidemic of sickness is considered, the more plausible becomes the theory that the Naugatuck ought to be held responsible for the more than ordinary amount of disease prevalent. The whole course of this river is a producer of poisonous germs. The atmosphere all the way from Waterbury to the Housatonic must carry from the vicinity of the river microbes of foulness. Between Waterbury and Naugatuck the stench arising from the river is such that passengers cannot ride in open cars on the electric line which follows the river bank, and closed cars have been in use most of the Summer. At Union City, Naugatuck, and Beacon Falls the water is absolutely filthy, and some days the odor arising from the river along through Ansonia and Derby is almost unbearable. It must be obvious that such a condition cannot exist without becoming a serious menace to health."

In the capital of the State there is a similar nuisance. The little stream which winds through a beautiful park in the heart of the city, on the edge of which the new Capitol Building stands, and which is clearly seen from the car windows by railway travelers as they pass through the town, is an uncovered sewer. The odor arising from it is very repulsive in the warm season. The Cities of Hartford and New-Britain unite in polluting this water course. New-Britain, nine or ten miles away, discharges sewage into a brook which is a tributary of this Park River, and this brook, owing to its very offensive odor, is a nuisance for miles of its course through a lovely rural region. One or two of the riparian owners have recovered damages from New-Britain for the injury thus caused.

The stream is foul enough when its waters enter the capital city, but Hartford pours more sewage into it before the Park is reached. We quote the following remarks from The Hartford Courant of the 2d ult.:

"The recent heavy rain changed somewhat the situation in Park River. Previous to these showers, the principal movement of the water had been perpendicular—up from the bottom to the top. As a result of the storm it has moved along a little horizontally, and the color has shifted from black, coated with green yellow, to an almost uniform chocolate tint. This is progress. If the river would always move, and if the water would cease to spoil and stink right under our very noses, the mere looks of things would be only a disagreeable incident. We pour into the stream vastly more of filth than its normal volume can take care of.

"People say that Park River is 'only an open sewer,' but we have more than once invited attention to the fallacy of such statements. A sewer intelligently planned and properly constructed does not hold the stuff poured into it. On the contrary, its very purpose is to carry this off quickly and before the processes of disintegration get in their work—to move it along until it reaches some body of water large enough to absorb it. So the Park River is not an open sewer—it's an open outrage.

"It threatens the life of the whole community, invites upon us the contempt of every traveler on the railroad that passes through here and every visitor that comes to Hartford, and turns away from this otherwise beautiful and attractive city many clean people who do not care to become fellow-citizens with this nasty creature."

In the course of time there is to be constructed an intercepting sewer which will divert Hartford sewage from this stream and pour it into the Connecticut, but that sewer will not extend southward ten miles to New-Britain. There is room for legislation in Connecticut—and many other States—for the prevention of such pollution, which sometimes crosses town and county lines, and may cause injury in many ways.

When a river or any other body of water thus polluted is a source of public water supply, the effect upon the public health is deplorable. The proof of this is abundant, and some of it can be found in the vital statistics of Albany, Chicago, Philadelphia, St. Louis, and several cities on the Ohio below the point where the sewage of Pittsburg and Allegheny City is poured into that river. Even when the contaminated stream does not furnish water for a city or village, there is always danger that persons will use it to their disadvantage, and there is evidence that the health of adjacent residents is sometimes affected seriously by the foul emanations which indicate the pollution.

November 4, 1895

THE WAR AGAINST DISEASE

Polluted Water Supplies.

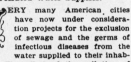

VERY many American cities have now under consideration projects for the exclusion of sewage and the germs of infectious diseases from the water supplied to their inhabitants. The effect of the pollution of public supplies of water upon the mortality due to intestinal maladies has been so clearly shown during the last few years that the demand for purification in cities marked by a high typhoid death rate is growing in force and soon will be irresistible.

The movement in Philadelphia for the construction of sand filtration works, for some time successfully opposed by an ignorant majority of the municipal legislature, will be vigorously supported by many prominent citizens during the next three or four months, and it is probable that an experimental filtration plant will be in use next year. On Sept. 1 it was reported that there were nearly 200 cases of typhoid fever in Indianapolis. The water supply of that city is to be purified, however, by the construction and use of filtration beds which will cover twelve acres.

In Chicago the typhoid death rate continues to be very high, the average number of deaths per week having been about fifteen for some time past. It is well known that the water taken from the lake is polluted by the city's sewage, and the health authorities urge citizens to boil it thoroughly. Chicago's water will be exposed to dangerous contamination until the great drainage channel shall have been completed.

Other cities on the great lakes, which procure water in the same way, suffer by reason of sewage pollution, as their vital statistics prove. In Milwaukee additional cause of complaint has recently been found in the contamination of the water by large quantities of garbage, which a contractor had undertaken to consume in a rendering factory, but which have been dumped into the lake from a pier. Forty-eight carloads of highly objectionable garbage were thus disposed of in four days.

No progress appears to have been made toward the purification of the water supplied to the people of Albany, the capital of the greatest of American States. Albany's water is polluted by its own sewage as well as by that which is discharged into the Hudson by neighboring cities just above it. The typhoid death rate is high in Albany at all times, and is frequently increased by epidemics of the fever. The legislators of the Empire State should use their influence to procure wholesome water for those who reside temporarily or permanently at the capital.

September 13, 1896

LAWFUL POLLUTION.

Justice BLACKMAR has rendered a decision which may be above criticism as a matter of law, but which he supports with singular reasoning. A Jamaica Bay oyster grower sued the city for damages caused by the discharge of sewage where oysters were growing for the market, making them inedible, or at least unsalable, since there is room for doubt of much which is sold for food. The Judge decided that the pollution of tidal waters is unlike the pollution of stream waters. All owners of the banks of streams are entitled to the use of the water, and the pollution of it is unlawful. Cities cannot lawfully empty sewage into streams, but they may into tidal waters. The plaintiff had no right in them except the right of access.

There is in this case no trespass by casting sewage on plaintiff's land, and there is no nuisance. The most that can be said is that the water has been rendered unfit for human consumption. No right of his is violated, unless a riparian owner on tidal waters has the right to have the salt water, as it is carried to and fro by the tide, kept fit for human consumption.

If there is no such law, why would it not be a good law to have? The Judge thought it unanswerable to say that, if the suit succeeded, "it would "render illegal every sewer in New "York City." The suit having failed, it is strictly lawful for the city to maintain such sewer discharges that the waters are unsafe for bathing, and even in such remote districts as Jamaica Bay it is unsafe to eat food taken from them. It is not a legal nuisance, the Judge says, but it is a threat to health.

The suit failed because it was impossible to give to every citizen the rights claimed, the court said, for the first time by the plaintiff. But why would it not be well for all to have the rights claimed, and to put the city upon prudent notice that its rights of pollution have been pressed to the point of wrongs by the public upon private citizens?

June 19, 1915

TAKE STEPS TO END BEACH POLLUTION

Acting on complaints that refuse swept shoreward from the sea was making bathing beaches a breeding place for deadly bacilli, Street Cleaning Commissioner Alfred Taylor and Captain A. Olmstead, U. S. N., Federal Supervisor of New York Harbor, conferred with Commissioner Taylor's office in the Municipal Building yesterday. The city and the Federal officials agreed to co-operate in an effort to trace the source of pollution.

After the meeting Dr. Alonzo Blauvelt, Sanitary Superintendent of the Department of Health, told Commissioner Taylor the Board of Health had received an unusually large number of complaints within the last few weeks about conditions at the beaches.

Long Beach, the Rockaways, Coney Island and Manhattan Beach have all reported that refuse was being cast up in greater quantity this year than ever before. Concessionaires at the resorts have reported many complaints from bathers. At the Brooklyn Eye and Ear Hospital four swimmers were recently operated on for mastoiditis, apparently contracted while bathing in polluted waters.

"There has been no change in our methods of disposing of the city's refuse," said Commissioner Taylor, "and I am unable to account for the refuse reported by residents at the beaches. The garbage is carried forty miles out to sea and is there dumped. On each scow, we have a Federal inspector, who sees that the scow actually goes out that distance and that all our requirements are faithfully performed. The scows are operated by the city and not by private contractors. We average three scows a day, each carrying about 2,000 yards of débris. A scow takes from nineteen to twenty-four hours to make the trip.

"We are co-operating with Captain Olmstead and it has been agreed that in the future two inspectors are to go out on each scow. We are also issuing orders that all paper or other material likely to float is to be carefully separated from the cargo of each scow.

"I am certain it is not our garbage that is being thrown up on the beaches. It may come from steamships, but I do not know just where it does comes from. We are going to make a thorough investigation and as soon as we find the source the nuisance will be stopped at once."

In support of Commissioner Taylor's theory that ocean traffic was responsible for the refuse was a statement yesterday by Eltinge F. Warner, publisher of Field and Stream, who is conducting a campaign to educate bathers to the dangers of polluted waters.

He pointed out that in 1911 the world's quota of oil-burning and oil-carrying ships was 364, with a total tonnage of 1,310,000, and that in 1921, only ten years later, the quota had increased to 2,536 vessels carrying or burning oil, with a tonnage of 12,797,000. These ships, said Mr. Warner, leave in cargo and come back in water ballast. Nearing shore on the return trip they throw and pump bilge, oil waste, sludge and slops into the ocean. This refuse collects on the water's surface twenty to thirty miles offshore, and winds or tide carry it landward. Mr. Warner charged that ear trouble developed by bathers was due to impurities in the water.

At the Brooklyn Hospital, Dr. R. H. Mitchell, chief of the house staff, confirmed reports that the institution had noted an increase of mastoiditis recently. He doubted that city sewage going to the sea had caused the increase, but admitted that uncontaminated water alone could not set up an inflammation. He reported four cases of mastoiditis foreign substance to pass through the eustachian tube and into the inner ear. He reported four cases of mastoiditis in the hospital yesterday and added that the history of the patient in each case showed he had been swimming within the last three weeks.

At the Manhattan Eye, Ear and Throat Hospital, 210 East Sixty-fourth Street, however, physicians reported no increase in mastoiditis or other ear ailments. Regularly, they said, a slight increase in mastoiditis was reported at this season, but this year it seemed lighter than in many years.

July 25, 1923

BIRDS AND OIL SLUDGE.

Although there is little new in the appeal of the English author, H. DE VERE STACPOOLE, that something be done to protect sea birds from the effects of oil on the water, his description of their suffering is moving. Reports from our own shores have told how the sludge pumped out by the tankers and oil-burning vessels clings to the feathers of birds at sea, sometimes making flight impossible and often causing raw places on the skin, which result in infection and death. According to British observation, slow death by starvation is also brought about. The birds are unable to clean either themselves or their fellows. As a result we find tangled on the sea wrack masses of black filth that are still living birds, and we have seen in oil when it is concentrated into long black lines the horrors of black creaking phantoms that were still birds.

This is off the coast of England and off France. The same condition exists off the coast of America. The problem can be solved only by some form of international agreement. Already the English and Americans have passed laws regulating the discharge of oil in coastal waters. But this does not affect the region beyond the three-mile limit. Oil has a long life afloat and can be carried far by wind or current. This means that until dumping at sea as well as near the shore has been prohibited it is hard to avoid damage whether to bird life or to fish or to the beaches. In England as here whole stretches of bathing sands have been rendered almost useless by the quantity of sludge washed ashore.

An international conference on the control of the use of oil at sea has long been contemplated. There is no reason to believe that any nation or group would oppose an agreement to end the destructive waste. The trouble comes from the fact that many shipowners claim that really satisfactory machinery for utilizing or storing oil waste has not yet been perfected. The cost of installation is undeniably high. Furthermore, those who live on the sea are more impressed with the capacity of that body of water to absorb waste than they are with the drift of oil. Many of them feel that it is simpler to empty their waste wherever they happen to be. It is hard not to believe that the present methods are uneconomic. Claims have been made in both Japan and England that it is possible to utilize 90 per cent. of what is now discarded, and to do so at a profit. Only by stopping in some way the practice of dumping oil at sea can we be sure to end the damage to animal life and to human pleasure caused by it.

June 19, 1925

ARCHAIC DRAINAGE SYSTEMS.

Chicago newspapers, loyally defending Chicago's case for diverting the waters of Lake Michigan into the drainage canal, are disturbed at the possible increase of typhoid fever should the city have to give up that practice. The Daily News goes so far as to express the fear that all the communities on lower Lake Michigan would find their waters contaminated. Presumably this reasoning is based on the fact that, if Chicago is forbidden to use the waters of the lake which have heretofore helped force Chicago's drainage into the Mississippi watershed, the canal will drain into the lake and all the near-by waters become foul.

But Chicago, like New York, now has to meet the cost of a modern sewage disposal system. Both cities have been notoriously slow in admitting this. Obviously, wholesale pollution of the lake waters is even more dangerous than New York's contamination of the harbor and ocean near by. The present system of forcing Chicago's refuse into the rivers that flow into the Mississippi is objectionable.

The difficulty lies in the enormous cost of establishing disposal plants for such large areas as Chicago or New York. When the Secretary of War granted the Chicago Sanitary District the right to continue to withdraw 8,500 cubic feet of water per second from Lake Michigan for the drainage canal, he made this conditional on the Chicago Sanitary District undertaking the construction of artificial treatment works covering a period of twenty years, the cost of which was estimated at $121,000,000. New York's plants are expected to cost even more.

In his book "The Principles and Methods of Municipal Government," Professor W. B. Munroe pointed out the paradox that while the United States has in many ways led the world in sewage disposal as a science, it has lagged far behind in sewage disposal as a practical art. Europe, by the greater age of her cities and the general density of population, was long ago forced to work out methods other than draining sewage into rivers, lakes or the sea. Paris and Berlin, for example, both have their systems of sewage farms which have worked satisfactorily for years. In various American cities other systems have been tried with success. January 14, 1927

CHICAGO IS ORDERED TO CUT DIVERSION

Special to The New York Times.

WASHINGTON, April 14.—The Supreme Court today entered its final decree in the Chicago water diversion case, incorporating in its decision, in the main, the recommendations of Charles Evans Hughes as special master.

The decision criticized the State of Illinois and the Chicago Sanitary District for "persisting in unjustifiable acts," and warned them that they "must find a way out at their own peril." The court ordered a gradual decrease in diversion of water from Lake Michigan until 1938, when the court expects Chicago will have completed her vast sewage disposal plant.

On and after Dec. 31, 1938, the defendants are enjoined from diverting more than 1,500 cubic feet a second, in addition to domestic pumpage. Chicago had insisted that the court should fix no limit of diversion, and the Lake States, among them New York, argued that the court should order all diversion stopped in 1938.

The Chicago Sanitary District is directed to file with the clerk of the court semi-annual reports on the progress of the disposal plant construction, and the quantity of water being withdrawn. The decree left the way open for either side to apply to the court later for action or relief, if the terms of the decree are found too stringent or too liberal. The court retains jurisdiction over the case.

Defendants to Pay Costs.

"We see no reason why costs should not be paid by the defendants, who have made this suit necessary by persisting in unjustifiable acts," said the decision, which was written by Justice Holmes.

Chicago's plea that it be given until 1948, or later, to complete its sewage disposal plant, because a rise in lake levels has demonstrated that there is no need for earlier completion, was dismissed by the court.

"Apart from the speculation involved as to the duration of the rise," the court said, "there is a wrong to be righted, and the delays allowed are allowed only for the purpose of limiting, within fair possibility, the requirements of immediate justice pressed by the complaining States.

"These requirements as between the parties are the constitutional rights of those States, subject to whatever modification they hereafter may be subjected to by Congress, acting within its authority. It will be time enough to consider the scope of that authority when it is exercised.

"In present conditions, there is no invasion of it by the former decision of this court, as urged by the defendants. The right of the complainants to a decree is not affected by the possibility that Congress may take some action in the matter."

Plea to Reverse River Denied.

The demands of the Lake States that Chicago be compelled to reverse the course of the Chicago River so it will flow into Lake Michigan as it originally did, and that Chicago be compelled to return to the lake its domestic pumpage after being purified in the sewage works, were dismissed by the court as "excessive upon the facts in this case."

The court did not touch one one of the chief points raised by the Lake States. This was their contention that water could not be legally diverted from one natural watershed to another without an interstate agreement.

The court merely pointed out that, in its original decision, it held that Chicago has no right to divert water "for the purpose of diluting and carrying away the sewage of Chicago."

Whether water could be diverted for navigation purposes also, apparently, is left open, for the court said that "all action of the parties and the court in this case will be subject, of course, to any order that Congress may make in pursuance of its constitutional powers, and any modification that necessity may show should be made by this court."

Chief Justice Hughes did not participate in consideration of the case because he acted as special master for the court prior to appointment as chief justice.

April 15, 1930

MOVES TO PREVENT STREAM POLLUTION

By JAMES W. WEIR.

Editorial Correspondence of THE NEW YORK TIMES.

ELKINS, W. Va., April 9.—The recent session of the West Virginia Legislature provided for research by a State water commission to develop practical plants for the disposal of industrial and domestic sewage. The law just enacted is regarded as constructive, inasmuch as it seeks a scientific solution of the stream pollution problem, instead of relying on the prosecution of industrial concerns responsible for such pollution. As if to illustrate the possibilities of a system of State research, announcement has been made by the State Geological Survey of the discovery of a process by which acid coal mine waters may be purified and made to yield valuable chemical by-products

The question of stream pollution has been a vexatious one for many years. There has been a desire, of course, to protect the purity of the streams, but public officials have been confronted with the fact that mere prosecution might drive many of the State's important industries to other localities. The question had almost resolved itself into one of fish or factories, when the idea was conceived by J. G. Prichard, secretary of the West Virginia Manufacturers' Association, of solving the problem by a system of research and regulation. The bill embodying the ideas of Mr. Prichard and others was drafted by Fred O. Blue of Charleston, former State Tax Commissioner.

Broad Powers Conferred.

The new law sets up a State water commission, consisting of the Commissioner of Health, the chairman of the Public Service Commission and the chairman of the Fish and Game Commission, to serve without additional compensation.

The commission has the power to cite any person, firm or corporation causing the pollution of any water to appear before it and to order offenders to use some specified means to eliminate such pollution.

More important still is a provision which directs the commission to study questions arising in connection with the pollution of streams, and, in cooperation with the College of Engineering of the State University and the director of sanitary engineering of the State Health Department, to conduct experiments in an effort to discover practical methods for the elimination of industrial wastes and stream pollution.

There was much opposition to the water commission measure from the radical wing of the Wild Life League, but other representatives of the league in Northern West Virginia, who had bitterly opposed a similar measure at the 1927 session of the Legislature, aided in the passage of the bill. Principal opposition came from sections of Kanawha County even after the passage of the bill, but Governor Conley declined to veto the measure on the ground that the new law was at least worth a trial. The claim was made by one wing of the Wild Life League that the new law would be conducive to greater stream pollution and that a large pulp mill would immediately be constructed at the headwaters of Elk River and thus pollute the Charleston water supply.

Research is expected not only to develop methods of eliminating stream pollution, but to provide for the utilization of waste water in line with the discovery of the use to which the sulphurous drainage water of coal mines may be put. Dr. D. N. Kaplan, chemist of the State Geological Survey, has developed this process conceived several years ago by David B. Reger, acting State Geologist, and it has been called the Reger-Kaplan process.

It consists of adding to the coal mine water a compound which, in combination with the chemical properties of the waters forms a blue basic pigment to be known as Monongahela blue and at the same time removes the acid from the water. The color of the pigment may be varied almost at will, producing beautiful shades of blue, green and red.

State Geologist Reger believes that "in the high sulphur regions the recovery of chemicals will prove so profitable that the exploitation of the process will be pushed by the larger mining and chemical companies. It is estimated that the mining drainage of the Monongahela Valley in West Virginia is approximately 50,000,000 gallons daily.

April 14, 1929

HIGH COURT ENJOINS DUMPING OF GARBAGE AT SEA BY NEW YORK

Special to The New York Times.

WASHINGTON, May 18.—New Jersey was upheld in its suit to prevent New York City from dumping garbage off the Jersey shores, but New York City shall have a "reasonable time" to construct an adequate incinerator system before the injunction takes effect, the Supreme Court ruled today in a decision from which there was no dissent.

The court's opinion followed the recent recommendations of Edward K. Campbell, special master, and in pursuance of his suggestion that a special master be named to determine the "reasonable time" for building the incinerators, Mr. Campbell himself was appointed for this task.

The decision, given by Justice Butler, denied the contention of New York City that the Supreme Court did not have jurisdiction.

Justice Butler reviewed Mr. Campbell's report to the court and his conclusion that New York City was maintaining a "public nuisance" on the property of New Jersey, adding that "the evidence abundantly sustains the findings of fact."

New York City, Justice Butler noted, maintained that Mr. Campbell erred in concluding that it has "unnecessarily delayed" providing incinerators.

Little Diminution Shown.

"The record shows that garbage gathered in the boroughs of Queens and Richmond has not been dumped at sea," Justice Butler continued. "The quantities shown to have been so dumped were taken from the boroughs of Manhattan, Bronx and Brooklyn."

Citing tables which showed that the amount of garbage dumped at sea was almost as much in 1929 as in 1924, he said:

"Further discussion of the evidence would serve no useful purpose. It is enough to say that defendant [New York City] has suggested no adequate reason for disturbing the findings. They are approved and adopted by the court.

"Defendant contends that, as it dumps the garbage into the ocean and not within the waters of the United States or of New Jersey, this court is without jurisdiction to grant the injunction. But the defendant is before the court and the property of plaintiff and its citizens that is alleged to have been injured by such dumping is within the court's territorial jurisdiction.

"The situs of the acts creating the nuisance, whether within or without the United States, is of no importance. Plaintiff seeks a decree in personam to prevent them in the future. The court has jurisdiction.

"There is no merit in defendant's contention, suggested in its amended answer, that compliance with the supervisor's permits in respect of places designated for dumping of its garbage leaves the court without jurisdiction to grant the injunction prayed and relieves defendant in respect of the nuisance resulting from the dumping.

"There is nothing in the act that purports to give to one dumping at places permitted by the supervisor immunity from liability for damage or injury thereby caused to others or to deprive one suffering injury by reason of such dumping of relief that he otherwise would be entitled to have. There is no reason why it should be given that effect."

After Mr. Campbell determines the "reasonable time," he must, "with all convenient speed, report to the court his findings and a form of decree."

May 19, 1931

CHICAGO IRRITATED BY POLLUTED WATER

Refuse From Indiana Mills Has Made It Disgusting in Odor and Taste.

Special Correspondence, THE NEW YORK TIMES.

CHICAGO, Jan. 12.—Interstate amenities are in danger of disturbance because of the pollution of Lake Michigan by waste from the steel industries on the Indiana littoral. For several weeks the drinking water of Chicagoans has been disgusting in odor and nauseating in taste.

Authorities have traced the cause of this offense to Hoosier industries, and specifically to quantities of phenol, a form of carbolic acid, escaping from industrial processes into the lake. When the wind blows from the southeast Chicago suffers. Even bathing is made unpleasant, since the unpleasant smell of the polluted water clings to the person after washing.

Indignation is developing to high pitch among householders, who find themselves forced to buy bottled water for drinking—an expense singularly unwelcome in these days.

Officials in charge of Chicago's water supply have had their lives made wretched by protests. They are powerless to act. Their one hope of relief lies in a change of the wind, but that, obviously, can bring only temporary escape. It may be necessary for the new State administration to take action by approach to the State authorities in Indiana, unless the offending industries mend their ways.

Jan. 15, 1933

CURB TAKES FORCE ON BAY POLLUTION

Special to THE NEW YORK TIMES.

ALBANY, Feb. 2.—Control of pollution of waters in and around New York City is assured by the tri-State compact which has just become effective between New York and New Jersey, and is to apply to Connecticut as soon as that State enacts similar legislation. It is believed to be the most comprehensive program ever entered into between two or more States.

The compact contemplates the making of all beaches in the tri-State area safe for bathing and recreational purposes and the return to shellfish culture of many of the areas that are now condemned.

It reads in part:

"Each of the signatory States pledges each to the other faithful cooperation in the control of future pollution and agrees to provide for the abatement of existing pollution in the tidal and coastal waters in the adjacent portions of the signatory States defined herein as coming within the district, and consistent with such object to enact adequate legislation which will enable each of the signatory States to put and maintain the waters thereof in a satisfactory sanitary condition, and particularly to protect public health; to render safe such waters as are now used or may later become available for bathing and recreational purposes; to abate and eliminate such pollution as becomes obnoxious or causes a nuisance; to permit the maintenance of major fish life, shellfish and marine life in waters now available or that may by practicable means be made available for the development of such fish, shellfish or marine life; to prevent oil, grease or solids from being carried on the surface of the water; to prevent the formation of sludge deposits along the shores or in the waterways; and with the fulfillment of these objectives to abate and avoid incurring unnecessary economic loss by safeguarding the rights of the public in its varied legitimate uses of the waters of the district."

Waters of the metropolitan area are divided into two classes, (a) those to be used primarily for recreational and shellfish purposes, and (b) those not expected to be used primarily for such purposes.

In 1931 the three States drafted a proposed compact, and the next year the New York State Legislature enacted a measure to become effective when New Jersey enacted its legislation. New York State commissioners are now to be named and the agreement will begin to function. The Federal Government has approved the compact.

February 3, 1936

OHIO VALLEY

Seven States Join in Program To End River Pollution

By GRADY E. CLAY Jr.

Special to THE NEW YORK TIMES.

LOUISVILLE, Ky., July 3—By signing a formal compact in Cincinnati Wednesday, the Governors of seven states marked a turning point in the history of anti-pollution efforts in the Ohio River Valley. The first Ohio River Valley water sanitation compact was signed by the Governors of Kentucky, Indiana, West Virginia, Ohio, Illinois, Pennsylvania and Virginia.

Gov. Thomas E. Dewey sent word that he would sign for New York promptly, although he could not attend the Cincinnati ceremony.

But whether the compact actually will reduce effectively the pollution of this giant waterway remains to be seen. Even its best friends admit that in dry weather the river is little better than an open sewer.

The compact is a goal toward which anti-pollution leaders in the valley have been working for thir-teen years. In 1940 President Roosevelt signed the Federal law creating and making the compact possible. It is being both praised and criticized in valley towns. Without doubt it is the first interstate compact of its kind on such a large scale. More than 20,000,000 persons live in the 200,000 square miles affected.

Its sponsors assert that it is the "first real step" toward pollution control, and predict Federal laws will not be needed to clean the river. But its critics say it is merely a smoke-screen to pacify the public and stave off future legislation to prevent pollution.

The compact creates a new Ohio Valley Water Sanitation Commission. This consists of three commissioners from each signatory state, and three representing the Federal Government. All serve without pay.

Under the compact's terms, all sewage flowing into the Ohio or its tributaries "shall be so treated" as to remove at least 45 per cent of suspended solids. It also says "all industrial wastes" flowing into these streams shall be "modified or treated to protect the public health or to preserve the waters for other legitimate purposes."

Critics of the compact, however, say it has no teeth. Article IX says the commission may take action against polluters. But the commission cannot act "unless and until it received the assent of at least a majority of the commissioners from each of not less than a majority (five) of the signatory states."

Also, no stop-pollution order against a town, corporation or person can be enforced without the approval of two of three commissioners from the state involved.

WHERE THE OHIO FLOWS

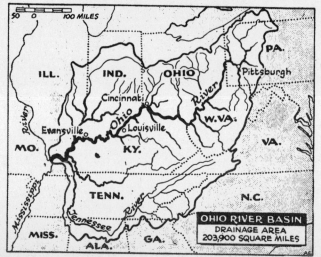

OHIO RIVER BASIN
DRAINAGE AREA
203,900 SQUARE MILES

July 4, 1948

POLLUTION NEEDS PUT AT 1.5 BILLION

U. S. Report on 19 States Says 724 Cities and 880 Plants in East Require Facilities

Clean rivers, lakes and bays in the nine-state area of the North Atlantic Drainage Basins will require sewage treatment facilities costing $1,500,000,000 at 724 municipal and 880 industrial locations.

This is the conclusion of a detailed Federal-state report on water pollution in one of the nation's key areas, a total of 103,000 square miles extending 510 miles from the northern boundary of New York to the mouth of the Potomac, and 440 miles from the western headwaters of the Po-

tomac to the eastern tip of Long Island.

Prepared by the Public Health Service of the Federal Security Agency on the basis of information supplied by sixteen state and interstate agencies, the report showed that waste discharges in the area were roughly equal to water consumption—more than twenty billion gallons a day. The report was released last night.

"Pollution abatement facilities needed by municipalities in the North Atlantic Drainage Basins are tremendous," the report declared, "since only 60 per cent of the more than 1,400 municipal sewer systems are provided with treatment works."

Deficiencies Affect Millions

"Nearly 7,000,000 people are served by facilities considered adequate for the present. About one-third of the municipalities, with a tributary population of over 9,000,000, need new treatment works, and enlargements or replacements are needed to properly serve about 8,000,000 people.

"The biggest task is to provide

facilities where none now exist. In many cases this will involve construction of intercepting sewers which may cost as much as the treatment plants."

On facilities in industry, the report disclosed that more than one-third of the industries listed in the area had treatment facilities, of which more than one-half were satisfactory.

A breakdown of the needs showed new treatment plants were required for 439 municipal sources of pollution and 683 industrial sources; enlargements and additions to existing facilities were needed to serve 224 municipalities and 184 industries, and replacements were required for facilities serving sixty-one municipalities and thirteen industries.

Seven Steps Proposed

The report recommended that:
1. Water pollution abatement be accelerated.
2. The municipalities, Federal and state institutions, industries and others responsible for pollution take positive action in accordance with the policies of local and interstate agencies to provide for pollution abatement.

3. Water pollution control agencies undertake investigations for additional information necessary for comprehensive programs.
4. Industries intensify present efforts and develop new approaches to solutions for unsolved industrial-waste pollution problems.
5. Land management improvement programs be supported vigorously to reduce silt loads now reaching streams.
6. State and local governments stimulate studies of financing methods for pollution abatement.
7. Public information programs on water pollution control be expanded.

The report covered the Hudson River drainage area, the Delaware, the Susquehanna, the Potomac, Upper Chesapeake Bay to the southern boundary of Maryland, and the many small coastal streams in New York, New Jersey, Delaware and Maryland, as well as that portion of the Great Lakes-St. Lawrence Drainage Basin that lies in New York, exclusive of the Lake Champlain drainage.

January 7, 1952

OHIO RIVER VALLEY REDUCES POLLUTION

Trend of 50 Years in 8-State Area Has Been Reversed by 5-Year Effort, Report Says

Special to The New York Times.

CINCINNATI, Dec. 12—"New pollution has been curbed, existing pollution is being decreased and the trend of half a century has been reversed," the Ohio River Valley Water Sanitation Commission announced last week.

Members of the commission are New York, Pennsylvania, Indiana, Ohio, Kentucky, West Virginia, Illinois, Virginia and the Federal Government.

Its fifth annual report since it was established on June 30, 1948, reviews anti-pollution achievements by states, cities, industries and by the commission itself.

"Five years is a short time in a campaign of interstate stream-pollution control when we consider that the problem has been more

than fifty years in the making," the report said. "Yet in the brief period during which eight states in the Ohio River Valley have been joined in a regional campaign to curb the degradation of water resources, a great transformation has occurred."

Results of Cooperative Effort

The analysis of cooperative effort between the commission and the municipalities in the eight-state area shows that 43 per cent of the sewered population of 9,319,000 now receives some form of (anti-pollution) treatment. Another 10 per cent is building new facilities. An additional 16 per cent has final plans ready for construction.

"Not only are streams receiving less sewage pollution than they had before, but the prospect for even clearer streams lies immediately ahead because of new treatment works reaching completion," the report said.

With regard to industries, the report added:

"The record shows that of the 1,247 industries discharging directly into Ohio Valley streams, 817 now are operating control facilities, thirty-one are constructing facilities and 117 are com-

pleting plans for installation of waste-reduction works.

"Substantial quantities of industrial waste are being treated in municipal disposal plants because the majority of industries are connected in city sewers."

Instances of Working Together

The eight states "are demonstrating they can work together to do a job that no one of them could do alone," the report said, offering the following instances:

¶Pooling of resources and police powers by means of a compact approved by Congress.

¶Support of an interstate commission to conduct investigations, hold public hearings, establish regulations and issue notices for compliance to municipalities and industries whose discharge of wastes may affect the quality of interstate waters.

¶Enactment of legislation to enable each state to carry out the pledge made in the commission agreement for "faithful cooperation."

In New York State, the report said, the total area that is tributary to the Ohio River, a section of the Allegheny River basin, is 1,955 square miles.

"New York has the smallest

sewered population of any of the compact states draining into the Ohio River watershed, 104,000," it was asserted. "Sewage-treatment facilities have already been provided for 94,000, about half of which are considered completely adequate. Of the twenty-nine industries discharging directly into streams, twenty-five have control facilities in operation but only one is considered adequate."

Discussing action by cities, the report said that "abatement of pollution from municipalities on the Ohio River has commanded primary attention." It also stressed that 66 per cent of the industries discharging directly into streams already had installed some form of control, an increase in a year of 18 per cent.

"Important gains have been made," but a tremendous amount of effort needs to be expended in development of treatment measures, methods of analysis and quality criteria by both industrial groups and regulatory agencies to achieve desired goals, the report continued, but it warned that "it would be a mistake to conclude" that industrial waste pollution represents a problem "well under control."

December 13, 1953

WATER CONTROL ACT IS MADE PERMANENT

GETTYSBURG, Pa., July 9 (AP)—President Eisenhower signed into law today a bill making Federal water pollution control a permanent program.

At the same time, in a statement issued at the temporary White House here, the President said the measure in one respect went beyond what he recommended—by providing for Fed-

eral grants to localities to pay part of the cost of constructing municipal sewage treatment works.

The bill as signed authorizes annual appropriations of $5,000,000 for grants to states carrying out water pollution control programs.

The President's statement said the bill "directs the Surgeon General of the Public Health Service, in every case, among other factors to give consideration to the propriety of Federal aid" in building sewage treatment plants.

"I have requested the Secre-

tary of Health, Education and Welfare and the Surgeon General to prepare criteria by which the propriety of Federal aid will be determined," he added.

I "Meanwhile, I urge that no community with sufficient resources to construct a needed sewage treatment project without Federal aid, postpone that construction simply because of the prospect of a possible Federal grant."

The first Federal Water Pollution Control Act was enacted in 1948.

July 10, 1956

Water

U. N. ISSUES DATA ON SEA POLLUTION

Special to The New York Times.

UNITED NATIONS, N. Y., Oct. 27—The vast damage resulting from the pollution of sea water by oil is indicated for the first time in a report just issued by the United Nations. It incorporates data from forty-two governments in all parts of the world, submitted in the form of answers to a questionnaire.

Major sufferers from such pollution, responses showed, are the industrial nations. The most specific listing of actual in-

stances was supplied by Japan, from investigation of twenty cases occurring since October, 1952. In seven of the twenty the loss was "unknown." Estimates of damage in the thirteen others totaled $70,000,000, and involved three fatalities, as well as a number of injured.

Twenty-seven of the nations replying to the questionnaire said that no studies had been launched in this field. Among those reporting that no problem existed on their coasts was Canada, "due to the relatively light density of shipping." New Zealand reported only minor difficulties because her coast "is little affected," possibly because

heavy fines are imposed for the discharge of oil from ships in port.

The chief types of damage have been to resort beaches, bird life and fishing grounds. Some countries mentioned the high cost of unsuccessful attempts to restore bathing beaches and beach homes to their normal cleanliness, and the loss of profitable tourist trade that ensued.

Norway, Italy, Belgium, West Germany, Argentina and Venezuela mentioned with varying degrees of emphasis the pollution of resort beaches. None included financial estimates of losses.

Western Germany cited "continuous, partially enormous

losses of sea birds" from the accumulation of oil on their wings that prevents flight and results in the birds being drowned. Norway, Italy, Japan, the United States and Egypt all referred to damage to fishing grounds or to fishing equipment.

Two fires that broke out on the surface of oil-covered waters were listed in the Japanese study. In one fire ten barges were burned.

The consensus appeared to be that the pollution originated mainly with oil tankers that anchored offshore, too close to the coast, to empty and clean their tanks. In this way they avoid heavy charges incurred by having this work done in port.

October 28, 1956

WATER POLLUTION CALLED DISGRACE

By BESS FURMAN
Special to The New York Times.

WASHINGTON, Dec. 12— The polluted state of this country's streams was called a "national disgrace" today by Surgeon General Leroy E. Burney of the Public Health Service.

"It is tragic for the world's richest, most powerful and most technologically advanced nation to foul its own nest, limit its own growth, and threaten the health of its people," Dr. Burney said in his opening address to the first Federally sponsored National Conference on Water Pollution.

With about half the expected 1,500 delegates present despite the snowstorm, the three-day meeting got under way at 2 P. M. instead of 10 A. M.

Mark D. Hollis, Assistant Surgeon General of the Public Health Service and its chief engineer, said that the prevention of further flow of pollution into the streams had become "big business, important and urgent."

"The trend of the times and the complexity of the problem outmodes the philosophy of postponement," he said. "The need is not tomorrow. It is today, maybe yesterday."

He said that the increasing population "has progressively degraded the recreation and aquatic life in most areas of the nation."

In materials prepared for the conference, the Public Health Service estimated that in the next ten years $10,600,000,000 would have to be expended to clean up the streams.

Reuse of Water Cited

Albert E. Forster, president of Hercules Powder Company of Wilmington, Del., who called himself a "spokesman for indus-

try" as director of the Manufacturing Chemists Association, said that the Department of Commerce made much higher estimates than the Public Health Service on possible reuse of water.

"By recirculation and reuse, total requirements for industrial water could be greatly reduced," he said. "Present reuse of water by industry approaches 100 per cent. This may be expected to rise to 400 per cent—that is, one gallon of intake water could be used five times."

President Eisenhower, who called the conference, sent a message saying that most important of all in the solution of the problem was "support of the individual citizen in cleaning up the waterways."

The speech of Dr. Ira N. Gabrielson, president of the Wildlife Management Institute, read by C. R. Gutermuth, vice president, said the chief deterrent to cleaning up the streams had been "a clash of philosophies." He said the best ap-

proach to date had been the current program of Federal aid to the building of treatment plants.

In speeches delivered at tonight's dinner, Representative John A. Blatnik, Democrat of Minnesota, said he would introduce a bill for stepped-up research, more Federal funds and an independent agency in the Health, Education and Welfare Department to handle the program.

Senator Robert S. Kerr, Democrat of Oklahoma, said the $90,000,000 bill vetoed by President Eisenhower last session was "too little." Senator Francis Case, Republican of South Dakota, said that disposal of atomic waste probably was the "ultimate threat"; and that progress was being made in taking salt out of sea and brackish inland waters, a possible solution that offers "real promise."

December 13, 1960

Antipollution Bill Is Signed; U.S. Water Control Widened

More Funds Provided for States and Cities—Federal Government Can Act to Curb Some Nuisances

By JOSEPH A. LOFTUS
Special to The New York Times.

WASHINGTON, July 20 — President Kennedy signed a bill today to provide more money and to extend Federal interest in preventing water pollution.

Federal action, however, is limited in the bill. Citizens who want purer streams and lakes for drinking water and for boating, swimming, fishing and other recreation will have to look to their state and local governments in the first instance.

The states and communities, for example, will have to put up $7 for every $3 in Federal funds. Where actual nuisances exist, in intrastate waters, the Federal Government can prosecute only when the governor of the state involved gives his consent.

One important provision makes the new law applicable to all navigable waters. At present, Federal action is restricted to bodies of water that cross state lines and to instances

where the pollution itself crosses state lines.

To help communities construct sewage-treatment plants, the Federal Government will make $80,000,000 available in the current fiscal year, $90,000,-000 in the following year, and $100,000,000 in each of the succeeding four years, an aggregate of $570,000,000. The previous ceiling on grants for a comparable period was $262,000,000.

The ceiling on grants to individual communities was raised from $250,000 to $600,000, or to 30 per cent of the building cost, whichever is less. The bigger cities had been penalized by the lower ceiling.

The law retains, however, the old provision that 50 per cent of the funds shall be used for grants to communities with populations of 125,000 or less.

Joint sewage treatment projects serving more than one

municipality may receive assistance up to $2,400,000 for each project.

The law increases the annual Federal matching grants to states for the administration of water pollution-control programs from $3,000,000 to $5,-000,000. This program was extended for seven years, through June 30, 1968.

Under the new law action to clean up Raritan Bay in New Jersey will start on Aug. 22 and 23 with conferences of Federal, New York, and New Jersey authorities in the United States court house in Foley Square, New York.

In thirteen actions taken since 1956 under the old antipollution law, Federal clean-up projects have embraced 131 cities and 220 industries. They have also established time schedules to purify 4,000 miles of streams.

July 21, 1961

136

SOAP MEN TACKLE SUDS NUISANCES

Industry Seeking to Solve Water Pollution Problem

By MARTIN L. GREEN

Cleanliness may be next to godliness, but cleanliness with synthetic detergents can be a nuisance, when resulting foam appears in the wrong place.

Such is the cry when froth shows up on the nation's streams and in glasses of household tap water.

There are those who also argue that the effects of detergent use is a lot worse—glutting with "waste" our once clean waterways and resulting, finally, in pollution and health menace. Others deny such effects.

The controversy has led members of Congress to consider legislation, manufacturers to appeal for respite, scientists to offer a solution and consumers to join the chorus of discordant interests.

The source of trouble, however, may soon be gone. Manufacturers recently announced plans to make a "soft" detergent that has all the cleaning power of present-day products, but none of the resistance to bacteria that break down troublesome suds.

The current industry zeal to come up with a product that is easily broken down in sewage treatment is considered a move to head off legislation that would limit or prohibit the use of hard detergents.

Representative Henry Reuss, Democrat of Wisconsin, is in favor of new laws, but has said he would withhold action until Aug. 1, while the industry negotiates for voluntary controls with the Department of Health, Education and Welfare.

Most detergents in use are made from a petroleum derivative called ABS, (alkyl benzene sulfonate). ABS has a complex molecule not readily decomposed by oxidation processes. Detergents made from it have survived the attacks of bacteria, gumming sewage systems and finally appearing as foam in water sources.

At times, the suds have backed up and poured out of bathtub and sink drains. More common is the glass of water with a "head."

The new soft detergents reportedly have a less complex molecule than the hard, and can be "de-sudsed" easily by sewage disposal facilities. An industry spokesman says. "They also are completely adaptable to use in present detergent-making equipment."

The Union Carbide Corporation and the Continental Oil Company (Conoco) recently have announced that they each will build two plants to produce the new detergent ingredient. The Conoco investment reportedly approaches $20,000,000.

The Monsanto Chemical Company is expected to announce production plans for a similar material, and the Esso Research and Engineering Company is at work on a different process, the bombarding of petrochemicals with gamma radiation.

Possibly the most confusing aspect of the whole affair is to determine the real objections to detergent waste. People are concerned with suds, pollution and all kinds of potential dangers to humans, animals and fish.

Research done by Prof. Ross McKinney of the Massachusetts Institute of Technology showed that bacteria in the presence of ABS continued to decompose organic matter present in sewage as effectively as they do in its absence, according to a report by the research council of the Association of American Soap and Glycerine Producers, Inc.

ABS is not toxic to bacteria, the report indicated.

In other words, the incidence of pollution is no different whether ABS is present or not. Industry and government sources say that ABS itself is not poisonous.

In regard to fish, the problem is much the same. An unusually high toll of marine life was reported on Great South Bay, L.I., in 1960, and detergents were investigated as the cause. A. F. Dappert, then executive secretary of the Water Pollution Board, said detergents contributed to the amount of phosphates and nitrogen on which algae feed, and that it was too much algae that caused the death of marine life, not detergents.

The real issue, then, is the suds problem. No one denies that it will become increasingly important, as sales of detergents continue to rise.

In the first quarter of this year, synthetic detergent sales accounted for 78.9 per cent of all soap and detergent sales, according to the Soap and Detergent Association. Detergent sales totaled 909,606,000 pounds, both record first quarter figures.

Interest by Government officials is also growing. Representative Robert Jones, Democrat of Alabama, also is examining the detergent-waste problem, and his Subcommittee on Research and Power on the House Government Operation Committee is making preliminary investigations.

Representative Reuss says that West Germany is outlawing hard detergents after October, 1964, and suggests that the United States move faster than it has. Anthony Celebrezze, Secretary of Health, Education and Welfare, has admitted the need for urgency, but would rather see voluntary action than legislation.

Representative Reuss questions whether voluntary action will be sufficient to eliminate the problem. He say there are no guarantees.

One local law already has been passed. Dade County, (Miami) Fla., recently enacted an ordinance making hard detergents illegal after next year.

A question often asked is, why use detergents in the first place?

One reason is that detergents seem to clean better than soap; they work faster to break down the surface tension of water. Also they are more resistant to water acids and hard water. They perform better.

It seems there is little doubt that detergents are here to stay; the thing not resolved is the form they will take.

July 7, 1963

Wisconsin Imposes Ban On Foamy Detergents

MADISON, Wis., Dec. 18 (AP)—Gov. John W. Reynolds signed a bill today that makes Wisconsin the first state in the nation to ban the sale of nondegradable household detergents, which are unaffected by normal sewerage treatment processes and cause foaming later in streams, rivers and occasionally in household water supplies drawn from such sources.

The measure makes the ban effective on Dec. 31, 1965, and requires detergent manufacturers to report periodically to the State Board of health on progress the industry is making to market soft detergents.

Supporters of the detergent ban contended that a deadline is necessary to force the detergent industry to market a new product by 1966.

December 19, 1963

LAKE POLLUTION IS CHICAGO ISSUE

Industries in Indiana Blamed for Peril to Water Supply

Special to The New York Times

CHICAGO, Jan. 2 — A new commotion arose this week over the pollution of Lake Michigan. The lake, which is the source of Chicago's water supply, has been called a "killer lake" because of the mysterious deaths of thousands of aquatic waterfowl.

Officials of Chicago blamed adjacent Indiana and its numerous lakeside industries for the threat to the purity of Chicago's water. Indiana officials said that they knew a problem existed and that they were making efforts to cope with it.

The issue erupted when Peter G. Kuh, a water-pollution expert with the Health, Education and Welfare Department, expressed concern in Washington Sunday about the situation. He said remedial measures should be started before a crisis developed.

Mr. Kuh, who was formerly with the Metropolitan Sanitary District of Chicago, said pollution at the south end of Lake Michigan — now endangering Chicago's water supply — had been brought to the attention of Federal officials by directors of the Great Lakes-Illinois Give Basin Project.

Three-fourths of Lake Front

Mr. Kuh said that the pollution at times had reached as far north as Wilson Avenue—about three-fourths of the distance along Chicago's lake front.

Anthony J. Celebrezze, Secretary of Health, Education and Welfare, earlier this month called a conference of Indiana and Illinois health officials for March 2 in Chicago.

The conference is the first step toward enforcement of the Federal Water Pollution Control Act. Violations of this act could lead to prosecution in Federal courts.

Under the act, the Secretary is empowered to call such a conference "whenever he has reason to believe that pollution arising in one state endangers the health or welfare of persons in another."

A study begun in 1960 disclosed previously that 35 municipalities and 40 industrial establishments at the south end of Lake Michigan in Illinois and Indiana caused "significant damage" to the water.

Mayor Richard J. Daley of Chicago said there should be no concern over the purity of Chicago drinking water because of excellent filtration facilities, which were extended this year to filter all of the city's water. But, he said, there could be a problem if pollution continues or increases in the lake in the next 10 or 15 years.

James W. Jardine, City Water Commissioner, said there was no pollution from Chicago or the Metropolitan Sanitary District except in severe storms. But he said that pollution from Indiana increased the cost of treating Chicago's water.

A Water Crisis Feared

Frank Chesrow, the sanitary district's president, and Vinton Bacon, the district superintendent, charged that "Indiana's failure to act not only is threatening us with a water crisis, it is negating millions of dollars worth of work that we are doing."

Mr. Bacon said that Chicago's water was endangered by "phenols, cyanides and other chemicals, and increasing amounts of bacteria that are coming out of Indiana."

Blucher Poole, technical secretary of the Indiana Stream Pollution Control Board, said Indiana knew it was a "sizable and probably major" contributor to pollution at the south end of Lake Michigan. The state intends to embark on an aggressive program to eliminate the problem, he said.

Mr. Poole said that more than a dozen communities and about 20 industries in northern Indiana had been asked by his office to participate in the meeting March 2 in Chicago.

Citing the mysterious deaths of birds, Dr. W. J. Beecher, director of the Chicago Academy of Sciences, said "this alarming situation is a symptom of the trouble with Lake Michigan, which we are polluting every day."

Gulls and Loons Killed

In November, 1963, about 10,000 gulls and loons were found dead or dying along the south and west shores of the lake, and in the fall of this year nearly 5,000 birds, most of them loons, died along the northern shore.

Dr. Beecher commented on this:

"Perhaps the bird catastrophe is concerned with increasing commercial use of Lake Michigan, and increasing numbers of people living on the lake. The entire biology of the lake has changed in the last 30 years. Lake trout are now a thing of the past.

"It may be significant that both of these happenings were in the fall, which is the time the lake begins to overturn. The surface water cools and sinks to the bottom, and the water at the bottom comes to the top. Perhaps the birds are being killed by something that has accumulated at the bottom of the lake. This matter needs a full research program."

January 3, 1965

POLLUTION STIRS MEMPHIS DISPUTE

Pesticide Found in Sewer— Health Hazard Is Seen

Special to The New York Times

MEMPHIS, Jan. 16 — Discovery this week of a deposit of 8,000 pounds of the agricultural pesticide endrin in a city sewer is the latest development in a year-long controversy here over stream pollution.

The endrin deposit is enough to pollute the Mississippi River, a United States Public Health Service official warned Memphis officials. The officer, George Putnicki, said the chemical "would create a tremendous health hazard" if they reached the Mississippi.

Memphis sewers dump into the Mississippi. The endrin was found caked in deposits up to three-feet thick in the sewer.

On June 3, 1963, the Memphis Health Department reported complaints from 20 persons living near Cypress Creek, an open stream flowing through the north side of the city. Nausea, vomiting and watering eyes were the symptoms produced by gas rising from the stream.

Company Refuses Blame

An official of the Velsicol Chemical Corporation denied responsibility. "Endrin could not have caused the symptoms," said Wilson Keyes, director of manufacturing.

On June 7, 1963, 26 workmen in plants near Velsicol were taken to five hospitals after becoming ill from chlorine gas fumes. Within a year law suits totaling over $5 million had been filed by more than 40 persons claiming injury.

Velsicol reacted in the first weeks of the trouble with a dinner for 150 political, civic and business leaders. "It came as quite a shock to us to discover that there was some question about whether we were welcome in the City of Memphis," said John Kirk, executive vice president, down from Chicago for the event.

Then Mayor Henry Loeb responded that "this plant is very much wanted by Memphis." His successor, William B. Ingram Jr., took much the same position a year later.

Endrin Pinpointed

The Public Health Service investigated early in 1964 and concluded that endrin was the contaminant causing fish deaths in the lower Mississippi. Velsicol continued its denials of responsibility.

Anthony J. Celebrezze, Secretary of Health, Education and Welfare, said last June that Memphis was the source of the poison. Velsicol is the only Memphis manufacturer of endrin. Bernard H. Lorant, Velsicol vice president, said from Chicago that he was "amazed that Secretary Celebrezze would accept such a report."

In July, Senator Abraham A. Ribicoff, Democrat of Connecticut, called a Memphis dump "a biological wasteland" and told a Velsicol official that the concern's "waste disposal system can be described as primitive and a dangerous nuisance." He wondered aloud "how the City of Memphis lets you get away with it."

The controversy subsided until it was renewed on Thursday by Mr. Putnicki's announcement. Mr. Putnicki told Memphis officials about the 8,000-pound deposit along a 3,400-foot stretch of sanitary sewer.

The city ordered the sewer sealed off and a bypass sewer built around the 3,400-foot length of contaminated sewer.

Velsicol, meanwhile, continues production of its agricultural pesticide.

The Public Health Service has opened an office in West Memphis, Ark., just across the Mississippi. From this and 13 similar stations it will study pollution of the Mississippi from Fulton, Ky., to the Gulf of Mexico. A $500,000 appropriation is financing it. January 17, 1965

N.A.M. SAYS CITIES POLLUTE WATERS

Finds Industry Spends $100 Million to Treat Waste

A report by the National Association of Manufacturers yesterday said most pollution of waterways comes from municipal sewage, not from industrial wastes.

A survey of nearly 3,000 companies using 90 per cent of the total intake of water by American industry was said to have disclosed a great increase in the installation of waste-treatment facilities.

The findings were made public by the N.A.M. at a news conference in the Belmont Plaza Hotel. According to the survey, 69 per cent of the manufacturing companies had installed water-treatment facilities as compared with 18 per cent in 1949.

The report, "Water in Industry," added that $100 million was being spent annually by manufacturers to curb pollution. The report was published in cooperation with the United States Chamber of Commerce and the National Technical Task Committee on Industrial Wastes.

Officials Are Concerned

Virtually all industrial plants being built today incorporate waste-treatment equipment in their designs, a panel of experts said at the conference.

The manufacturers' defense of their position in water pollution follows expressions of concern by President Johnson, Governor Rockefeller and other officials on the growing problems of protecting and conserving the country's water resources.

In December, Governor Rockefeller made an urgent recommendation for a $1.7 billion program to purify New York's polluted lakes, streams and rivers.

Almost two-thirds of the state's population, he said, lives in areas troubled by pollution, mainly from "1,167 communities pouring poorly treated and even raw sewage into our waterways."

On the problem of industrial pollution, the Governor proposed that the Federal, state and local governments adopt a system of tax incentives to stimulate private concerns to deal with the problem. He estimated that it would cost $67 million to eliminate industrial pollution in New York.

Cited Gloversville's Plight

A week earlier, Governor Rockefeller cited the plight of the Mohawk Valley village of Gloversville as exemplifying the consequences of water pollution.

After the village's reservoir went dry in last fall's drought, he said, the residents faced a crisis because they could not use the water of Cayudutta Creek "running by their very doorsteps." The creek, he added, "is so polluted with sewage and grease from tanneries that it would foul the pumps of fire engines."

The industry representatives at yesterday's conference voiced opposition to features of a water-pollution control bill passed last week by the United States Senate.

They particularly objected to the bill's proposal of a new agency to administer the anti-pollution program. Authority in this field, they said, should be left with the United States Public Health Service.

The N.A.M., the spokesmen added, also opposes the establishment of Federal standards of water quality.

February 5, 1965

THE MIGHTY, DIRTY HUDSON

By RONALD SULLIVAN
Special to The New York Times

ALBANY, Feb. 20 — The Hudson was once a magnificent river, and parts of it still are. But disgruntled fishermen, driven from the river by its almost unbelievable pollution, now tell the story that a nail dropped into the river near here or near New York City won't rust—because there is no oxygen left in the water.

It's not that the Hudson was ever crystal-clear or that it ever had miles of sandy shoreline. It didn't. But there was a time when it was clean, and fish and young people loved it.

Now even the fish can't stand it. The shad run in the lower river beneath the George Washington Bridge has all but vanished. And the only fish that lives in the river near Troy is a vicious kind of eel that thrives on human waste.

The police keep young people from swimming, and piles of garbage, miles of sewer pipes and mountains of debris have driven off the fishermen. What once was called the "great river of the mountains" is becoming a cesspool.

People Angry

Slowly — so slowly that it pains the conservationists and the guardians of the public health—uncontrolled pollution is beginning to make people angry.

Water pollution has become a hot issue in the United States; hot enough to become the first item of President Johnson's legislative program to pass the Senate this year.

And in Congress, in state houses and in city halls from Boston to San Diego, everyone suddenly wants to make water clean again—as long as it doesn't cost too much money. But it does.

The Federal Government says it will take at least $8 billion just to keep the country's cities from polluting water, just to catch up. It will take $20 to $30 billion more to install the necessary storm and sewer systems across the country.

A major goal of the President's Great Society is to re-create a land where a boy can jump into a river or pond without catching hepatitis or dysentery. Chemical and human pollution has befouled streams, lakes and shorelines, making magnificent rivers like the Hudson and the Mississippi unfit even to live near.

Governor Rockefeller estimates that two out of three New Yorkers live amid pollution. The statistics are frightening and disgusting.

New York City alone produces 1.3 billion gallons of raw sewage every day. More than 500 million gallons of it are pumped each day into the Hudson and East Rivers. About 140 miles upstream, Albany and its equally drab and dingy neighbors pour more than 60 million gallons of filth into the Hudson daily. State studies show that there is now six times as much pollution in the river as there was at the turn of the century.

As more people move to the cities all over the country, their waste and that of the industries that serve them threatens to engulf metropolitan areas in a sea of defilement unless they do things they have never been willing to do—put sewage disposal on a par with schools and highways.

It is almost impossible to talk about the Hudson without talking about the Mohawk River, too, because the Mohawk pollutes the Hudson. The two rivers join just north of here, compounding the worst of both into a pollution nightmare.

It is like this in most other places. Someone pollutes a pond. The pond then pollutes a stream, which, along with hundreds of other ponds and streams, pollutes a river or a lake. Like the ripples from a rock thrown into the water, large-scale pollution is generally the culmination of countless little pollutions.

Sources

Governor Rockefeller has taken an impressive lead in fighting bad water by proposing a $1.7 billion antipollution program in New York State. He pointed his finger at 2,100 pollution sources and 1,167 communities that don't seem to care what they do with their waste.

"Everyone has a stake in ending the pollution of our lakes and streams," he said. "Farmers, housewives, sportsmen, workers, businessmen, children — all want pure water."

The log is clear. The state estimates that it would cost nearly $1 billion by 1970 to clean up the Mohawk and the upper and lower reaches of the Hudson. The state argues that it will eventually cost the people far more than $1 billion if action isn't taken now.

State officials believe that $1 billion spent on cleaning the Hudson and the Mohawk would pay for itself in hard cash—not just in terms of swimmers swimming and fishermen fishing. But it is almost impossible to convince the voter who sees no visible return for the money put into sewage-treatment plants. His implacable opposition has stood in the way of state action for decades.

The toughest nuts for the state to crack have been the cities of Utica on the Mohawk and Troy, which hugs the west bank of the Hudson 10 miles north of here, at the junction of the Mohawk and the Hudson.

In 1936, the state Health Department sent a stiff note to Utica demanding that it stop pouring raw sewage into the Mohawk. The state gave Utica four years to stop. That was nearly 30 years ago and Utica is still pumping untreated sewage into the Mohawk at the rate of 15 million gallons daily.

Over the years Utica became more and more of an embarrassment to New York's efforts to clean its waters. Smaller cities told the state to do something about Utica before it moved against them. After several court battles and an implied threat to cut off state aid to the city, Utica has finally agreed to build a sewage-treatment plant.

On the Carpet

The Health Department had Troy on the carpet last month. But Troy officials contend they have been made "fall guys" by the state and point to the dozens of other Hudson communities that have no treatment facilities.

A 150-mile trip down the Hudson from Albany to New York City once was a joy. Now it begins amid unspeakable filth. As the accompanying map shows, this gross pollution persists before it begins to dissipate about 35 to 40 miles downstream. But pollution again sets in as communities farther down, almost without exception, contribute their share of raw sewage.

The map shows a 45-mile stretch midway to Manhattan that is relatively unpoluted.

But a Health Department doctor said this did not mean the water is clean. He said he would not swim in it nor would he dare to drink it. The river is so befouled by the time it leaves Albany that it would take 10 years, even with every sewer blocked and every river factory closed, before it could recover its original state.

Reluctance

And so it has gone. One city or town after another, each waiting for the other to do something and each insisting that schools and highways are more important than sewers. Local reluctance has spurred state and Federal initiative.

In his special message to Congress on natural beauty last week, the President asked for new Federal powers to clean up the country's water.

The Government has already made millions of dollars available to communities for sanitation studies and sewage-treatment plants. All told, 5,851 communities have received $500 million for treatment facilities since 1956. Under the Federal Water Pollution Control Act, a community can obtain Federal grants of up to 30 per cent of the construction costs, with a $600,000 ceiling.

Companion Bill

The Administration bill that cleared the Senate last month would raise the ceiling. The companion bill in the House raises it to $2 million a plant and provides $6 million for a joint community operation.

Governor Rockefeller urges an end of Federal ceilings. Under his plan, the state and the Federal Government would pay 40 per cent each and the locality would provide the remaining 20 per cent. But Administration officials doubt that Congress would ever remove spending limitations, as asked by Mr. Rockefeller.

If the limitations remain, the Governor's program would be crippled.

February 21, 1965

Atlanta Loses Ground in Fight on Pollution

By GLADWIN HILL
Special to The New York Times

ATLANTA, Feb. 27—A sign on Marietta Boulevard at the edge of town says, "Welcome to Atlanta — The Dogwood City." But no dogwood, or much other attractive vegetation, is evident along the banks of the Chattahoochee River where it courses under the boulevard.

What meets the eye is a nightmarish scene that might have been etched by Gustave Doré—skeletal saplings dying in filth-strewn mud, a pervading smell of decay and an endless torrent of sewage gushing into the river.

The once-crystalline Chattahoochee, flowing southward toward Alabama and Florida, now is opaque from upstream pollution, although it is virtually the entire water supply of Atlanta.

Into a Water Plant

Fifty yards wide, the river is about the color of coffee-with-cream a quarter-mile upstream from the Marietta Boulevard Bridge. There a sizable part of its flow detours into a city water plant for Atlanta's half-million population and the pipes, vats and boilers of the city's growing industrial complex.

The familiar community debate has been raging in Atlanta over whether fluoride should be added to this water to help teeth. It is doubtful if many of the debaters realize that the water they get now has had to be treated with alum, chlorine, ammonia, carbon and lime to make it clear and potable.

A few hundred yards downstream from the water plant, the same water, having done its work for Atlanta, surges into the river again from the R. M. Clayton Sewage Disposal Plant—much darker in color and containing what the engineers euphemistically call "solids."

The Clayton plant, the main such facility on the river, is supposed to give the sewage "primary treatment," meaning removal of solids. But for some years the plant has been over-

The New York Times   March 2, 1965
The polluted rivers are shown by the heavy black lines

loaded up to, currently, 57 per cent more than its capacity of 42 million gallons a day. The excess sewage flows in one side of the plant and out the other, unaltered.

Downstream for many miles, the pollution not only precludes recreational use of the river, but even is inhibiting the development of industry. This complaint was made by representatives of several downstream communities at a conference of the State Water Quality Board last week. The state's Health Department director, Dr. George Venable, called the Chattahoochee situation "the most complex pollution condition in Georgia."

Other Murky Waters

This was somewhat short of a complimentary reference, inasmuch as two other Georgia river systems are already the subject of formal proceedings of the Federal Department of Health, Education and Welfare to abate interstate pollution—the Coosa, which flows into Alabama, and the Savannah, running between Georgia and South Carolina.

The Chattahoochee is the next Federal target. The U.S. Public Health Service terms Atlanta's water and sewage needs "critical."

All this might be very embassassing to Georgians but for the fact that, according to the Public Health Service, the situation simply typifies in some degree conditions that are rife across the nation.

President Kennedy in 1961 called general pollution of the country's rivers "a national disgrace," and President Johnson reiterated the sentiment in January in calling for stronger Federal pollution controls. Governor Rockefeller of New York has proposed a $1.7 billion program, spaced over six years, to clear up water pollution in New York.

Atlanta itself has a $30 million secondary-treatment construction program in progress. This augmentation of primary sewage treatment can remove up to 90 per cent of the troublesome contaminants from sewage.

Federal abatement proceed-

ings like those in Georgia are under way in Maine, Rhode Island, the Ohio Valley, the Great Lakes and elsewhere.

In the West, where rainfall ranges down to only a few inches a year, the main concern is still about water supply, although there are incipient pollution situations even in San Francisco Bay.

In the East, where there is plenty of water (Georgia gets 50 inches of precipitation a year) the problem is contamination.

Here there is little actual "consumption" of water supplies in terms of physical loss. Most water, after use, returns one way or another into watercourses or underground sources. Once it was cleansed naturally by earth filtration and biological and chemical action in streams and lakes. The tremendous recent national growth in population and industry simply has overtaxed these natural processes almost everywhere.

Municipalities and other public agencies are partly to blame through their general lack of adequate sewage-treatment facilities. Of 600 municipal sewer systems in a 146,000-square-mile basin involving parts of Georgia, Florida, Alabama and Mississippi, the Public Health Service reports that 200 give sewage no treatment and 200 give it only primary, solid-settling treatment.

In Georgia, a big offender is the State Health Department itself. Its Milledgeville State Hospital, with 15,000 patients, pours 2.5 million gallons of raw sewage every day into the relatively small Oconee River, which runs into the Altamaha and then into the Atlantic. A corrective project has been started, but is some years short of completion.

Industry shares the blame. Into the Coosa River system, just before it enters Alabama, the communities of Trion, Summerville and Rome have been pouring the raw sewage of 3,000, 3,200 and 40,000 people,

140

respectively. But a textile mill on the tributary Chattooga River pours in contaminants equivalent to the sewage of 112,000 people, and a paper mill on the Coosa itself the sewage equivalent of 243,000 people.

Rivers Overloaded

Industrial waste differs chemically from municipal sewage, but is as bad or worse in its effects. Its chemicals can overload a river so that natural decomposition and purification processes come to a halt. Some industrial effluents, such as acids, tars and pesticide com-

pounds, are man-made synthetics that nature just didn't equips watercourses to assimilate.

The pollution in the last 28 miles of the Savannah River has gone a long way toward ruining the once-flourishing shrimp, shad and shellfish production of the area. The sewage of 146,000 residents of the Savannah area, 80 per cent of it discharged raw, accounts for most of the excessive bacterial count in the water.

But in the equally important

terms of exhaustion of oxygen in the water, most of the problem is attributed by the Public Health Service to two industrial concerns — a paper-bag plant whose effluents are rated as chemically comparable to the sewage of 810,000 people, and a can factory whose discharge is compared with a population of 130,000.

Atlanta's secondary-treatment project conforms with a new Georgia law requiring this of municipalities. Similar laws have been passed in other states.

Federal experts estimate that pollution calls for a nationwide expenditure of nearly a billion dollars a year to bring sewage facilities up to needs.

In 1963, with the Federal Government's accelerated-public-works program helping, the Public Health Service reckoned national expenditures of $820 million, representing a net gain on the backlog of needed facilities. But last year the total dropped to $590 million. This, in the opinion of qualified officials, meant a loss of ground in the battle.

March 2, 1965

POLLUTION REPORT BLAMES CONCERNS

By GLADWIN HILL
Special to The New York Times

DETROIT, June 16 — When pleas are made to industry to reduce the pollution of major national watercourses, some companies respond constructively. Others turn a deaf ear.

This was the gist of the formal report presented today by the Michigan Water Resources Commission and the State Department of Health to a Federal hearing aimed at cleaning up the Detroit River and Lake Erie, into which the river empties.

The river—actually the channel between Lake St. Clair and Lake Erie — is polluted by a daily flow of 1.6 billion gallons of industrial and municipal waste from the Detroit area, according to the United States Public Health Service. Each day, it is calculated, the river carries into Lake Erie some seven million pounds of alien chemicals.

The cumulative effect of these —an effect aggravated by similar pollution from Ohio, Pennsylvania and New York—has been to rob the lake of a good deal of its oxygen, and cause an abnormal growth of slime, seaweed-like algae, and inferior species of fish and worms.

A Swamp Is Feared

This has disrupted commercial and recreational uses of the shallow lake, adversely affected many community water supplies, and raised fears that if present trends are not arrested, the lake is fated to become a noxious swamp.

In the report today, the Michigan state agencies listed 19 industrial concerns—among them some of the nation's leading corporations—that had been notified over the last three years that they were discharging too many pollutants into the Detroit River.

Ten of the 19 were listed as having instituted remedial steps. In the case of the nine others, the report was:

"No significant changes in treatment and control methods."

These nine were the American Agricultural Chemical Company, the American Cement Corporation, the Firestone Tire and

The New York Times (by Gladwin Hill)

SOURCE OF POLLUTION: One of plants blamed for polluting Detroit River

Rubber Company, the Pennsalt Chemicals Corporation, Revere Copper & Brass, Inc., the Union Bag-Camp Paper Company, the Consolidated Packaging Corporation and the Monroe Paper Products Company.

Companies in both categories were among 20 cited last month by the Public Health Service, after a three-year study of the Detroit River situation, as continuing sources of excessive pollution. The agency's report recommended the installation, by the respective plants, of waste-treatment facilities involving an outlay unofficially estimated at $100 million.

One-Third Is Sewage

About one-third of the Detroit River pollutant volume is sewage from the city of Detroit. Unlike most of the nation's cities, which give sewage both primary treatment, through settling of solids, and secondary treatment, through bacterial "digestion," Detroit has only primary treatment, followed by chlorination.

Municipal officials have been resisting the installation of

secondary-treatment facilities, which they estimate would cost $109 million.

Today's Federal session was the second day of an expected week-long hearing by the Department of Health, Education and Welfare. It is the second phase of Detroit River pollution abatement proceedings started in 1962 at the request of Michigan officials.

The objective of the current sessions is to work out an agreement between Federal and state officials and pollution sources on a remedial program, as has been done in 34 other pollutions situations across the country.

The Federal department is also beginning a comprehensive pollution abatement action in respect to Lake Erie as a whole, involving all four states that border it.

The Michigan water pollution law—which some state officials have described as lacking strength—provides that when damage from excessive pollution is established, the offender can be called upon to desist or face penalties for contempt of court.

Well-informed state sources said the law had been much more conspicuously applied in outlying sections of the state than around Detroit, where the multibillion-dollar concentration of industry exerts pronounced influence on public affairs.

Today's state report presented no pointed challenge to the Federal cleanup recommendations. In effect, it said that Michigan officials had been doing their duty in trying to reduce pollution, but by implication passed to the Federal officials the job of clamping down on the most flagrant pollution sources.

Under Federal law, if a satisfactory remedial program is not pursued on schedule, those responsible for the pollution can be taken to court.

The most forceful sentiment on the problem today came from Clarence Eddy, a geologist representing the Michigan Department of Conservation. He said:

"We believe that the concept of disposing of wastes by committing them to lakes and streams ultimately must change."

June 17, 1965

Water

'SOFT' DETERGENTS FIGHT FOAM CLOGS

Industry Unveiling Product to Erase Drain Problems

By GLADWIN HILL

Special to The New York Times

LOS ANGELES, June 26— Next week will mark a major juncture in the battle against detergent foam, which has reportedly cost the public countless millions of dollars, and has irked housewives –along with helping them—from coast to coast.

July 1 is the date designated by the multibillion-dollar detergent industry for completion of a transition from chemically "hard" detergents to "soft" ones, whose bubbles will not remain so stubbornly intact after use.

The foam—harmless in ordinary concentrations — recurrently rises in tap water, and throughout the country it has deluged sewage treatment plants in great masses of foam that is difficult and costly to cope with.

The Soap and Detergent Association, the trade organization representing 90 per cent of the nation's "washday product" output, reported this weekend that the switch to more easily defoamed compounds had been substantially completed.

Shift Cost $100 Million

The shift was said to have cost the industry about $100 million over the last decade in research and new equipment and procedures. The work was spurred by nationwide vexation with foam problems that

precipitated many threats of Federal regulatory legislation and actual enactment of antidetergent laws by the State of Wisconsin and Dade County (Miami), Fla.

Unnoticed by most of the people, who buy four billion pounds of detergents every year, the central cleaning ingredient in packages on store shelves has changed in recent months from Alkyl Benzene Sulfonate (A.B.S.) to Linear Alkylate Sulfonate (L.A.S.).

The main difference is in the molecular structure of the compounds. The former is chemically "disgested" only with great difficulty by the microscopic organisms that neutralize the extraneous ingredients in waste water. This has necessitated the employment of all sorts of techniques — mainly tedious recycling of detergents through treatment processes— to subdue the froth. A plethora of foam, as most housewives have discovered, can actually stall a washing machine.

Easy to Digest

The new L.A.C. chemical is relatively easily digested. It works the same as old detergents, foams similarly and costs about the same.

However, detergent manufacturers, who ordinarily would use any pretext for the word "new," have eschewed the potential promotional innovation. One reason is that all competitors—63 manufacturers and 13 major chemical suppliers— have the same thing. The other is that they do not want to raise the public's hopes too high.

Although soft detergents alleviate much of the foam nuisance, they leave some major problems unsolved, a number of them problems people did not know existed. One is that, in many places, water supplies are polluted.

The advantage of soft detergents will be reflected mainly in the 50 per cent of the nation's communities that have comprehensive sewage treatment plants involving bacterial digestion processes.

The rest of the country has systems that give sewage only partial treatment or no treatment at all, or does not have any sewage collection. Nearly a third of the nation has only cesspools and septic tanks, whose contents gradually seep into the ground.

From partial-treatment and no-treatment sources, any detergent is still likely to percolate into ground water and thence into streams and lakes and wells. When reagitated, it may foam again.

In the past, when foam came out of water taps, it primarily reflected the stubbornness of detergents to defrothing. Now it is known, from extensive laboratory testing, that the detergents in general use are amenable in large measure to standard sewage treatment.

Evidence of Intrusion

Therefore, the appearance of foam from now on, according to sanitation experts in the United States Public Health Service, will be evidence of the intrusion of sewage elements into water supplies.

This is expected to happen on Long Island, where much of the domestic water supply comes from wells only vaguely separated, in the sandy, porous ground, from outlets of cesspools and septic tanks.

Nor will the new soft detergents do much to solve the problem of foam back-up in the drains of many city apartment houses. This, sanitation engineers say, is due to drain outlets not being built big enough: The bacteria that deflate detergent foam cannot go to work until it gets into

the sewer main. Since the new detergents will produce just as much foam as the old, the back-up problem is not changed.

Another major problem that will not be affected is the fact that the new detergents, like the old, contain up to 60 per cent of phosphate compounds. Phosphates are a prime fertilizer.

Most waste water eventually makes its way into steams and lakes. Here the phosphates cause an abnormal growth of plant life. This in turn can alter the amount of dissolved (free) oxygen in the water. This oxygen is what supports fish, and also gives the water its power to purify or neutralize sewage through oxidation.

In some major lakes and streams, the cumulative deposit of phosphates and other extraneous chemicals has become so great, and the oxygen content so small or nonexistent, that the water has lost its regenerative capacity and for all practical purposes has become inert. Any more sewage that goes into it simply drifts, its chemical and bacterial potential undiminished.

This had led to growing pressures from sanitation experts and conservationists for major reductions in the quantitites of phosphates and other oxygen-consuming chemicals that flow into water courses.

One line of research, strongly advocated by some sanitary engineering groups, is toward a "third generation" of detergents with quite different chemical bases. They point to the fact that there are such detergents, not involving foam at all, now in use in small quantities for specialized purposes in industry—employed in conjunction, for example, with other processes to clean metal.

June 27, 1965

States Agree to End Lake Erie Pollution

By GLADWIN HILL
Special to The New York Times

BUFFALO, Aug. 12—The five Lake Erie basin states reached an agreement with Federal officials today on a program to halt the lake's severe pollution, possibly within four years.

The program calls for an investment of billions of dollars in new municipal sewage works, industrial waste treatment installations and other facilities by New York, Pennsylvania, Ohio, Indiana and Michigan.

More than a ton of chemical contaminants now pours into the 240-mile-long lake each minute from such points as Detroit, Toledo, Cleveland, Erie, Pa., and Buffalo.

The pollution has impaired municipal and industrial water supplies and recreation facilities of a basin population of 10 million, and has disrupted the lake's normal biological processes.

It is possibly the worst case of large-scale waterway pollution in scientific annals, and has ominous implications in the face of recurrent water shortages in the East.

Lake Erie is the fourth largest of the Great Lakes, which contain one-fifth on the earth's fresh water supplies.

The interstate agreement culminated two weeks of hearings at Cleveland and Buffalo called by the Department of Health, Education and Welfare.

Murray Stein, Federal water pollution enforcement director and chairman of the hearings, siad the pact represented "tremendous" progress in Federal antipollution efforts. These have included 35 such abatement actions in the last decade.

One jurisdiction hitch arose during the three-hour closed conference.

Objection from Ohio

Ohio, which formally requested the

ed the hearings under Federal law, objected to the agreement's covering the state's intrastate waters as well as interstate. This issue was left for the incoming Health Secretary, John W. Gardner, to settle.

Mr. Stein said such a reservation might force Federal officials to laboriously establish proof of the interstate impact of pollution originating in Ohio.

However, Ohio's representative, Fred Morr, state director of natural resources, said no obstruction of the clean-up program was contemplated.

Normally, the Department of Health Education and Welfare, through the United States Public Health Service, collaborates in and oversees the execution of abatement programs that are directly administered by state health and water agencies. Pollution sources that do not conform are subject to Federal court action.

The participants agreed today to produce remedial construction schedules within six

months. A tentative timetable calls for completion of plans and xpecifications by next August, completion of financing by February, 1967, start of construction by August, 1967, and completion of construction by Jan. 1, 1969.

New York's part in the Erie clean-up—involving chiefly reduction of municipal and industrial pollution from the Buffalo area—is expected to be encompassed in the pending $1.7 billion state pollution abatement program. Voters in November will be asked to approve a $1 billion bond issue for this.

New York's representative at the hearings was Robert D. Hennigan, director of the State Health Department's Bureau of Water Resource Services.

The principal points in the agreement call for the following steps:

¶"Secondary" treatment of municipal sewage, or equivalent reduction of contaminants.

¶"Maximum reduction" by industrial plants of a dozen basic types of waste discharge ingredients, ranging from oil to heat.

142

¶Separation of municipal sewage and storm-drain systems in all new urban development.

¶Design of all new sewerage facilities to obviate bypassing untreated waste fluids.

'Secondary' Treatment

"Secondary" sewage treatment is a standard process by which up to 90 per cent of the main pollutants are removed biologically or chemically. The sewage of several million inhabitants of the Erie Basin—notably in Detroit and Buffalo

—does not receive secondary treatment.

Combined sewage and storm-drain systems, although common throughout the country, are considered undesirable. This is because even a light rain overtaxes the capacity of most sewage plants. The result is that part of the waste load is bypassed untreated.

The Mobil Oil Company, which was bitterly criticized at the hearing yesterday for polluting the Buffalo River at its Socony Vacuum plant, issued

this statement:

"Mobil Oil Company has always cooperated with the United States Public Health Service and with other governmental agencies in efforts to achieve a reduction of pollution in the Buffalo River. Where the weight of scientific evidence establishes the need, we will take prompt, sound and realistic corrective action.

"The impression left with the conference that the Mobil refinery is pouring oil and petroleum waste into the Buffalo

River is totally without foundation. It is equally untrue that all oil or oil waste that may be found in the Buffalo can be attributed to effluent from the Mobil refinery."

The United States Public Health Service report to the hearing listed the Socony Vacuum Oil Company as one of five industrial plants discharging waste into the Buffalo River. The company's discharge components were listed as oil, phenolics and cyanides.

August 13, 1965

New U.S. Agency and New Policy to Enter Fight Against Water Pollution

By GLADWIN HILL
Special to The New York Times

LOS ANGELES, Dec. 20—Early next year a team of experts from the Department of Health, Education and Welfare will descend on one of the nation's large cities to begin a Federal water pollution abatement action.

The city may be Atlanta. The polluted Chattahoochee River, which is the source of Atlanta's water supply and its sewage disposal channel, is high on the department's target list. It will be the 38th of such abatement proceedings, which have been instituted at an accelerating rate over the last decade. But in some respects it will be a landmark.

It will betoken the beginning of a new era in what President Johnson calls "the control of our environment." It will mark the debut of a new Federal agency, operating under a new policy, and in a radically changed climate of public opinion.

Department Shift

The new Federal Water Quality Act, which became law in October, shifted antipollution work from the United States Public Health Service to a new division in the Department of Health, Education and Welfare. Its head, who will be an assistant secretary of the department, is expected to be named soon.

The law also implicitly activated a new Federal approach to the pollution problem: prevention as well as cure. The act enjoined the 50 states to promulgate reasonable codes of water quality for their rivers and lakes by July 1967. Failure to comply will bring Federal authorities in to establish standards.

The law was enacted under some noteworthy circumstances. In signing it, President Johnson enunciated with sharp emphasis national policies whose previous obfuscation had aggravated the mounting contamination of waterways by municipal and industrial wastes.

One principle was that "no one has the right to use America's waterways, which belong

The New York Times (by Gladwin Hill)

Effluent from Mobil Oil Company plant spills into river near Niagara Falls, area plagued with pollution problem.

to all the people, as a sewer." A corollary was that individual sources of pollution bear the responsibility for extracting or neutralizing contaminants before wastes are dumped in the public domain.

Reversal In Attitude

The enactment of the law coincided with a remarkable reversal in attitude about water pollution on the part of many state officials and many industrial executives. A year ago a prevalent philosophy seemed to be that waterways had to be sewers and that attempts to end pollution should be firmly resisted as at best impractical.

By the fall, although there are many diehard defenders of pollution in both categories, leading corporations had swung to the clean-water cause and many state officials were taking a new look at possibilities of improvements.

The reason for the shift in attitudes was apparent: rapidly mounting pressures of public opinion.

Latent Dangers Seen

Resentment of dirty streams and lakes, which had long been voiced chiefly by conservationist groups, had spread to the public at large—to the extent that in Ohio alone, hundreds of thousands of citizens signed

petitions demanding a cleanup of Lake Erie.

Also giving impetus to the new antipollution law, which was shepherded through Congress by Senator Edmund S. Muskie of Maine and Representative John A. Blatnick of Minnesota, both Democrats, were some immediate implications in the areas of public health and water resources.

A summer outbreak of waterborne gastroenteritis in Riverside Calif., one of the worst localized epidemics in years, suggested the dangers that are latent in a poorly supervised national water system.

And the water shortage in New York City and many other Northeastern communities, in the face of large amounts of water made unuseable because of contamination, underscored the immense negative role that pollution can play in water supply.

Such developments in recent months as the mounting of an interstate effort to clean up Lake Erie, the detergent industry's mass shift to a chemical base that yields a less-stubborn foam, and the New York voters' ratification of a $1.7 billion statewide antipollution program are additional evidence of a new era in waterway conservation.

How soon can nationwide improvement be expected?

Murray Stein, water pollution enforcement director in the Public Health Service, which has handled the abatement program up to now, said in an interview in Washington a few days ago:

"Cleaning up the nation's waterways is essentially a matter of eliminating individual pollution sources—thousands of communities that don't give sewage adequate treatment, including some of our biggest cities, and thousands of industrial establishments.

"Our 37 abatement actions to date have involved more than 1,000 communities and 1,000 industrial plants in 40 states—but all that represents only a start on the problem. How fast improvement comes about will depend primarily on the states, and the attitude of individual pollution sources themselves.

"The Department of Health, Education and Welfare is au-

143

thorized by law to deal with situations of interstate pollution. and of intrastate pollution when the department's assistance is requested by governors. But this, again, is only part of the problem. We can only conduct so many abatement actions. These are complex protracted proceedings, and on our present scale of operation we couldn't initiate more than about one a month.

"The underlying cause of water pollution is that all over the country you have municipal sewerage systems that are inadequate for the loads that have been imposed upon them in the last few years. You have industries that give tremendous amounts of waste inadequate or no treatment at all.

"These conditions, in the final analysis, had to be rectified by the people immediately concerned. This can be gotten at best through states having adequate standards, and enforcing them. The new Water Quality Act is a big step toward obtaining these standards."

An Essential Process

Only about one-third of the nation's population is on sewerage systems with the "secondary" treatment processes now considered essential.

These remove up to 90 per cent of contaminants in wastewater by precipitation and biological "digestion." Detroit still has only primary treatment—removal of grosser solids. New York City discharges 400,000 gallons of untreated sewage daily.

According to the last comprehensive survey by the National Association of Manufacturers, American industry -- which is the source of about two-thirds of the fluid waste volume—was spending only about $100 million a year on waste treatment operations. The survey was made in 1959.

Outlays for equipment and operations have increased substantially since, but still probably total less than 1 per cent of industry's annual product value of some $150 billion. Two of the nation's industrial giants, the steel industry and the auto industry, are major pollution producers.

Danger of Rainfall

Needed municipal and industrial waste treatment facilities probably amount to upward of $100 billion—equal to a year's whole national budget. In 1964, construction contracts for municipal sewage collection and treatment facilities totaled less than $1 billion.

It has been officially estimated that correction of pollution sources in five states that are contaminating Lake Erie may cost $20 billion. The states are Michigan, Ohio, Indiana, Pennsylvania and New York.

The prospective cost for 1,900 communities to separate storm drains from sewer lines, so that rains do not regularly cause potentially lethal overflows of untreated sewage, is officially estimated at $20 billion to $30 billion. This may be greatly reduced by engineering innovations now under study.

Conservative quarters in industry, which regularly opposed Federal intervention in private enterprise and denounced Federal subsidies in principle, are now saying that Federal financial help will be necessary for industrial pollution cleanup.

A debate in coming months about the possibility of curbing pollution by imposing "effluent charges" on entities—municipal and corporate—that dump waste in waterways, is a certainty. This system is followed in Germany's Ruhr.

Senator Robert F. Kennedy of New York, among others, has suggested that the Ruhr system is worth consideration, and the President's Scientific Advisory Committee, in its big October report on "Restoring the Quality of Our Environment," recommended that the system be studied.

This year the Public Health Service initiated seven pollution abatement actions—on the Blackstone River (Massachusetts-Rhode Island), the Savannah in Georgia, the Mahoning in Ohio, the Red River (Minnesota-North Dakota), the Hudson, and Lake Michigan and Lake Erie. There were follow-up proceedings on actions that had previously been instituted, involving the Detroit River, the Columbia, the Raritan and the Missouri.

Federal Action Possible

In 1966, in addition to eight or 10 new actions, there will be follow-up proceedings in 14 situations. The actions involve hearings, formulation of remedies, and explicit scheduling of remedial programs. Failure to comply can bring Federal Court action.

In 1960, the Public Health Service began a series of long-term comprehensive studies of water pollution on the basis of 20 topographical basins over the whole country. The ninth and tenth of these studies—one covering a 10-state segment of the Missouri River, the other most of California and part of Oregon—are just being started. They involve one-sixth of the nation's areas and will take seven years.

It was the Great Lakes phase of this survey program that enabled Federal officials to come up with the critical facts and figures about Lake Erie this year. The Lake Erie portion of the study is scheduled to be completed by July, and the rest a year later.

'Tertiary' Treatment

The Department of Health, Education and Welfare is spending about $145 million on water quality improvement work in the current fiscal year. More than $100 million of this is earmarked for grants to municipalities covering up to 30 per cent of the cost of new sewerage facilities.

Substantial amounts of money are going into medical studies of pollution—on such matters as the still-mysterious dissemination of viruses by water—and into research on improved sewage treatment methods.

Among a dozen projects conducted on collaboration with municipal sewer systems are ones at Santee and Bijou, Calif., for "tertiary" treatment—a step beyond standard secondary treatment, aimed at completely renovating waste water.

Water pollution by pesticides from agricultural runoff and accidental industrial discharges has killed tremendous numbers of fish in recent years, but possible effects on people are still indeterminate. The Health, Education and Welfare Department will investigate this closely in 1966 in a 15-state monitoring program.

"We made a major breakthrough this year in public awareness of the water pollution problem." James M. Quigley, Assistant Secretary of Health, Education and Welfare, sumarized in an interview last week.

"Industry's shift from opposition to cooperation in water cleanup efforts, which emerged at abatement proceedings in the Middle West, could be described as a major philosophic breakthrough.

"The coming year will indicate whether we have the necessary talent and courage and imagination to follow through on these gains. Five years from now I'll tell you whether we grabbed the ball and ran with it, or fumbled.

"Unless American industry really adjusts to the fact that pollution control is a regular part of overhead, we're not going to get the kind of action we need.

"Also, this is the states' last big opportunity to do their part in this job. If they do it, water pollution control will be truly a Federal-state-local partnership. If the states fall, the trophy is going to the Federal Government by default. Nineteen sixtysix is the key year on this."

December 21, 1965

Pollution From Stricken Tanker Continues to Threaten Cornish Coast

Tanker Torrey Canyon yesterday on Seven Stones Reef off southwest coast of Britain. Craft went aground March 18.

By EDWARD COWAN
Special to The New York Times

LONDON, March 26 — An attempt to pull the tanker Torrey Canyon off the Seven Stones Reef failed today, and the 936-foot ship's hull cracked and buckled.

[The Dutch tug Utrecht, which was trying to pull her free, sent a radio message to the Netherlands saying that the Torrey Canyon was gradually sinking and must be considered a total loss, Reuters reported from The Hague.]

The Torrey Canyon, the biggest ship ever to run aground, was insured for nearly $16.5-million. As a total loss, she would represent the most expensive marine insurance claim in history.

The ship, which had been carrying 118,000 tons of crude oil from the Persian Gulf, was still in one piece when night fell.

Rocks had ripped open the Torrey Canyon's underside and protruded 17 feet into her hull, trapping her. The rocks acted as a giant pivot when the Utrecht tried to pull her free, foiling the attempt.

The tanker, owned by the Uniol Oil Company of California, ran aground about 18 miles off Land's End on March.

18. She was on charter to British Petroleum and bound for Milford Haven, on England's west coast.

Thousands of tons of oil from the tanker have flowed onto the resort beaches of Cornwall, and the cracking of the hull, at about midships, sent more of her cargo pouring into the sea. Royal Navy officials could only guess how much remained in the Torrey Canyon's tanks. Their best estimate was that 35,000 tons had escaped, but, as one official said, "It could be a lot greater than that."

Battle on Beaches

Even if half the original volume was left, the leaking oil could bring ruinous damage to the English coast and, depending on the tides and weather, foul the beaches of Belgium and France.

Local officials, civilian volunteers and 600 servicemen continued the battle to prevent the oil from soiling the resort beaches of Cornwall, but it seemed to be a losing struggle. At least 75 miles of coastline had received some oil and the affected zone was widening.

In small boats and on foot in the surf, the volunteers and servicemen spread detergent with pumps and spraying devices in an attempt to break up the thick oil.

Hotels reported cancellations. Residents and Easter weekend vacationers complained that they could smell the oil, and fishermen worried about the threat to marine life, including lobster and oysters.

Optimism Had Run High

Sea birds, their feet and feathers black with oil, struggled to get ashore and hundreds were said to have died.

United Press International Cablephotos

Detergent to combat fouling by oil leaking from tanker is poured into the harbor at Portleven. At least 75 miles of beach has received some oil and the area is widening.

Optimism about refloating the Torrey Canyon had been comparatively high this afternoon. Prime Minister Wilson flew to the mainland from his vacation retreat in the Scilly Isles for a four-hour meeting with several ministers.

Mr. Wilson said afterward that the Government's "present intention" was to deny the Torrey Canyon entry into territorial waters if she could be put afloat. Apparently, he feared that she would leak more oil, increasing the hazard to the beaches.

Lieut. Comdr. Mike Fournel, the helicopter pilot who took Mr. Wilson back to the Scilly Isles, reported on his return to the mainland that the tanker had split at about midships.

"She had gone deep in the water at the stern." Commander Fournel reported. "On a second circuit I thought I noticed a bend in the hull. She was sagging at both ends. During the next few minutes it became pronounced until there was a positive bend—a hump in the middle.

"The well decks were awash and oil was coming out in great quantities. I have never seen so much coming out on previous occasions that I have flown over her.

"She had not exactly split into two halves, but the back of the ship was broken."

A navy spokesman at Plymouth said it was impossible to know whether the Torrey Canyon had cracked because of the towing attempt this afternoon or whether her already weakened hull was simply succumbing to nine days of pounding by the sea.

March 27, 1967

Tanker Operators Strive to Avoid Pollution of the Sea

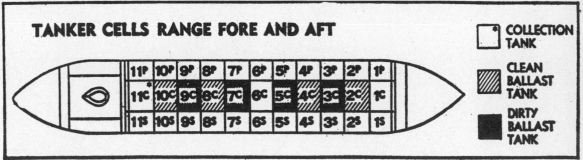

The New York Times

June 11, 1967

Tanker diagram shows compartment arrangement and how mariners select various tanks for ballasting, for cleaning and waste collection to avoid pollution of sea. Each tank holds approximately 10,000 gallons of fuel.

By GEORGE HORNE

Laden with a capacity cargo, the medium-sized tanker Texaco Maine moves in from sea to drop her 327,000 barrels of Arabian crude oil at the company's northernmost terminal at Portland, Me., for movement by pipeline to its refinery in Montreal.

Back and forth across the seas the Texaco Maine moves constantly, like hundreds of sister tankers owned by oil companies —homeward with oil, outward in ballast. All of them are blamed by the average layman for the oil wastes and oil-bearing residue that pollute the seas.

"Find a smear of oil on the beach and all hell breaks loose on the big oil companies," said one of a group of operators at a recent conference on pollution.

The Torrey Canyon disaster last March off the British Coast heightened public awareness of the increasing pollution problem. This was pollution on an Olympian scale, and it was the kind of spoiling against which there can be only minimum safeguards.

The real concern of maritime nations is the drop-by-drop pollution of the sea, the lone vessel pumping its tank washings overboard, the ancient freighters cleaning out bilges under leaky engineroom valves and the slow emissions that may come from scores of heavily loaded tankers sunk off the coast in World War II. There are 5 million barrels of oil lying submerged off the American coast.

Common Concern

The problem is a matter of much concern to the Maine's operators and other major companies—Humble Oil, Mobil, Gulf, Sinclair, Cities Service and others who help carry the 700-million tons of petroleum and products that move over the seas of the world each year.

They complain that they, too, get tarred when the gooey blobs show up occasionally on a South New Jersey beach or Cape Cod, and they insist that their restrictive orders to all ships, instituted years before the Torrey Canyon piled up on a reef, are the tightest in the world.

Present law implementing the international antipollution

treaty carefully delineates areas to which dumping must be restricted. The line is 100-miles off the coast in this area.

Every tanker carries a booklet showing exactly where the permissible dropping areas are, and an official Coast Guard oil record book is carried on every United States-controlled tanker. In this book the master must note any discharge.

But the oil companies say they do not put overboard any dirty tank washings, or slops, even in areas where they are allowed to pump, except in the most urgent circumstances, and that rarely.

Nonpersistent Oils

They do unload treated tank washings containing lighter products that are classified as nonpersistent oils and these, according to scientific research, are not pollutants because they evaporate or oxidate, completely disappearing in the sea.

Independent tanker operators also say "not guilty." They own far more tanker tonnage in the world than the oil companies. Both foreign and United States operators of dry cargo ships and passenger liners also say they have stringent rules against clearing oily bilges or oil emulsions in sea water used for ballast in fuel tanks emptied on long voyages.

Most passenger liners are designed with their own ballast tanks that are used for nothing but sea water. As for the traditional freighter bilges, a steamship company with more than two-score freighters replies:

"Nonsense; we don't have dirty bilges, and we'd better not have. Our ships are too taut."

But seamen on the ships say it happens frequently; they see the slicks on the ocean.

Who does it?

Commander J. A. King, the Coast Guard's chief enforcement officer in this district,

concedes that catching the culprit is almost impossible.

"It is dark out there at night, and there is an awful lot of sea," he says.

Every major oil company in this area claims a voluntary antipollution program. Saving up slops and carrying a tankful around the world until a shore disposal facility is available can cost considerable money in unearned carrying revenue.

On the other hand, tight scheduling can avoid much of the loss, and there is some return in the salvage of oils once freely dumped in mid-ocean.

Moreover, the tanker companies and the American Petroleum Institute, which has done considerable research on the program, believe that the world maritime nations will soon write a new pollution treaty banning dumping of any kind, even the nonpersistent oil products. They consider such a treaty necessary.

What are the precautions followed today? Take the Texaco Maine as a fair example of rigid operating instructions much like those of the other 18 companies that responded to a survey of what the companies are doing.

Panamanian Flag

Texaco Maine flies the flag of Panama.

She is 736 feet long and is registered at 46,442 deadweight tons, which means this is what she can pick up. She has 33 tanks in fore-and-aft rows, 11 tanks to port, 11 to starboard and 11 in the center. From bow to stern, they are numbered, beginning with 1P, 1S and 1C. At present she is operating in crude oil from Middle East or North African ports.

When her cargo has been thrust ashore in Portland by the big shipboard pumps and lines and the main pumps can suck no more, she will still have some 1,700 barrels of oil. The pipe-

lines themselves hold 500 barrels. About 1,200 barrels of oil remain coating the sides of the tanks, the underdeck surfaces, the bulkheads and tank bottoms.

More modern tankers have expensive inner epoxy coatings in the tanks, minimising the amount of oil left coating the surfaces, and simplifying the cleaning process. But the Maine was built by Bethlehem at Sparrows Point, Md., in 1944, and her problem, like those of the majority of tankers, is extremely difficult and ticklish.

So the process begins. She has been ordered to proceed to Ras Lanuf, Libya, to pick up a cargo of Libyan crude for delivery to a refinery at Eagle Point, N. J., on the Delaware, below Camden. Before leaving Portland, the oil remaining in the pipeline system is drained back into one tank, and smaller "stripping" pumps put it ashore.

Harbor Water Pumped

Now ballasting begins. The weather outside is fair, so the ship's skipper orders harbor water for center tanks 3, 5, 7 and 9. This is now dirty ballast. The Texaco gets under way. At sea, as she proceeds toward Gibraltar, the master orders four center tanks cleaned, selecting tanks 2, 4, 8 and 10 so that later they can take on clean ballast, to replace the dirty ballast.

Tank washing machines are let down into the tank tops, spraying hot water under high pressure. Now these four center tanks have a mixture of oil and water in an emulsion, and this stuff is pumped out into a collection tank, the aftmost center compartment.

When they are empty these tanks are filled with clean seawater. While this process was going on the dirty ballast tanks were settling, with oil floating to the top. The water below the oil is pumped into the sea from the bottom of the tank, under

supervision of a ship officer.

Using detection equipment, he stops the flow when the slightest trace of oil, as little as 100 parts per million is indicated. The remaining mixture of oil and salt water is also added to the collection tank, No. 11C. As the ship steams on, this tank also settles, and the clean water at the bottom is put overboard.

Oil in Collection Tank

Now the ship has traveled a thousand miles, her tanks have clean ballasting sea water and the 1,200 barrels of waste oil are in the collection tank. If her African port has a slops facility, the collection tank residue is pumped ashore and can be utilized. If there is no such facility, the Texaco Maine "loads on top."

This means that the ship takes on a fresh cargo of crude oil, creating problems back at Eagle Point, since the residue contains water. Every barrel of water means one less barrel of crude delivered.

A few of the companies are trying out one of several patented compounds that break down the oil-water emulsion (billions of droplets of oil, each imprisoning a lesser droplet of water) thus speeding and improving the separation process.

The ship that stays in the crude oil run is fortunate, because the problem is simply solved. When a variety of products are carried, some are incompatible and the "load on top" recourse is not possible.

A premium gasoline cannot be loaded over kerosene, for instance. The higher refined product would be downgraded. However, both are considered non-persistent oils that are dispersed and broken down by natural processes in the sea.

In addition to crude oil, the persistent products include diesel oil, furnace fuel and residual oils.

June 11, 1967

ALL STATES MEET WATER DEADLINE

U.S. Official Leaves Texas Hailing Pollution Efforts

Special to The New York Times

HOUSTON, July 1 — James M. Quigley, Commissioner of the Federal Water Pollution Control Administration, said today that all 50 states had met yesterday's deadline for establishing programs to clean up their waters.

Mr. Quigley flew back to Washington with the Texas plan to look at the 49 others. He said the "needling" he had given state pollution control officials might have produced a good majority of acceptable water quality criteria and plans for implementation.

The states had been ordered to submit plans by yesterday or face Federal standards on interstate waters within states. The order was incorporated in the Clear Waters Restoration Act, guided through the Senate by Senator Edmund S. Muskie, Democrat of Maine.

Notes Reluctance

Noting that there had been some reluctance, Mr. Quigley said:

"When they were given only 18 months to produce the criteria, there were cries of anguish from the states." They did not have the manpower, and legislatures were not in session to provide it. Even California memorialized Congress to give them another year. "For about six months, generally, the states sulked and said they couldn't do it. When they realized that Senator Muskie was not going to give them any more time, they started work.

"Most of the states have done a remarkable job, better than I thought they would and much better than they thought they

could. I will know in six weeks, but I am pretty encouraged."

Teams of experts in the agency will review the plans. Mr. Quigley's target date to send the state programs to Secretary of the Interior Stewart L. Udall is Oct. 2, the second anniversary of the signing of the law.

Mr. Quigley expects some state plans to be very good, a few pretty bad, and the bulk to fall in a category of "not good enough to approve and not bad enough to turn down." So a period of negotiation to strengthen the state programs is likely.

This will let the agency, just a year ago moved from the Health, Education and Welfare Department to the Interior Department, go into 1968 with water quality standards adopted in most of the states. Its next task will be to set up Federal water pollution controls in the states that do not produce satisfactory standards.

The agency also will set up a control system for discharges of oil in navigable waters, a job

transferred to it from the Army Corps of Engineers. Mr. Quigley expects this to be done in cooperation with the Coast Guard and with the port authorities.

Mr. Quigley's latest "needling" effort produced a stronger state plan of implementation than had been ready a month ago, when he had talks with Texas Water Pollution Control Board officials. He spent last week touring Texas pollution areas with the President's Water Pollution Control Advisory Board.

Mr. Quigley says states and local agencies and authorities must have the political courage to make bold attacks with Federal help, to solve problems for which there is now no known solution. He described the New York standards as the best in the nation, but the water as the worst, because no one had the courage to enforce them until Governor Rockefeller came up with his huge state fund for pollution control.

July 2, 1967

Soviet Moves to Halt the Pollution of Lake Baikal

By THEODORE SHABAD

Special to The New York Times

MOSCOW, Feb. 7—The Soviet Government, heeding appeals from conservationists, announced today a program of stringent measures to stop the pollution of Lake Baikal, the world's largest body, in volume, of fresh water.

It declared not only the Siberian lake itself but also its entire drainage basin a "protected zone," in which timber cutting and other industrial operations are to be severely restricted.

According to the summary of a decree published in Izvestia, the Government newspaper, ministers and other agency heads will be held "personally responsible" for implementation of the comprehensive conservation program.

The decree prohibits timber-cutting altogether in certain kinds of forests in the basin and places lumbering operations elsewhere under strict controls to prevent discharge of wastes into the lake and its tributary streams.

Pulp Mill a Culprit

A paper and pulp mill at Biakalsk, at the southwest end of the lake, was given until the end of the year to complete a purification plant to prevent untreated wastes from being discharged into the lake.

A second mill under construction at Kamensk on the Selenga River, which flows into the lake, was ordered to build its waste treatment plant before it starts production. The Baikalsk mill dumped untreated effluent when it started operation.

According to the decree, all other industries and municipalities in the drainage basin

The New York Times   Feb. 8, 1969

Lake Baikal, in Siberia, is the world's deepest body of fresh water. Program to stop pollution by industrial wastes and sewage was announced by Soviet Government.

are expected to build installations for the treatment of industrial wastes and municipal sewage. The construction of any new industries likely to pollute the lake will be prohibited.

The Ministry of Fisheries was charged with the conservation, replenishment and moderate exploitation of the lake's fish and wildlife, which include several rare species.

The Government decision came in response to a campaign among scientists here and abroad and in the Soviet press urging protection of the unusual lake against the discharge of industrial wastes and sewage.

The complaints began several years ago, when the mill at Biakalsk, began operations without an adequate treatment plant.

The sickle-shaped Lake Baikal, about 400 miles long, is the world's deepest lake—6,000

feet—occupying a rift in the earth's crust.

Although it is exceeded in area by other lakes, its great depth makes it the world's largest body of fresh water in terms of volume.

Despite its remote location, the lake represents a tremendous source of clean water in a nation that has become increasingly pollution conscious in recent years.

The pollution problem has become especially serious in the heavily populated industrial districts of European Russia.

A Soviet geographer has suggested that the time may come when the Russians may have to construct an aqueduct 2,000 miles long to carry clean water from Lake Baikal to the west.

Part of the Baikal drainage basin lies in the Mongolian Republic, to which Soviet conservation measures would not automatically apply. The Russians may, of course, expect Mongolian cooperation in the project.

Control over implementation of the new regulations was entrusted to three agencies.

The Ministry of Reclamation and Water Management will be expected to insure that untreated wastes and sewage will no longer be discharged into the Baikal Basin.

The Government Hydrometeorological Service will be expected to analyze regularly the chemical composition of Baikal waters and of any wastes being discharged into the lake.

Finally, the People's Control Committee, a watchdog agency with representatives in every factory, has been enlisted to organize "systematic control" of implementation of the decree and to call violators to account.

February 8, 1969

Santa Barbara Harbor Closed; Oil Fouls Beaches, Fire Hazard Feared

By GLADWIN HILL

Special to The New York Times

SANTA BARBARA, Calif., Feb. 5—Cleanup crews toiled along 16 miles of oil-smeared beaches tonight as the contamination of this resort community by an oil slick worsened. The Coast Guard, coordinating the effort to

plug a leaking underseas oil well, reported that shoreward winds over the next 24 hours threatened to aggravate contamination of the coast by the huge floating pool of black oil that has been accumulating in the Santa Barbara channel for the last week. The Union Oil Com-

pany, operator of the drilling platform six miles off shore, where the oil is originating, said tonight that a "massive effort" would be made tomorrow to plug the 3,000-foot-deep shaft and surrounding natural fissures. The Santa Barbara waterfront and its oil-drenched marina

were barricaded to the public as an unending succession of trucks drove up to the community's main pier, Stearn's Wharf, with 10,000 barrels of "mud," the viscous compound used to plug oil-well shafts. The marina, containing 700 pleasure boats, was befouled last night when oil

Workmen in Santa Barbara, Calif., pour chemicals on water to repel the oil, right, and spread straw to absorb it

broke through a floating barrier. It was newly barricaded today with a "boom" of logs and plastic floats running from the end of the pier to the tip of a breakwater.

"Live-aboard" boating people have been evacuated because of the danger of fire.

The stench of the oil wafted several miles inland, and the wrath of the community was even more pervasive. From Santa Barbara to Washington, legislators of the locality bombarded Federal officials with demands that oil drilling in the channel be halted permanently.

The well was among the first of hundreds contemplated in the Santa Barbara Channel under a $600-million leasing program executed by the Department of the Interior a year ago.

The unwanted gusher is the first episode of the sort among countless offshore drillings going back nearly a century. Most of them have been within the three-mile limit under the jurisdiction of the state.

The Federal leases are administered by the Department of the Interior's Bureau of Land Management and its Geological Survey office, which lay down detailed engineering specifications on how offshore drillings are to be performed.

A Union Oil Company spokesman said tonight that its entire drilling program for the runaway will had been approved by the appropriate Federal officials.

The oil slick now covers hundreds of square miles, extending beyond the Channel Islands —San Miguel, Santa Rosa,

Santa Cruz and Anacapa— which lie about 25 miles offshore.

It is estimated that 200,000 gallons of oil have gushed from the well so far.

Prisoners Help Cleanup

The Coast Guard said that the beach contamination extended from Carpinteria Beach State Park, 12 miles southeast of Santa Barbara, to Arroyo Burro Canyon, four miles up the coast from Santa Barbara.

Reports on the amount of oil on beaches ranged from a thin film to several inches. The Coast Guard reported "heavy concentrations."

Several hundred persons, many of them prisoners from state conservation camps, were engaged in the cleanup efforts.

This work consisted mainly in scattering straw and talc to soak up the oil.

The oil company said that it had 19 boats operating in the channel, spreading straw and dispersant chemicals.

Meanwhile, another drill was slowly boring into the ocean bottom toward the base of the existing shaft to provide a "relief" vent for the gas pressure that is causing the trouble. But it is not expected to reach its target for another week or 10 days.

Senate Hearing Held

By WARREN WEAVER Jr.
Special to The New York Times

WASHINGTON, Feb. 5—A California official pleaded today for Congress to halt offshore

oil drilling in the Santa Barbara Channel until the community had some reasonable protection against further oil pollution.

But the president of the Union Oil Company, whose well spewed up the oil slick now spreading along the coast, said that this would be "like shutting down the California education system because there's a riot at San Francisco State."

"Gentlemen, we need help and protection," George Clyde, a Santa Barbara supervisor, told the Senate Subcommittee on Air and Water Pollution. "Not just Santa Barbara and California, but all the areas where oil wells are being drilled. And we're not getting that help from the Interior Department."

Fred Hartley, the oil official, said that it was true that his company had not expected "this blow up of the earth's crust" that produced the pollution, but he maintained that "it was not reasonable for us to anticipate that."

Faced with these divergent views, Senator Edmund S. Muskie of Maine, the subcommittee chairman, tended to question the oil industry's position.

He asked the oil official whether the companies had "learned enough" in the 24 hours that Secretary of the Interior Hickel had halted drilling to permit them to resume safely now.

"A lot of decisions are made faster in industry than in some other places," Mr. Hartley responded.

Interior Agency Criticized

Mr. Clyde, who has been dealing with the Santa Barbara pollution as an official of county government, was critical of help provided by the Interior Department under the previous and current Administrations.

The Californian called Mr. Hickel's action "tokenism in its worst form" and said it "smacks of pure hypocrisy."

But he thought little better of the department under Stewart L. Udall, which he said "always led us to believe we had nothing to fear" from oil spillage when it licensed the offshore oil wells.

When Santa Barbara officials tried to keep the oil wells away in 1967, Mr. Clyde said, "everyone in the mineral resources division of Interior was hellbent to lease this land and get the drilling started."

"I can't say how much the oil industry influenced the Department of Interior," he said, "but I do know that whenever we said anything to Interior, within a matter of minutes the oil industry knew about it in detail."

While he conceded that the offshore oil drilling had produced an unfortunate result," Mr. Hartley insisted that the damage should be kept in perspective.

"I'm amazed at the publicity for the loss of a few birds," he said.

No human life has been lost in the pollution episode, he observed and added that that was more than could be said for the crime situation "in this fair city."

February 6, 1969

Coast Oil Leak Is Plugged; Udall Criticizes Own Role

Drilling Decision Blamed

By ROBERT H. PHELPS

Special to The New York Times

WASHINGTON, Feb. 8—Stewart L. Udall, former Secretary of the Interior, said today that he bore the responsibility for the decision last year to permit oil drilling in the Santa Barbara Channel.

Mr. Udall said in an interview that he was "sickened" by the vast damage caused by an oil leak in an undersea well in the channel, but added: "I bear responsibility and I take it."

Heads Consulting Firm

The former Secretary, now head of The Overview Group, an international consulting firm on environmental problems, said that he had asked experts throughout the Interior Department for their opinions about letting oil companies drill in the channel.

"There was no dissent," he said, and therefore he approved the leases that opened the channel for oil drilling.

Mr. Udall said that the question of tighter regulations governing the drilling had never come up, though geological conditions in the Santa Barbara area were known to be unstable.

Interior Department experts apparently had no doubts about the regulations, Mr. Udall explained, because 12 years of experience in drilling in the Gulf of Mexico, off Louisiana and Texas, had not led to any big leaks, even during hurricanes.

The state of California contends that its regulations governing drilling in state underwater lands are more stringent than the Federal Government's. The new Secretary of the Interior, Walter J. Hickel, has ordered his department to draft tougher drilling rules.

The first step in drawing up new regulations will be a survey of geological conditions in the earthquake-prone Santa Barbara area. Mr. Hickel met with top people in his department today and ordered the study expedited.

Although the leak has reportedly been plugged, Mr. Udall expressed some doubt that drilling should ever be resumed in the channel.

He said the disaster "causes me to wonder," particularly in the light of the discovery of new oil fields in Alaska, whether "we should not hold back" on further drilling.

The man who prodded the Nixon Administration to stop drilling yesterday, Representative Charles M. Teague, went further than Mr. Udall. Mr. Teague a Republican whose district includes Santa Barbara, said in an interview:

"This is not the place where oil exploration should occur. I don't want to put a 'forever' label on any subject, but I hope the drilling will be permanently stopped."

Although Mr. Teague appealed directly to President Nixon to stop the drilling—after Mr. Hickel suspended it, then let it be resumed—the Congressman praised the Secretary for moving quickly to solve the problem. Mr. Udall also said he would not criticize Mr. Hickel.

But both Mr. Udall and Mr. Teague were fearful of the ecological effects of the oil slick. Even after the water and beaches are cleaned up, they agreed, it will be years before the marine life can recover.

Torrey Canyon Recalled

Mr. Udall recalled the wreck of the oil tanker Torrey Canyon off the coast of England in 1967. More than 117,000 tons of crude oil poured from the tanker into the sea, fouling beaches from Cornwall to northern France.

A Government-sponsored study of the effects of the accident showed that it might take up to five years before many of the coves covered by oil recovered fully. Detergents used to clear the oil were found to be more harmful than the oil because they destroyed the oxygen in the water.

From the Torrey Canyon experience, Mr. Udall said, it will unquestionably be years before Santa Barbara recovers.

The oil and chemicals used to disperse the Torrey Canyon slick are thought to have disturbed the chain of marine life by killing the plankton and other small organisms, small fish and shell fish. While attention has focused on dead birds, naturalists are most concerned about the effect on marine life. Birds can fly from the scene, but marine life is trapped.

Mr. Teague expressed similar sentiments about the ecological effects of the oil leak, agreeing that in many ways a "dead sea" had been created—at least for a time—off the California coast.

February 9, 1969

WATER POLLUTERS WHO FAIL TO ACT FACE FEDERAL SUIT

Hickel Orders New Drive by Government to Identify and Prosecute Violators

Special to The New York Times

WASHINGTON, Sept. 3—The Interior Department plans to speed up its drive against water pollution by suing individual polluters if necessary.

Announcing a Government drive to "prosecute those who pollute," Interior Secretary Walter J. Hickel today ordered hearings before the Federal Water Pollution Control Administration. The City of Toledo, Ohio, four steel companies and a mining company have been charged by the Government with pollution, and their representatives will appear at the hearings.

The steel companies are the United States Steel Corporation, the Republic Steel Corporation, the Interlake Steel Company, and Jones & Laughlin Steel Company. The mining company is Eagle-Picher Industries, Inc., of Baxter Springs, Kan.

Those charged with water pollution are not required to attend the hearings, but they were sent official notification of the charges yesterday, according to an Interior Department official. If they are found guilty of the charges, and fail to take steps to eliminate pollution within 180 days, the Interior Department plans to bring suit.

Until today, states have had the responsibility for initiating court action against water polluters. Interior officials, however, feel that state court actions have been too slow in coming. The new drive at the Interior Department is designed to accelerate the cleaning up of the nation's waterways.

Act of 1965 Cited

In a statement, Secretary Hickel said: "This is just a beginning. We intend to continue the identification of polluters for prompt cleanup and pollution elimination."

Carl L. Klein, Assistant Secretary for water quality and research, said in a telephone interview that the Government for the first time would use the "abatement proceedings" provision of the Federal Water Pollution Control Act of 1965, which defined water pollution standards in various bodies of water.

Mr. Klein said the Interior Department's campaign procedure started with "fact finding" by scientists and engineers and called for voluntary hearings involving companies, municipalities or others charged with pollution.

Mr. Klein stressed that those found to be polluting waterways would be required to "come up with a time table to abate the pollution and a step-by-step plan to alleviate the situation."

Mr. Klein said if the offenders had not taken steps to eliminate pollution within 180 days after notice of the hearings, Secretary Hickel was empowered by the act of 1965 to ask the Justice Department to seek a mandatory court injunction for "pollution abatement."

Pollutors will now be required to meet a "real, tight schedule" in stopping pollution, Mr. Klein said, ranging from one to three years, depending on engineering factors.

Since 1965, the Government has used the "enforcement conference" provision of the pollution law to combat pollution in the nation's waterways. That provision, however, involved a process that Interior officials regarded as too time-consuming. Included in the procedure was the holding of joint Federal-state discussions to identify pollution problems and sources, with the states having responsibility for taking "remedial action."

But the main force behind the "enforcement conference" procedure was public opinion, according to Mr. Klein.

149

The first phase of Interior's drive against water pollution will center on the Great Lakes and certain rivers that supply many communities with drinking water.

The department's action today involves pollution of interstate waters in Kansas, Oklahoma, Ohio and the severely polluted Lake Erie basin.

Later actions will involve the Hudson, Missouri, Mississippi and Ohio Rivers.

Hearings on Sept. 23 in Joplin, Mo., will review charges against Eagle-Picher Industries, a mining company accused of violating water quality standards for the Spring River in Kansas and Oklahoma.

Hearings in Cleveland on Oct. 7-8 and in Toledo Oct. 9 will review charges of inadequately treated wastes discharged by the City of Toledo and Interlake Steel in the Maumee River, and of excessive discharges of pollutants by Republic Steel, United States Steel and Jones & McLaughlin in the Cuyohoga River. Both the Maumee and the Cuyohoga flow into Lake Erie.

Investigation was done by the Department of Interior study group on pollution enforcement that was created by Secretary Hickel July 25.

September 4, 1969

Waterways, Waste and Words

By GLADWIN HILL
Special to The New York Times

LOS ANGELES, Oct. 19—The deterioration of the nation's physical environment is linked in some degree to semantics. Elegant euphemisms are sometimes used to camouflage unpleasant facts, lulling the public into a false sense of security. The field of water pollution, particularly, has developed a wondrous vocabulary. It is no longer fashionable to call sewage sewage; it's "waste water." "Receiving waters" is the fancy term for "where we dump the stuff," Nobody dilutes sewage any more; he engages in "flow augmentation."

Some of the terms are as loaded with subtle propaganda as Marxist jargon or admen's references to "the expensive spread" and "flameless cooking."

Thus, "assimilative capacity" may be used by engineers purely in the technical sense of how much waste you can put in a river before it starts showing. But there is also the implication that you can dump a lot of stuff before it does show.

A Hopeful Sign

A hopeful sign of the times for antipollution forces, semantically speaking, emerged a few days ago in Dallas. Five thousand water pollution people — engineers, industrialists and public officials — were milling around in a convention of the Water Pollution Control Federation. It was a record turnout and there was relatively little use of the greatest cliché of them all: the innocent-sounding term "beneficial uses."

Most of the federation's membership could, logically, be presumed to be opposed to water pollution. But there are those who have manifested more interest in defending the status quo than improving it. With them, "beneficial uses" has long been a magic catch phrase.

"Public policies," a speaker will intone, "should be directed toward maximization of the beneficial uses of water."

It sounds innocuous enough. But that is just the point at which the average citizen simply interested in drinking a palatable glass of water, or fishing down by the old mill stream, might best start looking out.

"Beneficial uses" in itself is a bogus expression, so far as antipollution men are concerned. No one has ever pointed out any "nonbeneficial" uses of water, they observe. They say the word "beneficial" is just windowdressing, thrown in to mask and soften the pointed pointed pluralism of the word "uses."

The pluralism epitomizes, and is meant to emphasize, the philosophy of generations of water polluters, both in industry and in the operation of substandard community sewage systems.

Their doctrine is that waterways were divinely created for several purposes. One, to be sure, was as a drinking water source, a scenic attraction and a place of recreation. But equally hallowed waterway functions, in this philosophy, are as disposal channels for the fluid wastes of industry and communities.

Industry, under this doctrine, has a vested interest in the use of waterways and the upstream community has some inherent right to dump its effluents on the folks downstream.

Pollution-conscious quarters in industry and municipal sanitation are, increasingly, repudiating these concepts, as the decline of the "beneficial uses" phrase at Dallas indicated. Conservation lawyers increasingly are pointing out that there is no record of the public's ever having assigned its rights to clean waterways to anybody.

Kindred Cliches

But the abuse of public waterways has been bolstered down the years by the iteration of the term "beneficial uses," with the implication that all uses rank equally.

Kindred cliches are the righteous-sounding "legitimate users" (there being no illegitimate users) and "highest and best use" (translation: "In some situations, sewage deserves priority").

Conservationists, and an increasing number of average citizens, are viewing waterways as a source of public and industrial water supplies and, quite compatibly, as a source of enjoyment. They are rejecting the use of waterways to any detectable degree as channels for carrying away fluid wastes, which there is ample technology for handling in other ways.

And so it was that one of the Dallas conventioners remarked quizzically: "Things must be improving. I've hardly heard anybody talk about 'beneficial uses' this year."

October 20, 1969

U.S. Orders Iowa to Halt Pollution of 2 Rivers

By WILLIAM M. BLAIR
Special to The New York Times

WASHINGTON, Oct. 29—The Federal Government invoked for the first time today its power to set water quality standards for a state and told Iowa to clean up the raw sewage that it pumps into the Mississippi and Missouri Rivers.

Further, the Government urged Illinois, Missouri and Kansas to move faster on installing water clean-up facilities as a part of what Secretary of the Interior Walter J. Hickel described as an "over-all schematic plan to clean up the Mississippi, Missouri and Ohio Rivers."

The crackdown on Iowa, which has bucked the Federal prescription for clean waters, presaged similar action against other states. Iowa and several other states have challenged the Federal Government's right to dictate a policy of nondegradation of all waters.

The action against Iowa also covered 25 smaller interstate streams with pollution problems.

Iowa must provide secondary treatment of all wastes flowing into the Mississippi and Missouri by Dec. 31, 1973, under the Governments' fiat.

Secretary Hickey said that he was taking action against Iowa under the Water Quality Act of 1965 because "protection has to be provided for these major waterways upon which so many people, industries and municipalities depend for their livelihood, water supply and enjoyment."

It was his responsibility under the law, he said, to assure that state water quality standards protect public health, enhance water quality and otherwise carry out the purposes of the Federal act.

Furthermore, he said, Federal standards carry a provision that "dilution [of sewage] shall not be considered a substitute for proper waste treatment at any time."

Iowa's current standards were approved by the Interior Department, through the Federal Water Pollution Control Administration, last Jan. 16 except for treatment requirements and implementation plans for the Mississippi and Missouri Rivers and various guidelines on the 25 smaller streams.

The question of mandatory secondary treatment for sewage pouring into the two major rivers has not been resolved despite lengthy conferences with state officials. Secondary treatment is a process of "biological" digestion that eliminates most of the pollutants from sewage.

Iowa officials had maintained that when the Federal experts demonstrated that secondary treatment was necessary to maintain the rivers' quality the state would agree. Also they contended that Iowa contributed less pollution than other states on the waterways.

150

If Iowa fails to comply with the Federal standards, the Government could seek compliance in Federal Court.

Carl L. Klein, Assistant Secretary of the Interior for Water Quality and Research, said that thus far Iowa had relied upon primary treatment of wastes. He said that primary treatment removes only between 30 and 40 per cent of oxygen-consuming pollutants from waste water, compared to 85 to 95 per cent removal of impurities through secondary treatment.

Communities and industries will be given until Dec. 31, 1973, to achieve the secondary treatment objective, Mr. Klein said.

Mr. Hickel said he was asking Illinois to "update its secondary treatment target dates" for Rock Island, Moline, East Moline, East St. Louis and Chester, Missouri, he said, was being asked to set an earlier date than its proposed target date of 1982 for completion of secondary facilities. Kansas was being urged to "set a viable date for installing secondary treatment," he said.

Other pollution problems were being discussed with the cities of Omaha, Memphis and New Orleans, he said.

Iowa and other parties affected by the standards set for Iowa have 90 days in which to file written views and arguments on their position.

The Government's action came as Dr. Glenn T. Seaborg, chairman of the Atomic Energy Commission, blamed "unsubstantiated fear-mongering" for a major part of public concern over another aspect of pollution, the effects of nuclear power plants on the environment.

He and another member of the A.E.C., Wilfred E. Johnson, told the Joint Congressional Commitee on Atomic Energy that preliminary studies indicated that since the inception of nuclear power plants employes had suffered no measurable damage from radiation. The studies covered 170,000 persons in nuclear plants at Hanford, Wash., Oak Ridge, Tenn., and elsewhere.

October 30, 1969

Auditors Find U.S. River Clean-Up Effort Failing

WASHINGTON, Nov. 4 (AP) — The Federal Government's long and costly effort to clean up the nation's rivers had been hampered by poor planning, inadequate funds and unchecked industrial pollution, Government auditors said.

In a comprehensive report on the Federal water pollution control program, the General Accounting Office said that little or nothing had been accomplished despite the expenditure of $5.4-billion on waste treatment facilities since 1957.

Present funding levels are totally inadequate to meet the problem of waste-choked streams and rivers, the auditors said. But the agency added that before more money is spent a new basis for awarding grants should be developed.

The Government has contrib-uted $1.2-billion of the $5.4-billion that has been used to construct 9,400 projects.

The report is based on the study of eight rivers in various parts of the country where Federal, state and local forces have been battling pollution for years.

In every case the accounting agency found that the efforts of Government were being overwhelmed by continued outpourings of industrial waste.

Along a stretch of the Willamette River in Oregon, the agency said, the expenditure of $2.1-million by intergovernmental agencies on waste treatment facilities reduced the amount of municipally produced pollution by 20,000 units on a test scale. But during the same period, it said, two paper mills dumped between 500,000 and two million units into the same area of the river.

In another example cited by the agency, six cities along the Mississippi River in Louisiana used $7.7-million in Federal grants to build facilities that cut pollution by 147,000 units. But 80 industrial plants along the same stretch are putting 2.4 million units into it.

Similar examples were given for the Nashua River and Ten Mile River in Massachusetts, the Tualatin River in Oregon, the Pearl River between Louisiana and Mississippi, and Saco River and Presque Isle stream in Maine.

Rivers Called Typical

The accounting agency said the rivers studied for the report had been chosen not because they were unusual but because they were regarded as typical of the water pollution problem throughout the country.

The agency recommended that the present method of awarding Federal grants, which is generally on a first-come, first-served basis, be changed to consider the benefits to be derived from construction of facilities and the actions that might be taken by industrial polluters.

It also called for improved planning, involving systems analysis techniques to determine the requirements for controlling pollution in a particular area, the alternatives available and the establishment of priorities.

At the present rate of Federal support, the agency said, the backlog of cities awaiting funds is increasing constantly and no significant increase in the effectiveness of the program can be expected unless funding is increased.

November 5, 1969

'Dead Sea' in Harbor Linked to Pollution

By THOMAS F. BRADY

Dumping millions of tons of sewage sludge and dredging spoils just south of the Ambrose Light has created a "dead sea" that is spreading toward New York and New Jersey beaches, according to a report by the United States Marine Laboratory at Sandy Hook.

The report was released yesterday by Representative Richard L. Ottinger, Democrat of Westchester County, at a news conference at the Americana Hotel. He said the dumping was poisoning marine life and endangering the health of those who eat sea food caught in the polluted area.

The laboratory, a unit of the Bureau of Sport Fisheries and Wildlife, prepared the report at the request of the Army Corps of Engineers and submitted it Dec. 3, and since then, Mr. Ottinger charged, the corps has been "sitting on it."

A spokesman for the corps said the Coastal Engineering Research Center, to which the report was submitted, had turned it over to the Smithsonian Institution for evaluation to see if the procedures needed modification. He said that it was an interim report and that the study would continue.

For this reason, he added, the report was not published, but he said copies would be mailed Monday to those who had requested it earlier. A further report is expected in September.

Representatives of the Smithsonian Institution participated with the Corps of Engineers and a committee of scientists in March, 1968, to define the pollution problem for the study and set guidelines for determining the efects of waste disposal in the area.

Dumping in the area has been going on for at least 40 years, Mr. Ottinger said, but the noxious consequences have grown very markedly in the last five years.

Mr. Ottinger said the corps had issued permits for dumping five million cubic yards of sewage sludge and six million tons of dredging spoils a year at points five miles southeast and five miles southwest of the Ambrose Light.

The Key Pollutants

The sewage sludge comes from plants of 12 sewage authorities in the metropolitan area, including 10 plants of the New York City authority. The pollutants are augmented, the Representative said, by 365 million gallons a day of raw sewage that the city pours into the Hudson and East Rivers.

Another sewage sludge contributor is the Passaic Valley Sewage Commission plant in New Jersey, where sludge is now accumulating at a "critical" rate, according to Gov. William T. Cahill, because of the tugboat strike.

The "dead sea" area covers 20 square miles at the seaward entrance to New York Harbor,

151

according to the report, and officials of the Food and Drug Administration have recommended that all waters within a six-mile radius of the contaminated area be immediately closed to shellfish harvesting, Mr. Ottinger said. He added that no action had been taken so far, however.

More than a dozen species of fish captured in the polluted area were suffering from a disease known as fin rot, so virulent that sometimes the tails of the fish had simply disintegrated. The disease, produced by bacteria nurtured by the pollution, also resulted in lesions that destroyed the scales of the fish.

Lobsters and crabs exposed to the pollutants developed fouling of their broncial chambers and gills. Lobsters' gills were found covered with spots of oil and debris, and the fine integument covering the gills was perforated with small openings left by dead tissue.

Risk of Hepatitis

A risk was also reported by Mr. Ottinger that people who ate mollusks affected by the pollution could get hepatitis.

The report, signed by Dr. Jack B. Pearce of the Marine Laboratory, said bottom currents were probably carrying heavy pollutants, northeast. Estimates indicate that even if dumping were halted at once, it would take a decade for curernts to remove the pollutants.

The dredge spoil, the report said, is "even more impoverishing than the sewage sludge." The study shows that sewage sludge has spread out in a northerly direction from the

The New York Times (by William E. Sauro)

Representative Richard L. Ottinger at his news conference yesterday. Two darkened circles on map behind him indicate areas south of Ambrose Light where dumping occurs.

designated sewage dumping gounds over an area of 14 square miles.

Here the benthic (deep) macrofauna has become severely impoverished in contrast to that of the surrounding area," the report continued. "Several species which usually tolerate polluted conditions, such as nematode and rhynochocoelan rubber worms, were absent from the impoverished area."

Sewage sludge is the dense semiliquid residue left after plants, as distinct from raw, treatment of wastes in sewage or untreated, sewage. Dredging spoils are the bottom deposits, left from years of dumping industrial wastes, that are brought up mechanically to deepen commercial waterways.

Many of the pollutants are directly toxic, according to the report, and others exhaust the oxygen in the water and prevent penetration of light so that plant life cannot nourish itself by photosynthesis.

Mr. Ottinger said that if President Nixon did not put a stop by executive order to issuance of permits to dump, he would introduce legislation to halt the pollution.

If the President does not act, he said, "it will show his statements against pollution are just more words."

The Congressman told newsmen that he would announce his candidacy for the Democratic nomination for United States Senator within the next two weeks.

He said also that he shared the concern of two New York Congressional colleagues, Representatives Ogden R. Reid, a Republican, and Lester L. Wolff, a Democrat, who conducted a hearing Thursday on prospective pollution of Long Island Sound and the Hudson River by proposed atomic generating plants that would dump hot water and thus reduce the oxygen content of the waters there.

Mr. Ottinger said he believed that as a temporary solution to the pollution problem, the sewage sludge and dredging spoil should be dumped beyond the continental shelf, about 140 miles from Ambrose Light. New York City sludge carriers now make six trips a day to the dumping area 12 miles off Coney Island.

However, the congressman added, only new processing plants could provide a permanent solution to the pollution problem.

February 8, 1970

U.S. Making First Move Against Thermal Pollution

By GLADWIN HILL
Special to The New York Times

WASHINGTON, Feb. 21—Federal officials will make their first major move next week against a new and largely uncharted environmental problem —thermal pollution

At a Miami hearing starting Tuesday, the Federal Water Pollution Control Administration will reportedly ask that Florida's biggest electric company stop excessive discharges of hot water from a generating plant into Biscayne Bay.

The discharges, according to Federal investigators, have done "severe damage" to fish and marine plant life in the bay, where President Nixon has a winter residence and where the Government is projecting a national marine preserve.

The Miami hearing will mark the first time in 13 years of Federal water pollution abatement actions that a move has been based entirely on thermal pollution.

Seen More Widespread

The move presages a new era of Federal action against a problem that occurs in various degrees with many of the nation's 3,000 electric plants and many other industrial facilities. It is a problem that will become more widespread in the years ahead.

The nation's need for electric power is doubling every 10 years. Massive use of cooling water to condense steam is an integral part of the generating process. Almost half the water used in the United States is for industrial cooling, a Government engineer said this week, and, by the year 2000, the need will equal two-thirds of the nation's natural daily water runoff of 1,200 billion gallons.

Require More Water

The atomic power plants that are superseding fossil-fueled ones require even more cooling water than their predecessors.

"This conflict between electric power and the environment," the President's science adviser, Dr. Lee A. DuBridge, said recently, "is not apt to diminish but in fact is becoming increasingly acute."

Power plants are generally situated near waterways so as to have a big supply of cooling water. Having to get rid of heat from a plant that depends on heat to run steam turbines sounds paradoxical. But the waste heat is only a small fraction of the total heat used, and, in most situations, no constructive use for it has been devised.

Two oil-fueled generating plants of the Florida Power and Light Company near Miami, for instance, discharge about 10,000 gallons of cooling water a second at around 100 degrees.

This can raise ocean temperatures from around 85 degrees to 100 degrees or more. A peak of 103 degrees last June caused a sizable fish kill.

Florida Power and Light discharges exceed both Federal and state thermal limits. The company has been operating under a series of variances granted by Florida authorities.

In colder waters in other parts of the country heat is absorbed more readily. But while fish can tolerate sizable seasonal temperature ranges, a permanent temperature change of as little as four degrees may exterminate them.

Heat adversely affects fish in many ways while it reduces the amount of oxygen in water. Heat speeds up their metabolism so they have to breathe faster, slows them down in pursuing food, and inhibits reproduction.

It may also kill normal vegetation and stimulate the growth of algae, disrupting the plant-fish-insect-bird "food chain" underlying the ecology (the relationship between organisms and the environment) of an area.

Most of the nation's major rivers now have some thermal pollution. The Federal Government has prescribed limits on how much the temperature of waterways may be raised artificially.

For inland waters, the general limit is five degrees above the normal monthly average—to maximums of 86 to 96 degrees for short periods in different parts of the country. For trout and salmon streams, the ceiling is 55 degrees. For coastal waters, the general limits are a four-degree increase in winter and a one-and-a-half-degree increase in summer.

These limits have been incorporated in Federal-state water quality standards for all 50 states now in the final stages of adoption, which will open the way for general enforcement. The criteria have occasionally been cited in abatement actions centered on other kinds of water pollution.

Myriad of Species

Biscayne Bay is considered to be an exceptional ecological area, with 58 species of fish, more than 100 species of shellfish, and a dozen species of birds, including pelicans, cormorants, herons, egrets and the rare roesate spoonbill.

In the bay, researchers have discerned a "biological degradation zone" covering more than a square mile in the path of the power plant's effluent. The power company has been planning to add two nuclear generating units that would increase fourfold the current waste heat discharge.

Alternatives to discharging heated water into waterways are holding ponds and towers, where water is dribbled and cooled by air. At the generating plants near Miami, ponds would require an estimated 4,000 acres and towers would cost several million dollars. Florida Power and Light has proposed building a canal to cool the water somewhat while carrying it several miles to a sound south of Biscayne Bay. Federal

officials say this would also cause extensive biological damage.

Federal intervention in the Biscayne Bay situation was requested in December, 1967, by Gov. Claude R. Kirk Jr. under the Federal water pollution laws.

Next week's hearing is the first stage in a formal Federal pollution abatement proceeding aimed at producing a remedial program enforceable in the Federal courts. Technical advisers to Secretary of the Interior Walter J. Hickel, who is responsible for water pollution abatement, have recommended that the power company be required to confine discharges, by one means or another, to the Federal-state standards.

February 22, 1970

Countries That Touch the Fabled Rhine Take Steps to Clean Up the 'Sewer of Europe'

By ERIC PACE
Special to The New York Times

HOOK OF HOLLAND, the Netherlands—The pollution that has clouded the onceclear Rhine is spreading anxiety in the countries along its course, and they are increasingly moving to control the sources of the filth that the river picks up and spews into the North Sea.

Sewage and industrial effluent make up 20 per cent of the river's waters by the time it ends its 820-mile journey from Switzerland to the Dutch coast. The Dutch call it "het riool van Europe"—the sewer of Europe.

Over the decades the increase in contamination has stemmed mainly from population growth and industrial expansion and modernization.

More than a thousand pollutants are present in the Rhine's waters at any time, including detergents, pesticides and other chemicals. At least 30,000 tons of salts alone are carried into the Netherlands each day.

This pollution has already robbed the river of much of its fabled beauty. It has outstripped its natural capacity for cleansing itself. And it may someday prove harmful to human life.

New Plants Are Built

Accordingly, new effluent-treatment plants are being built by communities and private concerns along its course. There have been improvements in legislation, international cooperation and emergency procedures to prevent pollution or to minimize its effects.

Further measures are under discussion. They range from the creation of a European anti-pollution

The New York Times   March 22, 1970

The course of the Rhine is indicated by the heavy line.

fund to way of disposing of waste salts without flushing them into the Rhine.

Fears rose last June when millions of fish died mysteriously in the river's German segment. Their deaths, ascribed at the time to a pesticide in the water, have never been completely explained.

Concern about the Rhine has also been buoyed by a general anxiety about the deterioration of the environment in the countries of Western Europe.

Called "Vater Rhein" by countless German poets, it flows through or past six nations: Switzerland, Liechtenstein, Austria, France, West Germany and the Netherlands.

More than 30,000,000 citizens of those countries live within the Rhine's watershed, making it economically the most im-

portant river in Europe as well as the best known.

The Rhine remained relatively clean until heavy industry began sprouting along its banks in the last century. Since then pollution has gradually increased — particularly since World War II.

From the nineteen-forties to the nineteen-sixties the amount of chloride in the river's water increased by more than 20 per cent. The amount of ammonia grew by more than 7,000 per cent.

The amount of oxygen, important for a river's capacity to clean itself, fell by one-third. And impurities markedly reduced the clarity of the water in the Rhine's northern reaches.

Authorities differ on whether and how much the Rhine has deteriorated in the months since what the Germans call "Das grosse Fischsterben"—the great fish extinction.

Dutch experts contend that the amount of salts dumped into the Rhine, a major aspect of its pollution, still seems to be on the increase. They say there has been little abatement of other key aspects. Some German authorities are more optimistic, however.

A statistical study of pollution along the whole river for the 12 months ending last December is being prepared by the five-nation International Commission for the Protection of the Rhine Against Pollution. But it will not be made public until later this year.

Another international body, the council of Europe, reported last month that as many as two million bacteria can now be found in a single cubic cen-

timeter of Rhine water here in the Netherlands.

The council found that more than 20,000 bacteria per cubic centimeter were present at the midpoint of the Rhine, in Germany, while the river's water is "almost pure" at its source in Switzerland, where it is fed by mountain streams.

For purposes of comparison, the acceptable standard for drinking water in the United States is 50 bacteria per 100 cubic centimeters, (about 60 cubic inches) of water; that for waters where shellfish grow is 70 per 100, and that for bathing and for treated sewage is 2,400 bacteria per 100 cubic centimeters.

Waters Fresh in Austria

The waters are still sparkling and clear as they flow into Lake Constance from the Vorarlberg area of Austria, where Ernest Hemingway liked to ski. Their taste is cold and fresh.

But before they flow out again, past the ancient German city of Constance, they are exposed to a variety of modern sources of pollution, including the pleasure craft that ply the lake in warmer weather, dirtying it with their engine exhausts and other wastes.

In addition, the growth of the population in the surrounding countryside has increased the amount of sewer-water flowing into the lake, although several sewage-treatment facilities are planned or under construction nearby.

Still more pollution comes from the fertilizers in use in the area's farmland. The amount of phosphorous found in its waters more than tripled between the nineteen-forties and the nineteen-sixties.

Pollution has sped up the natural process that naturalists call eutrophication—the aging process by which a lake evolves into a marsh and ultimately solidifies and disappears.

Algae on Increase

Already the amount of algae in Lake Constance has increased by 1,000 to 3,000 per cent since 1920.

Downstream, a simpler source of pollution appears after the Rhine turns northward from Basel: the potash mines of French Alsace. They produce vast quantities of salt as a waste product. Dissolved into water, these are pumped into the river and left to drift northward at the rate of 7.5 million tons a year.

Other large quantities of salts are added from German mines and industrial concerns. At this season, the salts are not a particular threat since rain and melting snows have lifted the river's level. But at low water, later in the year, the concentration of salts sometimes becomes high enough to injure sensitive plants and otherwise complicate water problems in the Netherlands.

Another dumping problem worries the authorities, notably in West Germany. This is the possibility that one of the 10,000 boats abroad at any one time may accidentally dump overboard some substance that is poisonous to humans.

The result has been a growing awareness of what the prestigious Munich newspaper Die Süddeutsche Zeitung calls the dangers of carrying "death under the deck planks."

The West German chemical industry finds use of the rivers essential. For the sake of cheap transport and ample water, chemical companies have built enormous riverside factories at Ludwigshafen, farther downstream at Leverkusen and up the Main, a Rhine tributary.

Such factories have been heavy contributors to pollution, as have German metallurgical industries and mines. Flowing past these factories, the river picks up acids, bleach, cadmium, copper, dyes and a whole roster of other substances, including pesticides and detergents. (The production of synthetic washing materials doubled in West Germany between 1956 and 1966.)

In the Netherlands the problem is compounded by urban sewage and effluent from Dutch chemical factories and various agricultural processing industries.

Waters Grow Cloudy

All these sources of pollution have had wide-ranging consequences along the Rhine's path. Its waters become less clear as they move north. They are sometimes discolored by effluent and often have a bad taste.

In addition, the wildlife beneath the river's surface has been gravely affected, particularly "der Lachs," the salmon. Around 1885 the catch in the Rhine began to dwindle and salmon disappeared from the river several years ago. Pollution had driven them away.

Other, coarser varieties of fish are still found in the Rhine's murky waters.

Authorities in Bonn have largely written off the Rhine fishing industry as uneconomic, but other consequences of pollution are taken very seriously.

As the number of impurities in the Rhine's water has risen, the costs and complexity of making the water fit to use has risen, too, and the authorities fear that the costs will rise even further.

This is in part because both the Netherlands and West Germany expect to become more dependent on Rhine water as their populations and economies continue to grow. As it is, the Dutch rely on the Rhine for 63 per cent of their freshwater supply.

Stronger Action Called For

With so much at stake, governments and private businesses have taken a wide variety of measures to counteract pollution in recent years. But avid conservationists complain that stronger action is required.

There have been calls for stronger action by the West German authorities, although they have their hands tied. This is because under the West German Constitution measures against water pollution are mostly left up to the individual states.

Nonetheless, investment by Government authorities in effluent-treatment factories is now approaching $500-million a year in West Germany alone.

Notable among municipal effluent-treatment projects is one planned by the Swiss city of Basel. For lack of space, the city has arranged to have elements of the $85-million layout constructed on nearby German and French soil.

Private Battle Is Costly

The private sector has found pollution expensive to combat. In one West German state alone, North Rhine-Westphalia, industrial concerns have paid over $500-million to build effluent-purifying facilities since the early nineteen-sixties. In the Ruhr, $800-million have been spent on pollution-abatement measures since 1955.

At Leverkusen, local authorities and West Germany's largest chemical company are building a $55-million effluent-treatment facility that is to go into operation next year.

"This plant is vital for us—the life of a chemical firm depends now on getting rid of its wastes," said Paul Henkel, a director of Farben Fabriken Bayer A.G.

In general, German companies that are conscious of environmental problems are earmarking 6 per cent of their new construction budgets for facilities to reduce pollution.

Tough Law Planned

In the Netherlands, the Government plans to implement a tough new law against water pollution this year. The Dutch authorities also plan to anchor a testing barge in the Rhine where it flows from Germany across their country's border.

The vessel will be equipped with devices to detect any undue amount of pollution. Among them will be tanks of fish. If the fish die, the alarm will go out.

Teunis Verheul, the Netherlands director of water-planning activities, told a visitor to his office in The Hague: "We must have a continuous watch on the Rhine, on the physical and chemical data—and on the biological data, too. We must know if there is death."

Situated at the dirty end of the Rhine, the Netherlands was one of the main moving forces behind the creation of the international commission to prevent Rhine pollution. Its other members are Switzerland, France, West Germany

and little Luxembourg, which gets a cut rate on her dues. Although she is in the Rhine watershed, she does not adjoin the Rhine.

The head of the commission's secretariat, Jurrie Huizenga of the Netherlands, in an interview at his office in the West German city of Koblenz, said: "We are hopeful of improvement in the situation."

Set up in 1965, the commission collects fortnightly pollution statistics and carries out other studies. It holds periodic meetings of officials from its member governments and has helped to tighten up communications among them.

The procedures for passing the alarm in case of a pollution emergency have been worked out more carefully since the "great fish extinction."

Still No Solution

Yet all the commission's deliberations have not led to a solution of the salt problem, which worries the Dutch particularly.

Dutch officials say the French potash mines could simply let the waste salt pile up like mine wastes elsewhere in the world. The French have raised objections. Discussions on the subject have been going on, but there is no immediate prospect that the dumping will stop.

Also in the talking stage is the French proposal for the creation of a European anti-pollution fund. Putting the idea forward at a Council of Europe conference last month, the French agricultural minister, Jacques Duhamel, said the fund could be useful in financing large-scale measures in case of further pollution disasters.

Despite antipollution expenditures past and future, there is little hope that the Rhine will regain the purity of bygone centuries as long as Europe's industries continue to grow.

At this season, its waters seem particularly grey and greasy as they slide down a man-made estuary here and are carried out into the North Sea.

A public trash can stands on the estuary bank, but it is little used. As one hardy picnicker from The Hague remarked here the other day, "everything gets thrown into the Rhine."

March 22, 1970

Oil Called Peril to Food Supply in Sea

By WALTER SULLIVAN

Ever since the ships of Columbus first became entangled in its weeds, the Sargasso Sea has been known for its drifting plants, but today its coating of tar and oil is more extensive than its famous seaweed.

This is reported by a leading authority on the origin and fate of oils and other organic substances in the oceans. He says global spillages have reached the point where the entire world ocean is affected.

Petroleum constituents, some suspected of causing cancer, are entering the oceanic food chain, he says, and eventually could reach dining room tables. Thus, the world faces the possibility of losing the seas as a food source just as the population reaches the level where such protein becomes essential.

This warning has been issued by Dr. Max Blumer, a Swiss-born scientist formerly with Royal Dutch Shell who is now senior scientist in the chemistry department of the Woods Hole Oceanographic Institution in Massachusetts.

His views appear in his contribution to a book, "Oil on the Sea," published last month by Plenum, and in a recent issue of Oceanus, organ of the Woods Hole institution.

Similar experience

While no one else could be found yesterday who had made a similar study, oceanographers at the Lamont-Doherty Geological Observatory of Columbia University said they, too, had encountered lumps of oil and tar over much of the Atlantic.

"We were often amazed to find it in open ocean," said Dr. Allan W. H. Bé. They suspected at first that it might be coming from their own ship.

However, they now believe it to be the far-flung spread of spilled oil. As one Lamont-Doherty scientist pointed out, a beach on Barbados Island in the Antilles is fouled with tar and oil, although ships rarely pass. It is believed that the scum has been blown from afar by trade winds.

In a telephone interview yesterday, Dr. Blumer told how the chemical peculiarities of oil from a single source could be used to place the blame for oil damage. Petroleum, as it comes out of the ground, is an enormously complex mixture of substances, some, extremely poisonous, some probably long-term inducers of cancer, some

The New York Times   Jan. 16, 1970

OCEAN POLLUTION: Diagonal shading shows area in which the research vessel Chain of Woods Hole Oceanographic Institution found extensive oil and tar on the surface.

suitable as food for lower organisms.

However, since the mixture from any one source is unique, a relatively simple chemical analysis can show whence it came. Last Sept. 16, according to Woods Hole scientists, the barge Florida grounded off West Falmouth, Mass., spilling 65,000 gallons of No. 2 fuel oil.

Subsequently, oysters and scallops harvested along that shoreline were unfit for consumption because of their oil content, Dr. Blumer said. Furthermore, he added, it was possible, by chemical anaysis, to show that the damage had been done by No. 2 fuel oil of the same type as that spilled by the barge.

Whorl of Currents

The spread of oil over the Sargasso Sea is remarkable in that it lies in mid-Atlantic, far from heavily traveled shipping lanes. However, it is in the central whorl of the great circulating system of North At-

lantic ocean currents whose best-known element is the Gulf Stream.

Thus, as debris collects in the center of a whirling eddy, so, apparently, drifting algae, known as sargassum, have accumulated there. The oil and tar seem, likewise, to have come from afar.

The accumulation was observed by the Woods Hole research ship Chain during a sweep of 540 miles along the 67th meridian west of Greenwich. She systematically dragged a net designed to skim from the ocean surface its rich assortment of plant and animal life.

After two to four hours the net became so clogged with oil that it had to be cleaned with a strong solvent, Dr. Blumer reported. Oil-tar lumps more than two inches thick were brought aboard. In an area 500 miles south-southwest of Bermuda the oil was so thick that towing had to be suspended.

Dr. Blumer estimated that

three times more tarlike material was hauled aboard than sargassum.

Annual world oil production is about 1,800 million metric tons, he said, of which at least 60 per cent travels the seas. The total spillage, from leaks, accidents, flushing of bilges and tanks is, he estimates, about a million tons.

This, however, he believes to be a conservative estimate and it does not include production accidents, such as the recent leakage in Santa Barbara Channel, or industrial dumping into rivers and harbors that flow to the sea. The grand total, therefore "is likely to be 10 to 100 times higher," Dr. Blumer said.

While stringent preventive measures are the obvious remedy, where spills have occurred he favors burning or actual removal of the oil from the sea. The use of detergents, as in the great Torrey Canyon spill off England, breaks up the oil into tiny particles that are more easily taken up by tiny animals and thus enter the food chain.

Some of the hydrocarbons derived from oil are so stable that they can pass through a chain of animals without alteration. It is, in fact, possible to analyze the hydrocarbons found in a sea creature and determine whether or not they came from a spill.

The food chain of the oceans, from drifting plant to edible fish, involves far more steps than the simple one, on land, from field crop to edible cattle. But eventually, Dr. Blumer said, the hydrocarbons may reach human consumers, imparting an unpleasant taste.

"Far more serious," he added, "is the potential accumulation in human food of long term poisons derived from crude oil, for instance of cancer-causing compounds."

January 16, 1970

CURB ON OIL SPILLS IS SIGNED BY NIXON

New Law Raises Penalties and Expands Liability

By E. W. KENWORTHY
Special to The New York Times

WASHINGTON, April 3—President Nixon signed today the Water Quality Improvement Act of 1970, which sharply increases penalties for oil spills and extends the liability for the cost of cleaning them up.

The new law covers spills from onshore and offshore installations as well as from vessels.

The section dealing with oil spills had been the most controversial in the omnibus bill of amendments to the Federal Water Pollution Control Act. The conference committee that drafted the final version was deadlocked for weeks on differences between the Senate and the House on the question of liability for the cost of cleanup.

Perhaps the most obdurate opponent of the tougher Senate bill was Representative William C. Cramer of Florida, the ranking Republican on the

House Committee on Public Works. Mr. Cramer finally yielded, but only after an oil tanker went aground near St. Petersburg and spilled a large quantity of oil on the beaches of his district.

The principal author of the new law was Senator Edmund S. Muskie, Democrat of Maine, chairman of the Senate Public Works Subcommittee on Air and Water Pollution. Senator J. Caleb Boggs of Delaware, ranking Republican on the subcommittee, and Representative John A. Blatnik, Democrat of Minnesota, of the House Public Works Committee, cooperated with Mr. Muskie on the bill.

All four men were at the

signing this noon. Although the bill contained some provisions the President did not particularly favor, he was in good humor.

Turning to Mr. Muskie, who could be his opponent in the 1972 election, Mr. Nixon said with a grin, "It's your bill, isn't it?"

Mr. Cramer intervened, saying, "This is the Muskie-Blatnik-Boggs-Cramer bill."

Thereupon Mr. Nixon asked Mr. Cramer, "How did you get your name on it?"

Mr. Cramer replied, "I just snuck it on now."

Then, picking up the pen, Mr. Nixon said to all of them,

"If the environment doesn't improve, you're to blame."

Until now, any person found guilty of discharging oil from a vessel into navigable waters was subject to a fine of $2,500 for each offense, and the owners of the vessel were subject to a fine of $10,000 for each offense.

If the United States cleaned up the spill, the owners were liable for reimbursement of costs up to a limit set by the Brussels Convention of 1954. The limit was $5-million, or $67 a gross ton of the vessel, whichever was less.

But to assess these penalties or collect these costs, the Government had to prove gross negligence.

The new act introduces a new principle — absolute liability "without regard to whether any such act or omission was or was not negligent."

Except where a spill from a vessel was the result of an act of God, an act of war, negligence by the United States or an act by a third party, the owner of the vessel is liable for costs of cleanup incurred by the United States Government up to $14-million, or $100 a gross ton, whichever is less.

Subject to the same exceptions, owners of onshore and offshore facilities are liable for cleanup costs up to $8-million.

Where the United States can show that the owners of a vessel, onshore or offshore facility were guilty of willful negligence or misconduct, the owners are liable for repayment to the Government of the full cost of the cleanup.

Furthermore, any owner or operator of a vessel, onshore or offshore facility who knowingly discharges oil is liable to a fine of up to $10,000 for each offense, and is also subject to a fine of the same amount if he fails to notify the appropriate Federal agency that a spill has occurred.

It was on the issue of absolute liability that the Congressional conference committee bogged down. The House bill had provided that the owner must have "willfully or negligently" discharged the oil in order for the United States to collect cleanup costs, although it had said the burden of proving nonnegligence would be on the owner. The oil and shipping industries, maritime insurance companies and the Administration had favored the House version.

The new law authorizes a $35-million revolving fund to finance Government cleanup activities.

April 4, 1970

The Polluted Potomac: Sewage and Politics Create Acute Capital Problem

Mass of garbage and sewage floats on the Potomac in Washington while two young visitors to Capital play on shore

By GLADWIN HILL

Special to The New York Times

WASHINGTON—The heat of summer is enveloping the nation's capital, and with it has come the annual resurgence of a problem residents have come increasingly to dread: A stomach-turning miasma rising from the Potomac River.

The odor comes from sewage and its concomitants, chiefly algae, a scummy green plant that thrives in the filth and warmth, then rots, only to be succeeded by proliferating generations.

This year, the pollution of "the nation's river" has become more than an aesthetic affront. It poses pressing problems regarding not only sewage treatment but also housing, finance and Federal-state relations.

In Maryland and Virginia, flanking the capital, the authorities have had to ban home building for lack of sewage treatment facilities. Developers, caught in mid-construction, are muttering about bankruptcy.

Housing Shortage

Housing experts are talking about a severe shortage of dwellings for the capital's growing population, and the sewage situation is a major factor in this. State sanitation officials are at loggerheads with Federal officials to the point of preparing or considering lawsuits.

The situation on the Potomac has become the quintessence of water pollution problems that in various degrees plague communities from coast to coast.

Despite clean-up efforts, the river remains polluted because of "a hundred years of underestimates, bad decisions and outright mistakes," according to Eugene Jensen, regional director of the Federal Water Quality Administration.

This agency, an arm of the Department of the Interior, made the latest of many efforts at remedying the situation at an interstate hearing here May 22. The proceeding generated so much rancor about ways and means and blame that it had to be abruptly adjourned to let participants cool off.

Hearing to Resume

Resumption of the hearing is scheduled for later this month. Under Secretary of the Interior Fred J. Russell has admonished the parties—the District of Columbia, Maryland and Virginia—that they had better come up with some immediately applicable plans or face the possibility of prosecution as water polluters.

No one is optimistic about a quick solution. The clean-up effort has been dragging on for more than 30 years.

The Potomac originates 385 river miles northwest of Washington in the hills where Maryland and West Virginia abut Pennsylvania.

The upper reaches of the river have their problems, such as mine acid drainage, industrial discharges and silting. But the river is clean enough for swimming as it approaches Washington. It provides the capital's drinking water through an intake above Great Falls just north of the city.

A Tidal Estuary

But the lower 114 miles of the river, from the capital to Chesapeake Bay, are essentially tidal estuary, or an arm of the sea. There is an illusion that the river is flowing, but much of the time the water is just sloshing back and forth.

Under the bridges that link the capital with Virginia, the 1,500-foot-wide ribbon of water is a repellent, opaque gray-brown, so laden with silt, intestinal bacteria and other pollutants that an official of the water quality agency called it "a severe threat to the health of anyone coming in contact with it."

A report of the Water Quality Administration last year gave this description of the river:

"[Sewage] sludge deposits have blanketed fish spawning grounds and destroyed the bottom aquatic life on which fish feed. Along the margins of the estuary sludge deposits have released obnoxious odors when uncovered by ebb tide. Floating sludge masses, lifted by gases of decomposition, add to other debris on the water's surface."

Historically, the estuary in the capital area has been a pestilential sink. From 1881 to 1909, Washington had an average of 200 deaths a year from typhoid.

As recently as 1938, most of the area's sewage flowed untreated into the Potomac.

Then the Blue Plains sewage plant was built eight miles down the river's east bank from the capital.

For 20 years the Blue Plains plant gave sewage only primary treatment, which is basically allowing it to stand so that the lumps settle out. In 1958, this was stepped up to secondary, or a two-stage, treatment. After the primary stage, bacteria are added to "digest" pollutants. This process and oxidation remove a large part of the contaminants.

Today, the Blue Plains plant still handles about 80 per cent of the sewage from three million people in the 600-square-mile Washington metropolitan area, comprising the District of Columbia and parts of Maryland and Virginia.

Plant Overloaded

The Blue Plains plant is overloaded, handling about 250 million gallons a day against a designed capacity of 240 million. At this rate, it can give the sewage only cursory treatment, holding it for about two hours. The residue gushes into the Potomac dotted with four-inch-high clumps of undigested detergent foam.

To meet current standards, the plant should be removing upwards of 90 per cent of the principal pollutants—oxygen-absorbing matter and phosphates and nitrates, which stimulate algae growth. In reality, the plant is removing less than 75 per cent of the oxygen-absorbing matter, only 15 per cent of the nitrates and only 10 per cent of the phosphates.

This substandard treatment, in combination with the discharges of the other sewage plants, means that every day the Potomac is infused with wastes equivalent to about 80 million gallons of raw sewage.

To be clean enough for swimming, authorities say, water should not contain more than 1,000 coliform (intestinal bacteria) organisms per 100 milliliters. Coliform counts in the Potomac vary widely, and in the last couple of years systematic chlorination of sewage effluent has lowered the amount of live bacteria in the river by an uncertain degree. But last year there were counts as high as four million fecal coliform organisms per 100 milliters—4,000 times the safe level.

Pollutants Change

Since the Blue Plains plant instituted two-stage treatment, the regional load of oxygen-absorbing material on the river has been nearly halved. But between 1913 and 1969, while the Washington area's population was increasing eightfold, the discharge of nitrates into the water has increased ninefold and of phosphates twentyfold. This was caused primar-

WHITE HOUSE · CAPITOL · THE MALL · PENN. AVE. · LINCOLN MEM. · WASH. MON. · WASH. MON.

WASHINGTON, D.C.

PENTAGON · ANACOSTIA NAVAL AIR STA. · WASHINGTON NAT'L AIRPORT · BOLLING A.F.B.

VIRGINIA

SEWAGE PLANT

Blue Plains

MARYLAND

BELTWAY

The New York Times   July 12, 1970

ily by greater use of detergents.

The Potomac was one of the first targets selected after Congress in 1956 gave the United States Public Health Service the authority to proceed against interstate water pollution.

At a Federal interstate abatement hearing Aug. 22, 1957, representatives of the District of Columbia, Maryland and the Federal Government—Virginia refused to attend—agreed the Potomac's condition was deplorable and that reforms should be formulated.

At a follow-up hearing Feb. 13, 1958, with Virginia participating, agreement was reached on major goals. These included 80 per cent pollutant-removal efficiency for the Blue Plains plant that year and a major expansion of the plant to be completed in 1965.

There were supposed to be progress-report meetings every six months. But then an 11-year hiatus in Federal efforts to clean up the Potomac set in.

The river became a political shuttlecock, bandied about among conflicting aspirations of a half-dozen Federal and state agencies. The clean-up program was submerged in a sea of studies of the Potomac Basin.

Meanwhile, jurisdiction over water pollution was transferred from the Public Health Service, in the Department of Health, Education and Welfare, to the Department of the Interior.

In 1968, Secretary of the Interior Stewart L. Udall produced a 128-page report purporting to encompass all the previous proposals.

Just before leaving office in January, 1969, Mr. Udall called for the reconvening in April of the Potomac pollution abate-

ment proceeding that had been recessed 11 years before.

'A National Disgrace'

Opening the resumed hearing, the new Interior Secretary, Walter J. Hickel, said:

"We are here today, to put it quite simply, because the recommendations for collective action approved by the conferees in 1957 and 1958 have not been fulfilled. The lower Potomac River is grossly polluted. It is a shocking example of man's mistreatment of a natural resource—a national disgrace."

A survey at that time indicated that since 1957 only $165-million had been spent on sewage plants and some related piping in metropolitan Washington—an average of only $15-million a year in an area where municipal budgets aggregate more than $1-billion a year.

The water quality agency estimates that it will take about $500-million to clean up the Potomac.

The abatement proceeding went over the ground explored in 1957, and a construction program involving a dozen sewage treatment plants and extending into 1974 was agreed on. The broad goal was removal of as much as 96 per cent of the chief pollutants in sewage.

$29-Million Plant

Blue Plains now represents a total investment of $29-million. Most of this has been District of Columbia funds. The Federal Government has put in $5.4-million for experimental facilities on "advanced waste treatment" — three-stage processing that, for several times the cost of ordinary treatment, can produce effluent of virtually drinking-water quality.

Virginia pays $240 a million gallons for Blue Plains treatment of about four million gallons a day from a population of 80,000.

Maryland, a partner in Blue Plains from the beginning, contributed $5-million toward its construction. This, Federal officials calculate, gives Maryland a firm entitlement to the treatment of 67 million gallons a day. Maryland's actual load on the plant has been amounting to 114 million gallons a day.

Like a plug in a water pipe, the overload on the Blue Plains plant has had repercussions right back to the source.

For more than a year, Maryland and Virginia officials have had to impose constantly tightening limitations on new sewer connections in Montgomery and Prince Georges Counties in Maryland and Fairfax County in Virginia, where rolling rural countryside is being rapidly transformed into suburban subdivisions.

In recent weeks the problem has become acute. On May 20, because of "a menace and a nuisance to the public health, safety and comfort from discharges of raw and inadequately treated sewage," Maryland health authorities banned new sewer connections in sections representing 37 per cent of the Montgomery - Prince Georges area, where there has been $175-million worth of home building in the last two years.

A survey by the American University's real estate department indicated that in the next year, lost construction work, due to the sewerage restrictions will amount to $67.1-million in Montgomery County, $194-million in Prince Georges and $105.3-million in Fairfax.

Local officials are drawing much of the blame. Mrs. Hester McNulty, Potomac Basin committee chairman of the League of Women Voters, told a recent Federal hearing: "For too long sewage has been left to local officials who want to keep their budgets low and provide minimal treatment."

Into this situation the Federal Water Quality Administration threw a bombshell when the Potomac abatement hearings reconvened May 22, telling the outlying areas they could no longer count on the Blue Plains plant for relief.

Carl L. Klein, Assistant Secretary of the Interior for Water Quality and Research, served notice that the new Federal strategy for cleaning up the Potomac would be this: Blue Plains, instead of undergoing indefinite enlargement, would be held at its present capacity, and efforts would be concenterated on better processing. Virginia would be limited to its present input, and Maryland would be cut back from 114 million gallons a day to 67 million gallons.

The outlying areas, Mr. Klein continued, will have to develop their own treatment plants. He added that pending such construction they could resort to "package plants," compact portable units roughly the size of a freight car, which by sophisticated chemical and filtration processes produce an innocuous effluent with about 95 per cent of the principal pollutants removed.

State officials reacted apoplectically to the Federal plan. James Coulter, Maryland's normally mild-mannered deputy secretary of natural resources, who is a former water quality agency official, said the revised plan was "absolute drivel."

"It will take five years for Maryland to build an adequate treatment plant on the Anacostia River," he said, "and in the meanwhile this idea offers no relief from overloaded sewer lines.".

Maryland's Montgomery County prepared a lawsuit to prevent the Blue Plains plant from reducing its sewage intake.

June 12, 1970

U.S. SURVEY WARNS OF IMPURE WATER

Drinking Supply of Millions Described as Inadequate

Special to The New York Times

WASHINGTON, Aug. 17—A Government survey warned today that millions of Americans were drinking water of hazardous or inferior quality.

The study covered 18.2 million persons in Vermont and eight metropolitan areas across the country.

However it was found that the larger water systems, such as those in New York City and Cincinnati, were delivering mostly good water. The larger systems also showed adequate operation of treatment and distribution facilities, although improvements were urged by the investigators in many cases.

The survey found that 900,000 persons were receiving potentially dangerous water, some of it containing excessive arsenic, lead or fecal bacteria.

The United States Environmental Health Service defines a "hazardous" substance as one that may cause sickness or long-term illness. Levels of what are deemed excessive amounts in the water are established for each contaminant. For instance, one bacterial organism per 100 millimeters of water is considered hazardous.

In addition, 2 million persons were found to be drinking inferior water, which was described as safe but having a bad taste, odor or appearance. Most of the poor-quality water came from relatively small water systems serving communities of less than 100,000 persons, it was reported.

State health departments are not required by law to accept Federal definitions of inferior or hazardous water, but they "commonly follow" the Federal standards, according to Federal environmental officials.

The survey was conducted in 1969 by the Bureau of Water Hygiene of the Environmental Health Service, a division of the Department of Health, Education and Welfare. Results of the study were released today in the form of a statement from the director of the Water Hygiene Bureau, James H. McDermott.

"The survey was not expected to provide a perfect random sample of water supply systems throughout the country," Mr. McDermott said. "But the results are reasonably representative of the status of the water supply industry in the United States."

About 150 million people consume water from the same type of community water sources surveyed by the bureau, he explained, adding that the rest of the nation uses water from private sources, such as wells and springs. These sources were not covered in the survey.

In the nine regions covered, 969 public water systems were examined. Among the factors evaluated were the quality of water delivered to the consumer's tap, the adequacy of physical facilities and the effectiveness of the surveillance programs of health officials.

Widespread Defects

Health risks and other defects in drinking water were found in "both large cities and small towns irrespective of geographical location," according to Mr. McDermott.

Water systems serving communities of less than 100,000 persons, however, "evidenced a prevalence of the water quality deficiencies and health risk potential," he said. "Some of the very small communities were even drinking water on a day-to-day basis that exceeded one or more of the dangerous chemical limits, such as selenium, arsenic or lead."

A spokesman for the Water Hygiene Bureau said potentially hazardous levels of contamination were found in several areas of New York.

Excessive amounts of bacteria were found in the Tarrytown water supply, at the Mount Kisco Water Department, the Herod Point Estates at Wading River, and at the New York State Rehabilitation Hospital in West Haverstraw, he said.

Potentially hazardous levels of lead were found in water samples taken from faucets in some homes in Queens and in St. Dominic's Home in Blauvelt, as well as in the Croton-on-Hudson water supply. The spokesman said most of the lead had probably come from old pipes in the homes or in the water systems.

When investigators identified hazards in the water supply, they notified appropriate health officials, but the Water Bureau spokesman did not know if any of the conditions had been corrected.

The survey also found that 56 per cent of the 969 water systems showed physical deficiencies of some kind, such as poorly protected ground water sources.

Seventy-seven per cent of the plant operators were inadequately trained, the study said, and 79 per cent of the systems had not been inspected in the last full year prior to the survey.

Mr. McDermott concluded:

"There can be little doubt that this situation warrants major national concern. Most of our municipal water supply systems were constructed over 20 years ago. Since they were built, the populations that many of them serve have increased rapidly—thus placing a greater and greater strain on plant and distribution system capacity."

He recommended that water distribution methods be improved and used more intensively. He urged better training for persons responsible for operating water supply systems and also recommended regular analysis of water samples by state health authorities to detect contamination.

Specific sources of the contaminants described in the study were not named. Mr. McDermott said, however, that chemical pollution had generally increased over the last 25 years because of "the dramatic expansion in the use of chemical compounds for agricultural, industrial, institutional and domestic purposes."

The regions surveyed were Vermont and the metropolitan areas surrounding New York City, Charleston, W. Va.; Cincinnati, Charleston, S. C.; Kansas City, New Orleans, Pueblo, Colo., and San Bernardino, Calif.

August 18, 1970

Mercury Pollution in U.S. Found More Widespread Hazard to Health Than Expected

By GLADWIN HILL
Special to The New York Times

LOS ANGELES, Sept. 10—Mercury pollution, which suddenly emerged as a new environmental problem six months ago to worry officials, industry and the public, has proved a far more widespread hazard than was expected.

Evidence of abnormal amounts of mercury in water, fish and game birds has turned up in at least 33 states —almost twice as many as was officially foreseen even as recently as mid-July.

Reports from the 17 other states indicated no signs of mercury concentrations as yet. But the conclusion in some cases was admittedly uncertain and based on a lack of information rather than positive findings. Essentially, qualified officials have acknowledged, the problem is nationwide.

Mercury, traditionally considered an inert metallic element and casually released into the environment by industry and other users, was discovered last spring in potentially poisonous forms and quantities in the Great Lakes. There is a possible accumulation of millions of tons from past generations throughout most of the country.

The extent of this hazard is still undefined. But it is known that even minute quantities, under certain circumstances, can have dreadful physiological effects, and that, like some pesticides, mercury moves along the natural "food chain" from water and plants to fish, birds and humans in ever-increasing concentrations.

Federal and state officials have made impressive progress in reducing the current flow of mercury into the environment.

But their current efforts and the long-range eradication of the problem are hampered by unclear legal responsibility, outmoded laws and lack of information.

The reports from the 50 states also led to the following conclusions:

¶The only instance of mer-

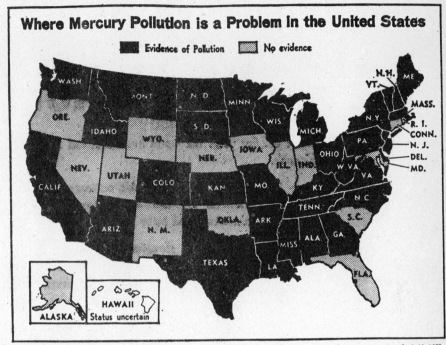

Where Mercury Pollution is a Problem in the United States

■ Evidence of Pollution   ▨ No evidence

The New York Times   Sept. 11, 1970

Thirty-three of the 50 states report waterways that have been hurt by mercury pollution

cury poisoning identified so far is still the case of the Huckleby family of Alamagordo, N. M., which ate pork from animals that had eaten grain accidentally treated with a mercury compound. Two of the family's children lost their sight and remained in comas for months.

¶While mercury-tainted fish and birds have been found in many states, the Federal Food and Drug Administration has found none moving in interstate commerce.

¶Reduction of the entry of more mercury into the environment from major industrial sources has proved remarkably quick and simple.

¶Control of mercury in agriculture has led—as with other pesticides—to an inconclusive administrative and legal tangle.

¶The mercury problem has cost some states and their citizens millions of dollars in lost revenues from commercial fishing, fishing licenses, tourism, retail fish sales and even restaurants.

The states where no evident mercury problems were found are Alaska, Connecticut, Florida, Illinois, Indiana, Iowa, Maryland, Nebraska, Nevada, New Mexico, Oklahoma, Oregon, Rhode Island, South Carolina, Utah and Wyoming. Hawaii also produced a negative report, but it was based on a

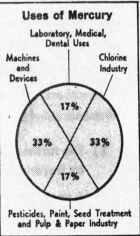

Uses of Mercury

Laboratory, Medical, Dental Uses

Machines and Devices

Chlorine Industry

17%   33%   33%   17%

Pesticides, Paint, Seed Treatment and Pulp & Paper Industry

The New York Times   Sept. 11, 1970

lack of information rather than affirmative findings.

In all the other states, manifestations of mercury have cropped up in water, fish or game birds. Concentrations of mercury in fish and birds is considered to be a threat even to "unaffected" states because of migration.

"All the states are somewhat involved," said Richard Ronk, director of the Food and Drug Administration's mercury investigation. "There are probably isolated mercury sources in every state. There are a lot of unanswered questions about this thing."

'National Scope'

Carl Klein, Assistant Secretary of the Interior, concurred. Commenting in Congressional testimony, he said that "mercury contamination of the aquatic habitat is of national scope."

Federal regulations set one-half part per million as the acceptable limit of mercury in food, and a half part per billion in drinking water. A half part per million has been compared in terms of proportion to a jigger of vermouth in a tank car of gin.

But these limits are admittedly arbitrary rather than realistic correlations with specific known effects on humans; the numbers were based on the smallest amount detectable some years ago.

Mercury concentrations of up to seven parts per million—14 times the allowance limit—have been found in fish and birds. There were no reports of excessive concentrations in drinking water, which in most places undergoes chemical treatment to neutralize a variety of common contaminants.

Sources of Mercury

The principal sources of mercury are the manufacture of chlorine-caustic soda, or "chlor-alkali," manufacturing plants, agriculture, lumber and paper-making operations and paints.

Water

But there are many secondary sources. Municipal sewage often shows appreciable amounts of mercury. It has been found in some samples of newsprint. And it was found recently in Tennessee moonshine liquor, apparently from old auto parts used in stills.

In some places the sources are a complete enigma. Fish in the northern reaches of Hudson's Bay were found to have excessive quantities in their systems. Lakes in Vermont, many miles from industrial plants, homes and automobile traffic, have shown mercury. Some experts think such occurrences may result from mercury vapor "fallout" from industry, incineration or even automobile fumes.

Mercury attacks the brain, causing neurological disorders such as blindness, paralysis and even insanity, and vital internal organs such as the liver. Swedish scientists reported recently in the American Medical Association's Archives of Environmental Health that mercury can also cause abortions and defective babies.

The human tolerance level and capacity for excreting mercury before this level is reached are still uncertain.

In the biggest use, the chloralkali industry, no mercury technically is "consumed." About 20 of the 69 chloralkali plants in the country use mercury in the process of converting brine into chlorine and caustic soda. The mercury does not end up in either product, but is washed away in residual brine—as much as 1,200,000 pounds a year, according to one Federal estimate.

Mercury originates in nature in an ore called cinnabar. Dr. David Klein, a chemistry professor at Hope College in Holland, Mich., who has studied mercury for several years, estimated for a Congressional inquiry last July that 5,000 tons of mercury move through the global environment every year by natural processes—erosion from ore, followed by drift to rivers and the sea—and that an equal amount now circulates annually from man's activities—about one-quarter of the latter in the United States.

The lethal effects of mercury compounds became known as far back as 1960, when reports were published of 111 persons being poisoned between 1953 and 1960 from eating fish at Minimata Bay in Japan, into which such compounds were discharged by a plastics factory.

Amount of Discharge

Since then there has been increasing evidence but neither the Government nor industry evidently inferred much from it. But on the basis of the Huckleby tragedy in New Mexico, the Department of Agriculture suspended the distribution of a major portion of mercury

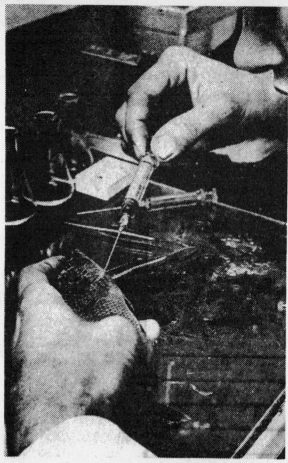

Allan Kain—Cincinnati Enquirer

MERCURY TEST: A blood sample is taken from a fish at the Taft Sanitary Engineering Center in Cincinnati. The impact of reports of danger from mercury in fish has been widespread, and devastating in some sportsfishing areas. However, Federal agents have found no fish in interstate commerce containing more than one-half part of mercury per million parts of fish, considered permissible rate.

seed-treatment compounds.

At Sarnia, Ont., the Dow Chemical Company was casually discharging 200 pounds of mercury a day into Lake St. Clair last spring when a 35-year-old Swedish chemist, Norvald Fimreite, doing graduate work at the University of Western Ontario, called the Canadian Government's attention to high mercury levels in commercial shipments of fish from the lake.

On March 31, the Canadian Government banned the sale of mercury-susceptible fish from the lake and the nearby Detroit and St. Clair Rivers. Michigan joined in the ban 10 days later.

Some quick calculations indicated that Canadian industry alone was discharging 250,000 pounds of mercury into the Great Lakes every year. The Federal Walter Quality Administration estimated that discharges on the American side may have been more than twice that, with 100,000 pounds going into Lake Erie alone.

The mercury problem found the Federal Government struc-

turally ill-suited to apply quick corrective measures.

Water pollution on an interstate basis is a responsibility of the Department of the Interior. The use of mercury in agriculture is a responsibility of the Department of Agriculture, which does not have a doctor on its staff. Food contamination on an interstate basis is a responsibility of the Food and Drug Administration in the Department of Health, Education and Welfare. Drinking-water standards are under another branch of that department, the Bureau of Water Hygiene.

The Federal agency that might logically be expected to be concerned with the over-all health threat of mercury pollution, the Public Health Service, lacked grounds for intervention because there have been no reported illnesses from mercury since the Huckleby case.

Congressional Hearings

The chairman of the Senate Commerce Subcommittee on Environment, Senator Philip A. Hart, Democrat of Michigan,

opened hearings on mercury pollution last May. But it was mid-July before Secretary of the Interior Walter J. Hickel, under Congressional pressures, began aggressive moves to curb the mercury problem.

He designated a task force of the Water Quality Administration and the United States Geological Survey to inventory mercury pollution sources, warned the Governors of 17 states of possible mercury problems and persuaded the Department of Justice to file suits against 10 chloralkali plants of eight chemical companies in seven states that allegedly were dilatory about abating discharges.

Many facilities acted speedily. The Dow plant at Sarnia cut its 200 pounds of mercury a day to only a few pounds in a few weeks. Other plants that were discharging 20 to 40 pounds a day reduced the emissions to a pound or less.

Temporary corrective measures were not complicated. At worst, a plant could divert fluid wastes from mercury processes into a holding pond until some sort of neutralization or recovery treatment for the mercury could be installed. Several chlorine plants serving adjacent paper and textile mills simply shut down.

Rapid Reduction

"I never had a pollutant where we got reductions this rapidly before," says Murray Stein, Federal water pollution enforcement chief. "The Erie Basin was 200 pounds a day. Now it's less than 10."

A spokesman for the Chlorine Institute estimates that the industry's mercury "losses" have been reduced by 95 per cent.

To get quick legal action, the Government had to invoke a seldom-used 1899 law making it an offense to discharge anything into a waterway without permission from the Army Corps of Engineers.

The lawsuits have drawn a lot of criticism from industry and state officials as an unnecessary and unfair "grandstand play." Many of the suits are under negotiation between the defendants and the Federal district attorneys in various states. If the companies clean up, the actions may be settled by a device such as the consent decree. Otherwise, they are subject to a $2,500 fine for each day's violation if they are found guilty.

The Federal Water Quality Administration has set as its goal the elimination of all mercury discharges. Establishments making "bona fide efforts" are being given three months to reduce discharges to a half pound a day and an additional three months to achieve a virtual zero level.

Systematic Check

Meanwhile, a systematic check is being made by the

160

Hickel task force of about 600 users of mercury throughout the country to pin down mercury "leakage" not accounted for by the chlorine industry. No additional Federal suits, according to informed sources, are imminent.

Internally, the F.D.A. has had its field force of 535 inspectors and 750 chemical analysts working on mercury problems, mainly helping state agencies in chemical analyses. The agency has found no fish or other interstate food products subject to seizure for containing more than one-half parts per million of mercury.

Governor Rockefeller of New York said Aug. 22 that reports by industrial concerns to the State Department of Environmental Conservation indicated that mercury discharges into New York's waterways had dropped from 71 pounds a day in mid-May to less than three pounds, and that the total would be down to less than 12 ounces by next summer.

In several states, the mercury problem has led to restrictions on fishing, resulting in millions of dollars in losses by the fishing industry, tourism and restaurants.

The closure of sections of the Tombigbee-Tensaw-Mobile and Tennessee River systems in Alabama to fishing has cost the state about $300,000 in commercial and sports fishing licenses, according to the Alabama Conservation Department. Fishing interests have filed damage suits totaling $200,000 against chemical companies allegedly responsible for the pollution.

In Michigan, the impact of restrictions on fishing in Lake St. Claire and the St. Claire and Detroit Rivers was described by some as a "disaster" to fish sales in the Detroit area and to the sports fishing business.

In Wisconsin, where fishing in the 430-mile length of the Wisconsin River was restricted, motels in some sections reportedly have had 80 and 90 per cent cancellations in reservations. There are estimates that the resort business has sustained losses of $1.5-million. Public hearings on mercury pollution control measures have brought out throngs of irate citizens, who described the mercury and DDT problems as an industrial outrage.

Scare in Kentucky

Although no official warning was issued in Kentucky, publicity about mercury pollution at Kentucky Lake, a popular fishing area, caused a widespread scare.

A fish market proprietor in Gilbertsville, Ky., said sales had dropped off $1,000 a week because of public apprehension. A nearby fish market that had been in operation 20 years closed down.

Pinpointing mercury pollution in many places has been arduous and slow. Although there is equipment now that can analyze parts per billion quickly, some states have such primitive facilities that it has taken as much as three weeks to process single samples of fish and water.

Minnesota alone has 10,000 sizable lakes. Montana officials are puzzled by mercury readings in the Clark Fork River near Missoula. In some places where chlorine plants cut off their discharge entirely, as much as three pounds of mercury has continued to flow out of waste channels every day from old accumulations.

Dr. Klein, the Michigan chemistry professor, said in an interview that a study he made of the ocean bottom at Los Angeles' main sewage discharge point had yielded mercury readings up to .8 parts per million. The original sources are a mystery.

Mercury, the only metal that is liquid at ordinary temperatures, evaporates readily. There is no fix as yet on how much mercury may be escaping directly into the air from industrial operations or coming out of smoke stacks and automobile exhausts and thence precipitated by rainfall.

There is a suspicion that unexplained mercury in northern New England may be residues of compounds painted on fallen timber after the 1938 hurricane to preserve it until it could be harvested.

Dr. Jesse Steinfeld, the United States Surgeon General, told the Congressional inquiry Aug. 27 that "the problem of health defects of toxic metals is a legitimate area of concern, but not a legitimate cause for hysteria."

"Toxic metals," he added, "must be placed in that growing collection of ubiquitous substances like pesticides about which we need to know much more, especially concerning the effects of low-level, long-term exposure."

"We've been so busy just trying to track the stuff down that we haven't had a chance to get into other investigations," one Federal official said.

Other qualified observers have a less sanguine outlook than Dr. Steinfeld has.

Dr. Henry A. Schroeder of the Dartmouth Medical School, an expert on toxic metal, said the mercury problem had opened up a whole area of problems of substances with hazardous physiological effects, including lead, cadmium, arsenic and some nickel compounds.

'Lagged Behind'

Robert E. Jordan 3d, a Defense Department counsel, got no argument when he remarked at a Congressional hearing: "Much of the country, including portions of the Congress and other agencies, have lagged behind the mercury problem."

Ralph Nader, the consumer advocate, told the same inquiry: "It's incumbent on us to ask why it took so long for industry and pollution authorities to recognize the danger."

The mercury contamination reveals how very weak is our system of food testing and monitoring."

While some quarters in the Administration have implied that the mercury investigation was a coordinated effort, inquiries to the various Federal agencies have adduced little evidence of effective liaison.

Federal Legislation

"Mercury, more than any other environmental problem presently known, exhausts the capability of government regulating agencies to respond to the problem," Dr. Albert Fritsch, a chemist and consultant to Mr. Nader, told Senator Hart's committee: "The tangle of jurisdictions means that control will be indefinitely delayed unless some prompt, extraordinary action at the Presidential level is forthcoming."

The idea of sweeping Federal legislation has occurred to many people, but shaping it is less obvious. There is at least one bill in Congress to provide severe penalties simply for any discharge of mercury. But it is given little chance of adoption.

Senator Hart has proposed appropriations to enable the Corps of Engineers to add 200 employes so that the 1899 water pollution law could be rigorously enforced, with specific permits for all discharges into waterways.

Senator Hart and Senator Warren C. Magnuson, Democrat of Washington, have introduced a commercial technology assessment bill to establish a surveillance agency to provide an "early warning system" of hazards such as mercury.

Some key officials and legislators forsee as the likeliest early answer to the "tangle of jurisdictions" the forthcoming establishment of an Environmental Protection Administration that might bridge existing gaps in those statutes and coordinate action.

September 11, 1970

A Study in Pollution Control: How Seattle Cleaned Up Its Water

By E. W. KENWORTHY
Special to The New York Times

SEATTLE—A few weeks ago, a Seattle newspaper carried a brief article under the headline: "Flow of Raw Sewage Into Elliot Bay Ends."

The gist of the item was that the last two sources of raw sewage on the bay, an arm of Puget Sound that forms much of downtown Seattle's waterfront, had been connected with a centralized metropolitan sewerage system.

The event, long anticipated here, attracted little attention among the nearly 400,000 users of the vast network of interceptor lines and treatment plants. But for people interested in pollution control, it was good news and it was important news.

For the forging of the last link in the sewerage system marks the culmination of a 10-year, $145-million program that is a case study in how a community willing to pay the price can clear up its water.

The ambitious program, carried out by an organization formally named the Municipality of Metropolitan Seattle but known localy as "Metro" for short, is regarded by environmentalists as exemplary in many ways.

Metro went to work long before saving the environment became a popular issue. It has succeeded in harnassing together the divergent interests and efforts of 18 separate metropolitan governmental units. And it has been paid for almost entirely by local bond issues.

The White House Council on Environmental Quality, in its first annual report, cited Seattle's Metro project as one of the two outstanding anti-pollution success stories in the nation. The other was the cleaning up of San Diego Bay.

Greater Seattle is 80 per cent surrounded by water, and its pollution problem was compounded by geography and political structure. On the west lies Puget Sound. To the east is the freshwater Lake Washington, 20 miles long and two to four miles wide, bordered by many incorporated towns.

161

Lake Washington, a freshwater lake to the east of Seattle, was made safe for bathers by an antipollution program

The outlet for Lake Washington is the Duwamish River, which empties into Elliott Bay.

This was the situation in 1958 when Metro was formed to deal with the sewage problem.

On the Seattle waterfront and the Duwamish, there were four primary treatment plants. (A primary plant settles out solids and removes about 35 per cent of the organic matter.) But there were also 46 "outfalls" dumping 70 million gallons of raw sewage a day into Puget Sound.

The deep sound, with its tidal flow, might have been able to handle this if the outfalls had been located off-shore in deep water with defuser attachments. But they were close to shore.

The result, according to Charles V. Gibbs, executive director of Metro, "was not only a terrible esthetic problem but a terrible bacteriological problem as well." The raw, stinking discharge floated around the docks and shoreline. Almost all beaches were closed as unsafe.

Around Lake Washington, where suburbia had proliferated since World War II, there were 10 secondary sewage plants discharging 20 million gallons of treated effluent into the lake each day from 23 cities and sewer districts. (A secondary plant removes 85 to 90 per cent of organic matter.)

The problem in Lake Washington was not bacteria; it was nutrients — phosphorous and nitrogen compounds in the effluent that stimulated the growth of algae.

When the algae died and decomposed, oxygen in the water was depleted. The result was a cloudy lake, foul-smelling and full of scum. The transparency of the lake—measured by the depth to which a white, eight-inch disc is visible below the surface—decreased from 12 feet in 1950 to about two and one-half feet in 1958. Many of the beaches on the lake were closed. Salmon suffered as the dissolved oxygen became exhausted in the deep water.

Dr. W. T. Edmonson, a zoologist at the University of Washington who began to study the pollution of Lake Washington back in 1952, described the irony of the situation in the late 'fifties. Noting that 10 treatment plants had been constructed since 1941 to deal with the "intolerable" contamination from raw sewage, he wrote that "the situation had been changed from one in which the lake was contaminated with organic wastes to one in which it was being fertilized with inorganic 'plant food.'"

It was the studies of Dr. Edmondson and others by Dr. Robert O. Sylvester, a professor of civil engineering at the University of Washington, that provided the spur for the creation of Metro. The principal civic force involved was the Municipal League of Portland. And the principal leader was James R. Ellis, who in 1953 was 32 years old, just three years out of law school and deputy prosecuting attorney for King's County, which includes Seattle and its environs.

In 1953, at the urging of Mr. Ellis, the Municipal League set up a Metropolitan Problems Committee, which spent two years digging into the problem of sewage pollution.

It was obvious to the committee that the problem could

The New York Times   Sept. 18, 1970

be solved only if a system was developed to serve the entire drainage basin. But such a system not only presented complicated design problems; it also raised jurisdictional problems on financing. Existing plants, some of them being financed by bond issues, would have to be abandoned. The suburban cities would have to be recompensed.

In 1956, the Municipal League recommended that the Mayor of Seattle appoint a citizens' committee to prepare a report and draft legislation to create a metropolitan authority to deal with areawide problems. The Mayor did so and named Mr. Ellis as chairman. The same year the state, county and city combined to hire Brown & Caldwell, a sanitary engineering firm, to prepare an area sewage plan.

The citizens' committee was ambitious. It wrote legislation permitting the formation of a "Metro" that would deal not only with sewage disposal but also with transportation and comprehensive land use planning.

The state legislature passed the bill in April, 1957. But the actual establishment of Metro required a majority vote in both Seattle and the county. In March, 1958 the voters of Seattle approved it, but it was defeated in the outlying county and thereby killed.

The citizens' committee thereupon redrafted the legislation to limit Metro's authority to sewage disposal, and city and county voters approved it in September, 1958.

Two Bond Issues Passed

As one of its first acts, the 21 members of Metro's decision-making body adopted the plan prepared by Brown & Caldwell. It provided for retention of one small existing primary treatment plant on Puget Sound, and the construction of four large new plants —three primary plants on the sound and one secondary plant on the Duwamish. A total of 110 miles of interceptor lines and 19 pumping plants would channel the sewage to the treatment plants. All 10 plants on Lake Washington would be abandoned, and no more effluent would go into the lake. All raw sewage discharges into the sound would be halted.

The original plan called for a bond issue of $125-million, but extension of the system to some suburbs not originally included has required a supplementary $20-million bond issue.

The bonds were to have been paid by additional sewer charges of $2.50 a month for each residential connection, with a proportionately higher charge for commercial and industrial connections. However, by careful planning, the residential fee was reduced to $2 a month before construction began in 1961 and has been kept at that figure.

Because it was begun long before passage of the Clean Waters Restoration Act of 1966, which provided Federal grants of up to 55 per cent

of the cost of a sewage project, the Metro system has received less than 6 per cent of its financing from the Federal Government.

The system is now in full operation. The beaches on the sound and on Lake Washington are now open. In Lake Washington, phosphorous, which was 70 parts per billion in

1963, has fallen to 29 parts per billion, and summertime transparency has increased to nine feet. There has been a 90 per cent reduction in the oxygen demand of the effluent released into the Duwamish River, and salmon can now migrate to their spawning grounds.

"Without achieving a miracle or a utopia," Mr. Ellis said, "Metro has nevertheless brought its civic activists some satisfying rewards. It has demonstrated the great potential of local initiative in the Federal-state-local framework."

September 18, 1970

Federal Officials Devise Plan to Set 'Heat Quotas' on Industrial Discharge Into Waterways

By GLADWIN HILL
Special to The New York Times

LOS ANGELES, Oct. 31—An unusual "heat quota" system has been devised by Federal officials as the most equitable solution to the growing problem of thermal pollution of waterways from power plants and other industrial facilities.

Under this plan—soon to be applied to Lake Michigan — a limit will be set on the total amount of heat that can be discharged into a particular area of water from all sources, including tributary streams and municipal sewage plants.

The amount of heat already going into a waterway from established discharge sources in a given area thus will determine what additional facilities can be put into that area without built-in water cooling systems. Once the "heat quota" is reached, any new heat-emitting installations would have to be built several miles away.

In the case of Lake Michigan; according to authoritative sources, the limit for any particular section of lake shore will be two billion British ther-

mal units per hour.

This, qualified officials said, will automatically preclude discharges from nuclear power plants—which use exceptional amounts of boiler-cooling water —and from large conventional power plants, fueled by coal, oil or gas. These plants would have to install cooling ponds, canals, or water-cooling towers.

Exemptions Provided

Federal experts estimate the cost of such amenities on Lake Michigan at from about 1 per cent to 10 per cent of present generating costs. This would amount to a monthly increase of 5 cents to 50 cents on the average $10 household electric bill.

Facilities emitting less than 500-million B.T.U.'s per hour would be exempt from restrictions in the present situation, while plants discharging between 500-million and two billion B.T.U.'s could continue under special requirements regarding dispersion of their hot water.

When a section of shoreline reached a thermal input total

of two billion B.T.U.'s per hour no additional discharge would be permitted closer than five miles.

There are now 24 power plants—one of them nuclear— discharging hot water into Lake Michigan. They account for about 75 per cent of the 40-billion B.T.U.'s per hour going into the lake.

Electric generating facilities are doubling every decade throughout the country, and more power plants—five of them nuclear—are scheduled to be in operation on Lake Michigan by 1974.

Without any restrictions, experts of the Federal Fish and Wildlife Service calculate, the heat input into Lake Michigan would increase 10-fold in the next 30 years, and ruin the lake's extensive fish and plant life.

The effects of the projected regulations on individual existing power plants and industrial facilities have not been detailed.

The "heat quota" plan, without any numerical specifics, was agreed to in principle by officials of Michigan, Illinois,

Wisconsin and Indiana at a closed conference with officials of the Federal Water Quality Administration and the Fish and Wildlife Service in Grand Rapids, Mich., Thursday.

Regulations Yet to Come

The conferees established a technical committee to work out details, and the foregoing heat limits are the ones the Federal representatives had prepared. The conferees are supposed to adopt regulations by next Feb. 15.

The "heat quota" plan in one sense is a compromise with a virtual ban on industrial thermal discharges propounded originally last May by Assistant Secretary of the Interior Carl L. Klein.

But from another standpoint the plan could be more rigorous, since it lumps a given plant's emissions in with other adjacent heat sources. Thus once the input quota along a segment of shoreline is reached no additional hot water dischargers, no matter how small, could set up in business.

November 1, 1970

SUFFOLK FORBIDS DETERGENTS' SALE

Action, Believed the Nation's First, Is Taken to Keep Drinking Water Pure

---

By CARTER B. HORSLEY
Special to The New York Times

HAUPPAUGE, L. I., Nov. 10 —The Suffolk County Legislature unanimously approved today a ban on the sale of virtually all detergents in the

county. The only exceptions are those used in such relatively minor products as dishwashing powder, shampoos and tooth paste.

The action, which was believed to be the first of its kind taken by any local government in the country, will take effect next March 1, following the expected approval of County Executive H. Lee Dennison.

The law bars the sale of most laundry detergent products, including many national brands, in Suffolk, a county of 1.1 million people.

The Suffolk County Board of Health and the Suffolk County Water Authority had warned

that continued widespread use of the detergents would seriously pollute the county's underground water supply and make it increasingly unpotable.

One spokesman maintained that there was no reason to "have water come out of the tap with a head on it." The agencies maintain that the water now is often foamy and frequently has a stench.

Mrs. Jane A. Stanley, the only woman in the 18-member legislature, said she had used soap products for some time and believed that other housewives would be "pleasantly surprised and will stay with them."

Mrs. Stanley predicted that neighboring Nassau County would soon follow suit in

barring the sale of specified detergents. The ban approved here affects only sales and does not prevent housewives from bringing detergents into the county from outside.

The Soap and Detergent Association estimated that Suffolk residents used 27 pounds of detergent for each person annually, as opposed to only five pounds of soap products, which will not be affected by the sales ban. The association estimated that detergent sales in the county totaled $6-million annually.

Industry spokesmen indicated that nearly 200 brand-name detergents would be affected by the ban. The county plans to inform storekeepers through the Health Department about specific brands affected.

Few Sewers in County

The ban is aimed primarily at compounds that do not break down naturally after use. Such substances, which are called nonbiodegradeable, involve alkyl-benzene sulphonates, alcohol sulphates and methylene blue active material.

The legislation does not concern itself with phosphates, which have been criticized as a source of pollution in surface water.

Almost all of Suffolk County is without sewers. A 10-year project to install sewers in the Towns of Babylon and Islip at an estimated cost of $275-million was approved in a referendum last year.

The legislation passed today was severely criticized by Charles G. Bueltman, the vice president of the Soap and Detergent Association, who termed the new law "irresponsible and bad technically."

Mr. Bueltman accused the Legislature of "attacking a pimple on a mountain" and maintained that the major problem was the lack of sewers, not the detergents.

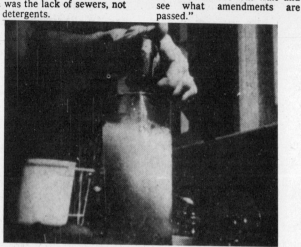

Tec Dannenberg

A HOUSEHOLD PROBLEM: As early as 1961, water from a private well in Suffolk County foamed with detergents.

He said the industry "will not take action against the law at this time but will wait and see what amendments are passed."

Monitoring Suggested

John V. N. Klein, the presiding officer of the Legislature, said that possible amendments might establish a "constant monitoring agency of the sale of detergents" and also prohibit nitrilo-triacetic compounds.

Legislator Louis T. Howard, whose district embraces most of Amityville, which is next to the Nassau County line, raised the question of homemakers' bootlegging detergent. Although he voted in favor of the ban, which would impose a maximum fine of $250 and imprisonment for up to 15 days for violators who sold the prohibited products, Mr. Howard declared that the ban would hurt two "innocent groups — the homemakers and the merchants."

He also said that the industry should be able to solve the problem of biodegradeability.

November 11, 1970

Polluters Have Members on Many State Antipollution Boards

By GLADWIN HILL

Most of the state boards primarily responsible for cleaning up the nation's air and water are markedly weighted with representatives of the principal sources of pollution.

This has been established in a nationwide investigation by The New York Times into the composition and operation of these boards and their role in environmental improvement.

The inquiry revealed that the membership of air and water pollution boards in 35 states is dotted with industrial, agricultural, municipal and county representatives whose own organizations or spheres of activity are in many cases in the forefront of pollution.

Top Companies Involved

The roster of big corporations with employes on such boards reads like an abbreviated blue book of American industry, particularly the most pollution-troubled segments of industry.

The state boards—statutory part-time citizen panels of gubernatorial appointees and state officials—are in most states the entities that set policies and standards for pollution abatement and that then oversee enforcement. They are the agencies that the Federal Government usually has to deal with.

The possibility that board members' personal connections could prejudice objective handling of pollution problems is deplored by Federal officials. They say privately that the composition of such boards is perhaps a major reason why abatement has not progressed faster.

These officials have no objection to spokesmen for special interests serving on boards that are purely advisory. In fact, most of them welcome it. But they point out that the state pollution boards have policing powers and they think that it is wrong for members to be responsible for policing their own areas of activity.

The widespread practice of putting individuals linked directly or indirectly with polluting on state pollution boards is defended by those involved on two grounds.

One is that such individuals bring to bear needed expertise and familiarity with pollution problems. The other is that such entities as industry, agriculture and local government, because of their civic importance, rate special consideration in the councils of government.

Brewery vs. Itself

Although there is no precise way to measure the impact of such boards on pollution problems because conditions vary so widely from state to state, there is abundant circumstantial evidence that they do not expedite pollution abatement.

One Colorado state hearing on stream pollution by a brewery was presided over by the pollution control director of the brewery. For years a board member dealing with pollution of Los Angeles Harbor has been an executive of an oil company that was a major harbor polluter. The Governor of Indiana recently had to dismiss a state pollution board member because both he and his company were indicted as water polluters.

Only seven states were found in The Times inquiry to have boards without members whose business or professional ties posed possible conflicts of interest.

Eight other states—among them New York and New Jersey—get along without such boards, dealing with pollution entirely through full-time state agencies.

Conservation organizations and citizen groups in many states are campaigning against what they call "stacked" boards, and a number of states are contemplating reforms. But

boards weighted with representatives from the pollution sector still dominate the national picture.

The controversial composition of most state pollution boards can be traced to their origins in state legislatures.

Many water pollution boards were created in the nineteen-fifties when water pollution first emerged as a nationwide problem and Congress passed laws giving Federal officials authority in interstate abatement. Air pollution boards were mainly formed in the last three years in response to analogous Federal legislation.

Familiar interest-group pressures in legislatures resulted in board seats in many cases being allocated by statute to such categories as industry, agriculture and municipalities.

Agriculture is a major source of pollution, from field burning and the drainage of animal wastes and farm chemicals. Counties and municipalities by the thousands have inadequate sewage facilities and noisome dumps and incinerators, and they are often as slow about remedying them as other polluters.

Some states even viewed pollution abatement as having a partisan aspect. The laws of Missouri, Utah and Ohio require that certain pollution board seats be split between Republicans and Democrats.

State Officials' Role

The presence of state officials on pollution boards does not always guarantee objectivity. Often they are from state Departments of Agriculture, Industrial Development or other agencies functionally allied with pollution sources.

Rarely, in the creation of the boards, were there any lobbyists for the general public. So it is unusual for more than one or two seats on a board to be earmarked for representatives of the public at large—if there are any at all—even though pollution is a problem distinctively affecting the entire public.

The arguments for composing boards largely of special-interest representatives are emphatically contradicted by the top Federal officials.

The Federal Water Quality Commissioner, David Dominick, said in an interview:

"Where a statutory board has responsibility as part of state government to establish standards for pollution abatement, the public is ill-served to have representatives of private vested interests passing judgment on such regulations.

"I think there's enough expertise in the public sector where no conflicts of interest would occur. The whole board should represent the public."

Opposes Present System

Dr. John Middleton, director of the National Air Pollution Control Administration, said:

"I think boards should represent disciplines that bear on air

pollution rather than economic interests. Industry can provide any helpful information on a nonmembership basis.

"The pattern of one or two seats on a board earmarked for representatives of the public doesn't make any sense. All the members of a board should represent the public."

Industry Aids Boards

In many instances, industry is demonstrably subsidizing in some degree the operation of state pollution boards. Typically, they meet monthly for a day or two. Members get only nominal compensation—$6.30 an hour in Ohio—or sometimes only travel expenses. That means they are serving on their employers' time, and even if they forgo their regular pay they are beholden to there employers for leaves of absence.

Critics of the "weighted" state boards do not contend that such boards are entirely unproductive. It is generally conceded that they have been the spearhead, however blunt, of much of the progress that has been made in pollution abatement. This applies particularly to air pollution, in which Federal regulatory steps to date have had little effect outside

the field of automobiles.

The critics' contention is simply that with disinterested boards, action would have been more decisive and progress faster. This applies particularly to water pollution, an area in which Federal authorities in the last two years have been impelled repeatedly to go around state machinery and bring actions directly against polluters.

The Times investigation, conducted over the last two months, disclosed no instances of corruption. Indeed, a Federal official commented: "As far as we know these are all upright people. Many undoubtedly strive to be objective. But if you were trying a case against the X.Y.Z. Paper Clip Company, would you want an official of the company on the jury?"

No two of these state panels are exactly alike in composition. They range from five to 15 persons. A typical pattern is a nine-member board composed of several state officials, one or more representatives of industry, a representative of agriculture, representatives of municipalities and counties, and perhaps representatives of "conservation" and of "the public."

Industry is the most ubiqui-

tous presence on such boards, with the steel industry, a big source of both air and water pollution, the most heavily represented.

The United States Steel Company, which has been cited as a polluter by Federal, state or local authorities in Ohio, Pennsylvania, Alabama and elsewhere, has executives on the air pollution boards of Alabama and Utah.

The company also had a man on Indiana's air board until a few months ago, when he was removed by Gov. Edgar Whitcomb because both he and his company had been indicted by a Federal grand jury for alleged violations of Illinois air pollution regulations.

Bethlehem Steel and National Steel have employes, respectively, on Indiana's air and water pollution boards. Bethlehem is also on the air board of Erie County, N.Y. (Buffalo, where it has been cited as an air polluter.

Other metal concerns also have a prominent part in pollution policymaking and enforcement among the states.

The Anaconda Company—recently sued in Montana for damages attributed to its fluoride emissions — has an executive on the air pollution board in Kentucky.

An Anaconda lawyer is on the water board in Utah. And the former head of an Anaconda subsidiary is chairman of Montana's Water Pollution Control Council.

A Reynolds Metals man is on the Alabama Water Commission. An Aluminum Company of America lawyer is an industry representative on North Carolina's Pollution Board and a staff doctor of the company is chairman of Iowa's Air Pollution Control Commission.

Role of Chemical Makers

The lead industry is well represented in Missouri pollution control. A former executive of the Eagle Pitcher Company (defendant in a recent Federal water pollution action in Kansas) is on the Missouri air board, and a National Lead executive is on the Missouri water board.

The next most active industry in providing expertise for state pollution boards is the chemical manufacturers—also a widespread pollution source.

Monsanto has men on the Arkansas Pollution Board and on the air boards in Tennessee and Idaho. Union Carbide which has temporized for more than a decade in controlling noxious fumes from its Alloy, W. Va., metallurgical plant, has an executive on the state air pollution board in Colorado.

The DuPont company, a recurrent water polluter in its headquarters state of Delaware, has abstained from pollution

---

COMPOSITION OF STATE POLLUTION BOARDS

Note: This table is not a classification of states' air and water pollution conditions

● Means state pollution board contains representatives of basic pollution sources (industry, agriculture, county and city governments).

○ Means state board is free of such representation

"No Boards" means air and water pollution regulation statewide is handled by a full-time state agency.

	Air Board	Water Board	Combination Air-Water Board		Air Board	Water Board	Combination Air-Water Board
Alabama	●	●		Montana	○	●	
Alaska	No Boards			Nebraska	●	●	
Arizona	No Boards			(3)Nevada	●		
Arkansas			●	New Hampshire	●	●	
(1)California	○	●		New Jersey	No Boards		
Colorado	●	●		New Mexico			○
Connecticut	●	●		(4)New York	No Boards		
Delaware			●	North Carolina	●	●	
Florida			○	North Dakota	●	●	
(2)Georgia		●		Ohio	●	●	
Hawaii			○	Oklahoma		●	
Idaho	●	●		Oregon		●	
Illinois	No Boards			Pennsylvania	●	●	
Indiana	●	●		Rhode Island	No Boards		
Iowa	●	●		South Carolina	●	●	
Kansas			○	South Dakota	●	●	
Kentucky	●	●		Tennessee	●	●	
Louisiana	●	●		Texas	●	●	
Maine			●	Utah	●	●	
Maryland	No Boards			Vermont	○	○	
Massachusetts			○	(5)Virginia	○	○	
Michigan	●	●		Washington	No Boards		
Minnesota			●	(6)West Virginia	●		
Mississippi	●	●		Wisconsin		●	
Missouri	●	●		(7)Wyoming		●	

(1) Pollution sources represented in regional branches of State Water Board. (2) Air pollution handled by State Board of Health.
(3) Water under State Board of Health. (4) State Environmental Board is advisory. (5) Interest conflicts banned by law.
(6) Water under State Division of Water Resources. (7) Water Pollution Control Council is advisory.

Dec. 7, 1970

board participation there—although another chemical company, Hercules Inc., is represented. But in Tennessee, a DuPont man is chairman of the air pollution control board and another DuPont man is on the state water board.

The Stauffer Chemical Company, whose fumes periodically tincture the air around Las Vegas, has a man on Nevada's Air Pollution Advisory Council, the source of panels that consider appeals from citations. The company also has an executive on Nebraska's Air Pollution Control Board.

And the Paper Makers

The paper industry, another big pollution source nationally, is also well represented in pollution control agencies.

A Scott Paper Company man is on Alabama's water commission. A West Virginia Pulp and Paper Company man is on Kentucky's water board. An International Paper Company executive is on the Alabama air board.

The Brown Paper Company long criticized by conservationists for pollution, is on New Hampshire's air board. The Weyerhaeuser Company is on North Carolina's pollution board. The Bowaters Southern Paper Company is on the Tennessee air board.

On Wisconsin's pollution control board are two lawyers whose clients have included the St. Regis Paper Company and Consolidated Papers, Inc.

On the Water Pollution Control Commission in Kentucky, where acid drainage from coal mines is a big water pollution

problem, is the president of the Kentucky Coal Operators Association.

These are only some of the bigger corporations from major pollution fields with representatives on the state boards. A complete list would run well over a hundred.

Impact of Influence

What are the effects of polluter influence within pollution boards?

Direct evidence that it is retarding pollution abatement is hard to find. No precise scales have yet been developed to gauge the degree of pollution in any particular state. Thus there is no way of comparing the relative progress of states with and without "weighted" boards.

There is no way of telling, moreover, how biased one board may be without prolonged observation and the effects of bias may be mingled with the effects of weak laws and regulations.

Evidence is available on all sides that from a national point of view efforts to reduce air and water pollution are making little headway under the prevailing system.

In four years, for example, air pollution has increased from an annual total of 142 million tons of contaminants to well over 200 million tons.

More than three years after the statutory deadline, as another example, only 18 states have adopted water quality standards satisfactory to the Federal Government.

What is more, Federal officials have information indicating that 32 states have extended

various abatement deadlines without the approval of the Secretary of the Interior—technically a violation of Federal law.

But, correlating general evidence of this kind with a particular "weighted" board in a given state will almost always be open to argument until more exact pollution measuring techniques are worked out.

'Drag Effect' Seen

What remains is circumstantial evidence. There is a wealth of it at hand and it clearly indicates the "drag effect" that such boards may have on attempts to get pollution programs moving. Here are some samples:

¶The Nebraska Water Pollution Control Council, notwithstanding widespread water pollution from cattle feed lots, went 14 years without issuing a citation to such offenders until last May, when Federal officials threatened to move against an aggravated case of pollution.

¶In Minnesota, where one statutory requirement on the composition of the nine-member State Pollution Control Agency is that it shall contain a farmer, air pollution regulations for cattle feed lots were proposed two years ago but have not been enacted yet.

¶Since May, 1967, the Connecticut Water Resources Commission has issued 863 orders regarding pollution abatement. Official records indicate that compliance has been obtained in less than half of these cases.

¶Wisconsin's Attorney Gen-

eral, Robert Warren, has publicly chided the state Natural Resource Board's enforcement arm for occupying itself with "small cheese factories and small fry polluters" rather than big offenders.

¶Ohio—where four of the five members on the Air Pollution Control Board have ties with the pollution sector (with industry represented by Procter & Gamble, a soap company with an acknowledged pollution record)—has the smokiest city in the country, Steubenville.

¶Colorado's Air Pollution Control Commission recently went along with industry suggestions that preliminary enforcement of clean-air standards not be started until 1973, although disinterested citizens have contended that the standards could be met by mid-1971. Full-scale enforcement is not scheduled until 1980.

¶On Alabama's 14-member Water Improvement Commission, all six "industry" seats are occupied by executives of companies now involved in pollution proceedings. Alabama recently was denied a $600,000 Federal pollution control assistance grant because its laws were adjudged so weak.

¶Louisiana's air and stream control commissions — composed of state officials and representatives of such groups as the Louisiana Manufacturers Association and the Louisiana Municipal Association — have never imposed a fine on anyone, and the air commission has brought only one corporate polluter into court in five years.

¶On Pennsylvania's 11-mem-

The New York Times (by Gary Settle)

Children playing in shadow of steel mill near Pittsburgh. Industry is most ubiquitous presence on many state boards primarily responsible for antipollution efforts, and steel industry, a major polluter, is the most heavily represented.

ber Air Pollution Control Board the lone "public" member is a former vice president of a steel company. Another steel executive left the board only recently. An executive of a third steel company is on the state's water board. Scranton, Johnstown and Pittsburgh are among the top 10 on the Federal list of smokey cities.

A confidential vignette of one board's activities was provided by a recent official in a Midwestern state where pollution problems are conspicious.

"The chief problem," he said, "was a general atmosphere of timidity [on the board] due to a hostile, lobby-ridden legislature and an apathetic Governor.

"We had money troubles constantly, so we didn't get much done. Some members would knuckle under if industry seemed to be getting to the

Governor. The Governor had some ties with the power industry, which restrained us from adopting tough emission restrictions."

Alternatives Tried

Virginia has been so conscious of the possible conflicts of interest that it has adopted a law to eliminate it: Even members of the Legislature are ineligible to be on pollution boards.

The few states that have panels composed of engineers, professors, pharmacists, housewives and other disinterested citizens and that obtain expertise from outside sources give every evidence of getting along just as well as boards with members from the pollution sector.

This is also true of the states that have no citizen pollution

boards. Several of these, such as New Jersey and Illinois, established professional environmental control agencies in the last year or two to supercede polluter-connected boards.

The Indiana Legislature next year will consider a proposal to supercede its present industry-oriented pollution boards with a full-time state agency like that in Illinois, which pays its five professional pollution control board members from $30,000 to $35,000 a year. Iowa and North Carolina are among other states considering structural revisions.

In Ohio, a "Breathers Lobby" of health, labor, church and conservation organizations has been pushing bills that would orient the state air pollution board more toward public interests by including an ecologist and an engineer among its

members.

Short of statutory changes, one remedial strategy is the "end run" around slow-moving boards. Pennsylvania's Department of Justice three months ago established an "Environmental Pollution Strike Force" of six young lawyers, who have filed 17 actions against polluters and already won nine.

Finally there is the power of citizen pressure. State pollution boards generally are required by law to hold public sessions, and citizens in some states are finding that a sedulous gallery of observers may change the tenor of boards' deliberations.

At a recent stormy meeting of Alabama's water commission, an irate woman conservationist hauled off and slapped a member of the board.

December 7, 1970

PRESIDENT ORDERS CURBS ON DUMPING IN U.S. WATERWAYS

Approval of Environmental Unit and Army Engineers Required of Industries

1899 LAW IS EMPLOYED

Discharge Must Meet State and Federal Guidelines— Congressmen Hail Move

By ROBERT B. SEMPLE Jr.
Special to The New York Times

WASHINGTON, Dec. 23 — President Nixon ordered today new procedures that will require companies to obtain Federal permits to discharge wastes into virtually any of the nation's waterways.

The Administration regards the action as a major move against industrial pollution.

The permits, which will be issued by the Army Corps of Engineers, cannot be obtained unless the industries receive certification from state and interstate agencies that the discharges meet existing water quality standards.

Agency Has Veto

In addition, permits will not be granted without the approval

United Press International

DISCUSS NEW ANTIPOLLUTION MOVE: Russell E. Train, at left, chairman of the Council on Environmental Quality, and William D. Ruckelshaus, head of the Environmental Protection Agency, briefing newsmen on a Presidential order that will require industrial concerns to obtain a Federal permit to discharge wastes into U.S. waterways.

of the new Environmental Protection Agency, which in effect will exercise final veto power.

Federal officials said today that the Environmental Protection Agency would make its judgment not only on whether or not an industry had received proper certification from state water quality agencies, but also on whether or not it conformed to new effluent guidelines,

which the Federal agency plans to issue shortly.

In effect, the new procedures would introduce Federal effluent standards long opposed by industry, which prefers to be measured against state standards.

'Most Important Step'

Russell E. Train, chairman of the Council on Environmental Quality, called the move "the

single most important step to improvement of water quality that this country has taken."

In a statement that accompanied the Executive order setting forth the new procedures, Mr. Nixon declared:

"The new program will enhance the ability of the Federal Government to enforce water quality standards and provide a major strengthening of our ef-

Water

forts to clean up the nation's water."

The President's action was based on the Refuse Act of 1899, which prohibits discharge into navigable waters without a permit from the Corps of Engineers. The law has been generally ignored, largely because the Corps of Engineers construed its responsibilities under the act to deal only with threats to navigation.

In March, however, a House subcommittee on conservation headed by Representative Henry S. Reuss, Democrat of Wisconsin, began publicizing the 1899 law. The panel urged the Administration to use the law to enforce water quality standards, instead of relying on the cumbersome administrative procedures available under more recent statutes.

The new order is expected to affect about 40,000 industrial facilities, which must submit applications for permits no later than July 1, 1971, and about 1,000 new plants built each year.

$2,500 a Day Penalties

The 1899 law provides for penalties of $2,500 a day against violators. The Justice Department could also seek an injunction against the company, which in effect would mean closing the plant. This step would probably be taken only as a last resort.

Initial Congressional reaction to the President's move was favorable.

Representative Michael Harrington, Democrat of Massachusetts, said the refuse law "has finally come of age as a weapon against the pollution of our waterways." Last April, Mr.

Harrington filed a list of 150 industries with the Attorney General of Massachusetts, asking for court action against them under the 1899 law.

A member of Mr. Reuss's staff said the Representative was pleased by the President's move but would withhold final judgment until more detailed guidelines spelling out the new procedures are published in The Federal Register in the next day or so.

Crucial Question

One crucial question in the minds of some environmentalists is whether the Environmental Protection Agency intends to check meticulously the judgments of state authorities or merely to spot-check applications for permits.

Under questioning by newsmen today, William D. Ruckelshaus, director of the new anti-

pollution agency, conceded that one problem would be monitoring the performance of companies after permits were granted.

He said permit holders would be required to provide periodic reports on the nature and amount of discharges and to remain in compliance with applicable water quality standards.

In his statement, Mr. Nixon said the new order would not mean any lessening of his efforts to persuade Congress to improve the existing Water Quality Control Act. In a message last February, on which Congress has taken no action, he asked for new water quality standards, stiffer penalties and the extension of Federal authority to intrastate as well as interstate waterways.

December 24, 1970

Flush Toilet a Growing Concern for Environmentalists

By DAVID BIRD

The flush toilet has been stirring increasing concern because of the damage it has done and is continuing to do to the environment.

"There are two crimes against humanity which at their inception seemed like real boons," Dr. Donaldson Koons, chairman of Maine's Environmental Improvement Commission, said in an interview. "They are the internal combustion engine and the flush toilet. Even with the best of treatment, they create serious problems for our lakes and waterways."

The internal combustion engine, in automobiles and motor boats, has been singled out for some time by ecologists as a key source of pollution in the environment.

But the dispute over the flush toilet has been relatively quiet. Now, however, there are suggestions that, in some cases, the outhouse may be better.

What concerns the environmentalists is that flush toilets use vast amounts of water, which is then released, contaminated, to pollute once-pure streams, lakes and even the ocean.

Outhouses for Camps

Dr. Koons says that "for summer camps, outhouses are the best disposal you can find."

What is feasible for summer camps, however, is not feasible for the cities, where there is not enough ground area to absorb all the waste. And while the environmentalist agree that it would be better for man to return his waste organic material to the soil to complete the ecological cycle, they differ on how best to achieve that aim.

Martin Lang, New York City's Commissioner of Water Resources,

agrees on the wastefulness of the flush toilet, but he does not see anything now to replace it.

He says that if the city changed, for example, to recirculating toilets, such as those used on planes, there would be a huge problem of solid waste collection because the units would eventually have to be emptied from each apartment building and home.

Mr. Lang and others believe that the most promising solution would be to return the treated effluent of sewage plants to the land outside the cities, such as areas worn out by strip mining or overcultivation.

In rural areas, the changing attitude toward outhouses is already opening the way for a revival of the privy.

Skylark, Inc., a recreational home community in Maine that is a subsidiary of the Scott Paper Company of Philadelphia, has already changed its "General Instructions to Lessees." The new instructions no longer prohibit the use of privies.

Moosehead Resort Corporation, which manages the Squaw Mountain Resort near Greenville, Maine, is investigating the use of the privy at campsites to be developed near that resort.

"The thing about the privy is that it doesn't work when there are a lot of people, but nevertheless man still belongs in the soil cycle," Dr. Barry Commoner, the ecologist who is director of the Center for the Biology of Natural Systems at Washington University in St. Louis, said in an interview.

Must Find a Way

Dr. Commoner said that with the flush toilet man had been dumping waste that was putting too much stress on the water system while at the same

time draining nutrients from the natural cycle of the land.

Nutrients added to water can lead to such dangers as the unchecked growth of algae, which chokes off all other life and leads to such situations as the dying of Lake Erie.

Although the outhouse may not be suitable for the city, Dr. Commoner said, technology will have to find a way to return waste to the soil.

When you break a key link in the ecological cycle by adopting a technological innovation like the flush toilet, Dr. Commoner explained, another technological innovation must be devised to complete the link and avoid the danger of ecological collapse.

So far only part of that technology has been devised. Sewage treatment plants now, in varying degrees make the waste less objectionable, but in almost every case some of the effluent from these plants acts as an excess nutrient in the water and some of it turns out as a smothering sludge. That is the case in New York City, where a vast "dead sea" of sludge has been created off the mouth of the harbor.

"It is insane to be taking all this organic matter off the land and dumping it into the water," Dr. Commoner said.

'Have Lagged Behind'

Dr. Koons said that "we have lagged dreadfully behind in devising systems for separating contaminated water from relatively clean water."

He said that because of the flush toilet we have to design water systems to provide 100 gallons of water per person per day, which he feels is extravagant in days of dwindling water supplies.

If the flush toilet were not

in general use, one person would, on the average, use only about 10 gallons of water a day.

Even those who are charged with building the billions of dollars worth of sewage treatment plants now being undertaken in a massive effort to prevent further degradation of the nation's waters agree that a second look should have been taken before the flush toilet was generally adopted.

Kenneth Walker, the regional director here of the Federal water pollution control office, said that "if we had to do it all over again we might do it differently, but we're kind of stuck with the system now."

He said the flush toilet system was inefficient because "it uses a whole lot of water to carry away a little bit of waste." He explained that sewage was actually about 99.9 per cent pure water and that the billions being invested in sewage plants were being spent just to remove that small fraction that makes it polluted.

New York City is now in the midst of a $1.3-billion program to upgrade its sewage plants and build new ones.

And although using the sludge from sewage plants to revitalize worn-out land outside the cities seems a good idea, landowners have resisted taking on sewage sludge.

Some of that resistance is being broken down with experiments in places such as southern Illinois, where sludge from Chicago's sewer system, shipped down by rail, is being put on farmlands.

Federal legislation requiring more complete rehabilitation of strip mines is also expected to help the process of returning the sewage waste to the land.

July 26, 1971

168

Detergents: Series of Health Controversies

By BOYCE RENSBERGER

Almost from the day they came on the market described as "washday miracles," household detergents have been embroiled in one environmental health controversy after another.

Over the years, the manufacturers, vying for a market in the billions of dollars, have tried to outdo each other with a succession of "new," "improved," and "amazing" ingredients.

Yet, in the eyes of scientists, government officials, and environmentalists, nearly every major change created a new problem or left standing an old one. In many ways, the problems resulted unavoidably from the trade-offs and compromises necessary to satisfy each of a variety of interests.

Although detergents first appeared on the market in the nineteen-thirties, they were used so little that their first problems did not appear until they had taken over a substantial share of the market in the fifties.

Soap Flakes Were Used

Until then, housewives relied on soap flakes and powders. Through the forties, however, the greater cleaning ability of detergents increased their popularity, especially in parts of the country where the water had a high content of dissolved minerals.

Because the minerals in "hard" water caused a soap-and-dirt scum, like the ring around the bathtub, to form and turn clothes gray, many housewives used washing soda in the machine to deactivate the minerals and make the water "soft."

Detergents were popular in these areas because they already contained an ingredient, phosphate, to soften the water.

In the early fifties, however, as detergents largely displaced soap, the first environmental problems appeared. Waterways in scattered parts of the country began to billow with suds. Sewage treatment workers were pushed out of their plants by heaps of foam from sewage water. Suburbanites found tapwater had a head on it.

Did Not Break Down

It soon became apparent that the detergent ingredient that made suds in the washing machine did not break down in the sewage plants. It continued to make suds everywhere it went.

Six years after the detergent industry recognized the problem, it found a way to correct it. It was another nine years, however, before the industry brought out products that would not foam up the nation's rivers.

The final action did not come until after Wisconsin and Dade County (Miami) banned detergents and Federal legislation seemed imminent. Then, in 1965, the entire industry converted to biodegradable detergents in which the sudsing ingredient could be broken down by decay bacteria.

Even as the industry switched one ingredient, however, problems with another— the only other basic ingredient — were already apparent. Phosphates were reported to be one cause of water pollution, especially in Lake Erie.

Activists in the environment movement, just gaining steam in the mid-sixties, complained that the detergent makers were doing only half the job by replacing one offending ingredient without replacing the other.

How the Ingredients Work

Understanding the detergent controversy, even as it continues today, depends on understanding how the ingredients of soap and detergent work.

Soap essentially contains one ingredient that serves the same function as one of a detergent's two basic ingredients. Chemists call that ingredient a surfactant, or surface active agent.

In soap, this chemical is sodium stearate, derived from animal or vegetable fat. In the first detergents it was ABS (alkyl benzene sulfonate) which was replaced in the biodegradable detergents by LAS (linear alkylate sulfonate).

In soap and detergent, the job of the surfactant is the same. It helps water soak into clothing fibers by reducing the surface tension that normally keeps water beaded up on the surface.

Although plain water may appear to soak into fabric, it may not always penetrate the individual strands of thread in which dirt is lodged. By making water, in effect, wetter, the surfactant helps to loosen dirt particles and flush them out.

Dirt Particles Drift

Once the dirt particles are loose, they drift in the wash water, surrounded by surfactant molecules that hang on like kernels on a corn cob. In this state the dirt cannot re-attach itself to the clothing.

However, if the water contains dissolved minerals such as calcium and magnesium, these minerals will combine with the surfactant-coated dirt particles to form a gray scum or curd that spin-dry washing machines can push back into the fabric.

It was to avoid these problems that detergent makers first put phosphates in their products. The phosphates, which the industry calls builders, chemically react with the minerals to produce a combined chemical that does not form scum. Thus, the surfactant-coated dirt particles remain separated in the water and are able to drain away with the wash water.

Before detergents, housewives in hard water areas used washing soda to achieve the same result. To make detergents effective in these same areas, many manufacturers sold detergents that were as much as 60 and 70 per cent phosphate.

Through the late sixties, however, environmentalists, both scientist and citizen, increased the pressures to remove phosphates because, as the Federal Water Quality Administration estimated, detergents contributed 65 per cent of the phosphate in the nation's waste water—phosphate that fertilized lake-choking quantities of algae.

During 1969 and 1970, local and state governments across the country passed laws banning or limiting the phosphate levels permitted in detergents. Again, as in the case of the foaming surfactant, public pressure forced the industry to change.

Procter & Gamble, the nation's largest detergent maker, had already been testing a phosphate replacement, NTA (nitrilotriacetic acid) developed in Sweden. In 1966 small quantities were test marketed with enough success to warrant a decision gradually to phase NTA in and phosphate out.

By 1968, some 10 per cent of Procter & Gamble's detergents had been converted from all-phosphate to one-fourth NTA and three-fourths phosphate. The company announced plans to eliminate all phosphates and convert everything to NTA over a period of years.

Health Hazards Cited

Even as the industry was building multimillion dollar plants to produce NTA in the quantities needed for the huge detergent market, medical researchers were turning in reports of possible health hazards. NTA, they said, breaks down into substances called nitrosamines and these have been shown to cause cancer. The enormous quantities that would be drained into the nation's waterways and, hence, back into drinking water supplies, could pose a serious hazard, they said.

Dr. Samuel S. Epstein, a toxicologist who has done research on NTA, told a Senate hearing, "We may well be jumping from the ecological frying pan into the toxicological fire."

Meetings were quickly arranged with detergent industry executives, some of whom insisted that NTA was safe, but all of whom agreed to abandon the substance.

Many manufacturers who had already lowered or eliminated phosphate, switched to an old reliable — washing soda, also called sodium carbonate, and increased the levels of other minor detergent "builders" that had been used, sodium silicate and sodium metasilicate.

Alarm Over Toxicity

Even this move, including the return to a substance that had been the housewife's choice in pre-detergent days, failed to correct all the problems. Within months of the switch from NTA, health experts became concerned over the toxicity of the new builders when used in such quantities.

Depending on the relative mixtures of the three sodium compounds and the use of other chemicals in the varying detergent formulations, some of the new detergents could be dangerously caustic when accidentally swallowed or gotten into the eyes.

Dr. E. William Ligon, of the Food and Drug Administration's division of hazardous substances and poison control, tested the new phosphate-free detergents on animals. "Some of them," he said, "did so much damage to animals that it was clear they would do more to people."

Few Poisonings Noted

Although few poisonings had occurred, the F.D.A. and other Federal agencies moved quickly to advise the public that phosphate detergents, though more ecologically harmful, posed less of a direct health hazard.

For the moment, it appears there is virtually no detergent on the market that can be considered safe in all respects. In the near future, however, several alternatives are likely to be tried.

Since June of 1970 the Gillette Company Research Institute has been working with a $344,000 Federal grant to find phosphate-free heavy duty detergents.

Dr. Anthony Schwartz, head of the project, said he had found several promising alternatives that met the demands of health officials and environmentalists but were more expensive and not quite as effective in cleaning.

His findings are expected to be made available to the entire detergent industry later this year.

September 18, 1971

Nations of Mediterranean Seeking to Clean Their Sea

Special to The New York Times

ROME, Oct. 22—The Mediterranean has been widely diagnosed as a dangerously sick sea; now the experts are seeking the therapy.

As a result of international conferences, the formation of numerous committees, and voluminous reports published on the deteriorating health of their sea, most Mediterranean nations have become aware of the need for cleansing their marine environments.

In the process they have also become aware that unless they act together, they are apt to drown separately in a sea of pollution.

The principal causes of the Mediterranean's poor health are its limited water circulation, slow rate of oxygen replenishment and lack of sufficient nutrients for marine life. These factors have been compounded by oil, chemical and sewage pollution that has already taken a substantial toll of public health, tourism and fisheries in the 18 Mediterranean countries.

An Ecological Conscience

Now the Mediterranean states, shaken by evidence that the cradle of civilization is turning into a graveyard, have begun to develop an ecological conscience.

The attention of most is now focused on the United Nations Human Environment Conference, set for June, 1972, in Stockholm. The countries hope the Conference will achieve one of its principal goals: viable solutions to international pollution problems.

The Mediterranean has always had a precarious ecological balance, according to a comprehensive report presented at a conference in Malta in July by Lord Ritchie-Calder, an associate fellow of the Center for the Study of Democratic Institutions at Santa Barbara, Calif. The center sponsored the conference, called Pacem in Maribus.

The situation is believed to have reached its present critical phase because of increased industrialization, the rise in population and the absence of adequate controls.

A group of 10 Mediterranean countries is currently working on a draft convention to be presented at Stockholm for a Regional Agreement on the Control of Marine Pollution in the Mediterranean. The 21 states bordering on the Mediterranean and the Black Seas have been invited to take part in this collective action.

At the Pacem in Maribus conference, more than 180 diplomats and scientists from 30 countries agreed on the urgency of multilateral legislation.

One obstacle to such action in the past has been the reluctance of governments to publish the details of their pollution problems for fear of hurting their tourist industries. Some scientists, according to authoritative sources, are afraid to publish certain information because they might lose their jobs.

The only real opening of the 970,000-square-mile Mediterranean is the Strait of Gibraltar. Surface water from the Atlantic flows in, bringing oxygen; underneath it is an outgoing current of dense, saltier, colder water.

Such a narrow opening means that it takes about 80 years for a complete turnover of Mediterranean water, according to estimates by Arthur R. Miller of the Woods Hole Oceanographic Institute in Massachusetts.

The ecological balance of the marine environment, particularly in a semi-enclosed system like the Mediterranean, can be destroyed in several ways. Some pollutants poison and kill plants and animal life directly. Others, such as oil, make so great a demand on the oxygen supply that marine life suffocates. Some pollutants, such as fertilizers and detergents, produce an excessive growth of certain plants or animals at the expense of others.

A fourth type of pollutant, DDT and other chlorinated hydrocarbons, accumulates in some types of marine life that have an affinity for this type of substance, and it eventually passes into the food chain with dangerous or lethal effects.

The Two Worst Areas

The two most polluted regions of the Mediterranean are the Provençal Basin off the French Riviera, and the Venice-Trieste area in the Adriatic Sea. The Rhône River flows through Marseilles into the basin and the Po flows into the Adriatic just south of Venice.

The two rivers had often been called the lungs of the Mediterranean because they and the Strait of Gibraltar were the sea's main sources of oxygen regeneration. The rivers have now become, in effect, sewers of debris, waste and pollutants and they contain relatively little oxygen.

There has been more detailed research on the extent of pollution in the areas fed by the rivers than in any other part of the Mediterranean. Yugoslav experts from the Center for Marine Research Institute at Zagreb, led by Joze Stirn of the Biology Department of the University of Ljubljana, report that pollution processes in the Venice-Trieste area have reached "very serious levels, probably the most disturbing in the whole Mediterranean."

The region of Marseilles was described as similar to that of Trieste. A large center of ports, industry and population pours waste, most of it untreated, into the sea. The bacterial load in the water was said to have gone "beyond safety limits" and marine sediments are impregnated with intestinal germs 100 yards from the shore. The Bay of Marseilles is contaminated by detergents, pesticides and an over-coating of oil. Some marine animal and plant species have decreased; others have grown abnormally.

15 Countries Cooperating

The first important step toward a comprehensive survey of the state of pollution in the Mediterranean was taken at the Conference on

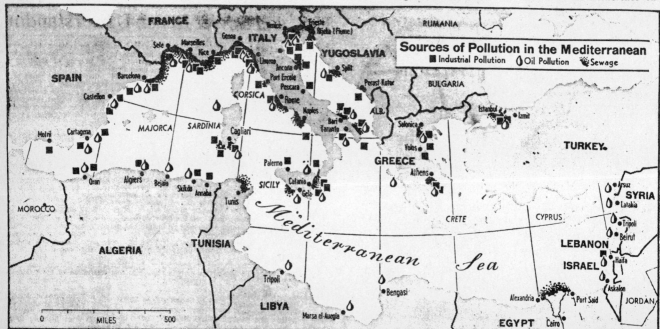

Sources of Pollution in the Mediterranean
■ Industrial Pollution ◖ Oil Pollution ≋ Sewage

The New York Times/Oct. 23, 1971

Pollution of the Seas, organized by the Food and Agriculture Organization in Rome last December. Fifteen countries agreed to cooperate in the project.

A report by experts of the sponsoring organization's General Fisheries Council for the Mediterranean, and the International Commission for the Scientific Exploration of the Mediterranean, provides a global view of the state of pollution for the whole region, specifying types of pollution and existing legislation.

Domestic, or sewage, pollution from large urban centers or areas of high population concentration is already a source of concern along the Spanish, French and Italian coasts where sewage goes directly into the sea with little or no treatment, according to the report.

The main sources of industrial chemical waste in the Western Mediterranean basin, which includes the Adriatic, were said to be metallurgy, industrial textiles, tanneries, pulp mills, fertilizers, chemical works, pesticides, detergents, docks and dockyards, metal industries, cement works, and plastic manufacture.

The widespread use of vegetal pesticides, particularly in estuaries and lagoons, was said to be responsible for the loss of marine life at all levels from zooplankton to fish. Aquatic birds have also suffered from the effects of pesticides. In the Eastern Mediterranean, pesticides are not widely used and the problem arises only in Israel, Cyprus and the Nile Delta.

Agriculture was included as a significant factor of industrial pollution because of the runoff into the sea of artificial fertilizers, pesticides and animal wastes off Spain, France, Italy, Yugoslavia, Greece, Turkey and Israel.

Legislation Is Summarized

The Food and Agriculture Organization's Conference on the Pollution of the Seas summarized water pollution legislation.

The summary showed that most of the industrialized nations along the Mediterranean coasts had begun to take action on mounting pollution in their coastal waters, but that there was still an urgent need for action at the international level.

As preparation for the Stockholm Conference, an Inter-governmental Working Group on Marine Pollution held its first session in London last June. Representatives of 33 states—oceanographers, marine scientists, and fishery experts — proposed regulation of the dumping of wastes and regional agreements on pollution in the North Sea and the Mediterranean.

October 23, 1971

The Mississippi: Highway of Pollution

By ROY REED

Special to The New York Times

NEW ORLEANS, Dec. 7— A government study to be published soon will report that waste dumped into the Mississippi River may be threatening aquatic life in the Gulf of Mexico and human health in southern Louisiana.

The report will be the latest and most urgent in a long series of warnings that Americans are destroying their greatest river.

"The father of waters" and "Old Man River"—those titles once came naturally to the mighty Mississippi.

Its floods once devastated entire regions. States have disputed each other for its islands and sandbars. People have lived out their lives on it, some in shanty boats and others in floating palaces. Many have died in it and because of it.

A Ditch for Waste

But now the Mississippi has lost much of its awe. Channeled and leveed, the river is now regarded by many as a highway for barges and a ditch for carrying off waste and poison.

The Mississippi is so dirty now that swimming and fishing in it are almost out of the question.

"I never serve Mississippi River catfish," remarked Alvin Pierce, owner of the Bon Ton Cafe here.

Like many of the city's more demanding restaurateurs, he buys catfish from cleaner streams.

It has been customary to think that the Mississippi River system, which drains 41 per cent of the continental United States, is too large to be destroyed by pollution. On an average day, the river carries more than 500,000 cubic feet of water a second under the Greater New Orleans Bridge.

This giant discharge is the greatest source of nutrients for life in the Gulf of Mexico. But the Mississippi has also become the Gulf's greatest source of man-made poison.

Some authorities now think that even the Mississippi is in danger.

'More Trouble Ahead'

"The view that the mighty Mississippi flows on unchanged just isn't so," officials of the Federal Water Pollution Control Administration said in 1968. "Old Man River is troubled and is showing signs of more trouble ahead."

The problem has grown worse since then. Cities in the river basin are slowly gaining in treating sewage, but industrial dumping and the runoff of agricultural pesticides and fertilizers are still major sources of contamination.

Louisiana Wild Life and Fisheries Commission

Discharge from chemical plant, near storage tanks, flowing into the Mississippi River below Baton Rouge, La.

Industrial pollution has increased as more water-using plants have been set up along the banks of the Mississippi and its tributaries.

Some industries, such as the Humble Oil Company at Baton Rouge, La., are spending large amounts of money to curtail pollution. But others are moving forward slowly, and only after insistent prodding by government agencies and environmentalists.

Federal and state authorities know surprisingly little about the specific dangers to man and other forms of life from the hundreds of man-made substances that are dumped into the river.

But some officials are aware of the general dangers and of certain isolated threats to the public or the ecological chain.

They know, for example, that a study in 1959 found that residents of New Orleans had three times as great an incidence of bladder cancer as those of Atlanta or Birmingham, Ala. No one knew why, but the New Orleans drinking water, which comes from the Mississippi, was suspected.

Pelican Exterminated

They know that Louisiana's state bird, the brown pelican, was exterminated in that state a few years ago. They suspect pesticides, especially the Endrin dumped for years from a single factory in Memphis.

But day by day and year by year, no government agency at any level tries to keep track of the wastes dumped into the Mississippi system and their impact on people, wildlife and plants.

Only minimal and erratic efforts are made to measure the amounts of oil, heavy metals, organic chemicals and other toxic substances that may be in the seafood eaten by thousands of people in this area each·day.

Scientists assume that at least some of those substances are absorbed into the flesh of the fish, crabs, oysters and shrimp that people in this region consume in great quantities. But the government's only concentrated efforts to identify and measure the substances come after publicized incidents such as the mercury scare of last year.

Little Testing

A spokesman for the Food and Drug Administration said here last week that the agency had done little testing of fish in this area in the last several months.

But he predicted that the agency would soon begin testing for other heavy metals besides mercury. It made a number of mercury tests last year and found varying levels of contamination.

Oysters are the only seafood tested continuously in the New Orleans area. "You can find everything in them, depending on where you get them," the spokesman said.

The absence of general and continuous testing to determine the possible hazards in food and water increases the significance of a study now being prepared for publication by the Baton Rouge staff of the Environmental Protection Agency.

The study raises more questions than it answers. For example, the agency has found 46 organic chemicals in the drinking water of New Orleans and Carville, a

small community up the river, after the water had been treated and presumably purified.

Two of those chemicals are thought to cause cancer. Four others have caused changes in the tissues of experimental animals. What the 40 other chemicals might do is largely unknown.

"The health and well-being of 1.5 million people who drink water from water plants using the Mississippi River as the source of raw water may be endangered by the discharge of industrial wastes containing materials known to have toxic, carcinogenic, teratogenic or mutagenic properties," the study concludes.

Southern Section

The study focuses on the southern section of the Mississippi. Sixty industries that manufacture mainly chemicals, paper and petroleum products dump their effluent into the river between Saint Francisville, La., just above Baton Rouge, and Venice, La., the last town of any size before the river reaches the Gulf.

New industrial sites are constantly being bought along both banks of the river, carved from land long used for sugar cane plantations.

The Environmental Protection Agency has found that the industrial plants already established are contaminating the river with cyanides, phenols, arsenic, lead, cadmium, copper, chromium, mercury and zinc.

Threat to Humans

The study says that 37 manufacturing plants dump

at least five pounds a day of at least one heavy metal, such as lead or mercury. These concentrations "may endanger human life and the life of the aquatic biota," the report says.

Lead, which is highly toxic in large amounts, causes the greatest concern. The agency has found that 27 industrial plants dump from 5 to 3,700 pounds of lead a day into this section of the river.

Industries have drastically reduced the dumping of mercury in recent months, but some still goes on. The Kaiser Aluminum and Chemical Corporation was granted permission recently by the Louisiana Stream Control Commission to continue dumping spent bauxite, containing small amounts of mercury, for at least three more years.

The Council on Environmental Issues, a student group at Louisiana State University, has tried unsuccessfully to persuade Federal officials to prosecute Kaiser for its pollution. A few days ago, the council filed a lawsuit in Federal court to try to force the officials to issue warrants for the arrest of Kaiser officials.

Some industries have only recently acknowledged that they had a problem. Kaiser, for example, insisted until a year ago that its spent bauxite did not harm the river. The company apparently has changed its mind. It is now studying alternative means of waste disposal.

James J. Friloux, acting chief of the Baton Rouge office of the Environmental Protection Agency, believes

that industrial pollution will be substantially solved in two or three years. He said most industries on the river planned to have secondary waste treatment plants either finished or under way by the end of 1972.

After that, he said, agricultural chemicals will remain the greatest problem and perhaps the toughest to solve.

Sewage Treatment

The thousands of municipalities in the Mississippi Basin have spent hundreds of millions of dollars to treat the sewage they dump directly or indirectly into the river. Progress has been steady, but much remains to be done.

Figures compiled by the Environmental Protection Agency's headquarters in Washington show that 374 communities, with a total population of 2,370,000, still dump raw sewage into the river or its tributaries. Four years ago, the untreated sewage of 3,895,000 persons went into the river. The total population of the basin is about 47 million.

The sewage of 736 other communities gets only primary treatment. Another 5,323 communities provide secondary treatment. Only 49 provide the more effective tertiary treatment that pours reasonably pure water back into the river.

The Mississippi may never again be as clean as it was when Mark Twain knew it. But, with enough money spent to correct past mistakes, neither is it likely to become a sewer.

December 8, 1971

States Curtailing Polluters On Pollution Control Units

By GLADWIN HILL

Across the country, a perceptible swing is developing away from what pollution fighters consider a major obstacle to environmental reform: the representation of polluting interests on state boards that deal with pollution.

In the last year, a 50-state New York Times survey shows, a score of states have taken or plan to take steps to eliminate or reduce such potential conflicts of interest from regulatory panels.

The number of states with air or water pollution boards reflecting pollution interests

dropped from 35 to 32 during the year.

The number with full-time state environmental agencies or other arrangements instead of the part-time citizen boards rose from eight to nine.

And more than a dozen states mapped such changes, or revamped their regulatory machinery in that direction.

The potential conflicts of interest on the boards in the 32 states arise from the membership of executives of polluting corporations; representatives of such major polluters as agriculture and local govern-

ment, and state officials who, in one degree or another, are spokesmen for agriculture or industry.

While there has been little evidence of outright misfeasance by such boards, many Federal and state officials, conservationists and laymen share a conviction that the pattern of "foxes guarding the henhouse" at best does not expedite pollution abatement.

The boards' defenders insist it is necessary to include members with links to pollution sources in order both to draw on their expertise and to achieve balance of a sort between clean-environment advocates and major economic interests.

But conservationists reply that expertise is available elsewhere; that environmental

cleanliness and economic prosperity are complementary rather than at odds, and that the general public interest deserves more than minority-group representation.

The classification of a state board is not a measure of its air and water pollution. States start out with widely different conditions. No two have identical administrations or rules. Corrective programs have been under way for different lengths of time and the changes in the administrative picture nationally have been too recent to have conspicuous effects. Still there is a repeated circumstantial correlation between persistent pollution and pollution-linked control boards.

The conditions reported in a survey a year ago impelled the nation's antipollution chief, William D. Ruckelshaus, head

COMPOSITION OF STATE POLLUTION BOARDS

Note: This table is not a classification of states' air and water pollution conditions

● Means state pollution board with regulatory authority contains members associated with basic pollution sources (industry, agriculture, county and city governments).
○ Means state board is free of such representation.

"No Boards" means air and water pollution regulation statewide is handled by a full-time state agency.

	Air Board	Water Board	Combination Air-Water Board		Air Board	Water Board	Combination Air-Water Board
Alabama	○			Montana			○
Alaska	No Boards			Nebraska			●
Arizona	No Boards			Nevada			●
Arkansas			●	New Hampshire	◐	●	
(1)California	○	●		New Jersey	No Boards		
Colorado	●	●		New Mexico			○
Connecticut	No Boards			(2)New York	No Boards		
Delaware			●	North Carolina			◐
Florida			○	North Dakota	◐	●	
Georgia	●	●		Ohio			●
Hawaii			●	Oklahoma			●
Idaho	●	○		Oregon			●
Illinois	No Boards			Pennsylvania			●
Indiana	●	●		Rhode Island	No Boards		
Iowa	●	●		South Carolina			●
Kansas			○	South Dakota	◐	●	
Kentucky	●	●		Tennessee	◐	●	
Louisiana	●	●		Texas	◐	●	
Maine			●	Utah	○	○	
Maryland	No Boards			Vermont	○	○	
Massachusetts	○	○		Virginia	○	○	
Michigan	●	●		Washington	No Boards		
Minnesota			●	(3)West Virginia	◐		
Mississippi	●	●		Wisconsin			◐
Missouri	●	●		(4)Wyoming			

(1) Pollution sources represented in regional branches of State Water Board. (2) State Environmental Board is advisory.
(3) Water pollution under State Division of Water Resources.
(4) Water pollution under State Department of Health and Social Services.

The New York Times/Dec. 19, 1971

of the Environmental Protection Agency, to urge the Governors of all 50 states that such potential conflicts of interest be eliminated.

Neither Mr. Ruckelshaus nor, in many cases, the Governors had the power to change conditions directly since generally the composition of the boards is prescribed by state laws. Nevertheless, there has been extensive movement toward change.

Some states in 1971 passed laws abolishing such boards. In others, boards were reconstituted to lessen pontential polluter influence and increase representation of the "public at large." In a number of states programs or proposals with the same objectives are scheduled to be put before the new legislative sessions opening in January.

The year's most pronounced change was in Alabama, whose air and water pollution boards had had heavy industrial representation and which, during the year, had had aggravated instances of air and water pollution.

Confronted by extensive mercury pollution of waterways and by an April air pollution "episode" in Birmingham — a foretaste of last month's temporary industrial shutdown there — the Alabama legislature brusquely replaced the old pollution boards with new ones structured to exclude polluter influence.

The state's new air pollution control board by law has seven members—the state health officer and six gubernatorial appointees: an engineer, a doctor experienced in respiratory diseases, and four representatives of the public.

Makeup of New Board

The new water pollution control board comprises the state health officer, the state conservation director, a doctor, a lawyer, an engineer and two public representatives.

Eliminated in the process were pollution-board seats that had been occupied by employes of United States Steel, Reynolds Metals and Scott Paper—all of which have been in court on pollution charges—International Paper, the Alabama Power Company and the Amoco Chemicals Corporation.

The legislature also enacted a law that no board appointee could be an employe or even a large stock holder (over 7.5 per cent) of any regulated corporation.

Connecticut, whose old Water Resources Commission and Clean Air Commission were by law heavily weighted with representatives of pollution areas, likewise abolished them.

In their place Connecticut put pollution control in the hands of a full-time professional Department of Environmental Protection. As its head, Gov. Thomas J. Meskill, a Republican, appointed Dan W. Lufkin, a Wall Street financier and conservation crusader who headed the financing drive for the original Earth Day observances in 1970. Mr. Lufkin has taken a furlough from his business interests to devote most of his time to direction of the department.

Action on Potential Conflicts

Montana created a new State Department of Health and Environmental Sciences, operating under the State Board of Health, a seven-member panel comprising three doctors, a dentist, a pharmacist and two housewives. The state downgraded its water pollution board—on which potential polluter influence had been strong —and its air pollution board

to advisory bodies.

Among the states that overhauled their boards to lessen potential conflicts of interest were North Carolina, Pennsylvania, Tennessee and Utah.

North Carolina, taking explicit cognizance of the possibility of conflicts of interest on pollution boards, revamped its State Board of Air and Water Resources—which had included an employe of the Aluminum Company of America, an executive of Carolina Power and Light, and a pulpwood executive—to reduce agriculture and industry's representation each from two seats to one. The public representation was increased from three to five.

The new law provided that the "public" members could not be officers or employes of either private or public organizations affected by the board's decision.

Tennessee replaced its nine-member Stream Pollution Control Board—on which industry and municipalities had had four seats—with a seven-member board with only one seat each for industry and municipalities. The State Air Pollution Control Board was nominally replaced also, although representation of special-interest groups was kept substantially the same.

Pennsylvania replaced its air and water pollution boards with an Environmental Quality Board and a new state Department of Environmental Resources. The board, which sets standards, is composed largely of public officials, eliminating some former corporate representation, although including a broad range of members in fields associated with pollution, such as agriculture, industry, commerce, transportation and public utilities.

In line with Mr. Ruckelshaus's recommendation, Utah's Gov. Calvin L. Rampton persuaded the Legislature to increase the "public" representation on Utah's air conservation commission from one to three members. He also replaced Miles Romney, a water pollution board member identified with the mining industry, with Robert Redford, the actor, who has been active in conservation.

Called 'Encouraging'

Commenting on the year's changes, Mr. Ruckelshaus said in an interview this week: "I think it's very encouraging that the states are moving in this direction. Governors seem to be responding to this problem very well, and should be encouraged by their constituents.

"I found in the past year that there is rampant mistrust of governmental institutions at every level. The only way to overcome that is to operate as openly as possible, and to eliminate any real or apparent conflict of interest, so that people can trust the decisions of these boards."

In some states, the situation remained unchanged. Conservation elements in Indiana, Nebraska, Nevada, North Dakota and Texas were thwarted in legislative efforts to replace pollution boards with full-time state regulatory agencies or to increase public representation on boards.

Nebraska's air pollution control director, Walter Franke, quit in disgust to take a position in Illinois' full-time pollution agency. In his letter of resignation, he told Gov. J. J. Exon, a Democrat, that Nebraska was "all talk and no action" on air pollution.

The state of Arkansas increased the representation of industry on its Pollution Control Commission, which deals with both air and water pollution. It enlarged the panel — which already included an employe of the Monsanto Company — from eight to 10 members, to include seats for a representative of the mining industry and the director of the State Geological Commission.

Action Planned in 1972

States where measures to change pollution control machinery will be considered by the 1972 legislatures include Arizona, Georgia, Idaho, Iowa, Kentucky, Maine, Michigan,

Mississippi, Ohio and West Virginia.

Arizona, Georgia, Idaho, Iowa, Maine and Ohio all are contemplating establishing full-time state departments to regulate pollution.

The lingering sentiment in defense of the status quo was expressed by Gov. George Guy, Democrat of North Dakota, who told a reporter:

"I do not believe pollution control boards should be divested of members from industry. Environmental management will always be a process of accommodating jobs and those concerned with environmental protection.

On the other side, Michigan's

Representative Raymond Smit of Ann Arbor, a Republican, a leading conservationist in the Michigan legislature, opposes even the inclusion of state officials from pollution boards.

"Basically third-level bureaucrats sit in for the department heads and at best they are very defensive about past activities of their departments," he said.

Of industry and municipal members, Mr. Smit said: "While the individuals can be of a very high caliber, they basically represent polluters on the board. They represent a constituency and the constituency includes the people on organizations the commission is set up to regulate."

December 19, 1971

GREAT LAKES PACT SIGNED IN OTTAWA BY NIXON, TRUDEAU

Antipollution Agreement to Cost U.S. Up to $3-Billion Over the Next 5 Years

---

By ROBERT B. SEMPLE Jr.
Special to The New York Times

OTTAWA, April 15—President Nixon and Prime Minister Pierre Elliott Trudeau signed a joint agreement today to begin the large-scale job of cleaning up the Great Lakes, the world's largest reservoir of fresh water.

Under the agreement, the United States plans to spend $2.7-billion to $3-billion over five years in Federal, state, local and private funds, and the Canadians will spend about one-seventh that amount.

At a signing ceremony here this morning, Mr. Nixon noted that in recent years "the quality of the Great Lakes' water has been declining, with ominous implications for 30 million Americans and seven million Canadians who live near their shores."

'Significant Step'

The new agreement, he said, "represents a significant step toward reversing that decline." The pact, known as the Great Lakes Water Quality Agreement, reflects six years of study and two years of bargaining between the two countries.

"This agreement," Mr. Nixon said, "bears witness to all the world of great concerns which unite our two countries: our common appreciation for the natural heritage which undergirds our national strengths, our common recognition that problems which cross international boundaries require international solutions, and our common confidence that our traditional relationships can grow to meet new demands."

The boundary between the United States and Canada runs through the middle of four of the five lakes—Superior, Huron, Erie and Ontario. The fifth, Lake Michigan, lies entirely within the United States, but at its narrow junction with Lake Huron it contributes much of the lake-system's flow.

Superior, Too, Is Affected

All five lakes are afflicted with some form of pollution, including relatively clean Lake Superior. Lake Michigan is befouled by sewage from innumerable industrial and municipal discharges, while Lake Erie is a virtual sump for the

sewage effluence of more than 12 million people and for industrial discharges from Detroit, Cleveland and other cities.

American officials who briefed newsmen here this morning conceded that the United States commitment depended heavily on Congressional appropriations and, in addition, the willingness of the Nixon Administration to persuade industry, through legal and other means, to provide up to $1-billion of the $3-billion total. They also acknowledged that other remedial steps to clean the lakes had consistently failed but said that they regarded the latest agreement as a "solemn commitment."

The agreement will not require the Administration to ask Congress for new funds. Instead the Administration will continue to spend, at the current rate, funds appropriated by Congress to finance the Federal share of municipal waste-treatment plants in cities bordering the Great Lakes.

Of the $2-billion in public funds the project will require, about half will be provided by the Federal Government and about half by state and local governments. Nearly all of the funds will be spent on new and improved municipal waste-treatment systems.

By contrast, the $400-million to $500-million that Canada plans to spend will be new funds — that is, expenditures above the current level for Great Lakes antipollution work. Negotiators for the two countries had originally hoped to

produce an agreement by last Christmas but were delayed by arguments on the American side over methods of controlling phosphate pollution, which is particularly serious in Lake Erie.

Canada agreed to force manufacturers of detergents to reduce the phosphate content to 5 per cent by the end of this year. The United States, reportedly under pressure from detergent manufacturers, has chosen instead to build waste-treatment facilities to neutralize phosphates. This decision has been severely criticized by Senator Edmund S. Muskie, Democrat of Maine.

At today's briefing, William D. Ruckleshaus, Administrator of the Environmental Protection Agency, defended the Administration's position by saying that only half of the phosphate problem is caused by detergents and that the other half is attributable to industrial wastes and other causes.

The Administration's commitment to the Great Lakes clean-up has also been challenged by critics who point out that the Office of Management and Budget earlier this year vetoed a demonstration program costing about $141-million recommended by Mr. Ruckelshaus's agency.

The signing ceremony this morning, held in the Confederation Room of Ottawa's Parliament building, represented the last official act of Mr. Nixon's two-day visit to Canada.

April 16, 1972

U.N. Parley Appeals for Curbs On the Release of Toxic Metals

By WALTER SULLIVAN
Special to The New York Times

STOCKHOLM, June 6—Delegates to the United Nations environmental conference approved today an appeal to all countries to minimize the release of toxic metals and chemicals into the environment.

While avoiding a judgment on the effects of supersonic air traffic on climate or on health, the delegates also voted to ask nations undertaking programs that might alter the climate to "carefully evaluate" the results. The fruits of such research would then be disseminated to the world "to the maximum extent feasible."

The phrase of qualification was inserted into the recommendation at the suggestion of the United States and was approved by a small margin. However, the 112-nation committee considering these recommendations then agreed to the final text without dissent.

The committee, one of three considering the agenda items, met as the United Nations Conference on the Human Environment embarked on its working sessions.

While all decisions made in the committees must be considered at the final full-scale sessions, representation in those sessions will be the same, so changes are unlikely. The recommendations being voted on one by one were drafted by a preparatory committee and are being amended in various ways.

To underline the urgency of the recommendation on toxic substances, the Japanese presented a report on three types of poisoning that have affected people in Japan. One had affected 1,081 people by February of this year, of whom 16 have died. It was traced to the leakage of a widely used chemical, polychlorinated biphenyl, or PCB, into rice bran oil widely used in Japanese cooking.

Officials of the World Health Organization at the conference say this is the first indication that PCB is lethally hazardous. Because the substance has been widely used in paints, in electric insulation, in carbonless paper and in other ways and because it is not appetizing to bacteria, it has been accumulating in the environment.

One cause for concern is the strong chemical similarity of PCB to DDT, which is even more widespread but has not yet been shown to be a serious threat to human beings.

Jack Davis, Canada's Minister for the Environment, told the conference delegates that substances like PCB have had a devastating effect on salmon runs and bird life. Use of PCB has been halted in Canada, he said.

Cadmium Poisoning Described

The Japanese also reported on the ailment known as the "itai itai disease," which for many years puzzled physicians in the area of the Jintsu River near the Sea of Japan coast of Honshu Island. "Itai," meaning "painful," or "sore," is a customary Japanese cry of pain, so the name of the disease can be translated roughly as "ouch, ouch."

It attacks the bones, which become painful and can break with even normal body movements. After following many false leads, researchers finally traced the ailment to cadmium poisoning. For years the metal had been entering the river in effluent from a silver, lead and zinc mine upstream. The river water was used for irrigation and the metal had soaked into the soil of rice paddies.

The disease has been reported from the area for many years. Since its nature was recognized in the late nineteen-sixties, 132 cases have been identified, and 32 of the patients have died.

Second Outbreak

The third and perhaps best-known of the poisoning outbreaks struck persons eating fish and shellfish from Minamata Bay on Kyushu Island. The effect has come to be known as the Minamata disease. It was traced to methyl mercury discharged by a chemical plant. As of March, 181 victims had been identified and 52 had died.

A second outbreak of Minamata disease occurred along the Agano River in Niigata Prefecture northwest of Tokyo, starting in 1965, and has been traced to fresh water fish contaminated by methyl mercury from another plant.

When a mother ingests mercury-contaminated food during pregnancy, her baby may become mentally retarded.

Delegates here were confronted this morning with photographs of such children and yesterday three adult victims of the disease were displayed by those seeking to force the chemical companies to provide larger compensation.

Informed sources said that possibly the most severe incidence of mercury poisoning on record may be one that has recently occurred in Iraq. While no final tally is yet available, the death toll is estimated at 1,000, with 10 times that number seriously affected.

Grain imported for planting had been treated with a mercury compound as protection against fungus growth. Although the bags were prominently marked with skull and crossbones, indicating poison, the grain was apparently fed to livestock. Then, when no immediate ill effect was observed, it was reportedly made into bread with disastrous results.

June 7, 1972

United Press International

DISCUSSES ENVIRONMENTAL DESTRUCTION: Premier Olof Palme of Sweden addressing yesterday's session of the United Nations environmental conference in Stockholm.

Clean-Water Bill Is Law Despite President's Veto

By E. W. KENWORTHY

Special to The New York Times

WASHINGTON, Oct. 18—The Senate and House representatives overrode today President Nixon's veto of the Federal Water Pollution Control Act of 1972, which thus becomes law and authorizes $24.6-billion over three years to clean up the nation's lakes and rivers.

The Senate vote to override, 52 to 12, came at 1:30 this morning, only about two hours after the President had sent up a veto message saying that the price tag on the bill was "unconscionable" and "budget-wrecking."

The President had delayed his message until 40 minutes before the bill would have become law without his signature, at midnight. His delay apparently was intended to give Congress time to accede to his request for a spending limit of $250-billion for this fiscal year. The limit was rejected and the President vetoed the bill.

The House vote to override the veto was 247 to 23. It came at 1:20 P.M. today.

A Warning Ignored

In overriding by such decisive margins, members of both parties ignored the President's warning that those who did so were "charge-account Congressmen" who were voting for inflation and higher taxes.

Anticipating the rejection of his veto, Mr. Nixon said that "even if the Congress defaults its obligation to the taxpayers, I shall not default mine." Noting that the bill gave him discretion in spending the funds authorized, he said, "I mean to use those provisions to put the brakes on budget-wrecking expenditures as much as possible."

That was taken here as a warning that he would not spend all the sums authorized, and particularly not those to pay the Federal share of waste treatment plants.

Senator George McGovern,

the Democratic Presidential candidate, said that Congress had acted "with great wisdom and courage" in refusing to sustain the veto.

"The Presidential veto," Mr. McGovern said, "reveals the Nixon Administration's record on behalf of the environment for what it is — hypocritical platitudes coupled with spineless inaction."

Cost Termed 'Staggering'

During nearly two years of Congressional deliberation on the bill, the White House had supported industry's opposition to many of its provisions. particularly the goal of no discharges of industrial pollutants by 1985 and the setting of limitations on effluents for classes of industry.

However. Mr. Nixon based his veto solely on what he called its "staggering" cost of $24.6-billion. Of that amount, $18-billion would be for the Federal share—75 per cent—of the cost of waste treatment works. The states and municipalities would pay the remainder.

In addition. $2.75-billion would be earmarked to reimburse states and cities for the Federal share on projects already completed or under construction that the states and cities have paid themselves in expectation of Federal reimbursement.

Of that amount, $2-billion would be for reimbursement for projects between 1967 and 1972—an amount that the Environmental Protection Agency agrees the Government owes and should pay. The remaining $750-million is for projects from 1957 to 1966, which E.P.A. insists the Government does not owe since "no significant Federal assistance program existed during this period and there was thus a lack of federal commitments."

$6-Billion 'Enough'

A year ago, Mr. Nixon had proposed a three-year program with $6-billion as the Federal share for waste treatment plants, only one-third of the amount contained in the bill. Furthermore, the Federal share would have been 50 per cent, rather than 75 per cent. Finally, there was no provision in Mr. Nixon's proposal for reimbursing the states and cities for the unpaid Federal share on past projects. Presumably any reimbursement would have had to come out of the $6-billion.

In his veto message, Mr. Nixon said that his proposed $6-billion was "enough to continue and accelerate the momentum toward that high standard of cleanliness which all of us want in America's waters."

In saying that, the President took direct issue with William Ruckelshaus, in a 33-page letter of the Environmental Protection Agency. On Oct. 11, Mr. Ruckelshaus, in a 33-page letter to Caspar W. Weinberger, Director of the Office of Management and Budget, "strongly" recommended that the President sign the bill. On the question of financing, Mr. Ruckelshaus made the following points:

¶The $18-billion for waste treatment plants provided by Congress was "the result of the Congress adopting a later E.P.A. needs survey than the one that provided the basis for the Administration's request" of $6-billion. E.P.A.'s revised estimate of needed expenditures over three years, Mr. Ruckelshaus said. was $18.1-billion. The 75 per cent Federal share of that figure would be $13.6-billion — almost $5-billion less than Congress provided, but $7-billion more than the President proposed.

¶Because actual expenditures for waste treatment projects would be spread over nine years, Mr. Ruckelshaus emphasized, the major outlays required by the Congressional bill "will not occur until the fiscal years 1976-1981," and "the total value of construction initiated in the near-term under th enrolled (Congressional) bill is expected to correspond close-

ly to the total value of construction that would have been initiated under the Administration bill."

¶"The major fiscal impact during the fiscal years 1973-1975 will result not because of obligations incurred for new construction . . . but as a result of reimbursement for projects already constructed or under construction."

On the last point, Mr. Ruckelshaus said the $2-billion owed "represents a commitment," and he recalled that in his 1971 environmental message, Mr. Nixon had said, "We must also assure that adequate Federal funds are available to reimburse states that advanced the Federal share of project costs."

Mr. Ruckelshaus also reminded the President that E.P.A.'s estimate of need provided to Congress ($18.1-billion) "was constructed to support the commitment of the President in his State of the Union message of Jan. 22, 1970, to 'put modern municipal waste treatment plants in every place in America where they are needed to make our waters clean again, and to do it now. '"

'It Seems Reasonable'

In conclusion, Mr. Ruckelshaus said he was "aware of the fiscal magnitude" of the bill. But, he added:

"It seems reasonable to me to spend less than 1 per cent of the Federal budget and two-tenths of 1 per cent of the gross national product over the next several years to assure future generations the very survival of the gross national product."

Mr. Nixon rejected Mr. Ruckelshaus's recommendation and reasoning.

In urging that the veto be overridden today, Representative Robert E. Jones Jr., Democrat of Alabama, who was in charge of the bill, said that Congress knew the bill would be costly.

"But we also know," he said. "that the people are prepared to pay the price of this undertaking."

October 19, 1972

Protecting Water: New Law Aims at Bias

By GLADWIN HILL

Special to The New York Times

WASHINGTON, Oct. 22 — A noteworthy side-effect of the so-called Environmental Revolution has been the increasing currency of the expression ". . . foxes guarding the hen house . . ."

Governmental efforts to deal with environmental problems have repeatedly led to the dis-

News Analysis

covery that the requisite regulatory authority lay, under traditional governmental structures, with agencies ill-fitted to exercise that authority impartially.

A major reason for the creation in 1970 of the Environmental Protection Agency was to center in an independent Federal agency powers over

such things as pesticides and radiation safety that had been, exercised ambivalently and uncomfortably by the Department of Agriculture and the Atomic Energy Commission.

Similar problems have existed in state governments. One of the most conspicuous is the fact that in most states, water pollution and air pollution have been under the jurisdiction of boards composed solely or part-

ly of citizen appointees who, in many cases, had ties with the major pollution sources — industry, agriculture and municipal, county and state governments.

State laws have often earmarked a certain number of seats on these boards for representatives of industry, agriculture and interest-oriented government agencies — the rationale being that such members would provide special "expertise."

Surveys by The New York Times in 1970 and 1971 found that a majority of states (the 1971 total was 32) had such potential conflicts of interest among members of one or more pollution boards. No instances of outright malfeasance were found. But it seemed a fair presumption that a board with built-in bias might not deal so effectively with pollution problems as would a completely impartial board.

One of William D. Ruckelshaus's first official acts as administrator of the Environmental Protection Agency was to write to all 50 governors urging that such potential conflicts of interest be eliminated. Some states took steps to do so.

Now a provision of the water Pollution Act enacted last week has struck a sweeping blow at built-in bias on the state boards.

Impact of New Law

Referring to the new program under which all nonmunicipal dischargers of fluid waste into waterways will have to obtain state permits (subject to Federal veto), the law says that "no board or body which approves permit applications or portions thereof shall include as a member any person who receives, or during the previous two years has received, a significant portion of his income directly or indirectly from permit holders or applicants for a permit."

This requirement, according to Representative John D. Dingell, Democrat of Michigan and a leading sponsor of the provision:

"Is intended to wipe out all industry representation on any water pollution control board or similar body that has anything to do with issuing, denying, or recommending permits. It is a condition precedent to any state obtaining the power under Section 402 to issue permits.

"Even one [industry] representative shall not be allowed," he said "because of the potential that the board will consider permits in which he has an income interest."

While the new requirement deals specifically with industrial connections of members of state water-pollution boards and does not cover board-member affinities with agriculture and local governments (which are generally big polluters), it will force a number of states to alter board memberships, existing statutes and particularly their philosophies of board representation.

The new law also poses an obvious question: If this is desirable for state water pollution boards, what about parallel state air pollution boards, which generally are similarly composed, but covered by no such requirement?

The latter boards are expected to be an early target of the "environmental lobby," one of whose members remarked whimsically: "Pretty soon we'ell be having a problem of what to with unemployed foxes."

October 23, 1972

91 Nations Agree on Convention To Control Dumping in Oceans

By JULES ARBOSE
Special to The New York Times

LONDON, Nov. 13 — Representatives of 91 countries, including all of the world's major maritime nations, agreed today on a global convention to end the dumping of poisonous waste matter at sea.

Under the convention, the dumping of high-level radioactive waste, biological and chemical warfare agents, crude oil, some pesticides and durable plastics is prohibited. Other, less harmful substances and materials, such as arsenic, lead, copper, scrap metal and fluorides, can be discharged only with special permits.

The convention, hammered out by 250 delegates during 14 days of discussions, was termed "a historic step toward the control of global pollution" by Russell E. Train, head of the 14-member United States delegation.

The agreement came after a weekend of talks that threatened to collapse because a sizable bloc of countries raised the issue of offshore territorial limits by insisting on the establishment of a "pollution zone" ranging from 50 to 200 miles off their coasts.

The delegates ultimately agreed to shelve the issue. It

Associated Press
Russell E. Train

will be taken up again by the United Nations Law of the Sea Conference, which may begin in the next year or two.

This conference will deal with territorial limits that are now confused by claims ranging from the traditional three miles to more than 200 miles.

Mr Train, chairman of President Nixon's Council on Environmental Quality, said that the convention "had achieved substantially all of the objectives which the United States had been seeking." He added that it was a strong and effective measure and gave "practical evidence of the increasing priority the nations of the world are giving to environmental problems."

Article 1 of the 22-article convention calls on the signatories to "individually and collectively promote the effective control of all sources of pollution of the marine environment."

Specifically, the countries pledge themselves "to take all practical steps to prevent the pollution of the sea by the dumping of waste and other matter that is liable to create hazards to human health, to harm living resources and marine life, to damage amenities or to interfere with other legitimate uses of the sea."

The Convention on the Dumping of Wastes at Sea — its official title — will take effect as soon as it is ratified by legislatures of 15 of the signatory countries.

Only 57 of the 91 countries had signed the convention when the conference ended today at Lancaster House. However, the fact that the other participating countries did not sign the document immediately was not considered vital.

Twelve of the countries attended the conference as observers, and the others are expected to sign after the English text of the convention is officially translated into the three other working languages: Russian, French and Spanish.

On the 'Gray List'

While following the general lines of American antidumping legislation signed by President Nixon six weeks ago, the convention goes much further in the number and types of substances whose discharge into the oceans by aircraft or vessels is prohibited.

The American legislation prohibits specifically only high-level radioactive waste and chemical and biogolical warfare agents. Other substances on the "black list" of the convention, in addition to those already mentioned, include mercury and cadmium and their compounds, fuel oil, heavy diesel oil, lubricating oils, or generally just about anything that finds its way into the food chain or does not rapidly convert into substances that are biologically harmless.

Also on the "gray list" of substances and materials requiring special permits are zinc, silicon compounds, cyanides and waste containing large quantities of beryllium, chromium, nickel or vanadium. The convention stipulates that any substance or material not mentioned in the first two categories will require a general permit, giving governments effective control over anything dumped into the oceans.

The enforcement of the antidumping measures and sanctions is left to individual countries. There is no attempt in the convention to coordinate penalties.

The convention calls for the creation of a secretariat, to coordinate and disseminate the latest technical and scientific data on oceanic pollution and dumping. One of its duties will be to advise countries on how to get rid of waste matter that cannot be dumped at sea. The convention did not advise on methods of safe disposal of poisonous matter on land.

November 14, 1972

E.P.A. Is Curbing Sewer Main Subsidies

By GLADWIN HILL
Special to The New York Times

WASHINGTON, Oct. 14—The Federal Government is tightening up on its billion-dollar program of grants to localities for sewer mains because of findings that the projects are encouraging unsound community growth.

A federally sponsored study by an independent research organization has indicated that communities, in apparent zeal to obtain Federal subsidies, are building sewerage facilities on a scale calculated to handle their expectable population increases for as much as the next 2,000 years.

In less extreme instances, communities are installing up to twice as much sewerage capacity per household as some experts estimate are needed. And in some cases they are contracting debt loads for sewerage that impel them to seek more residents who can contribute toward paying off the debt.

To counteract these tendencies, the Environmental Protection Agency is instituting new criteria for its grants, particularly in regard to analysis of their impacts on land use and community growth.

This reorientation of Federal policies was outlined last week by Russell W. Peterson, chairman of the Council on Environmental Quality, in reporting on the results of a survey made for the council by Urban Systems Research and Engineering, Inc. of Cambridge, Mass. The council advises the President and Congress on environmental policies.

The study scrutinized sewerage projects in 52 communities, including some in New York City and New Jersey.

Mr. Peterson, a former Governor of Deleware, said the findings were "not a criticism of the Environmental Protection Agency, but a criticism of the way we have become accustomed to building sewers in this country."

He said Russell E. Train, the E.P.A. administrator, "is aware of the results of the study and has begun a process to modify the Federal waste water treatment grant program so as to offset those undesirable stimulants to undesirable land use development."

Under 1972 water pollution legislation, the Federal Government may pay up to 75 per cent of the costs of new community sewage treatment facilities and related construction.

Of $3.4-billion obligated in grants to date, $1.2-billion has gone for "interceptor" lines—collection mains leading to sewage plants.

"The study shows that the pattern of interceptor sewer construction now supported by the grant program is encouraging low-density development in urban fringe areas and thereby exerting significant adverse impacts on land use patterns and efforts to conserve energy," Mr. Peterson said. "These land use and energy impacts are not being identified and evaluated as part of the grant award process.

"In a typical case in the [52-community] sample, over half the land to be served by the interceptor sewer is currently vacant. Nearly $145 per capita is being expended to build capacity beyond that required by the existing population."

Typical projects were designed for 50 years ahead, Mr. Peterson continued, with one unspecified project scaled to 2,000 years of growth at present rates, while the study indicated that contruction looking ahead only 25 years was "more economic."

"The study shows that most interceptor sewers are being designed for 100 or 125 gallons per capita per day, even though current per capita water use averages only between 60 and 80 gallons per day. No valid basis could be found for the higher consumption figures.

The Environmental Protection Agency, Mr. Peterson said, is considering limiting funding of interceptors beyond those needed for current population; cutting design capacity to 25 years; using the lower per capita use figures in planning; reviewing financing processes; requiring much more detailed information on land development prospects; and requiring that "full environmental impact statements be prepared on all major grants and circulated to the public and other agencies for comment early in the project review process."

Sewerage facilities have been the limiting factor in hundreds of communities with growth problems. Conversely, residents of many communities have resisted enlargment of sewerage facilities as paving the way for unwanted growth.

In one major current community growth battle, involving Petaluma, Calif., a United States District Court—in a decision that is being appealed—ruled that a community could not refuse to finance sewerage facilities as a device for excluding additional population.

One community covered in the study was Oakwood Beach on Staten Island in New York City, where two interceptor lines costing $40-million are projected.

Commenting that the study "found no evidence that any serious land use impact analysis had been performed," the report said:

"Despite the tremendous size and cost of the Oakwood Beach project and the serious problems of high-density development in other sections of New York City, a negative environmental impact declaration was accepted and the project was approved without preparation of an environemntal impact statement."

Another place surveyed was Ocean County in New Jersey, where the E.P.A. last spring advised officials they should scale down their growth projections to meet federally acceptable limits on the discharge of treated sewage effluent.

However, the survey found, "since the municipalities responsible for land use control have been unable to shape or to curtail new development in their areas; and the developers, in the absence of a regional sewage system, can and will install other less desirable sewage disposal facilities, the local planners believe the construction of sophisticated regional sewerage systems is the only sensible way to proceed."

October 15, 1974

Life vs. livelihood By Wade Greene

DULUTH, Minn.—"We drink it, our children drink it, we bathe our babies in it, we believe it's the purest, best-tasting, most unpolluted water found anywhere," said Wayne Johnson, attorney for the town of Silver Bay, Minn., home of the Reserve Mining Company. That was two years ago, and Johnson was voicing the usual lakeside pride in the purity—innocence, you might even say—of Lake Superior. In a society that has judged itself pervasively guilty of ecological sins of emission, Lake Superior has stood as a vast, rare bulwark of the unpolluted. The ultrablue waters of this magnificent inland sea, whose 31,820 square miles make it the largest freshwater lake in the world, have been admired as being so clean, so clear that you could see 50 feet deep into them.* Lakeside dwellers have boasted that they draw and drink their water directly from the lake, unfiltered, unsanitized.

And that's just what Wayne Johnson did to demonstrate his faith in the lake's purity—he downed a glass of Lake Superior—straight, as they say.

It may have been a deadly drink. In an injunction issued last April at the end of what has been by far the longest and most expensive environmental trial in United States history, a Federal District Court found not only that Superior's waters were impure in some sectors but that they contained a cancer-causing agent—asbestos, or some-

*Insofar as it has the greatest surface area, Lake Superior is the world's largest fresh-water lake. Its total area is 31,820 square miles. The fresh-water lake with the world's greatest volume, however, is the Soviet Union's Lake Baikal, with an estimated volume of 5,750 cubic miles. Baikal is the world's deepest lake, measured as 6,365 feet deep in certain places.

Wade Greene, a frequent contributor to this Magazine, specializes in social and environmental subjects.

thing virtually identical to asbestos—which for 18 years has been flowing into the waterpipes and gastrointestinal tracts of a number of communities on the southwestern tip of the lake, including one of the nation's major Great Lakes ports, Duluth.

This potentially horrendous contamination was laid directly at the sluices of the Reserve Mining Company, one of the few large employers in the region, one of the principal iron miners in the country. Submicroscopic particles of an asbestos-like substance were found by Federal Judge Miles Lord to be a major component of Reserve's massive discharge into the lake—a daily 67,000 tons of ground rock. As a result of his findings, Judge Lord ordered the Reserve plant closed down.

For 32 hours last spring, the giant lakeside plant, through which 12 per cent of the nation's domestic iron-ore supply normally flows, stood nearly empty and noiseless, a mute brooding symbol, for a moment, of the triumph of environmental interests over economic interests. The plant was reopened when a Federal Court of Appeals stayed Judge Lord's injunction pending development of an alternate on-land disposal system by Reserve. Yet the Court of Appeals branded the original decision to allow the dumping into Lake Superior a "monumental environmental mistake" and has told Reserve, in effect, that it must move to halt its dumping. Therefore, it seems to be only a matter of time and selection of an on-land disposal site before the Reserve operation will be radically overhauled. Prodded by the courts, Reserve and Minnesota authorities are currently negotiating for a site.

The sheer dimensions of the case are certain to make it a major chapter in the modern epic of industrial society's efforts to harmonize environmental and economic values. Clashes between these often antithetical forces have become increasingly frequent in recent years with the heightening of environmental sensitivities. But nowhere have the stakes and symbols of the confrontation been so stark and so immense: possible cancerous contamination of an extraordinary fresh-water lake—and some 200,000 Duluth-area residents who drink from it—on the one hand, and the fate of a major portion of a major industry, with regional, national and even global economic implications, on the other. For Reserve itself, the lakeside plant represents an investment of about one-third of a billion dollars, and the company says that it will cost more than that to convert it to a land disposal system. Thus, the fate of some 3,000 Reserve employes is directly at stake, too.

The Reserve Mining case may also come to be regarded as a landmark in the nation's efforts to grapple legally with environmental health threats that medical research is bringing to light, many of which involve carcinogenic, or cancer-causing, agents that may be affecting the lives of thousands—even millions—of people. A profound issue of law is at stake in the Reserve Mining case and may well end up being resolved by the Supreme Court. This is whether full scientific and legal proof of a suspected health threat to a large population is needed before a court can halt the hazard, or whether the probabilities that are the rough terrain of medical science's frontiers will legally serve to avert a potential calamity before irrefutable, corporal proof is established.

"To us, there are neither heroes nor villains," the Appeals Court philosophized about the Reserve Mining saga in its stay of Lord's injunction. Perhaps not. This epical chapter begins—partially, at least—

in the morally neutral and impersonal terrain of classroom economics and geography. Northeastern Minnesota's Mesabi Range, as most grade-school pupils once were instructed, was where much of America's iron ore came from, and iron was what much of America's industrial economy was built on. But the rich, smeltable range ores, which have been described as the "raisins in the pudding," began to run out in the late nineteen-forties. Little but the pudding remained—a dark, hard, flintlike rock called taconite, which contained less than 30 per cent iron, or barely half of what was needed for steel smelting.

So it seemed an unequivocal triumph of what was then called "American ingenuity" when engineers, centered mainly at the University of Minnesota, developed a process to refine, or "benefact," the low-grade taconite into high-grade iron ore ready for the blast furnace. The process amounted largely to pulverizing the taconite rock into smaller and smaller pieces until some of it was flour-fine, and using huge magnets along the way to extract the pieces with high-iron content. The fact that two-thirds of the original rock had to be discarded somewhere was considered an incidental, purely technical problem.

As defenders of Reserve now emphasize, the government of Minnesota was a prime mover in getting Reserve to set up shop; Reserve was the first to arrive and the only one to use the lake for its refuse, but it is no longer the largest of what are now seven taconite processors in Minnesota. Minnesota passed special tax legislation and later even a constitutional amendment aimed at luring taconite processors into the Mesabi. State agencies also readily acquiesced to Reserve's request to cart the raw taconite rock by private rail line from its mining area in the range, 50 miles from the lake, to a refining facility on the lake, and dump its tailings, as the taconite residue was called, into the lake. The question of just what happens to the tailings has come to be a central issue in the Reserve Mining Case. In its original 1947 hearings for permits to discharge into the lake, Reserve asserted that even the finest tailings would flow down to, and remain at, the bottom of a wide 900-foot deep trough off the shore near the Reserve plant. In any case, Reserve contended, the tailings were just "inert sand."

The plant was formally opened in 1956. At the same time that it went up, so did Silver Bay, a brand-new town of prefabricated houses and one-car garages built in the hills inland for Reserve's workers and their families, now some 3,000 inhabitants in all. In St. Mary's Catholic Church in Silver Bay, the altar and baptismal font were hewn out of taconite rock.

The symbolism was apt. Yet downcurrent from Reserve, skepticism about the plant began to grow, slowly, as Reserve's production, and discharge, increased and as the effects of the tailings began to be noticed more widely. Milt Mattson, whose family has long been in the fishing business near the Reserve site, recently recalled that a milky gray sediment appeared from time to time around Beaver Bay, just south of Reserve, after the plant opened, and that it "really began to show" after production doubled in 1961. Over the mewing of herring gulls, he described one of the problems for fishermen. "The herring won't tolerate any dirty water," he said. "And they wouldn't go near the murky areas."

Fishermen also began to complain that their nets came up covered with slime—evidence, they contended, that the tailings increased algal growth. By the mid-sixties, some conservationists were also beginning to object to the "green water"—in contrast to Superior's famous ultrablue—that streaked out at times from Reserve's plant in a long plume.

Grant Merritt, director of the Minnesota Pollution Control Agency, has described the Reserve case as "one classic situation above all others." As he observed, it contains "all the elements an environmental case could have: problems of pollution, economy, politics and law." Not least of all, politics.

A good deal of local- and state-level politics was involved in establishing Reserve and the taconite-mining industry in Minnesota. But once it was there, Reserve and its absentee co-owners, Republic and Armco Steel, proved highly impolitic. The iron-and-steel industry is not used to dealing directly with the public. Reserve was no exception. Its only public reactions to growing complaints were to shrug them off as unfounded or misdirected. The "green water," for instance, had nothing to do with Reserve's tailings, the company said; it did not know just what did cause the "illusion." (Since the trial began, Reserve has declined to discuss anything to do with charges against it except in court.)

But Reserve ignored politics and public pressures at its peril. In the late nineteen-sixties, a grassroots movement began to take hold against Reserve, and while at first it seemed almost pathetically inadequate to the challenge, this movement is above all responsible for bringing the industrial behemoth on Lake Superior to its knees. And grassroots it truly is; it has been led by such improbable giant-tamers as a prim, United States Government secretary in Washington, Verna Mize, and a former hair stylist in Duluth, Arlene Lehto.

Mrs. Mize, who grew up on the Michigan shore of Lake Superior, has been gently badgering Washington officialdom since 1967, when a visit to the lake convinced her it was being sullied by Reserve. She has written thousands of letters, picketed and petitioned and accosted the influential with ever-ready bottles of Superior water—shake any bottle and it turns murky gray-green

with Reserve tailings. She has talked to nearly every Great Lakes Congressman or Governor by now.

Arlene Lehto's parents own a small lakeside resort down-current from Reserve. That is where she grew up and where her visions of a more crystalline lake stem from. At about the same time that Verna Mize was becoming an antagonist of Reserve in Washington, Mrs. Lehto organized the Save Lake Superior Association in Duluth. By approaching various groups and government entities around the southwestern end of the lake, she and her organization have concentrated their fire on Reserve. Recently, Mrs. Lehto proposed that Duluth post billboards on highways leading into the city to warn visitors not to drink the city's water, a suggestion that was not received warmly by the city fathers. "To me, it's sort of a simple thing," she philosophized not long ago. "I think people's health is more important than economic matters, and the environment is the basis of their health."

Another unlikely, early challenger of Reserve was a mild-mannered regional coordinator for the Department of the Interior named Charles Stoddard. It is largely because of Stoddard that the cause grew into a prickly national political issue.

Stoddard headed a governmental study group that looked into Reserve's discharge, and in late 1968, shortly before the Johnson Administration and his own employment came to an end, Stoddard "released" or "leaked" the study group's report. The effect of Reserve's dumping, said the report, "is to increase the turbidity of this once clear lake, accelerate eutrophication by enrichment of its water, raise certain chemical constituents to levels beyond established limits and to decrease fish food and habitat through deposition of sediment." Mild enough stuff in light of subsequent findings, but enough to lead to a recommendation in the report that Reserve be given three years to "investigate and construct alternate on-land waste disposal facilities." Predictably, that multi-

million-dollar suggestion caused no small stir around Duluth and Washington.

Stoddard is retired now and spends much of his time in his country home in the evergreen forests of Wisconsin's Indian Head region about 50 miles from Duluth. I drove down to see him during a visit to the Duluth area.

In a small, cluttered study, and swatting at mosquitoes as he talked, he spoke a little about the fuss and politics of the Stoddard Report. He said there had been considerable pressure to disavow the report, much of it stemming from Representative John A. Blatnik, in whose district the Reserve Mining Company is located. Stewart Udall, then Secretary of the Interior, was Stoddard's boss. "Udall," Stoddard said, "told me Blatnik had come to him almost with tears in his eyes, saying his reputation was at stake unless Udall said the report was just a preliminary inner-agency memo" and did not, in effect, bear the imprimatur of the Interior Department.

Udall did just that in mid-1969: He asserted that the report was only a "preliminary staff report." He also denied allegations that Blatnik had attempted to suppress the report. A few weeks after talking to Stoddard, I talked to Representative Blatnik in his office in Washington, in a room dominated by a wall-sized photograph of the John A. Blatnik Bridge that connects Duluth to Superior, Wis. Blatnik, who is retiring himself this year, was clearly sensitive to the charge, which is gospel among anti-Reserve activists, that he had tried to suppress or discredit the Stoddard Report. When the document came to his attention, he explained, he considered it "very, very preliminary, a small mimeographed report by an on-the-spot chief." He called up Udall, he said, and asked, "'What about this Department of Interior report?' and Udall said, 'What report?' He didn't know it existed. I said, 'Why don't you wait another season so you can tell whether effects are increasing or decreasing?'"

Later, I asked Udall about the Stoddard Report. "I don't come out with too much glory on this whole thing," he said.

He recalled that Blatnik had been "very upset" by the report, that "we needed his cooperation on a lot of things . . . and we sort of muffled our oars a little bit as a result of his remonstrations. Everyone agreed to kind of fudge it by saying it wasn't an official report. But it was all double-talk. It was an Interior Department report. Blatnik and his people, in their effort to blunt it, were playing a lot of little games."

Blatnik emerges as a highly ambivalent actor in the Reserve saga. More than anyone, he seems to personify the profoundly conflicting values at issue—namely, the environmental vs. the economic tensions. On the one hand, Blatnik, who is now chairman of the House Public Works Committee, has gained his principal national reputation as the House's main author and champion of water-pollution control legislation, stemming back to the mid-fifties. He is now known to some as "Mr. Water Pollution Control."

This interest in clean waters stems naturally enough from the large amount of it in Blatnik's thinly settled district—a considerable portion of Minnesota's 15,000 lakes of more than 10 acres in size are contained in the district, as is 161 miles of Superior's shoreline. On the other hand, so are the Mesabi Range, the taconite industry and the thousands of jobs it provides. Accordingly, "Mr. Water Pollution Control" was once glad to have been known at home as one of "the Taconite Twins." Blatnik was one of two Minnesota state legislators who, in the forties, were mainly responsible for getting tax legislation approved to attract taconite mining companies to Minnesota.

Blatnik, however, denied having anything to do with the Superiorside site that Reserve chose. And as I talked to him, the Representative produced a large framed painting of the National Water Quality Laboratory on Lake Superior just 50 miles south of Reserve's plant, and he suggested that he had been instrumental in getting the lab built at that particular site in order to keep a monitoring eye on Reserve. "Blatnik's lab," as it is called locally, has indeed played a critical role in building the case against Reserve. So the political currents and crosscurrents in which Blatnik figures clearly have been complicated.

One major finding of the lab was that the mineral cummingtonite was turning up in water samples in the southwest part of the lake, and that Reserve's tailings were the only significant source of this mineral—a direct challenge to Reserve's insistence, stemming back to 1947, that virtually all of its discharge settled immediately offshore from the plant.

This and other evidence against the plant were put forth as Federal pure-water laws began to close in on Reserve in a series of enforcement conference sessions in Duluth in 1969 and 1970—sessions held to establish abatement schedules for various Lake Superior polluters. Reserve alone refused to come up with an abatement plan, even in the face of a final 180-day deadline set by law. Instead, as subpoenaed notes and memoranda have confirmed, Reserve and its co-owners, Armco and Republic Steel, practiced their own brand of counterpolitics. "By vigorous political activity, primarily in Washington," advised one memo, "it may be possible to amend or modify the 'conclusions' and 'recommendations'" of the enforcement conference.

The political activity included meetings between company representatives, including heads of Armco and Republic, both active in Republican party financing, and high-level Government officials.

Nevertheless, William Ruckelshaus, who was then head of the Environmental Protection Agency, stood off White House intrusions and, after some delay, asked the Justice Department to take Reserve to court.

The legal case against Reserve up to this point was hardly an open-and-shut one. As John Hills, the Justice De-

partment lawyer who headed the trial forces against Reserve—he has since moved to the President's Council on Environmental Quality — explained after the trial: "You had very subtle effects from the discharge. You had mammoth discharge, and yet the effects were not of the same kind as if you were putting cyanide in the water."

But then, in the final months before the trial was set to begin, the possibility arose that the effects of Reserve's discharge might not be very subtle after all, that in fact the tailings contained, in effect, asbestos and might be slowly killing thousands of people. With this possibility came a major shift in the case against Reserve. The case moved a giant step to the very borders of scientific knowledge. The carcinogenicity of asbestos has been scientifically established only in recent years, largely as a result of tracing specific cancer cases to asbestos inhaled in certain occupational settings. Yet medical scientists are still not certain how asbestos causes cancer. This recent and incomplete understanding of the asbestos peril set the framework of much of the Reserve trial; it made scientific proof as much of an issue as legal proof. And scientific proof in court would be complicated by the fact that it takes 20 to 30 years, or more, from the time of initial exposure for asbestos-caused cancer to begin to take its toll. Reserve had been operating for only 17 years —and only 12 at its current rate of output.

The original connection between Reserve's tailings and asbestos was, significantly enough, stressed not by any of the growing body of scientists who were working on the Reserve case, but by Arlene Lehto, the former hair stylist who became an environmental activist.

In the print-and-rubberstamp shop she and her husband operate on the Duluth hillside, Mrs. Lehto described how she happened to put forward the connection. She said she was casually talking with a geologist the night before she was to speak at a Canadian-American conference on Lake Superior in late 1972. He

mentioned the similarity between Reserve's discharge and asbestos, and somewhat rashly she worked it into her speech. Members of both the National Water Quality Laboratory and the Minnesota Pollution Control Agency were there, and the rash observation stirred their thinking, at least in the back of their minds.

One night, shortly afterwards, Gary Glass, a young scientist at the lab, had a nightmare. "My dream was that I should not drink the water," he recalled recently in his office overlooking Lake Superior. The next morning, he said, he and Philip Cook, who was the lab's technical coordinator for the upcoming trial, began poring over books in the lab's library and making calls to experts around the country. Within a week, they established the possibility of a real-life nightmare—the nearly complete identity of Reserve particles with asbestos.

Actually, Glass said, the identity had literally been staring him in the face for three years—in the form of blow-up of a microscopic photograph of a lake-water sample that had been taped up in his office. "I didn't know what it was, not having a mineralogical background," Glass confessed.

He pointed to a corner of the picture, at a black blotch about two inches long. "This is the one that really turned us on here," he said with the sort of jolly exuberance of scientific discovery that transcends the potentially horrible implications of the discovery. "This has got the very same morphology—a bundle of fibers—as some of the papers on asbestos we saw right down in our library. Zing!"

Many of those originally aware of the asbestos finding were worried that Duluth-area residents would panic upon learning that their water may be carcinogenic; that given the long latency period of cancer from asbestos, the water may already have planted cancerous seeds in older residents of the area. But the reaction of Duluthians was, and still is, remarkably calm. For a while, there was a spurt in the sales of bottled water, but that leveled out rapidly. Duluth's

public-school system by now has installed a filter on at least one water tap on each floor of each school. But even that is too much precaution for one member of the local board of education. "Filtered water doesn't taste good," he has complained. "We ought to have unfiltered water for those who prefer it."

When I visited the Duluth area a year after the first announcements of the cancer risk, I was eerily reminded of "On the Beach," the novel in which people are going about their usual ways knowing about, but seemingly ignoring, the inescapable radioactive death engulfing them. The difference being, of course, that a good deal could be done to avoid Duluth's cancer peril, although very little apparently was being done.

Gene Kaari of Silver Bay, a filter operator for Reserve with 14 years' seniority, seemed to be expressing the general attitude—or, at least, a widely enunciated one: "What's the matter with the water? We've been drinking it for all these years, and we're healthy. If there is a health hazard, it should have shown up before this."

Was there more than mere ignorance of the basic chronology of asbestos cancer—of the 20-year minimum dormancy period—beneath the apparent indifference? "Let's face it," said Jeno Paulucci, a flamboyant and independent-minded food processor and Duluth's best-known businessman, "I would imagine that a good share of our population, whether it's banker or printer, survives on the mining industry. So what they have in their hearts, they don't say in their minds. So there's damned few that speak about the potential horror of Reserve's pollution because there's fear. There's fear of losing business. They are afraid of the iron-mining business going away. I say, Jesus Christ, how stupid can you be?"

Those most fearful about the economic consequences, of course, are the residents of Silver Bay, Reserve's employes. They have been living for more than a year now under the cloud of the possibility that their jobs will be swept from under them—a possibility made intensely real by the fact that the plant actually was shut down for a moment.

At that point, according to newspaper interviews at the time, Reserve's workers seemed to be almost unanimous in condemning the shutdown and Judge Lord, even though their own health may have been most imperiled of all by the plant's operation. (Reserve's smokestacks are said to be spewing asbestos fiber in the air around Silver Bay. So both the town's air and water supply may be carcinogenic.) Since the temporary shut-down, however, there has been a growing resignation to the likelihood that some major change in the plant's operation and in the lives of the workers is inevitable. "We just wish the courts would hurry up and decide what to do," said Silver Bay's Mayor, Melvin Koepke, who works at Reserve himself.

A slight, soft-spoken man, Koepke was wearing unfaded blue jeans and looking after three potatoes in aluminum-foil wrapping as they baked in a circular barbecue rack when I talked to him. Wasn't he worried about getting cancer, I asked, trying to elicit some sense of what it was like to be earning a living at a job that might be killing you. "Well, you know, you hear about getting cancer from cigarettes and food additives, and people don't stop smoking or eating," he said, shrugging. Was anyone in Silver Bay bothered? Yes, he said, but they wouldn't discuss it with an outsider.

The next day, however, I arranged a clandestine meeting with a Reserve worker several miles from the town. The worker acknowledged he was concerned about the health threat *and* about his job. He was a bachelor, he explained, but if he had children, he would be less worried about his health and more worried about losing his job.

T he Federal Court case against Reserve began on a warm August day last year in the modern Federal building in Minneapolis. The trial proceedings lasted almost

without interruption until April of this year. The proceedings cost by one estimate more than $12-million and produced 20,000 pages of transcript, including an incredible amount of complex scientific testimony and countertestimony.

On one side were arrayed the U.S. Government; the states of Minnesota, Wisconsin and Michigan; the cities of Duluth and Superior, and half a dozen environmental groups. On the other: Reserve Mining, supported by the Northeastern Minnesota Development Association, the Duluth Area Chamber of Commerce, the villages of Silver Bay and Babbitt (where Reserve's mine is located) and others. In the course of the trial, Reserve's co-owners, Armco and Republic Steel, were named as co-defendants by Judge Lord.

Both sides marshaled scores of expert witnesses whose data and analysis reached into many arcane corners of science, ranging from oceanography to epidemiology to microscopy. Some of the testimony revolved around the old question as to just where Reserve's tailings went after they were dumped into Lake Superior. Reserve acknowledged that cummingtonite was a major ingredient in its discharge and that the mineral appeared in water samples taken far from the plant.

But Reserve witnesses testified that cummingtonite was also found in the suspended sediments of 60 rivers that emptied into the lake. Government witnesses said that the X-ray analysis upon which this contention was largely based was faulty; they said that Reserve was the only major source of cummingtonite and that microscopic particles of the mineral found in the water supplies of Duluth and other downcurrent communities were therefore definitely traceable to Reserve's outpouring.

The focus of the scientific arguments, however, was the asbestos issue, which revolved around several major scientific premises that were extensively debated:

■ *That a substantial portion of Reserve's cummingtonite discharge is similar or identical to amosite asbestos, chemically and morphologically—*

in form and shape. Reserve witnesses stressed what they said were morphological differences: A crystal of amosite, they testified, was composed of bundles of fibers, whereas a cummingtonite crystal was more like "one coherent fragment." Not so, said Government witnesses; Reserve's tailings appeared in electron microscope photography as bundles of fibers "indistinguishable" from amosite asbestos.

■ *That small fibers are as dangerous as larger ones.* The Reserve particles in question were under five microns in length — less than one five-thousandth of an inch — whereas Federal safety standards covering occupational exposure to asbestos restrict only the level of particles over five microns long. The Government said that this restriction existed because particles shorter than five microns were difficult to measure until the development of electron-microscope techniques. Reserve said this restriction existed because particles under five microns are not hazardous.

■ *That ingesting, as well as inhaling, asbestos fibers could cause cancer.* The great bulk of the concrete evidence on asbestos-related cancer comes from industrial cases in which workers have inhaled asbestos particles. But Government witnesses, including Dr. Irving Selikoff, a leading authority on asbestos-related health problems, testified that high gastro-intestinal cancer rates were found in asbestos workers and theorized that such cancer was caused by workers coughing up inhaled particles and then, in effect, ingesting them. Reserve witnesses referred to animal studies which suggested that ingestion of asbestos did not cause cancer.

■ *That there is no safe level of exposure to asbestos.* Government witnesses cited cases in which exposure to asbestos for only a short time or in relatively small concentrations—in an asbestos worker's home, for instance — was enough to cause cancer. The question of a safe or "threshold" level of exposure was important because of both the difficulty of measuring fiber levels in water supplies

and the certainty that asbestos levels in Superior water were considerably less intense than in occupational settings upon which epidemiological evidence of a hazard was based. The Government presented evidence that there was no safe level; Reserve presented evidence that there was.

"The available scientific and medical evidence," Reserve stated near the end of the trial, "compels a conclusion that no possible health hazard can be attributed to Reserve." The Government, on the other hand, argued that it was "probable" that ingestion of Reserve's fibers "will produce cancers and an increased rate of death." Missing from the wide gap between the impossible and the probable were "the dead bodies," as they have been inelegantly referred to—the presence or absence of cancerous deaths clearly attributable to Reserve's discharge.

A tissue test of Duluth residents who had been drinking Superior water for most of the time Reserve had been operating was conducted during the trial, but the test results were inconclusive. Nor did one court-appointed expert, Dr. Arnold Brown, a research pathologist associated with the Mayo Clinic, sustain either the probability or the impossibility. From his review of the scientific literature, he said, "I would be unable to predict . . . that there will be an adverse effect" from Reserve's discharge. Yet he also testified that "until we know what safe levels are, as a physician, who would rather see well people than sick people, I have some sort of compulsion to protect ourselves against known agents that produce cancer. . . . "

T hat same compulsion evidently was what moved Judge Lord. He also seems to have been moved by the mounting suspicion that he and the legal process were being used by Reserve—that whatever the merits in either side's case and whatever the eventual outcome, the long and lengthening trial was serving Reserve's commercial purposes by prolonging its use of its present equipment.

A turning point in the trial came with the surfacing of concrete evidence that Reserve was using the trial for just such an end. Alongside the scientific arguments and counterarguments was waged an almost equally complex battle over the question of how feasible, financially and technologically, it was for Reserve to alter its operation. Reserve officials insisted that the company could not afford to quit dumping in the lake, that on-land disposal of its tailings would involve a vast overhaul of its operation and cost upwards of $500-million and that the company could not run a profitable operation with on-land disposal.

The company also argued that it would take from three to five years for it to convert to on-land disposal and that, if it were forced to discontinue dumping in the meantime, it would have to close down and lay off its 3,000 workers. The Government, on the other hand, presented evidence that it would cost only half as much to convert as Reserve claimed, and that it would be less costly for Armco and Republic to overhaul Reserve's operation than for the steel companies to turn to the open market for their iron.

The Government, however, seems to concede that Reserve's operation would be less profitable. That concern, along with a desire to get as much use as possible out of its present equipment, appears to be the driving force behind Reserve's resistance to mending its water-polluting ways.

In order to weigh the economic aspects of the case, Judge Lord, from early in the trial, tried to find out from Reserve officials what on-land

disposal plans had been worked out in the event that he prohibited them from further discharging in the lake. The Reserve officials insisted that their only alternative plan was to pump the tailings out, still into the lake, through a deep pipe. But in mid-February, after the trial had been under way for nearly half a year, the Government subpoenaed documents from Armco and found a thick report of an "engineering task force" titled "Tailings Disposal, Alternate Studies, Final Recommendations, July 17, 1972." In the report, it was concluded that the deep-pipe system was "not technologically or economically feasible," and that an on-land disposal system, one which had been thoroughly looked into, appeared to be the only method that would satisfy regulatory agencies as well as technological requirements.

"What should I do to the sons-of-bitches?" Lord asked a visitor in his chambers. And in court he was not much more restrained. "This court has been misled for five months on a plan you weren't even serious about," said Lord, who had once been a Golden Gloves boxer and now sounded as though he were tempted to take up the sport again. Lecturing C. William Verity Jr., chairman of the board of Armco, near the final moments of the trial, he went on: "The people of Duluth had that unwelcome addition to their diet for five months while you made $5-million profit." A few days later, Lord closed down the Reserve plant.

A basic difference in legal philosophy, as well as a dif-

ferent reading of the testimony and evidence, separated Judge Lord's decision to close the Reserve plant immediately from the Appeals Court's stay of that decision. Lord weighed the possibility that further discharge endangered the lives of thousands of people against the economic damage to Reserve, its owners and its employes. He evidently also put on the scales Reserve's and its owners' apparent duplicity — and ruled in favor of halting the health danger. The Appeals Court, on the other hand, found that "although Reserve's discharges represent a *possible* medical danger, they have not in this case been *proven* to amount to a health hazard" (italics added), and decided in effect that the degree of risk was less weighty than the "urgency" it found in the economic dislocation caused by the shutdown.

Environmental lawyers are worried about the legal precedent the Appeals Court's ruling, if upheld, may set concerning the need to establish scientific certainty about a health hazard before relief is granted. Still, Reserve's challengers take comfort in the fact that both District and circuit courts have said that Reserve must halt its pollution. Even the United States Supreme Court, while it has rejected Government appeals to overturn the circuit court's stay of Judge Lord's plant-closing injunction, has indicated it might be willing to uphold Lord's injunction if an on-land site is not soon agreed upon and if plans are not set in motion to stem the daily flow of the 67,000 tons of refuse into the once pristine lake.

In surveying the quarter-century course of Reserve,

from the company's welcome on Lake Superior to the current tide sweeping it off, the Appeals Court's neutral moral verdict — "neither heroes nor villains" — may stand the test of further history and of greater detachment. Others, however, find the moral lines quite clearly and oppositely drawn; and some would include the Appeals Court itself in the cast of villains.

By any reckoning, Reserve has hardly acquitted itself heroically. It has spent at least $6-million so far in legal costs in order to keep dumping in Lake Superior. It is still spending and still dumping. Thus, it is hard to dispel the suspicion, which provoked Judge Lord, that Reserve looks at such legal costs as simply another operating outlay, like the expense of new linings for its rock pulverizers. Nor can Reserve find moral relief, even if it has found legal relief, in the distinction that its discharge may be only a *possible*, rather than a proven, threat to the lives of thousands.

One's moral, or amoral, weighing of the case at this juncture must ultimately depend to a great degree on how great a health risk Reserve's discharge is seen to be. But given the long dormancy period involved in any cancer that may result, it may be another five or 10 years, or more, before irrefutable evidence is in. At that point, finally, the "monumental environmental mistake" on Lake Superior will reach a conclusive denouement, either in an epidemic of cancerous death or — the happy ending — in a relieved glance backward at a dreadful hypothesis that for some reason never materialized. ■

November 24, 1974

HIGH COURT KILLS NIXON'S BLOCKING OF WATER FUNDS

By WARREN WEAVER Jr.
Special to The New York Times

WASHINGTON, Feb. 18— The Supreme Court ruled

unanimously today that President Nixon did not have the right to impound $9-billion in water pollution funds approved by Congress because the legislation authorizing the program had not given him that authority.

The case, the first on impoundment to reach the high court, did not involve the more complicated question of

whether a President has implied power under the Constitution to refuse to spend money that Congress has appropriated. That issue is still being debated in the lower courts.

Although they did not say so directly, the Justices appeared to order the immediate release of $5-billion of the funds that Mr. Nixon impounded, and President Ford

has since classed as "deferred," or indefinitely postponed. Mr. Ford released $4-billion on Jan. 27.

Allocation Process Slow

As a practical matter, however, the decision may have little immediate impact. A spokesman for the Environmental Protection Agency said that the $5-billion had been considered available for some time, and that the allocation process was so slow that it would not be tapped for many months.

In New York and Connecticut, environmental officials estimated that the two states would eventually receive at least $690-million as a result of the Court's ruling.

What the ruling might mean for the remaining $8-billion or more that the Ford Administration is still withholding in construction, health and other environmental funds was anybody's guess.

White Writes Decision

In programs based on statutes like the Water Pollution Act of 1972, the money may now have to be released, following today's precedent. In other programs where a different sort of authority was invoked by the Administration, the money may remain impounded, awaiting the outcome of further legal action.

The decision was a victory for New York City, which had challenged Mr. Nixon's order to restrict Federal aid for sewers and sewage treatment works. The former President had impounded the $9-billion immediately after Congress overrode his veto of the Water Pollution Act.

Writing for the Court, Associate Justice Byron R. White said that the legislation "was intended to provide a firm commitment of substantial sums within a relatively limited period of time in an effort to achieve an early solution of what was deemed an urgent problem."

"We cannot believe," he continued, "that Congress at the last minute scuttled the entire effort by providing the executive with the seemingly limitless power to withhold funds from allotment and obligation."

Associate Justice William O. Douglas concurred in the result

of the decision, but did not offer any reason for not also endorsing the legal reasoning supporting it. The high court presumably decided the case shortly after it was argued in November, long before Mr. Douglas was hospitalized by a stroke on Jan. 1.

New York City, joined by several upstate cities and later Detroit, won its case in Federal District Court, and the United States Court of Appeals for the District of Columbia Circuit affirmed the ruling, holding "the act requires the administrator to allot the full sums."

When the Government asked the Supreme Court to review the case, the Justice Department argued in its petition that Congress could not impose this kind of restriction on the President under the doctrine of sovereign immunity. This argument was dropped, however, in the Government's subsequent briefs and oral argument.

Position Is Changed

In the lower courts, the Government maintained that the President had unlimited

power to impound funds. In the Supreme Court, however, the Administration conceded that the full $18-billion approved by Congress for water pollution projects would have to be spent eventually, but not necessarily on the Congressional schedule of $5-billion in 1973, $6-billion in 1974 and $7-billion in 1975.

In his Budget Message this year, President Ford listed the $9-billion at issue as "deferred," a classification set by the 1974 Congressional Budget Act for money that will be spent eventually but not during the coming fiscal year.

Mr. Ford said at that time that "release of all these funds would be highly inflationary, particularly in view of the rapid rise in non-Federal spending for pollution control."

Of a half-dozen similar lawsuits brought to win release of the water pollution funds, the Government lost all of them in the lower courts except for one in Federal District Court in California.

February 19, 1975

Study Finds Chemical Pollution Of Drinking Water in 79 Cities

By HAROLD M. SCHMECK Jr.
Special to The New York Times

WASHINGTON, April 18—

The Environmental Protection Agency has found the drinking water of 79 American cities polluted with traces of organic chemicals, including some that are suspected of being causes of cancer.

The agency has concluded that the problem exists throughout the country. The health consequences of the pollutants, if any, are unknown but under study.

Cities sampled in the survey included New York, Buffalo, Rhinebeck, Waterbury, Conn., Passaic Valley, N.J., and Toms River, N.J. New York City is one of 10 for which more detailed studies are to be made but the analyses are not yet complete.

"People should not react with any sense of panic, but they should know there is a problem," said Russell E. Train, the agency's administrator, at a news conference today.

The survey concentrated on six chemicals, two of which—chloroform and carbon tetrachloride—are considered potential causes of cancer. Traces of chloroform were found in water from every city tested.

In a few cities, where test were done in greater detail, traces of other hazardous chemicals were found. These included dieldrin, a pesticide, and vinyl chloride, which has been linked to cancers in persons heavily exposed to it.

The quantities of pollutants found in the survey were extraordinarily small. Chloroform, for example, had a range of less than one-tenth of one part per billion to 311 parts per billion.

"Even at these low levels, the chemicals are a matter of concern that warrants the diligent carrying out of our safe drinking water plans," said Mr. Train.

"These plans include determining how widespread the problem is how significant these low levels are to human health, what is the source of these chemicals and what can be done most practically to solve the problem," he said.

The six chemicals were chosen for the survey because they were prevalent, could be analyzed rapidly and some were thought to be byproducts of the chlorination process by

which drinking water is purified for human use. In addition to chloroform and carbon tetrachloride, they include bromodichloromethane, dibromochloromethane bromoform and 1,2 dichloroethane.

Organic chemicals are compounds that contain carbon, hydroben and, often, other elements. The organic chemicals are associated with living things and include many manmade compounds.

"Today's survey was designed to provide a cross-section of the national problem," said Mr. Train. "In the U.S. today there are approximately 240,000 drinking water supply systems of significant size. Today's results indicate that we will probably find one or more of the six chemicals in most of the nation's chlorinated water systems."

Chlorination Method

Although chloroform is among the compounds thought to be a byproduct of chlorination, Mr. Train emphasized that chlorination remains the single most effective method of preventing serious water-borne bacterial diseases such as typhoid and cholera.

"We continue to believe that the benefits of chlorine used to prevent immediate, acute biological diseases far outweigh the potential health risks from chlorine-derived organic compounds," he said.

There is now simply no single proved method for dealing with all aspects of the problem of organic pollutants in water, Mr. Train said.

The broad survey of water

in 80 communities throughout the United States began after the discovery last year that drinking water supplies for New Orleans, originating from Mississippi water, contained traces of 66 different chemical pollutants, including the six covered in the survey.

Hopewell, Va., one of the 80 communities, will not be sampled until later this month. For that reason, the survey results today listed only 79 cities.

Ten cities were selected for more detailed study. The results for five of these were made public today. These were Miami, Seattle, Philadelphia, Cincinnati and Ottumwa, Iowa. Traces of 36 compounds were found in water from Cincinnati and Philadelphia, 35 from Miami and lesser numbers, so far, from the other cities.

The survey was based on samples from water supply systems, but did not necessarily cover all the sources of supply for a given city.

In most cases the specific source was not listed. In other cases, such as Passaic Valley, it was. An E.P.A. spokesman identified this as the Passaic Valley Water Commission plant serving Passaic, N.J.

In the survey of six chemicals, the sample from New York City showed traces of only three: 22 parts per billion of chloroform; 7 parts per billion of bromodichloromethane and nine-tenths of one part per billion of dibromochloromethane.

April 19, 1975

ALLIED CHEMICAL GETS A FINE OF $13 MILLION IN KEPONE POLLUTING

PENALTY IS MAXIMUM ALLOWED

Judge Gives Company 'No Credit' for Not Contesting 940 Counts of Dumping Into James River

By BEN A. FRANKLIN
Special to The New York Times

RICHMOND, Oct. 5—The Allied Chemical Corporation was fined the maximum sum of $13,375,000 in Federal district court today for polluting the James River for nearly four years with the persistent, highly toxic insecticide Kepone.

Judge Robert R. Merhige Jr. said that he had given the company "no credit" for having pleaded no contest—which has the same effect as pleading guilty—to 940 counts of knowingly dumping Kepone-laden process water and other chemical discharges into the river, and thence into the Chesapeake Bay, from its chemical plant in Hopewell, Va., just south of here.

The United States attorney here, William B. Cummings, charged in a series of indictments last summer that Allied officials in Hopewell had conspired to withhold from the Federal Environmental Agency information about the toxic discharges—data that would have alerted the Government earlier to what the prosecutor called "this greatest disaster of the environmental decade."

Acquitted of Conspiracy

Judge Merhige acquitted the corporation of the conspiracy charge at the end of a four-day trial, concluded last Thursday.

What had to be done today was the sentencing of Allied and the sentencing of a former Kepone-manufacturing subcontractor of the chemical corporation and of two former Allied executives, Virgil A. Hundtofte and William P. Moore. They had pleaded nolo contendere, or no contest, to a variety of Federal pollution violations between 1971 and 1974.

The separate criminal actions completed before Judge Merhige today were brought under the recently invoked antidumping provisions of the Federal Refuse Act of 1899 and under the Water Pollution Control Act of 1972.

The $13.3 million fine levied against Allied, which has 90 days to pay, was believed to be the largest ever imposed on a corporation or an individual for polluting the nation's waterways. Allied lawyers here said that none of the fine could be covered by insurance.

Other Fines Recalled

A spokesman for the Environmental Protection Agency said that the largest pollution penalty he could recall was $7-million levied against the Ford Motor Company for not complying with emission control regulations.

He said that Exxon had been fined $100,000 for illegal discharges into a waterway but that the Kepone fine was apparently the first fine for toxic chemical discharges into a waterway.

Noting the severity of the sentence, Judge Merhige told a courtroom at 6 P.M., after eight hours of argument by corporate lawyers for mitigation of the penalty, "I hope, after this sentence, that every corporate employee who has any reason to believe that pollution is going on will say to himself, 'I'd better do something about this if I want to keep my company, if I want to keep my job.'"

After giving much lighter sentences, virtually the minimum, this morning to Mr. Hundtofte and Mr. Moore for their roles in operating the Life Science Products Company, the Kepone polluting subcontractor of Allied, Judge Merhige tonight offered Allied, which had $2.5 billion in sales last year, a possible reduction of the massive fine.

He was making no promises, the judge said. But if, in the 120 days Federal courts allow for a convicted defendant to apply for reduction of sentence, Allied Chemical "takes some voluntary action to help the people who have been directly hurt" by the widespread Kepone contamination here, the judge said he would not have "a closed mind" about a motion to mitigate his sentence.

Fishing on River Stopped

He referred, in mentioning those "directly hurt," to about 80 former Life Science Products workers who suffered neurological and other disturbances as a result of exposure to the chemical, and to scores of fishermen on the James River who have been denied their livelihoods because the state has closed the poisoned river to fishing.

Sentencing today ended the criminal aspect of Virginia's 18-month Kepone scandal, but the disabled former Life Science workers and the James River fishermen have filed nearly $200 million worth of civil damage claims in separate suits against Allied Chemical. Those cases are also to be heard by Judge Merhige—if they come to trial.

The judge's suggestion to Allied today

that its sentence might be reduced added to the speculation here that the corporation was more likely to seek out-of-court settlements with the civil plaintiffs than to undergo another long trial, with Kepone-affected, trembling witnesses, that might be more damaging to the company's repute than the criminal proceedings.

Judge Merhige seemed to be suggesting that the corporation be generous in any such settlements. He said today that he had explored the possibility of channeling some of the fine money to the Kepone victims but had discovered that that "cannot be done." The $13.3 million will be paid directly to the United States Treasury.

Commenting indirectly on the pressure he had placed on Allied with his suggestion, Judge Merhige quipped, "People think that if you are a Federal judge you must not have any good commercial sense." There was laughter in the courtroom, except from the defendant's table.

Deliberately Took Time

It was one of the few moments of wry humor provided today by an otherwise grave judge.

"The environment belongs to every citizen, from the lowest to the highest," the judge told the corporation's lawyers, as he opened his sentencing statement. "I took so long today because I wanted to be sure that my feelings about pollution were not so strong that I was forgetting to temper justice with mercy.

"But I am satisfied that we, as a nation, are dedicated to clean water. I disagree with the defendant's position that this was all done innocently. I think it was done as a business necessity, to save money. I don't think we can let commercial interests rule our lives."

Judge Merhige could have sentenced Mr. Hundtofte and Mr. Moore to long prison terms—81 years and 153 years, respectively. Also, fines could have totaled nearly $2 million for Mr. Hundtofte and $3.8 million for Mr. Moore. Neither is a wealthy man.

Instead, the judge suspended for each of them all but $25,000 of their fines and gave them five years under court probation but with no prison terms, to pay them. Both men had pleaded guilty in pre-trial bargaining with the Federal prosecutor and had cooperated fully with the government.

Judge Merhige's $3.8 million fine against Life Science Products Company, the dismantled Kepone plant in an old Hopewell gasoline filling station, was even more a token. Mr. Cummings, the prosecutor, noting that the defunct company now had assets of only $32, said "that is one fine that will never be collected."

Kepone, the chemical that led today's sentencing is a nonbiodegradeable chemical cousin of DDT that is retained in the liver and fatty tissues of animals and humans. It has been used chiefly in the tropics for the control of fire ants and banana pests. The product was introduced at Allied Chemical in the 1950's and manufactured intermittently at its Hopewell plant until 1974.

October 6, 1976

The Great Lakes Have A New Enemy: the Air

By GLADWIN HILL

"The Great Lakes are dying. . . ." "The Great Lakes are getting cleaner. . . ." "The Great Lakes are worse. . . ." The disparate reports depend on where the pollution studies are made.

Indeed much of the community and industrial waste that has been befouling the lakes is being abated. But new sources of pollution are now being discovered that raise questions as to how affective conventional abatement efforts will be in restoring the lakes to something like pristine quality. There have been reports in recent weeks, for example, of the danger of Mirex, a toxic compound, to Lake Ontario. The New York State Department of Environmental Conservation has called the contamination an environmental tragedy and has said it would cancel a salmon-stocking program in the lake.

The new cry of the limnologists, however, is the words of Chicken Little: "The sky is falling"; it now develops that a significant portion of the Great Lakes' contamination is coming out of the air.

The five lakes—Superior, Michigan, Huron, Erie and Ontario—are a truly elephantine work of nature. They contain one out of every five gallons of the world's fresh water. They cover 95,000 square miles, an area nearly twice as big as the state of New York. Extending for 800 miles, from Minnesota to the St. Lawrence River, they drain an area of 300,000 square miles that includes some of the most densely populated and heavily industrialized sections of the United States and Canada.

The fluid wastes of 47 million people and hundreds of industrial establishments drain into the lakes. Despite their vast water volume (65 trillion gallons) the waste load began a generation or more ago to exceed the lakes' natural capacity to neutralize extraneous material, mainly by oxidation.

Lake Erie began to stagnate, its celebrated whitefish replaced by great smelly masses of seaweed-like algae, whose proliferation is blamed to a considerable extent on the detergent phosphates in sewage. There were signs that the other lakes were heading in the same direction.

Four years ago the United States and Canada which share ownership of all the lakes except Michigan, launched a major joint effort to halt this systematic pollution.

Billions of dollars have been committed to installing hundreds of elaborate municipal and industrial facilities for cleaning up fluid wastes before they gravitate into the lakes. The implementation is impressive. At Niagara Falls, finishing touches are being put on a huge, fortress-like concrete structure that will be one of the world's biggest sewage treatment plants, a $60 million system to eliminate Niagara River pollutants that have long assaulted the sensibilities of visitors to the Falls. At Muskegon, Mich., as part of a $44 million multi-community project, 50 giant sprinklers spray purified effluent on miles of cornfields to sidetrack phosphates.

No one has determined precisely how much pollution has been abated. But Ohio beaches on Lake Erie are reopening, and trout and salmon have returned to the Detroit River, the lake's biggest tributary.

The U.S.-Canada International Joint Commission, in its latest annual assessment, has reported that 94 percent of the Great Lakes Basin population on the Canadian side and 60 perce-- -n the American side now have adequate sewage treatmer

But the commission said the lakes still have 63 "water quality problem areas," involving 66 municipal and 104 industrial pollution sources, notably because of still-incomplete sewage treatment plants, handling both community and industrial discharges, at Duluth, Minn., Detroit, Gary, Ind., Cleveland and Tonawanda, Buffalo and Syracuse in New York.

Meanwhile, the perceived dimensions of the problem have grown, through many research projects, far beyond the original conception of cities and industries and familiar chemicals as the basic culprits.

Examinations of fish and aquatic plants show excessive levels of such persistent man-made chemicals, as the pesticide DDT and Mirex, and PCBs (poly chlorinated biphenyls), a family of industrial compounds, Both Mirex and PCBs are suspected of having cancer-causing properties.

The Worst Jolt of All

But the worst jolt is the new finding that much of the contamination comes not through pipes, where it can be controlled, but from the air.

Dr. Thomas J. Murphy, a chemist at DePaul University in Chicago, got to reflecting on the fact that much of the Great Lakes' water—50 percent in the case of Lake Michigan—comes not from surface flows but from rain and snow. He decided to find out just how dirty that 32 inches of annual precipitation is.

His initial inquiry focused on phosporus, the elemental parent of phosphates. Containers were placed, around the Lake Michigan shore in Illinois and Wisconsin. After storms the contents were analyzed. The phosphorus fallout was calculated at 1,100 tons a year. This is 18 percent of the estimated amount of phosphorus in the lake.

When current goals for reducing the phosphates in municipal and industrial sewage are achieved, Dr. Murphy said, the unreduced atmospheric component will be even more significant, amounting to 30 percent of the total amount.

The experiment was repeated with PCBs. The finding was that rain and snow precipitate more than 1,400 pounds of PCBs into the lake in a year, nearly as much as the amount flowing in from known surface sources.

Where these atmospheric contaminants are originating is not known. Winds regularly transport dust particles hundreds and perhaps thousands of miles. Because there is no way of controlling this atmospheric transport, the discovery magnifies the importance of dealing with the surface flows of pollutants.

The extent of other contaminants descending from the sky remains to be determined. But a new chapter has opened up in the ambivalent saga of the Great Lakes' restoration.

No steps have yet been taken by any governmental body to deal with this problem, primarily because few appear to know how to approach it, having learned only recently of the extent of the air pollution on the Great Lakes. But the recent passage of the Toxic Substances Control Act is expected to help in the cleanup.

Gladwin Hill is the New York Times national Environment correspondent.

October 10, 1976

Chemical, Paper and Metal Industries Say Economy Will Be Hurt by Costs of Achieving Zero Pollution

By GENE SMITH

The chemical, paper and pulp, metal and mining industries are stepping up their opposition to strict pollution controls.

These industries, in particular, have in recent months released detailed surveys aimed at showing that the costs would be prohibitive if they are forced to comply with the goal of the Water Pollution Control Act of 1972—zero discharge of pollutants into water by 1985.

U.S. Steel Warning

Judson Hannigan, president of the International Paper Company, said at a recent meeting of the Syracuse Pulp and Paper Foundation that, if such laws were enforced, "the economic impact of the capital and operating costs involved during the next 10 years will drastically change the competitive posture of many of the major basic industries of the United States, like steel, aluminum, chemicals, petroleum and pulp and paper."

Edgar B. Speer, chairman of the United States Steel Corporation, said last month at the annual luncheon of the Pennsylvania Chamber of Commerce that the nation's air is much cleaner than required by Federal law and that, unless the Clean Air Act of 1970 is modified, industrial expansion could be adversely affected.

Mr. Speer also warned that, if the law was not amended, the steel industry might not spend some $36 billion for expansion and modernization and an additional $12 billion for "nonproductive air - pollution control equipment." He added:

"It would be difficult to find a more effective method for bringing economic growth to a halt."

United States Steel's chief executive said his industry had already committed close to $3 billion for pollution control equipment and had eliminated or was controlling more than 90 per cent of its emissions.

'A Natural Reaction'

Peter B. Lederman, acting director of the industrial and extractive processes division of the Environmental Protection Agency, says he feels that the concerted efforts of businessmen against environmental control laws is "just a natural reaction."

In an interview at the 35th Exposition of Chemical Industries at the New York Coliseum last Monday, Dr. Lederman said he was "not surprised" to see the protests coming from the various industries.

"In the present crunch, it's easy to use environmental controls as a scapegoat," he said. "Personally, I have found that implementation of our regulations does not necessarily cost as much as industry estimates.

'Of course, there are some definite problem areas, but we're flexible. We hold a carrot out to industry and it's up to them to come in with their ideas. Personally, I find that we've done a good job when we get complaints from both sides—industry and the environmentalists."

Speaking at a seminar at the chemical show, Robert Schaffer, director of E.P.A.'s permits division, said there were "about 60,000 industrial dischargers of waste into water in the United States." He reported that to date permits had been issued to 43,500 and that all were supposed to have them a year ago.

In previous talks, Mr. Schaffer said the E.P.A. would not extend 1977 deadlines for meeting industrial water-pollution control standards. Under the Water Pollution Control Act of 1972, all industry must have the "best practicable" pollution control technology in operation by July 1, 1977, to meet waste water discharge standards. (A National Water Quality Commission report has estimated that 10 percent of industry would not meet that deadline.) The act requires that industry, by 1983, use the "best available technology economically achievable."

Ten leading chemical companies—Allied Chemical, Dow, DuPont, Exxon, Hercules, Hooker, Monsanto, PPG, Stauffer and Union Carbide—undertook a study of the impact of these requirements on their operations.

They found that they had already installed water-pollution abatement equipment worth $473 million and would have to invest $695 million more to meet the 1977 requirements. To meet the 1983 standards, they would have to invest an additional $941 million, or a total of $2.1 billion, just for existing chemical manufacturing facilities.

Assuming a reasonable rate of growth and the addition of new facilities, they estimated that their total investment in water-pollution control equipment by 1983 would be $4.5 billion and that the annual operating costs for such equipment would increase from $550 million in 1974 to $1.4 billion after 1983.

They also said: "Substantial reductions in pollutants discharged have already been achieved, and equally significant reductions are expected as a result of 1977 controls. By comparison, meager reductions will result from meeting 1983 requirements."

Estimate of Added Cost

Richard E. Chaddock, executive coordinator of environmental affairs for Hercules Inc., said it would require about an 85 percent increase in investment to meet the 1983 standards, plus a 75 percent increase in annual operating costs and a 130 percent jump in energy usage over 1977 standards.

The chemical industry's study called for the 1972 Act to be amended after an evaluation of what has been accomplished under the 1977 requirements.

The Kennecott Copper Corporation estimated that it would have to spend $240 million on top of the $345 million it has committed to build air-pollution control facilities at its four copper smelters. The company said the additional equipment would consume energy equivalent to 22 million gallons of oil a year. This would be in addition to the equivalent of 26 million gallons for the original equipment. Kennecott urged that the Clean Air Act of 1970 be amended to require the E.P.A. to consider the economic and energy impact before implementing air pollution regulations.

Cornell C. Maier, president of the Kaiser Aluminum and Chemical Corporation, reported that his company had spent $110 million since 1973 to meet environmental control standards and estimated that the mining and metal industry had spent or committed about $1 billion to comply with the Clean Air Act.

"This same industry is currently spending one out of every four dollars of capital expenditures for air control purposes. Clearly this represents major support and accomplishment under the Clear Air Act," he said.

Alexander Calder Jr., chairman of the Union Camp Corporation, said, "Environmental legislation that has been passed in the last five years is among the worst drawn up anywhere in the world." He said that he knew of quite a few paper mills that would have been built but were scrapped because of the "zero discharge" requirements.

Earlier this year, as chairman of the American Paper Institute, Mr. Calder led the industry's battle against the E.P.A. described existing legislation as "too foggy" and added that "we'll be in the courts for years trying to settle this issue."

The American Paper Institute released in September a detailed study on the economic impact of pollution control costs. The study, prepared by the URS Research Company of San Mateo, Calif., concluded that adherence to clean water and air legislation would force consumers' outlays for paper to nearly double by 1983.

December 8, 1976

E.P.A. Bans Discharge of PCB's Directly Into the Nation's Waters

By BAYARD WEBSTER

The Environmental Protection Agency yesterday ordered the ban of the direct discharge of PCB's, a highly toxic industrial chemical, into United States waters.

The chemical, a close relative of DDT, has been found in scientific studies to cause deformities in fetuses, changes in liver function, nervous disorders and cancers in animals. Widespread in the environment, it is found in almost all major bodies of water in the world. Significant amounts have also been monitored in the air.

The ban follows results of recent studies that show that its levels in water and fish exceed by several factors those standards set by the E.P.A. and the Food and Drug Administration.

The only plants covered by the ban are some 20 factories that manufacture electrical transformers and capacitors and discharge their PCB's, used in electrical insulation, into bodies of water.

The E.P.A. noted in announcing the ban late yesterday that "past widespread use of the chemicals in the production of lubricant additives, hydraulic and compressor fluid, carbonless copy paper, plasticizers, paints and other products has resulted in PCB's being present throughout the environment.

"Although most of these uses have now been substantially curtailed, PCB's which have entered the environment cannot in most cases be recovered and will require many years to degrade. The public will be alerted to potential hazards by careful long-term monitoring of PCB levels in food."

One of the most prominent PCB contamination cases involved the General Electric Company, which had been dumping its PCB wastes into the Hudson River, causing fish to accumulate many times the permissible level of the chemical. A negotiated settlement between the company and the State Department of Environmental Conservation resulted in G.E.'s agreeing to cease its dumping and to pay $3 million toward cleansing the river and $1 million for research toward ending the problem.

The ban did not cover the PCB wastes of transformer plants that discharge their wastes into municipal sewage systems. Such "indirect" discharges, the E.P.A. said, would be taken care of by additional regulations being prepared by the environmental agency.

The ruling, which in effect calls for the zero discharge of PCB's into main bodies of water, follows the recent studies that indicated that the recently proposed permissible standard of one part per billion of PCB's in transformer plant waste waters could not be met.

"The present problem of PCB contamination in the environment is so severe that in many waters throughout the United States PCB loads are already in excess of the criteria," E.P.A. Administrator Russell E. Train said in announcing the ban. His action came on the final day under the law for a decision in the PCB case, which had been argued most prominently by the Environmental Defense Fund, a Washington-based environmental law firm principally responsible for the banning of DDT in 1972.

The industrial chemical is widely used for insulating electrical equipment, in the recycling of wastepaper and in metal casting plants. Because of its widespread distribution in the environment, its resistance to biological degradation with a half-life of more than 25 years, PCB's are considered one of the most serious of the many environmental contamination problems prevalent today.

Monsanto Industrial Chemicals Company, the only American maker of the chemical, announced several months ago that it would quit production of the substance by October 31 of this year. There are no restrictions on the importation of PCB's. The Food and Drug Administration has already banned use of the chemical in the processing of food and feed, where it was sometimes used as cloth and paper insulation in containers and cartons.

January 20, 1977

Flag-of-Convenience Oil Tankers Magnifying Concern About Spills

By JOHN KIFNER

The wreck of the Agro Merchant, spilling 7.5 million gallons of oil off Nantucket Island last December, and the rash of tanker mishaps that followed, have aroused national concern over oil pollution and over the ships flying "flags of convenience" that bring in much of this nation's oil.

Senate hearings are under way and Brock Adams, the new Transportation Secretary, has announced a set of safety rules, which have been long in the making.

But the Coast Guard has had the authority to make and enforce safety regulations since 1972 and has not done so. Possible Congressional action includes a bill to fix liability for oil spills and their damages, but similar legislation failed under oil industry pressure last year.

Thus, largely because of the flag of convenience arrangement, shipping has become an industry without effective regulation.

The series of accidents is not so much a sudden phenomenon as part of a growing problem, atributable to the increasing imports of oil and the economic organization of the shipping industry, a five-week examination has found.

The examination included interviews with government officials, Congressional sources, industry figures, conservationists, fishermen, Coast Guard officers and scientists, as well as the study of governmental, academic and industry reports.

The Agro Merchant—an aged, rusting Liberian-flag tanker of murky ownership, chartered out to an oil company, manned by a polyglot crew and captained by a Greek officer who read his radio direction finder backwards—was not particularly unusual among the fleet that brings oil into ports along the Atlantic, the examination found.

Central to the situation is the flag of convenience arrangement under which countries like Liberia, a small, West African nation with no natural harbor, yet which in the last 30 years has accumulated the world's largest merchant and tanker fleets, offer shippers freer and more profitable operation.

Third Is American-Owned

About a third of the Liberian-flag fleet is owned by American interests.

Defenders of the system, who like to call it a "flag of necessity," contend that it is the only way American shipowners can stay in the competitive market. Its critics contend that the arrangement allows dangerously decrepit and illmanned ships to ply the seas.

Last year, a record 19 tankers were lost, according to preliminary figures kept by Arthur KcKenzie, director of the Tanker Advisory Center in New York. The lost ships' tonnage, 1,129,000 deadweight tons, was nearly 50 percent greater than the losses in 1975, when the previous record was set.

Of the 19 tankers, 11 flew the Liberian flag. Two of the ships were registered in Cyprus, a flag of convenience with even less restriction. Four of the remaining lost tankers were Greek, one was Spanish and one was East German.

Mr. McKenzie said he expected the final 1976 figures to show two or three more tankers lost and noted that the first nine months of the year had already set a record before the Argo Merchant had the largest spill in American waters.

The figures indicate the growing dimension of the problem.

45 Percent Imported

The energy-hungry United States is now importing 45 percent of its oil—6.3 million barrels a day, the equivalent of 35 Argo Merchant loads. The imports, already up from 36 percent before the Arab oil embargo in 1973 and 1974, are expected to continue rising.

Of the imported oil, 94 percent is brought in aboard foreign-flag ships, and 40 percent of that in ships of Liberian registry, according to former Secretary of Transportation William T. Coleman Jr.

Amid this flow of traffic, Mr. McKenzie told the Senate Commerce Committee, which is looking into the shipping industry, the rash of accidents was "not unexpected" and was "part of an evergrowing phenomenon."

Some of the incidents involving Liberian flag ships were commonplace and relatively harmless groundings. But they also included a major oil spill in the Delaware River, a breakup off Hawaii and an explosion in the Los Angeles Harbor that killed 11 sailors.

The severity of the pollution situation is underlined by the fact that the oil spills and wrecks like those that have aroused current concern account for only 10 to 15 percent of the two million tons of oil spilled into the oceans during transit, according to a 1975 study by the National Academy of Sciences.

Most of the spills occur in almost routine operating procedures, primarily the flushing of tanks, which again reflects the lack of effective regulation. While there are international rules against flushing tanks near shore, the practice, often done at night, is common.

International regulations calling for seregated ballasts have been approved for all new ships over 70,000 tons. But the rules, drawn up in 1973 but not in effect since not enough countries have ratified them—the United States is among those that have not—would exempt both the existing fleet and new smaller ships of the size that make up four-fifths of those in American waters

Not in Heavy Seas

There are technical problems, too. While the industry says it can control spills is seas of up to five feet, this proved of little avail in the seas so heavy they picked up the fenders placed around the Argo Merchant to contain the oil and flipped them onto the deck.

The two major factors involved in ship accidents, according to studies by Mr. McKenzie and others, are the age of the ship and human error, which often results from unseamanlike conduct of the crew.

Shippers who operate under the Liberian flag are able to make huge savings by hiring cheaper crews and do not have to meet the stricter safety requirements or undergo the inspections and crew-licensing procedures required by the United States and other traditional maritime nations, such as Britain.

The Liberian shippers contend that their safety record is comparable to those of other fleets, noting that between 1964 and 1967 the percentage of their tanker fleet lost—half of 1 percent—was about in the middle of the maritime nations. The high was Greece, with 0.76 percent of its tanker fleet lost. The United States record was 0.15 percent.

Biggest Number Lost

Because the Liberian fleet was larger, however, the percentages mean that the Liberian registry accounted for the highest number of ships lost, 68. The Greeks lost the next highest, 26. The United States fleets lost nine ships.

Philip J. Loree, the director of the Federation of American Controlled Shipping, an organization of 20 American oil companies and other shippers using the Liberian and Panamanian flags, contended that the ships represented by his group, averaging 100,000 tons, represented a substantial capital investment and were therefore treated carefully.

similarly, Guy E.C. Maitland, the executive director of the Liberian Shipping Council, an organization of 62 major international shipholders, says that the Liberian flag-fleet includes some of the most modern, largest and best-equipped tankers now in use.

But Robert J. Blackwell, the assistant secretary for maritime affairs of the Maritime Administration, told the Senate Commerce Committee that, while many of the Liberian-flag vessels were modern, "we have never seen that fleet."

Older Ships Here

For geographic and economic reasons, the ships that bring oil into American ports are, by and large, older ones.

Since the early 1960's, shipowners have been building bigger and bigger tankers, or Very Large Crude Carriers, as those above 200,000 deadweight tons are known.

But there are no deepwater ports on the Atlantic and Gulf Costs, and only a few on the West Coast, that can accommodate the draft of these large ships.

Thus, according to a recent study by the Coast Guard, 80 percent of the ships arriving in American ports are under 70,000 deadweight tons.

At the moment, there is a glut of tanker tonnage, which means that oil companies can choose ships offering to work at cheaper rates. And that means older, less well-maintained ships run by marginal operators are often the ones chosen. Further, there are economic pressures to cut costs on the crews or to steer shorter but more dangerous courses to save time and money.

Bad Ships Prevail

The wreck of the Argo Merchant, Mr. Loree conceded, "highlighted a problem" where,

particularly in the shallow waters of the North Atlantic, the economic pressures made for a "kind of Gresham's Law" under which bad ships drove out good.

Studies by Mr. McKenzie and others have found that tankers over 10 years old become increasingly subject to metal fatigue, making them more likely to break up.

More importantly, while the older coastal tankers flying the American flag are subject to Coast Guard inspections, those flying foreign flags have not been. So ships can be put under a Liberian or other flag and kept going for a few more voyages with little incentive to keep up their maintenance.

"Because of the shallow waters and the economics of oil transportation, what you get is smaller tonnage, which is older tonnage," Howard F. Casey, deputy assistant secretary for maritime affairs of the Maritime Administration, said in a telephone interview.

Reports Usefulness Ended

"They were probably registered under another national flag for part of their life, and frequently these are old crocks that have outlived their usefulness," he added.

"We are the garbage dump for the tankers of the world," Jessie Calhoun, the president of the National Marine Engineers Beneficial Association, an American union that has long opposed foreign flag shipping on the ground that it takes away jobs, told the Commerce Committee.

Human error was a major factor in more than 80 percent of the accidents at sea, according to a National Academy of Sciences study.

The Liberian authorities accept the papers of most other nations in licensing seamen and officers, although few nations accept theirs. A number of accidents involving Liberian-flag ships have been found to involve unqualified men.

While the Liberian flag in maritime practice is essentially the creation of interlocking economic interests, oil and shipping, it is treated as a sovereign nation.

Adm. William O. Siler, commandant of the Coast Guard, told the Senate Commerce Committee that his agency did not enforce safety standards on Liberian and other foreign flag ships, as it had been empowered to do by 1972 legislation, because he did not want to cause trouble abroad.

The Coast Guard, which has defended the

safety record of the Liberian-flag ships, has come under increasing criticism from members of Congress and environmentalists, who accuse it of footdragging in drawing up safety regulations under the 1972 Ports and Waterway Act.

Finds Lack of Enthusiasm

"It has proven less than enthusiastic in carrying out its mandate," Representative Norman D. Dicks, Democrat of Washington, said.

One area of criticism, for example, has been the Coast Guard's method of drawing up possible regulations for two proposed new tanker safety requirements, double bottoms and segregated ballasts. Every nongovernmental member of the Coast Guard's study group came from the oil or shipping industries. The regulations that were drawn up fit all the ships now under construction.

Last week, Secretary Adams announced a new set of regulations, long in preparation by the Coast Guard, requiring for the first time that ships coming into American waters have such basic equipment as magnetic compasses and radar.

Rules Don't Always Work

But in the traditionally risky and free-wheeling shipping industry, even the regulations that are in existence are not always effective.

One-way shipping lanes, for example, are voluntary and there have been a nnumber of near collisions of ships in the wrong lane. Tricks are often used, experienced captains say, to load tankers beyond their safety limits.

Jacques Cousteau, the underwater explorer, says that his men and equipment were almost run over several times off Greece last summer and were only saved when flares and blank pistols were fired at Liberian-flag ships to wake up helmsmen who were apparently asleep at the wheel.

Thus, it was in this kind of industry that the Argo Merchant, flagged as a dangerous and defective ship, banned from the port of Philadelphia and cited for deficiencies on a recent pilots' report in Panama, had its final rendezvous with a well-marked shoal off Nantucket.

February 13, 1977

High Court Upholds E.P.A. Curbs On Industrywide Water Pollution

By LESLEY OELSNER
Special to The New York Times

WASHINGTON, Feb. 23—The Supreme Court unanimously upheld today the Federal Government's authority to impose binding, industrywide regulations controlling the amount of pollutants that factories may dump into the nation's waterways.

The ruling, written by Justice John Paul Stevens, was seen as giving a substantial stimulus to the Federal effort to combat water pollution.

The acting administrator of the Environmental Protection Agency, John R. Quarles, hailed the action as "a very important victory," and said it strengthened the agency's ability to "maintain the water cleanup momentum."

The environmental agency has consistently maintained that it has broad authority to control dumping of industrial wastes under the Federal Water Pollution Control Act Amendments of 1972. How-

ever, the courts have reached somewhat differing views on the extent of that authority.

The issue came before the Justices as the result of a challenge initiated by eight chemical companies, including E.I. du Pont de Nemours & Company, which contended that the E.P.A. did not have the authority to issue industrywide regulations.

The United States Court of Appeals for the Fourth Circuit rejected the companies' challenge to the agency's authority to issue precise limitations on the discharge of pollutants by existing "sources," or concerns. But it said that these rules, and those for new plants, were only "presumptively applicable" to individual plants.

Both the Government and the companies sought review of aspects of the ruling.

The 1972 statute provided that a series of steps be taken to achieve the goal of eliminating all discharge of pollutants into waterways by 1985. In part, it set out two stages for limiting the discharge of pollutants, called "effluents," some to be taken by mid-1977, some by mid-1983.

The environmental agency, in an effort to implement the statutory goal, issued regulations with three sets of limitations —the first two imposing progressively

higher levels for existing "sources" of pollutants, meaning factories and plants, for mid-1977 and mid-1983; the third setting limitations on plants to be constructed in the future.

Considering these today, the Court made the following specific rulings:

¶The statute authorizes the 1977 limitations as well as the 1983 limitations to be set by regulation, "so long as some allowance is made for variations in individual plants."

¶The E.P.A. is not required by the statute to provide variances from compliance with regulations issued for new plants.

The Supreme Court also held that the Federal appeals courts rather than the Federal district courts were the appropriate forums for direct review of the agency's regulations regarding discharge of pollutants.

Eight Justices participated in consideration of the case and all joined in the Stevens opinion. The opinion noted that Justice Lewis F. Powell Jr. did not participate, but did not give a reason.

The decision was titled E.I. du Pont de Nemours & Co. v. Train, encompassing the petitions in Nos. 75-978; 75-1473; 75-1705.

February 24, 1977

Reserve Mining Wins Court Battle Over Disposal Site for Ore Wastes

ST. PAUL, April 8—The Minnesota Supreme Court today unanimously affirmed the decision of a lower court and ordered state environmental agencies to issue permits for the Reserve Mining Company to dump ore wastes at a disposal site in northeastern Minnesota that the company prefers.

The State's high court said that the permits should be subject to stringent conditions designed to prevent environmental harm and to alleviate a potential health threat from the wastes, called taconite tailings.

Reserve Mining has discharged about 67,000 tons of tailings daily into Lake Superior since 1955 from its taconite processing plant at Silver Bay, Minn., about 55 miles northeast of Duluth. Federal and state environmental officials contend that the wastes contain microscopic asbestos fibers that could cause cancer and other serious ailments if inhaled or ingested.

Such fibers have been found in the Lake Superior drinking water used by Duluth and several other lakeshore cities in Minnesota and Wisconsin, although it has not been shown that they have caused any detrimental health effects.

For the last eight years, the State and Federal Governments have taken several

legal and administrative actions in an effort to halt the discharge into the lake.

The Federal courts have ruled that Reserve must halt the discharge and switch to a land disposal site in northeastern Minnesota. In addition, a Federal judge has ordered a halt to the discharge into the lake by midnight July 7.

Because of the State Supreme Court's ruling, it is expected that the Federal courts will now lift that deadline to allow Reserve time to prepare the disposal site, called Milepost 7.

Reserve has said that it could have a disposal system in operation at Milepost 7 in about 33 months, thus ending the discharge into Lake Superior.

The State Supreme Court had heard the arguments in the case just yesterday. Its swiftness in issuing a decision surprised state attorneys and environmental officials, who had generally expected that it would take several weeks to get a ruling.

Milepost 7 Opposed

Those attorneys and officials oppose the use of Milepost 7 as a disposal site, and they had asked the Supreme Court to reject it. They contend that the site would be environmentally unsound and that asbestos fibers could be blown off it and on to nearby Silver Bay.

The Minnesota Attorney General, Warren Spannaus, declined to comment on the Supreme Court's ruling, in part because it was not accompanied by a formal opinion. That opinion, which will contain the court's rationale for approving Milepost 7, will be issued later.

The issue facing Mr. Spannaus is whether to appeal the ruling to the Federal courts. He and the state's top environmental officials favor an alternative site, called Milepost 20. The two sites are so named because of their location along Reserve's railroad, over which raw taconite ore is hauled from an open-pit mine to the lake shore processing plant at Silver Bay. Milepost 20 is farther inland from Lake Superior than is Milepost 7, and state environmental officials had argued that because of that, Milepost 20 would be a safer site.

Reserve is owned jointly by two Ohio concerns, the Armco Steel Corporation and the Republic Steel Corporation.

The Supreme Court's decision was hailed by Reserve Mining and by the Mayor of Silver Bay, Melvin Koepke.

"It's indeed a good Friday, to say the least." Mr. Koepke said. "The people of Silver Bay feel relieved. We've lived under tension and uncertainty for the last eight years, wondering if the plant would be shut down and we would be put out of our jobs."

Mr. Koepke is a machinist for Reserve and has worked at the company's Silver Bay plant for 21 years.

April 9, 1977

CITIES HELD LAGGING ON SEWAGE PROJECTS

By GLADWIN HILL
Special to The New York Times

LOS ANGELES, May 13—Contrary to the hopes of Congress, and despite the expenditure of billions of dollars, two-thirds of the nation's cities have not improved their sewage plants sufficiently to meet Federal specifications by a statutory deadline of July 1, according to a new expert analysis.

Of some 12,806 municipal sewage facilities in the country, according to the new analysis, only 4,244 will meet the July 1 requirement.

Degrees of compliance, the analysis found, ranged from 57 percent in the E.P.A.'s Region I, New England, down to

18.6 percent in Region VII, comprising Nebraska, Iowa, Kansas and Missouri. The ratio in Region II, comprising New York, Connecticut and New Jersey, was 38.2 percent.

The cities' cleanup record contrasts considerably with that of industry, which is loosely credited by Federal officials with being something like 90 percent "on schedule." That means that 90 percent of the major industrial pollution sources either have installed the necessary corrective equipment or are on firm schedules for doing so.

The analysis, based on data obtained by the Environmental Protection Agency in its annual municipal needs survey, was carried out by the Water Pollution Control Federation, a Washington-based organization of public officials, engineers, scientists and industrialists.

'Nonpoint' Sources Stressed

The problem of municipal delay in compliance has been highlighted in two

other recent ways: an E.P.A. suit, filed May 6, over Detroit's continued sewage discharges into Lake Erie, alleging that the city had been unduly slow in enlarging its treatment facilities, and a May 11 vote by the House Merchant Marine Committee in support of a plan to bar the dumping of sewage into the oceans after Dec. 31, 1981.

The Water Pollution Control Act of 1972 set July 1 as a deadline for both cities and industry to install the "best practicable" fluid waste treatment equipment.

This was recognized as an initial, and only partial, cleanup phase. Mid-1983 was set as a second deadline, for installation on "best available" treatment facilities. With these in operation, Congress hoped, by 1985 objectionable discharges could be reduced virtually to zero.

The slow pace of the cities' cleanup program—to some degree unavoidable—points to a delay of several years in abatement of pollution of the nation's waterways that Congress, in 1972, hoped might be accomplished by 1985.

However, there is a sort of wry consolation to the cities' deficiencies. That is the fact that discharges from sewage plants, while often locally deleterious, have been found to be a much smaller factor in the overall degradation of rivers and lakes than originally thought.

When the pollution act was passed, the widespread, if undocumented, notion was that sewage plants and industry shared most of the responsibility for water pollution.

But later studies have indicated that both sources are probably dwarfed by contaminants from nonspecific ("nonpoint") sources, such as runoff from cities, farms and forests—a new focus of stepped-up attention from Federal and state water pollution authorities.

Obviously far more difficult to define and regulate, the nonpoint sources represent a big imponderable in the question of when people will start seeing waterways generally restored to something like a pristine clarity. But it evidently will be long after 1985.

The problems that have developed have produced a barrage of proposals that Congress, which is considering amendments to the water pollution act, revise the deadlines. An opposing view is that the schedule should be adhered to, and variances from it granted only in individual cases with demonstrably difficult problems.

Two Stages in Deadline

In the case of municipal sewage systems, the 1977 deadline for "best practicable" facilities was interpreted to mean two-stage sewage treatment. This removes about 85 percent of the principal contaminants, leaving a fluid residue objectionable mainly in the fact that it contains nitrates and phosphates that promote unwanted vegetation in lakes and rivers.

Failures to meet the deadline are variously attributed to financing problems, engineering, hardware and construction delays, and unforeseen difficulties in carrying out the $18 billion program of subsidies to communities that Congress voted in 1972. Under the program the Federal Government pays 75 percent of the cost of improvements.

Some muncipalities were slow in applying for grants; the E.P.A. did not have enough people to process the applications expeditiously; there were widespread complaints from state and local officials of excessive red tape; and there have been countercharges from Federal investigators that many millions in grants were used unwisely or dishonestly.

A decade ago it cost a community only about $50 per resident to install two-stage sewage treatment facilities. Today, according to the E.P.A., the cost is around $200.

May 14, 1977

SOLID WASTES

MAJOR U.S. CITIES FACE EMERGENCY IN TRASH DISPOSAL

By GLADWIN HILL
Special to The New York Times

LOS ANGELES, June 15—An avalanche of waste and waste-disposal problems is building up around the nation's major cities in an impending emergency that may parallel the existing crises in air and water pollution.

Some features of the situation are being described by conservative Federal officials as "a national disgrace."

Experts feel that resolution of the problems may require radical changes in people's patterns of consumption and disposal, major shifts in municipal administration and sweeping revision of the nation's attitude toward its environment.

These are the conclusions and implications that emerge from interviews with leading authorities in the field and a 13-city check by The New York Times of what is becoming known as "the third pollution": the problem of solid wastes and their disposal.

Refuse Burden Grows

Every man, woman and child in the country, on average, is now generating more than five pounds of refuse a day—household, commercial and industrial —ranging from garbage to iron

filings but excluding vastly larger amounts of uncollected trash such as agricultural waste.

Collection and disposal facilities are staggering under the load, and the load is increasing rapidly.

"The major metropolitan areas are standing in front of an avalanche, and it is threatening to bury them," says Karl Wolf of the American Public Works Association, which has been conducting research projects on waste disposal.

Where It Comes From

Since 1950, the nation's population has increased 30 per cent, but the waste load has increased 60 per cent and is expected to increase 50 per cent more in the next decade.

The increase results from increased population, increased consumption of commodities and affluence, which has brought the regular discard of things that once were saved.

"We used to get few newspapers — people would save them to make money in paper drives," says Detroit's Public Works Commissioner, Robert P. Roselle. "Now we collect them all. The telephone company used to collect old phone books for their paper value. Now we collect them."

In San Francisco, even with no population increase, waste has increased by a third in the last decade. In Chicago it has been rising 2 per cent a year despite a slight decline in population. Cities across the country are also grappling with mounting numbers of automo-

biles, a physical and fiscal nuisance, although their mass is minuscule in the total waste picture.

"Half the communities in this country with populations of 2,500 or more," said Wesley Gilbertson, the first Federal Solid Wastes Program director, "are not doing even a minimally acceptable job of solid waste collection and disposal."

Only a Partial Answer

Once community trash could be disposed of simply by dumping it on the outskirts of town or burning it. Now with cities growing, outskirts have become fewer and farther away.

And incineration can only be a partial answer, because of the 20 per cent ash residue, the large volume of noncombustibles and air pollution problems.

So, from New York to San Francisco, the cry is going up that cities are running out of disposal space. A year's rubbish from 10,000 persons covers an acre of ground seven feet deep. Philadelphia has been short of space for years and has had to shift largely to incineration. But it now is near the end of its rope because of incinerator obsolescence and high replacement costs.

New York City has been using up many acres of dumping space a year, with little space left. Even Tucson, Ariz., surrounded by desert, estimates that it will run out of disposal space within the next three years.

The problem is economic. Close-in urban land has become too valuable for dumping; other local dump sites are too far away because of trucking costs. And costs are rising.

Boston's refuse collection costs jumped last year from $2.6-million to $3.9-million just because of payroll increases.

New York City's sanitation department, concerned primarily with solid wastes, has 14,000 workers, and its budget has climbed to nearly $150-million a year.

In Milwaukee, annual rubbish-removal charges for a household have risen in a decade from $26.40 to $35.25. Albuquerque's rate went up from $30 to $36 this year. In Portland, Ore., the monthly rate goes from $2 to $2.25 next month.

Alternative Systems Sought

These trends have touched off a scramble for alternative refuse disposal systems somewhere in the range of present costs.

The average community outlay for refuse collection and disposal, according to a survey made in 1968 by the Public Health Service's Solid Wastes Program, is $6.80 a person a year.

But in some Washington suburbs, household rates run as high as $46.20 a year. New York City figures it costs $30 to dispose of a ton of trash.

San Francisco, where the current household charge is $22 a year, is planning to ship its refuse 375 miles by railroad to a desert disposal area in Lassen County.

Philadelphia is close to completing an arrangement for railroading its rubbish to distant abandoned mine pits. Chicago is considering a 250-mile haul.

Milwaukee has a rail-haul plan, but has not been able to find an amenable disposal locality. Philadelphia figures it now costs about $7.50 a ton to get rid of refuse by incineration; under the railroad plan it would pay a contractor only $5.35 a ton.

Chicago and Detroit are experimenting with trash compac-

The New York Times

Westchester County's only garbage dump, at Croton Point Park on the Hudson River south of Peekskill. The dump must be closed by the end of this year; there will be no more space left. Throughout the nation, refuse collection and disposal facilities are staggering under the load of millions of tons of solid wastes each year.

tion into blocks; Japanese researchers say that the blocks can be used as building materials.

Pneumatic tube waste-dispensing systems have been tried in Europe, and the Walt Disney organization is planning such equipment for its new Florida development.

Some organizations have been working on ways of macerating household trash, in a device like a garbage grinder, and forcing it in liquefied form to treatment centers. However, this would take care of only part of a community's waste.

Similarly, there has been a lot of talk lately about devising more easily destructible packaging, particularly bottles and cans. But if all packaging were abolished completely, it would reduce the waste load only about 15 per cent.

'Back Yards' Studied

Despite all the quests for solutions, some experts believe that the most serviceable short-range answer may lie in cities back yards.

Around 75 per cent of the nation's trash, by tonnage, still goes to open dumps, with about 15 per cent going into incinerators.

The Public Health Service found that 94 per cent of the dumps and 75 per cent of the incinerators were inadequate in respect to sanitation and pollution and termed this "a national disgrace."

Only 5 per cent of refuse is disposed of by the "sanitary landfill" method, in which each day's deposit is covered with six inches or more of compacted dirt, making it rodent-proof and odorless. The process, according to the Public Health Service, in a typical situation costs only $1.27 a ton, against 96 cents for obnoxious dumping.

No Simple Solution

There is no simple answer to the nation's refuse problem, experts say, because it is a composite of myriad problems that hinge on local economics.

These can vary greatly even in different sections of the same city. In New York, scrap metal contractors pay the city from 21 cents to $4.03 to pick up abandoned automobiles in the Bronx, Brooklyn, Queens and Staten Island.

In Manhattan, because of logistic difficulties, they pay nothing.

Across the country, scrap metal prices vary sharply, depending on transportation and the nearness of metal works.

If Milwaukee could find a disposal area, it is estimated that railroad removal would cost from $5.45 to $6.23 a ton, $2 less than its present dumping-and incineration system.

Yet Denver backed off from a plan to haul its trash by train 75 miles to a point near Colorado Springs because estimates were that within a decade it would cost $419,000 more than local disposal.

Apart from such variables, the quest for new disposal methods starts from the most disorderly of economic bases. Across the country, trash systems have evolved from the primitive town dump, and there is no uniformity or consistency in methods or financing.

Roughly half the nation's refuse collection is by public agencies and half by private collectors, either franchised or dealing directly with customers.

In many places, householders are billed specifically for the service; in others the service is financed from general tax

funds.

Sometimes the charge is calculated at cost; sometimes it is pegged to yield a profit. Sometimes the service is given below cost with the differential coming from taxes. The profits of franchised collectors may or may not be subject to regular public scrutiny.

Inefficiency Found

Thus, in many cases citizens have little way of knowing whether they are getting their money's worth and whether the cost of an alternative system would be justified.

Inefficient approaches to waste disposal are the rule rather than the exception.

"Most of what is wrong with solid waste management in the United States," says one official, "can be attributed to fragmenting responsibility down to small political jurisdictions which lack sufficient resources for the job."

In Los Angeles County, 70 communities share rubbish collecting depots and disposal sites. But elsewhere in California, Federal researchers found one area where 80 public agencies were running 70 separate disposal systems.

The hazy economics of some municipal systems, experts have observed, provide a field day for officials and agencies to juggle funds—not necessarily into their own pockets, but to the detriment of efficient refuse disposal.

What is being done about the waste problem nationally?

Congress got the word on the impending waste glut as far back as 1965 and passed the Solid Waste Disposal Act. It gave the Public Health Service primary responsibility for a program of research and technical assistance and matching grants to states and localities to help in the development of disposal programs.

Most of these states have availed themselves of these grants, the average being about $50,000. But total appropriations for the solid waste program have remained less than $20-million a year.

No Cure-Alls Found

While much valuable basic information has been amassed —(nobody knew anything about the national waste disposal picture before)—and a number of experimental projects launched, the program has not yet come up with any cure-alls.

The general line of thinking is that, as with air and water pollution, each area will have to work out a solution fitted to local circumstances and that waste collection and disposal will have to be handled regionally, with localities pooling efforts.

Ultimately, coping with solid waste, it is believed, should be part of an integrated system covering also liquid and gaseous effluents.

"One of the most significant items of progress," says Richard D. Vaughan, current director of the Solid Wastes Program, "is that state budgets are starting to show specific items for waste disposal work.

"There is public concern about this, where three years ago there wasn't any."

Senator Edmund S. Muskie, the Maine Democrat who fostered basic water pollution legislation, has turned his attention to solid waste and has a bill pending that would increase financing of the solid waste program tenfold, to a level of more than $200-million a year.

Seemingly there are only two things that can be done with rubbish—burn it or bury it, in the ground or in the ocean.

But there is an often overlooked third possibility, and some scientists think it is the only one for the long run. That is to reclaim refuse—to break it down into its main constituents for reuse.

That has been the dream of many people, but it has not been realized, except for a small amount of scrap metal and paper salvage, because it has been cheaper to proccure new materials.

A number of ingenious experimental factories for converting trash into garden compost have been established around the country, but most of them closed down because there wasn't a market for that much compost.

Use and Discard

So the nation's economy has continued on a "use-and-discard" pattern. But there comes a point where the cost of getting rid of used material gets so high that the cost of renovating it would represent an economy.

That point is already being approached with the commonest commodity, water.

The same tactic is even more applicable to solid waste. Dirty water, if not renovated, disposes of itself one way or another. Solid waste does not; it just piles up. And scientists foresee the day when accumulation of rubbish, and gases from its incineration, will be as troublesome over the face of the globe as are the wastes in a spacecraft.

The latter are systematically "recycled." The same recycling eventually will be imperative with everyday wastes, it is felt, and may as well be undertaken soon as a decisive answer to the refuse problem.

A scientific report prepared for the Senate Public Works Committee last year said:

"It is now evident that the industrial economy of the United States must undergo a shift from a use-and-discard approach to a closed cycle of use and salvage, reprocess and reuse . . . or else faces the alternative of a congested planet that has turned into a polluted trash heap, devoid of plant and animal life, depleted of minerals, with a climate intolerable to man."

Short of this long-range solution, the public is confronted with some early large-scale expenditures just to get relief from current rubbish bugbears.

Mr. Vaughan estimates it will take an outlay of more than $2-million a day—$835-million a year—for five years, just to "upgrade existing collection and disposal practices to a satisfactory level."

This is a sizable sum, but only a fraction of the $4.5-billion a year the Public Health Service estimates is being spent to deal with all types of solid waste.

Relief may be some time in coming. Frank Stead, a former official of California's Department of Public Health and one of the nation's leading environmental experts, said recently that it would take 16 years to bring about basic changes in solid waste management.

"Two years to convince the public that new concepts will be successful, four years to put through the necessary legislation and 10 more years to put the changes into effect," he said.

June 16, 1969

PACKAGING LINKED TO WASTE ISSUE

Rosenthal Cautions Industry of Possible U.S. Curbs

By DAVID BIRD

The chairman of a Congressional consumer subcommittee told representatives of packaging companies yesterday that their industry's practices exploited the consumer and were creating an unmanageable solid-waste problem for the nation.

The Representative, Benjamin S. Rosenthal, Democrat of Queens, warned that if the industry did not change its packaging "the Federal Government may be forced to act sooner and more drastically than you ever thought possible."

Mr. Rosenthal spoke at the American Management Association's National Packaging Conference and Exposition, which is being held here through tomorrow. The conference sessions are in the Americana hotel. The exhibition of packages and packaging machinery occupies four floors in the Coliseum.

Costs Seen Multiplied

Apart from Mr. Rosenthal's speech, much of yesterday's meeting emphasized the promotion of one-way "convenience" packaging that many persons contend has added sharply to the country's waste burden.

"Our great economy and our soaring standards of living," R. A. Miller, a packaging executive, said in his address to the meeting, "are based on the rapid movement of enormous numbers of attractively packaged, easy-to-find, easy-to-buy products."

But Mr. Rosenthal maintained that the consumer was really a multiple loser as a result of today's packaging. First, he said, the consumer pays more for a disposable container than for a reusable one. The buyer then has to pay to have the used container collected for disposal. Finally, he said, "the consumer must pay again" when the "container does not degrade, but lives on to foul environmental quality."

Mr. Rosenthal, who is chairman of the Government Operations Committee's special consumer inquiry, said that discarded packaging "produces about 13 per cent of the nation's solid-waste output, which has become a major pollution problem."

He added that containers have become an increasing waste problem because they were made of tougher materials that do not decompose readily.

Mr. Rosenthal said that the several hundred packaging executives at the meeting here this week were showing "only minimal concern about the problems of solid waste disposal that increased package use will create."

He said that corporations "must establish an effective mechanism through which the public interest can be represented and considered." "The election of 'public interest' representatives from outside the company to its board of directors may be the best solution," Mr. Rosenthal said.

The packaging conference will end tomorrow—Earth Day-Antipollution forces in that nationwide protest against degradation of the environment have announced that they plan to demonstrate against the packaging convention.

April 21, 1970

Program to Recycle Waste Products Is Found to Be Lagging

By ROBERT A. WRIGHT
Special to The New York Times

LOS ANGELES, May 6—Recycling of castaway bottles, cans and papers so far is failing to make much headway around the country despite the efforts of thousands of ecology-minded people who are hauling the waste products to collection points for re-use.

In most cases, collection groups depend on subsidies, and many have been forced to close down as they are caught in the ebb and flow of volunteer interest and action. And even those that are successful have failed to channel more than a minute proportion of wastes back into the manufacturing process.

The reason, according to reports from several major cities, is that recycling so far is not paying its own way. And there is a growing belief that it will be years—if ever —before it makes any major dent in the nation's mounting piles of solid wastes.

While some experts increasingly have come to believe that recycling offers a long-range solution to waste disposal problems, the scrap industry warns that the supply of many waste materials is far greater than industry can consume and that virtually no markets exist for others.

Nonprofit volunteer groups that collect waste materials may reduce litter, which accounts for less than 10 per cent of all solid waste, but cannot operate without subsidies. Many have disbanded their efforts because of a lack of markets or waning volunteer interests.

Meanwhile, massive public relations and advertising campaigns by industry hail progress in reducing litter through collections and conservation of resources through the recycling of glass, steel, aluminum, plastic and paper waste. But they omit some pertinent facts.

A press release by the Glass Container Manufacturers Institute, Inc., announces that used bottles and jars redeemed from the public are being recycled at the rate of 912 million a year. While it notes that the program is "only a first step toward our long-range goal," it fails to state that the American industry produces about 36 billion glass containers a year. Thus, recycled botles and jars account

The New York Times

Youths pouring cans into container at a reclamation center operated here. Nation's effort to collect refuse for recycling has, in some instances, failed or moved slowly.

for only 2.6 per cent of the total number of glass containers produced in the United States annually.

Recycling of Cans

Similarly, the Aluminum Association reported a four-fold increase in the collection of aluminum cans for recycling last year. The increase—to 770 million cans—amounts to about 35 million pounds of metal and contrasts with aluminum shipments for container production last year of 929 million pounds, or 3.7 per cent of the total used by American container manufacturers.

The American Iron and Steel Institute announced that it retrieved about 1.5 billion cans for recycling last year through magnetic separation from municipal dumps, but the institute failed to note that some 65 billion cans are manufactured each year. Thus, the 1.5 billion cans recycled in 1971 amounts to 2.3 per cent of the total manufactured.

A typical example of the ebb and flow of recycling efforts occurred late last year in Minneapolis. In the middle of a campaign against a proposed ordinance in suburban St. Louis Park that would prohibit cans for soft drinks and beer, the

Theodore Hamn Browning Company and Coca-Cola Bottling Midwest, Inc., announced that they would sponsor "the most comprehensive, full-time recycling center in the country."

Center Keeps Profit

The St. Paul center, called the Metropolitan Bottle and Can Recycling Center, redeems nonreturnable bottles at half a cent a pound for clear and amber glass and one-fourth a cent for green glass. Steel or bimetal cans bring one-fourth cent a pound and all-aluminum cans 5 cents. These rates are about half of those paid by glass and metal companies, and the profit is being retained to help make the center self-supporting.

About a month after the center opened, Robert Schmitt, who had collected about one and a half million pounds of bottles in 14 months as a private hauler, went out of business because, he said, he could not collect enough bottles fast enough. He said he found that he had to haul 44,000 pounds of bottles a week, getting a cent a pound, to break even in his three-man operation.

Confusion reigns in the emerging science of recycling, as an in-depth study by the

Bank of America notes. The bank, which announced that it would step up its use of recycled paper, released its report to aid other companies considering the move. The report said, "Any corporation today that seriously investigates the feasibility of expanding its use of recycled paper almost certainly will find itself bogged down in a morass of conflicting claims and opinions."

'Claims' and 'Disclaimers'

The bank added that it encountered "expansive claims" from "zealous conservationists" and "equally consistent disclaimers" from the paper industry."

The bank's research found that recycling of 35 per cent of the nation's annual paper consumption by 1985 was technically feasible and, "theoretically at least, economically feasible, too."

Even some of those who regard recycling as the long-range answer to the solid waste problem express concern about the clamor for expensive municipal systems to recover recyclable materials. The Institute of Scrap Iron and Steel, Inc., whose 1,300 member companies have long been in the recycling business, warns of the "danger of establishing recycling centers on the theory that the recovered materials will result in income over cost when sufficient markets for available recyclable materials simply do not exist."

The institute notes that planned municipal recovery plants are based solely on supply concepts. "The recycling center, as it is referred to, does not recycle," the organization says. "It merely increases the supply of waste that is available for recycling."

The institute contends that materials recovered from municipal refuse are of lower grade than other scrap already in oversupply, and suggests that increasing the supply could cause "a more thorough depressing of the entire recycling industry with stockpiles of increased accumulation of metallics throughout the country."

The institute points to the untold numbers of abandoned automobiles and household appliances despoiling the countryside and estimates that obsolete cars alone represent more than $1-billion worth of re-usable metal that is not being recycled.

Carl Sexton, president of Los Angeles Byproducts, says that "there is no question about a viable market for glass and metals in the West." The company, which recovers ferrous metals from plants at four municipal landfill operations in Northern California, sells its recycled materials to mining companies that use it to extract.

copper from sulfite solution. The company is making a marginal profit, according to Mr. Sexton.

Los Angeles Byproducts is building a separation machine in the plant of the private companies that collect the refuse of San Francicso and expect to recover at least 90 per cent of the ferrous steel cans. Mr. Sexton believes the project will prove economically successful. But he added, "Just how successful such programs will be on a nationwide basis remains to be seen" because of the lack of markets in many areas.

M. J. Mighdoll, executive vice president of the National Association of Secondary Industries, Inc., says: "It's difficult to tell people it's not desirable to collect any kind of materials that seemingly are environmental no-no's. It sounds as if you're being anti-environment. The fact is, however, that while this is good as far as reducing litter—a minor part of the problem—unless you expand markets for these materials, you're not really recycling."

Depend on Subsidies

The American Iron and Steel Institute, for one, has announced its commitment to recycle into steel every old can it gets. And the Glass Container Manufacturers Institute is similarly commited. But volunteer groups still depend on subsidies to survive.

The San Diego Ecology Centre, which has collected 130 tons of glass since it started last August, "does a little better than break even," according to James Jacobson, president of the center. "We sometimes make $100 on a shipment," he added.

The center is subsidized by the Sanitainer Disposal Com-

pany, which donates large collection containers and trucks at cost for the transportation of the glass to the nearest glass plant, 75 miles away in Santa Ana. The Kerr Glass Company pays $20 a ton, a $5 subsidy "in the interest of ecology" over the going commercial price of used glass. Kerr officials say recycling is "a very marginal thing economically."

Recycling Experiment

The city of Los Angeles, conducting a six-month experiment in recycling, opened six reclamation centers, consolidating volunteer efforts in a program subsidized by business. Greg Barron, coordinator of the program, says that the volume of the first nine weeks of collection is about half what is needed for the project to break even.

The $20 that glass companies are paying for each ton of used glass "is not entirely public relations," according to Richard L. Cheney, president of the Glass Container Manufacturers Institute. "We also are subsidizing research and development for new markets," Mr. Cheney said. "It is the only way we have of getting large quantities of cullet [used glass] back in the plants."

Mr. Cheney concedes that the technology of glass recycling is just emerging, but he contends: "We know enough to say it can be economically sound and is the long-range solution. Not that you will ever recover enough from municipal waste to turn a profit for a city. But the negative economics of it—the elimination of the alternative cost of disposing of it—are obvious. In 10 years, we will have make real strides in the direction of full recycling."

Mr. Cheney does not believe

that returnable bottles would be an alternative solution. "A study by the Midwest Research Institution showed that, if the single-trip bottle was eliminated, you would reduce solid waste by just 1.37 per cent, and the volume is growing at 4 per cent a year." Mr. Cheney said. "Experience has shown that, particularly in urban areas, people do not return them. In New York, returnable bottles are only making three or four trips."

A comprehensive study conducted by the National Association of Secondary Industries for the Federal Environmental Protection Agency found that the recycling of solid waste could be greatly expanded. But the study, made by the Battelle Memorial Institute and the Conference Board, found that substantial amounts of recoverable secondary materials of all types are being "lost" and that discriminatory tax and transportation rates present obstacles to wide-scale recycling.

Mr. Mighdoll said the study showed that 60 per cent of available secondary copper was not being recycled. "We need a combination of depletion allowances and tax incentives that do not exist for recycled materials if new markets are to be created," he said. "Another inequity is transportation rates. Virgin copper moves at 50 to 60 per cent cheaper than scrap copper, waste paper 100 per cent higher. Also, the Government continues procurement specifications that preclude recycled materials in a lot of things it buys."

From June, 1967, to last June, the Environmental Protection Agency expended $7-million in grants for research, demonstration projects and internal study of recycling. One of its grants

is helping finance a municipal separation system in Franklin, Ohio, for steel, aluminum, glass and paper.

In San Francisco, where municipal refuse is collected by private companies under contract to the city, a privately funded project to separate ferrous metals is being installed by Los Angeles Byproducts. Leonard Stefanelli, president of the Sunset Scavenger Company, which collects more than 60 per cent of San Francisco's rubbish, says he expects the system to extract more than 275 million steel and bimetal cans a year.

Mr. Stefanelli says his company will get a royalty of $2 a ton for ferrous metals from Los Angeles Byproducts. "We will also save $3.57 for each ton of metal we don't have to transport to the landfill site," he added. "We save another $2 on each ton of metal that doesn't have to go into landfill. The can reclamation system will pay for the grinding up of the garbage, which reduces its bulk by 25 per cent. It means that a 20-year landfill site can now be used for 25 years."

On other systems, Mr. Stefanelli is not as sanguine. "A salvaged material doesn't become recyclable until there is a market for it on an economic basis," he said. "We could supply tons of paper and corregated salvage if we knew someone was ready to buy it."

"In the past, Sunset hand-separated bottles, rags and other scrap and we couldn't make a profit on this even though our volume was $2-million," Mr. Stefanelli added. "Home separation and recycling centers, while they are good for keeping people aware, don't make economic sense."

May 7, 1972

DEBATE WIDENING ON 'BOTTLE LAW'

By GLADWIN HILL

One afternoon last spring, two buses pulled up at the state Capitol at Lincoln, Neb., loaded up with state legislators and public officials, and took them 60 miles to Omaha for a festive evening, courtesy of the Falstaff Brewing Company.

A few days later a bill to ban nonreturnable beverage containers, opposed by brewing and other interests, was killed in a legislative committee.

Nebraska was one of 15 states — among them New York — where, a New York Times survey shows, the "beverage lobby" so far this year has fended off proposals to emulate Oregon's widely ac-

claimed "bottle law."

The Oregon measure, instituted in 1972 to reduce litter, requires deposits on all beverage containers, bans detachable-tab "pop-top" cans and contains other provisions aimed at encouraging reuse or a systematic recycling of containers.

The law has been pronounced a success by state officials and the Federal Environmental Protection Agency. Its efficacy is challenged by beverage and container makers and distributors, who consider it an expensive nuisance and who have mounted farflung, overt lobbying efforts to prevent its spreading.

U. S. Measure Studied

Congress is considering a Federal version, endorsed in principle by the E.P.A., that would apply to all the states. The influence of the anti-

regulatory forces is suggested by the fact that Senator Jennings Randolph, West Virginia Democrat, is sponsoring a measure that would deny Federal solid-waste grants to any state with a "bottle law."

South Dakota is the lone state where efforts to enact a "bottle law" have been successful this year.

In Vermont, previously the only state to emulate Oregon, efforts this year to repeal a 1973 mandatory-deposit law were thwarted.

In Delaware, a tax on nonreturnable containers was passed by the House of Representatives, but was not acted on by the state Senate.

A container-regulation measure is under consideration by a legislative committee in New Jersey. In Virginia, a bill was put over to the 1975 legislative session for interim study by a

commission on solid waste.

In both Florida and Idaho, bottle bills were blocked for the third successive year, and in Louisiana several container-regulation bills died in committee.

New York Bill Dies

In New York, a measure on the Oregon pattern died in committee, to the pronounced satisfaction of the Empire State Chamber of Commerce and other business interests.

In Ohio, labor organizations were among other interests that mounted what was described locally as "awesome pressure" to kill a bottle bill.

Other states where bottle bills came to naught included California, Kansas, Michigan, Minnesota, Tennessee, West Virginia and Wyoming.

The new South Dakota law, which takes effect July 1, 1976,

calls for a complete withdrawal from the market of nonreturnable or nonbiodegradable cans and bottles.

It also provides for stricter penalties for littering, a public education program on litter problems, and $115,000 in state matching grants for antilitter campaigns by local governments.

The two-year interval before the law takes effect was injected to permit legislative review of its potential impact and of the educational programs, and as a transitional period for the beverage industry.

"If that is regarded as a club over the head of national suppliers," commented state Senator E. C. Pieplow, one of the measure's sponsors, "then so be it."

To Press Nebraska Move

Despite the defeat of the legislation in Nebraska, its sponsor, state Senator Steve Fowler, said he would reintroduce the bill next year and, if necessary, again in 1976.

The proposed Federal law, on which the Senate Commerce Subcommittee on Environment held hearings in May, would mandate a two-cent deposit, or refund evaluation, on beverage containers usable interchangeably by various bottlers, and a five-cent deposit on other containers. The measure would ban detachable-tab cans on which the rings are considered to be, among other things a hazard to animals.

The aim of container controls is to arrest a "throw-away explosion," in which, according to the E.P.A., the nation's use of beverage containers has more than tripled, from 15.4 billion to 55.7 billion in the 1959-72 period when per capita beverage consumption increased by only one-third.

"This dramatic increase," says John R. Quarles Jr., E.P.A. deputy administrator, "can be traced in large part to an increase in the use of the nonrefillable container."

August 3, 1974

COINING TRASH
GOLD STRIKE ON THE DISASSEMBLY LINE

By Boyce Rensberger

There are two ways of getting steel—out of the ground, in the form of iron ore, and out of garbage and trash, in the form of rusty cans, bent paper clips, broken toys, old refrigerators and the like. The steel is the same. But, in a world of dwindling natural resources and growing environmental pollution, the two methods differ in ways that have profound implications for the future of industrialized societies. For every ton of steel produced from recycled municipal solid waste instead of ore, the following things happen:

☐ Enough electricity is saved to power the average American home for eight months—a 74 percent saving in the amount of energy consumed to produce that one ton of steel.

☐ Two hundred pounds of air pollutants, of the kind produced in making steel from ore, are not produced—an 86 percent decline in air pollution.

☐ About 6,700 gallons of fresh water are not used—a 40 percent saving.

☐ As the water that is used is returned to streams and sewers, 102 pounds of water pollutants are not discharged—a 76 percent reduction.

☐ And 2.7 tons of mining wastes are not heaped on the landscape around the mine. (Figures based on calculations by the Environmental Protection Agency.)

The same garbage from which steel can be extracted can be combed—with comparable environmental benefits—for aluminum, copper, paper, glass and other materials. The 134 million tons of refuse Americans throw away each year contain an estimated 11.3 million tons of iron and steel, 860,000 tons of aluminum, 430,000 tons of other metals (principally copper), more than 13 million tons of glass, more than 60 million tons of paper, and burnable organic materials containing energy equivalent to 150 million barrels of crude oil, or 70 percent of the expected annual yield of Alaska's North Slope. What it takes to extract these materials from the unsightly waste that piles up on city sidewalks is high technology—and that technology is already with us. The benefits of recycling are already beginning to be achieved. Major "resource-recovery" programs are spreading rapidly across the country.

St. Louis processes 325 tons of garbage a day to recover iron and steel; organic matter is burned to make enough electricity to power the equivalent of 25,000 homes, and the system is being expanded to handle upward of 6,000 tons per day by mid-1977. Baltimore has completed a plant that handles 1,000 tons of refuse each day, extracting ferrous metals and glass and burning the balance to make steam. San Diego County is building a 200-ton-a-day plant that will recover ferrous metals and glass and turn the organic matter into fuel oil. By late next year, Milwaukee will be getting some of

Boyce Rensberger is a science reporter for The New York Times.

its electricity from garbage as part of a private resource-recovery system that will also reclaim steel, tin, aluminum, paper and glass. The system will take in 1,200 tons a day and convert the organic portion into fuel for sale to the Wisconsin Electric Power Company. Similar systems are operating in half a dozen smaller towns, and are planned or under construction in a score of larger cities.

Interestingly enough, recycling has acquired a champion in Dr. Glenn Seaborg, who, during his many years as chairman of the Atomic Energy Commission, had stood for full speed ahead on development of nuclear power—and had thus been a villain to the environmentalists. Seaborg has declared that the industrialized nations must evolve in the direction of a "recycle society," in which virtually all materials would be reused indefinitely. "In such a society," he said, "the present materials situation is literally reversed; all waste and scrap, what are now called 'secondary materials,' become our major resources, and our natural, untapped resources become our back-up supplies."

Another proponent of recycling is Dr. Rocco Petrone, the former National Aeronautics and Space Administration official who headed the Apollo moon-exploration program, a venture that to many symbolized a failure to attend to problems on earth. Petrone now heads the National Center for Resource Recovery, Inc., a Washington-based group established five years ago by several industries and labor unions. Working with private and Government funds, the group conducts re-

search on solid-waste recycling methods and promotes their wider adoption. Neil Armstrong, the first man on the moon, is on the board. "Sometimes people are afraid of technology," Petrone says. "However, it's not the use of technology but the abuse of technology that causes problems. I think that if we can get good technology developed and applied to recycling, we'll be able to put less demand on nature to maintain our society."

The "recycle society" may be arriving faster than most people realize:

☐ In 1974, about 2.3 billion aluminum cans were recycled. That amounted to 17 percent of the total manufactured. And it represented a 40 percent gain over the number recycled in 1973, which, in turn, showed a 28 percent increase over 1972. The rate for 1975 shows another sizable increase. There are now 1,300 aluminum reclamation centers around the country.

☐ In 1970, only two cities had separate collection programs for paper, which constitutes about half the weight of all municipal waste; by 1974, the number had grown to more than 100. Last year, 22 percent of all paper manufactured in the United States was of the recycled variety.

☐ In 1970, "throwaway bottles" were thrown away. Recycling them was virtually unknown. Today, there are about 100 glass-recycling centers in 28 states, collecting about 250,000 tons of used glass annually. That is still only 2 percent of a year's production, but the proportion is growing.

By almost every accepted criterion, solid-waste recycling has become a major new growth industry. It has its own trade journal, called Resource Recovery & Energy Review. A number of engineers and scientists have switched from the limping aerospace industries to resource recovery. A number of American corporations have begun developing equipment for the new industry; the better-known ones include Monsanto, Union Carbide, Grumman, Raytheon and Allis Chalmers, and dozens of smaller companies, including many established specifically for resource

recovery, have joined the competition.

At the heart of the new movement is a system of machines capable of taking truckloads of municipal waste at one end and processing them into steel, aluminum, glass (sorted by color, if need be), burnable fuel containing five times as much energy as it takes to run the entire operation, and a variety of other materials. The requisite technology — from mammoth shredders that can reduce an automobile to fist-sized chunks in 30 seconds to a sophisticated "aluminum magnet" capable of extracting that valuable metal—is already in existence. Much of it is fully operational. There are at least half a dozen patented, commercial techniques for turning organic wastes, such as plastics or table scraps or paper, into various kinds of solid, liquid or gaseous fuels. (Most of the paper thrown away becomes too soiled and intermingled with other wastes for reuse as recycled paper.)

It is estimated that a resource recovery system of average size, capable of handling the 750 tons of waste generated daily by a city of 250,000 people, would recover each day 24 tons of paper, 44 tons of steel, 5.3

The boom in waste recycling is driven by recognition that there is money to be made in the business.

---

tons of aluminum, 44 tons of glass and 3.9 tons of other metals. If New York City were to process the 30,000 tons of refuse it generates each day, these numbers would be multiplied by 40. All this is material that, until now, was used once and thrown away, never to be used again.

☐

The age of recycling may be said to have begun on Earth Day, of 1970, when thousands of young people, and many who were not so young, staged a variety of events to impress upon the public that mankind lives on a planet of finite resources and vulnerable life-support systems. "In North America, the recycling of wastes from municipal refuse was almost unheard of before Earth Day, 1970," says Dr. Jack Milgrom, a chemical engineer for Arthur D. Little, Inc., who specializes in application of recycled materials. "But that day it became an important issue." In hundreds of communities around the country, volunteer centers were set up to accept paper, glass and metal waste. For all the impact of those and subsequent educational efforts, however, public awareness of the praiseworthiness of recycling was not enough, by itself, to bring about the present boom. It took two other factors to do that. Both were economic.

There was growing recognition that conventional waste-disposal methods —incineration, land filling and ocean dumping—were causing health hazards and destroying fishery potential through pollution of the air and seas. Yet the new and improved pollution-control regulations proved very costly. Also, many municipalities were running out of suitable land to fill and were having to pay increasingly large sums to haul their garbage to outlying sanitary land-fill areas. Recycling came as the answer.

Ordinarily, it can cost a city from $1 or $2 a ton to $10 or $12 a ton to dispose of wastes. A modern materials and energy recovery system, operated privately at a 15 percent profit margin, brings the cost down to $3.45 a ton, making recycling economically attractive for most urban areas, which are paying more than that, or soon will be. Economists forecast that the rising costs of energy and of virgin material will gradually make solid waste so valuable that cities may some day be selling their garbage to private resource-recovery systems.

It is this rising cost of exploiting natural resources that provides for the recycling boom. Aluminum offers the best example. It takes 20 times more energy to produce a pound of aluminum from bauxite ore than from urban solid waste.

In a report prepared for the Energy Policy Project of the Ford Foundation, scientists of the Midwest Research Institute estimated that if the aluminum, copper and ferrous metals now being lost through ordinary disposal of solid wastes were recycled to the fullest practicable extent, the country would save 403 trillion B.T.U.'s of energy annually—energy that would have to be used, at an increasingly higher cost, to make these metals out of natural resources. That is the equivalent of 3.22 billion gallons of gasoline, or a 10-day supply for all the motor vehicles in the country. It is also estimated that the energy in a year's worth of urban waste is equivalent to $1.5 billion worth of imported crude oil, and the metals and glass in the same waste are worth another $1 billion. Americans are now paying on the order of $1 billion a year to throw these resources away.

□

A resource-recovery system is a combination of machines operating in sequence as a kind of "disassembly line." The machines sort out and process the different ingredients of municipal waste into various kinds of relatively pure materials. In a typical system, garbage trucks dump their contents onto a conveyor belt that carries everything into a shredder. There, a hammer mill—a spinning horizontal shaft studded with swinging hammers — bashes everything about, breaking large objects into smaller pieces. Some shredders are capable of taking anything up to the size of a refrigerator and reducing it to baseball-sized chunks. Automobiles are handled in separate, larger hammer mills.

Another conveyor belt carries the pulverized refuse to an "air classifier." The waste is dumped into a shaft with a strong updraft. Light objects,

Aluminum cans enter a recycling machine in a New York plant. Below, shredded trash like this is used to generate electricity in St. Louis, where recycling has gone furthest.

including much of the organic matter, are blown upward and shunted off to other devices, either for incineration or to be converted into fuel. Heavy objects fall down the shaft to a third conveyor belt and move past a magnet that pulls out the steel and iron items, mostly cans. Nonferrous metals move on to yet another device, called a trommel.

Imagine an oil drum on its side, the wall perforated with holes; if the drum is rotated about its axis, objects inside it that are smaller than the holes will fall through. So, of course, will dirt and small stones. That is how the nonferrous objects are separated by size. Some trommels consist of a series of drums, with each succeeding drum having larger holes than the one before; objects that do not fall through the smaller holes roll on to the next stage.

Near the end of the disassembly line, some systems have one of the most technologically advanced components of the system—an electrodynamic separator, or "aluminum magnet." Under ordinary conditions, aluminum does not respond to a magnetic field. However, if an alternating current is induced in aluminum by passing it through such an electrical field, the aluminum becomes temporarily magnetic and can be drawn out of the waste stream. Other specialized devices may be added—for extracting lighter objects by making them float on flowing water, for instance; or for controlling the density of a vat of water by adding pulverized minerals, so that some objects, such as aluminum and glass, may float while heavier metals sink; or for separating glass from stones

One company, for example, markets all the forms of R.D.F. under the brand name of Eco-Fuel.

While the emphasis in resource recovery has been on materials with established uses, significant research is done on new uses for recoverable products that today can be used only as land fill. One example is glass. Because virgin glass is still cheap (it is made chiefly of silicon, the most abundant mineral on earth), it has not proved economical to convert discarded bottles and broken glass into new glassware.

solid wastes. The reason, of course, is that the metals are so finely dispersed in garbage that the cost of reclamation would exceed the metals' value. If the price of silver and gold continues to rise, reclamation may some day be economical; but, for the moment, recycling of these metals is confined to industrial scrap. Jewelry manufacturers, who account for 60 percent of the country's gold consumption, employ a wide range of devices to prevent loss from casting and grinding. These range from more effective sweeping of work-

of the countryside this year than they otherwise might have. Similar reprieves for other aspects of the natural environment are being won through increased recycling of other materials. Although the favorable environmental impact of the current boom in resource recovery is still negligible, there is every reason to believe that it will grow.

The city where recycling of garbage has gone furthest is St. Louis. Within two years, virtually all of the refuse generated by the 2.5 million people of the St. Louis metropolitan area—8,000 tons a day —will be recycled through a $70 million plant now being built by the Union Electric Company. The company is doing this because a much smaller experimental plant, mostly funded by the Environmental Protection Agency, has proved so successful.

That experiment began in 1968, before the creation of the E.P.A., when the city of St. Louis applied to the Federal Government for a grant to set up a demonstration plant to recover energy and ferrous metals. The request was approved, and in 1972 a $3.9 million facility began processing 325 tons of refuse daily, shredding all of it and extracting the light organic matter with an air classifier. The heavy portion went through a magnetic separator to recover steel and iron. The organic matter was then trucked 18 miles to the electric company's generating station and mixed with coal to fire the boilers.

Because of the plant's experimental status, St. Louis pays nothing to dump its garbage there. However, studies have shown that under normal financial arrangements the city would be paying roughly $2 a ton to dispose of its wastes this way, instead of about $5 a ton for the normal method of incineration and land fill. Each ton of garbage, it was found, yielded $1 worth of ferrous metals and $2 to $3 worth of energy. Assuming a saving of $3 a ton, this comparatively small facility would theoretically have saved St. Louis $1 million during the past three years,

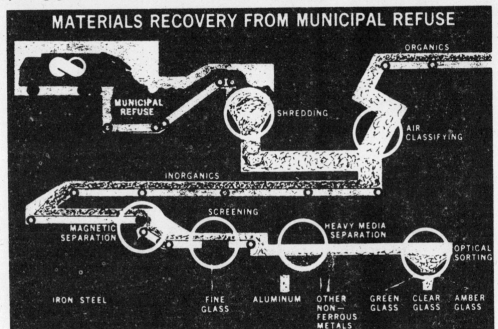

MATERIALS RECOVERY FROM MUNICIPAL REFUSE

ORGANICS

MUNICIPAL REFUSE

SHREDDING

AIR CLASSIFYING

INORGANICS

SCREENING

MAGNETIC SEPARATION

HEAVY MEDIA SEPARATION

OPTICAL SORTING

IRON STEEL

FINE GLASS

ALUMINUM

OTHER NON-FERROUS METALS

GREEN GLASS

CLEAR GLASS

AMBER GLASS

Garbage as growth industry: Systems, such as this one, for extracting iron, steel and other valuable materials from municipal refuse are being installed in cities around the country.

—and even for sorting out glass of different colors—by means of a photoelectric detector. Various combinations of these devices are planned in systems capable of handling up to 8,000 tons of solid waste each day.

The organic wastes, as distinct from metals, can be converted to any of several types of "refuse-derived fuel," or R.D.F. Pound for pound, R.D.F. has about half the energy value of coal and a third of that of oil. In the simplest method, the organic matter is shredded into fragments that can be mixed with coal for burning in a coal-fired electric generator. More sophisticated methods reduce the organic wastes into more convenient forms, such as coallike briquets, oil or gas.

However, broken glass is being used in the manufacture of "glassphalt," a road-paving material with high durability; mixed with portland cement, waste glass is being used to make building panels and blocks; glass is replacing as much as 60 percent of the costlier marble chips used in terrazzo flooring; and a mixture of 90 percent glass and 10 percent synthetic polymer is being used to make sewer pipe that is more resistant to corrosion than clay or concrete.

As yet, little if anything is being done to reclaim silver and gold, although it is estimated that somewhere between 2.7 million and 6.7 million troy ounces of these metals are discarded each year with the nation's municipal

benches to chemical treatments that scavenge gold molecules from equipment and fluids.

The doubling in the price of silver over the last two years has prompted many silver-using industries to employ similar measures. According to the Silver Institute, the refining of scrap silver is up 15 percent over last year and 30 percent over the year before. In 1973, some 45 million ounces of silver (out of a total consumption of 196 million ounces) were recycled; the projection for 1975 is 62 million ounces. That means that about $76 million worth of silver that would have been thrown away two years ago is now being saved.

It also means that silver mines will scar a little less

199

as against the costs of the normal method.

Impressed with these figures and with the usefulness of shredded garbage as a fuel, Union Electric decided to go into the resource-recovery business through a full-time subsidiary. The huge new facility, now under construction and scheduled for completion by mid-1977, will get its garbage from a vast new collection network that will have trucks picking up refuse and taking it to one of four indoor truck-to-rail transfer terminals. Trains will then haul the garbage to the plant adjacent to Union Electric's generating station. The new plant will be able to extract ferrous metals, aluminum, copper, brass and glass. Several scrap-metal processors are planning to build facilities adjacent to the new plant; one, for example, will extract the tin plating from steel cans.

"We went into this," says a spokesman for Union Electric, "because we were impressed by the prototype plant. We're going into it strictly as a private business because we think we can make money." Although inflation and more sophisticated technology are certain to make the new plant's costs much higher than the experimental plant's, company officials expect to quote an at-

tractive price to the city and to private suburban haulers.

There is one major drawback to reclaimed iron and steel — it costs two and a half times as much to ship by rail as an equivalent weight of iron ore, a differential that is enforced by the Interstate Commerce Commission.

Freight rates are supposed to be the same for competing commodities. But, in the days before recycling, scrap steel and iron were not regarded as being in competition with virgin ore for the steelmaker's dollar. The total cost of making steel from ore—including the cost of transportation from mine to mill — averages about $70 a ton. The total cost of making steel from scrap metal is 6 percent higher—and the difference comes from the higher freight rate. That difference, according to analysts, has inhibited the growth of iron and steel recycling. "In essence, the Federal Government is condoning a freight structure that makes it more economic to strip ore from a hillside than to 'mine' the metallic solid waste defacing the hillside," says Herschel Cutler, executive director of the Institute of Scrap Iron and Steel, Inc.

Despite many challenges

and legal efforts to change the freight-rate inequity, the I.C.C. has refused to modify its position. Curiously, however, it has taken a different stand on other recycled materials—and, citing a need to encourage recycling, has held down proposed rate increases for the transport of waste paper and nonferrous scrap metals.

The freight-rate problem is not the only stumbling block to increased recycling. Another is what many view as lack of experience in this field on the part of political leaders who must make decisions paving the way for resource recovery. The current situation in Dade County (Miami), Fla., is a prime example.

The county solicited proposals for a major resource-recovery plant, and 10 firms responded with such a wide variety of proposals that the county had difficulty evaluating them. It called in a private consulting firm, which found that the county had grossly miscalculated the probable costs of the various systems; had they gone ahead to make a decision on their own, the county officials would have committed themselves to spending $20 million more than they needed to. The proposals, which cost the competing firms an estimated $2

million to draw up, have been tabled, and county officials are in Europe studying recycling systems in operation there.

In the long run, however, inexperience on the part of public officials is not likely to slow down the growth of recycling very much. For the boom in resource recovery is driven by the recognition that there is money to be made in the business—money that is drawing private enterprise to regard solid waste not as a city's headache but as a natural resource. As Dr. Helmut W. Schulz, director of Columbia University's Urban Technology Center, puts it, "The brightening economics for resource recovery offer the opportunity to municipalities and regional authorities to shift the burden of solid waste disposal to private industry, at considerable savings to the taxpayer."

Environmentalists may still extol the beauties of an unspoiled landscape and urge Americans to stop polluting for the love of nature. Their battle, however, is now being waged most effectively not by ecologists but by engineers and industrialists. And, though the environmentalists may hate to admit it, that provides them with their best hope that their own cause will prevail. ∎

December 7, 1975

'Silent Spring' Is Now Noisy Summer

By JOHN M. LEE

The $300,000,000 pesticides industry has been highly irritated by a quiet woman author whose previous works on science have been praised for the beauty and precision of the writing.

The author is Rachel Carson, whose "The Sea Around Us" and "The Edge of the Sea" were best sellers in 1951 and 1955. Miss Carson, trained as a marine biologist, wrote gracefully of sea and shore life.

In her latest work, however, Miss Carson is not so gentle. More pointed than poetic, she argues that the widespread use of pesticides is dangerously tilting the so-called balance of nature. Pesticides poison

not only pests, she says, but also humans, wildlife, the soil, food and water.

The men who make the pesticides are crying foul. "Crass commercialism or idealistic flag waving," scoffs one industrial toxicologist. We are aghast," says another. "Our members are raising hell," reports a trade association.

Some agricultural chemicals concerns have set their scientists to analyzing Miss Carson's work, line by line. Other companies are preparing briefs defending the use of their products. Meetings have been held in Washington and New York. Statements are being drafted and counter-attacks plotted.

A drowsy midsummer has sud-

denly been enlivened by the greatest uproar in the pesticides industry since the cranberry scare of 1959.

Miss Carson's new book is entitled "Silent Spring." The title is derived from an idealized situation in which Miss Carson envisions an imaginary town where chemical pollution has silenced "the voices of spring."

The book is to be published in October by the Houghton Mifflin Company and has been chosen as an October selection of the Book-of-the-Month Club. About half of the appeared as a series of three articles in The New Yorker magazine last month.

A random sampling of opinion among trade associations and

chemical companies last week found the Carson articles receiving prominent attention.

Many industry spokesmen preface their remarks with a tribute to Miss Carson's writing talents, and most say that they can find little error of fact.

What they do criticize, however, are the extensions and implications that she gives to isolated case histories of the detrimental effects of certain pesticides used or misused in certain instances.

The industry feels that she has presented a one-sided case and has chosen to ignore the enormous benefits in increased food production and decreased incidence of disease that have accrued from the development and use of modern pesticides.

The pesticides industry is annoyed also at the implications that the industry itself has not been alert and concerned in its recognition of the problems that accompany pesticide use.

Last week, Miss Carson was said to be on "an extended vacation" for the summer and not available for comment on the industry's rebuttal. Her agent, Marie Rodell, said she had heard nothing directly from chemical manufacturers concerning the book.

Houghton Mifflin referred all questions to Miss Rodell. The New Yorker said it had received many letters expressing great interest in the articles and "only one or two took strong objection."

In an interview, E. M. Adams, assistant director of the biochemistry research laboratory of the Dow Chemical Company, said he would be among the first to acknowledge that there were problems in the use or misuse of pesticides.

"I think Miss Carson has indulged in hindsight," he said. "In many cases we have to learn from experience and often it is difficult to exercise the proper foresight."

Benefits Against Ills

Emphasizing that he spoke as a private toxicologist, Mr. Adams said that in some procedures, such as large-scale spraying, the possible benefits had to be balanced against the possible ills.

He referred to the extensive testing programs and Federal regulations prevalent in the pesticides industry and said, "What we have done, we have not done carelessly or without consideration. The industry is not made up of money grubbers."

Tom K. Smith, vice president and general manager of agricultural chemicals for the Monsanto Chemical Compay, said that "had the articles been written with necessary attention to the available scientific data on the subject, it could have served a valuable purpose —helping alert the public at large to the importance of proper use of pesticide chemicals."

However, he said, the articles suggested that Government officials and private and industrial scientists were either not as well informed on pesticide problems as Miss Carson, not professionally competent to evaluate possible hazards or else remiss in their obligations to society.

He said "the preponderance of evidence" indicated that chemical pesticides had not taken a significant toll of wildlife.

P. Rothberg, president of the Montrose Chemical Corporation of California, said in a statement that Miss Carson wrote not "as a scientist but rather as a fanatic defender of the cult of the balance of nature." He said the greatest upsetters of that balance, as far as man was concerned, were modern medicines and sanitation.

Montrose, an affiliate of the Stauffer Chemical Company, is the nation's largest producer of DDT, one of the pesticides that Miss Carson discusses at length. She also discusses the effect of malathion, parathion, dieldrin, aldrin and endrin.

"It is ironic to think," Miss Carson states at one point, "that man may determine his own future by something so seemingly trivial as his choice of insect spray." She acknowledges, however, that the effects may not show up in new generations for decades or centuries.

A spokesman for the National Agricultural Chemicals Association said, "We are quite concerned over the misrepresentation of an industry which has tried to do right.

"We don't intend to answer directly. We don't want to be on the defensive. But we are expanding our public information program and making available a number of new brochures."

Termed Disappointment

The Manufacturing Chemists' Association said Miss Carson's work was "a disappointment." Various courses of action are being considered, a spokesman said, but no decisions have been made.

Chemical Week, a trade magazine, said in an editorial, "Industry must again take up the Sisyphean task of repeating— again and again—that its research is aimed at profit through knowledge—not the sale of more and more pesticides whether they kill us or not."

The Department of Agriculture reported that it had received many letters expressing "horror and amazement" at the department's support of the use of potentially deadly pesticides.

The industry had a favorite analogy to use in rebuttal. It conceded that pesticides could be dangerous. The ideal was to use them all safely and effectively.

The public debate over pesticides is just beginning and the industry is preparing for a long seige. The book reviews and publicity attendant upon the book's publication this fall will surely fan the controversy.

"Silent Spring" presages a noisy fall.

July 22, 1962

SCIENTISTS URGE WIDER CONTROLS OVER PESTICIDES

President's Panel Calls for Stiffer Rules to Protect Health of the Nation

By ROBERT C. TOTH
Special to The New York Times

WASHINGTON, May 15— The President's Science Advisory Committee cautioned the nation today on the use of pesticides.

In a critical report that stirred agencies to action even before its official release, the scientists called for changes in laws and regulatory practices to guard against the hazards involved in widespread use of the chemicals.

The committee said that the use of pesticides must be continued if the quality of the nation's food and health is to be maintained.

But it called for more research into potential health hazards and, in the interim, for more judicious use of the substances in homes and the field.

43-Page Document

The 43-page document said application of certain "specially hazardous and persistent materials now registered" by the Government should be modified and new compounds "rigorously evaluated" before being approved.

It urged that the Food and Drug Administration review "as rapidly as possible" the current residue tolerances in foods, particularly those based on "inadequate" evidence.

The scientists all but told the Department of Agriculture to stop its controversial mass sprayings to eradicate gypsy moths, fire ants and two types of beetles. The aim should be control, not eradication, the report suggested.

President Kennedy said without qualification that he had "already requested the responsible agencies to implement the recommendations in the report, including the preparation of legislative and technical proposals which I shall submit to Congress."

The Food and Drug Administration has already begun a reassessment of the tolerance limits of pesticides in foods. The study is largely the result of facts turned up by a special panel of the Science Advisory Committee. The chairman of that committee, which took up the presticide problem, was Dr. Colin M. MacLeod, professor of medicine at New York University Medical School.

Among its recommendations, the report released today called for Government programs of public educaticn on the poisonous nature of 500 pesticides now registered. These include chemicals to kill insects, weeds, fungi, herbs, and rodents. More than 700,000,000 pounds of the chemicals were produced in the United States in 1961.

Found in Food and Clothes

The substances are detectable in many food items, in some clothing, in man and animals and in natural surroundings, the report said. They travel great distances and persist for long periods of time. One of every 12 acres of land in the 48 contiguous states has been sprayed with pesticides in the last year.

The report said that about 45,000,000 pounds of the pesticides are used annually in urban areas, much of it by home owners. On the average, one "bug bomb" per household is sold each year.

Dishes, utensils and food may inadvertently be contaminated. Citizens come into contact with the chemicals, which can be absorbed through the skin, in their mothproof clothing and blankets as well.

In a tribute to a popular author, the report said that "until publication of 'Silent Spring' by Rachel Carson, people were generally unaware of the toxicity of pesticides."

The book, a study of the dangers inherent in the unchecked use of pesticides, was published last fall by the Houghton Mifflin Company.

The report was critical of both Government agencies with the principal responsibility for pesticides—the Food and Drug Administration and the Department of Agriculture. He said the effectiveness of the chemicals was better proved than their safety.

The scientists said that the Department of Health, Education and Welfare should develop a comprehensive system for gathering data on pesticide residues, as well as a national network to monitor residues in air, water, soil, wildlife and fish.

Members of the special committee on pesticides, in addition to Dr. MacLeod, are:

H. Stanley Bennett, Dean of the Division of Biological Sciences, University of Chicago.

Kenneth Clark, Dean of the College of Arts and Sciences, University of Colorado.

Paul M. Doty, professor of chemistry, Harvard University.

William H. Drury Jr., director of the Hatheway School of Conservation Education, Massachusetts Audubon Society.

David R. Goddard, Provost, University of Pennsylvania.

James G. Horsfall, director of the Connecticut Agricultural Experiment Station.

William D. McElroy, chairman of the Department of Biology, The Johns Hopkins University.

James D. Watson, professor of biology, Harvard University.

The chairman of the full advisory committee is Jerome B. Wiesner, special assistant to the President for science and technology.

May 16, 1963

DDT Detected in Aquatic Life
In Both the Atlantic and Pacific

By WALTER SULLIVAN
Special to The New York Times

KANSAS CITY, Mo., Nov. 14 —The insect-killer DDT has "invaded the water environment of the world," according to Dr. Luther L. Terry, Surgeon General of the United States Public Health Service.

In ways that are still mysterious, he said, it has spread over the planet. It is found in fish caught off Iceland and Japan and in the aquatic plants and wildlife of the Arctic.

Dr. Terry gave the Bronfman Lecture tonight to the annual meeting of the American Public Health Association here. He also hailed the contributions made by the three winners of this year's Bronfman Prizes for Public Health Achievement.

The prizes provide $5,000 to each recipient.

The winners were:

¶ Dr. Harold Delos Chope, director of public health and welfare in San Mateo County, Calif.

¶ Marion B. Folsom, chairman of the National Commission on Community Health Services and former Secretary of Health, Education and Welfare.

¶ Dr. Herman E. Hilleboe,

DeLamar Professor of Public Health Practice at Columbia University.

Accumulates in Fat

Among the peculiarities of DDT is its slowness to react chemically. It thus persists in the environment. Furthermore, it tends to accumulate in fatty tissue and has been found in surprisingly large amounts in the fats and oils of deep sea fish, such as tuna.

"We do not know," Dr. Terry said, "whether DDT has been carried by air currents or whether it has entered the water cycle through the movement of plant or animal life."

He told of Public Health Service plans to analyze water obtained from the glacial ice of both polar regions in an effort to find clues to this puzzle. Layering of the polar ice makes it possible to determine the year in which the ice, at various depths, was laid down as snow. It was not until the mid-nineteen forties that DDT came into general use. The Public Health Service plans to extend its ice study back to before this period.

Dr. Terry said that the use of insecticides and other chemicals was unavoidable. He stressed, however, the many public health problems that have been raised, particularly in assessing the danger of longterm exposure to very small amounts of such substances.

Earlier in the week-long meeting, which ended today, Dr. Wayland J. Hayes Jr., of the Communicable Desease Center of the Public Health Service in Atlanta, argued that DDT was "fully as safe as the older insecticides" when used according to instructions. More than half of those who have died of pesticide poisoning in the United States, he said, were children who should never have had access to such chemicals.

Beneficial Effects Cited

He pointed to the beneficial effects of DDT in controlling disease-bearing insects and in reducing crop losses. In India, he said, the incidence of malaria has been cut from 75 million cases in 1953 to fewer than 5 million in 1962.

Another participant reported on the widespread appearance of DDT and dieldrin in American rivers. Dieldrin is another potent pesticide.

He was Dr. Gordon E. McCallum, Assistant Surgeon General. A study of 38 water samples from 10 rivers between May and December of last year disclosed these substances in all of them.

The levels are extremely low—comparable, he said, to a teaspoonful dropped into the Potomac River.

Dr. McCallum also discussed the invasion of the nation's water by detergents, primarily alkyl benzene sulfonate. This chemical, he said, "is now showing up in ground and surface water in every populated area in this country and in Europe."

He questioned whether widespread use of such a chemical should have been permitted when it was introduced after World War II. The soap and detergent industry, he said, is now developing a substitute that is appetizing to certain microorganisms. This substitute should be in general use by the end of next year, he said.

November 15, 1963

U.S. FINDS ENDRIN KILLS RIVER FISH

By JOHN W. FINNEY
Special to The New York Times

NEW ORLEANS, May 5 — On the banks of the Mississippi River, where fish have been dying by the millions for the last four years, the Federal Government began today its first formal action against the pollution of waterways by modernday pesticides.

For the first time in unequivocal terms, the United States Public Health Service said endrin, used widely in the South to control insects in sugar cane and cotton fields, was responsible for the fish deaths.

The service said an endrin manufacturer in Memphis was a major contributor to the pesticide pollution of the Mississipi River.

The charge was promptly denied by the Velsicol Chemical Corporation. It accused the Public Health Service of reaching scientifically "unsound" conclusions in indicting endrin in the fish kills.

The health service presented its indictment at a four-state antipollution conference orderly by the Department of Health, Education and Welfare.

Pollution Officials Attend

The water pollution control agencies of Arkansas, Louisiana, Tennessee and Mississippi were represented.

Until now, in its public statements, the health service has described endrin as the "most likely cause" of the fish deaths. Today it stated flatly that the pesticide "was responsible for the fish kill observed in the Mississippi and Atchafalaya Rivers in Louisiana during the fall and winter of 1963-64," a conclusion it contended was confirmed by laboratory studies of

10 dead catfish taken from the Mississippi.

Initially the health agency suggested that the pollution was coming from normal agricultural use, with the pesticides being washed off the fields into streams and rivers.

Today it suggested that industrial wastes were a major cause of the pollution.

There must be other sources than the Velsicol plant, however, the service said, to account for the fact that the level of endrin in the Mississippi at New Orleans is two or three times the level observed in the river between Memphis and Vicksburg, Miss.

Possible additional sources, it said, are fabricating plants that package the endrin, sugar cane factories in Louisiana, surface run-off from endrin-treated fields in states upstream from Louisiana, and sediments of endrin in the river bottoms that are swept up with high waters.

The fact that "minute con-

centrations" of endrin have been found in the treated water supplies of Vicksburg and New Orleans, it said, "is a matter of concern."

"It is practically certain that no one has suffered acutely from this condition thus far," the service said.

However, it pointed out that the accumulative effects of long, continued ingestion of extremely small amounts of the toxic pesticides "have not been defined."

"Either these compounds must be eliminated from the water supplies by control at the sources or by water treatment," it declared, "or positive assurance must be forthcoming from some authoritative source that continued ingestion of a certain tolerance level in the water supplies will not result in ultimate damage to health and well-being of the one million Mississippi and Louisiana water consumers."

May 6, 1964

Ban on DDT Sales Voted in Michigan

Special to The New York Times

LANSING, Mich., April 16 —Michigan moved today to outlaw the sale of DDT. It is the first of the 50 states to take such action.

The State Agriculture Commission voted to cancel all

registrations for the sale of the controversial pesticide. The commission acted under a procedure that requires hearings if demanded, but the circumstances left little doubt that these would be at most a formality, delaying the effect of the order for a few weeks.

Michigan, where tourism and recreation are an industry second only to auto man-

ufacture, has been shaken by discoveries pointing to pesticides as a menace to commercial and sport fishing.

Gov. William G. Milliken reportedly may ask neighboring states to follow Michigan's lead in an effort to protect life in Lake Michigan from potential peril. A large volume of chemicals used on crops runs off into streams and rivers and, finally, the lake.

Soaring concentrations of DDT in the pesticide dieldrin has turned up in the fat of salmon, whitefish, trout and perch. There has been talk of need to close down the state's commercial fisheries.

A shutdown of Michigan's fisheries would directly affect 1,000 people, state officials have estimated. Other economic effects would be far-reaching.

The state is caught up in a new fishing craze for a variety of salmon called the Coho, which was introduced into Lake Michigan a few years ago. The scrappy commercial and game fish, imported from Oregon, has stimulated a boom along Lake Michigan's shoreline of 400 miles.

The Agriculture Commission's decision means that licenses will no longer be issued to chemical companies wishing to distribute DDT-bearing products.

Nearly a year ago the commission halted the sale of DDT for mosquito control applications, including retail distribution of aerosol bomb preparations.

"This will end DDT use altogether in Michigan," commission spokesmen said.

Most agricultural uses to control pests in orchards, dairying, field crops and the like have already been phased out by users in favor of more expensive but safer chemical substances.

Agriculture Director Dale Ball said the commission acted on recommendation of Michigan State University's agricultural experiment station.

Dr. Gordon Guyer, chief of the university's pesticide research center, welcomed the decision as a "commendable step forward." At the same time, Dr. Guyer spoke of "hysteria" in public controversy over what, if any, hazard DDT posed to humans.

"I very, very seriously question if there is any medical significance to findings on DDT build-up in animal and fish tissues. Still, the residues have been relatively high and this is a very persistent material," he said.

A deputy in the State Department of Natural Resources has reported that the Federal Food and Drug Administration might limit pesticides in the flesh of fish eligible for interstate shipment to 3.5 parts per million.

One lot of seized salmon in Michigan contained up to 19 parts per million in its dorsal fat and entrails.

April 17, 1969

WIDE CURB ON DDT SCHEDULED BY U.S. IN NEXT 2 YEARS

3 Departments Set to Join in Phasing Out Nonessential Use of the Pesticide

COMMISSION IS HEEDED

Step May Be Most Important by Administration So Far in Environment Control

Special to The New York Times
WASHINGTON, Nov. 12—The Nixon Administration plans to ban all but "essential" uses of DDT during the next two years, Robert H. Finch, the Secretary of Health, Education and Welfare, announced today.

Mr. Finch, in a news conference, said that the Departments of Agriculture and the Interior would join with his department in a plan to phase out uses of the pesticide in the United States.

He offered no details on when or how the phase-out would begin, but said that even if the use of DDT stopped "tomorrow, it would take 10 years or longer for the environment to purge itself" of the chemical.

The far-reaching decision may be the most important made to date by the Nixon Administration in health and environmental control.

Earlier actions to ban use of the cyclamate sweeteners and to curtail oil pollution had broad impact on the nation's health, ecology and economy.

From Home to Farm

But a curb on DDT affects everyone from consumers of foods that may contain undue amounts of the chemical to the farmers who spray or dust it on their crops in an effort to protect them from insects.

In calling for sharply limited use of the chemical, Mr. Finch did not spell out what acceptable uses might be. He said that this would be subject to careful determination.

Concern over the use of DDT on food crops in particular has grown since the National Cancer Institute completed a study early this year showing that cancerous tumors could be produced in laboratory mice with a diet of large doses of DDT.

Although there is no scientific information on the tolerance level of humans to DDT, Mr. Finch said, "prudent steps must be taken to minimize human exposure to chemicals that demonstrate undesirable responses in the laboratory at any level."

Mr. Finch said that he and Clifford M. Hardin, Secretary of Agriculture, and Walter J. Hickel, Secretary of the Interior, had signed an agreement to put into effect the recommendations of a special study commission on pesticides.

The commission, appointed in April by Mr. Finch, urged elimination of all but "non-essential uses of DDT over the two-year period. Its chairman is Dr. Emil M. Mark, former chancellor of the University of California at Davis.

Asked if the interdepartmental agreement meant that the Administration would sharply curtail the use of DDT, Secretary Finch replied:

"Absolutely. We made a commitment."

Secretary Hardin said in a statement that he thought the commission's recommendations were "sound" and confirmed that he was in "substantial agreement, if not complete agreement," with Mr. Finch.

The decision to impose the ban was a harder one for the Secretary of Agriculture to make, Mr. Finch said, because his "constituents" are farmers.

Abundant Signals

Mr. Finch said that there was no hard evidence to guide the Government in determining which uses would be suitable. But he added that there were "abundant signals" to justify the ban.

He said the Administration would not seek to halt manufacture of the pesticide because most of the DDT produced in the United States was exported and it would be up to other nations to decide whether benefits of its use outweighed harmful effects.

The commission also recommended in its 44-page report that the Administration develop guidelines for the control, review and study of a variety of similar "hard pesticides," socalled because they remain in the environment for long periods of time.

The report warned that curtailment of the use of DDT could cause increased use of the other chemicals, including aldrin, dieldrin, endrin, heptachlor, chlordane, benzene hexachloride, lindane, and compounds containing arsenic, lead or mercury.

At the same time, the commission cautioned against overreacting. It said that "consideration must be given to both the adequacy of the evidence of hazard of human health and possible consequences to human welfare that flow from the imposition of restrictions."

Pesticides are so important to modern society, said one section of the commission's report, that "we must learn to live with them."

But the thrust of its recommendations was for tighter governmental control over the use of the chemicals and more regular study of their effect.

Conservationists have sought a ban on DDT amid growing criticism — and governmental concern — about its potential harm to humans and their surroundings.

The month before Mr. Finch named the 17-member study panel, the Food and Drug Administration seized 28,150 pounds of Coho salmon from Lake Michigan containing a high residue of DDT.

The Canadian Government acted earlier this month to ban some 90 per cent of the uses of DDT. At least two states, Michigan and Arizona, also have enacted legislation to curb its use.

Mr. Finch said the imposition of a ban on DDT would proceed after discussions with the other two departments. Although he said he did not wish to be tied to a specific timetable, he added that "I wouldn't want to see us wait six months" to begin phasing out the pesticide.

The interdepartmental agreement was necessary, he said, because he does not have legal authority to act unilaterally. It would not require new legislation to curb DDT, he said.

Mr. Finch forwarded the commission report to the President's Environmental Quality Council. He said the council would consider the report at its meeting Nov. 20.

November 13, 1969

HICKEL EXTENDS PESTICIDES CURB

Virtual Ban Is Imposed in One-Fourth of the U.S.

By RICHARD D. LYONS.
Special to The New York Times

WASHINGTON, June 17—Interior Secretary Walter J. Hickel imposed today a virtual ban on the use of pesticides on more than 500 million acres of Federal lands, one-quarter of the land area of the United States.

Those chemicals prohibited include DDT, Aldrin, 2, 4, 5-T, dieldrin, endrin, DDD, mercury compounds and nine other lesser known agents.

Mr. Hickel's order put 32 other chemicals and classes of chemicals on a "restricted list," allowing them to be used under limited circumstances but only with the approval of the Cabinet Subcommittee on Pesticides.

The Interior Department had previously banned some of the chemicals, but the prohibitions were widely ignored by the agency's 75,000 employes, a spokesman said.

DDT, for example, was officially banned six years ago, but it was still in use up until last year. Bans on other agents were effective in some areas, but not in others, said.

"Mr. Hickel's order was issued to dispel any notion by department employes that the pesticides could continue to be used," the spokesman said.

Mr. Hickel's order was viewed by some Federal sources as serving to goad other Federal organizations managing large tracts of Government land, such as the Departments of Defense and Agriculture, into adopting stronger policies against the unrestricted use of pesticides.

Clifford M. Hardin, the Secretary of Agriculture, banned the use of DDT in residential areas seven months ago and set an almost total prohibition on its use by farmers, gardeners and homeowners by the end of this year. Yet many other presticides, herbicides and economic poisons that have been deemed harmful to the environment remain in use.

Mr. Hardin announced today that his department had called a working conference on the disposal of sprays and powders containing chemicals harmful to human life. The parley, to be held here on June 30 and July 1, is expected to attract conservationists, state agriculture officials and members of other groups concerned with the environment.

In a related action today, the Agriculture Department's director of science and education, Dr. Ned D. Bayley, told members of the Senate Commerce Subcommittee on Energy, Natural Resources and Environment that the current laws governing environmental chemicals were inadequate to protect public health.

Dr. Bayley said many agriculture officials believed that an entirely new law might have to be written to clear up discrepancies in the old statutes, mainly the Pesticide Act of 1964.

Further legal power may be needed, Dr. Bayley testified, to stop the sale of products while administrative investigations are under way, to require new warnings on packages, to require quality control during manufacturing with increased inspections, and to impose liability on the users of some chemicals.

Senator Philip A. Hart, Democrat of Michigan, the subcommittee chairman, said he believed that the current law was adequate but that the Department of Agriculture was interpreting it too literally.

Harrison Wellford, a lawyer connected with Ralph Nader's Center for the Study of Responsive Law, complained to the subcommittee that the Department of Agriculture had been ineffective, tardy and overly secret in administering the pesticide laws.

Mr. Wellford said the partial bans on DDT and the herbicide 2, 4, 5,-T were a "sham" and that much of the chemicals were still in use. He said laws were so vague that manufacturers could continue to sell these chemicals for a year or longer after they were officially prohibited.

Mr. Hickel's new order requires all Interior personnel to "consider safety and environmental quality as the primary factors in making the decision whether or not to use a pesticide."

The chemicals banned also include Heptachlor, Lindane, Toxaphene, Amitrol, Azodrin, Bidrin and Strobane, as well as inorganic arsenical compounds and thallium sulphate.

June 18, 1970

Science
How to Turn Insects Against Themselves

The future of pest control will depend less and less on nonselective chemicals that pollute the environment and poison desirable organisms and more and more on such selective approaches as seduction with sex lures, involuntary birth control, jammed radar, enforced celibacy and germ warfare.

This is the future that Rachel Carson insisted upon eight years ago in "Silent Spring" when she forecast inevitable doom resulting from overdependence on lethal chemical. This is the future that the current wave of environmentalists sees as essential to our survival.

It is a future that every day is coming closer to realization. Last week, a scientist with the United States Department of Agriculture told the annual meeting of the American Chemical Society that traps baited with a synthetic sex lure may effectively control the gypsy moth, which has defoliated much of the Northeast since the mid-1960's, when spraying with DDT was discontinued throughout most of the region.

The powerful lure, synthesized following an analysis of the sex attractant extracted from more than 78,000 female gypsy moths, is used in traps to attract unsuspecting males to what they think are sex-hungry females.

Curb on Sprays

If the man-made love potion proves sufficiently potent, it might be used as the sole method of control, with the males destroyed inside the traps and few left to fertilize females. Otherwise, the traps could serve as a means of detecting infestations, with chemicals used only when trapped insects show they are needed.

Thus, instead of a farmer spraying his crop once a week in July to prevent an infestation, he could wait until the trap showed that the pests were present and cut his spraying to, say, once for the month or perhaps not at all.

In addition to sex attractants, which have been isolated for a dozen or more important agricultural pests, scientists are experimenting with traps baited with lights, ultra sound waves and electromagnetic waves that are particularly attractive to insects.

This biological approach to pest control originated some 40 years ago with Dr. Edward F. Knipling, director of the Entomology Research Division of the Agriculture Department.

Dr. Knipling conceived the idea that sterilizing insects might be a more effective way to control them than killing them outright. Years later, he proved his point by ridding the southern United States of the screwworm fly, which was causing $120-million damage to cattle each year.

Billions of male screwworms were reared in a Texas fly factory, sterilized by radiation and released. After several generations of barren females, the screwworm was gone.

At the same time, a long-term search for natural parasites and predators of important agricultural pests is beginning to yield promising results. In one test, the release of 200,000 aphid lions per acre was shown to be as effective in controlling the cotton bollworm as are insecticides.

And a parasitic ant imported from South America is being studied as a possible control measure for the South's fire ant, a serious pest of people, crops, livestock and wildlife. The parasite lives in fire ant mounds and tricks its host into feeding and rearing its young, to the neglect of the fire ant's own colony.

Scientists in the Agricultural Research Service call this the era of "integrated control," when biological and chemical methods are used in concert, to the detriment of agricultural pests and the benefit of mankind.
—JANE E. BRODY

September 20, 1970

204

DDT, the First Domino

By NORMAN E. BORLAUG

ROME—The current vicious, hysterical propaganda campaign against the use of agricultural chemicals, being promoted today by fear - provoking, irresponsible environmentalists, had its genesis in the best-selling, half-science-half-fiction novel "Silent Spring," published in 1962. This poignant, powerful book—written by the talented scientist Rachel Carson—sowed the seeds for the propaganda whirlwind and the press, radio and television circuses that are being sponsored in the name of conservation today (which are to the detriment of world society) by the various organizations making up the environmentalist movement.

The moving forces behind the environmental movement today include The Sierra Club, National Audubon Society, Isaac Walton League, The Boone and Crockett Club, and the new legal arm of the movement: The Environmental Defense Fund, with its scores of lawyers baptized into the movement with the motto, "Sue the Bastards." The principal individual supporters of the movement are wilderness explorers, bird-watchers, wildlife lovers, ill-informed press and television personalities and confused youth and older members of society who have been frightened so badly by the doom sayers that they have joined.

Although the collective active membership in these organizations is perhaps less than 150,000, their superb organization and tactics make them an extremely effective force in lobbying for legislation to ban pesticides, and for brainwashing the general public.

Previously both the Environmental Defense Fund and the National Audubon Society had stated that DDT causes cancer, even though the Surgeon General of the U.S. Public Health Service has stated: "We have no information on which to indict DDT as a tumorigen or carcinogen for man and, on the basis of the information now available, I cannot therefore conclude that DDT represents an imminent health hazard."

The evironmentalists would now like to have a legislative ban placed on DDT so as to prohibit it for any use in the U.S.A. Almost certainly as soon as this is achieved, these organizations will begin a worldwide propaganda barrage to have it banned everywhere. This must not be permitted to happen, until an even more effective and safer insecticide is available, for no chemical has ever done as much as DDT to improve the health, economic and social benefits of the people of the developing nations.

The World Health Organization in 1955 launched a worldwide campaign against malaria. In summarizing the progress of the campaign on Feb. 2, 1971, organization officials made the following statement:

Tomi Ungerer

"More than 1,000 million people have been freed from the risk of malaria in the past 25 years, mostly thanks to DDT. This is an achievement unparalleled in the annals of public health. But even today 329 million people are being protected from malaria through DDT spraying operations for malaria control or total eradication.

"The improvement in health resulting from malaria campaigns has broken the vicious circle of poverty and disease resulting in ample economic benefits: increased production of rice (and wheat) because the labor force is able to work; opening of vast areas for agricultural production: India, Nepal, Taiwan and augmented land value where only subsistance agriculture was possible before.

"The safety record of DDT to man is truly remarkable. At the height of its production 400,000 tons a year were used for agriculture, forestry, public health, etc. Yet in spite of prolonged exposure by hundreds of millions of people, and the heavy occupational exposure of considerable numbers, the only confirmed cases of injury have been the result of massive accidental or suicidal swallowing of DDT. There is no evidence in man that DDT is causing cancer or genetic change."

Although more than 1,400 chemicals have been tested by W.H.O. for use in malarial campaigns, only two have shown promise and both of these are far inferior to DDT.

It is now obvious that the current aim of the Environmental Defensive Fund and its affiliated environmentalist lobby groups is to ban DDT first in the U.S.A. and then in the world if possible. DDT is only the first of the dominoes. But it is the toughest of all to knock out because of its excellent known contributions and safety record. As soon as DDT is successfully banned, there will be a push for the banning of all chlorinated hydrocarbons, then in order, the organic phosphates and carbamate insecticides. Once the task is finished on insecticides, they will attack the wood killers, and eventually the fungicides.

If the use of pesticides in the U.S.A. were to be completely banned, crop losses would probably soar to 50 per cent, and food prices would increase fourfold to fivefold. Who then would provide for the food needs of the low-income groups? Certainly not the privileged environmentalists.

These are excerpts from a speech by Dr. Norman E. Borlaug, winner of the 1970 Nobel Peace Prize for his work in developing wheat strains, delivered at a U.N. conference in Rome.

November 21, 1971

'Environment Is Not a Motherhood Issue'

Bruce Roberts/Rapho Guillumette

By BARRY COMMONER

BOSTON—Environmental quality is supposed to be an issue which is so innocuous and so free of dangerous controversy that, like motherhood, it can be readily discussed without offense to anyone. Sometimes those of us who emphasize the environmental crisis are accused of diverting attention, even deliberately, from really serious and controversial issues, into this benign channel.

In fact, the environment is not at all a motherhood issue. Quite the contrary; pursued to its source every environmental issue generates a confrontation with the grave, unsolved, intensely contested issues of the world—war, poverty, hunger and racial antagonism.

Striking evidence is the recent attack by the distinguished agronomist, Norman Borlaug, on what he terms the "current vicious, hysterical propaganda campaign against the use of agricultural chemicals being promoted by fear-provoking, irresponsible environmentalists."

Dr. Borlaug's complaint is that "No chemical has done as much as DDT to improve the health, economic and social benefits of the people of the developing nations." The facts simply do not support Dr. Borlaug's position. First, concerning the need for DDT to control insect-born diseases, especially malaria, in developing countries: While it is true that DDT has largely eliminated malaria in a number of countries, this success is limited to relatively dry areas, such as Greece, or to islands such as Ceylon. In contrast, campaigns intended to "eradicate" malaria in Central America through spraying with DDT and other insecticides, have largely failed. Now, after some

12 years of effort the malaria incidence is as high or higher than it was at the start of the campaign, for the mosquitoes have become resistant to insecticides.

Secondly, contrary to Dr. Borlaug's claims, the elimination of pesticides would apparently have only minor effects on United States agricultural productivity (and therefore probably elsewhere as well). This is evident from a recent Department of Agriculture symposium at which agricultural experts presented a series of careful analyses of precisely what would happen to U.S. agricultural production if pesticide use were restricted.

(1) In the case of wheat, banning the use of herbicides (2-4D and 2-4-5T) would lead to a loss of 55.3 million bushels—about 4 per cent (not "50 per cent") of the total production. The computed effect on the price of wheat would be a 12 per cent (not a "fourfold to fivefold") increase.

(2) A ban on the use of the same herbicides on 62 million acres of treated crops would increase the cost of production by $290 million—which is only 1.5 per cent of the value of all crops and 6.6 per cent of the value of crops now treated with herbicides.

(3) A 70-80 per cent reduction in insecticide use could be accomplished, with no reduction in agricultural output, simply by increasing harvested acreage by 12 per cent. This represents the amount of land diverted from agricultural production in 1967 by various government land retirement programs—land which, according to one agricultural expert, ". . . is suitable for regular cultivation with no additional investment."

Given these data, I am unable to accept Dr. Borlaug's claim that a severe reduction in the use of agricultural chemicals would sharply reduce

agricultural output.

The chief cause of the environmental crisis in the United States is the counter-ecological nature of the new industrial technologies which have transformed the American economy since World War II.

Most of these transformations, such as the substitution of fertilizers and pesticides for land, crop rotation and cultivation, have greatly increased the environmental impact of production. In other words, to produce the same amount of goods and transportation as we did in, say, 1946, we now pollute the environment much more than before.

These transformations are irrational, because industrial and agricultural production is itself dependent upon the integrity of the environment which is being destroyed by the new technologies. The only "explanation" which can be offered for the irrational, counter-ecological direction of the United States economy since World War II is that it has been guided by the drive —natural in our economic system—to increase the rate of economic return, for there is, for example, more profit in the manufacture of detergents than of soap.

If we accept as inviolable the present economic inequities which restrict the access of poor nations, and of poor people in every nation, to the resources of the earth—then, and only then, does a conflict arise between the integrity of the environment and the welfare of the people who inhabit it.

Barry Commoner is director of the Center for the Biology of Natural Systems at Washington University (St. Louis) and author of "The Closing Circle."

December 7, 1971

DDT BANNED IN U.S. ALMOST TOTALLY, EFFECTIVE DEC. 31

Ruckelshaus Decides After 3-Year Fight That Risk to Environment Is Too High

---

COURT APPEAL IS FILED

Farmers Are Given Time for Instruction in the Use of a Substitute Pesticide

---

By E. W. KENWORTHY
Special to The New York Times

WASHINGTON, June 14 — William D. Ruckelshaus, administrator of the Environmental Protection Agency, banned today almost all uses of DDT, the long-lived toxic pesticide that lodges in the food chain of men, animals, birds and fish.

After almost three years of legal and administrative proceedings, reports by scientific bodies and public hearings, Mr. Ruckelshaus declared in a 40-page decision that the continued use of DDT over the long term, except for limited public health uses, was an unacceptable risk to the environment and, very likely, to the health of man.

"The evidence of record showing storage [of DDT] in man and magnification in the food chain," said Mr. Ruckelshaus, "is a warning to the prudent that man may be exposing himself to a substance that may ultimately have a serious effect on his health."

Three Crops Affected

Mr. Ruckelshaus's order is effective Dec. 31, 1972. In the meantime, he explained, growers of cotton, peanuts and soybeans — the three crops that account for almost the total domestic use of DDT—will get instruction in the handling of a substitute pesticide, methyl parathion. The substitute is toxic, but unlike DDT, it degrades quickly.

Samuel Rotrosen, president of Montrose Chemical Corporation, sole United States manufacturer of DDT, immediately asked the United States Court of Appeals for the Fifth Circuit in New Orleans to set aside the order. Montrose is jointly owned by Chris-Craft Industries, Inc., and the Stauffer Chemical Company.

Formulators Appeal

A parallel appeal in the same court was filed by Robert L. Ackerly. a Washington lawyer who represents a group of formulators of DDT, that is, manufacturers of various commercial pesticides containing DDT.

Mr. Ruckelshaus's order represented a major victory for environmental groups that in October, 1969, petitioned the Secretary of Agriculture to begin the proceedings that would lead to cancellation of DDT for shipment in interstate commerce. At that time the Department of Agriculture had authority to register "economic poisons" for such shipment. The authority was transferred in December, 1970, to the Environmental Protection Agency.

The organizations, which later took their cause to the Federal courts, were the National Audubon Society, the Izaak Walton League, the Sierra Club, the Western Michigan Environmental Council and the Environmental Defense Fund, which also supplied legal counsel for the plaintiffs.

The decision came 10 years after Rachel Carson, the biologist, set off the controversy over DDT with her book "The Silent Spring," in which DDT was called "the elixir of death" for birds, mammals, fish, insects and perhaps for man if it proved to have cancer-causing properties.

The order represented a defeat for the maker and the 31 formulators of DDT; for the Department of Agriculture, which had entered the case on their side, and for Representative Jamie L. Whitten, Democrat of Mississippi, chairman of the House Appropriations subcommittee that handles funds for the environmental agency. Mr. Whitten, a cotton-state Congressman, has maintained that DDT is necessary to combat the boll weevil and boll worm.

Discovered in 1939

The insecticidal properties of DDT were discovered in 1939. During and after World War II it was hailed as a miracle chemical because of its ability to control typhus and malaria in tropical and semitropical countries and the boll weevil and many other agricultural pests in the United States.

In recent years, domestic use of DDT has steadily declined, partly in response to warnings by scientists about its persistence when taken into the food

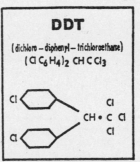

The New York Times/June 15, 1972
DDT, as shown in the formula and diagram, is a compound of carbon, hydrogen and chlorine.

chain, partly as a result of Federal restrictions and partly because of the discovery of substitutes.

According to the Environmental Protection Agency, domestic use has declined from about 79 million pounds in 1959 to 12 million to 14 million pounds in 1970.

Roughly 86 per cent is used for cotton, 9 per cent on peanuts and 5 per cent on soybeans. Minimal quantities are used on some other vegetable crops.

Exports of DDT, mostly for malaria control by the World Health Organization and under the foreign aid program, total about 26 million pounds a year.

Mr. Ruckelshaus's order will not affect manufacturing for export.

Also, it will permit use for protection of public health — as in outbreak of disease following a flood—and for health quarantine purposes.

The order permits the use of DDT on three minor crops, green peppers, onions and sweet potatoes in storage, for which no effective alternatives are now available. However, certain conditions must be met by applicators.

These three uses now account for less than 1 per cent of domestic sales.

Mr. Ruckelshaus said in his findings that DDT could persist in soils "for years and even decades" and also in aquatic ecosystems, that it could attach to eroding soil particles and so contaminate water supplies and that it could get into the air by vaporizing from soil and crops.

Further, he said, it is toxic to fish and birds and impairs their reproductive processes. The thinning of eggshells through ingestion of DDT, he said, has accelerated the decline of many birds such as eagles and ospreys.

He also said that experiments on laboratory animals had shown that DDT causes tumors and that this raised the possibility that it may cause cancer in human beings.

The chronology of the DDT controversy is as follows:

October, 1969. The environmental groups petitioned the Secretaries of Agriculture to give notice of cancellation of all uses of DDT and to suspend its use as an "imminent hazard" to health pending the cancellation proceedings.

November, 1969. The Secretary, Clifford M. Hardin, issued cancellation notices for use on shade trees, tobacco, around homes and in marshes except to control disease carriers but refused suspension. The environmental groups went to court.

May, 1970. The United States Court of Appeals for the District of Columbia circuit ordered Mr. Hardin to issue cancellation notices on all other uses and suspend all uses pending cancellation proceedings or say why he could not do so.

August, 1970. Mr. Hardin issued cancellation notices on more than 50 uses, but not on cotton spraying, the most important. He still refused to order suspension.

January, 1971. The Court of Appeals ordered Mr. Ruckelshaus to issue cancellation notices on all remaining uses and to consider immediate suspension.

January, 1971. Mr. Ruckelshaus issued the cancellation orders and thus set in motion the long administrative proceedings allowed by law. The companies requested the appointment of a scientific advisory committee to examine the question and also public hearings.

March, 1971. Mr. Ruckelshaus declined to suspend use, finding DDT was not "an imminent hazard."

September, 1971. The scientific advisory board recommended accelerated reduction in use of DDT "with the goal of virtual elimination of any significant additions to the environment."

April, 1972. After seven months of hearings and 8,900 pages of testimony, Edmund S. Sweeney, an Interior Department officer selected by the Civil Service Commission to preside over the hearings, said that benefits of DDT outweighed the risks and recommended reversal of the proposed ban. The Environmental Protection Agency said it would contest the findings.

June 15, 1972

E.P.A., Citing Risks of Cancer, Bars 2 Widely Used Pesticides

By E. W. KENWORTHY
Special to The New York Times

WASHINGTON, Aug. 2 — A halt in the manufacture of aldrin and dieldrin, the two most widely used pesticides in United States, was ordered today by the Environmental Protection Agency on a finding that their use poses an "imminent hazard" to the public health.

The highly toxic pesticides are suspected, on the basis of tests in mice and rats, of causing cancer.

The order to the Shell Chemical Company, the sole manufacturer of the two pesticides, becomes effective in five days unless Shell or a company using one of the pesticides in a trade-name product requests a hearing. But that hearing must be held as quickly as possible and cannot last longer than 15 days.

For more than a year hearings have been going on as a result of an E.P.A. order in June, 1972, canceling many of the uses of the pesticides, and soliciting views of the public on whether as the law permits, there should be an immediate suspension of some uses, because of an "imminent hazard" to health.

Without waiting for the conclusion of those hearings, Russell E. Train, E.P.A. Administrator, decided to order suspension of manufacture. The effect will be to prohibit use of the pesticides after current supplies, which are not large, are used up. This also means that the pesticides could not be imported.

The company today called Mr. Train's action "blatantly unfair, a violation of due process and an insult to the integrity of E.P.A.'s administrative proceeding." It said, "There is no evidence whatsoever that associates this chemical with cancer in man."

Dieldrin and aldrin are closely related, persistent, long-lived chlorinated hydrocarbons that accumulate in fatty tissues when taken into the body through the food chain.

Although used on a number of garden crops, for termite control and in mothproofing, more than 90 per cent of the two pesticides manufactured in the United States are applied to the soil in corn fields to control the "corn insect complex."

After application, aldrin breaks down into dieldrin and, as a result, it comprises most of the manufacture of the related pesticides, although dieldrin is produced and marketed separately.

Mr. Train said he acted because "I find that a situation exists in which the continued manufacture of aldrin and dieldrin will likely result in unreasonable adverse effects on man and the environment."

The protection agency said that pathologists had discovered "significant increases" in tumors, many diagnosed as malignant, in two kinds of mice and two kinds of rats when over a brief period they were fed a diet that included as little dieldrin as one-tenth part per million.

Furthermore, the agency reported, in tests made in fiscal 1973, the Food and Drug Administration found "measurable amounts" of dieldrin in 83 per cent of all dairy products sampled, in 88 per cent of all garden fruits sampled, in 96 per cent of all samples of meat fish and poultry and in 12 to 42 per cent of grain and cereal products, potatoes, leafy vegetables, oils, fats, shortening and fruits.

Finally, the agency said that samples of human tissue taken during surgery and autopsies in 1970 showed that 96.5 per cent of the individuals tested had detectable residues of dieldrin in their fat tissues. By 1971, similar tests showed that 99.5 per cent of those tested had such residues.

The agency said that "based on National Cancer Institute methods for estimating human cancer risk, the present average human daily dietary intake of dieldrin subjects the human population to an unacceptably high cancer risk."

Moreover, the E.P.A. stated, "children, particularly infants from birth to one year, because of their high dairy product diets, consume considerably more dieldrin on a body weight basis than any other age segment of the population."

Mr. Train's order today represented a notable victory for the Environmental Defense Fund, which has headed a three-year fight to suspend the manufacture of the two pesticides aldrin and dieldrin.

A spokesman for the group said today that it was "very pleased" with the ruling and that the ban would not affect corn production because there were good alternative pesticides available.

August 3, 1974

World Use of DDT Feared on Rise

By HAROLD M. SCHMECK Jr.
Special to The New York Times

WASHINGTON, Feb. 13—Worldwide use of DDT is probably as high today as it was more than a decade ago when United States production was at its peak and may be higher, a scientist said here today.

Use of the long-lived pesticide was almost totally prohibited in the United States in 1972 because of its potential hazards to men, animals, birds and fish. By that time it was so widely distributed throughout the world that traces could be found in virtually all types of living things.

DDT is still a worldwide problem and North America cannot be isolated from it, said Dr. G. M. Woodwell, director of the Ecosystems Center of the Marine Biological Laboratory, Woods Hole, Mass., one of the world's foremost research centers.

In recent years, Dr. Woodwell said, the bulk of DDT use has shifted away from the Temperate Zone to the tropics, but it spreads through the air and oceans and continues to be a scientific and political challenge to the world.

In this country, use of the toxic pesticide had been declining throughout much of the nineteen-sixties even before the ban by the Environmental Protection Agency.

Restrictions were put on DDT use, Dr. Woodwell said, because it is toxic to many animal species, persistent in the soil and virtually uncontrollable once released into the environment. It is also soluble in fat and therefore accumulates in animals and man.

The scientist spoke at the Smithsonian Institution at a symposium on marine biomedical research sponsored by the National Institute of Environmental Health Sciences.

Dr. Woodwell said there were, currently, powerful efforts to bring about a reversal of the United States ruling on DDT. Some scientists are known to favor its use under controlled conditions, contending that the advantages would outweight the hazards.

Dr. Woodwell said, however, that the pesticide was uncontrollable if used on a major scale because it is so persistent and because it vaporizes from the ground and thus gets into the atmosphere where it can be widely distributed.

He said that studies have shown measurable amounts in the air over land that was last sprayed with the pesticide several years ago. The relatively new evidence of its potential for vaporization, he said, makes much less mysterious than it once seemed, the finding that DDT had spread to all regions of the earth including the poles.

One favorable sign in the situation, the scientist said, is evidence that DDT used in the tropics is broken down chemically more rapidly than in the Temperate Zone. He suggested that this might result from atmospheric patterns that carry the pesticide aloft to where it can be attacked by radiation from the sun, which is much more powerful over the tropics than over more northerly and southerly regions.

A questioner in the audience, however, asked the scientist if it were not possible that chlorine loosed from molecules of DDT high in the atmosphere might contribute to destruction of the planet's protective ozone layer. Dr. Woodwell said the answer to this was unknown.

Dr. Woodwell said the available evidence had shown that the concentration of DDT in ocean waters of the Temperate Zone appeared to be declining and that over-all concentrations in living animals of the Temperate Zone were also declining.

On the subject of worldwide use of DDT, he said that Dr. Edward Goldberg of the Scripps Institution of Oceanography, in a recent anlysis of DDT use, estimated the current total use at about twice the total of United States production in the early nineteen-sixties.

Dr. Woodwell said United States production was about 81,000 metric tons in 1963 and that roughly the estimated world use now is about 150,000 metric tons, the bulk of it being in the tropics. A metric ton is 2,205 pounds.

Although DDT is little used in the United States now, a substantial amount is still exported. A specialist at the Department of Agriculture said today that 1973 exports totaled about 73,-712,000 pounds—about 33,000 metric tons—valued at about $16-million. The export total was the highest so far in the nineteen-seventies. Figures for 1974 are not yet available.

February 14, 1975

A Ban on Most Pesticides With Mercury Is Ordered

By United Press International

WASHINGTON, Feb. 18 — The Environmental Protection Agency ordered today an immediate ban on the production of virtually all pesticides containing mercury.

Russell E. Train, chief of the environmental agency, cited cases of nervous system disorders caused by mercury poisoning in Japan, Iran and the United States as evidence that unchecked use of mercurial pesticides would pose an unreasonable human health hazard.

The new order will halt the production of all mercurial pesticides used as bacteriacides or fungicides in paints, varnishes and lacquers. It also puts an end to the use of such pesticides on turf, including golf course greens and other golf course areas, and in the treatment of seeds.

"Economic, social and environmental costs and benefits of the continued use [of mercurial pesticides] are not sufficient to outweigh the risk to man or the environment," Mr. Train said.

He said that registered and recommended pesticides existed and were available to replace the hazardous substances at a reasonable cost.

Mr. Train's order stopped short, however, of banning all mercurial pesticides.

He allowed the continued production of fungicides containing mercury compounds to treat fabrics such as awnings and tarpaulins intended for continued outdoors use, for the control of brown mold and for the control of Dutch elm disease.

A spokesman said Mr. Train's action should reduce the mercury now entering the environment through pestcide use by an estimated 98.5 percent.

Mr. Train said that most mercury compounds used in pesticides were not dangerous by themselves. But, he said, relatively innocuous mercury compounds are converted in the environment to toxic methylmercury, a substance that builds up to ever higher levels in the food chain.

Ingestion or inhalation of enough methylmercury causes disorders to the central nervous system in animals and humans that can cause illness or death, the environmental agency said.

A spokesman for the E.P.A. said that the paint industry accounted for almost 90 percent of the mercury used in pesticides.

The 359,000 pounds of mercury used in pesticides in 1973, he said, included about 321,000 pounds for paint, 26,000 pounds for turf treatment, 7,000 pounds for seed treatment and 5,000 pounds for such things as treating fabrics, protecting wood

and controlling Dutch elm disease.

[The Associated Press reported that a spokesman for the environmental agency in Washington said the "preventive measure" had been sparked by a case in Alamagordo, N.M., in which members of a family became blind and suffered damage to the nervous system after eating meat from pigs that had eaten seed contaminated by a mercury-treated pesticide.]

Mr. Train's order resulted from a notice of cancellation issued in 1972 for all pesticides containing mercury because of the "substantial question" about their safety even when used according to label restrictions. An E.P.A. administrative law judge issued a ruling last Dec. 12 after hearings on that notice.

Mr. Train said all mercurial pesticide products produced before the issuance of his order may be sold. He said the use of those products represents the most practical means of disposing of them.

February 19, 1976

Farmers Turn to Pest Control in Place of Eradication

By JANE E. BRODY

In fields and orchards throughout the country, a small but growing number of farmers are abandoning the heavy use of pesticides.

Instead of trying to eradicate pests—a strategy that has proved to be ineffective and also an economic and environmental disaster—they are turning to an ecologically based concept of pest control that is expected eventually to dominate world agriculture. The concept is 50 years old, but has long been disregarded.

These farmers have learned that a pest need not be annihilated to prevent costly damage to crops, and that, in fact, maintaining a small population of pests in the field may be the most effective means of reducing crop losses.

Savings Are Cited

Farmers who are using this approach, called pest management, are saving money, getting higher yields and achieving more effective control of the insects and plant diseases that are the bane of modern single-crop agriculture.

At the same time, the farmers have greatly reduced their

The New York Times/Andrew Sacks

Royal Kline, a Michigan apple grower, uses magnifying glass to inspect leaves and fruit. He relies on scientific advisories to plan his battle against pests.

use of hazardous pesticides, which in the last three decades have become widespread environmental contaminants that threaten the health and in some cases the survival of many animals, including man.

Using pest management techniques, a Michigan potato farmer saved $25,000 in one summer by not using aerial spraying to control a fungus called late blight.

In Illinois, where pesticide residues had contaminated milk produced by alfalfa-fed cows, alfalfa farmers participating in a regional program of pest management have cut pesticide applications from eight or 10 a season to one or two.

And in Israel, a simple, inexpensive pest management strategy—the importation from Hong Kong of a pest parasite—has resulted in control of Florida red scale, Israel's principal citrus pest, without any pesticides and with an annual saving of $1 million.

Key Concept Explained

"In pest management," said Dr. Robert L. Metcalf, professor of entomology at the University of Illinois, "pesticides are used as a weapon of last resort, rather than the first thing you reach for."

Pest management is a system of restricting the numbers of pests so the injury they cause will not result in significant economic losses. This often means that sizable pest infestations can be tolerated without reducing crop yields. The pest population may have to be suppressed for only brief parts of the growing season, when it is able to cause economic injury to the crop.

The basic concept is "one of optimizing control rather than maximizing it," Dr. Metcalf and his collegue, Dr. William Luckman, state in their book, "Introduction to Insect Pest Management."

Rather than relying solely on pest-killing chemicals, management uses one or more cost-effective measures that wherever possible take advantage of nature's own controls on pest populations.

These include cultivation practices to discourage infestation and reproduction of the pest, crop varieties that resist insects and diseases, natural bioogical controls such as insects or diseases that prey on or parasitize the pest and the monitoring of pest invasions so pesticides are used only when they do the most good.

Parasites and Computers

Pest management can be a single tactic, such as introducing a parasite that attacks the pest, or it can be a network of interrelated options so complex that a computer is needed to integrate them to the farmer's best advantage.

Farmers using pest management strategies have abandoned the old "crop insurance" approach of spraying pesticides on a set schedule during most of the growing season, whether the target pest is present or not. Instead, they spray "only when needed"—that is, only when the pest is there in sufficient numbers and in the pest's life stage that can cause economic injury.

With suppression of pest populations rather than eradication of them as the goal, farmers have shown that they can achieve pest control with no more than half the usual amount of pesticides. Such a reduction has been achieved in a simple pest management system for cotton, which accounts for 40 percent of all the insecticides used for agriculture in the United States.

A spray-only-when-needed approach has been facilitated by the development of early warning systems—methods of predicting and detecting a pest's presence long before it causes crop damage and when minimal control is needed to prevent its further spread.

Predictions Made

The early warning systems may involve computers that collect and analyze weather data around the clock for large regions and make predictions as to when and where a pest is likely to reach economic significance. Sometimes a four-week warning can be obtained, long enough to achieve effective control.

The warning system may involve scouts who periodically check fields for infestations or insect traps that herald impending invasions.

Pest management relies heavily on understanding complex biological and environmental relationships that together determine a pest's ability to cause crop damage. These include the following:

¶How the reproductive cycles of the pest mesh with the maturation of the crop, and how planting time and weather affect this interaction. If, for example, cold weather delays the emergence of a pest, the injurious life stages may not appear until the crop has matured beyond the point where it could be seriously damaged.

¶How the life cycle of the pest mesh with those of the pest's natural enemies. If natural enemies could destroy the pest at the time it reached potentially damaging levels, the farmer would not need chemicals to do the job. Sometimes small populations of the pest are added to the field to feed the parasites and predators and keep them from dying out.

¶The effects of pesticide spraying on beneficial insects as well as on target pests. "When we kill a pest's natural enemies, we inherit their work," said Dr. Carl B. Huffaker of the University of California at Berkeley and Riverside. When a pesticide must be used, pest management specialists prefer one that is selective to one that attacks a wide range of insects.

¶Structuring the crop environment to encourage natural controls or to reduce the amount of pesticides needed. The introduction of some crop diversity into the monoculture of modern agriculture can often restore a favorable balance in the insect population.

For example, by planting some evergreen blackberry bushes near their vineyards, California viticulturists gave a winter home to a parasitic wasp that can control the grape leafhopper.

¶Determining the "economic injury level," or how many of each pest must be present before the crop is significantly damaged. For soybeans, researchers figured out how much food a single insect of each pest species consumed in its

development. They concluded that much larger numbers of soybean pests could be tolerated than anyone had realized before costly damage would be done.

¶Recognizing unusual interactions in the crop environment. For example, the use of the herbicide 2,4-D on corn greatly increases the damage done by insects. Annual rotation of corn and soybeans prevents damage by the western and northern corn rootworms, but aggravates problems with white grubs and the black cutworm.

"Pest management is like the intelligent practice of medicine," Dr. Metcalf said in an interview. "You study the patient and the disease and then figure out what to do, instead of giving antibiotics just because the patient says he's sick."

Although the research to develop effective pest management plans can be difficult, Dr. Metcalf maintains that the plans are not hard to carry out. "The adoption of pest management is the only way modern agriculture can survive, both in the developed and in the developing countries," he said.

Before 1945, when, as Dr. Edward H. Smith, Cornell University entomologist, put it, "DDT led us down the primrose path" of increasing dependence on chemical pesticides, agriculture used successfully several ecologically based pest control systems that were the harbingers of modern pest management.

In the 1920's, Prof. Dwight Isely of the University of Arkansas evolved a system for controlling cotton pests that focused on preventing economic injury rather than on eradication. He integrated the use of an insecticide with insect scouts who searched plantations for infestations. Often, as a result, only 5 percent of the total cotton acreage needed insecticide treatment.

According to Dr. L. Dale Newsom, entomologist at Louisiana State University, "Professor Isely was the father of pest management."

"The system he developed more than 50 years ago," Dr. Newsom said, "included most of the components of the most sophisticated systems that are currently being demonstrated in cooperative state-Federal pest management programs throughout the cotton-producing areas of the United States."

The 'Miracle' of DDT

Dr. Dean Haynes, entomologist at Michigan State University, said that "the best papers on biological control of pests were published in the '30s."

But seduced by the "miracle" of DDT, which seemed capable of killing every creature that walked on six legs, farmers and entomologists alike focused on

broad spectrum chemical insecticides.

For the next two decades, instead of studying the biology and ecology of insect pests, entomologists spent nearly all their time testing potential new insecticides. And farmers, impressed with the quick, easy, cheap kill afforded by pesticides, used them with abandon.

For cotton, for example, farmers adopted what Dr. Newsom calls a "womb-to-tomb" program of insect control—"weekly applications of heavy rates of broad spectrum insecticides from the time cotton emerged to a stand until the crop matured beyond the stage of susceptibility to insect attack."

Although at first these chemicals produced the desired insect control and consequent increases in yields, not a decade passed before the problems began to multiply. One by one, economically important insect pests became resistant to the effects of heavily used chemicals.

With the balance of power in the insect world disrupted by the destruction of certain major pests or of natural predators and parasites, new insect species emerged and caused problems as bad or worse than the original ones.

With less than 10 percent of the pesticides applied hitting their targets, the general environment quickly became contaminated with persistent, potentially harmful chemicals.

But the worst blow has finally struck agriculture. According to a report released last February by the National Academy of Sciences, because of pesticide abuse, yields of a number of major crops, including corn and cotton, have begun to level off and in some cases decline. Pesticides are becoming counterproductive, the academy said, and the losses are expected to accelerate.

Pesticides Add to Problem

The more pesticides the farmer uses, the more new insect problems emerge and the more pesticides he then has to use to control them. Dr. David Pimentel of the New York State College of Agriculture and Life Sciences at Cornell said that despite a 30-fold increase in the use of pesticides on the American corn crop, insect losses have increased threefold in the last 20 years.

"We simply cannot maintain agricultural productivity if ecologically unsound methods of pest control are used," said Dr. Waldemar Klassen of the Depart of Agriculture.

Since 1972, the Department's Cooperative Extensive Service and Animal and Plant Health Inspection Service have been supporting pilot pest management projects covering 19 crops in 29 states. In addition, the National Science Foundation and the Environmental Protection Agency have sponsored research to develop integrated pest management systems for six major crops.

A simple pest management system has already resulted in a 50 percent reduction in the amount of insecticides needed to control cotton pests, Dr. Newson said.

Despite the rising costs and declining effectiveness of pesticides, it is proving extremely difficult to wean farmers away from the concept of chemical crop insurance. This year, farmers are expected to use more pesticides than ever before—treating 70 percent of the record 333 millions acres they plant — and pesticide production is expected to exceed 800 million pounds, according to industry sources.

Trained professionals who can guide growers in successful pest management are in short supply, although extension agents are now being schooled in its principles.

"Today in the Central Valley of California, there are about 200 people selling integrated pest management, while 2,000 or more are selling chemical pesticides," according to Dr. Louis Falcon, insect pathologist at the University of California, Berkeley.

"Pest management," Dr. Newsom said, "is like growing old. The alternatives are highly unsatisfactory."

August 1, 1976

The Lasting, But Partial, Influence of 'Silent Spring'

By BAYARD WEBSTER

A lovely country town, once alive with plant and animal life, had become severed, as it were, from nature. Its streams were lifeless, its vegetation withered, and no birds sang. And on the fields, in the gutters of houses and between the shingles of the roofs lay the root of the trouble—an evil white powder.

It was 15 years ago that a shy, small, retiring woman wrote such an allegory as a preface to a book that has become a legend: The author was Rachel Carson and the book was "Silent Spring."

It was a documented report on the effects on the environment of modern chemical poisons, notably the synthetic organic insecticides and herbicides—the "white powder"—that had been developed since World War II.

She pleaded, not for abolition of all pesticides, but for more restrictions and control and care in their use, for searches for alternative methods of pest control, and for an understanding of the nature of the chemicals.

The book, frightening in its implications of danger to life from long-term chronic exposure to even minute levels of chemicals, created a hubbub. There were whispers of fear from doomsayers, blasts of anger from industry, praise from conservationists, charges of unscientific research from chemists, and a controversy over almost every one of its 368 pages.

Now, after a decade and a half, most, but not all, of the controversy has died down as more and more people came to support her arguments and found her premises to be true. She did not, however, single-handedly solve the world's environmental dilemmas.

Recent ecological disasters such as the Kepone casualties in Virginia, the PBB contamination of Michigan farms, and the discovery of the worldwide presence of PCB's are evidence of the kind of problems that will undoubtedly plague the world for years to come.

But "Silent Spring" is now credited by most environmentalists with being a consciousness-raising event that has resulted in a long-lasting environmental movement in the United States. This movement, concerned not with pesticides alone, encompasses a spectrum of interests ranging from air and soil pollution problems to radioactivity to urban affairs to aesthetics and the quality of life.

As the movement first gained momentum in the sixties, a series of hopeful events occurred. Within weeks following publication of "Silent Spring," various sections of the ponderous Federal apparatus began to move. President Kennedy put his science advisory committee to work investigating pesticide hazards and Congress called for inquiries into life-threatening chemicals in the environment.

Further aroused by an increasing number of environmental horrors—fish kills in the Mississippi, strip-mining scars in Appalachia, the Torrey Canyon oil spill in the Atlantic—Congress passed the National Environmental Protection Act in 1969. This required environmental impact statements for all major Federal construction projects and called for the creation of the President's Council on Environmental Quality, an advisory body.

Then, in rapid succession, came the creation of the Environmental Protection Agency and the passage of the Clean Air Act in 1970, and the Water Pollution Control Act of 1972—all landmark events that for the first time established environmental regulations and standards and provided an agency to implement and enforce them. The Environmental Protection Agency is now the nation's largest regulatory body. Meanwhile, on the international scene, the United Nations in 1972 organized its Environmental Program and set up headquarters in Nairobi.

Zeroing in more specifically on the problem of hazardous chemicals in the environment, the Congress last year passed the Toxic Substances Act, giving the Environmental Protection Agency a firmer measure of control over the manufacture and use of the 50,000-odd pesticides registered for use in the United States.

As the chemical industry reacted to pollution problems only as it was forced to act by law, young attorneys and scientists around the country formed more than a score of environmental public interest law firms, seeking action in the courts that they could achieve in no other way.

These law firms, led by the Environmental Defense Fund in the field of pesticides and the Natural Resources Defense Council in other sectors, succeeded in pressing legal actions that stopped or slowed the degradation of many forests, streams and lakes; helped to keep the air relatively clean and provided stricter regulation of toxic chemicals.

Miss Carson, who died in 1964, had focused on 12 pesticides as being particularly hazardous to animals and humans. These were DDT, malathion, parathion, dieldrin, aldrin, endrin, chlordane, heptachlor, 2,4-D, toxaphene, lindane and BHC (benzene hexachloride).

All 12 of these compounds, unregulated 15 years ago, are now either banned, restricted or proposed for possible phase-out by the Federal Government, mostly as the result of actions by environmental law firms, principally the Environmental Defense Fund.

Citizen action for environmental improvement has been generally amorphous but Earth Day celebrations sprang up in 1970 and increased environmental activity was initiated by such organizations as the Sierra Club, National Audubon Society, Nature Conservancy and Friends of the Earth.

Most environmental experts feel that the slow pace of action and the lack of leadership at the Federal level have been the main hindrances to the environmental movement in general.

But many of them are optimistic, even among the wreckage. Dr. George Woodwell, head of the Ecosystems Center at the Marine Biological Laboratory in Woods Hole, Mass., cites the banning of DDT in 1972 as one of the most significant events in recent history.

"It was an extraordinary step for a technological society like ours to take," he said. "I think we're now at the point where there's a real glimmer of hope that we can handle substances like this in the future."

And the banning of DDT, of course, was the legacy of Rachel Carson.

January 9, 1977

State Noise Control Regulation Signed Into Jersey Law by Cahill

By RONALD SULLIVAN
Special to The New York Times

TRENTON, Jan. 24—The nation's first statewide comprehensive noise-control legislation, designed to muffle the clamor in what has been called the country's noisiest state, was signed here today by Gov. William T. Cahill.

The new law gives the New Jersey Department of Environmental Protection the power to regulate all noises it regards as harmful to physical health and mental serenity. Fines up to $3,000 can be levied for each violation.

In signing the measure this morning, Governor Cahill, a Republican who is gaining a reputation as a tough protector of New Jersey's badly damaged environment, said the state was the first to legislate in this field.

As a result, instead of excessive noise remaining a local violation of disturbing the peace, for which the fine would be roughly $25, it will now become a state offense of disturbing the environment.

Because New Jersey is the most densely populated state in the nation, because it has the most heavily traveled highways, and because it is one of the most industrialized states in the country, state environmental officials say it also is the noisiest.

The noise has prompted one community after another to adopt local ordinances aimed at reducing it. But state officials say most of the communities simply do not have the technology to deal with the problem. Moreover, they say that most towns have treated noise as a local nuisance rather than as an environmental pollutant that some psychologists regard as just as injurious as air and water pollution.

Car Standards Due

The new law empowers the Commissioner of Environmental Protection to establish statewide noise-level standards for automobiles as part of the state-operated inspections; to restrain industries from disturbing surrounding residents; set curfews for specific kinds of noises; prohibit the use of machines that do not have mufflers; and bar the use of machines and other noisy equipment unless they meet state-established noise-level specifications.

For example, the state would no longer allow a sleeping suburbanite to be jarred from a

Associated Press
Gov. William T. Cahill with the new bill in Trenton.

Saturday morning's sleep by the unmuffled racket of a power mower or by the ear-grinding clatter of a garbage truck grinding up refuse. However, there are no state provisions for barking dogs.

Factories Already Covered

State and Federal regulations already govern limits of noise in factories and in other industrialized facilities. The new law essentially extends this protection to the average citizen in his home or on the move.

Staffing arrangements for the Office of Noise Control have not been completed.

Industry won a concession by having a 13-man council that would be empowered to review any regulations adopted by the state. The council will have nine public and four state members.

According to state officials, a noise level of more than 70 decibels has a disturbing impact. They say that physical damage can occur with prolonged exposure to sounds between 85 and 95 decibels.

State officials do not anticipate that New Jersey will ever be as quiet as Vermont or Montana. But they hope that by keeping down the existing racket as much as possible they can achieve some measure of tranquillity.

Wilentz Sponsored Bill

The noise-control act was originally conceived by Robert N. Wilentz, a Middlesex County Democrat who left the Assembly two years ago. It was sponsored by Kenneth Wilson, a former Republican Assemblyman from Essex County.

Governor Cahill also signed two other environmental bills. One appropriates $20-million for acquiring land for recreation and conservation, while the other creates a 15-man council to protect the environment in the Pine Barrens of central New Jersey.

January 25, 1972

Clamor Against Noise Rises Around the Globe

By GLADWIN HILL
Special to The New York Times

WASHINGTON, Sept. 2 — Noise, long tolerated around the world as an inevitable concomitant of accomplishment and progress, is reaching a historic peak of unpopularity.

Belatedly recognized as the most ubiquitous and most annoying, if not the most deleterious, of all the pollutions, it is under new attack on many fronts.

The current debate in New York City over a proposed new antinoise ordinance is symptomatic of the growing clamor against din—a movement now involving the United Nations, Federal, state and local governments, science, industry, the legal profession and citizens.

Their common goal is to revitalize a public asset so fundamental that, ironically, it is cited in the first sentence of the Constitution: ". . . to insure domestic tranquility . . ."—a reference that courts have held covers noise. At stake in the effort are countless billions of dollars and possibly the mental and physical health of millions of people.

Evidences of the mounting concern about noise include the following developments:

¶Only last Wednesday, the University of California reported that children attending schools near Los Angeles International Airport ran the chance of suffering permanent hearing damage and were threatened emotionally because of jet aircraft noise.

¶The United Nations environmental conference in June pinpointed noise as an important area for international study and control.

¶Congress is now processing noise control legislation.

¶The states have begun adopting comprehensive antinoise laws.

¶Cities are abandoning ancient ineffectual "nuisance" laws on noise in favor of more enforceable scientific standards.

¶Courts have been handing down an increasing number of rulings granting citizens physical or monetary relief from noise.

¶The medical profession, long preoccupied with the specialized problem of noise within industry, is giving more attention to the effects of noise on ordinary citizens.

¶Industry is giving quietness new emphasis in the design of many kinds of machines.

Start Being Made

To date, no one has noted any marked increase in quietude across the nation. But many signs suggest that a start is being made on stemming the steady ominous increase in background noise in recent decades, and that an actual rollback of the country's cacophony level may not be far in the offing.

At the same time, the conquest of noise gives indications of being the most intricate and difficult of all efforts against pollution. On the troublesome side are such considerations as the following:

¶A certain level of noise, probably in an objectionable degree, is inherent in present patterns of urban life—although these can be changed.

¶Enforcement of antinoise laws is difficult because noise is intangible and so often fleeting.

¶Amelioration of a major noise source, airplanes, is a legal mare's nest that will be

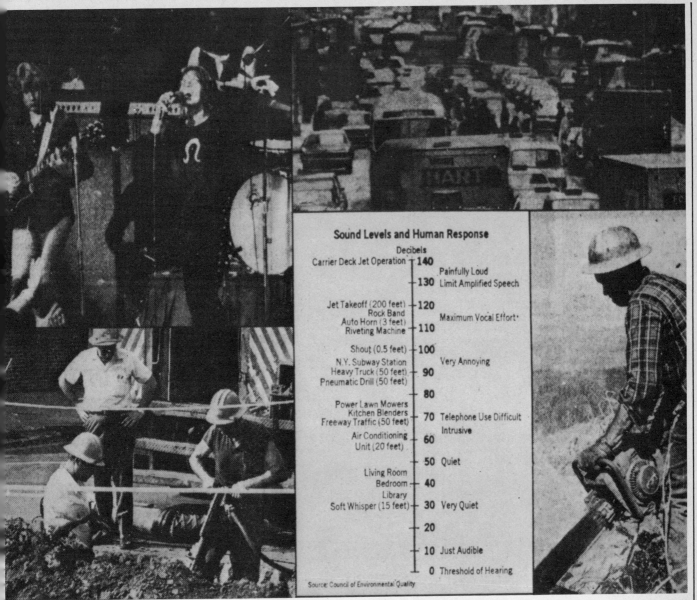

Sound Levels and Human Response

Decibels

Carrier Deck Jet Operation	140	
	130	Painfully Loud — Limit Amplified Speech
Jet Takeoff (200 feet)	120	
Rock Band		Maximum Vocal Effort
Auto Horn (3 feet)		
Riveting Machine	110	
Shout (0.5 feet)	100	
N.Y. Subway Station		Very Annoying
Heavy Truck (50 feet)	90	
Pneumatic Drill (50 feet)		
	80	
Power Lawn Mowers		
Kitchen Blenders	70	Telephone Use Difficult
Freeway Traffic (50 feet)		Intrusive
Air Conditioning		
Unit (20 feet)	60	
	50	Quiet
Living Room		
Bedroom	40	
Library		
Soft Whisper (15 feet)	30	Very Quiet
	20	
	10	Just Audible
	0	Threshold of Hearing

Source: Council of Environmental Quality

From the intrusive bustle of traffic to annoying rattle of the pneumatic drill to the ear-splitting sounds of a rock band, noise assails man

years in resolving.

¶Although persistent noise has irritated people to the point of murder, and there is evidence it can cause mental illness, noise's long-term effects generally are inconspicuous and scientifically imprecise.

¶Mobilizing public opinion against noise is difficult because people have become accustomed or even addicted to it (subconsciously, for instance, equating the thud of car doors and the roar of vacuum cleaners with solidity and power).

Measured in Decibels

Noise is measured in decibels, on a scale that runs, for practical purposes, from zero, the threshold of human hearing, to around 140, in the area of loudness that causes pain and permanent ear damage. Each increase of 10 on the scale represents a tenfold increase in sound intensity.

Thus, while 30 decibels is approximately the quietness of a library, 60 decibels represents a thousandfold increase in noise, and is about the point where it becomes objectionable. This is roughly the equivalent of big-city traffic noise.

Ninety decibels is the Federal limit for all-day exposure of factory workers, and constant exposure to more than 90 decibels can cause permanent hearing impairment.

The noise of jet planes is in the 90-to-120 decibel bracket. The volume of some rock concerts has been measured at over 130 decibels, where even short-term exposure can cause ear damage.

Measurements of environmental noise inevitably fluctuate from place to place and from moment to moment. Therefore there is no ready yardstick of the "noise level" of a community. However, experts concur that the noise level in American cities has been steadily rising rather than diminishing. The Environ-

mental Protection Agency and other authoritative observers have estimated the increase at at least one decibel a year, equivalent to a tenfold increase in a decade. At this rate, they have said, by the year 2000 the general din would be deafening.

Extent of Danger Debated

How grave this danger is is a matter of debate among scientific observers. But most of the emerging evidence tends to substantiate concern about it.

The Environmental Protection Agency's latest figures are that about one out of 20 persons has some hearing impairment, and that in about 25 per cent of these cases, the impairment is "noise-associated."

"We are becoming a nation of auditory cripples," according to Dr. Moe Bergman of Hunter College in New York. "Environmental noise is one of the most serious public health problems, urgently requiring solutions and

public controls."

Dr. Jack Westman, a University of Wisconsin Medical School psychiatrist, reported at a recent scientific meeting that housewives' increasing complaints of headaches, stomach upsets and nervous tension "are related to exposure to noise, which brings to the surface submerged tensions and results in emotional outburst."

Study Made in England

A recent two-year study of 124,000 persons in two communities in England disclosed a significantly higher rate of admissions to mental institutions from the group that lived near London's Heathrow Airport, with recurrent exposure to 100 decibel noise.

Dr. Lester Sontag of the Fels Research Institute at Yellow Springs, Ohio, reported in 1969 that his studies of unborn babies "justify our concern about the possibility of fetal damage from such violent sounds as sonic booms."

And Dr. William F. Geber, a pharmacologist at the Medical College of Georgia, reported that rabbits and rats exposed to urban noise levels only 10 per cent of each day had produced 25 times as many defective fetuses as animals kept in a quiet environment.

Apart from bodily trauma, noise's toll is extensive. The World Health Organization has reported that in the United States excessive noise cost: upward of $4-billion a year in compensation payments, accidents, inefficiency and absenteeism.

Noise's depreciation of real estate values also undoubtedly runs into the billions. School buildings in Los Angeles and elsewhere had to be deactivated because airplane and traffic noise cut down effective classroom time by as much as 30 per cent. In Mesa Verde, Colo.; Bryce Canyon, Utah, and other national parks in the West, sonic booms have damaged ancient dwellings and caused landslides. The Air Force has had to pay out about $1-million in sonic-boom damage. The noise element was a big factor in the termination last year of the multibillion-dollar supersonic transport development program.

Two Types of Noise

There are two kinds of environmental noise: the noise inherent in 20th-century machines, from kitchen blenders to airplanes; and entirely untoward manmade racket, from the snorting hot rod to the wild neighborhood party.

Four obvious avenues of noise abatement are available:

¶Making and enforcing laws against unnecessary manmade tumult.

¶Building quieter machines.

¶Muffling noise, through better building construction, landscaping and use of soundabsorbers such as trees and shrubs.

¶Through land use planning, segregating unavoidable community noise sources, such as industrial and commercial activity, heavy traffic and airports, from residential and recreational areas.

Generalized laws against undue noise go back to ancient Rome. Alvin G. Greenwald, a Los Angeles legal expert on noise, has counted more than 12,000 community noise codes in the United States. But until lately there has been little effective use of them.

Enforcement Difficult

Equipment that will scientifically measure sound and record it can now be obtained for about $1,000. But catching of excessive-noise makers in the act, getting conclusive evidence and prosecuting them is a task most local law enforcement agencies have despaired of. The Los Angeles Police Department, when que-

ried recently, could not produce statistical evidence of a single noise arrest last year despite a municipal antinoise ordinance.

Public apathy has contributed to noise increase. A recent visitor to Stockholm, a city with heavy traffic, heard an automobile horn only three times in three weeks. Yet of all the cities in the United States, only Memphis has achieved a comparable reputation.

Memphis in 1938 simply banned unnecessary horn blowing and began issuing tickets for it. This reduced offenses to a current rate of only about 150 a year. This has won Memphis numerous "quietest city" awards, although some Memphis residents say that in other respects it is not notably quieter than other cities.

Stress on Vehicle Noise

Until recently the most explicit effort to abate din was the action of a number of states in limiting vehicle noise on highways to around 85 decibels. But here also enforcement has been sketchy.

Federal officials say California has the most comprehensive vehicle noise law. Its state highway patrolmen handed out 18,000 tickets last year for noisy cars. But with only six two-man teams to cover 162,000 miles of highways, the level of enforcement is admittedly low.

The first comprehensive state noise legislation was enacted by New Jersey last January. The law made excessive noise a state offense, with fines up to $3,000, and directed the state's Department of Environmental Protection to draw up antinoise regulations.

The agency is now in the process of implementing that legislation. A 13-member citizen council provided for in the law to review regulations is just being appointed.

"We're trying to frame a model ordinance for communities," the Environmental Commissioner, Richard J. Sullivan, said, "so the state won't be in the business of trying to deal with noisy neighborhood parties.

"Airplanes we can't do anything about, because the Federal Government has pre-empted jurisdiction. But we've got to formulate standards and enforcement methods for major noise areas like traffic and industry. A year from now we'll be able to tell you whether we're making any progress."

Illinois, Colorado, and some other states are in similar preliminary stages of noise regulation.

Chicago Antinoise Program

In July, 1971, Chicago put into effect the most compre-

hensive program to curb noise of any American city.

Its 3,000-word ordinance sets noise limits for a dozen categories of sources, from bulldozers to garden tools. The limits range from 94 decibels for heavy machinery down to 55 decibels as the maximum that may emanate from a residence. Progressive reductions bring the limits on vehicles and machinery down as low as 65 decibels by 1980. The law carries a penalty of up to a $500 fine and a six-month jail term.

The law is administered by the city's Department of Environmental Control, under a novel technique designed to overcome the classic obstacle in noise law enforcement: the fact that police officers do not have the time or technical wherewithal to issue citations, while technical people generally do not have police power.

In Chicago three-man teams comprising two noise inspectors and a police officer cruise the city. When violations are spotted, citations can be issued on the spot.

Under this system, 1,649 cases were brought to court in the year ending last June. Of about 1,000 cases completed to date, convictions were obtained in 809 and compliance was obtained in most of the others. A $5,950 fine was imposed on the Grand Trunk Western Railroad for 40 violations at one of its loading docks.

Since Chicago instituted its program, another group of cities has adopted or moved toward similar legislation. The group includes New York, Washington, Baltimore, Kansas City, Mo., Dallas, St. Paul, Minneapolis and Grand Rapids, Mich.

The proposed New York City ordinance, like Chicago's, sets decibel limits for practically every sort of noise source, including subway trains.

It limits the blowing of auto horns to "emergency use" and provides for the muting of horns starting in 1974. A 76-decibel limit is set on car noise, as perceived at a distance of 25 feet, in the city, and this maximum is to be reduced to 70 decibels in 1978.

Enforcement of the ordinance is assigned to the Police Department, with criminal penalties ranging from a $50 fine to $2,000, and 60-day jail terms.

The proposed ordinance has been challenged by people in the construction industry as entailing undue increases in building costs. It is expected to come to a vote in the City Council around the end of September.

Airplanes Cited as Source

The most acute single source of noise is airplanes. About one out of every 10 persons in the country lives close enough to airports to be bothered by plane noise and the number of airports and the amount of air traffic are expected to multiply

in the years ahead.

Abatement of airplane noise is a legal puzzle that has lawyers and public officials, as airport area residents, in a quandary.

The Federal Aviation Administration has jurisdiction over all civilian air traffic and over many aspects of airport design and operation. In 1968 Congress also gave the agency the authority to set noise limits on planes from a design standpoint.

In November, 1968, the F.A.A. promulgated limits of 102 to 108 decibels, as "perceived" from nearby points, for the new "generation" of jumbo passenger planes—the 747's, DC-10's and L-1011's.

The older passenger jets produce from 110 to 120 decibels. Argument has been raging for two years about quieting aircraft engines, with the air transport industry saying "retrofitting" is impractical because it would cost a billion dollars. The F.A.A. is expected to issue some modification requirements within the next few months.

Airplane Noise Continues

Meanwhile, the F.A.A.'s design limits on plane noise do not necessarily match the amount of racket a plane may make flying over a community, and the F.A.A. does not profess to monitor or police individual flights' noisemaking.

This appeared to leave a jurisdictional gap in which communities could set noise limits for airplane operations. A number of communities have tried this. But the Federal courts have repeatedly invalidated such ordinances as an intrusion on a Federal regulatory area.

A Burbank, Calif., "curfew" banning jet traffic between 11 P.M. and 7 A.M. is before the United States Supreme Court and a contested Inglewood, Calif., regulatory ordinance is headed there.

The jurisdictional gap has left airport operators in the middle. California courts have awarded several million dollars in property devaluation damages to residents around the Los Angeles International Airport and the city of Los Angeles is faced with nearly $5-billion in additional suits. Hundreds of similar suits have been filed in other parts of the country.

The jurisdictional bind was made particularly acute in May when a California court ruled in a case involving the city of Santa Monica that an airport was liable not only for property devaluation but also for compensation for personal annoyance.

This ruling moved Los Angeles officials to exclaim that on that basis they might have to close down the Los Angeles airport, second busiest in the country, lest they incur astronomical damage claims. The problem remains unresolved.

Meanwhile, the state of California, which contends it can legally promulgate aviation regulations as long as they do not conflict with existing Federal enactments, is preparing to put into effect in December flight restrictions aimed at reducing noise. The state fully expects its regulations will be challenged by the Federal Government and the airlines.

The mounting concern about airplane noise convinced Congress it should do something about noise generally.

In December, 1970, it created an Office of Noise Abatement and Control in the Environmental Protection Agency, and directed it to study the problem.

The agency turned in a massive report last January. The House of Representatives in February passed a noise control bill (HR 11021) drafted by Representative Paul G. Rogers, Democrat of Florida.

It directs the environmental agency to establish national noise emission limits for four kinds of machinery: transportation equipment, construction equipment, motors and engines, and electrical equipment. It authorizes the agency to assess civil fines of up to $25,000 for violation of these standards by manufacturers and distributors.

In regard to airplanes the measure gives the environmental agency only an advisory role, leaving authority with the F.A.A.

Tougher Bill in Senate

The Senate has been consid-

ering a more stringent bill (S 3342) sponsored by Senators Edmund S. Muskie of Maine and John V. Tunney of California, both Democrats.

The chief difference in the Senate bill is that it would give the Environmental Protection Agency comprehensive jurisdiction over aircraft noise—even though the agency has demurred at accepting this responsibility on the ground it lacks technical expertise.

The Senate has completed committee hearings and the next step will be to reconcile House and Senate versions of the legislation.

Both bills give the states leeway to formulate their own noise control regulations as long as they do not conflict with Federal standards. The laws would also provide states with technical assistance from the Environmental Protection Agency in setting up organizations to administer noise-control regulations.

Efforts by Industry

Industry began sensing the public unhappiness about noise several years back, and doing something about it.

The auto makers have been trying to make cars quieter. New York City last year completed replacing its old fleet of 1,480 clanking refuse trucks with quieter hydraulic-compaction trucks.

Inspired by European progress, American manufacturers have been designing quieter air compressors, a major racket-

maker on construction projects. Research is under way to tone down the noise of diesel trucks, whose snorting often reaches the noise level of jet planes.

New York City is experimenting with such refinements as recorded siren noises for emergency vehicles, which can be focused at street level, rather than actual sirens, which project a needless barrage of sound in all directions.

The "leisure time equipment" industry, involving everything from snowmobiles to hedge-clippers, advanced last year through the National Industrial Pollution Control Council a noise-reduction program for machines. Under it, equipment noise now as high as 92 decibels at a distance of 50 feet would be reduced over the next decade to a maximum of 77 decibels.

Land Use Planning

The least-used tactic to date to lessen noise has been land use planning, because most of the nation's communities are locked in, at least for the time being, to archaic layouts in which noise problems were not considered.

Congress has before it several proposals for Federal-state collaboration in more rational land use, in which noise would be a factor. But the measures have been bogged down in debate, and there is no telling when, if ever, legislation will emerge.

Meanwhile, the chief influence in this direction has been the Department of Housing and Urban Development, which can

control many things through its construction financing.

A year ago the agency set noise limits in construction specifications and in the location of buildings financed by H.U.D. As a result plans for a number of big projects have been altered.

A planned residential development near the new Dallas-Fort Worth airport in Texas was relocated. Noise-reduction features were superimposed on the design of a nursing home in New York City. H.U.D. is a party in the environmental-impact delay in construction of the new outlying Los Angeles airport at Palmdale.

Perhaps the nation's most venerable antinoise campaigner, is Dr. Vern O. Knudsen, an internationally noted acoustics expert and chancellor emeritus of the University of California at Los Angeles.

As far back as 1927, in an era of relative silence, he said: "Americans today are paying in shortened tenure of life and reduced efficiency for the noise amid which they must work and live."

Dr. Knudsen is naturally pleased with the blossoming of public awareness of his favorite problem. But he does not think it's any too soon or too fervent.

"I always used to say, quoting Victor Gruen, that noise and smog are slow agents of death," he said recently. "But in a few years we can change 'slow' to 'sure' if this continues. If everybody gives up, we'll be like the dinosaur."

September 3, 1972

CONCORDE FLIGHTS TO U.S. APPROVED FOR 16 MONTHS; FOES RENEW CHALLENGES

Supersonic Jet Would Give Limited Service Here and in Capital

By RICHARD WITKIN
Special to The New York Times

WASHINGTON, Feb. 4 — Transportation Secretary William T. Coleman Jr. ruled today that France and Britain could operate limited service by the Concorde supersonic jet airliner to New York and Washington on a 16-month trial basis.

While the long-awaited decision could mean flights to Washington's Dulles International Airport by mid-April, most key officials still viewed it as unlikely that the supersonic transport would be allowed into New York for many months, if ever.

This was because Mr. Coleman's Federal approval now has to be supplemented, in the case of New York, by permission from the Port Authority of New York and New Jersey, which operates Kennedy International Airport.

Governor Carey, who in effect can veto Port Authority actions,

is on record as saying that commercial Concorde operations to Kennedy "must be denied." He repeated his opposition today. Expected lawsuits challenging a Carey veto could drag on for a long time.

No local permission is needed in the case of Dulles because that airport is federally owned and operated. But environmentalists and Congressional opponents immediately initiated legal action to bar flights to Dulles.

Congressional legislation to delay or bar Concorde flights is also being pushed. However, it is uncertain whether Congress

will act before flights get under way.

In a 61-page opinion, Mr. Coleman specifically approved requests of Air France and British Airways to make two flights a day each to Kennedy and one a day each to Dulles.

Drawbacks Conceded

The Secretary acknowledged the environmental and economic drawbacks of the 1,350-mile-an-hour Concorde, emphasizing that "the most serious immediate consequence of limited Concorde operations is noise."

He concluded, nevertheless, that the disadvantages seemed to him outweighed by the advantages of a very limited test operation that could be canceled forthwith for safety or other overriding reasons.

Among the advantages of the plan that could almost halve over-water flight times, he cited the following:

¶Facilitation of commercial and cultural exchange.

¶Determination whether a quieter, more efficient future supersonic transport (SST) was

215

feasible.

¶Prevention of serious economic harm to two close allies.

¶Maintenance of international fairness and equity.

* "It may well be," Mr. Coleman argued, "that further development of this technology is not economically sensible in the energy and environmentally conscious period in which we live. If so, then the Concorde will fail because it is an anachronism, and its failure will be recognized as such rather than attributed to an arbitrary and protectionist attitude of the United States out of fear that our dominance of the world aeronautical manufacturing market is threatened."

In 1971, Congress canceled this country's SST program under pressure from the environmental movement and economy-minded officials who argued that there were much more pressing uses for the funds. About $1 billion had been spent on the project.

The French and British, despite soaring costs that they could afford even less than the United States in the short run, pressed ahead.

Passenger Service

Two weeks ago, the Concorde went into regular scheduled passenger service from London to Bahrain, and Paris to Rio De Janeiro.

The Russians have a plane very similar to the Concorde. But it is making only scheduled cargo runs so far, within the Soviet Union.

Mr. Coleman's decision, which he announced at a brisk news conference down the hall from his office here, was followed promptly by an announcement from the White House that "the President will stand behind the Secretary's decision."

A large question on the horizon was whether President Ford might veto any of several varieties of Congressional legislation to delay Concorde operations, at least to New York, or ban supersonic operations anywhere in the country.

On Dec. 18, the House voted to ban the graceful elongated craft for six months from all United States airports except Dulles while more more data were obtained on its effects, including noise and the much-disputed peril it poses for increasing the incidence of a non-fatal form of skin cancer.

A Senate committee is scheduled to make a similar proposal tomorrow. In addition, a group of Congressmen have introduced, or plan to, various anti-Concorde bills or riders to other bills.

The six-month delay provisions are amendments to bills authorizing significant sums for airport development. Congressional observers think a more significant threat to the Concorde lies in a number of appro-

The British Concorde making first commercial flight, Jan. 21, from London to Bahrain

Associated Press

The New York Times/Feb. 5, 1976

priations bills to which anti-Concorde riders could be attached—riders that would block flights to Dulles as well as to other gateways.

In this connection, one of the strongest of many criticisms of Mr. Coleman's action today came from the chairman of the Senate Appropriations Committee, Birch Bayh, Democrat of Indiana.

Last year, Mr. Bayh sponsored an appropriations amendment that would have kept the Concorde out of the country if its noise level exceeded that of subsonic planes, which it does to a degree that is a matter of continuing dispute. The vote defeating the Bayh amendment was only 46 to 44, while

a similar amendment was defeated in the House by a 214-to-196 vote.

The big question remains whether President Ford will veto any anti-Concorde legislation.

Criticism by Bayh

Today, Mr. Bayh, one of many who are seeking the Democratic Presidential nomination, said of Mr. Coleman's decision:

"I regard this as a very serious mistake. The Concorde is wrong economically, wrong environmentally, and wrong in terms of energy. It is a means of transportation available only to a very few rich people. I will push strongly for legislation to overturn the Secretary's decision."

The fare for current flights to Bahrain and Rio in the 100 passenger Concorde has been pegged at 15 to 20 percent above first class. British Airways said today the round-trip fare between London and New York would be $1,168, and between London and Washington $1,240. This comes to 17.7 percent above today's first-class fares.

Mr. Bayh's outrage at the Coleman decision was echoed by many long-time opponents of the Concorde, including numerous other Congressmen from areas not facing Concorde operations.

The Coleman decision is effective March 4. He explained that the 16-month period had been picked so that tests could encompass operations in all seasons and then leave four months for the results to be analyzed.

Mr. Coleman set down a number of expected conditions for any Concorde flights, including confining operations to the period between 7 A.M. and 10 P.M. local time and authority for the Federal Aviation Administration to impose any added noise-abatement procedures it deemed safe and technically feasible.

The Secretary directed the F.A.A., which is in his department, to proceed with a program to measure high-altitude pollution and to set up thorough going noise-measurement systems at Kennedy and Dulles.

The New York Times/Teresa Zabala

William T. Coleman Jr., Transportation Secretary, holding a copy of his opinion on the Concorde in Washington.

February 5, 1976

Technology Seeking to Solve the Problems of Progress

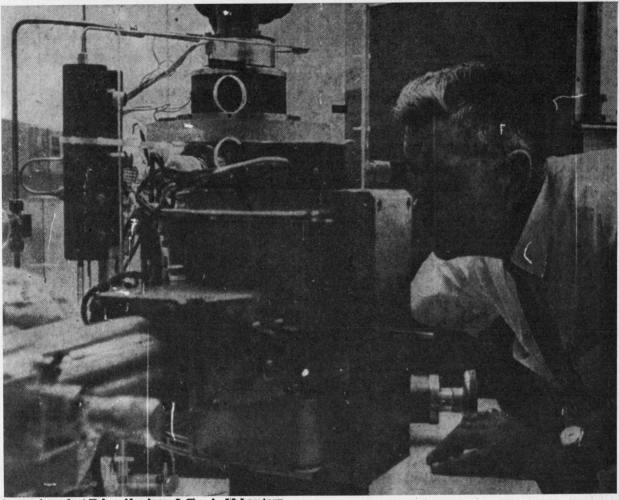

Research work at Kaiser Aluminum & Chemical laboratory

"In the 1970's R&D won't be fashionable; it will be an absolute necessity."

Industrialists Get Word: Environment

By GLADWIN HILL

IN California, three major power-plant projects are stymied by public objections to their prospective smog emissions.

In Minnestoa a mining company faces a multimillion-dollar outlay to change its mode of disposal of iron-ore tailings.

Citizens of New Mexico recently gave the cold shoulder to a big paper mill project.

And in South Carolina a community is divided over the desirability of a $100-million chemical plant.

The message of 1969 for industry sounded loud and clear. "Environment" is the new watchword. Its impact promises to be tremendous—to some extent in terms of dollars, but also in terms of procedures and philosophies.

●

The more progressive segments of industry were just starting to catch up with problems of water pollution and air pollution when the blow fell.

The blow was the Santa Barbara off-shore oil-well blowout of a year ago. The once-abstruse word "ecology" entered popular parlance. From water and air pollution, the screen suddenly magnified to include the effects of chemicals, solids, wastes and noise; urban beautification and the exploitation of natural resources.

Not since the trust-busting days of Theodore Roosevelt has the force of public opinion instruded so emphatically on the business community's patterns of operation. People are palpably fed up with filth, noise, ugliness and contamination of the—there's that word again—environment.

The sentiment has impinged on more than just the historically dirty heavy industry. Manufacturers of both DDT and cyclamates, recently considered unalloyed boons to man, felt the sudden crunch. The gamut of affected businesses ranges from airlines sued for excessive jet noise to detergent makers whose phosphates, however kind to housewives' hands, are under fire for

The Cost of Pollution Abatement

The New York Times

New York harbor and smog-shrouded skyline
"Environment" is the new watchword. Its impact promises to be tremendous.

propagating obnoxious vegetation in lakes and rivers.

Some executives have loudly deplored public "emotionalism" and "hysteria" about restoration of a clean environment. But the sentiment isn't likely to go away, judging by strategic decisions in some of the nation's biggest industries.

•

"For over half a century, emission control wasn't even among our criteria in making cars," a Detroit executive remarked the other day. "Now it's become the No. 1 criterion."

What started as perennial complaints of women's clubs and the Izaak Walton League about dirty water has suddenly assumed the shape of an economic revolution whose ultimate dimensions still are indeterminate.

The coming year promises to be a period of further impact, agonizing reappraisal and altered bookkeeping.

The divergent reactions to the environmental crunch were reflected in a colloquy at a recent national water pollution conference. A steel company head, invoking an old shibboleth, said plaintively: "We can't put any money into pollution control that we haven't first made as profits." Another company head hastened to correct his accounting. "Pollution control isn't an out-of-profit item," he said. "It's part of the cost of production."

The "We've got to do it" school of thought seems in the ascendancy. When one of the nation's top corporation executives, namely Robert O. Anderson, chairman of the Atlantic Richfield Company, says in a major public forum that we had better start shifting our sights from quantity to quality, a historic corner plainly has been turned.

How much the cost of environmental reform may be to industry appears incalculable. Formerly a company could figure that for a new facility, adequate water pollution control equipment might run 10 per cent of capital costs. Par for air-pollution control equipment has been less, with more like a 5 per cent maximum—except for some test situations such as power plants, where to date a shift to nonsulphurous fuels may be the only answer.

The big problem area has been plants that are not new, where the cost of superimposing controls may be a marginal proposition. This has been perplexing to many corporations, but what is it going to be in an era when communities are rejecting lucrative new industrial installations out of hand because of their environmental implications? And when not only outright contamination but sheer esthetics have become an unignorable consideration?

These are some of the questions industrialists see themselves grappling with in 1970.

Beyond all this, there is another dimension. Dominating discussions of the "environmental crisis" is the problem of population—the prospect that inevitably at some point its unbridled growth has to stop, and with it the business predicated on growth. At the same meeting of the United States Commission for UNESCO where Robert Anderson broached the quality-instead-of-quantity concept another businessman said: "We should start preparing right now for a no-growth economy." And nobody laughed.

Industry Hurting in Pollution Costs and Bethlehem Knows It

Blast furnace thickener at Bethlehem Steel's complex at Burns Harbor, Ind., during construction. Sludge is reclaimed from bottom of tank and is upgraded. Steelmaker says it has little more money for pollution control.

By GERD WILCKE

Special to The New York Times

BETHLEHEM, Pa.—"We have just about reached the end of the line as far as spending large sums for pollution control is concerned. Something has to be done with the tax laws as they now apply."

The statement by Lewis F. Foy, the president of the Bethlehem Steel Corporation, underlined a problem that many industry executives were trying to get across to the public last year as the pressure for a cleaner environment continued unabated.

Mr. Foy was quite serious. The economic aspect of efforts to remain good corporate citizens in this age of environmental concern was brought out strongly by the nation's second largest steel producer late last year when the company announced that it might have to cut back its operations at Lackawanna, N. Y., by close to 30 per cent.

The company offered three reasons for the cutback: local taxation, employe productivity and the "lack of a realistic approach to the cost of environmental improvement."

Bethlehem officials acknowledge that the recent public pressure on regulatory agencies has in turn brought pressure on industry. However, they insist that the company's concern over the environment dates back several decades.

Mr. Foy and Dr. A. D. Brandt, who is manager of the environmental quality control division of the company's industrial relations department and who came to Bethlehem in 1946, estimated in an interview at corporate headquarters here that Bethlehem had spent about $200-million so far on pollution abatement.

For example, when the company opened new steel-making facilities at its plant at Burns Harbor, Ind., in late 1969, more than $47-million had been invested in the environmental quality program. Steelmaking and related facilities required about $13-million in new water treatment equipment. This was in addition to the existing $25-million waste-water treatment system.

Also, more than $9-million had to be spent to equip the new facilities with air pollution control equipment.

In the last three years, sums invested in pollution control equipment accounted for about 5.6 per cent of the total investment in property, plant and equipment.

Contemplated pollution-control expenditures in the next five years run to about $40-million annually, or about 11 per cent of the planned total investment.

Visitors to this eastern Pennsylvania city of 76,000 cannot help notice the work going on at a 23-story structure known as the Martin Tower, named after Edmund F. Martin, former chairman and chief executive officer at Bethlehem.

What most visitors do not know is that the company gave serious consideration in the wake of a long steel haulers' strike last year to not completing the future corporate headquarters.

In October, the company decided to move ahead on the $18.5-million structure but announced that it would be sold and leased back to raise cash for other capital purposes.

The episode illustrates that corporate size does not exempt Bethlehem from being hard pressed for funds.

"Over the past 20 years," Mr. Foy said, "we have spent as much money on pollution control as we could prudently afford.

"Since this type of investment is not productive, it puts a tremendous load on our earnings. One of the most distressing things is that many of the dollars already spent are now useless because of increased stringencies by enforcement agencies."

Mr. Foy continued: "With the general economy the way it is, with the loss of the 7 per cent tax investment credit, we at Bethlehem have just about reached the end of the line as far as continuing large capital pollution expenditures are concerned."

Mr. Foy explained that it would be "tremendously helpful" if industry were permitted to consider capital expenditures for pollution control as an item of operating cost for deduction in the year they are made, instead of treating them as an averaged depreciation cost.

"Until this happens," he said, "you won't get the full empathy of industry because the money just isn't here. If we could run a deficit like the Federal Government, it would be a different story."

Dr. Brandt noted that, from the technological point, one of the most troublesome aspects of pollution control in the steel industry was the problem of coke ovens.

By way of emphasis, Dr. Brandt showed a visitor a copy of the April 27, 1970, report by the Secretary of Health, Education and Welfare, which said: "Although there are a number of new plant design features which hold promise for emission control, current technology is not satisfactory."

Dr. Brandt denied that Bethlehem in the past had any serious conflicts with the law over environmental issues.

However, he acknowledged that the plant manager of the Steelton facility near Harrisburg, Pa., was fined $100 for taking out an air pollution control installation without reporting it to authorities.

Dr. Brandt also said that the company paid a fine of $100 and $25 in costs in San Pedro, Calif., where it runs a shipyard, after being notified of a violation of the Los Angeles County Air Pollution Control Regulation. The control authorities said that Bethlehem, in the process of sandblasting, had created a dense dust cloud.

Late last year formal hearings took place in Erie County in connection with the company's refusal to close down a vessel at one of Lackawanna's basic oxygen furnaces, which the county contended had been in violation. The company said it had recorder transcripts showing that no violation of the pollution code had taken place.

To a visitor, the city of Bethlehem didn't look any dirtier than towns of similar size that are not the homes of major industries, although no fisherman was seen trying his luck in the Lehigh River.

However, Mayor Gordon Payrow, who was interviewed by telephone, insisted that fish had returned to the river. "You wouldn't have found any fish in that water 10 years ago," the Mayor said.

Asked whether his office received many complaints about pollution coming from the huge steel facility here —the plant has a capacity of 3.3 million tons of steel annually, and extends 4.5 miles along the river—Mr. Payrow said that "complaints are at a minimum," and were mostly in relation to the plant's coke ovens.

Was Bethlehem a good corporate citizen? "Absolutely," his honor said.

January 10, 1971

Cost of Cleanup

Or, a Myth Of Factory Closings Is Exploded

By GLADWIN HILL

LOS ANGELES—A prominent industrialist told a large Western college audience a few weeks ago: "U.S. Department of Commerce figures show that 219 plants last year were forced to shut down primarily because of environmental pressures."

His words, which were untrue, were part of a rising clamor that one Federal official remarked had been taking on the proportions of "a national myth": the idea that current pollution - abatement regulations are crippling substantial segments of American industry.

Another industrialist said recently that environmental controls "need to be placed back in perspective before the country is faced with economic ruin."

Headlines such as "Pollution Laws Closing Plants by the Hundreds" and "A Drive to Find Jobs for Victims of the Pollution War" have appeared in both business and lay periodicals. A Department of Commerce publication said recently, "More plant closings are being reported daily from countless small communities throughout the nation."

However, a nationwide check by The New York Times, corroborated by Government reports, provides little substantiation for such assertions and apprehensions.

To the contrary, the survey yielded indications — supported by a number of officials, economists and other observers — that the costs of pollution control, while they may be causing dislocations in a few specialized situations, could also be a constructive element in terms of plant modernization and increased efficiency.

Highlights of the findings were the following:

¶Among the more than 10,000 business enterprises that fail every year in the United States that fail every year, in only a few cases have pollution-control requirements been a major factor.

¶Plant closings in which pollution-control costs did figure almost invariably involved marginal facilities, some of them a century or more old.

¶Repeatedly pollution regulations have been blamed for industrial shutdowns when the basic reasons were otherwise.

¶Unemployment caused by such shutdowns, according to the latest Federal data, has aggregated fewer than 1,500 jobs.

¶Recent studies made by independent consulting organizations for Federal agencies indicate that the cost impact of present pollution-control standards in the years immediately ahead will be small on both industry and the economy as a whole.

Nearly all the states now have water-quality regulations approved by the Federal Government. The Federal Environmental Protection Agency has

> *Among the more than 10,000 business enterprises that fail every year in the U.S. in only a few cases have pollution-control requirements been a major factor.*

promulgated general air-quality standards relating to the major air pollutants. In addition, the states have just submitted to the Federal Government "implementation plans" for regulating their particular pollution sources.

The Council on Environmental Quality in its last annual report projected the total national outlay to meet existing air, water and solid-waste pollution-control standards at $105-billion for the period 1970-75. Of this total, industry's share, as distinct from governmental and private expenditures, was placed at $28-billion, an annual average of less than $4.7-billion.

Annual sales of all American industrial concerns total about $750-billion, and 127 companies have sales of more than $1-billion each.

Industry's outlay in 1970 of $2.5-billion for new pollution-control facilities, the council said, was only 3.1 per cent of its over-all expenditures for new plants and equipment.

"The cumulative expenditures over the six-year period," the council's chairman, Russell E. Train, has said, "are expected to be less than 1 per cent of gross national product (the nation's total output of goods and services in one year). Industry air and water pollution-control expenditures will generally be less than 1 per cent of the value of shipments."

Nevertheless, the thrust of many expressions from industrial and financial quarters has been that existing regulations impose inordinate economic burdens.

The Chamber of Commerce of the United States, a business group, last September said of pollution-control costs: "All existing firms will be adversely affected, but in some cases the economic impact will be severe."

C. Howard Hardesty, vice president of the Continental Oil Company, said the nation's pollution-control bills were going to be "staggering."

Maurice H. Stans, recently Secretary of Commerce and industry's foremost exponent of the "go slow" attitude on pollution control, said abatement costs might "throw thousands of people out of work" and cause "whole communities to be run through the economic wringer."

Lewis F. Foy, president of the Bethlehem Steel Corporation, said 18 months ago: "We have just about reached the end of the line as far as spending large sums for pollution control is concerned."

And the National Industrial Pollution Control Council said early in 1971: "Increasing public concern with the pollution consequences of our society has inspired responses at some levels of government which are incompat-

ible with the economic health of our society."

Contradictory views have been equally emphatic.

"The environment pretext," said Ralph Nader, the consumer advocate, "promises to become a convenient scapegoat for some of the 10,000 businesses that go under each year at the hands of the marketplace or predatory corporate practices."

And I. W. Abel, president of the United Steelworkers of America, said last August: "We are increasingly being confronted by claims from industry that the demands made on it by citizens and government will not control pollution but rather will bring complete stoppage of plant operations.

"This is a throwback to the antiquated escape route of 'smoke means jobs.' It is environmental blackmail of the worst sort. Many obsolete plants use environment control as a convenient public-relations tool to justify a production decision to terminate proved on inquiry to be a second factor.

The New York Department of Environmental Conservation, for instance, noted four plants that had closed with pollution-control costs cited but said that in each case other economic factors appeared to be the real determinant.

In New Jersey, the State Department of Environmental Protection reported that three small chemical plants had shut down in 1970 after receiving orders to stop polluting the Raritan River, but officials said the abatement cost was not a clear-cut determinant.

In Montana, where there has been much controversy over smelter fume-emission standards, the chairman of the State Board of Health, Mrs. John C. Sheehy, said, "Pollution controls have not cost a single job."

These and many similar reports conformed with the findings of a detailed study of the prospective impact of pollution controls on 14 major industries between 1972 and 1976 recently made for the Council on Environmental Quality, the Environmental Protection Agency and the Department of Commerce by 11 independent consulting firms.

Of 12,000 plants in the 14 industries, the study report said, about 800 would close "in the normal course of business" between 1972 and 1976 anyway. Pollution abatement requirements, it was estimated, might cause an additional 200 to 300 to close, but many of these would be economically weak units that would probably close anyway in a few more years.

These possible environmental closings, it was said,

might involve a theoretical loss of 50,000 to 125,000 jobs —1 to 4 per cent of the employment in the industries involved.

Mr. Train and others have operation and rationalize past failures to modernize facilities."

Leonard Woodcock, president of the United Auto Workers called it "corporate irresponsibility" before a Senate hearing.

The latest Federal survey on the pollution-control impact came out in March under a newly established "early warning" system. In this the Environmental Protection Agency advises the Department of Labor every three months of situations where pollution requirements may have adverse economic effects. The survey deals only with operations with more than 25 workers.

This initial report, which went back to 1970, said that throughout the country only eight plant closings had been observed "where environmental regulations were a factor."

(A spokesman for the Department of Commerce said it had never issued any such figures as the "219 forced shutdowns" cited by the anonymous industrialist. A departmental inquiry indicated the speaker had gotten the number from a Washington business - news syndicate that had misconstrued figures from a Federal study that did not relate to actual plant closings.)

The eight closed establishments listed in the Federal survey were a woolen mill in South Grafton, Mass.; a chemical plant at Saltville, Va.; a vegetable oil concern in Durant, Okla.; smelters in Superior, Ariz., and at Selby, Calif.; a copper mine at Tucson, Ariz.; a pulp mill at Coos Bay, Ore., and a lumber mill at Philomath, Ore.

Thirty-eight more "threatened" closings or production curtailments were reported, but in only 18 cases did Federal investigators assess these eventualities as "highly probable."

Of the 38 plants, 13 were in the pulp and paper industry, seven were in the metals industry, six were sugar mills and the rest were scattered among various fields. The plants were in 16 states.

A canvass by The New York Times of 20 states, from Maine to Hawaii, produced a scattering of additional cases of actual or threatened plant closings. But generally there was no evidence of any great impact.

In each state, officials and industrial associations were asked to specify any plant shutdowns in which pollution-control costs were a major factor. Few such instances

could be cited and, in most of these, pollution abatement also noted that national pollution-abatement cost projections are seldom balanced against the existing costs of pollution itself—at least $16-billion a year for air pollution, by Federal estimate, and $13-billion a year for water pollution.

Other factors that tend to make industrial pollution-control cost figures larger are:

¶Such outlays are not an all-at-once expense, but can be spread over 20 years or more — the lifetime of the facilities—even though their depreciation can be compressed for income-tax advantages to as little as five years.

¶In many cases pollution-control-equipment expenditures are accorded other Federal and state tax advantages. Facilities can often be financed through tax-exempt bond issues sponsored by public agencies, affording industries lower financing rates than on ordinary construction.

Joseph Kivel, an officer of the E.P.A.'s planning and evaluation division, said: "I think almost all the plants affected by pollution-control regulations so far are marginal. They're not making much money or they aren't making any. I don't think any industry that is healthy will be closed down as a result of pollution legislation."

The idea that pollution control is unduly costly to industry was challenged recently by two economists — Joseph H. Bragdon Jr. of H. C. Wainwright & Co. and John A. Marlin of the City University of New York— who studied the financial records from 1965 to 1970 of 17 pulp and paper companies covered in a recent pollution analysis by the Council on Economic Priorities.

"The evidence," they said, "suggests not only that this hypothesis [of undue expense] is entirely untenable but also that the reverse—pollution is unprofitable—is the case in the long run. The earnings performers during the 1965-70 period seem to include a disproportionate number of companies with strong records in the areas of pollution control." They suggested these concerns had good management that was conducive to both low pollution and high profits.

"In some cases," William D. Ruckelshaus, head of the Environmental Protection Agency, said recently, "it happens that by prodding industry to clean up we're encouraging modernization and development of more efficient plant facilities, which is a net benefit to our general economy."

Conservation Lawyers Move to Defend the 'Quality of Living'

By GLADWIN HILL
Special to The New York Times

WARRENTON, Va., Sept. 12 — Do people have a constitutional right to freedom from air pollution and other environmental hazards and annoyances?

This and other legal defenses against the increasing number of assaults on "the quality of living" were explored by 75 of the nation's leading conservation lawyers at an unusual meeting in Warrenton this week.

For two days, lawyers concerned with the new field of "environmental law" discussed their special problems, and possible strategic responses, in a closed-door conference designed not for public consumption but, rather, as orientation for themselves.

The principal conclusion of the lawyers was that some radical changes in the traditional patterns of jurisprudence are necessary to accommodate growing public dissatisfaction with the deteriorating environment.

The meeting, with only invited participants, was sponsored by the Conservation Foundation — a Washington-based organization financed by the Ford Foundation and other donors — and the Conservation and Research Foundation of New London, Conn.

Nader in Attendance

The participants included Ralph Nader, the consumer advocate; Victor J. Yannacone Jr., who has been in the forefront of a nationwide campaign against persistent pesticides; lawyers for prominent conservation organizations, faculty members of leading law schools, and scientists.

Until the last few years, they noted, mass public complaints against environmental conditions have had a hard time in the courts, because the law was geared to the old concept of offenses by one individual against another, the tradition that governmental discussion was unchallengeable, and a presumption that public inconvenience was acceptable in the cause of private enterprise.

A landmark in a new direction was New York's Storm King Mountain case, involving a proposed power plant on the Hudson River. The United States Court of Appeals for the Second Circuit ruled in 1965 not only that conservationists had a right to challenge a Federal Power Commission permit granted for the plant, but also that the commission had erred in not considering the environmental impact of the plant.

Since then, environmental suits have proliferated. But the conservation lawyers agreed, they still face a formidable array of problems, including the following:

¶In general, the burden of proof is still on plaintiffs that a glue factory is obnoxious or that a beautiful valley should be preserved rather than flooded for power generation.

¶Expert witnesses are hard to get. In the Santa Barbara, Calif., oil slick case, it was noted, corporations and other opposing interests pre-empted the talent.

¶Environmenttal cases usually develop only after the damage is done or when it is imminent; characteristically, the initiative is on the other side.

¶While many governmental administrative agencies are now under formal orders to take into consideration the environment and ecological consequences of their actions, there is still some question about the extent to which such orders are enforceable by the courts.

¶Money, depositions, trial transcripts, expert-witness fees and other unavoidable expenses —along with standard lawyers' fees—put the cost of a full dress suit beyond the resources of most conservation groups.

"I think that capitalism and conservation are essentially incompatible," said William M. Bennett, a former California public utilities commissioner, who has been leading a drive against large pipeline interests. "Corporate management's prime responsibility legally is to make money for stockholders, and the management can be challenged on any 'nonproductive' expenditure, such as for pollution controls."

Mr. Nader, whose Center for the Study of Responsive Law in Washington is extending his attacks on automotive hazards to other consumer and governmental fields, questioned whether the prevailing American system was capitalism.

"I think it's more corporate Socialism," he said.

Mr. Yannacone chided his colleagues for being naive in trying to fit conservation action into traditional legal patterns.

"It's about time the legal profession got some ecological sophistication," he said. "We have to invent causes of action. We have to find new legal rules to overcome traditional government secrecy. Administrative agencies are the abortive offspring of modern legislation.

"Every bit of progressive social legislation of the last 50 years has come about only after litigation," he continued. "It's the highest use of the courtroom—even when we lose—to focus public attention and disseminate information about intolerable conditions."

There was a discussion about establishing a nationwide conservation legal organization, patterned after the American Civil Liberties Union, with a national center coordinating regional branches where talent could be systematically mustered on a semivolunteer basis.

Sidney Howe, president of the Conservation Foundation, was selected as chairman of an ad hoc committee to proceed with this and other suggestions for group action.

The conferees agreed that the following methods of legal strategy would be pursued in conservation causes:

¶The fundamental concept of suing to abate a "nuisance."

¶The mass-action suit.

Another line of action is based on the doctrine that much, if not all, land and other resources are essentially held in trust by government and the delegated representatives, to be used for the public good. Therefore, they are legally protected from transitory and arbitrary abuses of their pristine condition.

In addition, the conferees agreed, there is the concept that several sections of the Constitution — particularly the Ninth Amendment, reserving to the people all powers not explicitly given to the Government— guarantee the people protection from the mounting number of encroachments on their privacy, peace and pursuit of happiness.

September 14, 1969

Environment May Eclipse Vietnam as College Issue

By GLADWIN HILL
Special to The New York Times

LOS ANGELES, Nov. 29 — "We want to stop the war, end pollution—and beat Stanford!" yelled a Berkeley pep leader at last weekend's big football rally.

The mention of pollution brought a roar of approval from a University of California crowd of 5,000 that almost drowned out the reference to the big game.

Rising concern about the "environmental crisis" is sweeping the nation's campuses with an intensity that may be on its way to eclipsing student discontent over the war in Vietnam.

This is indicated by interviews with students and faculty members from many campuses and with leading conservation authorities around the country.

There is a strong feeling on the campuses that the war will be liquidated in due course. Meanwhile, it is physically remote. And, in the wake of the big protest marches, many students feel Vietnam offers only limited scope for student action.

But the deterioration of the nation's "quality of life" is a pervasive, here-and-now, long-term problem that students of all political shadings can sink their teeth and energies into. And they are doing it.

A national day of observance of environmental problems, analogous to the mass demonstrations on Vietnam, is being planned for next spring, with Congressional backing.

From Maine to Hawaii, students are seizing on the environmental ills from water pollution to the global population problem, campaigning against them, and pitching in to do something about them.

"A ground swell of concern is starting, on everything from population and food supply to the preservation of natural areas," commented Dr. Edward Clebsch, assistant professor of botany at the University of Tennessee.

"I've been floored by the intensity of their actions and feelings," said Dr. Vincent Arp, a Bureau of Standards physicist close to the University of Colorado at Boulder. "The student group is going like a bomb."

"They can see it, they can feel it, they can smell it. And they think they can change it," said William E. Felling, a program officer of the Ford Foundation, which contributes to many conservation activities.

In Los Angeles a fortnight ago, a student bloc stole the spotlight from 1,000 older participants in a gubernatorial environmental conference. Last week in San Francisco, at a meeting of the United States National Commission for UNESCO, something similar happened.

Words and Deeds

In Massachusetts last week, Boston University students put on a two-day campaign of public education in ecology. In Seattle, the University of Washington Committee on The Environmental Crisis was staging a similar "learn-in".

Words are only the surface of the iceberg. University of Minnesota students, fresh from a mock funeral demonstration against the fume-belching automobile engine, were planning to dump 26,000 cans on the lawn of a beverage manufacturer to protest use of such packaging. Northwestern University students were campaiging against a controversial regulatory proposal of the Chicago Sanitary District, and against the waste discharges of a big drug manufacturer.

At Stanford and the University of Texas, law students were researching new courtroom stratagems against despoilers of the environment. University of Arizona students in semisecrecy, were collecting data on the fume emissions of copper smelting operations.

Efforts Get Results

Already the student environmental front can point to many accomplishments. Student activists had significant roles in the campaigns to "save" San Francisco Bay and the northern California redwoods, and to block new dams on the Colorado River.

The University of Wisconsin's Ecology Student Association was active in the campaign against the recently truncated Project Sanguine, the Navy's high-power communications development; and provided important logistical support for the Environmental Defense Fund in the months-long Madison hearings on DDT.

At the University of Illinois at Champaign-Urbana, Students for Environmental Control sallied forth in freezing weather 10 days ago and extracted six tons of refuse from nearby Boneyard Creek. They persuaded city officials to follow up the effort, and are working on a beautification plan for the creek.

A University of Texas student is launching a state environmental newsletter. University of Washington students, on their own time, are preparing an 80-page report on ecological problems of Puget Sound. At the California Institute of Technology. students organized an intercollegiate summer research project in environmental

Daniel Dmitruk for The New York Times

University of Minnesota students conduct a mock funeral for the gasoline engine in a protest against air pollution. Concern for "environmental crisis" is sweeping campuses.

problems that already has attracted nearly $100,000 in foundation financing.

On some campuses—Vassar, the University of Oklahoma, and the University of Nebraska are examples — there are no evidences of organized environmental concern. But they are far outweighed by the ferment elsewhere.

On the University of Texas campus at Austin there are at least six environmental groups, with interests ranging from water pollution to conservation law. One group, in the College of Engineering, has filed 58 formal complaints against the University itself for pollution of a nearby creek. At the University of Hawaii, there are close to two dozen groups, each organized around a particular cause.

Action Is Keynote

Some groups, like Boston University's Ecology Coalition, have as few as a dozen members. Others have hundreds. But with causes on every hand,

mass membership and parliamentary formalities mean less than action, which can be initiated by a handful of people. Then the causes gather their own following.

A few groups cherish the designation of "radical" and are indirect offshoots of the leftist movements like the Students for a Democratic Society and California's Peace and Freedom Party.

"Capitalism is predicated on money and growth, and when you're only interested to maximize profits, you maximize pollution. We need a system that takes maximum care of the earth," said Cliff Humphrey, the 32-year-old leader of Ecology Action, one of several groups at Berkeley.

But generally the aura of the environmental "new wave" is conservative, with coats and ties as conspicuous as beards and blue jeans. "There's a role for everybody in ecology," said Keith Lampe, a cofounder of the

Yippie movement, who puts out an environment-oriented newsletter from Berkeley. "People with widely different styles and politics can talk to each other with no more tension than a Presbyterian talks with a Methodist."

Few 'Anarchists'

"I doubt if you'll find many anarchist ecologists," commented Steve Berwick, a 28-year-old Yale environmentalist. "Ecology is a system, and anarchy goes against that."

A typical group is Boston University's Ecology Action, whose 75 members are led by Bruce Tissney, a 20-year-old junior geology major. Edwardian rather than hippie in appearance, he has a trimmed red beard, wire-rimmed spectacles, and affects such sartorial accoutrements as a blue plaid vest and matching bow-tie, white shirt, and gold watch and chain.

Ecology Action's two-day educational program last week included "friendly" picketing of

PROTEST: A group of girls demonstrate against the smog in Los Angeles. Conservation groups are erupting at colleges all over country and may eclipse discontent over the war.

Associated Press

the state capitol, a pollution film festival, pamphleteering and lectures, and a mock award of a pollution prize to a local power company. The group has been conferring with state water pollution officials about doing spare time "watchdog" work, and is planning to set up dust-catching devices to monitor air pollution.

There have, across the country, been incidents, but mostly minor—such as the arrest last month of 26 University of Texas students who tried to block the felling of some trees for a campus building extension.

Local Orientation

Some of the campus groups are branches of national organizations such as the Sierra Club (which has just installed a campus coordinator at its San Francisco headquarters), the Wildlife Federation, and the newly established Friends of the Earth. But most of them are spontaneous local movements. Many tend to shun the established national organizations as being dedicated to old-line "conservation" rather than the environmental crisis. They also feel the older groups are wary of "direct action" for fear of losing the tax-exempt status that is their financial base. Ad hoc student groups don't have this problem.

"We don't want to be labeled

as 'conservationists' or 'antipollution'," said Wes Fisher, a 26-year-old ecology student at the University of Minnesota. "Pollution and overpopulation are like a web, and pollution is just the symptom."

The students are employing the gamut of communications and political-pressure techniques—meetings, lectures, rallies, picketing, research, pamphleteering, letter-writing, petitions, legislative testimony, collaboration with public agencies and contacts with politicians.

Last month, Illinois' representative William Springer, Republican, felt student heat when conservationists from the University of Illinois picketed a testimonial dinner for him because he backed a controversial dam project.

Impetus Is Recent

The environmental "new wave" gathered in California as far back as 1965, when Berkeley students staged a sitdown protest against a freeway and Stanford students became involved in campaigns for San Francisco Bay, the redwoods, and Point Reyes National Seashore.

But most of the organizing is recent, and is proceeding unabated. A Boston University group was sparked by a recent Ramparts magazine article by Stanford's Dr. Paul Ehrlich, the

"population bomb" crusader. San Francisco State College students were galvanized by a speaker from the Planned Parenthood organization. Bob Hertz, an organizer of the University of Minnesota's Students for Environmental Defense, said his inspiration came from Zen Buddhism and its emphasis on the interrelationship of man and nature. A student group gathering strength at Ohio State was motivated by concern over the Army Engineers' Clear Creek Dam project in southern Ohio, which threatened to flood a pristine natural area used by science students.

In more instances than not, students are welcoming faculty collaboration and counsel. In some places, faculty members have taken the lead. At the University of Arizona in Tueson, a philophy professor, David Yetman, and a recent law graduate, William Risner, organized "GASP" (Group Against Smelter Pollution) to do battle with the copper companies. The group now includes students and townspeople.

Academic Growth

A University of Illinois engineering instructor, Bruce Hannon, has been a leader of the Committee on Allerton Park, opposing a $70-million Army Engineers dam project near

Decatur. Students joined in a campaign that lead to the University's commissioning of an engineering firm to produce an alternative plan.

The environmental ferment caused Ohio State to establish a School of Natural Resources last year. Its original involvement of 180 has grown quickly to 300. An introductory conservation course that had 147 students last fall had 210 this fall. The college's perennial Biologists Forum, which used to draw 20 persons to its meetings, has been attracting hundreds. The University of Tennessee reports an enthusiastic reception for a new course in "Biology and Human Affairs." Colby College in Waterville, Me., has organized two special seminars in January and February on pollution problems and conservation law.

Students are taking the initiative in some environmental teaching. At Stanford, Jeff Bauman, a 22-year-old senior majoring in biology, this fall has been attracting 20 to 40 students to an informal after-dinner dormitory seminar.

Overshadowing Vietnam

There are differing indications on the campuses about how soon environment may overshadow Vietnam in student interest, but the trend is evident.

"A lot of people are becoming disenchanted with the anti-war movement," said Boston University's Bruce Tiffney. "People who are frustrated and disillusioned are starting to turn to ecology."

"I think environment is a bigger issue than the war, and I think people are beginning to sense its urgency," said Robert Benner, a 22-year-old geology student in the University of Colorado conservation movement.

"The country is tired of S.D.S. and ready to see someone like us come to the forefront," remarked Alan Tucker, a member of Ecology Activists at San Francisco State.

"Environmental problems will obviously replace other major issues of today," said Terry Cornelius, president of the University of Washington's committee on the environmental crises. "This is not just a social movement for Biafra or Vietnam, but for everybody and our closed system, Earth."

"Environment will replace Vietnam as a major issue with the students as the Vietnam phase-out proceeds," commented A. Bruce Etherington, chairman of the University of Hawaii's architecture department. "And it will not be just a political lever to be used by radicals."

Many of the "over-30" environmentalists see the student movement as the catalyst, if not the main driving force, that will get environmental improvement rolling and overcome the older generation's tacit resignation to the status quo.

"These kids are really remarkable in their understanding and maturity," said 52-year-old Dr. Barry Commoner, the prominent Washington University ecologist who has been addressing many student groups.

Campuses are seen as representing a greatly broadened base for the "conservation constituency" needed to jog bureaucrats and support the politicians through whom environmental reforms generally must clear.

Conservation lawyers look to campuses for the scientific expertise vital in pressing environmental battles in the courts, and for the energy necessary to raise funds for the usually expensive legal proceedings.

Indications are that coming months will see the student conservation tide swelling and manifesting itself in an arresting variety of ways.

Already students are looking forward to the first "D-Day" of the movement, next April 22 —when a nationwide environmental "teach-in", being coordinated from the office of Senator Gaylord Nelson, Wisconsin Democrat, is planned, to involve both college campuses and communities.

Given the present rising pitch of interest, some supporters think, it could be a bigger and more meaningful event than the antiwar demonstrations.

November 30, 1969

Some Troubled by Environment Drive

By JACK ROSENTHAL
Special to The New York Times

PHILADELPHIA, April 21— Some 10,000 people are expected to swarm into a local park here tomorrow for an Earth Week rally, but the 2,000 members of the Young Great Society won't be among them. They will boycott the rally because, like ghetto residents and urbanists elsewhere, they are troubled by the nation's new infatuation with the environment. They think it concerns the wrong kind of pollution.

"What about the pollution of the mind, the pollution of the houses, the pollution of the dirty, uncared-for systems left to the poor?" asks Herman Wrice, the head of the Young Greats, who have won wide respect for their work in such fields as rehabilitating slum houses and delinquent orphans.

Mr. Wrice is by no means the only skeptic about the environmental cause. At a recent dinner in Washington, Richard G. Hatcher, the black Mayor of Gary, Ind., observed:

"The nation's concern with environment has done what George Wallace was unable to do: distract the nation from the human problems of the black and brown American, living in just as much misery as ever."

A Widely Shared View

That view is widely shared by students of urban and minority problems, but they are deeply divided as to how best to reassert the urgency of those problems.

One strategy is to use the environmental cause. This "piggyback" strategy is openly pursued by Saul Alinsky, the noted community organizer, working in Chicago.

"The environmental thing represents a nice, good middle-class issue," Mr. Alinsky says. "More than that, it is a basis for organizing the middle class. As a consequence of that organization, things can start rolling on every other issue of political pollution."

Last fall he helped organize a 3,000-member Committee Against Pollution, which has fought air pollution in the Chicago area. Mayor Richard J. Daley complained that Mr. Alinsky greatly exaggerated the extent of air pollution. "The Mayor can say that," Mr. Alinsky responded, "because he breathes through his ears."

Common Roots

A second strategy sees natural and urban environments as having common roots. This combination is what environment really means, says Donald Canty, a Washington urban writer and editor.

Mr. Canty edited City, the publication of Urban America, until the group's recent merger with the National Urban Coalition. Now he is developing a new publication for the coalition, City—A Magazine of Urban Life and Environment.

"Urban problems are not problems that result from what people do to slums in cities," he says. "They come from what the slums do to people. The environment of an inner-city child is not merely bounded by the walls of his schoolroom, not merely the demeaning neighborhood, but also the quality of the teaching he gets, the respect he is accorded.

"We need to help people broaden their concern over environment to cover the combination of urban as well as the physical environment."

Clifford L. Alexander, one of the Johnson Administration's highest-ranking black officials, called attention to a more explicit combination—of political power.

"The balance of inner-city political power is shifting fast to blacks," he recently told a civil rights group in Durham, N. C. "If environmentalists want their little ordinances and their big ones, they must show they have an alliance with blacks.

"Unless environment is cleansed in black enclaves," Mr. Alexander said, "it will never greatly improve in white suburbia. Even the most energetic bigot cannot build a wall to the stars that keeps his air free from the effect of the inner city."

Still another strategy, that adopted by Mr. Wrice and the Young Great Society in Philadelphia, is indifference.

"Can we really solve anything with a big outdoor rally?" Mr. Wrice asks, referring to the Earth Day meeting here. "How many weekends are those college kids going to go out with their boats and nets to fish for trash? Meanwhile we've still got sewers stopped up with rats.

"The best way to let something temporary die is not to mess with it."

April 22, 1970

Millions Observe Earth Day Across Nation

The New York Times (by Patrick A. Burns)

Throngs jamming Fifth Avenue yesterday in response to a call for the regeneration of a polluted environment.

Activity Ranges From Oratory to Legislation

By GLADWIN HILL

Earth Day, the first mass consideration of the globe's environmental problems, preempted the attention and energies of millions of Americans, young and old, across the country yesterday.

Congress stood in recess because scores of its members were participating in Earth Day programs.

The activities ranged from huge demonstrations to the passage of environmental legislation.

Rallies involving up to 25,000 persons took place in New York, Philadelphia, Chicago and other big cities. The National Education Association estimated that 10 million public school children participated in "teach-in" programs.

Organizers of Earth Day said

more than 2,000 colleges, 10,000 grammar and high schools, and citizen groups in 2,000 communities had indicated intentions of participating.

Ten thousand persons joined in a rally at the Washington Monument that was embellished with a rock-music concert and the distribution of litter bags.

There was a minimum of disorder, despite the fact that the unprecedented event owed its format in some degree to the fractious antiwar protests of recent years.

In one of the day's few disturbances, 15 young people were arrested at Boston's Logan International Airport for blocking a corridor in a protest against the development of supersonic transport planes and their threat of "noise pollution."

The Earth Day idea originated with Senator Gaylord Nelson, Democrat of Wisconsin, and other conservationists in Congress. It was organized by Environmental Acton, Inc., a small cadre of young people based in Washington, and by ecologically minded persons in

thousands of schools, colleges and communities.

The purpose of the observance was to heighten public awareness of pollution and other ecological problems, which many sicentists say urgently require action if the earth is to remain habitable.

Amid a spate of oratory in hundreds of places, one prominent voice was that of former Vice President Hubert H. Humphrey. In a speech at a high school in Bloomington, Minn., Mr. Humphrey urged that the United Nations establish a global agency to "strengthen, enforce and monitor pollution abatement throughout the world."

"We can do things internationally and we must," Mr. Humphrey said amid repeated bursts of applause. "We've got to do it. That's what this nation must lead toward."

President Nixon informally expressed approval of the Earth Day program but took no active part in it, spending a routine day in his White House o'ce.

But nearby, in front of the Department of the Interior, about 2,500 young people staged a demonstration that was keyed to the department's controversial oil leases. They

chanted, "Off the oil!" "Stop the muck!" and "Give earth a chance!"

Secretary of the Interior Walter J. Hickel, the principal Administration official to endorse Earth Day, returned to his home state of Alaska for an appearance at the state university.

'Environmental Revolution'

Senator Edmund S. Muskie, a frequent critic of President Nixon's environmental proposals and a leading contender for the Democratic Presidential nomination in 1972, said that Earth Day indicated the need for "an environmental revolution."

Addressing a crowd of 25,000 in Philadelphia, the Maine Democrat said: "A cleaner environment will cost heavily in forgone luxuries, in restricted choices, in higher prices for certain goods and services, and in hard decisions about our national priorities.

"We are spending 20 times as much on Vietnam as we are to fight water pollution, and twice as much on the supersonic transport as we are to fight air pollution."

In a speech at Georgetown University in Washington, Senator Birch Bayh, Democrat,

226

The New York Times (by Jack Manning)

POSITIVE ACTION: Students from John Dewey High School, Brooklyn, cleaning up and painting benches on Plumb Beach

of Indiana, called for the creation of a "National Environmental Control Agency to conquer pollution as we have conquered space."

In no two communities in the country were the patterns of the day's activities just alike.

In Tacoma, Wash., 100 high school students rode down a freeway on horseback, demonstrating against automobile fumes.

A San Francisco group, calling itself "Environmental Vigilantes," dumped oil into a reflecting pool at the offices of the Standard Oil Company of California in a protest against oil slicks.

In West Virginia, five tons of trash were picked up along a five-mile stretch of U. S. Route 50 and dumped on the Harrison County courthouse steps in Clarksburg.

In Buffalo, most of the members of the Common Council paraded through the square at City Hall with brooms, shovels and a sanitation cart, symbolizing a community clean-up campaign.

Stewart L. Udall, former Secretary of the Interior, spoke at Michigan State University in Lansing and endorsed his $1,000 fee to the sponsoring campus ecology group.

No Observance in Earth

One of the few communities where Earth Day passed without observance was Earth, Tex., where the Chamber of Commerce said the occasion "just

slipped up on us" before there was time to plan an observance.

The Reynolds Metals Company sent trucks to colleges in 14 states to pick up aluminum cans collected in "trash-ins," with a bounty of one-half a cent for each can.

At the University of New Mexico in Albuquerque, students collected signatures on a big plastic globe to present as an "enemy of the earth" award to 28 state Senators accused of weakening a recent anti-pollution law.

Earth Day enthusiasm even overflowed across the border to Canada. Observances in the Place Bonaventure in downtown Montreal included a fashion show that featured putative feminine antipollution garb of 1984. Girls wore jumpsuits, heavy vinyl gloves and plastic face masks.

Not all the day's activities were in a negative vein.

In Ohio, Gov. James A. Rhodes lifted a partial ban on commercial fishing in Lake Erie.

The ban had been imposed because of the discovery of concentrations of mercury in the water. Governor Rhodes said the move was warranted by "new and more complete tests of fish samples."

Gov. William T. Cahill of New Jersey signed a law creating a state environmental protection agency, and Governor Rockefeller signed a measure coordinating pollution

abatement and conservation activities.

The Michigan House of Representatives overwhelmingly approved a bill assuring citizen groups of legal standing in court to press environmental grievances.

The Massachusetts Legislature enacted preliminary approval of a state constitutional "environmental rights" amendment to facilitate citizen action against pollution. The measure is subject to approval by the next legislature and the electorate.

Gov. Marvin Mandel of Maryland signed 21 bills and legislative resolutions dealing with environmental controls.

A House Commerce subcommittee approved a bill that would nearly triple the current annual Federal spending of $45-million on clean air research and bolster the now skimpy factory testing of auto fume control equipment.

Besides antipollution activities, attention also was focused on the question of population control, to which many environmentalists give paramount priority.

"Even a relatively pollution-free technology will be swamped by an unchecked birth rate," Senator Marlow W. Cook, Republican of Kentucky, said in a speech in Louisville.

Some caution also was expressed in the day's oratory. Calling for "rational and

thoughtful" activity, Senator Gordon Allott, Republican of Colorado, suggested that "some extremists want to use the environment issue as one more club with which to beat America."

Cites Recent Oil Spills

An official of the Department of the Interior warned oil industry executives that they would have to demonstrate conclusively that marine oil spills would be averted or "the pressure on an aroused public will make it virtually impossible to continue exploration and development of the petroleum industry on the continental shelf."

The official, Hollis M. Dole, Assistant Secretary of the Interior for mineral resources, told the Offshore Technology Conference at Houston that recent large oil spills made it certain that regulations governing offshore operations would be more stringent, enforcement more rigorous and penalties more severe.

Summarizing the implications of the day's activities, Senator Nelson said:

"The question now is whether we are willing to make the commitment for a sustained national drive to solve our environmental problems."

Remedying pollution, he said, will cost $25-billion to $30-billion a year "over and above what we're now spending at the national level."

April 23, 1970

Environmentalists Are 'Irrational'

By JEFFREY ST. JOHN

"The man who does not do his own thinking," observed the 19th century orator-essayist Robert G. Ingersoll, "is a slave and is a traitor to himself and his fellow-men."

The ecological movement in America has become a herd of hysterical intellectual sheep. The consensus and conformity of the movement about what should be done about our deteriorating environment leaves the impression they are the New Luddites.

The long-range goal of the environmentalists seems to be a novel process of nationalization of the means of production in the name of survival. However, the disappearance up to 90 per cent of the sturgeon and salmon in Russian rivers because of Soviet industrial pollution is a startling illustration that Socialism is no insurance against pollution. The naive philosophy of some environmentalists that we must "commune with nature" is charming. Except that the half-million or more human beings who perished in Pakistan's recent natural disaster offer a rude and dramatic illustration that nature is not always a friend of man.

The banning of DDT due to environmentalist hysteria is an ominous indication of possible man-made disasters to come. Two hundred and ten separate species of insects and reptiles are controlled by the pesticide. Yet in state after state that has banned DDT we find mounting reports of a rise in ravages to forests and crops from such pests as the gypsy moth. Maine banned DDT in 1967 and to save its forests from destruction by the spruce worm it was forced to reintroduce its use. Malaria is again on the rise in states which embargoed DDT. Overseas, the story is the same. The World Health Organization reports Ceylon's banning of DDT a few years ago resulted in "more than a million cases of malaria" having reappeared.

What the environmentalists purport to know about the pollution problem has created a form of intellectual pollution. Their mindless approach is all the more startling since the so-called "friends of the earth" continue to claim that our planet will perish unless we alter our current allegedly suicidal course. Since the mind and reasoned thought is man's only tool for survival, the irrationality of some environmentalists suggests that Mother Earth doesn't have a chance. This revolt from reality is further evidenced by the standards proposed by the environmentalists.

They put a premium on government coercion to "get the polluters." The vengeful vocabulary of the activists has taken on the tone of a punitive expedition pursuing political power rather than pollution problem-solving. The executive director of the Environmental Clearing House uses such revealing rhetoric as "putting stronger weapons into the hands of the public." When one talks of "weapons" it smacks of substituting the whip for human reason and wisdom.

Rejection by some of the environmentalists of the tools of technology for solving pollution problems is in reality a rejection of the mind. This ecological ethic ignores the three cardinal precepts of pollution problem-solving: technology, money, and politics. Dominant environmental politics seems to rule out the first two, premising most of its coercive proposals on the grounds that "greed" is at the root of pollution. And the nation's elected political leadership is listening more to the TV stars of the environmental movement than to the research and development people.

Objectively, we have much more reason to be optimistic than gloomy over the environment evidences. Despite Malthusian doomsday predictions over population, for example, recent U.S. census figures indicate strongly that our population is beginning to stabilize voluntarily without any help from Big Brother.

We need to turn over a new ecological leaf in 1971. Mother Earth was irrationally represented in 1970. Efforts at a new and reasoned ecological ethic will be seriously sabotaged if last year's ecological totalitarians prevail in the years ahead. It is their philosophy of coercion and rejection of the mind which has fouled the nest of human freedom.

Jeffrey St. John is a radio commentator and self-identified libertarian.

January 1, 1971

Polluting the Environment

To the Editor:

Where has Jeffrey St. John been? His basic contention in his Jan. 1 Op-Ed article is that "dominant environmental politics seems to rule out" the application of "technology" and "money" to solve pollution problems. He is wrong. He doesn't know what he is talking about.

Environmentalists are only too aware that the technology exists to control most forms of pollution. They are only too aware that it takes money to install pollution control equipment in new and old industrial plants and to build municipal waste treatment plants. And for years environmentalists have been urging industry and government to spend the money to apply this technology. Environmentalists have been supporting state and local bond issues for pollution control facilities and enforcement programs.

Both industry and government, with too few exceptions, have had to be pushed into spending money to control pollution. Only when environmentalists have resorted to what Mr. St. John calls the third "cardinal precept" of "pollution problem-solving" — politics — have they been able to move industry and government to action.

Human reason, wisdom and freedom are indeed precious, as Mr. St. John indicates. But where were reason and wisdom as industries and municipalities proceeded to foul our air and water? Was it reason and wisdom that led the Federal Government virtually to ignore the 1899 Refuse Act as an instrument to control water pollution until the past year or so? Or were industry and government simply exercising their "freedom" — the freedom of industry to pollute and the freedom of government to look the other way?

It is no "rejection of the mind" to recognize that self-regulation and voluntary approaches to pollution control have failed to stem environmental degradation. The public needed new legislative weapons — yes, "weapons," for "reason and wisdom" had not prevailed on a voluntary basis.

The environmentalists derided by Mr. St. John are primarily responsible for the antipollution weapons enacted in recent years. And the environmentalists will keep the pressure on to assure the fullest possible use of these laws.

The environmentalists derided by Mr. St. John deserve major credit for articulating a new public commandment to industry and government: "Thou shalt no longer muck up our environment."

Too bad Mr. St. John didn't know what he was talking about before assailing the environmental movement as a threat to freedom. And too bad he didn't read further in the works of Robert G. Ingersoll; he might have found Ingersoll's warning that "ignorance is the only slavery."

MARVIN ZELDIN
Washington, Jan. 4, 1971
The writer was formerly director of information services of the Conservation Foundation.

January 21, 1971

5 States With Citizen-Suit Laws Find Fears of Abuse Unfounded

By GLADWIN HILL
Special to The New York Times

LOS ANGELES, March 23—Last July a group of citizens in Detroit became annoyed at the smoke belching from a power plant. They brought a suit in State court as citizens of the community to stop the emissions.

There was nothing remarkable about that—except that in about 40 states they might not have been able to get into court. Under traditional legal ground rules, plaintiffs must present evidence that they personally have suffered specific, measurable damage from whatever they are trying to stop.

But Michigan is one of six states that, since 1970, have passed laws guaranteeing citizens the right to sue against environmental abuses simply because they are environmental abuses.

Fears Unwarranted

The laws encountered opposition on the ground that they would swamp the courts with frivolous suits. Similar objections have been raised against a bill before Congress to ac-

cord citizens the same environmental litigation rights in the Federal court system.

A research organization now reports that such apprehensions so far have proved unwarranted.

The Washington-based Consumer Interests Foundation said that a survey of five of the six states with "citizen suit" laws disclosed that only 50 such actions had been brought in the last two years.

Of these, 33 were in Michigan, the first state to give explicit sanction to citizen suits. The other states were Connecticut (one suit), Florida (3), Massachusetts (6) and Minnesota (7). No compilation was available from the sixth state, Indiana.

Of the 50 cases, 17 involved land use issues, 16 water pollution, 13 air pollution, two wildlife, one solid waste and one pesticides.

A number of other states are considering citizen-suit laws.

State legal and environmental officials queried in the survey were unanimous in stating that the laws had not overburdened the courts.

"The idea that there would be a flood of cases is a myth that has been exploded," said Gregor McGregor, Assistant Attorney General of Massachusetts.

Michigan's Attorney General, Frank J. Kelly, said the law had proved "an extremely important asset in the effort to abate pollution."

Prof. Joseph L. Sax of the University of Michigan Law School, who has been a prime mover in drafting and advocating state and Federal legislation on the subject, said:

"Plainly, a statute's influence is not limited to lawsuits actually instituted. Industrial and administrative agency behavior may be mollified by the fear of a lawsuit and its attendant publicity, and developments in one suit may bring about institutional changes of behavior."

Citizens' rights to sue are stipulated in the Clean Air Act of 1970, the Water Pollution Control Act of 1972 and other Federal environmental laws.

The National Environmental Policy Act of 1969 permits citizen suits where Federal agencies have failed to issue en-

vironmental impact statements on pending projects. About 350 suits have been brought in Federal courts under this provision.

A pending Federal measure, sponsored by Senator Philip A. Hart, Democrat of Michigan, and others would permit private citizens to bring Federal court suits against individuals, business organizations, or Government agencies in connection with interstate activities tending to impair the environment.

A similar bill in the last Congress accorded the right to "any person." The revised version grants the right to "any persons who are adversely affected . . . or who speak knowingly for the environmental values asserted."

The Consumer Interests Foundation is a nonprofit organization established last year by Consumers Union. Its trustees include Bess Myerson, New York City's departing consumer protection head; Ralph Nader, the public-interest lawyer, and Betty Furness, a former Federal consumer interests official.

March 24, 1973

Who Started Pollution?

By Barbara Garson

A very nice couple in Portland, Ore., donated the downstairs of their house for our day-care center. Imagine, use of the ground floor, (from eight to five) for fifteen little children! Naturally we were anxious to keep the place in good condition and follow any rules.

Unfortunately, these generous people were recyclers.

After snacks we parents had to peel the labels off the juice cans and flatten them. After lunch we had to wade through the kiddie litter, separating organic from inorganic. At the end of the day we collected the accumulated apple cores and orange peels to bury in the compost heap.

The other mothers seemed to think this was very moral behavior, quite in keeping with our concept of co-op. I did too, I guess. But deep down I always resented the idea of being so good.

One day, as I was running hot water over a grape juice bottle so that I could clean off the label before putting it in the proper receptacle, the whole thing shattered. (Not the bottle, but my sense of why we were doing it all.)

"What kind of nonsense is this?" I shouted. "I've got fifteen kids to take care of, and here I am washing labels off bottles one by one, while a machine is slapping them on faster than all the good women of Portland can possibly peel them off!"

I considered taking Juliet out of the nursery school.

"Either it's right to paste paper labels on bottles or it's wrong," I reasoned. "If it's wrong you don't sit here scraping them off. You go down to the factory and make them stop. Rip out the evil root and branch!"

Maybe I'm just a hard-liner, a law-and-order type, but I don't think that the *victims* of crime should pay.

I held on to these convictions; in fact they grew stronger. But I never could convince anyone in the nursery school.

In the natural course of events Juliet graduated into public school.

When she came home from kindergarten and told me, "The Jews killed Jesus Christ," I didn't feel any need

to defend myself. But when she came home from the first grade and told me, "People start pollution; people can stop pollution," I was furious.

"You didn't start pollution! I didn't start pollution! You're six years old, young lady, and it's time you realized you're living in a class society. Some people profit from pollution. The rest of us could stop them if. . . ."

"But, mommy. . . ."

"And furthermore, I forbid you to join the Brownie Scouts."

"But. . . ."

"I forbid you to go on neighborhood cleanups and paper-recycling campaigns. . . ."

"But. . . ."

"If you want to clean up the environment, join a revolutionary socialist organization."

"But. . . ."

"And if there isn't a nice one, start one. You've got a long life."

"But, mommy, all the girls I like in the second grade are in the Brownies."

"I know," I sighed, "I know. And all the nice grown-ups are recyclists. That's why we have pollution."

Barbara Garson is author of the play "MacBird." This is reprinted from WIN magazine. December 30, 1974

ECOLOGISTS SEEK TO 'EXPORT' CURBS

Groups Use '69 Law to Fight Pollution by U.S. Abroad

By GLADWIN HILL
Special to The New York Times

WASHINGTON—Almost imperceptibly, environmental activist organizations have begun a campaign to "export" the United States' burgeoning ecological standards to foreign countries.

The twin "secret weapons" in the conservationists' campaign are the National Environmental Policy Act of 1969 and the massive transfusions of money and materials the United States is accustomed to dispensing overseas.

The act, in its requirement that analyses of the impact on the environment be made before major Federal projects are begun, says this requirement applies to "all Federal agencies."

Public Is Informed

The impact-assessment process cannot stop the pursuance of any project if an agency ultimately decides it is worthwhile. The process was designed by Congress as a way of forcing agencies to consider previously ignored factors in their undertakings and to put the statements of proponents and opponents of the agencies' projects on the record for official and public comment that might alter policies and plans.

Some environmental lawyers could not see why—particularly when the United States is committed by the 1972 Stockholm Declaration to uphold sound environmental standards internationally—Federal agencies should follow double standards, adhering to certain environmental criteria at home but espousing inferior conditions abroad.

Accordingly, three environmental groups — the Sierra Club, the Environmental Defense Fund, and the National Parks and Conservation Association—filed a little-noticed suit late in 1973 against three Federal agencies involved in the program of exporting billions of dollars worth of atomic power generating equipment to foreign nations.

The defendants were the State Department, as policy maker; the United States Export-Import Bank, which lends foreign countries much of the money to buy the nuclear equipment, and the Atomic Energy Commission, as the technical arm of the program.

A United States District Court in Washington ordered the A.E.C. (and now its successor agency, the Energy Research and Development Administration) to prepare an environmental assessment. This is scheduled to be completed in a few weeks.

Pesticides Were Curbed

Last April a new front in the conservationists' campaign was opened in connection with the export of pesticides banned or restricted in this country. In this case the State Department's Agency for International Development was sued. The agency since 1970 has provided more than $50 million to help 20 foreign countries buy American pesticides.

Under another Federal Court ruling the agency agreed to prepare an impact assessment by next September, and meanwhile not to finance any further foreign procurement of long-lasting pesticides such as DDT, Aldrin, Dieldrin, Heptachlor, Chlordane, and two, four, five-T,, except in public health emergencies.

Meanwhile last June the environmentalists went after the Department of Transportation, challenging its channeling to Panama and Colombia, without any formal impact assessment, of funds appropriated by Congress to help build a key 250-mile segment in the international highway system planned to link Canada and Latin American countries.

Sued to Stop Project

The project, running from Panama City eastward across the Colombia border, spanning jungle in the vicinity of Panama's mountainous Darien Gap, has been questioned as a development that might help spread the hoof-and-mouth disease of cattle to North America, needlessly disrupt the lives of people in undeveloped regions and harm flora and fauna.

This time the Sierra Club, Audubon, Friends of the Earth, and the International Association of Game, Fish and Conservation Commissioners won an injunction to have the Department of Transportation stop the project. Judge William B. Bryant in Washington ruled that although the department had made a quasi-study of environmental effects, it had failed to circulate the findings according to requirements of the 1969 looked into the disease problem and had failed to look for less disruptive routes.

The Panamanian Government hit the ceiling—in the words of one diplomat — over the stoppage, complaining that it would direly affect employment.

In response to a plea from the State Department, the environmental groups agreed to a modification of the injunction that permitted work already contracted for to proceed, but held up work on the gap segment and the Colombian segment and forbade solicitation of further funds from Congress until a full impact assessment was completed.

March 17, 1976

Tensions Increase Between Labor And Environmentalists Over Jobs

By PHILIP SHABECOFF
Special to The New York Times

WASHINGTON, May 27—The United Automobile Workers, long considered by many to be one of the nation's more progressive labor organizations, joined with the auto industry this week to help defeat stringent controls on car exhaust fumes.

The union has long supported efforts to preserve and protect the environment. But in this case stricter emission controls, the union believed, could mean fewer jobs.

The union's role in defeating stronger provisions is only one of a number of recent examples of increased tension between organized labor's concern over jobs and the environmentalists' demand for clean air, clean water and the protection of other resources.

Neither labor nor those who put a high priority on environmental issues desires a conflict or believes one is inevitable. A number of unions have been finding that environmental protection efforts have meant more jobs for their members.

Carter Sees No Conflict

President Carter has insisted that "environmental protection is consistent with a sound economy" and says that antipollution programs have created and will continue to create more jobs than they destroy.

But the day after the President sent his energy message to Congress last week, a group of loggers from California, bearing an 18,000-pound redwood peanut on a flatbed truck, arrived in Washington to protest the Government's decision to expand the Redwood National Park in Northern California, contending that the enlarged park would mean fewer jobs cutting down the big trees.

Recent examples abound of clashes between labor's demand for jobs and environmentalists' demands for clean air and water and protection of other resources.

For example, George Meany, president of the American Federation of Labor-Congress of Industrial Organizations, issued a statement asking that United States landing rights be granted to the Concorde, the Anglo-French supersonic jet, on the grounds that to ban it would hurt workers here and abroad. The jet is opposed by residents near airports and others who object to its noise pollution and other possibly negative environmental effects.

Differ on Nuclear Power

Construction unions, backed by the A.F.L.-C.I.O., have been strong supporters of a growing nuclear power industry, including the development of breeder reactors, a position considered anathema to the environmental movement. The building trades also support the water construction projects that President Carter has opposed, to the applause of the environmentalists.

Miners, tuna fishermen, bottle makers and a wide spectrum of other workers are finding that their way of earning a livelihood is conflicting with environmental problems.

In the abstract, the trade union movement is a strong supporter of the nation's environmental goals. The A.F.L.-C.I.O. was an early supporter of strong clean air and clean water legislation.

When it comes to the work place itself the trade unions are insistent that the environment be safe and healthful.

In some areas where their interests run parallel, the unions are in the forefront of conservation efforts. The Sheet Metal Workers Union, for example, is strongly pushing the development of solar energy because a new solar industry would provide many jobs for its members.

But when real jobs of union workers are threatened, or even when the future growth of jobs that will add to union rolls is threatened, unions often become adversaries of environmentalists and allies of employers.

"We don't think our positions are irreconcilable with the environmentalists," said Albert Zack, chief spokesman of the A.F.L.-C.I.O. "But we do think that some of the environmentalists are too far out. You can have all the pristine air and water in the world and nobody will be eating."

More Jobs Won than Lost

Environmentalists in and out of government insist that the threat to jobs posed by environmental protection activities is exaggerated. Industry, they say, often tries to "blackmail" workers and people in surrounding communities into supporting their resistance to pollution laws, by threatening to close down and move away.

One frequently cited example of such tactics is the Reserve Mining Company's resistance to demands that it stop dumping potentially hazardous ore wastes into Lake Superior. The Federal district judge in the lawsuit against the company commented a couple of years ago, "In essence, defendants are using the work force at Reserve's plants as hostages."

The Environmental Protection Agency monitors the economic impact of antipollution programs and has found that the growing antipollution industry has created many more jobs than have been lost when plants closed because of environmental laws.

The agency reported that since 1971 fewer than 20,000 jobs were lost by plants closed for environmental reasons. In the same period, more than 250,000 workers were hired on clean water projects alone.

Small Comfort to Jobless

A study by Chase Econometrics indicates that while the antipollution industry is growing there will be an increase in employment, but that starting in 1979, when the growth of the industry declines, employment will, too. Over the long run, according to the study, the pollution control programs will have no appreciable net effect on employment.

But the fact that more jobs may be created than lost by pollution programs is small comfort to the workers who lose jobs, commented Anthony Mazzocchi, legislative director for the Oil, Chemical and Atomic Workers Union.

"Corporate America has painted everyone into a classic dilemma," said Mr. Mazzocchi. "Now its jobs versus the environment. The worker has a choice between his livelihood and dying of cancer."

Environmentalists are aware of the resistance to what they deem to be vital programs, and many are starting to pay serious attention to social and economic implications of their goals.

"One of the problems," said Gus Speth, a member of President Carter's Council on Environmental Quality, "is that we don't have any program to respond to this problem."

Mr. Speth proposed in a recent speech that the nation adopt new economic priorities, moving away from "undifferentiated expansion of gross national product per se" and toward an approach that focuses on providing jobs that are "environmentally benign."

May 28, 1977

CHAPTER 3

A Fuel and Energy Crisis

Wyoming feels the effects of the fuel crisis. Note alternate transportation on the right.

Nunley/NYT Pictures

THE GREAT EPIC OF POWER: IN SEVEN AGES

After Super-Power Era, Now at Hand, Scientists Predict an Even More Wonderful Period — In Humanity's Amazing March Each Stage Grows Shorter

Super-Power: Spun in the Waterfall and Woven Over a Tracery of Towers.

By EVANS CLARK

THE age of "super-power" is upon us. Its sign and symbol are the long, drooping lines of wires with their tracery of towers, which have been spun, like cobwebs and almost over night, over vast reaches of the country. Each day adds to the web and to the wonder.

Thus the world enters upon the sixth age of civilization—upon the sixth age of power. What the seventh will be man can only envision dimly, for this sixth is hardly yet upon him. But in this amazing march of humanity into the future each stage is shorter than the one before. Civilization advances by geometrical progression.

FIRST came the age of the power of individual man. It began when man began—and that date no one knows. Scholars who have delved into the dim origins of human life, digging up bits of evidence here and there among scratched stones, buried axe-heads and pottery, hazard a guess that man evolved into his present form about 35,000 years before the birth of Christ. For at least 20,000 of this procession of years man apparently had no power except that of his own body. He could walk and he could run, he could brandish crude weapons, chipped by his own hand from the stones he picked up. But he could summon to his aid no power other than his own.

His day was taken up with serving his immediate needs—food, clothing and shelter. The fire, kindled only with vast effort, by hacking bits of iron pyrites with flint, had to be kept going at all costs—constant watching and refueling was necessary. What clothing he had was of skins made into the crudest covering. For food he had a wide variety, but it took most of his time to collect it: nuts, fruits, berries, birds' eggs, snails, frogs, shellfish and such other fish as he could capture in his hands or with the aid of simple traps, the dead bodies of larger animals which he happened to come upon, and such of the small mammals as he could kill with clubs or stones.

In this first age there was no leisure, no surplus, no comfort—save for a bask in the sun snatched before the next hunt; only necessity met with bare hands.

SECOND was the age of the power of the earth and domesticated animals. "Man came to the threshold of civilization," says Morgan in his "Ancient Society," "when he harnessed the ox to an iron plow for the purpose of cultivating the cereals." Somewhere between 10,000 and 12,000 years ago man began his first attempts to use forces outside himself to help him cope with life. The primitive savage took nature as he found it; he snatched what he needed from the world, but did not make that world work for him.

The early Neolithic man who first planted seeds to reap a later crop seized the first outpost of power in the long effort to capture nature's forces and tame them to his use. He made nature grow his food for him instead of trusting to his own right arm to pick where he could find it, or kill it as it ran. It saved his energy and it saved him time.

The first use of animals in the service of man was a conquest of power that marked an epoch. It came at about the same time as the first conscious utilization of the power of growth. The Neolithic man had domesticated cattle, sheep, goats and pigs. "He was a huntsman turned herdsman of the herds he once hunted." He made them pull the plow and the cart. He made them carry him from place to place. He made them give him food and drink. He had reached out into the world and made another force outside himself labor in his behalf. The power of animals could be immediately applied and at any place desired. In this respect its capture was an even greater benefit to man than that of the power of plant growth.

For the first time in history man won a little leisure—a bit of surplus energy and time. Not every moment had to be given to meeting the immediate demands of food and warmth. The energy which crops and draft animals save man put into perfecting shelter and tools. Pottery, houses, plows, implements of a dozen kinds—first of wood or stone and then of metal, bronze and, later, iron—he fashioned during the leisure he

achieved. While none of these tapped new sources of power outside of man, they saved his own energy by making its application more effective—and released it for still other pursuits.

THIRD was the age of the power of slaves. The great limitation of animal power is the human supervision it entails. Animals must be guided, but the slave is the most self-directing power that man has ever used. He can not only do work but within certain limits can himself plan and guide it, and the overseers needed are few.

Man was not slow to perceive this, in comparison with the time it took him to discover the power of domesticated animals. In the fertile valleys of Mesopotamia civilization took root in Neolithic days. With the development of agriculture and the use of animals and tools life became less hazardous. Population grew and village communities sprang up. Then came cities, and finally—even as early as 6,000 B. C.—came evidences of conquest and empire—kings and armies, priests, written documents and a tradition already ancient. With conquest came captives. At first the men were killed—tortured or sacrificed to the victorious god—the women and children assimilated into the tribe. But it was not long—in this perspective of years by the tens of thousands—before conquerers began to make of captives living sacrifices to themselves instead of dead sacrifices to the deity, and slavery was born.

Civilization is surplus—the marginal stock of supplies, actual and potential, of the body and of the mind which man is able to build up beyond his immediate needs. The great pre-Christian peoples proved it. In the vast empires of the Sumerians, the Babylonians, the Assyrians and the Egyptians a few men had a great deal of surplus. They could trade, read and write, produce works of art, weave intricate webs of religious myth and ritual, make beautiful clothes—or have it all done for them. The earliest of them, the Sumerians, had time to build the Tower of Babel, and the Egyptians built the Pyramids. Hero of Alexandria constructed a primitive sort of steam turbine 130 years before Christ. And then came Greece and Rome, civilizations whose flower is fragrant today.

The surplus which made the great pre-Christian civilizations possible was the surplus created by slavery, the harnessing of the power of the many in service of the few. But these civilizations were affairs of but one-tenth of the population—the life of the nine-tenths was still the old weary round of immediate necessity. The needs of decay were in their midst. The few had too much leisure and the many did not have enough.

FOURTH came the age of the power of serfs. Were the Middle Ages the dark ages because they used less power? An interesting question, yet to be answered with finality. It is certain that serf labor is less mobile, less centralized, less controllable than slave labor. The serf was half slave, half free. Perhaps there were too many feudal lords, each with too few serfs to produce the concentrated power and the accumulated surplus that made the civilization of Greece and Rome. The serf owed fealty to his lord. He must fight for him in the wars and deliver his tithe of grain; he was even bound to the land. But on his land he was comparatively free. There was lack of cohesion about the whole feudal machine that could not drive either deep or far.

What monuments, what learning, the feudal days produced originated, for the most part, outside the feudal circle. The revival of learning came from out of the monastery; the great cathedrals were built in the cities where serfs had escaped into freedom.

THEN, fifth, came the age of the power of steam, working a greater change in the life of man in fifty years than had the use of animal and man power in 5,000. We call that change glibly the "industrial revolution," but it is difficult to realize the immensities of what had befallen the world. In our absorption with what the industrial revolution has produced, it is easy to overlook its cause—the unleashing of new power, power such as man had never known in all history, power that could be generated wherever it was needed, power in unlimited abundance, power that could be harnessed in a thousand ways.

Steam and the machinery that was devised to put it to work soon made all other forms of power obsolete. One man firing a boiler and another tending the machine could do the work that hundreds of men could never have done before. The thing was irresistible.

Steam is a great centralizer. Hundreds of shoemakers scattered through half a hundred villages were displaced by one big factory; one huge steel plant that blotted out the sun with smoke took the place of ironmongers scattered through ten counties round. With a cumulative effect that swept everything before it, steam tore people loose from the country and the village and drew them into the city.

Steam created a surplus beyond the most fantastic imaginations of man. In a period equal to about one tick in twelve hours since the clock of man began to run, the world has been literally flooded with a profusion of goods and chattels—an incredible range of material things, from baby carriages to battleships, and, by a variation of the steam engine theme, the gasoline motor, with automobiles and airplanes.

If steam has revolutionized the material life of man it has wrought, directly or indirectly, no less a change in government, property, morals, religion, health, comfort and education. Even though steam has given us a world in which ownership is highly concentrated and the majority of mankind has little in comparison with the wealth of the few, the general level of civilization for the masses of mankind is something that the savage, the slave or the serf would consider far better than had been their estate. The concentration of population has tended to break down family ties. Supported by the findings of science—made again possible by surplus and leisure —customs, traditions and morals have swung loose from the moorings of the past.

SIXTH is the age of the power of electricity. Just as man begins to find himself in this new world which steam has made, he taps a new reservoir of power. Electricity floods the world today with a torrent of development more swift and deeply channeled than that which steam released a hundred years ago. Steam is highly specialized, but electricity is universal. Steam must be produced at the point of consumption, in separate individual boilers. Electricity can be generated in giant power stations where costs of production are low—near the mine mouth or beside the waterfall—and carried on transmission lines of wire to a hundred thousand individual consumers.

Steam can be made only by burning some fuel—coal or oil; electricity can be made by water too. To get the coal or oil from the ground requires an expenditure of power, and to consume it subtracts from the world's increasingly valuable supply of natural resources. Water automatically delivers itself to the dynamo, once the plant is built; and there is no end to its flow. None of it is consumed in the process. In its very generation electricity can utilize a scarcely tapped and inexhaustible reservoir of energy.

Steam is wasteful, electricity economical. The steam locomotive, to take only one example, for all its lithe and pulsing strength, is inefficient beyond belief. Engineers have calculated that it costs many times as much per ton mile to operate a railroad by steam than by electricity —even when the electricity itself is generated by steam. Consider the waste of thousands and tens of thousands of separate boilers, each with its attendant crew, serving as many separate factories with steam power when they might all be served by electricity from one central powerhouse with a battery of half a dozen boilers at the most; or no boilers at all, just running water.

The logic of electricity is conclusive. It is only a matter of time— another tick of the historical clock— when steam power, except as it is used to generate electricity, will be as obsolete as the horse car. Twenty-two years ago the generating stations of this country produced 3,000,000,000 kilowatt hours of electrical current. Last year they produced 54,000,000,000—a growth of more than 1,300 per cent. in less than a decade. What they will bring forth in the next decade can only be guessed.

What changes does this "superpower age" hold for mankind? Those who live in the midst of a revolution can see only dimly some of its objectives—the final results wait upon their own fulfillment. Like every other accession of power electricity will release a vast amount of human energy formerly thrown into mere routine.

In this instance, women will be the principal beneficiaries. What women will do with the leisure which electricity will bring them remains to be seen. But the sudden elimination of a large amount of household drudgery leaves a gap to be reckoned with in terms of social consequence.

Some see in electricity the great decentralizer. As steam produced the city and the slum, they say, electricity may resurrect the country and the small town. The possibilities are prodigious, the probabilities difficult to define. At the least, electricity holds the promise of a cleaner, swifter, vastly more productive age.

SEVENTH is the age of the power of—what? Already the scientists have raised the question, and some have tried to answer it. An answer must be found. If there were enough water power in the United States to meet the electrical needs of the nation it would be another matter; but the available water power is very far from enough. And already the end of our oil and coal is in sight. The world has used up more of its mineral resources in the past twenty-one years than in all preceding history. At least one-third of the high-grade coal deposits of the United States has been dug up and there is only enough oil left to last for twenty years. Even if every available source of water power were utilized to the fullest possible extent it could furnish only about one-third enough current to meet our needs. Some other reservoir of power must be tapped.

March 29, 1925

Investigators Warn President of Waning Oil Supply for Future

OIL SUPPLY IN SIGHT FOR ONLY SIX YEARS SAYS COOLIDGE BOARD

Special to The New York Times.

WASHINGTON, Sept. 5.—Serious concern over the future oil supply of the United States, both for military and industrial purposes, is shown in the report by the Federal Oil Conservation Board, dealing with national petroleum conditions, which was made public today by Secretary Work, who announced that the report had been sent to President Coolidge in the Adirondacks.

The total present reserves in pumping and flowing wells in the proven sands, according to the report, has been estimated at about 4,500,000,000 barrels. This is theoretically only six years' supply, the board emphasizes, although it cannot be extracted from the ground that quickly.

Since the first oil well was drilled in Pennsylvania more than 68,000 oil wells have been drilled in the United States, more than a fifth of which were failures, and to June 30 last over 9,000,000,000 barrels of crude oil had been produced. But 3,000,000,000 barrels, or one-third of this great total, has been produced in less than five years, and the production last year alone exceeded 750,000,000 barrels.

The United States is producing and consuming 70 per cent. of the world's oil production, and the total investment in the business today is given as $9,500,000,000, while for the last year for which figures are available the total wholesale value of the products was $1,793,700,027.

Conservation Declared Imperative.

In what it considers a concise picture of true conditions, as determined by the exhaustive inquiry in which it has been engaged more than a year, the board, consisting of Secretaries Work, Hoover, Wilbur and Davis, asserts that with the current production coming from about 4 per cent. of the producing wells, most of them only a year or so old, and from fields discovered in the past five years, the future maintenance of current supplies implies constant discovery of new fields and new wells. The board holds that the situation renders it imperative for the welfare of the nation that every effort be exerted to obtain the maximum amount of oil from known fields and the most vigorous efforts at conservation along various lines.

Future supplies, the board holds, must depend on reserves, new fields, improved methods of recovery, better utilization of control, consumption economies, supplies from distillation of shale and coal, and even from foreign oil fields. The board urges better control of production, better mechanical devices for use of oil products, and that American oil companies should acquire and exploit foreign fields.

"While the production of oil upon our own territory is obviously of first importance," the report says, "yet in failure of adequate supplies the imports of oil are of vast moment. The present imports from Latin-American fields amount to about 62,000,000 barrels annually of crude oil, against which we export about 94,000,000 barrels of products.

Points to Southern Fields.

"The fields of Mexico and South America are of large yield and much promising geologic oil structure is as yet undrilled. That our companies should vigorously acquire and explore such fields is of first importance not only as a source of future supply but supply under control of our own citizens.

"Our experience with the exploitation of our consumers by foreign-controlled sources of rubber, nitrate, potash and other raw materials should be sufficient warning as to what we may expect if we shall become dependent upon foreign nations for our oil supplies. Moreover, an increased number of oil sources tends to stabilize price and minimize the effect of fluctuating production."

The provision of future supplies of essential oil products for the American people, the board asserts, must arise from the following sources:

1. The reserves already mentioned.
2. The possible discovery of new sands in the known areas by deeper drilling.
3. The possible discovery of new fields.
4. Improved methods which will recover a larger proportion of the oil out of the sands.
5. Better utilization of crude oils by diversion from less essential to more essential uses, such as conversion of fuel oil into gasoline.
6. Better control of the flush flow from newly discovered fields.
7. Economies in consumption by improved mechanical devices.
8. Supplies from distillation of oil shales and coal.
9. Foreign oil fields.

The board says the major part of the measures to be taken to protect our future supplies "must rest upon the normal commercial initiative of private enterprise." The field for Government action is declared to be "considerable, but to formulate the broader by-laws of the industry in the sense of conservation and to concentrate thought upon them is the major part of the board's task in cooperation with the industry."

How the Industry Can Help.

The directions in which industry can contribute to assure future supplies are set forth as:

1. Continued exploration for extension of known sands and deeper sands in known fields.
2. Continued exploration for new fields.
3. Systematic research and experiment upon methods of securing a larger proportion of the oil from the sands.

4. Systematic research and experiment in new methods and cheapened costs in refining and cracking oils and waste elimination.
5. Cooperative methods in sane development of new fields to prevent wasteful flush flow and overproduction.
6. Research and application by engine builders of more economical use of petroleum products.
7. Expansion of American holdings in foreign oil fields.

What the Government Can Do.

The contributions which the Government can make are given as follows:

1. Continued and expanded research by the Geological Survey in geologic studies of the accumulation of oil and structure of oil-bearing areas; by the Bureau of Mines into methods of producing and refining, including oil shales, and by the Bureau of Standards into questions of constitution and utilization of oil products.
2. The more intellectual handling of Government-controlled oil sources on public and Indian lands.

Of the fundamental conservation measures thus mentioned, that of cooperative methods in development of new fields to prevent temporary gluts is declared to merit more exhaustive discussion, "as it is a promising field for important action by both industry and the Government."

With respect to the demand by those who appear to think that Federal legislation is necessary to deal with the oil situation staring the nations in the face in the matter of future supplies, the President's board takes a stand against Federal legislation and in favor of State action, except in the case of Federal oil lands—unless national defense is threatened by waste or exhaustion, or the States fail to act.

Power of the Government.

"The power of Federal Government to regulate oil production is doubtless limited to its own oil land," the board holds, "unless the national defense is imperiled by waste or exhaustion of the oil supply.

"Here the policy of reservation to meet future Federal requirements has already been established and is being perfected, as prompted by increased appreciation of the need and as guided by better understanding of the means to that end. It has been suggested to the board, however, that the Government's jurisdiction as a sovereign owner of the oil under its land may justify and authorize Federal legislation to prohibit adjoining owners from appropriating the oil by means of wells drilled on their property. This suggestion involves many interesting implications concerning State and Federal power, but State legislation for the protection of all owners would be preferable, and the Federal authority, if any, should be invoked only when it is clear that the State is unwilling or fails to act, or when naval reserves are threatened with depletion."

The investigation conducted by the four Cabinet officers constituting the board was ordered by President Coolidge on Dec. 19, 1924, when he wrote them a joint letter saying he was advised that the "failure to bring in producing wells for a two-year period would slow down the wheels of industry and bring serious industrial depression." The President requested that the board "study the Government's responsibilities and enlist the full cooperation of the oil industry in an investigation to determine actual conditions."

More than 100 leaders of the industry met with the board in Washington in February and discussed all phases of the oil problem. Questionnaires were directed to the leaders of all branches

of the industry, with the result that the heads of the large producing, refining and marketing companies of the country supplied confidential information.

Two additional reports will be issued by the board, dealing with foreign oil conditions and with possible substitutes and the development of oil shale.

Says Question Has Two Sides.

Secretary Work, in making the report public, emphasized that there were two sides to the question.

"This report," the Secretary said, "is intended to present certain facts contributed by the oil industry or gathered by Government scientists. The facts and opinions received from these sources have been weighed with open minds, without conscious prejudice or thought of confirming theories already conceived.

"There are two sides to this nationally important question that must be weighed wholly free from individual bias and both sides are entitled to respectful consideration.

"Many producers claim the supply of petroleum, having heretofore been equal to the demand, probably always will be, and should time prove the contrary that substitutes for both lubricants and gasoline will be devised by science.

"Other producers argue that almost every natural resource is limited and may be exhausted by wasteful use. To illustrate, they cite depletion of soil fertility, timber and even fish in the sea, the exhaustion of old oil fields and the diminishing flow of all wells. They argue further that without discovery of new fields the present rate of production cannot be maintained and therefore urge that, as new discoveries are uncertain, improved methods of recovery, less waste, more ground storage and checking of competitive haste in drilling are economic provisions that should be enforced.

"These conflicting opinions are based partly upon facts and partly upon conjectures. Such opinions are valuable only in proportion to the logic of their reasons, which therefore must be understood and analyzed. It is hoped that this preliminary report, conscientiously prepared under the direction of this board by scientific men burdened by exacting daily routine duties, may furnish a concise picture of the true conditions as determined by the inquiry of this board."

The report itself, while 7,000 words long, is only preliminary, and the supplemental reports that are to follow will deal extensively with the different phases of the national oil problem touched upon in this short summary.

Concern Over Future Supply.

"There must be natural concern over our future supply of oil," the report declares, "because of the manifest dependence of so large a part of our industrial life, national defense and domestic comfort upon continued adequate supplies of lubricants for all machinery and fuel for automotive engines.

"Our future resources in coal and iron have been so determined by geological evidence and exploration that we can measure our proved supply by centuries. But the local character of oil deposits, their geological uncertainty, and the vast amount of unproductive capital required for long advance exploration, prevent any such advance proof in the case of oil. The very nature of all minerals, in that they do not reproduce themselves, of course renders their extraction a depletion of assets. And therefore the resources of oil which are proved at any given moment cannot, from the nature of things, be otherwise than a reserve against current production. They do not imply future production of oil.

"The total present reserves in pumping and flowing wells in the proven sands has been estimated at about 4,500,000,000 barrels, which is theoretically but six years' supply, though, of course, it cannot be extracted so quickly.

"Another addition to this natural cause of anxiety for future supplies lies in the fact that the maximum rate of production from all fields is in their early days, before gas pressures which expel the oil are diminished, and thus of the current production more than one-half is coming from about 4 per cent. of the producing wells—for the most part only a year or so old—and from fields that have been discovered within the past five years.

New Fields Required.

"Therefore, future maintenance of even current supplies implies the constant discovery of new fields and the drilling of new wells, and thus the maintenance of this large ratio of flush production. Hitherto there has been no failure to discover such new fields as required.

"However, this dependence upon fortuitous discovery of new fields renders it imperative that every effort shall be made to secure the maximum amount of oil from the known fields, and the most beneficial use of the oil that is produced."

The board recalls that the first successful oil well was drilled in Pennsylvania in 1859, and then says, respecting development and production:

"During this period of sixty-seven years over 680,000 wells have been drilled and to the end of June, 1926, over 9,000,000,000 barrels of crude oil have been produced, one-third of this in less than five years, with the production for 1925 exceeding 750,000,000 barrels. The United States produces about 70 perr cent. of the total world production and consumes about the same percentage of the total."

Natural Gas Wasted.

Besides crude oil, the board says, the same sands produce vast quantities of natural gas. No figures were available on the total amount of gas produced by wells that also produce oil, but the volume is declared to be large, and in some fields a comparatively large percentage is "wasted" into the air.

"This waste of natural gas," the report continues, "incident to the past and present methods of capturing oil at the surface, is occasioned by blowing the gas into the air to hasten the flow of oil from the wells. It has been estimated that in the Cushing Field in Oklahoma at one period there was an average daily waste of 300,000,000 cubic feet of natural gas, or more than 100,000,000,000 cubic feet in the course of a year, the equivalent of 5,500,000 tons of coal. Vast wastage of gas has occured at Cromwell, Okla.; Burkburnet, Texas, and Eldorado and Smackover, Ark., and at several of the fields in the Los Angeles basin.

"The dissipation of this gas by letting the wells produce wide open was a triple waste, containing, as it did, large quantities of casinghead gasoline, with the fuel value in addition, and being capable of much more effective use for recovering oil from the sands.

"About 300,000 wells were producing in 1925, the oil being transported from the wells by more than 40,000 miles of trunk pipe lines and 40,000 miles of gathering lines. The industry also employs some 142,000 tank cars and about 400 tank steamships for crude transportation and for distribution of refined products. There are some 500 refineries. The total investment in producing wells, transportation, refining and marketing equipment exceeds $9,500,000,000. The total wholesale value of the products was $1,793,700,027 in 1923."

The distribution of use is set forth in a table giving the fractions into which the 1925 production was split by refinement, showing that the largest percentage was 15,279,000,000 gallons which went into gas oil or fuel oil, and 7,294,000,000 gallons, or 23.4 per cent, into straight distillation of gasoline.

Tells of "Known" Fields.

With respect to "known" oil fields, the board reports:

"The producing and proved area in the United States is asserted to be in excess of 3,000,000 acres, the fields varying in production in 1925 from a daily average of 197,650 barrels for the Smackover field in Arkansas to only a few barrels daily for a number of the older fields in the country.

"As the oil flows into the wells from the sands, due to the pressure of natural gas, the flush flow is always larger and tends to diminish with the release of gas pressure. The annual production during many years has been sustained by the discovery of sufficient new fields each year to maintain a very large contribution from this flush flow. Therefore, supply equal to our present demands hangs definitely upon the rather precarious basis of repeated new and important discoveries."

As to future reserves the report says:

"From the most recent estimates it would appear that the reserves of oil available by flowing and pumping wells from present producing and proved fields amount to about 4,500,000,000 barrels.

"Theoretically, this is but six years' supply, and it, of course, could not be produced rapidly enough to in itself maintain the present annual production of about 750,000,000 barrels. These reserves do not in themselves prove a shortening of supplies; nevertheless, as stated above, their life, as compared to proven supplies of iron and coal, is such as always to maintain concern."

Extension of Reserves.

In addition to the "proven" reserves at any one time, the report continues, the known fields in many cases have proved of larger extent than at first estimated, due to the extension of "fringes" of such fields, particularly as the result of opening new sands and, in some instances, the extension of known sands by deeper drilling. The capacity of machinery for deeper drilling has steadily developed, until in 1925 an oil well 7,591 feet deep was completed in Southern California.

"At various stages in development it has usually been asserted that no greater depths could physically be attained," the board asserts, "yet almost every year demonstrates the penetration of still greater depths. As many of the oil sands slope into the earth, deeper drilling of known sands will bring still further production, as well as the discovery of deeper sands underneath those now being exploited."

"Certain parts of the country are known by their geology to be impossible of appreciable oil production," the report says in discussing new fields. Such positively barren areas are estimated to aggregate 43 per cent. of the total area of the United States. But this does not warrant the assumption that the remaining 1,100,000,000 acres of the country, or any large part of them, will be found oil bearing.

"Considerable portions of this area have already been drilled for oil or water. It is a certainty that we are learning each year more of geologic structure at the hands of a large body of public and private geologists, but the percentage of dry holes in new exploitation is increasing. To assert that no new fields will be found would be to deny a very strong law of probabilities, and we may conclude that such fields will be found, but obviously no forecast of their importance can be given.

Varying Estimates of Oil Left.

"There is a wide variation in estimates of the amount of oil left underground in the sands after production ceases with ordinary methods of flowing and pumping which have been hitherto employed. The evidence before this board shows the general belief of oil experts that not more than 25 per cent. of the oil can be recovered by ordinary methods.

"Some leading authorities consider that less than one-sixth of the oil is so

recovered. During recent years a considerable amount of investigation and experimentation has been made with different methods of forcing out the contained oil with water, air or gas pressure—either directly from the surface or through the proposed method of sinking shafts and driving galleries underground.

"Some of the authorities on these methods believe that a second crop from known sands can be obtained by these means as great as that already recovered and available in the proved reserves. Such a result, if attained, would add a total of over 13,000,000,000 barrels to our supply from known fields.

"While no positive assurance can be given from the results so far attained, yet it is the impression that developments have proceeded so far as to give strong belief in considerable further recovery from the known sands over and above the estimates which have been made as to the supplies available by present methods."

Would Restrict Uses of Oil.

Asserting that the most essential products from our crude oil are lubricating oil, gasoline and other oils for internal combustion engines, the report says the other uses for oil could be dispensed with, without "an industrial revolution, as other fuels and substitutes could be applied without prohibitive differentials," in economic costs.

"At present about one-half of our crude oil production is burned either as crude oil or as fuel oil to generate steam and heat—the remainder is used as kerosene, gasoline, lubricating oil, &c.," it continued.

"Up until thirteen years ago, the amount of gasoline which could be recovered was limited to the natural fraction of gasoline in the crude product, but the discovery of the cracking processes, by which the heavier oils can be broken down into gasoline, has opened an entirely new vision as to the gasoline supply.

"At present probably 64 per cent. of our gasoline comes from straight distillation of crude oil, 31 per cent. from cracking and 5 per cent. from natural gas gasoline. It seems theoretically possible to crack 80 per cent. of the oil now going into fuel and heat purposes if the gasoline were required.

"The discoveries in methods of cracking oil are of fundamental importance in considering the future of our essential oil supplies, and this conversion in itself comprises a potential of assured gasoline supplies.

"It may be contended that a control of production by which supplies would be reduced would so increase the price as to force the production of gasoline instead of so large a fraction of use as fuel oils. This program, however, raises impossible difficulties in artificially increased prices to the consumer, inability to regulate the flow accurately to the demand and the ultimate necessity to regulate profits in order to protect the consumer."

Production Control Is Stressed.

Better control of production is recommended.

"There are subsidiary phases of overproduction which deserve attention, as they lead to economic waste," the report declares. "At the initial opening of new fields the gas pressure is strong and the flush flow of wells is very large, rapidly diminishing to more settled production, and the opening of new fields is in most instances followed by a fever of drilling.

"Due too often to divided ownership in small areas, the drainage of which is threatened by adjacent wells, a rush of drilling leads to enormous flush flows which temporarily glut the market and force much oil into fuel consumption and, through over-release of the gas, diminish the amount of oil that can be ultimately obtained by flow and pumping. This question is discussed later on more fully."

Waste of Gasoline.

Urging better mechanical devices in the use of oil products, the board says many gasoline engines are inadequately designed for the most economic use of gasoline and estimates that 25 per cent. of the gasoline "could be saved and the same mileage of movement provided if there were more adequate designs of carburetors and engines."

As to supplies from oil shale and coal distillation the report says in part:

"The by-product distillation of coal for coke and gas purposes produces a certain amount of oil. The most adaptable coals produce about four and one-half gallons of light oils and fifteen gallons of tar per ton. This supply for 1925 amounted to about 146,000,000 gallons of light oils and 258,000,000 gallons of tar.

"The tendency in industry is strongly to increase the distillation of coal and the recovery of by-products; the whole of the coals adapted for this purpose, however, would not furnish an appreciable substitution of our liquid-oil supplies. They obviously could not be distilled for production alone, but must depend upon the market for the major products—that is, coke and gas.

"The oil shale and oil sand deposits of the country are of more promise in the future outlook. Very large areas of such shales exist, many of them containing as much as a barrel of oil per ton of shale. Their utilization is solely a question of price. There can be no doubt that these shales will some day be brought into production. They form an almost unlimited reserve and may, therefore, be taken as the final protection of our people in the matter of essential supply."

The right of the State under its police powers to prevent the action of one owner from working a deprivation of the rights of other owners of a common property and to prevent waste or destruction of the common property by one of the owners "seems reasonably clear," the board holds.

"The right of the State to prevent the waste of natural resources," the report continues, "is rendered more important in this matter by the newly discovered or at least more widely recognized facts regarding the role of gas in the oil sands. Gas is more than a commodity of smaller commercial value associated with oil; it is the efficient agent provided by nature for bringing the oil within the reach of man. Dissolved in the oil, the gas makes the oil flow more freely to the well and there forces it upward, and the longer the gas is retained in solution the larger is the recovery of oil.

Protection of Underground Supply.

"Waste of gas is therefore a double waste and the impairment of the gas pressure in an oil sand by one owner may prevent his neighbors from recovering any of the oil beneath their land and himself from securing more than a small part of the oil underlying his own land. * * * If the several oil-producing States should protect property rights in oil produced from a common underground supply, it undoubtedly would have some effect in the direction of stabilizing production, of retarding development whenever economic demand does not warrant and of making the business of oil production more economical.

"Such legislation, although not directly regulating production, would in part accomplish this by freeing owners and operators from the present pressure of a competitive struggle.

"The formulation and adoption of remedial measures permitting a better correlation of underground property rights is all the more urgent in view of the present intense activity of the engineers in perfecting better methods of oil recovery. To accomplish something in relieving the pressure of competition before it is too late, it is suggested that the lawyers familiar with the evidence of the oil operators take the lead in developing the needed line of action in accord with the discoveries of science and changes in engineering practice."

Another avenue of relief the Board thinks would be "restriction of development of operation through voluntary agreement of owners."

"Operators of experience have expressed to the board the opinion," the report says, "that duplication in drilling and the consequent unnecessary reduction of gas pressure constitute the cause of all waste worth mentioning on the production of oil. The danger that much of the remaining oil will be brought to the surface before either courts or lawmakers can be expected to remedy the situation has also prompted the expression of opinion to the board that voluntary cooperation offers the only practical solution.

"The question of the legality of cooperative agreements has been frequently raised in the recent discussions of remedial measures."

Wants Uncertainty Removed.

"The uncertainty as to whether the economic betterment through substituting cooperation for competition runs counter to Federal and State laws has served as an actual, or imagined, or pretended barrier to cooperative action, and the removal of that inhibition is asked, although the suggestion comes from the industry that, 'to protect the public, approval of such agreements by some commission or board should be required.' This doubt should be removed by appropriate legislation.

"The voluntary cooperation proposed would need to include the land owners and operators in a single field or pool, which is a relatively small unit of production, so that the possibility of monopolistic control need not be feared. Indeed, cooperative regulation of either the development or the operation of a single pool could control only a small percentage of the country's production.

"The largest flush pool in recent years, Santa Fé Springs in California contributed 11 per cent. to the output in 1923, and no pool contributed more than 8 per cent. to the output of either 1924 or 1925. Indeed, the three exceptional pools last year, Smackover, Ark.; Long Beach, Cal., and Tonkawa, Okla., together accounted for only 16 per cent. of the country's production. Even the flood of oil from the Cushing Pool at the time of its maximum yield in 1914 and 1915 is to be credited with only 17 per cent. of the country's production in those years, when the total yield was only a third of that of 1925.

"In the one instance of cooperative control by the Salt Creek Conservation Committee, the prorating of production in 1922 and 1923 reduced the output to perhaps one-third of the productive capacity of the 600 to 760 wells then producing. The effect of the committee's restrictions on production was a matter of only 8 or 9 per cent. of the country's production of that period.

"The question of the country-wide influence of such cooperative action on either supply or price would, moreover, under any legalized procedure, be always subject to 'appropriate and adequate governmental scrutiny,' quoting from counsel of the American Petroleum Institute, "to the end that these owners might not be stimulated to undue haste and a wasteful competition in the development of their properties and trade, but might have a greater liberty to consult the economic conditions of the industry from time to time'."

Lack of Cooperation Alleged.

The report cites various examples of lack of cooperative action, among them that of the Santa Fé Springs field in California, concerning which case the board says:

"In this small but highly productive field forty-five operating companies competed, drilling more wells than necessary, letting loose an unneeded flood of oil on the market with resultant disaster to the price structure, and creating conditions of increased operating costs.

"The total losses attributed to this competitive struggle in a single field are estimated as in excess of $900,000,-000; but even greater perhaps than this economic waste, involving a loss to the industry, is the loss to the country involved in the dissipation of gas pressure, leaving in the upper sands oil that might well have been recovered to an amount equal to that actually taken out by the rush methods of competitive drilling and producing."

The Dominguez field in California is cited as an example of development by cooperative agreement among a few large oil operating companies.

As to the Federal Government's own problem, the report points out that the Government is still the owner of vast areas embracing several hundred million acres of land, the title to which is held "really in trust for the public generally." There are proved oil fields in all these areas, and much unexploited territory holds promise of oil in commercial quantities.

"All these areas," the board reported, "except some special reserves, may fairly be said to be wide open to private appropriation through lease or otherwise.

Increase in Production Cited.

"There seems to be no discretionary power anywhere to resist the exploitation of these lands and the dissipation of the Government's oil resources. The leasing of the Indian lands is progressively mandatory until exhaustion."

Production from these Government lands jumped from 2,153,794 barrels of crude petroleum in 1921 to 12,371,473 barrels in 1925 from the naval oil reserves; from 9,214,792 barrels in 1921 to 29,153,768 in 1925 from the public domain.

"While it is true that the present leasing system of the Federal Government provides for prompt and continuous development with competitive drilling," the board asserted, "the administrative policy of late has fully recognized the advantage to land owner, oil operator and public alike in delaying production in periods of acute overproduction and, as far as the statutes permit discretion, extreme leniency has been the rule in granting applications for relief from the full drilling requirements as set forth in the terms of the lease.

Conservation by Holding Back.

"The desire of far-seeing operators to shut in production until the oil from their wells was needed has been fully met with sympathetic action by the Federal officials. Few, if any, private landlords have been so liberal. The field engineers representing the Department of Interior have also practiced conservation in their enforcement of every possible measure to prevent physical waste and to increase recovery, and in this the cooperation of the lessees has become more and more evident. Federal supervision of this type commends itself in the industry.

"The story of the development of the Federal oil lands is not unlike that of privately owned lands, except that a remedy has been available, that of legislation by Congress.

"But legislation even when free from constitutional questions does not evoke popular support when this purpose is avowedly to put on brakes on development of natural resources. This stumbling block is recognized whether Federal or State legislation is proposed."

Provisions for National Defense.

Dealing with national defense requirements the board continues:

"Under its constitutional power to provide for the common defense, the

Federal Government should continue to make and execute plans for an adequate supply of petroleum for all military and naval needs of the future. Tank storage sufficient to meet initial demand should be built and maintained intact against a wartime emergency.

"Underground reserves should be preserved to supplement the commercial supply as the next line of defense and, in the administration of these reserves of oil in the ground which form 'an important part of the national insurance,' future security not present economically should be the sole guiding principle.

"Current peacetime requirements of those branches of the Government responsible for the national defense are approximately 20,000,000 barrels of petroleum products a year. These requirements are adequately provided for under the present normal rate of production.

"In case of war, the national defense requirements would immediately increase many fold. This larger quantity would include the direct requirements, that is, the products actually used by the agencies of the Government engaged in national defense operations, and the indirect requirements—the amount which will be needed industrially to carry out the munition program or other similar programs of these agencies.

"The production from oil wells within the boundaries of the United States at present is in excess of the estimated maximum requirements for national defense in time of war."

Military Needs May Decline.

"It is barely possible that future discoveries may reduce, or possibly entirely eliminate the need for petroleum fuels in the national defense. It is also conceivable that substitutes for mineral lubricants may be developed on a scale sufficient to meet major requirements. With the development of the Diesel engine and its adaptability to motor vehicle use, the military consumption of petroleum as fuel will be reduced per horsepower."

The report of the Naval Oil Reserve Commission was made available to the Federal Oil Conservation Board, which concurs in general with the conclusions and recommendation of the Reserve Commission. The board believes steps should be taken to put into effect the principal recommendations therein, with such modifications as may be necessary.

"The wartime oil requirements of the navy in any overseas campaign would probably include the major portion of the whole deep-water tonnage under the United States flag," the report says in conclusion. "The increasing use of internal combustion engine drives on commercial carriers makes liquid fuel more and more necessary for wartime water transport. The logistic services of the army and navy and many of its combat weapons, such as tanks, tractor-drawn artillery and airplanes, are dependent upon petroleum products for fuel and lubrication. Should the oil supply accessible to the United States become exhausted and no satisfactory liquid substitute be developed, it would be necessary to resort to coal for propulsion.

"Our entire wartime reserve should not be in the form of refined products placed in tanks, for two reasons. First of all, the future needs of the army and navy for petroleum products may be in a ratio quite different from that of present use, and, in view of the natural tendency of gasoline or even crude petroleum to waste when held in storage, a better policy is considered to be the storage of the higher grades of fuel oil or topped crude, from which the needed products could be derived.

Proposes Underground Reserve.

"Further, it is important that there should be an underground reserve in the event that our commercial supply becomes exhausted before that of other nations. This underground reserve should obviously not be drawn upon unless and until other sources become insufficient."

Declaring the Government, "as the largest land owner is committed to practical conservation of irreplaceable raw materials by the protection of the public estate and the guidance of its development," the report says it will continue its search for "pertinent" facts, and that, while the Federal Government's obligation is "sacred and inescapable," State Governments should promptly study the advantage of cooperative action by land owners and oil operators, looking to sane development of new fields.

The States, says the board, should also consider legislation authorizing cooperative agreements, under proper safeguards, and State legislation "with the declared purpose of conservation should be enacted to stop the waste of gas, the loss of its content of gasoline, and the even greater loss incident to reduction of gas pressure in oil sands."

The President's board recommends "active cooperation" between oil producing States in the study of the proposed legislation, with a view to uniform laws, and says that "even more pronounced" should be the cooperation between State and Federal officers having authority in the regulation of oil and gas production, and the Federal bureaus whose investigative fields are essentially coterminous. Cooperation within the industry in both research and action, as well as field cooperation between neighboring operators, is urged, with cooperation of the industry and the Government in planning and carrying out the necessary research "to make full use of all facilities, resources and personnel available."

September 6, 1926

MORE ELECTRICITY USED.

Average Increase Was 51 Per Cent. Between 1920 and 1925.

WASHINGTON, Dec. 21 (AP).—During the five-year period 1920-25 there was a 51 per cent. increased consumption of electricity in the United States, and in some States, a report by the Geological Survey asserted, the increase ranged higher than 85 per cent.

In actual increase the Middle Atlantic States led with 5,950,000,000 kilowatt hours. The increased production in all States aggregated 22,315,000,000 kilowatt hours.

The percentage increases by sections were as follows:

New England 42, Middle Atlantic 48, East North Central 58, West North Central 34, South Atlantic 58, East South Central 78, West South Central 79, Mountain 21, and Pacific 58.

December 22, 1926

SUPERPOWER LINKS NORTH WITH SOUTH

10,000,000 Horsepower System From Pensacola to Boston Put in Operation.

PROVIDES A DOUBLE CIRCUIT

Hydroelectric and Steam Plants Joined in a 1,000-Mile Triangular Hook-Up.

ASSURES AGAINST TIE-UPS

Philo (O.) Plant, the 'Load Centre,' Cost $17,000,000—Eighteen Companies in Chain.

Electric generating stations with a combined capacity of 10,000,000 horsepower effected loop interconnection yesterday and established a double guarantee against interruption of service when a 132,000-volt transmission line between Kingsport, Tenn., and Saltville, Va., was put into operation, connecting the great hydroelectric resources of the South and the steam-generated power plants of the North. The first connection was completed on Sept. 6, between Roanoke, Va., and Raleigh, N. C.

The gigantic "power pool" is described by the men who have organized it as the world's greatest interconnection of electric generating stations. It does not include the metropolitan district of New York, but it includes other great cities of the country, such as Boston, Philadelphia and Chicago, Atlanta and Birmingham, and will very likely soon include Detroit.

Insures Against Tie-Up.

The American Gas and Electric Company has built the last links in the line which roughly forms a 1,000-mile triangle, from Boston to Chicago to Pensacola, Fla. George N. Tidd, President of the American Gas and Electric Company and the Appalachian Electric Power Company, in discussing the big double circuit hook-up, effected by completion of the Kingsport line, said:

"The completion of this line gives loop, or double circuit, facilities for the transmission of electric power as needed between systems covering the Southeastern States from Tennessee and the Carolinas to the Gulf and the extensive interconnected transmission lines reaching through the Virginias. Ohio and Indiana into Illinois and Michigan and to the northeast through and beyond Pennsylvania.

"In the contingency of an accident affecting a line, service can now be looped around the trouble by way of the other line, and even the line in trouble can be fed from both ends to the vicinity of the interruption. The importance of this new interconnection in the need of industries for reliable electric power is very evident.

"Reliable electric power attracts industry, and industry means population. Interconnection has made power available to hundreds of towns and villages which could never afford to put in a similar supply for themselves Now even the small village with transmission line service is in position to say to prospective industries: 'You can have all the electric power you need.'

"The larger the territory served by interconnected electric systems and the more consumers on the line, the greater the diversity of demand for power and the more efficiently and reliably each individual consumer can be served."

Balancing of Load.

"This latest interconnection makes for a balancing of load between hydroelectric and steam-electric plants which is of great importance to power users. In the event of lack of rainfall, with consequent power shortage in the hydroelectric territories, the steam-electric generating plants of the American Gas and Electric Company and other interconnected systems can, by relaying, be made available to supply the shortage, and at all times of surplus hydroelectric energy available in Southern districts it can be used in the territories of the fuel-burning companies instead of going to waste. This means conservation of coal resources to a degree which cannot readily be calculated at the present time.

"Electrically, the power companies must always be in advance of the communities they serve. They cannot afford to let one catch up or pass them—nor can the cities afford it. The companies must foresee city growth years before it takes place, and must begin raising new capital to carry out plans which will not yield returns for two years or more. When the new industry comes, the power must be ready. The power company cannot say: 'Yes, we will start construction now, and in a year or two your new industry can get power.' On the contrary,, the company must say: 'Yes, we foresaw the growth of this city; two years ago we made our plans for this very contingency, and today your power is ready.' Only by constant vigilance, by constant study, by constant planning, by constant action which can never be taken as quickly by any Governmental body can the electric companies be always ready with power at low cost.

"Interconnection has done more than any other one element to enable us to keep ahead of the growth of the communities without undue cost—and undue cost would reflect itself in impracticably high rates. Cities in our territory are growing more rapidly than the country's average. We feel we are partly responsible for that growth and prosperity."

Power Shifting System.

The generating stations included in the double loop were compared as to methods of operation to the Federal Reserve System by Mr. Tidd, who described them as a "power pool." Through loop interconnection they shift power back and forth over the map to the places where it is needed to meet the stress of peak loads.

The idea was worked out in a small way years ago by Mr. Tidd when he was directing the affairs of a comparatively large company in Scranton, Pa. He was on the point of enlarging the facilities of the plant at great expense when it occurred to him that a connection with the Delaware-Lackawanna plant would obviate the necessity for plant enlargement.

The Lackawanna plant was practically idle after 4 o'clock in the afternoon, when hauling coal from the miner stopped. But at that hour the American Gas and Electric plant began to handle its peak load. Interconnection enabled the mines plant to furnish the additional current needed by Mr. Tidd's plant. In like manner electric energy from the American Gas and Electric plant was shifted during its period of idleness in mid-afternoon to the mines plant, so that it was enabled to meet its peak load requirements without further plant extension.

"That arrangement has continued practically unchanged for fifteen years up to the present," said Mr. Tidd.

Of the 10,000,000 horsepower represented by the double circuit power pool, 6,000,000 horsepower is generated by steam plants and 4,000,000 by hydroelectric plants.

The companies included in the superpower system are as follows: South: Tennessee Power and Light Company, Carolina Power and Light Company, Southern Power Company, Georgia Railway and Power Company, Central Georgia Power Company, Columbia Electric and Power Company, Alabama Power Company, North-Appalachian Power Company, Ohio Power Company, American Gas and Electric Company, Cleveland Electric Illuminating Company, Toledo Edison Company, Commonwealth Edison Company (Chicago). East: West Penn Power Company, Penn Public Utilities System, Niagara Lockport and Ontario Power Company, Mohawk-Hudson Power Company, and the New England Power Association.

One of the plants in the system is the Philo, Ohio, plant of the American Gas and Electric Company, which will have a capacity of 256,000 kilowatts when its new addition is completed in October, 1928. The plant will have cost $17,000,000. Philo was chosen as the location for the largest plant in the system, and its builders describe it as the largest in the world, because Philo is the "load centre" of the system regardless of the fact that it has only 900 inhabitants. According to E. H. McFarland of the American Gas and Electric Company, who supervised construction of the plant, many of the oldest inhabitants of the village are still wondering why the corporation decided upon their town as the site for a $17,000,000 plant, but they are gratified, among other things, because the weekly payroll of the plant's builders amounts to about $15,000, which is spent mostly in Philo.

USE OF ELECTRICITY BIGGEST ON RECORD

Power Consumed Last Month in Manufacturing Plants Establishes New Peak.

VARIOUS GROUPS IMPROVED

Steel, Metal-Working, Rubber, Automotive and Textile Lines Are Among Them.

Consumption of electrical energy for power purposes by manufacturing plants of the country indicates that the rate of operations in February was 6 per cent greater than in January, and 10 per cent higher than in February last year, Electrical World reports. The February rate established new record figures at 0.3 per cent in excess of the previous peak reached in September, last year.

New high rates of productive activity were recorded in five primary industrial groups—rolling mills and steel plants, metal-working plants, rubber products, automobile manufacturing, including parts and accessories, and chemicals and allied products.

The automotive industry rose to new heights in February, with a rate of operations 8.1 per cent over January and 8.6 per cent over February of last year. Rolling mills and steel plants showed a rate of operations that was slightly more than 22 per cent greater than in February, 1928. Metal-working plants registered a gain of 20.5 per cent over the same month last year.

Conditions in the textile industry also improved, according to the index of electrical consumption, the gain over February of last year amounting to approximately 4.5 per cent.

All sections of the country reported a rate of productive activity greater than that of February last year. Increases over last year, by sections, were: New England, 6.1 per cent; Middle Atlantic, 3.2 per cent; North Central, 17.1 per cent; South, 2.2 per cent; the Western States, 12.6 per cent. In New England and the North Central districts the February rate, corrected for the number of working days, was of record proportions.

Manufacturing activity in the United States in February, compared with January this year and February last year, all figures adjusted to twenty-six working days and based on the consumption of electrical energy, with the monthly average of 1923-25 taken as 100, was as follows:

	Feb. 1929.	Jan. 1929.	Feb. 1928.
All industrial groups	140.4	132.5	127.7
Metal industries group	157.3	142.6	130.7
Rolling mills and steel plants	163.3	153.5	133.8
Metal working plants	153.7	135.8	127.6
Leather and its products	102.1	94.3	118.4
Textiles	133.0	129.4	127.5
Lumber and its products	107.6	107.4	113.3
Automobiles and parts	161.5	149.4	148.7
Stone, clay and glass	148.7	137.4	127.3
Paper and pulp	125.2	126.2	124.2
Rubber and its products	154.7	148.2	137.0
Chemicals and allied prod.	138.7	129.2	129.1
Food and kindred products	127.5	128.0	115.6
Shipbuilding	98.5	108.2	98.8

March 17, 1929

POWER FROM SUNLIGHT.

We are constantly hearing of some new accomplishment of the photo-electric cell, more picturesquely known as the "electric eye." It counts the number of vehicles that pass through the Holland Tunnel, levels high-speed elevators automatically at floor stops, tells the engineer of a power house or steamer when too much smoke is issuing from his stacks, sorts cigars, turns electric lights on and off, matches colors, measures the intensity and the variation of the light of stars, judges the turbidity of solutions and makes the transmission of photographs over wires possible. Its uses are legion. Yet its principle is simple. Light falls on a tube lined apparently with silver, but actually with an alkali metal—sodium, caesium, potassium or rubidium. Let light fall upon the lining and electrons fly out. We have a flow of electrons, a feeble current. Light has

actually been converted into electric energy. New types of cells are appearing, a little more efficient, a little simpler than the old. From Berlin, Pittsburgh and Newark come three separate announcements that tubes are unnecessary, and that a plate of metal coated with copper oxide, silver selenide or some other form of selenium gives stronger currents when illuminated. Surely this is not the end.

On high blazes the sun. What if its light were converted into a Niagara of electric energy? The physicists who have given us photo-electric cells suggest guardedly such possibilities. They jot down staggering figures. Some 200 billion horsepower fall on that portion of the earth not too obliquely exposed to the sun's rays, or, in more manageable figures, four million horsepower to the square mile. What if it now takes many million photo-electric cells to light even a

single incandescent lamp? Out of the spark that THALES drew from rubbed amber came radio and television. In the light of electricity's past is it foolish to imagine that we shall have vast areas given over to light-catchers? Only the other day Professor PUPIN told the American Women's Association that we are only on the threshold of a more rational and brilliant utilization of electric energy.

For fifty years we have been turning electric currents into light. At last a hopeful beginning has been made in turning light into electricity. Sunbeams carrying us in queer railway trains, driving airplanes and automobiles, lighting cities and homes, excavating holes to be filled by skyscrapers, cooking meals in millions of homes—why not? It would be no stranger than turning a waterfall into the glow that now illuminates many a Western city.

February 21, 1932

POWER FROM SEVERN TIDES.

At the mouth of the Severn, England, the waters rise and fall 47 feet at the equinoxes. Even the Spring tides have a range of 40 feet and the neap of 22. Why not build a dam to hold back the water and pay it out gradually through turbines which would be coupled with dynamos to generate electric energy? Moon and sun would turn the wheels of British factories.

English engineers have been toying with the idea for generations. One government report after another has examined it from the technical, industrial, navigational and economic angles. The latest, most thorough and authoritative of these examinations, representing six years of work by a

committee of engineers and much experimenting by Professor A. H. GIBSON with large-scale models of the Severn estuary to determine the effect of various types of dam on tidal levels and navigation, has been received tepidly by the British technical press. Not the feasibility of harnessing the tides but the commercial success of an enterprise which in the end may cost not much less than £40,000,000 is questioned.

When it is considered that each of the 2,000 million electrical units made available for the national network would cost 0.2372 pence and that a modern coal-fired station can generate energy at a cost of only 0.3 pence the margin in favor of the Severn project

is small—too small, perhaps, when we must reckon with still lower costs for steam-fired stations in the near future. Contingencies that cannot be foreseen may easily wipe out the slim margin of six-hundredths of a penny per electrical unit that now stands on paper between commercial success and failure if the Severn scheme is ever carried out. It is a pity that economics must thus spoil a romantic picture. There is something captivating in the thought of London lit by the moon as it swims overhead, of 706 tides produced in a year by lunar and solar attraction, of industry geared, as it were, to the solar system.

April 16, 1933

OIL INDUSTRY AIDED BY WIDER DEMANDS

Multiplication of Consumer Services Levels Off Former Seasonal Let-Downs

---

'DEPRESSION PROOF' TREND

Association of Products With Almost All of the Nation's Activities and Enterprises Cited

---

By J. H. CARMICAL

Largely through the increased demand for its products in essential fields, including the greater use of fuel oil for heating homes, the petroleum industry has solved most of its complex problems. Not only have the wide seasonal fluctuations, which in former years characterized the demand for oil products, been eliminated, but the oil industry now is so clearly interwoven with every form of human activity that it may go forward even in a general business recession.

No other important industry is currently considered so "depression proof" as the oil group. Despite the drastic decline in general business starting in the Fall of 1929, the demand for all oil products in 1930 and 1931 was larger than it was in 1929. It was only in the depth of the depression in 1932 and 1933 that the demand for petroleum products showed any serious recession. The constantly expanding demand for oil products in essential fields since then has placed the petroleum industry in a position where the return of conditions similar to those of 1932 and 1933 probably would not seriously disturb it.

The most influential factor in the progress of any industry is the stability of operations. Because of the former drastic decline in the use of oil products in Winter, the oil industry did not enjoy such stability until in the last year or two. With the development of oil burners for homes, the construction of better roads, the greater use of trucks and buses and the longer motoring season, the demand for oil products from November to May in the last few years has been larger than formerly.

New Trend in Demand

An analysis of the statistics of gasoline and fuel oil consumption discloses that, beginning with the Winter of 1934-35, the demand for these products has been in excess of that in the preceding Summer. From 1929 to that period the demand for these products showed declines ranging up to 11.5 per cent.

A detailed study of the figures in the following table supports the conclusion that the seasonal variations in the demand for these two major oil products have been eliminated.

INDICATED DOMESTIC CONSUMPTION
(In Thousands of Barrels)

Period.		Motor Fuel.	Fuel Oil.	Total Motor Fuel and Fuel Oil.
May 1929-Oct.	1929	213,792	204,523	418,315
Nov. 1929-Apr.	1930	178,037	204,080	382,117
May 1930-Oct.	1930	218,851	176,192	395,043
Nov. 1930-Apr.	1931	176,735	180,089	356,824
May 1931-Oct.	1931	229,492	160,990	390,482
Nov. 1931-Apr.	1932	175,755	170,892	346,647
May 1932-Oct.	1932	206,438	136,358	342,796
Nov. 1932-Apr.	1933	166,278	165,270	331,548
May 1933-Oct.	1933	212,439	152,809	365,248
Nov. 1933-Apr.	1934	178,930	183,541	362,471
May 1934-Oct.	1934	225,029	152,416	377,445
Nov. 1934-Apr.	1935	189,170	191,208	380,378
May 1935-Oct.	1935	241,627	168,171	409,798
Nov. 1935-Apr.	1936	204,213	214,427	418,640
May 1936-Oct.	1936	267,955	187,922	455,977
Nov. 1936-May	1937	228,978	235,655	464,685

With the possible exception of coal, which may be used instead of oil for heating purposes and generating steam, there is no substitute for oil products. In fact, at current levels for coal and oil, there is virtually no economic substitute for oil. With more than 1,000,000 homes equipped for burning light fuel oil, with the increased demand for oil in the airplane and shipping industries and with the trend toward the greater use of the Diesel engine in several directions, when these are considered, along with other essential consumers of oil products, it can be seen readily that the oil industry comes nearer to being independent of general business conditions than any other major industry.

Since it is generally known how many installations there are consuming oil in one form or another, the industry is able to estimate consumption with a fair degree of accuracy for months ahead. While it is admitted that general business conditions may recede to a level where oil consumption will be reduced somewhat, the fact that the trend in the demand is steadily upward, makes almost as good as certain that consumption will not decline below the preceding year, unless there should be a general stagnation in all industrial activity.

Auto Registrations Turning

In the decade from 1920 to 1930, automobile registrations in the United States increased from 9,-231,941 to 26,545,000, a gain of 185 per cent. For three years in the depression, registrations declined and despite the increased buying of motor cars in 1934 and 1935, there actually were 250,000 fewer automobiles registered at the beginning of 1936 than in 1930. However, in 1936, automobile registrations increased by about 2,000,000 and a further increase of about 1,800,000 or about 7 per cent, is expected this year. This will lift total motor registrations in the United States at the close of this year to a new peak of about 30,000,000.

Granting that gasoline consumption might be moderately less for each automobile next year if the recession in business continues, it is estimated that the increase will amount to at least 5 per cent. For the current year, domestic motor fuel consumption is placed at roughly 525,000,000 barrels. An increase of 5 per cent would mean the consumption of an additional 26,-250,000 barrels next year. Provided gasoline exports approximate the 35,000,000 barrels shipped abroad this year, the total demand for motor fuel in 1938 will be about 586,000,000 barrels.

In view of the fact that the demand for certain oil products, such as heating oils probably will show a larger gain, it is generally held safe to assume that the demand for all oil products will increase by 5 per cent in the next twelve months. However, in order to supply the gasoline demand, it will require about 56,000,000 additional barrels of crude oil since it takes about 2.15 barrels of crude to make one barrel of gasoline. In other words, in order to maintain stocks of all oil products at current levels crude-oil production next year must average 126,000 barrels daily more than for 1937.

In the week of Aug. 28, this year, crude oil production in the United States established a record at 3,735,-000 barrels daily. Total production of crude oil this year also will establish a record.

However, with the expected increase in the demand for oil products materializing next year, it is likely that crude oil output for some weeks during the Summer months next year will reach a figure of around 4,000,000 barrels daily.

The following table shows the indicated domestic demand for motor fuel, residual fuel oil, gas and heating oils and the combined demand for these products from 1926 through this year. The figures are in thousands of barrels:

Year.	Motor fuel in bbls	Residual fuel oil, gas oil and heating oil in bbls	Total in bbls
1926	267,128	339,572	606,700
1927	305,367	339,265	644,632
1928	338,881	383,974	722,855
1929	382,878	415,156	798,034
1930	397,770	368,531	766,301
1931	407,843	334,668	742,511
1932	377,791	308,157	685,942
1933	380,494	323,705	704,199
1934	410,339	340,371	754,710
1935	434,897	365,514	800,411
1936	481,591	408,991	890,582
1937 (1st 9 mos)	391,276	324,660	715,936
1937 (estimated)	525,000	450,000	975,000

Although stocks of crude oil in storage at the end of this year will be about 15,000,000 barrels larger than on Dec. 31, 1936, they are not considered to be any larger than is needed for refineries to carry on economical operations. From a level of 428,446,000 barrels at the close of 1929, they had declined to 288,184,000 barrels at the end of 1936. Currently, they are around 304,000,000 barrels, but with the shut-down of the East Texas field for four consecutive Sundays, which started on Nov. 21, and the pinching-back of production in other areas, it is estimated that there will be a further drop in crude oil stocks before the close of this year.

With the exception of gasoline, the statistical position of all other oil products is considered sound. Gasoline stocks are about 10,000,000 barrels more than at this time last year. In only one year have gasoline stocks shown such an increase and that was in 1929 when they were increased by 10,129,000 barrels. However, when taken in connection with the increased demand for gasoline, total motor fuel stocks are equivalent to only about forty-five days' requirements.

Although prices of other oil products have been showing strength in recent weeks, there has been a definite weakness in gasoline prices. The weakness has not spread to crude oil and the belief in the industry is that it will not, provided the State regulatory authorities do not permit a gross overproduction of that commodity. The weakness is due largely to the fact that, with anti-trust proceeding having been instituted against a large number of the major companies for purchasing jointly gasoline from the independent refineries, the purchase of "homeless" gasoline from the small refiners by the large companies has virtually ceased. As a result, profits from refineries largely making gasoline have been materially reduced, and in some cases it is non-existent. However, the situation is expected to adjust itself, probably with the return of the seasonal increase in the demand for gasoline next Spring.

The oil industry is closing the most satisfactory year, from an earnings standpoint, since 1929. For the first nine months of this year the companies that issue interim reports had an increase in net income of about 47 per cent over 1936. The increase for the final quarter is not expected to be so large as that for the first three quarters, but it is estimated that the industry as a whole for the full year will show an earnings increase of about 40 per cent over 1936.

December 5, 1937

NATURAL GAS RULES MADE

Power Commission Calls for Investigation and Reports

Special to THE NEW YORK TIMES.

WASHINGTON, July 7.—Three important orders affecting the natural gas industry were issued today by the Federal Power Commission as it put into effect the administration of the Natural Gas Act approved on June 21.

Under the law authorizing regulation of transportation and sale of natural gas in interstate commerce, the commission's orders institute an investigation of natural gas companies and direct filing of reports, promulgate and prescribe rules of practice and instruct the gas companies to file schedules of rates and charges and contracts and agreements for sale and transportation of gas.

Schedules for deliveries in Delaware, Maryland, New Jersey, New York and Pennsylvania are to be filed by Sept. 6. July 8, 1938

ELECTRIC CAPACITY 24% ABOVE DEMAND

Nation's Supplies of Power Found Ample for War Work and Civilian Use

NEW RECORDS BEING MADE

Seasonal Peak of Needs Near —Anticipated by Additions to Generating Plants

By THOMAS P. SWIFT

America's electric power and light industry is one of the few that is having little difficulty in meeting all demands for war production and civilian requirements as well. Since Pearl Harbor there has not been a single instance were a major demand for power by a war industry has not been met, and as our second year of war draws to a close the nation's utili-

ties, both private and public, find themselves in the enviable position of having a 24 per cent reserve capacity.

According to the latest available figures, the peak demand for electric power by all of the nation's consumers aggregated 39,500,000 kilowatts in October. The total installed generating capacity to meet this demand, exclusive of the numerous industrial plants having their own power facilities, totaled 49,000,000 kilowatts, resulting in a "spare" generating capacity of 9,500,000 kilowatts, or 24 per cent of the peak demand. This country's installed power capacity is now well in excess of the generating capacities of the Axis countries combined.

Peak Demand Approaching

In the next three to four weeks, as is customary with the industry, the nation's utilities will reach the point of maximum demand for essential war and civilian kilowatts. For the current month, November, the peak, it is estimated, will run between 40,000,000 and 40,500,000 kilowatts. The December peak demand, or the yearly high record, will be approximately 41,000,000 kilowatts, according to the industry's leading engineers.

By the end of the year, however, an additional 1,000,000 kilowatts of generating capacity are scheduled to be in operation, thus bringing the country's installed capacity contributing to the public supply to an even 50,000,000 kilowatts.

This would leave a "spare" capacity of some 9,000,000 kilowatts over peak demand for the month, or about 22 per cent—a comfortable margin when measured against the damaging effects of the last year's bombing raids over Germany and the occupied countries of Europe. The additional 1,000,000 kilowatts of capacity scheduled for operation before the end of the year will come from the completion of numerous power plants, both public and private, throughout the country.

On a kilowatt-hour output basis, the industry, according to statistics released by the Edison Electric Institute, turned out 4,482,665,000 in the week ended Nov. 13—the official highest thus far on record. For the week ended yesterday, preliminary estimates place output at 4,500,000,000 kilowatt-hours, or 19 per cent over the same week a year ago. With the high peak of the year expected for the week starting Dec. 12, the industry is anticipating an output of approximately 4,600,000,000 kilowatt-hours, or slightly higher depending on weather conditions.

Distribution of Power

The manufacture of planes, ships, tanks, guns and materials directly related to the war effort currently utilizes about 70 per cent of the nation's industrial power output, which, in turn, accounts for approximately 60 per cent of the total electric sales of the utilities. The remaining 40 per cent is taken by householders, commer-

cial establishments and miscellaneous users, such as street traction properties and municipal street lighting.

One of the largest consumers of electric power in the nation is the electro-metallurigical industry. It is estimated that this industry will have used some 35,000,000,000 kilowatt-hours this year, contrasted with only 10,000,000,000 kilowatt-hours in 1940. The aluminum, magnesium, electric steel and alloy manufacturers, going full blast on war orders, constitute the largest individual class of power users in the country.

While frequently in the last five years Federal power officials have predicted that power shortages were in the offing, nothing has happened to date to bear out such forebodings. The current power conservation program is in no way related to a shortage of power-generating facilities.

In formulating the recent "brown-out," or conservation program, J. A. Krug, head of the Government's Office of War Utilities, said reassuringly:

"In the electric utility industry the installed generating capacity, together with capacity now under construction, is ample to meet all foreseeable electric needs. It is essential, however, to save the use of electricity wherever possible so as to reduce directly or indirectly the demands for materials, fuel, transportation and manpower."

November 21, 1943

EARLY USE OF ATOM AS FUEL PREDICTED

Duc de Broglie, French Expert, Sees Energy by Splitting Particles 'Fairly Soon'

By HAROLD CALLENDER
By Wireless to THE NEW YORK TIMES.

PARIS, Aug. 7—The fabulous energy released by the invention of the atomic bomb seems destined "fairly soon" to replace power produced by coal, oil and water, the Duc de Broglie, eminent French physicist, said today in commenting on news of the bomb, which dazed and fascinated Paris today.

The Duke added that naturally this was far more important than the war or the defeat of Germany. He thought that it could be compared only with the discovery of fire by primitive man.

He estimated that the energy obtainable from a disintegrated atom would be about 200,000,000 times as great as that in the most powerful explosive hitherto known. The Paris newspapers seemed

uncertain today whether the announcement of the new missile exceeded the Pétain trial in importance. Some thought not; others quickly began hunting experts to tell them whether President Truman's statement on the bomb could really be true. The answer given was yes and that Mr. Truman did not exaggerate for propaganda purposes.

At the Foreign Office, where they had long been racking their brains over the future of the Ruhr, the Saar and the Rhineland, officials concluded that they might as well go fishing, too. Why acquire gray hairs over the Ruhr and the Saar if "fairly soon" coal is to become useless and abundant heat and energy are to be derived from systematically disintegrated atoms?

A few weeks ago Gen. Charles de Gaulle said that it was necessary that the Rhine, "from one end to the other," should be in French hands, and he claimed Cologne as part of the French zone of occupation on the ground that invasions had started there. It is now asked, however, whether it is worth while to make a fuss about the Rhine as a barrier against an enemy crawling along the ground when future wars, if any, will be fought with atomic bombs.

In official offices are men who

have spent years studying the relations between oil fields and the power of nations, all lately hunting feverishly for new oil supplies. They asked themselves today whether the time had come to roll up their geological maps and to revise many a policy since the quest of oil seemed to become in the future purposeless.

If a few atoms could be pulverized between now and winter and the released energy distributed in heat, France would need no longer to worry about getting 6,000,000 tons of coal from abroad.

"The invention of the atomic bomb is a great event for the destiny of the world, an event that in augmenting immensely the power of man also terribly increases his responsibility for the use he makes of it," he said.

"Twenty-five years ago a kind of super-chemistry was born, a chemistry of artificial transmutations which prolong and generalize radioactivity. This new chemistry concentrated on not groups of atoms, as did its predecessor, but the grouping of infinitely small bodies that form the nuclei of atoms.

"Invulnerable to chemical reactions, these bodies correspond to Lavoisier's conception of simple

bodies 150 years ago, which still remains true except when intervention takes place by methods that twentieth-century science has taught.

"Today we can penetrate inside the atom. This operation exacts and can liberate amounts of energy 200,000,000 times as great as that of the most powerful explosives. Yet we could so proceed only with atoms isolated in experiments in the most delicate laboratories. Only recently we did not know how to set in action in this way a large quantity of matter though we foresaw that the peculiar properties of the atoms of the heaviest elements, like uranium, would soon make this possible. Now we know that this final obstacle is overcome.

"It is too early to measure all the consequences. But here is a source of energy for all uses. It will be the motive force of tomorrow. Plans for machines to use this force have already been patented. In industry it will bring a fabulous revolution.

"Doubtless we shall no longer have to seek far and wide the raw materials coveted by nations. For now the new alchemists will have tools for the transmutation of materials not only theoretically but practically."

August 8, 1945

MIDWEST STATES

Use of Electric Power by Farmers Is Increased

By HUGH A. FOGARTY
Special to THE NEW YORK TIMES.

OMAHA, May 17—Despite the closing of many war industries, consumption of electric power has risen steadily throughout the Midwest in the last two years.

And production men who must plan for the future are gambling that the trend will continue upward—sharply and almost indefinitely.

Reasons for the increase are numerous. But among the most important must be listed the steady growth of power use on the farms —for pump irrigation systems, for the home and even for the barns and machine shops.

The experience of Nebraska is typical. It holds additional interest because the state is unique in the fact that all its power production facilities are publicly owned. During 1946 the state used 1,200,000,-000 kilowatt hours of electricity. This was a 10 per cent increase since the end of the war. The amount was five times what was consumed in 1920.

Even so, the number of farms in the state still without the benefit of electricity can be counted in the thousands. With production capacity nearing the maximum the three Nebraska hydro-electric districts have banded together to build a $10,000,000 steam electric plant on the Missouri River at Bellevue, a few miles south of Omaha.

This plant, expected to be ready late in 1948, will supply additional power for the Rural Electrification Administration's branches and for the Omaha public power district.

The story is largely the same throughout the Midwest. For that reason the billion-dollar-plus program envisioned under the Pick-Sloan plan calls for power production as an important by-product of the reservoirs planned to fill up behind 105 dams in the ten Missouri River basin states. In addition to providing flood control and dependable navigation the Federal-state program calls for producing ten billion kilowatt hours of electricity annually. This would be roughly comparable to the yearly output of the Tennessee Valley Authority.

That is a long-range program, however, and is subject to curtailment if Congress goes all the way toward rigid economy.

An important factor in the planning of the Midwest is a belief that additional power will attract additional industry and make the basin boom with new activity.

News Notes: Farmers have been making giant strides between frequent showers, many working around the clock to get corn into the ground. . . . Supreme Court action in upholding an ICC order granting more favorable rail freight rates to the South and West was greeted by one Midwest observer as "the best news in forty years."

May 18, 1947

NEW RECORDS SET FOR OIL INDUSTRY

Nation Becomes Net Importer for First Time—New Fields Sought in Hemisphere

The American petroleum industry has just closed its most momentous year.

Records in every division were shattered and production and consumption both established new highs. Capital expenditures were at a new peak, and earnings, reflecting the increased demand and the higher prices for both crude oil and refined products, were at new high levels.

The United States again became a importer for the first time since the early Nineteen Twenties when the Mexican fields were at their peak. Imports exceeded exports by an average of nearly 100,000 barrels daily. However, the nation's production was more than sufficient to meet the demand. The excess of imports over exports were added to stocks. Oil from the Middle East also made its appearance and at the year-end was being brought in at a rate of about 100,000 barrels daily.

Crude oil production in the United States for 1948 averaged about 5,500,000 barrels daily, an increase of almost 8 per cent above that in 1947. Natural gasoline recovered from gas production averaged about 400,000 barrels daily, making the total domestic output 5,900,000 daily.

The demand for petroleum products for domestic use was about 7 per cent higher than in 1947, or approximately domestic production for the year. Exports, however, were down about 50,000 barrels daily from the average of 450,411 daily in 1947. During the year inventories were increased at an average of about 235,000 barrels daily, resulting largely from the excess of imports over exports.

There were comparatively few price adjustments during the year. Crude oil prices held unchanged until the Phillips Petroleum Company, on Sept. 28, announced an increase of 35 cents a barrel, to $3, in the price it would pay in the fields in the Southwest. On Nov. 24 the Sinclair Oil Corporation met the advance but no other important purchasing companies made any change. Near the year-end, both companies rescinded their price advances.

With stocks of fuel oil becoming rather unwieldy, a reduction of 20 to 25 cents a barrel was made in heavy fuel oil on Dec. 1 by the leading companies on the East and Gulf Coasts. This was followed on Dec. 22 by an additional cut of 25 cents a barrel. A slight reduction in the price of heating oil also was made on the Atlantic seaboard. Gasoline prices at refineries also were shaded a bit. For the first time since the close of the war the oil industry was interested in getting new business and was making price concessions as an inducement.

Despite a slight excess of supply last year over the demand, there was little chance that the price of oil products will suffer any sharp decline. The demand continues to expand and indications are that the oil companies will have to expand their facilities in order to meet the requirements for oil products. An increase of 5 per cent in demand is looked for this year, which, if it materializes, will mean that an additional 300,000 barrels a day of petroleum products must be supplied.

Since the war the American oil industry has been engaged in its greatest expansion program in history. Capital expenditures last year were estimated at more than $2,000,000,000, a record, but only slightly in excess of such expenditures in 1947. For the three years 1946 to 1948 the total was about $5,500,000,000.

When it is realized that these expenditures have amounted to more than twice the average for the war years and about three times those for the five years immediately preceding, the financing problem that the industry has encountered becomes apparent. Financing through the sale of stock was out of the question because of the condition of the market. Consequently, the industry had to rely on loans from banks and insurance companies and from earnings.

Profits Rise Steadily

During the period, however, profits from operations rose steadily. Last year, only about 24 per cent of the net earnings of the industry were passed on to stockholders in the form of cash dividends, compared with about 35 per cent in 1947 and 45 per cent in 1946. In the pre-war years, the industry usually passed on to stockholders about 60 per cent of the net earnings.

Since such a large percentage of earnings were being taken for reinvestment in the business, many companies, particularly those in the major group, adopted the policy of supplementing their cash disbursements with stock dividends. The value of the stock dividend, plus cash disbursement, generally was not in excess of the estimated earnings of the particular company for the full year.

Such a method of financing the expansion program has its benefits. The stock dividend was subjected only to the capital gains tax in the event the recipient wished to sell the additional shares. To holders of large blocks, this was much less than income tax payments. In addition, holders could choose the time at which they wanted to convert the dividend shares into cash.

The efforts of the oil industry since the war, plus the large discoveries in the Middle East, have resulted in the United States being assured of adequate supplies of petroleum products in peacetime for many decades. However, in the event of a war in the near future, with the United States dependent only upon the petroleum that can be produced in the Western Hemisphere, there probably would be great difficulty in meeting both military requirements and essential civilian needs.

The oil industry and defense authorities in Washington are aware of this situation and it is quite likely that, in cooperation with the Federal Government, a determined effort will be made this year to solve the problem. Such a step would involve first an arrangement with the Latin-American countries by which private American industry could develop further oil resources.

Some of the Latin American countries already have seized the petroleum properties of United States interests and have excluded United States nationals from operating within their borders. These countries are now requesting loans from the United States Government in order to extend their operations of the seized properties. Without having the "know-how," in the operation of petroleum properties, oil officials say, the granting of such requests might not result in the proper development of such properties.

Oil requirements in another war may amount to 3,000,000 barrels a day, or about double the peak in the last war. The United States Government is interested in developing in the Western Hemisphere an output of at least 2,000,-000 barrels daily in excess of current demand. With such an excess supply, it is estimated that by rationing, the necessary oil products would be made available for any emergency.

The development of an excess supply of 2,000,000 barrels a day in this country virtually is out of the question. At present there is no excess production. To meet expanding consumption over the next five years is about all that can be hoped from the domestic industry. Thus in order to be sure of adequate supplies accessible in another war, the United States must rely on Latin America.

January 3, 1949

Expansion in Synthetic Fuels Called Vital to U. S. by Krug

Interior Secretary Says Rapid Development of New Resources and Processes to Fight Depletion Is Necessary for Security

By JAY WALZ
Special to The New York Times.

WASHINGTON, March 27—National security and the country's standard of living depend on an all-out program of resource development, especially in the fields of synthetic fuels and water power, Julius A. Krug, Secretary of the Interior, declared today.

Urging rapid development of synthetic liquid fuels from coal, and of extraction of fuels from oil-bearing shale, the Secretary said the United States would face a "transportation collapse" should our foreign supply of oil be shut off. He recommended the development be conducted by private industry with "Government encouragement."

As for water power, Mr. Krug advocated a twenty-year development program costing up to $15,-000,000,000, including the St. Lawrence power and seaway project.

The Secretary's plea for resource discovery and expansion, coupled with most rigid conservation of all existing natural resources, came in an annual report to the President.

"Unless future generations are to face a declining standard of living, we must rely more heavily on our inexhaustible supplies, stop the waste of irreplaceable materials and find and develop additional resources as fast as possible," Mr. Krug told President Truman.

While the report laid special stress on petroleum and water development, it also recommended restrictions on the use of such scarce basic minerals as copper, lead and zinc "to protect the nation's economy from the effects of critical shortages."

Petroleum, however, was the most critical energy source problem because "it powers the nation's transport system and is the key to national defense to a far greater extent than in any other nation," said the Secretary.

He reported the country had an estimated 21,000,000,000 barrels of "proved" crude oil reserve, while unproven reserves might be four times as great.

Oil Supplies a Problem

"Even if this were true, we have only about two generations of domestic crude oil supply left," he continued. "It is time for us to move with the greatest possible speed to develop alternate supplies.

"It may take us fully ten years to develop them sufficiently to keep our automobiles, airplanes and trains moving in a future continuing emergency, not to mention keeping many factories running."

Mr. Krug noted that the country was using its scarcest fuel, oil, "more liberally" than the more plentiful fuels, coal and water.

Department geologists reported that while petroleum production now was more than five and one-half times what it was at the end of the first World War, "still the demand cannot be fully met." Even with the extensive research to produce substitute liquid fuels, our needs for some time to come would have to be met by oil from the ground, it was noted.

Conservation Advocated

Mr. Krug said the United States could not count on development of atomic or solar energy in the foreseeable future to provide the vast power needed to keep the country's transport rolling, and added it was "dangerous" to depend on foreign sources of oil, "particularly half-way around the world."

He recommended that oil-producing states adopt conservation practices, and that oil reserves in submerged lands off the continental mainland be explored.

The Secretary's twenty-year water development program set a goal of at least 40,000,000 kilowatts of hydro-electric power. He said the Federal Government should build facilities to produce 30,000,000 of this total at a cost running from $12,000,000,000 to $15,000,000,000.

He estimated the combined cost of the St. Lawrence Seaway and Power Development at $966,678,-000, of which the United States' share would be $605,203,000 and Canada's $361,475,000. The projected output is 6,500,000 kilowatt hours.

The St. Lawrence project, Mr. Krug said, was needed not only for power, but also "to bring the newly important iron ore from Labrador and South America to American steel plants."

The power and seaway projects, rejected by a number of previous Congresses, are under consideration again by a Senate committee.

While stating that hydro-electric power production and irrigation projects in the West must be improved, the Secretary reported that in the last year the Bureau of Reclamation let 1,400 contracts, totaling nearly $160,000,000, in construction work. He said farmers supplied with irrigation water from Reclamation works produced crops valued at $555,000,000, and that 15,000,000,000 kilowatt hours of electric energy were sold from Reclamation power projects.

March 28, 1949

GIANT WINDMILLS FOR POWER URGED

P. H. Thomas at U. N. Parley Says Practical Designs for Generators Are Ready

By ROBERT K. PLUMB
Special to The New York Times.

LAKE SUCCESS, Aug. 31—The construction of giant windmills to produce cheap electric power was urged at the United Nations Scientific Conference on the Conservation and Utilization of Resources here today.

Percy H. Thomas, a member of the Federal Power Commission, said practical designs for such "areogenerators" had been completed. A machine capable of producing 7,500 kilowatts, according to the plans submitted, would be 475 feet high, with two rotors each 200 feet in diameter, mounted on a vast turntable.

Mr. Thomas submitted this proposal to a meeting of experts concerned with solving the engineering problem of providing increasing quantities of power in the face of expected shortages of high-quality fuel. French and Dutch scientists presented similar plans for harnessing the wind.

Would Combine Generators

The United States speaker recommended that aerogenerators be used in conjunction with conventional electric generating stations, hydro and heat power, to provide usable energy on a dependable basis. When the wind is blowing, the wind turbines can take the load off present generators. Wind energy can be saved up, he asserted, by pumping water into local reservoirs and withdrawing it on calm days as it is needed through water generators.

Mr. Thomas emphasized that he was presenting a realistic view of how to obtain power from the wind; that the practical problems had been solved and that many areas now "may look hopefully to this source of low-cost energy" as soon as machines had been erected.

"The statement that wind energy is now available for electric utility service is intended to be taken literally," he commented, "the only reservation being the allowance of a reasonable amount of time and of a sufficient preliminary design effort to produce the first full-sized physical structures.

"The appropriate design procedures, both aerodynamic and constructional, are now known. There remains, however, for designing and operation engineers the inevitable educational state so characteristic of all new developments. No untried types of structures of mysterious principles are involved."

A pilot machine on a small Vermont mountain known as "Grandpa's Knob" has provided essential engineering data for future aerogenerator design, Mr. Thomas said. The Vermont turbine was built in 1941 and ran continuously for three weeks before mechanical failure forced a shutdown. Necessary parts could not be replaced during the war, he added.

Cost Is Estimated

Mr. Thomas estimated that the installation of a giant wind machine would cost somewhat less than an inexpensive steam plant of comparable capacity and that labor and maintenance costs would also be less.

Studies of prevailing winds indicate that locations favorable for the erection of aerogenerators are world-wide, he said. If the wind machine is to be tied in with a hydroelectric system to "firm" the supply, however, the number of possible sites may be smaller.

There are industries that can use electric power in large quantities without requiring a steady supply, he added, and many nations that are handicapped "in the vital matter of low-cost power." The economic justification of the aerogenerator, he declared, is in the simplicity of installation and operation and the savings in fuel transportation and storage costs. And it utilizes a resource not now touched, Mr. Thomas reported.

September 1, 1949

COAL IS POOR THIRD IN FUEL FIELD RACE

Petroleum Imports Ruled Out as Factor, With Domestic Oil, Natural Gas Held to Blame

While a West Virginia coal producer, testifying before a Senate Labor subcommittee hearing in Washington, blamed the loss of tonnage on oil imports, he failed to mention inroads made by natural gas.

R. M. Davis, of Morgantown, W. Va., told the committee that 200,-000 miners might be out of work within a year-and-a-half if foreign oil imports continue at their present rate. While Mr. Davis was testifying, one of the major gas companies in the nation, the Columbia Gas System, was busy in his home state drilling wells from which millions of cubic feet of natural gas are flowing throughout a seven-state area. In many of the communities to which the fuel is being sent it is displacing bituminous coal at a rate that would seem to minimize the effect of oil imports, regardless of whether they carry a tariff.

Crum Development Cited

Significant of the national picture in the fast growing gas industry, was the dramatic occurrence at Crum, W. Va., recently. At 11 o'clock in the morning of May 19, the United Fuel Gas Company, a subsidiary of Columbia Gas, shot a 5,733-pound-time bomb into a 2,950-foot well to start a million cubic feet of natural gas flowing into the pipelines of the world's largest integrated natural gas distribution system. Similar wells are being shot throughout the Southwest in an effort to supply a steadily increasing demand for this fuel. Columbia, alone, serves about a million homes and industries directly. In addition, it sells gas wholesale to independent utilities who in turn serve another million customers through 7,000 miles of pipe in the mountain state, Pennsylvania, Maryland, New York, Ohio and Washington, D. C.

While West Virginia is first in the production of soft coal, industry and investors in that State are joining the rest of the nation watching the balance of popularity between coal and gas swaying toward the latter.

Since 1946, for example, the use of soft coal has declined from 48 to 38 per cent last year. In that same period, the use of natural gas has risen from thirteen to eighteen per cent, according to the United States Bureau of Mines.

Another significant development recently was the signing of what probably was the largest single natural gas supply contract. Texas Eastern Transmission Corporation has completed a contract for the purchase from the United Gas Pipe Line Company of 134,000,-000,000 cubic feet of gas a year for delivery to the Algonquin Gas Transmission Company. This latter company is now before the Federal Power Commission for permission to deliver gas to New England.

264,000-Mile Network

More than 264,000 miles of gas pipe link the abundant gas fields of the Southwest with thirty-five states that are using the fuel. About 7,400 miles of line were approved during 1949, and about 12,500 miles of line will be added if the F. P. C. acts favorably on applications before it. Leaders of the gas industry foresee a national web of pipelines by 1955, before which, they estimate, there will be about 300,000 miles of pipe carrying gas. The pipeline division of the $6,000,-000,000 gas industry represents about $2,500,000,000 of investment in the dozen or so major natural gas transmission systems.

Rather than signs of depletion, the gas industry today has more reserves than it ever had. Latest estimates show that every major city in the nation probably will have natural gas available by 1954. As to reserves, H. Leigh Whitelaw, managing director of the Gas Appliance Manufacturing Association, said recently a study of Federal Power Commission data and information received from the American Gas Association showed proved recoverable reserves of natural gas in excess of 180 trillion cubic feet at the end of 1949; the figure on Dec. 31, 1948 was about 175 trillion cubic feet. In 1919, it was 15 trillion cubic feet.

To refute the contention of Mr. Davis, one of the nation's top oil industry men, Eugene Holman, president of the Standard Oil Company (New Jersey), made the statement that coal has lost a very small fraction of its business to imported oil. He said the greatest loss by the coal industry last year was to industrial users other than electric power utilities and railroads, a drop from 247,000,000 tons in 1948 to 191,000,000.

In the meantime, while the coal industry continues to lose customers not only to the oil industry, but to gas pipelines and gas utilities as well, Senate and House committees ponder bills introduced to help the coal industry. The likelihood of any action this session, according to reliable Washington sources, is very slight.

June 5, 1950

ELECTRICITY MADE BY ATOMIC REACTOR

Heat Removed by Liquid Metal Gives Power for Lights and Pumps of Building in Idaho

Special to THE NEW YORK TIMES.

IDAHO FALLS, Idaho, Dec. 29 —Scientists have produced useful electric power from atomic energy for the first time, spokesmen for the Atomic Energy Commission announced here today.

Heat energy was removed from a breeder reactor by a liquid metal of a type not revealed and this energy produced enough steam pressure to drive a turbine.

The turbine, in turn, generated more than 100 kilowatts of power, which supplied a lighting system and operated pumps and other equipment.

This successful experiment, which wrote a new chapter in the history of the atomic age, took place in a modest brick and concrete building on the Snake River plains near Arco in southeastern Idaho.

The initial demonstration took place the night of Dec. 21-22 when, at the site of the National Reactor Testing Station, operated at Arco by the Argonne National Laboratory of Chicago, an engineer threw a switch while anxious colleagues watched the results.

There was restrained excitement but no handshaking nor backslapping as the ten or a dozen men observed this manual operation that cut in on the "house circuit" at the station and brought about the operation of the entire equipment of the station by electric power generated from atomic energy.

The first test lasted more than an hour. The reactor and equipment in its building, including lights, pumps and other devices, were operated by a normal load of 100 kilowatts of power fed through the circuit. This means that more than 100 kilowatt hours of electricity were used during that period.

Tests Will Be Continued

H. V. Lichtenberger, the project engineer for Argonne, threw the switch. Dr. Walter H. Zinn, director of the Argonne Laboratory, and other members of the Argonne staff were participants or observers at the momentous occasion.

A spokesman for the Atomic Energy Commission's Idaho operations, with headquarters here, said that the Argonne staff members had worked through most of the day and that by the time they were ready to shift the energy production to the house circuit they regarded it "pretty well as a routine sort of thing."

Mr. Lichtenberger agreed that what emotions were present "were fairly well suppressed."

"We knew, or thought we knew, what was going to happen," he related, "and we were too worried about such things as circuit breakers to show much excitement."

No specified time has been set for the test, he added, and it was late before everything was ready. He would not hazard a guess as to the next step in application of the findings but said that tests would be continued after the first of 1952. A logical step, he suggested, would be experimentation with more power and a bigger unit.

A commission spokesman said that adjustments were being made in the set-up at present with a view to resuming the tests.

"The whole purpose of the power take-off," he went on, "is first, to obtain information about handling liquid metals at high temperatures under radioactive conditions, and, second, getting experimental data on extracting heat from the reactor in a useful manner."

Station Is One of Four Projects

"This is not a production job. The main job of the reactor testing station is to explore the feasibility of obtaining more fissionable material from the reactor than is put into it. The task is to prove or disprove the concept that atomic fuel may be 'bred' in the reactor.

"We are very well pleased and happy over the result of this test, since this is the first reactor constructed at the Arco project."

The spokesman said the identity of the liquid metal used to remove heat energy from the reactor, which in turn generated steam pressure to drive a turbine that generated the useful power was "classified information" that could not be made public. In a similar category, he said, were several other items relating to the test, such as cost and efficiency factors.

The reactor testing station, almost hidden amid a 700-square-mile sagebrush tract of southeastern Idaho, is one of four projects under supervision of the Idaho operations office of the commission, directed by L. E. (Bill) Johnston. The station was turned over to the Argonne Laboratory on April 10. The reactor was installed and the build-up of the project has been "gradual." Dr. Zinn has been a frequent visitor from Chicago during the period.

An announcement in September last year that Chicago scientists had developed a practical method for the direct conversion of atomic energy into electricity through use of a "neutron thermometer" was called wholly untrue by the Atomic Energy Commission, which said the device never had been and never could be used for the production of useful amounts of electricity.

Great Hopes Held for Use

The scientists attach great hopes for peace-time development of atomic power to the experimental breeder oven. This is because it is essential, they say, to find a way to create new nuclear fuel faster than it is consumed.

For example, in the production of the atomic fuel known as plutonium at the Hanford (Wash.) works, the amount of Uranium 235 consumed in the chain reaction exceeds the amount of plutonium produced. Scientists hope the Idaho experiments will show it possible to "breed" more fuel than is used in such chain reactions.

Last month Britain announced that she had developed the first atomic house-heating system.

Using atomic energy instead of coal, the statement said, enough heat was generated to warm an eighty-room building equipped with a water pipe system.

December 30, 1951

NEW ATOMIC PLANT 'BREEDS' OWN FUEL; CIVIL USE IS NEARER

Dean Says Reactor Can Create Fissionable Material Out of Common Form of Uranium

LOWER COST NOW IS GOAL

Commission Head Tells Edison Institute Time Has Come to Let Industry Enter Field

Special to THE NEW YORK TIMES.

ATLANTIC CITY, June 4—The Atomic Energy Commission has developed a new atomic power plant that can produce new fuel "at least" as rapidly as fuel is consumed in operating the plant.

This new "milestone in the development of atomic energy in this country" was announced here today by Gordon Dean, retiring chairman of the Atomic Energy Commission, in a speech at the closing session of the Edison Electric Institute.

"It is a development," Mr. Dean said, "which holds out the promise of making a civilian atomic power industry even more feasible and attractive in the long range than it has hitherto appeared to be."

The fuel that operates an atomic power plant is uranium 235, the fissionable form of uranium, which constitutes only about 1 per cent of the total uranium in natural deposits. Part of the enormous cost of producing atomic power is in isolating this natural fuel from the elements with which it occurs.

In the new power plant, as Mr. Dean described it here, the burning of uranium 235 not only produces the heat to operate the power plant, but changes non-fissionable uranium into fissionable plutonium at the same time "at a rate that is at least equal to the rate at which uranium 235 is being consumed."

The significance of this, he explained, "is that it is now possible for mankind ultimately to utilize all of the uranium that can be extracted from the earth's surface for atomic fuel, whether it is fissionable or not in its natural state."

Process Is Called 'Breeding'

Scientists have known for a long time, the commission chairman said, that the process, known technically as "breeding," is possible not only for non-fissionable uranium but also for thorium, another relatively plentiful element. But they were never sure, he explained, that as much or more new fuel would be produced as was burned.

The demonstration that this was possible, Mr. Dean said, was made by Dr. Walter Zinn, Dr. Harold Lichtenberger and other scientists of the Argonne National Laboratory near Chicago, using the atomic reactor at the Reactor Testing Station in Arco, Idaho. That is the reactor that in December, 1951, produced the first useful atomic power by using the heat produced by burning uranium 235 to operate a steam turbine.

The success of the experiment with uranium "suggests," the chairman explained, that similar breeding may be possible with thorium, as the scientists have suspected. Thorium was not used in the experiment, however, and therefore he said "I do not wish to imply that its susceptibility to breeding has been proved."

"We must take care," he warned, "to see that this encouraging development is kept in its proper perspective. This news does not mean that economic power from atomic fuels is here. It does not mean that over night we have suddenly obtained all the fissionable material we want or need. It does not mean that uranium can now be regarded as a virtually costless fuel."

Mr. Dean specified two important technical limitations of the new process:

1. "Before the newly created fuel can be extracted and put to use, it must go through a chemical separation process which is currently one of the most expensive aspects of the atomic energy business.

2. "Breeding is a slow process, and a reactor may have to operate for five years or longer before it succeeds in yielding as much new fuel as was initially invested in it."

Policy Is at Crossroads

Nevertheless, the result of this and other developments in the field of atomic energy, Mr. Dean said, "is to bring us to the crossroads in atomic power policy."

"The last remaining technical obstacle," he asserted, "is to learn how to build atomic power plants so cheaply that the power they produce will be competitive with that from conventional fuels. The policy problem that faces us is how this cost-cutting job can best be done."

The Atomic Energy Commission has come to the conclusion, he said, that the time has arrived when "the present Federal monopoly should be relaxed to permit wider participation in the power reactor program." He based this position on two factors:

1. "There is good reason to believe that a demand outside of the Government's own sphere of operations is developing for atomic reactors. It is a commercial demand, an economic demand and a civilian demand. It seems only reasonable to expect that people outside of the Government be given a chance to work toward the development of reactors to meet this demand.

2. "The job ahead is a developmental one—a cost-cutting one—the kind of a job that can be done best by skilled people competing with other skilled people who are working toward the same or a similar goal."

As a start in this direction, the Atomic Energy Commission, he explained, has recommended that the Atomic Energy Act of 1946, which established the Government monopoly in the atomic field, be amended to permit:

1. The ownership and operation of nuclear power facilities by groups other than the commission;

2. The lease or sale of fissionable material under safeguards adequate to assure the national security, and

3. The use and transfer of fissionable or by-product materials by the owners of reactors, subject to purchase by the commission or regulation by the commission in the interest of health and safety.

Curb on Hard Practices Urged

The chairman urged that those legal changes be accompanied by changes in the practices of the Atomic Energy Commission to permit:

1. The granting of more liberal patent rights;

2. A progressively more liberalized information policy in the power reactor field; and

3. The performance in commission laboratories of the research and development work in the power field that is deemed warranted in the national interest.

"A few people," Mr. Dean said, "have already labeled these policy recommendations as 'the atomic giveaway program.' That is simply not true.

"This all boils down to a question of fair play and common sense. I think it is obvious that we cannot expect private concerns to come in and spend millions of dollars without getting some benefits, and I think it is obvious that private concerns as well as the Government must be in this power program. Otherwise we can never have real competition, the catalytic agent of progress, and we will always have a program limited in size and scope by the range of the Federal Government's imagination and vision, which is not always as broad as it might be."

In an interview before his speech the chairman said that when he retired on June 30 he planned to take a two months' vacation, and then do "some writing."

He said he was in complete agreement with the statement made to the convention's 3,000 delegates by Walter H. Sammis, president of the Ohio Edison Company, who declared he would "endeavor to further the cooperative efforts between the electric utility industry and the Atomic Energy Commission in the development and utilization of atomic power for the general welfare of all the people."

June 5, 1953

HIGH COURT VOIDS F.P.C. GAS DECISION

Special to The New York Times.

WASHINGTON, June 7—The Supreme Court ruled today that sales of natural gas by the Phillips Petroleum Company to pipelines that distribute it in interstate commerce are subject to regulation by the Federal Power Commission.

The 5-to-3 ruling of the court affirmed a decision of the Federal Court of Appeals here. It reversed an earlier decision by the Power Commission itself that it did not have jurisdiction over rates charged the pipelines by companies engaged solely in the "production and gathering of gas."

In consequence of the ruling, the Power Commission will be required to determine whether the rates charged by Phillips Petroleum to five interstate pipeline companies that distribute natural gas in fourteen states are "just and reasonable." Costs to several million consumers presumably could be affected.

Minton Delivers Opinion

Justice Sherman Minton delivered the opinion of the court. He was joined by Chief Justice Earl Warren and Justices Stanley F. Reed, Hugo L. Black and Felix Frankfurter, although the latter wrote a separate concurring opinion.

Justice Tom C. Clark was joined by Justice Harold H. Burton in a strong dissent. Justice William O. Douglas also wrote a dissenting opinion.

Justice Clark said that the majority ruling, "on its face," brought "every gas operator, from the smallest producer to the largest pipeline, under Federal regulatory control."

"In so doing," he said, "the Court acts contrary to the intention of Congress, the understanding of the States, and that of the Federal Power Commission itself. The Federal Power Commission is thereby thrust into the regulatory domain traditionally reserved to the States."

Justice Douglas held that this was a "question the court has never decided" and "involves considerations of which we know

247

little and with which we are not competent to deal."

He thought that the decision of the Power Commission, "made by men intimately familiar with the background and history" of the Natural Gas Act of 1938, should have been sustained.

The technical question involved was whether Phillips Petroleum was a "natural-gas company" within the meaning of that Act. The Power Commission held that it was not: the courts held that it was.

Phillips is an "independent" producer. It gathers gas from its own wells and from other producers. It operates five networks of pipelines in Texas, one in Oklahoma, one in New Mexico and two that extend into both Texas and Oklahoma. The gas from these pipelines is processed to remove extractable products and impurities. From the processing plants the gas flows through outlet pipes to delivery points where it is sold and delivered to the interstate pipeline companies.

Exempt From Regulation

The Power Commission held that this was "production and gathering," exempted by Congress from regulation under the natural gas act.

"We do not agree." Justice Minton stated. "In our view, the statutory language, the pertinent legislative history, and the past decisions of this court, all support the conclusion of the Court of Appeals that Phillips is a 'natural-gas company' within the meaning of that term as defined in the Natural Gas Act."

"Regulation of the sales in interstate commerce for resale made by a so-called independent natural gas producer is not essentially different from regulation of such sales when made by an affiliate of an interstate pipeline company," Justice Minton continued.

"In both cases the rates charged may have a direct and substantial effect upon the price paid by the ultimate consumers. Protection of the consumers against exploitation at the hands of natural gas companies was the primary aim of the Natural Gas Act.

"Attempts to weaken this protection by amendatory legislation exempting independent natural gas producers from Federal regulation have repeatedly failed, and we refuse to achieve the same result by strained interpretation of the existing statutory language."

One of the most recent attempts at the type of "mandatory legislation" referred to by Justice Minton was the so-called Kerr bill, sponsored by Senator Robert S. Kerr, Democrat of Oklahoma, that was rejected by the Eighty-second Congress.

In his separate concurrence Justice Frankfurter emphasized that Congress had passed the Natural Gas Act for the basic purpose of occupying a field in which the states might not act, namely, the regulation of sales for resale, or "wholesale sales" in interstate commerce.

Three Divisions in Industry

Justice Clark pointed out that the "natural gas industry, like ancient Gaul, is divided into three parts." These are production and gathering, interstate transmission by pipeline, and distribution to consumers by local companies. Phillips' natural gas operations, he said, are confined exclusively to gathering and production.

"By today's decision," said Justice Clark, "the court restricts the phrase 'production and gathering' to the 'physical activities, facilities and properties' used in production and gathering. Such a gloss strips the words of their substance. If the Congress so intended, then it left for state regulation only a mass of empty pipe, vacant processing plants and thousands of hollow wells with scarecrow derricks, monuments to this new extension of federal power."

Phillips Petroleum appealed the lower court ruling to the Supreme Court. The respondents were Wisconsin and its Public Service Commission, Detroit, Kansas City, Mo., Milwaukee and the Federal Power Commission.

Three States Join Appeal

Texas and its railroad commission, the Corporation Commission of Oklahoma, and New Mexico and its oil conservation commission joined Phillips in the appeal.

Lawyers representing Phillips were Rayburn L. Foster and Harry D. Turner of Bartlesville, Okla., and Hugh B. Cox of Washington.

Attorneys General Vernon W. Thompson of Wisconsin, John Ben Shephard of Texas, Mac I. Williamson of Oklahoma and Richard H. Robinson of New Mexico were among the lawyers who appeared for the respondents.

CONGRESS HAS LAST WORD

Phillips Executive Says Verdict Was Not Unexpected

Rayburn L. Foster, vice president and general counsel of the Phillips Petroleum Company, commenting on the Supreme Court decision, said: "I am not too surprised at the decision. I am, however, surprised at the line-up of the justices. The ultimate decision will be up to Congress."

The economic aspect of the decision, he added, will be that, rather than submit to regulating the commodity (natural gas) the independent natural gas companies will sell their gas in the states in which they produce it and not in interstate commerce. He made these statements at a meeting sponsored by the Practicing Law Institute in the Statler Hotel here. The institute is holding a week-long meeting devoted to the legal problems of the gas industry.

W. E. Torkelson, chief counsel for the Wisconsin Public Service Commission, indicated pleasure with the decision. His commission, along with other Midwest cities, had fought to have Phillips placed under jurisdiction of the F. P. C.

Nelson Lee Smith, an F. P. C. commissioner, said it would appear the Federal regulatory agency is now in the business of regulating well-head sales of natural gas. The decision is an important one, he added, although not as momentous as it might seem on the surface since the commission already has many regulatory powers over the natural gas business. June 8, 1954

Upstate Area Gets Atomic Electricity

By WARREN WEAVER Jr.
Special to The New York Times.

WEST MILTON, N. Y., July 18—At 2:57 P. M. today the chairman of the Atomic Energy Commission threw a giant copper switch and sent atomic-generated electricity coursing into the lines of a private utility. Government authorities said it was the first time in this country that nuclear energy had been sent into commercial channels to light lamps, heat stoves and turn fans. Before he threw the switch, Lewis L. Strauss, commission chairman, declared that the ceremonies provided "a moving demonstration that the atom can indeed be stripped of its military casing and adapted to the arts of peace."

Right after Mr. Strauss threw the switch, Francis K. McCune, vice president and general manager of the atomic products division of G. E., said:

"Ladies and gentlemen, the Free World's first commercial atomic-electric power is on the line."

The power, 10,000 kilowatts of it, came from the experimental reactor built here two years ago by the General Electric Company. It was designed initially to test a prototype of the power plant installed in the second atomic submarine, the Sea Wolf.

Lights a Giant Bulb

The power went into the private lines of the Niagara Mohawk Power Corporation. Where it went from there was anybody's guess, although some of it presumably wound up in local farmhouses and area industry.

Symbolically, the first commercial atomic electric power lit up a giant bulb on the platform before the 200 guests at the ceremonies. It also provided power for the routine needs of the personnel who run the reactor station in this rolling rural area.

The speakers' platform stood in the shadow of the 225-foot steel sphere that houses the reactor. Sharing it with Mr. Strauss were Douglas McKay, Secretary of the Interior; Senator Clinton P. Anderson, chairman of the Joint Congressional Atomic Energy Committee, and Ralph J. Cordiner, president of the General Electric Company.

The power distributed commercially is that left over after submarine experimental requirements are filled. Niagara Mohawk is paying three mills a kilowatt hour for it. The money goes to the Atomic Energy Commission, which owns the reactor.

General Electric owns the turbine-generator, which converts the heat of atomic fission within the reactor into usable electricity. This cost $1,250,000, but the company is generating the electricity without charge to either the Government or to the private utility.

The 10,000 kilowatt capacity of the atomic-powered generator is a small amount to Niagara Mohawk, which operates a system with a 2,500,000 kilowatt capacity. However, it is enough to serve a city of 20,000 to 30,000 population.

No one can tell who actually used atom-generated power today, except for those receiving it directly at the reactor installation. Trying to trace it would be like guessing whether a bucket of water poured into the Hudson River at Albany would wind up in a teacup in Tuxedo Park or a bath tub in the Bronx.

General Electric officials declined to say how much it was costing the company to produce the electricity it was presenting to the utility. Admittedly the cost is more than the three-mill rate Niagara Mohawk pays the A. E. C.

Speakers at the ceremonies repeatedly emphasized that the demonstration did not mean that atom-generated electric power was economically practical. They said, however, that the West Milton project was a forerunner of future peaceful atomic programs.

The switch-throwing ceremony itself was symbolic, as Mr. Strauss pointed out. The switch could have been closed in two ways. One would have activated the propeller shaft of the experimental submarine motor within the reactor shell. The other way, the one actually used, sent the excess power into commercial channels.

Swords of Plowshares

"This switch is a symbol of the great dilemma of our time," Mr. Strauss declared. "I throw it now to the side of the peaceful atom and by that choice we of the United States mark the beginning of a fulfillment of the scriptural injunction of Isaiah: 'They shall beat their swords into plowshares and their spears into pruning hooks.'"

The arrangement under which Niagara Mohawk markets the power is temporary. This is necessary because Federal law requires that municipal power plants and cooperatives in the area be given a chance to buy power generated in public projects. This is an Atomic Energy Commission project.

Two municipal power systems —in Illion, N. Y., and Holyoke, Mas.—are close enough to be eligible, as is a rural cooperative in Delhi, Delaware County.

Senator Anderson declared that it was "a day of vindication for all those whose faith in the peacetime promise of atomic energy had been unswerving."

He said it was also "a day of challenge" to those wishing to develop the potential of the atom still further.

Secretary McKay said that the nation, as the result of projects like the one at West Milton, was "merely at the beginning of a great new advancement in the atomic age."

July 19, 1955

TIME RUNNING OUT

HEAT WAVE POINTS UP RISING POWER DEMANDS

By WILLIAM M. BLAIR
Special to The New York Times.

WASHINGTON, June 22— When fans stop spinning, air conditioners slow down or fail, escalators stop and electric trains halt in the middle of a tunnel, dwellers in metropolitan centers become acutely aware of the electric power situation.

The "situation," which the informed and uninformed deplore from time to time, has been creeping up on the United States for twenty years. It was compounded by the industrial demands of World War II and doubly compounded by unleashed consumer demands and the expanding economy of the postwar period. It raises the question of how we are going to keep from running out of electricity.

The situation encompasses the need for prime power at peak periods of demand, such as during the winter heating season and during summer hot periods. The high demand of this week's hot spell caused the power trouble in New York, Chicago and elsewhere.

New York Projection

An indication of what such peak demands mean may be seen in projections the Consolidated Edison Company of New York has made for the coming winter's high demand period in late December.

Its projection on file with the Federal Power Commission shows that Con Ed and three companies associated with it in a metropolitan power pool will have a generating capacity of 4,856,000 kilowatts, the same as now available. But the December peak demand is expected to reach 4,625,000 kilowatts, leaving what is regarded by many engineers as too small a margin of reserve.

Figures on electric power production since 1946 also serve as a measure of the power problem. Total production of electric energy in 1946 by electric utility and industrial producers was about 206,609,000,000 kilowatt hours. In 1956 total production

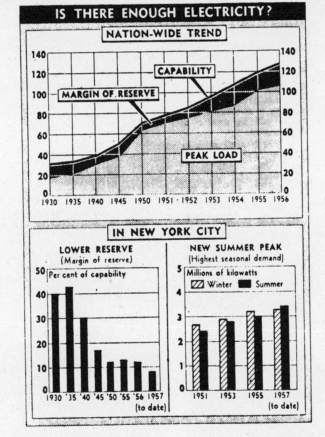

IS THERE ENOUGH ELECTRICITY?

NATION-WIDE TREND

CAPABILITY
MARGIN OF RESERVE
PEAK LOAD

IN NEW YORK CITY

LOWER RESERVE (Margin of reserve)
Per cent of capability

NEW SUMMER PEAK (Highest seasonal demand)
Millions of kilowatts
Winter   Summer

climbed to a peak of 683,894,-050,000 kilowatt hours.

The use of energy by the average residential user has climbed a record 250 kilowatt hours in two years. A prime factor has been the air conditioning of office and industrial buildings, not to mention home room conditioners and complete house conditioners.

Eliminating industrial energy output and limiting the energy production to utility production, because utilities are the source of power for United States homeowners, the growth of such output moved from about 220,-000,000,000 kilowatt hours in 1946 to 600,591,000,000 kilowatt hours in 1956.

While utility production and consumer demand have been increasing markedly, so has the cost of almost everything that goes into making the kilowatts surge through the lines and run the appliances. This fact the industry does not hesitate to point out to anyone who questions its production and its efforts to meet growing demands.

The pride industry takes in its record is not a whit diminished by its critics who contend that its interest lies more in profits than in expanding to meet new goals. In brief, these critics contend that the pace of industry is maintained according to the profit scale.

Quite aside from the contention between the industry, its critics and the inevitable public versus private power fight, the fact remains that a power shortage continues to loom large on the horizon of American mechanization.

This was amply demonstrated last week when New York and Chicago suffered energy letdowns that many attributed to the boom in air conditioners as a result of the heat. Undoubtedly air conditioners are a prime factor but in terms of total energy they are only a part of the picture. Homes use more than air conditioners. And there are more homes going up all the time, equipped with such modern aids as dishwashers, garbage disposals, clothes dryers and electric heating units.

Planning Needed

A key factor in industry calculations is the question of how best, for example, to shut down a few boilers or a generator for repairs, not to mention storm interruptions or just plain unexpected breakdowns. The industry must plan for its major repairs at a low season to keep its plants operating at the most efficient capacity. Thus, a boiler or two out for repairs in New York might make the difference between supplying the power needed when a heat wave strikes.

Nevertheless, the public, which pays the bills, assumes its asserted right to ask why the fans stop running and the elevator doors won't open. The public has heard brave words from the industry through an extensive advertising campaign that it is helping to make the country great and safe and can do a better job through private enterprise without Government intervention.

While the sparring goes on, there stares out from the figures on file the knowledge that added electrical capacity is needed.

The problem now is how best it can be done, with whose money and at what cost. Will the needed capacity come from adding to conventional power plants or from atomic energy? The answers are still not clear.

June 23, 1957

ENERGY SOURCES FACING DEPLETION

Conservation Termed Only Way to Avert Exhaustion of U. S. Fuel Supply

By RICHARD RUTTER

Americans, whatever else may be said, are a notably energetic people.

Sociologists may seek the reasons, but it may be no coincidence that the nation's economy—its very standard of living—is founded on the use of energy. It comes chiefly from the so-called fossil fuels—oil, natural gas and coal. It has come in prodigious abundance and has been consumed and wasted on a vast scale.

No one will gainsay the wonders that the use of energy has wrought, but one hard fact should not be overlooked: the sources of this energy are irreplacable. They are gradually being depleted and, in time, will be exhausted. It may be a very long time, as in the case of coal, or it may be comparatively short, as in oil. Conservation practices can postpone the day of reckoning, but they cannot prevent its arrival.

Take oil—the chief source of our energy. Some experts have predicted that the United States will reach its peak production within five years, ten at the most. This country will become increasingly dependent on foreign supplies, transported at great distance, and, in time of war, at great hazard. In fact, imports of crude oil have been steadily increasing since World War II.

Domestic Output On Rise

This has happened despite steadily growing domestic productic

The United States has been the world's largest petroleum producer for so long that it is hard to realize that supremacy is drawing to a close. With proved reserves of about 30,-000,000,000 barrels, the nation has only about 15 per cent of

the estimated 200,000,000,000 reserve barrels throughout the free world. Yet it has been poducing 48 per cent of the world's output, meanwhile consuming 56 per cent at home.

By 1966, according to a study by the Chase Manhattan Bank, America's share of world oil production may be down to 39 per cent.

The reserves of 30,000,000,000 barrels may seem large. They must be related, however, to future demand. And that is expected to swell to more than 14,000,000 barrels a day within the next decade. To fill that requirement and maintain the present reserve position would call for the discovery of 57,-000,000,000 barrels in new oil. Some 1,200,000 wells would have to be drilled by 1966. Obviously, a formidable problem looms.

It is extremely doubtful that it can be solved by discoveries in this country. The rate at which new supplies have been added through the discovery of new fields and of new pools in old wells has actually been declining since 1953.

The most optimistic forecast is that the United States has ultimate reserves of upward of 300,000,000,000 barrels of crude oil. This includes offshore fields. Granted this estimate, production would still level off some time between 1970 and 1980.

Another Complicating Factor

There is another complicating factor. The demand for petroleum products is growing more rapidly in the rest of the free world than in America. In 1956 it jumped 12.7 per cent, compared with an increase of 3.5 per cent in the United States.

Where will the oil come from? Assuming political stability—a rather big assumption —most of it will come from the Middle East. That area has 70 per cent of all proved reserves and yet it has been producing only 24 per cent of the free world's total. Venezuela is another major source; it now accounts for about 17 per cent of supplies. Canada is coming to the fore as an important producer. The Far East is expected to become another source through exploration and development.

The search for oil is unending. There have been recent discoveries in scattered areas, such as the Sahara, France, Israel and Italy. Perhaps the most dramatic search has been in the off-

shore lands in the Gulf of Mexico. More than eighty oil and gas fields have been found along the coasts of Texas and Louisiana. It is estimated that they contain about 1,300,000,000 barrels of crude oil and natural gas liquids.

Conservation came fairly late to the oil industry. Waste and overproduction were common until the early Nineteen Thirties, when the situation got out of hand. It showed up in the form of huge stocks in storage that were far in excess of requirements. Regulation was badly needed and it finally came in action by the oil-producing states.

States Established Controls

Various state agencies were established to control production. Operators were told how many wells they could drill and where, what kind of pipe they could use, the amount of oil that could be produced each day and just how it should be produced to reduce waste. In all, state conservation laws have resulted in the recovery of 50 per cent or so more oil than would have been recovered without regulation.

Much more remains to be done. Only about half of the oil in the average reservoir can be economically recovered by present methods. The industry is spending hundreds of millions to improve recovery techniques. Eventually perhaps as much as 75 per cent of oil-in-place may be recovered.

If America's oil reserves should run dry, there will still be oil. It can come from oil shale or coal. The known supply of oil shale in the country is vast; large and rich deposits have been found in Colorado, Utah and Wyoming. The Bureau of Mines has put at 500,000,000,000 barrels the recoverable amount of oil from shale. The problem, however, is to effect the recovery at a cost competitive with natural petroleum. Research projects are under way and it is believed the solution will be found. The same applies to making synthetic oil from coal.

As oil goes, so largely does natural gas. About a third of the natural gas supply comes from oil wells and the rest from gas wells that are usually discovered in the process of looking for oil. The growth in the use of natural gas, however, has been far more spectacular in recent years.

Production Has Doubled

In 1956 the marketed pro-

duction of natural gas in the United States was about 10,-100,000,000,000 cubic feet, almost double that of ten years earlier. In that same decade its share of the nation's over-all energy supply rose from 13 per cent to 24 per cent. Reserves at the end of last year totaled about 237,700,000,000,000 cubic feet. This country is the only major producer and consumer of natural gas at present, although Canada's output is on the rise.

The consensus is that there will be ample supplies of natural gas at least until 1975 and perhaps well beyond. As reserves dwindle, however, there will be no sizable imports to take up the slack.

There are some indications that the rate of growth in demand for natural gas is already leveling off. It now heats about 25 per cent of all homes, mostly in urban areas. Conversions from coal to gas are becoming fewer.

Coal is one prime source of energy that presents no supply problem in the foreseeable future. The United States has almost half of the world's recoverable reserves — about 948,000,000,000 tons. Since consumption has been running at about 420,000,000 tons annually, it takes only a little figuring to see that supplies will be ample for a long, long time.

Those who keep a close watch on the natural resources situation are generally agreed that coal will be the energy fuel of the future, as it has been in the past. It will fill the gap as the output of oil and natural gas declines. America, already a leading exporter, will have to help supply Western Europe's growing coal needs. These are expected to rise by at least 50 per cent over the next twenty years.

The coal industry is not taking anything for granted. Mechanization has been introduced into the mines and productivity has climbed sharply. Power-cutters and drilling machines have replaced the pick and hand-operated drill. Mechanical loaders have made the hand shovel obsolete. Belt conveyors carry the coal out of the underground. Waste has been reduced. All of this, too, spells conservation.

The sources of energy may be perishable, but there is no reason why they can not be nourished into a long, productive life.

December 11, 1957

NEW GENERATORS USING OLD IDEAS

By GENE SMITH

The search for improved ways of generating power seems to prove that there's really nothing new in science.

Four methods today seem the most promising. These are fuel cells, thermoelectricity, thermionic conversion and magnetohydrodynamics. These have been under intense investigation by the people who make equipment to produce power.

Yet, not one is really new. In fact, the concept of the fuel cell, which is said to be the most direct means of generating electric power, dates back to

1802. Thermoelectricity was first observed in 1821. The idea underlying magnetohydrodynamics (usually referred to as M. H. D.) goes back to 1831 and the basic phenomenon involved in thermionic power was reported in 1878. The key to all of them, the electron itself, was discovered in 1898.

"Actually, the basic principles underlying these new methods of power generation, or energy

conversion, have been known for many years," Dr. S. W. Herwald, vice president for research of the Westinghouse Electric Corporation, said recently. "So, perhaps it might be more exact to describe this new area of scientific endeavor somewhat facetiously as: conventional energy conversions by non-conventional means," he added.

Dr. Herwald explained that

the search for these methods ranged through all levels of science and technology, including basic research, applied research, development and engineering. He continued:

"Today one finds the mathematician, the chemist, the physicist, the electronics engineer working together in this broad and significant new area of science. Almost without exception, these new methods of power generation involve the conversion of less versatile, less controllable, less broadly useful forms of energy into electricity."

Westinghouse recently demonstrated these approaches and explained them this way:

The fuel cell is an electrochemical device that converts the "free energy" of a chemical reaction directly to electrical energy. It contains no moving parts and is, thus, silent in operation. Theoretically, it can operate at efficiencies as high as 70 to 90 per cent, against the maximum of 42 per cent for today's most modern power plants. Westinghouse personnel feel that high-temperature fuel cells (operating above 1,500 degrees Fahrenheit) might become competitive with conventional large-scale power sources in ten to twenty years. Special low-temperature cells operating at below 250 degrees Centigrade are expected to find special applications even today where capital and fuel costs are not of primary concern.

A German Discovery

The thermoelectric principle was discovered by the German physicist, Thomas Seebeck. He learned that the flow of heat through a metal segment could produce a voltage difference between the hot and cold ends. With the advent of new thermoelectric materials, practical generation of electricity is now possible. For example, Westinghouse was working a year ago with devices that could produce only slightly more than a single watt.

Today, the company has such generators rated at 100 watts and "very soon" will complete a generator rated at 5,000 watts. A 100-watt thermoelectric generator at the company's research laboratories changes the heat of a gas flame into enough electric power to drive a bench grinder and light a pair of headlights.

Again, there are no moving parts and no noise, so that in military power plants heat would be converted into electricity without noise. In space vehicles and missiles, the absence of moving parts would aid guidance and stability in orbit.

"We are confident that thermoelectricity will be a very important element of the technology of the Nineteen Sixties," a company spokesman said.

Thermionic Principle

Thermionic generators produce electrical power by using the electrons emitted from the surface of a material when it is heated to a high temperature. The heated electrons must, however, be emitted into a vacuum, thus posing certain structural problems. The biggest roadblock is materials that must operate for long periods at temperatures up to 4,500 degrees Fahrenheit.

Such generators are expected to be particularly useful to the military, where compactness, light weight, simplicity and high efficiency are required.

They also are expected to be used as boosters with nuclear power stations.

M. H. D. is really an adaptation of Faraday's discovery that a conductor moving in a magnetic field can be used to generate electric current. A flowing liquid, such as mercury, is substituted for the conventional solid copper bars in a generator. The problem here is to find the best conducting gas and devise materials that can withstand "low" temperatures of 4,000 to 5,000 degrees Fahrenheit. Westinghouse already has demonstrated a new M. H. D. generator that, operating at about one-fourth its full power rating, has produced two and one-half kilowatts of power and has run continuously for four minutes.

The reasons for this experimentation in direct conversion systems for electric utilities are obvious. Dr. J. A. Hutcheson. vice president for engineering at Westinghouse, pointed out that "if we could save one-tenth of a cent per kilowatt-hour in the generation of electricity, at the present rate of its production this would amount to a saving of $500,000,000 a year. •

"No single one of these new methods is likely to be 'the one way to do it.' We will find no one answer to our varied growing needs for electric power. That is why we think it is important in our research to have a broad program, to look at all methods that seem feasible and to use their unique characteristics in the best way we can to do the best over-all job. At this stage of the game, we think it important that we do not overlook any 'sleepers.' "

April 3, 1960

A.E.C. WINS TEST ON PLANT SAFETY

Backed by Supreme Court on Michigan Installation

Special to The New York Times.

WASHINGTON, June 12—The Supreme Court, in an important test case for the atomic energy program, upheld today the procedures of the Atomic Energy Commission for authorizing the construction and operation of nuclear power plants.

The court rejected, 7 to 2, arguments of three labor unions that the commission had not properly followed the safety regulations in the Atomic Energy Law in authorizing the construction of an experimental atomic power plant at Lagoona Beach, Mich.

Justice William J. Brennan delivered the majority opinion upholding the commission's procedures in granting a construction permit for the Lagoona Beach reactor and overturning a decision by the United States Court of Appeals for the District of Columbia.

A minority opinion was given by Justice William O. Douglas, who was joined by Justice Hugo L. Black.

The case was the first contested licensing proceeding to be decided by the commission. It was regarded within the commission and the industry as an important test of the procedures under which atomic power plants are being developed.

If the commission had been overturned, it would have necessitated a complete reshaping, and possible delays, in the program for development of atomic power.

At issue was whether the commission has to make a definitive finding of safety at the time a permit is issued for construction of a reactor or can defer this safety finding until a permit for operation is issued.

Following its established procedures, the commission in 1956 authorized the construction but not the operation of the $80,000,000 reactor now nearing completion at Lagoona Beach, with Government research assistance. The plant, utilizing a breeder type of reactor which produces more fuel than it consumes, is being built by the Power Reactor Development Company, a non-profit group of twenty-one companies, to demonstrate the feasibility of a breeder reactor for producing electricity.

3 Unions Opposed

The United Automobile Workers, the International Union of Electrical Workers and the United Papermakers challenged this procedure. They argued that for reasons of safety and under the requirements of the law the commission would have to make a finding of safety of operation of a reactor before authorizing its construction.

Justice Brennan said the commission had "good reason" to defer a safety finding until operation was actually licensed.

"Nuclear reactors are fast developing and fast changing," he said, and "what is up to date now may not, probably will not, be as acceptable tomorrow. Problems which seem insuperable now may be solved tomorrow, perhaps in the very course of construction itself."

The Lagoona Beach plant is about thirty miles from Detroit and the same distance from Toledo, Ohio. The area has 2,000,000 residents.

In the minority opinion, Justice Douglas said the legislative history of the Atomic Energy Law of 1954 "makes clear that the time when the issue of safety must be resolved is before the commission issued a construction permit."

The construction given the law by the commission, and now approved by the court, he said, is "a lighthearted approach to the most awesome, the most deadly, the most dangerous process that man has ever conceived."

The Government's case was argued before the court by Archibald Cox, the solicitor general. W. Graham Claytor Jr. represented the Power Reactor Development Company and Benjamin C. Segal the labor unions.

June 13, 1961

COAL STRUGGLING TO KEEP POSITION

Mechanization of the Mines and Lower Rail Costs Buoy the Industry

LABOR LENDS BIG HAND

But Nuclear Power Systems Are Joining Gas, Oil and Electric Competition

By J. H. CARMICAL

The bituminous coal industry appears to be well into the competitive fight with other fuels. Although badly battered, and with several markets definitely lost, the soft coal industry continues to hold a prominent position in the nation's energy picture, but only through its meeting of great challenges.

Through the mechanization of its mines and other adjustments, the coal industry in the postwar period has pretty well licked the problem of high production costs. Now its efforts are being concentrated largely in solving the problem of high transportation costs to bring coal prices in line with other fuels.

Since the end of World War II, the coal industry has undergone radical changes in its consumer markets. The switch of the railroads from coal-burning to diesel locomotives, the steadily increasing use of fuel oil both by home-owners and industry and the expansion of natural gas pipelines have resulted in the loss of several traditional coal markets.

Electricity Use Grows

With the gradual disappearance of these markets, the electric utility industry continued to expand in a spectacular manner because of the many new homes built and the greater use of electrical gadgets. Realizing that these utility companies offered coal a steadily increasing market, attention was directed at first toward lowering mining costs and later to reducing transportation charges to the utility plants.

Although these efforts met with success with respect to the utility companies, the use of coal in all other domestic markets has declined steadily since 1951. In the export field, where there was a sharp increase in 1956 and 1957, largely because of the Suez crisis, present shipments are much below the level of 1951.

During this period of declining markets, the productivity in the coal-mining industry has gradually been improved. From

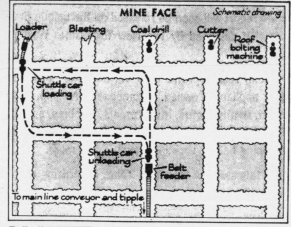

MINE FACE   *Schematic drawing*

Loader   Blasting   Coal drill   Cutter   Roof bolting machine

Shuttle car loading

Shuttle car unloading   Belt feeder

To main line conveyor and tipple

The New York Times   July 28, 1963

USE OF MACHINERY has made coal mining a nearly continuous process. In diagram, automatic loader feeds cars which shuttle to conveyor. When the loading is done, bolting machine will move to spot occupied by loader and other units will move one place to right. Rotation keeps coal cut, drilled and shot down for constant loading.

an average output of 7.47 tons a day for each miner in 1952, the rate had been raised to 15.41 tons in 1962, by far the highest in the world.

Productivity Raised

In the most efficient underground mines, the productivity rate is as high as 35 tons a day for each miner. In the strip mines of Illinois and Kentucky, the average last year was 27 tons a day for a man. In some of the more efficient strip mines, an output of 80 to 100 tons a day is being achieved.

With this increase in productivity, the operators have been able to make slight reductions in prices at the mines and at the same time increase the wage rate, which averaged $120.96 a week in April, sharply above the $74.69 average for 1951.

Currently, the mine price is around $4.50 a ton, down almost 10 per cent from the 1951 level, when it was $4.80 to $5 a ton.

The main contribution to increased productivity and lower costs in the industry has been the attitude of the United Mine Workers of America, which has not insisted on make-work and other featherbedding practices. During this period, almost all the cost reduction in the production of coal had to be passed on to the consumer in the form of a price cut in order to meet the competition with other fuels. Consequently, most of the coal industry has been standing still as far as profits were concerned.

The bottleneck in the coal-mining industry continued to be in the heavy transportation charges. Although it is being offered at a lower price at the mine-head, the value of coal is judged in the competitive market on its delivered price and heat content. With railroad freight rates constantly rising,

those interested in coal realized that economies in transportation costs must be brought about if the industry was to benefit from the lower production costs.

Coal Hauling a Big Job

Moving coal to market is one of the nation's biggest transportation jobs. About 75 per cent of the coal loaded at the mines moves by rail. Excluding some 3 per cent used at mines and moved out through high voltage transmission lines in the form of electricity, the rest is moved equally by barge and motor vehicles.

Recently, there has been a trend by the electric utility companies to build generating plants near the mines. Also, an experiment program is underway to increase the voltage of these lines, thus raising their efficiency.

In the transmission of electricity over long distances, even with high-voltage lines, there is a considerable loss of power. In addition, the cost of constructing these lines have risen steadily in recent years and the acquisition of land for a right of way offers another problem.

The movement of this 25 per cent of the coal output by other than rail has resulted largely from the efforts to lower transportation costs. Previously, the railroads almost had a monopoly on the movement of coal.

Realizing that they stood to lose considerably more of the coal business unless rates were made competitive the railroads started a movement about a year ago to adjust tariffs on the basis of train-load lots rather than single car rates. By merchandising their services on a competitive basis rather than the old artificial single car rates, it is estimated that coal freight charges will be reduced up to 50 per cent and that it will prove beneficial to the car-

riers, the coal-mining industry and the public.

Probably the most efficient method so far employed is that worked out between the Commonwealth Edison Company of Chicago, the New York Central and the Gulf, Mobile & Ohio railroads in the train-load movement of coal from the Illinois and Indiana mines into the Edison plants in the Chicago area.

Utility Provides Cars

Under the arrangements, the cars of 100-ton capacity each will be furnished my Commonwealth Edison. The coal will be moved in train-load lots of 100 cars each and around trip is made each 48 hours, including four hours each to load and to unload.

The rate of the New York Central for the 285-mile route from southern Indiana to Hammond, Ind., will be $1.45 a ton, and that of Gulf, Mobile & Ohio from Perry, Ill., to Joliet, Ill., a distance of 300 miles, will be $1.30 a ton. Added to these rates must be the estimated cost of, 25 to 30 cents a ton to cover car ownership, including amortization, maintenance and other charges.

The present rate on single-car shipment of coal between these points on the Gulf, Mobile & Ohio is $3.67 a ton, and the special rate on shipments of 1,000 tons or more is $3.17 a ton. Taking into consideration the car ownership cost, the train-load rate is just about one-half that of the 1,000-ton rate and $2.07 cents a ton below the single car tariff.

Most of the coal coming into the New York City area comes from mines in northern West Virginia, a distance of 461 miles. Based on the Gulf, Mobile & Ohio rate to Chicago of 4.3 miles a ton-mile, the railroad charge of bringing coal to the New Jersey piers would be $1.97 a ton. Because of the longer haul, the car ownership charge would be about 45 cents a ton, or a total of $2.42.

Costs Here Outlined

With the recent reduction, the present rail freight rate charge to the New Jersey piers is $3.12 a ton. To get the coal to Consolidated Edison Company's plants in New York City involves an additional barge charge of about 50 cents a ton. The cost of the coal at the mines is about $4.40 to $4.45 a ton, which makes the total cost of coal delivered to Consolidated Edison's plants here a little better than $8 a ton.

Residual fuel oil is coal's principal competitor for the electric utility business along the East Coast. A ton of coal is equivalent in heat units to 4.167 barrels of residual fuel oil so $2 a barrel for this oil now could make it competitive with coal.

The railroads are understood to be well pleased with their Commonwealth Edison deal and believe that it will prove to be reasonably profitable. By increasing tonnage a trip to 20,-000 or probably more through the installation of "slave" loco-

COAL-RAIL-UTILITY COMPLEX: Appalachian Power Company's $95,000,000 steam plant at Carbo, Va. It burns coal from nearby Clinchfield Coal Company installation. Norfolk & Western Railroad has also invested in the complex.

motives in the center of the train, which would be controlled by the engineer on the front locomotive, it is held that costs could be further reduced. Of course, this will depend upon the elimination of the certain present work rules on the railroads.

The proposal to build a pipeline connecting the West Virginia coal fields with electric generating plants along the Eastern seaboard has been under discussion for years. The taking of the coal slurry—a mixture of powdered coal and water — directly to the plants would save the electric utility plants serving Manhattan and adjacent areas the barge charge they now must pay.

Transporting coal by pipeline has been found commercially feasible, although the nation's only such carrier has now been placed on inactive duty. It has been delivering coal from a mine near Cadiz, Ohio, to a Cleveland utility for six years without interruption. By forcing the railroads to cut their rates from the mines to Cleveland from $3.54 to $1.88 a ton,

a reduction of 47 per cent, it has served the purpose originally intended.

When the pipeline started operating, the railroads reduced their rates on coal to Cleveland by $1.05 a ton. Early this year, the railroads proposed a further cut of 61 cents a ton. Since the reduction applied to the 4,000,000 tons of coal a year used by the utility and the pipeline could deliver only 1,300,000, the pipeline was shut down and all the coal-hauling business to the utility plant was turned back to the railroads.

Business Grows

Realizing that there is not much chance of stimulating greatly the demand for coal by other major users here other than the electric utility companies, both the coal industry and the railroads are centering their efforts in that direction. This business has grown since 1954, when the utilities consumed more coal than any other consumer group. They now are using an estimated 53 per cent of total output. By 1975, utility consumption is expected to grow to 475,000,000

tons a year.

The use of electric power in the United States is expected to continue to expand. A relatively new and expanding market for electricity is for space heating in homes and other buildings. At present, coal is used as the fuel to generate more than one-half the power produced by the electric utility industry and the coal interests hope to increase this rate over the years by meeting head-on the competition with other fuels.

So far, coal is not being used as a utility fuel in the Pacific Coast area and the industry is taking definite steps to make coal competitive in that area where the projected utility growth rate is even greater than the national average. There are large coal reserves in Montana, Wyoming and other Western states that could be shipped economically to the West Coast under train-load rates.

The biggest threat to coal in expanding its electric utility market is the Government's policy with respect to the building

of nuclear energy power plants. At present, there are two groups in California and one in Connecticut seeking $12,000,000 of Federal funds to construct nuclear reactors. In addition, the Niagara Mohawk Power Company is seeking free nuclear fuel for five years for a reactor it wants to build on Lake Ontario.

Joseph E. Moody, president of the National Coal Policy Conference, Inc., an organization sponsored by the United Mine Workers, the coal industry and the railroads, said that there was evidence that these lowering of costs will continue and that coal delivered to the customer will be even cheaper in the future.

Mr. Moody noted, however, that, if the Government is determined to continue to pay subsidies amounting to millions of dollars a year in an effort to make electricity produced from atomic plants as cheap as that produced from coal-steam plants, coal will be driven from more and more utility markets, no matter how much costs are lowered.

Coal to Compete

"If private industry wants to build atomic plants, and operate them in full, free competition with coal-fired plants, it has every right to do so," Mr. Moody said, "and the coal industry will beat any competition they offer. But it doesn't make sense for the Government to use public money to try to create a new source of electric power, which isn't needed, in order to drive coal and other conventional fuels out of the market."

The second largest customer of coal is the steel industry. With the more efficient techniques being introduced into the making of steel, including the oxygen process, fewer pounds of coke will be needed to produce a ton of steel than now. However, the increased use of steel might offset the decline in the amount of coke needed.

Exports of coal overseas this year will be about 10 per cent more than in 1962, and are anticipated at some 30,000,000 tons. Since only a small amount of this moves under the Government's foreign aid program, it is estimated that it will bring in some $275,000,000 in foreign exchange.

Japan is the biggest single buyer of American coal, last year taking about 6,500,000 tons. Most of this is coking coal for use in making steel. But in 1962, the United States Government sent some engineers to Indonesia to help that country develop its coking coal industry and raise its sales to Japan.

July 28, 1963

A.E.C. CHIEF TERMS ATOM PLANTS SAFE

Dr. Seaborg Backs Location in Populated Sections

By JOHN W. FINNEY
Special to The New York Times

WASHINGTON, Nov. 7—Dr. Glenn T. Seaborg, chairman of the Atomic Energy Commission, said today that he would have no fear of living "next door" to a nuclear power station.

Publicly joining the debate over such plants, Dr. Seaborg said it would be safe to locate them in populated areas. His statement marked a shift from the previous commission policy, which required that the plants be located in isolated areas.

Dr. Seaborg said that atomic plants would present less hazard to the general public than many other technological developments.

"The probability of a serious accident is extremely low" and the likelihood that such an accident would have a dangerous consequence" for the general public is "even lower," he said.

In a speech in Norfolk, Va., for the national convention of Sigma Delta Chi, a journalistic fraternity, Dr. Seaborg sought to counter public "misunderstanding" and "unreasoning fear" over the safety of atomic power plants. The speech was also intended as a rebuttal to David E. Lilienthal, the first chairman of the commission. In lectures, Congressional testimony and magazine articles in the last year, Mr. Lilienthal has emerged as a leading opponent of locating nuclear power plants in population centers.

Dr. Seaborg's speech, made public by the commission's office here, assumed the proportions of a policy statement. It was reported to have been given the general approval of the four other commissioners.

This spring, before the Joint Congressional Committee on Atomic Energy, Mr. Lilienthal said he "would not dream of living in Queens" if Consolidated Edison of New York built its proposed plant in Ravenswood, just north of the Queensboro Bridge.

Finds Fear 'Unreasoning'

Dr. Seaborg used the same expressions as Mr. Lilienthal to argue his case.

"Perhaps I can best summarize my feelings about the safety of these power reactors by saying that I would live next door to the atom," Dr. Seaborg said. "I would not fear having my family residence within the vicinity of a modern nuclear power station built and operated under our regulations and controls."

Dr. Seaborg said he appreciated that "many have an unreasoning fear of the unknown, and radioactivity appears as such an unknown."

He said, however, that, on the basis of what was known about radioactivity, "we are able to proceed with assurance in assessing the safety of nuclear power stations."

Ultimately, the commission will decide whether the million-kilowatt power plant should be built in Ravenswood. The proposal is now being reviewed by the commission's regulatory staff and the advisory committee on reactor safeguards prior to a public hearing by a safety and licensing board.

Economic Factors Noted

Dr. Seaborg did not specifically mention the Consolidated Edison proposal. But he indirectly touched on the issue by citing the economic considerations that prompted utilities on the East and West Coasts to build atomic plants closer to metropolitan areas.

Dr. Seaborg emphasized that the nuclear industry "has one of the best safety records in the country." As evidence he pointed to the fact that "in about twenty years of operation of reactors of various types, there has not been a single accident that has caused any known injury to the public" as distinct from A. E. C. personnel.

Within commission laboratories, however, there have been occasional reactor accidents, some leading to fatalities.

In January, 1961, for example, three enlisted men were killed at the reactor test station in Idaho when a still unexplained accident developed in a partially dismantled reactor.

Abroad, one of the most dramatic accidents occurred in October, 1957, when an experimental British reactor at Windscale, England, suddenly overheated and spewed radioactive debris over some of the countryside.

The Windscale reactor, however, was of a basically different type from those being developed for power stations in the United States and did not have the thick containment shell required on American atomic power stations.

November 8, 1963

A.E.C. CRITICIZED ON PLANT POLICY

Lilienthal Says Commission Takes Advocate's Role

By GENE SMITH

David E. Lilienthal charged here yesterday that the Atomic Energy Commission "has disqualified itself" to act on the Consolidated Edison Company's application to build a one-million-kilowatt nuclear power plant in Long Island City.

Mr. Lilienthal, who was the first chairman of the commission, spoke to a luncheon gathering of the American Nuclear Society in the New York Hilton Hotel. He said that the A. E. C. acted now as both advocate and judge of nuclear power plants.

"As advocate," he went on, "the A. E. C. appears before itself as judge. Under our traditions of due process, the roles of advocate and judge are not entrusted to the same hands, however honorable or competent those hands may be. And yet, no one raises the slightest question about the judges in this case."

Question From the Floor

Asked from the floor who should judge the safety of the proposed atomic plant if not the A. E. C., Mr. Lilienthal replied: "There are a number of possible answers, but the criteria are independence and competence. Congress could set up an independent body or some agency like the Federal Power Commission might take over the job."

Joseph C. Swidler, chairman of the power agency, who was in the audience, said he "could not offer any comment on that proposal."

Following the meeting, he said in a private interview that this was the first time he had heard Mr. Lilienthal advocate that the power commission enter the scene.

Also after the meeting, the press was invited to meet with a "truth squad" from the A. E. C. This group sought directly to dispute most of Mr. Lilienthal's statements.

It included in its membership: James T. Ramey, the agency's commissioner; Emerson Jones, special consultant to the general manager of Consumers Public Power District of Nebraska; Robert E. Ginna, chairman of the Rochester Gas and Electric Corporation, and Dr. Joseph Howland of the University of Rochester.

Lack of Evidence Charged

Mr. Ramey said that Mr. Lilienthal "has really changed his tune" from past public utterances. He carried with him the text of a talk that Mr. Lilienthal gave to the Detroit Economic Club on Oct. 6, 1947, in which he backed many of the points that he disputed today.

Mr. Ramey said that the former A.E.C. chairman's remarks had been "entirely uncalled for" and he charged him with "impugning the judgment of the [present] Commission."

In his talk, Mr. Lilienthal argued that a recent A.E.C. announcement of a new test reactor to be built "in the Western desert" was an admission by the commission that "not only has the evidence about the New York reactor's safety, not yet been put in the public record—apparently a test reactor is only now being built to provide some of the very kind of evidence that is needed."

The "truth squad's" answer to this was a complete outline of the safety procedures required for approval of a nuclear power plant. Mr. Ramey pointed out that the A.E.C. at present spends "about $25 million a year on reactor safety alone."

Mr. Jones charged that Mr. Lilienthal had taken "completely out of context" statements attributed to Dr. Glenn T. Seaborg, chairman of the A.E.C. Mr. Lilienthal argued that "when Dr. Seaborg, with his colleagues' concurrence, says it is safe enough for him and his family [to live next door to the plant], he is in effect stating a conclusion; he is informally prejudging the New York application in the most forceful possible language."

Dr. Ramey and others in the "truth squad" were strong in their defense of the safety record of the atomic energy industry as a whole, terming it one of the safest of all industries.

The leaders of the atomic industry were in town for the second day of the four-day "Atom Forum"—a combination of the Atomic Industrial Forum and the American Nuclear Society. An "Atom Fair" including exhibits of the industry is being held in the Americana Hotel. Mayor Wagner has designated this as "Atomic Energy Week."

Other speakers in sessions at the Americana, though they preceded Mr. Lilienthal, took strong issue with most of his comments. Representative Chet Holofield, Democrat of California, said the atomic industry was still an "adolescent that shows signs of maturing into a healthy adult." He took issue directly with previous utterances of Mr. Lilienthal.

Representative Holofield warned that private industry must take more decisive steps than it has since the Atomic Energy Act was passed in 1954.

Walker L. Cisler, president of the Detroit Edison Company and first president of the Atomic Industrial Forum, said that the private atomic industry would "not be building and operating nuclear power plants if there were any question as to their safety."

November 20, 1963

Paradise Is Stripped By HARRY M. CAUDILL

DESCENDANTS of the people who settled in the shallow valley of the Green River founded a small hamlet called Paradise, Ky. It was well-named, for the countryside was green and pleasant and the stream teemed with fish.

HARRY M. CAUDILL, a lawyer, is a lifelong resident of the coal country of eastern Kentucky. He is the author of "Night Comes to the Cumberlands: A Biography of a Depressed Area" and of several articles dealing with Appalachian problems. He is a former state Representative and has been a driving force behind Kentucky's recently passed strip-mine control law.

Game abounded and a man could live an unworried life, the tedium broken by an occasional visit to the county seat to listen to trials and swap yarns with friends from other parts of Muhlenberg County.

But times have changed. There is still a dot called Paradise on the map of Kentucky but last year Muhlenberg produced 17.6 million tons of coal—more than any other county in America—and the production record was achieved at a staggering cost. Paradise is isolated and shrunken, huddled in an appalling waste. Thousands of acres of earth are piled high into ghastly ridges, sometimes black

with coal, sometimes brown with sulphur. The streams that wind through this dead landscape are devoid of life.

Here in western Kentucky, part of America's Eastern Interior coal field, the mineral lies near the surface, and the region has fallen prey to strip mining on an immense scale. Strip mining is as easy as it is ruthless. It simply tears the earth apart stratum by stratum in order to rip out the minerals. Conventional tunnel and pillar mining leaves the surface relatively undisturbed, but stripping totally disrupts the land and its ecology.

255

IF YOU HAVE NEVER SEEN A STRIP MINE —Strip mine near Paradise, Muhlenberg County, Kentucky. Here game in green hills once

In a typical Appalachian operation the development may proceed in two or more seams of coal at different levels in a mountain. The uppermost seam may be laid bare by the violent expedient of blasting away the entire overlying crest—a process known to the industry as "casting the overburden." Lower down, the seam is exposed or "faced" by bulldozing and dynamiting the timber and soil away from the coal. The uprooted trees, loose dirt and shattered stone are pushed down the slope. The coal is loosened by light explosive charges, scooped up with power shovels and loaded onto giant trucks for hauling to the nearest railroad loading docks.

Where the terrain is level or gently rolling, the bulldozers and power shovels scrape away the dirt to expose the rock layer roofing the coal. The stone is then shattered with dynamite and lifted by immense shovels or draglines onto the spoil heaps, accumulations that sometimes rise almost sheer to a height of 200 feet. Several acres of the fuel may be bared in this manner before smaller shovels begin loading it onto the trucks.

One gigantic shovel owned by the Peabody Coal Company is as tall as a 17-story building and has become a major tourist attraction. Thousands of people drive out of their way to watch it devastate the American land.

A TRAVELER comes in along the

The U.S. uses ever more electricity.
To make it we use ever more coal, which comes
increasingly from strip mines that devastate
our most valuable national heritage—the land.

abounded and the Green River teemed with fish; last year Muhlenberg County produced more coal than any other county in America.

new Kentucky Turnpike from The Bluegrass, where the lawns and fields are manicured and the miles of wooden fences gleam with fresh white paint. There the influence of the early German immigrants lives on, and in the counties around Lexington it is easy to conclude that the nation's heritage in its land is being safely guarded.

The turnpike carries the traveler through a line of lovely low hills into the west

Kentucky plain. The land thins as it flattens. Muhlenberg County was never high-quality farm land, but it was adequate, and with constructive farm practices and enlightened forestry it had a substantial and permanent potential. By no stretch of the imagination did it warrant the destruction to which it has been subjected.

Aggravating the shock that accompanies one's visit to Paradise is the realization that

this is T.V.A. country and T.V.A. is the nation's bench mark in land conservation. Two billion dollars have gone into T.V.A. projects and a vast amount of favorable propaganda has accompanied its every venture. Millions of Americans assume that the T.V.A. territory is in good hands.

At Paradise, T.V.A. assumes tangible form in an enormous coal-burning electric power

BEHEMOTH—This power shovel, the biggest mobile land machine in the world, here strip-mining coal in Illinois, stands 20 stories high, has a single operator and can scoop up 250 tons of earth with every bite.

plant, built at a cost of $183-million. It consumes 12,000 tons of coal daily. Modern, automated, gigantic, the plant towers above a desolation created by its insatiable appetite for fuel. Just beyond the steel fences which surround it the desert begins. Within sight of the plant Peabody's machines rip the tortured earth while gargantuan trucks rush the coal to the voracious "cyclone" furnaces.

T.V.A. has two faces. One is composed of the green hills around Knoxville, enriched with cheap Government fertilizer and green with pines planted with Government subsidies. It sparkles with T.V.A. lakes and hums with profits from a multitude of new industries attracted by a pleasant climate, abundant water, flood control and dirt-cheap electricity. But T.V.A.'s other face is less pleasing to contemplate. The agency generates much more electricity from coal than from its hydroelectric dams and fuel-buying policies have long been the subject of bitter controversy. By insisting on rock-bottom coal prices for its growing string of huge steam plants it has stimulated strip mining enormously.

T.V.A. is the nation's biggest coal consumer and its purchasing policies have set the pace for the market elsewhere. Despite a general inflation, coal prices have remained stationary for 15 years. Squeezed by rising costs of machines and labor, countless underground pits have been forced to close. Strip mines have been able to hold the price line and meet T.V.A.'s bid requirements. Therein lies the tragedy of Paradise. And therein lies similar tragedy for hundreds of other communities elsewhere in America — and, ultimately, enduring tragedy for all Americans.

A FEW years ago, oil and gas interests confidently assumed that theirs were the modern fuels. Coal was "old-fashioned." In burning, it left a residue of ashes, soot and grit. The industry was archaic —fragmented into hundreds of small companies, undercapitalized and plagued with labor troubles. Obviously the future belonged to other fuels.

But coal has staged an amazing comeback. The demand for electricity has grown enormously—and for most of America coal is the best fuel for generating it. The most optimistic proponents of atomic power estimate that nuclear generating plants will be able to meet no more than 20 per cent of the nation's electric needs by the end of the century.

From a postdepression low in 1954, coal production had climbed 5 per cent by 1960. Then the market zoomed another 5 per cent in 1961 and nearly as much in 1962. In 1963, the gain was 6.5 per cent and in 1964 it was more than 7 per cent. Economists now predict that by 1970 the coal industry will be producing at 100 per cent of present capacity.

According to the United States Bureau of Mines, more than 1.3 billion tons of coal, valued at about $6-billion, have been mined in Appalachia alone in the past three years. This year production is expected to reach 500 million tons. New pits are being opened, and most of the new production will come from "surface mining"—a euphemism dreamed up by the industry's public-relations firm.

The Appalachian coal field extends through Pennsylvania, West Virginia, eastern Kentucky, western Virginia, eastern Tennessee and northern Alabama. This mountain range is one of the richest resource areas in the continent — rich with coal, oil, natural gas, sandstone, limestone, low-grade iron ore, water, timber-growing potential and marvelous scenery. The hunters and wilderness scouts who first penetrated the gaps never beheld a land more enchantingly beautiful than the wooded Appalachian hills and hollows on a misty morning.

With wise management of its resources, Appalachia could have been the richest part of America today. Instead, it has become synonymous with poverty of land and people. But Appalachian destitution did not occur by accident. It is the result of nearly a century of remorseless exploitation.

The timber stands were bought by Eastern lumber companies, and the forests were cut down, sawed up and shipped away in a barbarous manner which totally disregarded the capacity of the land to regenerate the stands. Few healthy seed trees were spared, and today the woods consist mainly of low-grade stock which the lumbermen have culled many times.

But the disastrous exploitation of timber never equaled that which characterized the coal interests. Traditionally, America's industrial muscle has rested on Appalachian coal seams, and mining has garnered huge fortunes for Philadelphia, Boston, Detroit, Cincinnati, New York and Chicago.

From the beginning, coal companies displayed a reckless contempt for the earth. Near their tipples they piled up tremendous culm heaps which they never troubled to vegetate. They poured—and continue to pour — immeasurable quantities of mine wastes into streams. Hundreds of them permitted the inhabitants of their towns to use waterways as dumping grounds for garbage and trash.

Their investment in schools and other local facilities was held to an absolute minimum. All proposals to impose severance taxes on minerals for support of local facilities and services were beaten. As successive generations of extractors prospered, they withdrew from the region, taking their money with them and leaving new accumulations of ugliness and poverty behind.

THE coal industry's lack of responsibility has culminated in today's strip mining. A flight along the Appalachian crest from Pennsylvania to Alabama reveals the awesome scope of the depredations. For hundreds of miles one passes over lands churned into darkening death.

In Pennsylvania alone, 250,-000 acres have been left as bleak and barren as the Sahara. As the hills steepen to the southward, contour strip mining begins. In West Virginia and in eastern Kentucky, whole valleys have been wrecked. Gigantic cuts are made at the face of the coal seam, the "highwalls" sometimes rising 90 feet straight up. The broken timber, shattered rock and dirt are shoved over the steep slopes, to be carried by the rains into streams and onto farm lands.

In Ohio, 202,000 acres in a half-dozen counties have been churned by the machines. Ohio's Senator Frank Lausche, a stanch friend of business, has repeatedly denounced the strippers who are ruining so much of his state. So complete is the devastation that in some areas once-valuable farmlands can now be bought for as little as 25 cents per acre.

In Wise County, Virginia, the coal lies near the hilltops and many mountains have been decapitated—turned into flat-topped mesas. In eastern Tennessee, where the seams are thin and the mountains a bit less steep, the damage subsides somewhat. Nonetheless, the hills which T.V.A. was established to save lie scarred as if by a monstrous whip.

In the Eastern Interior coal field the creeping ruin has spread across western Kentucky and far into Indiana and Illinois. Seventy-five thousand acres in Indiana have been ravaged. In Illinois, 105,000 acres embracing some of the world's best cornland have been turned upside down.

And the ravages of strip mining are spreading to other coal fields farther west. In North Dakota, the Truax-Traer Coal Company is strip-mining three million tons of lignite coal annually for the giant new power plant of the Basin Electric Power Cooperative. Using an ingenious device called the "launch-hammer mining wheel," the operation is incredibly swift and efficient. With the growing demand for electricity on the West Coast and the spread of extra-high-voltage transmission lines to carry the product from generator to consumer, the vast lignite fields in the Dakotas are likely to be wrecked on a gigantic scale in the next two or three decades.

In addition, there is talk of cooking Colorado shales for their petroleum. If this occurs, the shales will be recovered by strip mining, and the beautiful hills of Colorado will face the extinction that now threatens so much of the Appalachian range.

T.V.A.'S OTHER FACE—A Tennessee Valley Authority electric-generating plant at Paradise, Ky. Each day it consumes 12,000 tons of coal, torn from the adjoining strip mine, where a dragline is at work.

WHAT kind of corporations commit this murder of the landscape? One might suppose them to be obscure entities whose managers have not yet learned that in the 20th century it is good business to preserve that which cannot be replaced.

Not so. Many of the great strippers are subsidiaries of well-known American corporations — Bethlehem Steel, Republic Steel, Inland Steel, Interlake Steel, Weirton Steel, Youngstown Steel and Tube and United States Steel. Their advertisements proclaim an enlightened concern for the perpetuation of the American way of life, but in Appalachian valleys they ruthlessly kill the land on which future generations of Americans must depend. The most brutal example of corporate irresponsibility lies on the Poor Fork of the Cumberland River in Harlan County, Kentucky, where the United States Coal and Coke Company has shattered the Big Black Mountain for more than 20 miles, reducing much of this noble terrain feature to a rubble heap.

The coal companies' profits are fantastic. Pittsburgh-Consolidation Coal Corporation earned $19,420,000 in 1955 and $44,470,000 in 1964. The profits of Glen Alden Corporation rose from $40,000 a decade ago to $6,000,000 a year ago. Peabody Coal enjoyed a profit increase in the same decade from $9,430,000 to $30,470,000. Ayrshire Collieries increased its net income from $2,520,000 to $5,790,000. The profits of Pittston Coal Company zoomed from $2,190,000 to $13,720,000 in the same decade.

The land companies that own the mineral wealth and lease it to strip miners show the most profitable balance sheets in American industry. In 1964, the Virginia Coal and Iron Corporation retained as net profits 61 per cent of its income and paid out 45 per cent of its gross income as dividends. This record was equaled by another huge land company, the Kentucky River Coal Corporation. By contrast, in the same year, General Motors cleared a dime out of each dollar received and paid out a nickel in dividends.

AND what of the people whose communities are shredded for cheap coal? Obviously, ruined lands must be abandoned, and the strip-mined counties all have shown sharp population declines. A million people have moved out of Appalachia in the past 10 years, and the Eastern Interior field has fared little better. As subterranean mining declined, thousands of families passed onto the public assistance rolls. Today, welfare, not mining, provides most of the money spent by families in the nation's coal fields.

In Appalachia, exploitation has reduced a once-proud and even violent people to the most passive and trampled-upon part of the American population. Passivity has reached its ultimate depths in eastern Kentucky, where whole communities have been impoverished and debased with scarcely a protest. There, coal companies often own the minerals underlying the lands of farmers and the state's highest court has ruled that the companies have the right to destroy the land in order to get out the minerals.

With this license to wreck, many operators have proceeded with complete abandon. They have rolled rocks through some homes and have pushed others off their foundations. Many have been demolished by avalanches from the spoil banks. In Knott County, a one-armed coal miner came home from a retraining program conducted as a part of the war on poverty to find his house and all its contents buried beneath a mammoth landslide.

When a group of mountaineers calling themselves The Appalachian Group to Save the Land and People visited Gov. Edward T. Breathitt of Kentucky last June, an 80-year-old woman told him that she had stood on the front porch of her little home and watched the bulldozers invade her family cemetery. She said:

259

> **Most Americans have long assumed that the waste of resources was curbed and that victory over greed and wantonness was achieved in the days of Theodore Roosevelt. Nothing could be farther from the truth.**

"I thought my heart would break when the coffins of my children come out of the ground and went over the hill." This situation prompted one mountaineer to comment that the coal industry digs up the dead and buries the living.

But even in the Kentucky Mountains, suffering and patience have their limits. In the Clear Creek community, women copied the tactics of civil-rights demonstrators and stopped the strippers by lying down in front of their bulldozers. They were promptly hauled off to jail, but the machines are still stalled by recalcitrant mountaineers determined to save their lands from ruin — and themselves from total impoverishment.

AMERICA is not alone in its craving for coal or in facing the problems growing out of strip mining. In Europe, postwar reconstruction has seen a Socialist Government in Great Britain and a Communist Government in Czechoslovakia authorize the demolition of whole towns to get at the coal beneath them. In England, strip miners have gone to a depth of 700 feet — 500 feet deeper than anything yet attempted in America.

But European national coal boards have demonstrated that complete reclamation of strip-mined sites is possible. In England, for example, the topsoil is carefully scraped off and saved in separate heaps. Next, the subsoil is pushed aside and similarly isolated. So is the rock above the coal. At the end of the operation, the rock goes back into the pit first; then the subsoil is pushed in; finally, the topsoil is restored to its original site. The restored land is compacted as it is replaced. After it is contoured, it is fertilized and treated with limestone. It then undergoes an intensive five-year agricultural restoration treatment. Most of the land is returned to agriculture; the rest is planted in timber. Total cost averages about $1 a ton of coal recovered. These achievements demonstrate that "total reclamation" on the right terrain is within easy reach of American industry.

IN 1875, California outlawed hydraulic gold mining, on the ground that it silted streams and caused the flooding of valuable farm lands. This type of mining was comparatively harmless when compared with coal stripping, but it is unlikely that any of the states will face up so squarely to our modern problems. In the coal states, legislatures are composed in the main of little men who yield easily to blandishments and enticements, and the coal industry is rich and ruthless.

For years, Pennsylvania had the strictest reclamation law. It required operators to obtain stripping permits and to post bonds to assure reclamation of the land in accordance with a state-approved plan. But when the stripped land is very steep, a bond guarantees only the impossible — a situation comparable to authorizing rape on assurance that the rapist will afterward restore his victim's virginity.

Early this year, conservationists all over America watched with admiration the efforts of Kentucky's young Governor Breathitt to tighten his state's reclamation law. The act, as amended, authorizes the Kentucky enforcement agency to require strippers in mountainous eastern Kentucky to drag part of the soil off the spoil banks and use it to cover the perpendicular highwalls, reducing them to a slope not exceeding 45 degrees. In western Kentucky, the mined land must be shaped so that it can be traversed by farm machinery. All disturbed lands must be seeded to a prescribed vegetative cover or, in the mountains, planted with approximately 800 seedling trees for each acre.

•

To these modest and reasonable requirements the coal industry responded with outraged bellows. Scores of lobbyists descended on the State House. Huge funds were collected from operators and their suppliers for the avowed purpose of defeating the bill. Governor Breathitt spoke out against improper pressures upon legislators and the bill passed by a comfortable margin — the severest public setback ever suffered by any segment of the coal industry in Kentucky.

While Breathitt remains Governor, strict enforcement is to be expected, but his administration is likely to be exceptional. The state has had a reclamation statute since 1954, but it has been generally ignored, and it is to be feared that future state administrations will emulate this leniency.

Elsewhere in Appalachia, Virginia, Tennessee and Alabama have never bothered even to enact reclamation statutes. In the Eastern Interior field, only Illinois has any sort of reclamation law. In the Western field, such legislation has never been seriously considered.

If governmental power is to save the American land, it must be Federal power backed by a strong national will and conscience. The American population is growing rapidly; estimates of the United States Bureau of the Census indicate a population of 300 million by the year 2000. The nation's land base cannot grow by a single inch, but it can be effectively diminished by industrial processes which include not only strip mining for coal but quarrying, borrow pits, opencast iron mining and similar operations. Under any enlightened philosophy, the present occupants of the land hold it in trust for future generations and are under a positive obligation to pass it on in a tolerable state.

The blight of strip mining does not stop at the edge of the spoil banks. When freshly exposed, the soil is hot with sulphuric acid: for years nothing can grow on it. In the meantime, the sulphur and mud wash into streams, killing aquatic life and piling up in horrible weed-grown banks. The long-range cost of dredging the Mississippi and its tributaries of this coal-flecked debris will be astronomical — a burden all Federal taxpayers will share.

In my opinion, the Great Society should enunciate a clear-cut policy relative to extractive industry and its distorted practices of social accounting. Based on need and historical experience, strip mining should be permitted in those areas where terrain and weather permit complete reclamation — that is, in flat or gently rolling country. In West Kentucky, Illinois, Indiana, Ohio and most of Pennsylvania, coal can be extracted cheaply and efficiently by this method, and the same machines which rip the earth can heal the scars. Following British practice, the land can be restored to its original condition and, perhaps, even improved. If funds are made available for the purpose, and if the state and Federal Governments make certain it occurs, intensive treatment with fertilizers, limestone and leguminous plants and trees can restore the land to beauty and usefulness. Such costs should be borne by industry as part of the price of coal.

At the same time, it should be recognized that in most of Appalachia the land is simply too steep, too rough, too rugged, and the rainfall is too heavy, to permit restoration. When a hill is decapitated or flayed, the topsoil vanishes first. There is no feasible way to separate the rock, subsoil and topsoil and then to restore them to the pits in their natural order. In such terrain, and with the precipitation, freezing and thawing natural to the region, to strip mine is to destroy. Unless technological advances make possible a complete reclamation of mountainous land, national policy should outlaw strip mining in such terrain.

The technology of subterranean mining has made fabulous strides in the last two decades. Continuous mining machines, roof-bolting devices, battery-powered coal cars, improved ventilating techniques and strict enforcement of the Federal mine safety code have not only made the miner's life easier and safer but increased his productivity more than twofold. Coal from tunneled mines must sell for a little more per ton than that from strip mines, but the difference is a small price to pay for the preservation of the land.

The third feature of the national policy should provide an effective program for the reclamation of "orphan banks" — old strip mines worked out and abandoned long ago. Congress should appropriate funds for the purchase of these lands, and they should be smoothed and revegetated. When so restored, they could be sold, and the proceeds applied in payment of the Government's investment.

The same policy should apply to lands pitted by taconite mining in northern Minnesota, and to other areas where industries extract minerals by open-cast mining. Such rules would require only that the generation which damages the land — and benefits thereby — would pay for restoration, instead of future generations which, in all probability, will have enough problems of their own making.

To date, in the mining of all minerals, 1.75 million acres have been destroyed or severely damaged by surface mining. Some 400,000 of these acres are officially classified by state agencies as "reclaimed land," though often there is not a blade of grass or a seedling tree to support the claim.

THE hour is late and the agony of the land is intense. Most Americans have long assumed that the waste of resources was curbed and that victory over greed and wantonness was achieved in the days of Theodore Roosevelt. Nothing could be farther from the truth. Shocking as were the mass slaughters of the American bison and the passenger pigeon, they were no more grotesque than the present destruction.

As wealth multiplies, hordes of Americans will purchase country retreats and seek quiet areas for recreation and leisure. Someday, every acre will be needed for its food and fiber. Unless we act now our grandchildren may inherit vast man-made deserts, devoid of life, polluted with acids, hideous to the eye, baked by the sun and washed by the rains. If this is their heritage, they will curse us so long as the deserts remain to monumentalize our greed and folly.

Continued silence by the national Administration on this urgent issue is inconsistent with the dream of a Great Society. It is, in fact, inconsistent with simple patriotism and basic common sense, for unless the land lives the people must perish.

March 13, 1966

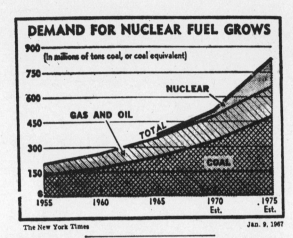

DEMAND FOR NUCLEAR FUEL GROWS

The New York Times   Jan. 9, 1967

Coal and Hydroelectric Areas Find Atomic Power Moving In

Reactors Are Planned by T.V.A. and Northwest Power Group—Utility, Long a Holdout, Joining Trend

By GENE SMITH

Two major events marked the emergence of atomic power last year as a real competitor against conventional fuels and methods of power generation. The first was the announcement last June that the Tennessee Valley Authority had decided to build two 1,065,000-kilowatt nuclear reactors near Decatur, Ala., at a cost of $247-million.

The second, announced last month, was the decision of the American Electric Power System to seek bids by Feb. 1 for a 1-million to 1.1-million-kilowatt atomic power plant on Lake Michigan, nine miles south of St. Joseph, Mich.

The importance of these two plans stemmed from the fact that T.V.A. is situated in the heart of low cost coal supplies and that American Electric has long been the technological leader in the industry and had, until now, resisted the move to atomic power.

Add to this the fact that earlier in December a Pacific Northwest power group had made public plans for a 1 million-kilowatt nuclear plant to be ready by 1973. This would be in the heart of the traditional home of hydroelectric power.

First In Northwest

Owen Hurd, managing director of the utility group, said this plant would be the first of 11 stations in the 1 million-kilowatt range to be required in the region by 1985. It is believed that most of these will be atomic.

Over-all, the most significant aspect of nuclear power in 1966 was the fact that more than half of all new steam-electric generating capacity ordered by the nation's investor-owned electric utilities last year were of the nuclear type.

William J. Clapp, president of the Edison Electric Instiute, the industry's trade association, said this meant that by the year's end some 130 electric power companies were participating in one or more of 53 nuclear power projects involving outlays of more than $3.5-billion.

Mr. Clapp, who is also president of the Florida Power Corporation, reported that "at least 18 orders were placed in 1966 by investor-owned utilities for the construction of large nuclear units with a total capacity of about 13.4 million kilowatts, or more than 12 times the capacity of the 13 nuclear units now in operation."

13 Plants Operating

He said that the 53 projects included 13 units now in operation; 33 under design, construction or in the planning stage and scheduled for operation by 1973, and seven that involve research, study and development of nuclear energy. The capacity of all these nuclear units totals 24.5 million kilowatts and does not include T.V.A. or other strictly public power activities.

While none of these new nuclear plants has been situated in an urban area, a proposed plant in Burlington, N. J., represents the closest approach to a city since Consolidated Edison Company two years ago cancelled plans for proposing atomic energy in its Ravenswood station in Queens on the East River.

Burlington, which will have a capacity of more than 1 million kilowatts, will be owned jointly by the Public Service Electric and Gas Company of New Jersey, the Philadelphia Electric Company, the Delmarva Power and Light Company and the Atlantic City Electric Company. Its site, on a 140-acre tract on the east bank of the Delaware River, is about 11 miles southwest of Trenton and 17 miles northeast of Philadelphia.

Sixteen cities or towns with populations in excess of 25,000 will be within 25 miles of the site, including the Philadelphia-Camden region with more than 2 million people and Trenton with 114,000.

The problem of locating nuclear reactors near the user of electricity seemed to be such an obstacle for the Boston Edison that it gave up plans to build a 612,000-kilowatt nuclear power station within the city limits for completion by 1971. The decision was based on the belief of reactor manufacturers that without years of operating experience with similar types of plants, it might be impossible to meet the target date. A less populated area was selected for the site.

Just now successful was the nuclear power effort last year may best be seen in a report from the Southern Interstate Nuclear Board, the agency in the nuclear and space fields for the 17 states of the Southern Governors' Conference. It showed that four Southern utilities, plus T.V.A., had ordered 7.66 million kilowatts of nuclear power reactors valued at $990-million during the last 12 months.

"Until a year ago there was not a single commercial nuclear power reactor in the 17-state area," the report said. "In fact, the only power reactor was a 65,000-kilowatt thermal experimental prototype reactor built and operated by Carolinas-Virginia Nuclear Power Associates at Parr, S.C."

It also noted that more of the announced units would be built in South Carolina than in any other Southern state. These units included a 700,000-kilowatt plant at Hartsville costing Carolina Power and Light Company $70-million, and the Duke Power Company's twin units with a total capacity of more than 1.6-million kilowatts near Clemson. The award for this $207-million project went to Babcock & Wilcox Company for its nuclear division at Lynchburg, Va.

Robert Gifford, executive director of the Southern nuclear group, said that last year's effort was the result of 20 years of study on the part of the Atomic Energy Commission, American industry and Southern utilities.

"Because the Southern states have been blessed with outstanding hydroelectric sites and low-cost coal, oil and natural gas, we did not expect competitive nuclear power to become a commercial reality this fast," Mr. Gifford said. "But the reduction in costs, particularly over the last two years, made it possible for Southern utilities to proceed with their commitments this year."

Wilfred H. Comtois of Westinghouse Electric Corporation's atomic power division joined a growing body of opinion that has advocated a combination of nuclear power plants that would also provide sweet water from salt or brackish water. Speaking at a meeting in November of the American Society of Mechanical Engineers, Mr. Comtois said the only question was one of economics.

He added that it appeared that the estimated cost of producing water from nuclear combination plants would be "competitive with conventional sources of water in many areas of the world."

January 9, 1967

Liquefied Natural Gas Gaining Role in World Oil Trade

SUPER COOLER: Natural gas, pumped from Zelten and Raguba fields in Lybia, is cooled at the Esso facility at Marsa el Brega to a temperature that changes it to a liquid and is then stored in tanks at the upper left.

By WILLIAM D. SMITH

In the history of the petroleum industry, 1969 will likely go down as the year of LNG.

Liquefied natural gas, to give the commodity its seldom used but proper name, will come into its own in the remaining months of this year as an important aspect of the world oil trade. LNG is the gas found at the well head but is reduced to liquid form by extreme pressure and cold.

The first significant event will be the shipment sometime this spring of LNG by Standard Oil Company (New Jersey) from Marsa el Brega in Libya, to Barcelona, Spain, bringing to fruition the largest single investment in the history of the international oil industry.

This will be followed by shipments to La Spezia, Italy, completing a $350-million project that Jersey began eight years ago.

Jersey, however, will not have a monopoly on the LNG news or business this year. In July, the Marathon Oil Company and the Phillips Petroleum Company hope to begin shipments of LNG 3,-250 nautical miles from Cook Inlet in Alaska to Japan. This project is expected to cost more than $200-million.

Facility in Algeria

The only existing facility

for international LNG trade is at Camel, Algeria, which ships the product to both France and England.

Another plant is being built at Skikda, Algeria, by Sonatrach, the Algerian national hydrocarbon company and Erap, a French company. Construction will begin in 1971.

At least three other LNG projects are in the advanced planning stage. The buyer in each case is Japan.

Royal Dutch Shell is talking about shipping gas from Borneo.

Russia, not to be outdone by capitalist initiative, is discussing shipments from Sakhalin or Vladivostok and British Petroleum Company, Ltd., and Compagnie Francaise des Petroles have plans for shipment from Abu Dhabi, the Persian Gulf sheikdom.

Start-up data for all three projects is sometime in the 1970's.

The initiative for international trade in LNG, strangely enough, began with a Midwest meat packer, Union Stockyards of Chicago.

The company felt it was paying too much for its locally produced fuel and began experiments with refrigerated barges designed to move natural gas, which was liquefied near its source on the Mississippi River.

A liquefaction plant was built and barges were insulated in a new way to carry the liquid gas. But the river authorities and insurance companies raised difficulties. Meanwhile, the company had managed to get a better price from its local suppliers.

This might have been the end of the story but for the fact that a much bigger customer became interested — the British Gas Council. The council put up capital with Union Stockyards and the Continental Oil Company to outfit a small tanker to carry the liquefied gas.

The first international shipment of LNG took place 10 years ago between Lake Charles, La., and Canvey Island in the Thames River in England.

The original plans called for the first commercial shipments to move from Venezuela but the discovery in Algeria of large gas fields changed the situation. This project has matured to where it now ships 1.5 billion cubic meters a year to the British Gas Council and 0.5 billion cubic meters to France.

When it reaches full contract level in 1970, Jersey's Libyan production will amount to 3.6 billion cubic meters a year.

The Algerian natural gas

is mainly produced from fields where there is little or no oil output. Jersey's operation, on the other hand, will utilize gas produced as a byproduct of oil.

In 1961, Jersey organized a task force to study and recommend the best means of utilizing this associated gas. Not only was the company looking for a way to make profitable use of this potentially valuable product, but the Libyan Government was putting pressure on all the oil operators to do something about the waste of this valuable national asset.

Use of the gas in the manufacture of petrochemicals and ammonia fertilizers was suggested but rejected for lack of ready markets for products made from the large amounts of gas available. Even the possibility of an underwater pipeline across the Mediterranean to Italy was considered.

Advances in cryogenics, or the technology of extreme cold, suggested the possibility of liquefying the natural gas from Libya's giant Zelten and Raguba oil fields so it could be shipped to Mediterranean ports in Europe.

The economics of the situation were carefully studied and sales contracts were signed in November, 1965. When the operation is in full swing, the gas will be brought by pipeline from the separations plants in the fields to the port of Marsa el Brega.

Complicated Process

There it is liquefied in a giant cryogenic facility and stored in insulated tanks until it is loaded on one of four specially constructed 250,000-barrel LNG ships.

At the oil fields, the gas-oil separations units already existed. Now, instead of flaring the separated gas, it is processed and fed into new compressors to increase the pressure to 720 pounds per square inch for transporting through a 36-inch pipeline.

After reaching Marsa el Brega, the gas is subjected to a complicated process and finally becomes liquid at atmospheric pressure.

The basic design of the liquefaction plant in Libya was prepared by ESSO Research and Engineering Company. The plant uses a refrigeration cycle developed by Air Products and Chemicals, Inc.

The Bechtel Corporation handled the detailed engineering of facilities and the plant was erected by SNAM Progetti S.p.A. and Compania Italiana Industriali, both of Milan, Italy.

The liquefaction plant stands on 10 acres and consists of two separate but identical units. It incorporates four cryogenic exchangers, each about 200 feet high.

After liquefaction, the LNG is stored above ground in two 300,000-barrel double-walled storage tanks.

Four ships are being built to transport the LNG to Spain and Italy. The cost of the four ships — which in many ways are floating thermos bottles — is about $80-million. Two have already been built.

To handle present and near-term ocean LNG contract trade, three tankers are now operating out of Algeria, two have been completed for the Libyan venture and another four are under construction. Bids are out on at least five more vessels.

While the Esso-Libyan venture is now running a little behind schedule, the Phillips-Marathon operation in Alaska appears to be a little ahead of the game in its race to become functional by the July deadline.

In March of 1967, the two companies signed contracts with the Tokyo Electric Power Company, Inc., and Tokyo Gas Company, Ltd., for delivery of 1.4 billion cubic meters a year under a 15-year contract.

Phillips Owns Share

Phillips owns 70 per cent of the project and will provide 70 per cent of the gas from its Cook Inlet gas field.

Marathon owns 30 per cent of the project and will provide 30 per cent of the gas from its Kenai field. Marathon is the operator.

Despite the massive scale of the Libyan and Alaskan projects, LNG will most likely continue to be an interesting, profitable yet minor contributor to world energy supplies, according to a study by Arthur D. Little, Inc.

Estimated LNG movements for 1970 will be at a rate of approximately 635 million cubic feet a day but will account for only a little more than one per cent of the natural gas used in the Free World at that time.

All the rest of the natural gas used will be in gaseous form through pipelines.

Viewed against total energy needs, LNG will supply a little more than two-tenths of one per cent of the estimated Free World demand in 1970.

Increase Predicted

Even with the predicted tenfold increase by 1980, LNG will still supply no more than slightly over one per cent of Free World energy demand.

The market for LNG is created by a unique set of circumstances. The consuming country must be deficient or totally lacking in indigenous gas and also be remote from a supply source capable of being brought to market economically by pipeline.

In addition, LNG is a very capital intensive product and for that reason there must be a willingness to accept long-term supply contracts.

Nonetheless, LNG appears to be on the world energy scene to stay and 1969 will likely go down as the year it made its mark.

February 9, 1969

Supply of Natural Gas Is Debated

By GENE SMITH

Is there really a shortage of natural gas?

The answer to this question tends to vary greatly. The Potential Gas Committee, an all-industry group financed by the American Gas Association, the American Petroleum Institute and the Independent Natural Gas Association of America, released last June 11 updated figures on gas supplies and concluded: "There is plenty of natural gas in the United States—but somebody must drill the wells to get it."

W. Morton Jacobs, president of the American Gas Association and of the Southern California Gas Company, told a New York Society of Security Analysts meeting: "Experts estimate there is enough gas in the ground to meet the nation's requirements to the turn of the century, based on known geology and the present state of the art in production."

Gordon C. Griswold, president of the Brooklyn Union Gas Company, in response to a question on gas supplies at a hearing by the New York Public Service Commission, said: "I feel that this concern for the availability of supply is essentially a short-term problem and that the supply in the mid- and long-term future will be adequate to meet

Forty-foot sections of 3-foot pipe being welded into a third main natural gas transmission line for the Trunkline Gas Company. The new line, part of concern's Texas-to-Michigan transmission system, will add 1.5 billion cubic feet of gas daily in network.

the future needs for Brooklyn Union."

Although the problem of gas supplies has been getting public notice for at least six months, the threat of dwindling supplies dates back to at least 1960 when the Federal Power Commission made its basic decision that rates at the wellhead should be fixed on a regional, or area, price formula.

When the Permian Basin rates were set for all gas in that producing area of Texas and New Mexico, there were those in the industry who immediately warned that drillers would slow down their efforts because there would no longer be sufficient incentive to locate and to develop new gas fields.

The Permian Basin formula was eventually appealed all the way to the Supreme Court, which decided that area pricing should be applied to all of the gas-producing regions of the country. Producers viewed this as a new philosophy of "mass control based on averages" and vowed to seek a way out of such rigid controls.

One obvious way has been to sell gas only for intra-state use, which is not subject to F.P.C. regulation. The other was only too simple: Just don't spend money for costly drilling operations.

The results have been equally obvious. In the 10 years ended 1967, proved reserves of natural gas dropped from more than a 22-year supply to one of less than 16 years. In mid-1968 Dr. Paul Mc-Avoy of the Massachusetts Institute of Technology, predicted that by 1970 the supply of natural gas will fall short of demand by 1.8 trillion cubic feet.

Earlier this year, Ralbern H. Murray, director of marketing for the Consolidated Natural Gas Service Company, Inc., of Pittsburgh, told a meeting of engineers that projected domestic consumption of natural gas will exceed the anticipated discovery rate by 10 trillion cubic feet a year by the year 2000. He stressed that the use of natural gas since World War II gas been growing faster than the amount of proved reserves.

Mr. Murray also found that 1000 trillion cubic feet of natural gas would have to be discovered between now and 2000 but that discovery rates can only be expected to rise to 25 trillion cubic feet a year during the next 30 years.

F.P.C. commissioners are gradually adopting the view that some changes will have to be made in the area pricing concept if gas supplies are to be adequate to meet demands.

Flexible Pricing Urged

Carl E. Bagge, vice chairman of the F.P.C., told a seminar at Oklahoma State University in May that even the Supreme Court had referred to the Permian formula as "an experiment rather than in inflexible guide to the future."

"The method of pricing natural gas, it seems to me, must be made more flexible," Mr. Bagge said.

Lawrence J. O'Connor Jr., another commissioner, pointed out that in the 1970's attention may be focused more on the cost of new gas supplies to the customer than on the over-all price level received by the producer.

Albert B. Brooke Jr., a third member of the F.P.C., submitted that the commission consider exempting from price controls all but the largest gas producers.

Producers Defended

The president of one of the nation's major pipelines told an industry-executives meeting in Colorado Springs last June 11 that between 1963 and 1967 the percentage increase in marketed production of gas was five times greater than the increase in net reserves added to supply. He told the meeting that it was completely untrue that major producers were withholding large, undisclosed gas reserves from sale in order to create an emergency.

He then outlined a plan of action to obtain more gas, including: streamlining or eliminating much of the formal hearings before the F.P.C.; increasing rates for new supplies of gas; giving producers "current dollars" to meet inflated expenses and investments; exemption from controls for minor supplies; and creation of premium prices for special exploratory projects.

July 15, 1969

Wide Power 'Brownouts' Likely in East in Summer

By BEN A. FRANKLIN
Special to The New York Times

WASHINGTON, June 6 — Much of the Eastern half of the nation is almost certain to have some dislocation of electrical service this summer.

Government and utility industry officials say that in large areas of this "land of electrical living," as the advertising men put it, there is simply not enough power to go around during the sultry periods when air conditioner use is at its peak.

Pressed to meet air conditioning power demands, a number of utilities have already resorted to a systemwide voltage reduction to spread the available electricity around.

The situation differs from those of 1965 and 1967, when blackouts darkened huge areas of the populous Northeast for hours and then were repaired. The problems now are chronic and systematic. They are apt to have nagging widespread effects for months and perhaps for years.

The new worry in the electrical industry, accordingly, is the "brownout," a disruption of less than total proportions. But along the way, there may be scattered blackouts as well — some of them unavoidable but deliberate.

In most areas — if major equipment failures do not bring more total outages — the public evidence of overtaxed generating capacity may be picture shrinkage and loss of intensity in color television sets. Utilities usually do not announce that brownouts are in progress, and in many jurisdictions they need not report them later.

A brownout is the signal of an energy-rationing decision by power companies to reduce line voltage. A voltage drop leaves more power to spread around.

According to engineers at the Federal Power Commission, even small voltage reductions cut the efficiency of such devices as electric ranges and toasters, shorten the life of fluorescent tubes and raise the operating temperatures of electric motors.

Effect of Voltage Drop

The engineers say a temporary drop of no more than 5 per cent should not damage "significantly" refrigerator or air conditioner motors. But overheating may trigger automatic cut-off switches.

Some electrical motors will restart automatically after cooling off. On others, a "reset" or "overload" button, often difficult for the unpracticed to find, must be pushed to restart when the protective device turns them off.

When the hot weather arrives, millions of consumers, who may already have unwittingly experienced voltage cuts, will be asked to make voluntary reductions in air conditioner use. And they will experience "cycling"—the intermittent shutdown of overheated air conditioner motors starved for voltage.

In their first joint action on any such matter, the Public Service Boards of Pennsylvania, New Jersey, Delaware, Maryland and the District of Columbia—agencies that monitor utilities in so-called P-J-M Pool—last week ordered the 100 power systems they regulate to send cautionary letters to all their customers by July 1.

Warning to Consumers

P-J-M consumers will be warned that some of the region's companies "will be unable to meet the need [for power] from their own resources" and requested to conserve power by setting air conditioners at "no less than 75 degrees."

Government officials say that isolated blackouts may occur even without equipment failures when power companies decide to "shed" portions of their system overloads at peak hours. In such cases, a power company simply pulls switches to cut off certain substations from overtaxed generating plants in the hope of preserving service on the rest of its system.

There may be economic repercussions as well. Some factory shifts may have to be rescheduled by large industrial consumers of electricity. The installation of electrical machinery may have to be postponed. Some home appliances —particularly air conditioners

—may not be sold.

Many utility companies have already suspended their institutional air conditioning promotions and are concentrating on electrical heating advertisements instead. Winter power loads are smaller. So are summer sales of heaters.

But if summer brownouts come—and such a generating giant as the 19-million-megawatt Tennessee Valley Authority is calling the present situation "an emergency"—a winter of serious disruptions may not be far behind. According to anxious Government and industry officials, the trend of sharply rising power consumption and a long-term lag in generating capacity is casting long brownout shadows for months ahead.

The reasons for the situation are as complex as the country's vast, interconnecting power grid itself.

Experts endlessly debate the causes. But nearly all agree that there is blame enough to go around to all concerned— the Federal Power Commission; other top Government energy and fuel planners, particularly the Atomic Energy Commission; the private and public electric utility industries, the coal industry and the railroads.

S. David Freeman, the top Federal energy policy planner, says they have all been "living in a dream world."

Based on interviews with spokesmen for each of these interests, the explanations and excuses are as follows:

¶The coal industry is unable to deliver enough of the basic fuel of steam-electric power— more than half the total electrical energy is generated by burning coal—and it attributes the situation to the Atomic Energy Commission. In the late fifties and early sixties, its critics say, the commission persuaded the country—and therefore most coal executives and most investors in coal— that cheap, nonpolluting atomic electric power was just around the corner. It was not. Most atomic plants have experienced unexpected, recurring technical delays. At the same time, coal apologists say, the mining industry was being mechanized, but it suffered from a shortage of capital, and not enough new mines were developed. Hundreds of mines are being opened now, but it takes two to three years to begin significant production. And there is a shortage of trained miners and mechanics.

¶The coal railroads, also in

a slump, failed to order enough coal hopper cars and, on some lines, even enough locomotives to transport the coal to markets. The rapid rise in coal exports, particularly to Japan, further disrupted the rail transportation system. Hundreds of 100-ton coal cars have stood full—and idle—for days and weeks at Hampton Roads, Va., docksides, awaiting off-loading to ships still on the high seas. And the inter-line competition for hoppers—they can also carry beets, wheat, gravel and other bulk cargoes—has left some major coal hauling lines, like the Louisville & Nashville, with thousands of its coal cars "off line," tied up elsewhere in the service of other railroads.

Government officials say the utility industry, by consistently underestimating its own annual sales growth and assuming a 7.5 per cent yearly increase in power consumption when the average advance has been 9 per cent or more, has laid itself open to the extreme difficulties it now confronts. Federal power economists believe the utilities also have been too cautious in adding new high-voltage transmission lines that could help meet surge demands in one

region with excess power from another.

¶The utilities assert that they have been blocked in expansion efforts at nearly every turn by the activists and lawsuits of the new "ecological revolution." There is aggressive resistance now even by state and local governments to the air, water and radioactive pollution threats of both conventional and nuclear power plant construction proposals. Esthetic protests have blocked or delayed transmission lines. The trend of environmental opposition is up.

But the Federal Power Commission has never advised the industry that it would need the 25 per cent margin of generator capacity that many now agree is essential to withstand sudden air conditioning loads. And while the coal industry has been publicly advertising its failures, the railroads and most utility spokesmen have been minimizing theirs.

Six months ago, A. H. Aymond, president of the biggest utility trade association, the Edison Electric Institute, described as "sheer nonsense" warnings by top coal executives that brownouts and blackouts lay ahead.

June 7, 1970

SCIENTISTS WARN ATOM POWER FOES

Say Public Must Back Gains or Give Up Conveniences

By NANCY HICKS
Special to The New York Times

UNITED NATIONS, N. Y., Aug. 15—An international gathering of scientists and businessmen concerned with atomic energy in effect told an increasingly hostile public this week that it would have to choose between air-conditioners, dishwashers, television sets and garbage disposals on the one hand and, on the other, greater responsibility for the effects— notably environmental pollution—of increased power demands.

This view emerged from an international symposium, Environmental Aspects of Nuclear Power Stations, at the United Nations.

About 400 representatives of science and industry from 25 countries—all committed to the

future of atomic energy as a source of power—attended the meeting, which was sponsored jointly by the International Atomic Energy Agency and the U. S. Atomic Energy Commission.

They met daily, exchanging papers that asserted the relative safety of nuclear energy. They were quick to reply to charges that they excluded any opposing points of view from the meeting.

Power Plant Dispute

Over the last year and a half, in a time of rising demand for electric power, public opposition to nuclear power plants skyrocketed as the future sanctity of the environment became a national issue. Critics have charged that such plants involve hazards of radiation and thermal pollution of waterways. At the same time, resistance to power plants using conventional fuels has risen because of concern over air pollution.

Many scientists involved in research contend that much of the opposition to nuclear power production is based on "stirrer-uppers," as one man called them, who do not have facts. Others concede that the public has genuine concern that should

be handled honestly. All agree, however, that the future of nuclear power generation as well as that of fossil fuel (oil or coal) or water will not progress without active public participation in planning.

"I'm personally glad that the public is so interested in the problem of nuclear power generation and the environment," mused Dr. John Dunster of the United Kingdom Atomic Energy Authority. "I hope this means more money and more enthusiasm to be more heavily taxed to pay to protect the environment."

In one paper, Dr. Chauncey Starr, dean of the School of Engineering and Applied Science of the University of California at Los Angeles, developed a philosophical approach to deal with the problem of risk-versus-benefits in future atomic energy production.

'Trade-offs' of Risks

He said that in all of American activities there are "trade-offs" between risks and acceptability.

"There are contradictory assumptions in the operations of our society," he said. "First, it is commonly accepted that everyone should have the opportunity for a natural death. Second, it is commonly ac-

cepted that every individual should have an opportunity to use and enjoy the fruits of our centuries of technological development.

"Third, it is the philosophy of an egalitarian society that where the activities of an individual infringe on others in an undesirable way, the society may intervene to control individual activities in order to achieve a balance between group well-being and the privileges of the individual. It is evident that these inherent assumptions are not compatible."

He added that the risks that the American public was willing to take in sports and transportation was about statistically equal to the death rate caused by disease. He suggested this might be a yardstick to use in determining the probable safety of controversial, risk-benefit questions involved in atomic energy safety, contraceptive pills and the like.

Some scientists, however, have said that these problems of long-range risk are not realistic because nuclear power is just a stopgap measure until some method can be found to harness solar energy, a finite source, for electric energy production.

August 16, 1970

Hazards of Nuclear Power

To the Editor:

The current discussions, pro and con, on the dangers or relative safety of radiation emissions from nuclear power plants should, by the diversity of beliefs expressed, cause thinking people to be deeply concerned.

The Atomic Energy Commission has been entrusted with the responsibility of promoting and controlling nuclear energy. In examining its many publications, the "Understanding the Atom" series in particular, it becomes apparent to the reader that this huge, bureaucratic monster, the A.E.C., has developed a philosophy of life and death in sharp contrast to the accepted beliefs of all modern religions. The A.E.C. subscribes to the tenet that we must accept a certain amount of deaths, cancers, leukemias and gene-

tic malformations in order to "enjoy" the "benefits" of nuclear power.

How do they sustain this premise? Quite simply; by publishing reams of verbiage designed to placate an unsuspecting public.

We are told that the advantages of nuclear power use in our society are such that we must be prepared to pay some price. We are told that we must "balance lives against lives." I'm certain those who must be sacrificed would like to know in advance, to face their executioner. But such is not the case.

It would seem that any technology so replete with dangers should be avoided rather than coddled. Any technology admitting there is no threshold in the genetic effect of radiation and there is no "safe" amount of radiation insofar as genetic effects are

concerned but continuing to pursue a course destined to elevate the radiation levels throughout our entire environment is a technology totally unneeded by mankind.

We can produce electricity from other power sources. We do not have to accept the inevitability of a radiation-saturated world just to satisfy the egos of a group of well-intentioned but morally misdirected scientists.

The presumption of any governmental agency to literally decide the life or death of anyone for the supposed benefit of others far exceeds the authority intended them or any earthborn mortal.   JOHN K. MUSTARD
Executive Director, Delaware Valley
Committee for Protection of
the Environment
Moorestown, N. J., Sept. 25, 1970

October 4, 1970

MORE NATURAL GAS IS SOUGHT FOR USE IN EASTERN STATES

By ROBERT D. McFADDEN

The utility commissioners of eight Eastern states have appealed to Secretary of the Interior Walter J. Hickel to replenish the nation's dwindling reserves of natural gas by fostering a more rapid of development of new supplies.

The growing shortage of natural gas has prompted three New Jersey utilities to restrict new large customers, and the New York State Public Service Commission has said it is considering an order to impose similar restrictions on gas companies throughout the state.

In a meeting with Mr. Hickel last Tuesday, the eight commissioners urged tighter requirements to insure the prompt development of outstanding leaseholdings by producers, more frequent lease sales and economic incentives to permit producers to share the risks of developing leases with the Federal Government.

A Study Is Promised

Reporting yesterday on the meeting, Joseph C. Swidler, chairman of the New York State Public Service Commission, and Anthony J. Grossi, Public Utility Commissioner of New Jersey, said Mr. Hickel had agreed to study long-range

proposals for offshore drilling and to carry out some programs to increase onshore development.

A spokesman for Mr. Swidler explained last night that the commissioners had decided to report on the substance of the closed-door meeting because they felt it was "important for the public to know about it." The spokesman asserted that Mr. Hickel had declined earlier in the week to make any public statement about the meeting, or even to identify the participants.

Several Groups Represented

Participants in the meeting included representatives of Maine, Connecticut, Rhode Island, Pennsylvania, Delaware and North Carolina and of the National Association of Regulatory Utility Commissioners, the United States Conference of Mayors and the New York Conference of Mayors.

The scope of the meeting reflected the growing concern over the dwindling reserves of natural gas. Last month, the New York commission announced that it was contemplating new-customer restrictions on gas companies after receiving a staff report that said supplies for state utilities were "rapidly reaching a critical situation."

Last week, Consolidated Edison said it could not build a controversial addition to its Astoria power plant if it had to comply with a city requirement that natural gas be used exclusively as fuel for the plant.

The company said it would be economically unsound to build the $370-million addition if it were forced to lie idle because uncertain supplies of gas were shut off.

The demand for natural gas, a relatively clean and adaptable fuel, has soared in the last few years with the nation's growing awareness of pollution. Because its price is regulated by the Federal Power Commission, natural gas competes economically with oil and coal in most market areas.

Between 1954 and 1968, gas reserves were growing at a rate of 2.1 per cent a year, against a consumption rise of 5.3 per cent a year. Consumption's current rate of increase is 6 per cent a year.

As a result, the reserve ratio fell from 29 years of gas supplies in 1954 to 14.6 years in 1968. Each passing year is trimming about another year's reserves from that figure, so that the reserve is now down to about 11 years' supplies.

The problem, however, is a question of economics — not that the world is running out of natural gas.

Large sources of natural gas exist in Louisiana, Texas, Alaska and in offshore areas, but oil companies who hold about 70 per cent of the leases have not found the development of gas supplies as profitable as their oil holdings. The result has been that many gas lease holdings have not been developed.

In an effort to force such development, the utility commissioners last week suggested a series of revisions in the Government policies regulating the producers.

For one thing the commissioners suggested a requirement that leases be developed immediately after they were granted. Another proposal called for

more frequent sales of potential gas acreage in the offshore areas of the Gulf of Mexico, where reserves are estimated to total 48.6 trillion cubic feet. According to Mr. Swidler, total leases under the United States Geological Survey supervision have grown only 3 per cent over the last five years.

The commissioners also suggested that the Government encourage lease sales in the Atlantic offshore areas, which have not been developed and which are closer to the heavy demand in the East.

Proposed also were economic incentives that would enable producers to share their economic risks of developing leases with the Government.

One such incentive would combine the so-called "bonus plan," in which the Government gets a "bonus" by requiring immediate and full payment for leases, with a royalty-bid system, in which payment for leases is delayed and tied to returns on the development of leases. The latter would in effect have the Government share the producers' investment risks as well as windfalls and would eliminate the large capital barrier to entry into the production field, enabling smaller companies to get involved.

Noting that "considerable gas reserves are available on shore in Louisiana, Texas and even upstate New York," Commissioner Grossi of New Jersey said yesterday:

"With proper economic incentives, that gas can be developed to meet our crisis while the long-range plans for offshore drilling are discussed with ecological experts. The point is that the Federal Government has the power to alleviate the crisis right now by proper encouragement of onshore lease development."

August 16, 1970

POWER SHORTAGE LEADS TO AD CUTS

But Some Companies Still Seek New Customers

WASHINGTON, Dec. 26 (AP)—The threat of fuel and power shortages this winter has prompted several states to call for a moratorium on utility company advertising drives for new customers, but some companies have declined to go along.

A study indicates that, while some electric and gas utilities have voluntarily cut back or changed the thrust of their advertising, others continue to seek new customers despite the possibility of not being able to service them.

Earlier this year, the Federal Power Commission reported that 17 states had taken action to restrict or prohibit utility promotional practices and that 26 states had investigated the promotional practices of utilities in their jurisdiction.

Warned In 1964

The Oregon Public Utility Commission warned electric and natural gas utilities as early as 1964 against imprudent statements for their products. Two years later, the presidents of the Northwest Natural Gas Company, the Portland General Electric Company and the Pacific Power & Light Company were told in person to stop making excessive statements.

However, according to Public Utility Commissioner Sam Haley, extensive promotional advertising by Oregon gas and electric utilities continues.

Citing a utility company report "that there will not be sufficient gas available during the next few years to meet all the requests by present and potential customers," Mr. Haley said, "There appears to be no reason for gas utilities to promote sales that will increase peak loads in the face of such a bleak supply situation."

Appeal In Michigan

Willis Ward, chairman of Michigan's Public Service Commission, has asked gas and electric utilities "to avoid the solicitation of business which could not be served because of the gas shortage."

Two of Michigan's largest utility companies have taken opposite positions on the issue.

The Consumers Power Company, which has a million electric customers and 830,000 gas customers in the state, says it has dropped nearly $1-million in gas and electric promotional advertising this year. But the Michigan Consolidated Gas Company says it has no intention of cutting back.

The inability of a gas pipeline company to deliver promised supplies and a delay in the completion of a nuclear generating plant prompted Consumers' action, a spokesman for the company said.

But a spokesman for Michigan Consolidated said: "Since we have sufficient gas to supply all anticipated new residential and commercial customers as well as essential industrial users, we believe it is necessary to inform the public of the advantages of natural gas over other fuels, particuarly for domestic use and to encourage the purchase of gas appliances."

In Illinois, gas shortages outside of Chicago are affecting only large commercial customers. Gas utilities have established priorities approved by the Illinois Commerce Commission that guarantee full service to residential users but allow restricted and interruptible service to new commercial and industrial customers.

The Central Illinois Light Company is continuing normal promotion levels, but has changed the thrust of its advertising. Harold Haig, the marketing manager, says the winter ads show consumers how to conserve heat by using new air filters and storm windows.

Acute Situation

In Chicago, the situation is more acute. Spokesmen for the Peoples Gas Light & Coke Company say they believe they can serve all existing customers this winter if supply and weather conditions are normal. But the company has stopped accepting new gas customers of any kind, and all types of advertising are down.

At last count, Peoples had 11,740 applications for gas service that could not be filled—the equivalent of 185,000 single-family homes.

Throughout the rest of the country, the story is much the same.

Commonwealth Natural Gas, a wholesale distributor in Virginia, is still participating in national advertising campaigns to stimulate sales and spokesmen say the company will continue to do so.

Heavy utility advertising continues in California. The State Public Utilities Commission has asked major utilities to voluntarily submit a statement on promotion practices and amounts spent by the end of the year and to submit voluntarily any new promotion programs or budgets 30 days ahead of their use.

Minnesota's largest power supplier, the Northern States Power Company, has not cut back its advertising but has changed its focus—from attempting to attract new customers to emphasizing the costs involved in building new generating systems and protecting the environment.

December 27, 1970

5-YEAR OIL ACCORD IS REACHED IN IRAN BY 23 COMPANIES

6 Persian Gulf States Gain More Than $10-Billion in Additional Revenue

SHUTDOWN IS AVERTED

By JOHN M. LEE
Special to The New York Times

TEHERAN, Iran, Feb. 14—Twenty-three Western oil companies agreed today to additional payments of more than $10-billion to six Persian Gulf states for a five-year oil agreement intended to stabilize the crisis-prone industry.

The settlement increases payments by more than $1.2-billion this year, rising to $3-billion dollars in 1975. Without the new payments, oil income in the Gulf area has been $4.4-billion a year.

The agreement, reached after a month of maneuvers and only under threat of government-dictated terms beginning tomorrow, averts the danger of a halt in oil supplies from the Persian Gulf, the major source for Western Europe and Japan.

However, European and Asian consumers will be presented with a huge bill. The price of gasoline, fuel oil and other petroleum products will almost certainly be raised.

Importers to Pay More

In addition, importing countries will pay more for their oil to the detriment of their balance of payments. Less-developed countries such as India are expected to be hard hit.

The immediate effect on American consumers should be limited since the United States imports only 3 per cent of its oil requirements from the Middle East. However, since these imports are large, fuel oil, these prices could be raised. The world oil picture is so interrelated, however, that other increases could show up in time.

February 15, 1971

'Clean' Reactors Delayed in Drive For Atom Power

By WALTER SULLIVAN

The White House has decided that, instead of pressing for the development of "clean" fusion reactors, it will throw heavy budgetary support behind a new type of atomic power production that, like today's reactors, manufactures hazardous radioactive by-products.

The decision has been made to meet critical power shortages anticipated for the next few decades. The projected reactors are the so-called liquid metal fast breeders.

They use liquid metal, such as sodium, to transfer heat from the reactor to steam generators. They will be breeders in that they will make more fuel than they consume, thus helping relieve present dependence on limited uranium supplies.

While investment in such reactors, which derive their energy from atom-splitting, or fission, is being sharply increased, spending for fusion research is being somewhat reduced as part of the general budget tightening.

The reasoning is that the technology for breeder reactor production is largely in hand whereas many uncertainties remain as far as fusion is concerned. The Atomic Energy Commission also argues that, contrary to the fears of some, breeder reactors and methods for disposal of their by-products can be made safe.

According to A.E.C. sources, the planned budget for technical development of breeder reactors in the fiscal year beginning in July will be $103-million, compared with $81-million in the current year. For work on a demonstration plant the budget will rise from $10-million to $36-million.

The increase bears out a recent statement by President Nixon's science adviser, Dr. Edward E. David Jr., that the development of breeder reactors had become one of the country's chief technological goals.

The decision runs counter to the wishes of those who fear that heavy dependence on breeder reactors and other power plants driven by fission will endanger the environment.

Leaders of the fusion program fear that a lack of funds will considerably delay the "golden age" of fusion power. Nevertheless, a survey of the field indicates that, at best, such power could not be generated in quantity much before the end of this century.

Fusion reactions are those in

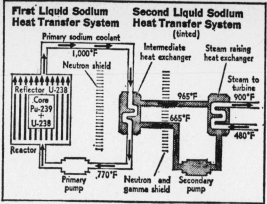

First Liquid Sodium Heat Transfer System
Primary sodium coolant
1,000°F
Neutron shield
Reflector U-238
.Core
Pu-239
+
U-238
Reactor
Primary pump
.770°F

Second Liquid Sodium Heat Transfer System (tinted)
Intermediate heat exchanger
Steam raising heat exchanger
Steam to turbine
965°F   900°F
665°F
480°F
Neutron and gamma shield
Secondary pump

The New York Times   March 8, 1971

In a typical fast breeder reactor, as indicated schematically in diagram, liquid sodium in the first circulation system carries energy in the form of heat from the reactor core to a primary heat exchanger. There the heat is transferred to a second circulation system, filled with sodium that is not radioactive. The heat is then transferred to a water system, which generates steam to drive a turbine and produce electricity.

which the nuclei of light atoms, such as those of hydrogen, helium or lithium, are forced together to form heavier atoms. In the process, a small amount of the starting material is converted into a large amount of energy.

Such processes are essentially safe in that they produce little or no radioactive by-products and they are virtually foolproof against runaway reactions. Furthermore, their fuel is almost as abundant as water itself.

But such reactions are extremely difficult to achieve in a controlled manner. They occur in a hydrogen bomb, but only under the momentary conditions of high pressure and temperature induced by an atomic explosion, whose energy comes from fission.

The breeder reactors that are now to be the main focus of the American developmental effort are an elaboration of the fission-style reactors that now dot the country. At the end of last year, 108 such reactors were either in operation or on order as part of the country's power-generating complex.

As in fusion, the fission process converts a residue of matter into energy. The conditions necessary for a self-sustaining fission reactor were relatively simple to establish, once the principles were known. The first was built secretly at a squash court at the University of Chicago in 1942.

In one way or another, a fusion reactor must establish conditions of high temperature and pressure like those in the core of a star, where fusion reactions occur continuously.

The trend toward atomic power is strong. More power-generating capacity is now on order for atomic plants than for the conventional type. Those that generate energy by fission are springing up all over the country.

Some atomic plants, like the

one at Indian Point on the Hudson River, are on a vast scale. But they are intrinsically inefficient. This inefficiency will largely be overcome by the new breeder reactors.

In the first place, the latter make more fuel than they consume. Second, their fuel cycles mesh nicely with those of the "old-fashioned" reactors now in general use. The technology for their construction is now well in hand, whereas that for fusion reactors is still at an uncertain stage.

And they will produce less thermal pollution in the waters used for cooling than today's reactors. (Some fusion reactors, if they prove feasible, would produce no thermal pollution at all.)

It is hoped that, before the end of this year, one of three rival contractors will have been chosen for construction of the demonstration breeder plant. They are Atomics International, General Electric and Westinghouse, each with its own supporting team of public utilities. The A.E.C.'s goal is for operation of the plant to start by the late nineteen-seventies.

In the projected breeder reactors, a fissionable fuel — plutonium 239—will be split, releasing neutrons that then split other atoms of plutonium 239 in a chain reaction. In ordinary reactors, the excess neutrons plunge into the surrounding material, generating radioactive by-products.

These surplus neutrons in the breeder reactor will be used to convert uranium 238 into plutonium 239 (the numbers refer to the total of protons and neutrons in the nucleus). The plutonium 239 can then be used as fuel.

Thus, when a mixture of oxides of uranium 238 and plutonium 239 is "burned" in a breeder reactor, the amount of fissionable material increases with time, instead of the other way around. A starting mixture

of about 15 per cent plutonium is required.

Uranium 238 cannot be used as fuel. Only uranium 235 can be split; it serves as the fuel for most of today's power plants. Less than 1 per cent of natural uranium is of the 235 variety; the remaining 99 per cent is largely wasted.

However, some plutonium 239 is manufactured in these conventional reactors and, when the new breeders come into operation, it can be used to drive them. Also, because the breeders make possible the ultimate use of virtually all uranium, instead of less than 1 per cent of it, the drain on the world's uranium reserves will be greatly reduced.

The new fast breeders will be more efficient in that they will use liquid metal, such as sodium or potassium, to carry heat from the reactor to generate steam. The advantage of such metals is that they can be heated to very high temperatures without producing pressures so great as to risk rupture of piping.

In this way, energy can be carried away from the reactor at high speed. The reactor, therefore, can be run "faster" and one of a given size will be able to generate much more electric power than its current counterpart.

While designers of atomic plants go to great lengths to provide double protection against accidents, skeptics fear that, although accidents are rare, those that occur could be disastrous. They note that, if the molten sodium or potassium used in breeder reactors comes in contact with water or steam, a violent reaction occurs.

Thus, if there is a leak in that part of the system where the liquid metal is used to heat steam, it could be catastrophic. To isolate the reactor from such a danger, typical designs provide double liquid metal systems. One carries heat from the reactor and heats another liquid metal system, which then heats the steam.

In any production of energy from fission, there are potential hazards. They involve those who mine the fuel and, at the end of the process, the disposal of radioactive by-products.

The plants themselves are subject to "excursions" if the chain reaction gets out of hand. This can melt reactor components and release highly radioactive fumes. One of the challenges in reactor design has been to prevent such excursions and provide containment of the fumes if they do break loose.

If they prove technically and economically feasible, fusion plants would eliminate most, if not all, of these hazards. However, because even at best they cannot fill the energy requirements of the next two decades, the emphasis is on breeders and on efforts to minimize any danger of accidents.

March 8, 1971

Coal Rush Is On as Strip Mining Spreads West

The New York Times/Ben A. Franklin

The Navajo strip mine of the Utah Construction and Mining Company near Farmington, N. M., is the largest single coal producer in the country. The strip mine's output from Indian coal lands was more than six million tons last year.

By BEN A. FRANKLIN

Special to The New York Times

WASHINGTON, Aug. 21—A new stage in the development of the American West is beginning on the arid plains and badlands that flank both slopes of the Rocky Mountains.

On thousands of square miles of vacant land west of the Mississippi—much of it in Federal ownership or in Government land grants to Indian tribes and railroads—a feverish coal rush is on.

The scramble is for coal leases and rights that will open an enormous and virtually untapped reserve of cheap Western fuel to strip mining.

On a scale far larger than anything seen in the East, where acreage totaling half the area of New Jersey has been peeled off for coal near enough to the surface to be strip mined, portions of six Western states – Arizona, Colorado, Montana New Mexico, North Dakota and Wyoming—face a topographic and environmental upheaval.

It is being brought on by the nation's apparently insatiable demand for energy, by the air pollution crisis in urban centers, by new technology in the conversion of coal to clean fuels, and by the economies of bulldozing rather than tunneling for coal that are available in the West.

In resolving the energy and air pollution problems, however vast areas of isolated open spaces in the West may be drastically altered.

The visual impact of strip mining is invariably stunning. On flat or rolling terrain, mammoth power shovels crawl day and night through great trenches, lifting, wheeling and depositing the unwanted strata above the coal seam into thousands of uninterrupted acres of geometrically perfect windrows of spoil banks.

In mountain coalfields where one, two or as many as five seams may lie horizontally through timbered slopes far above the valley bottom, the contour strip mines are notched in continuous, sinuous strips around the mountainsides. Trees and earth and rock are cast down the mountain flanks to expose the strippable edge of the coal bed.

The legacy of upheaval remains. Silt fills streams for thousands of miles. Sulphur-bearing coal, left in place and exposed to the elements, yields a long-lasting trickle of sulphuric acid which chemically burns streams and kills aquatic life. From the air over a "hot" acidic strip mine, pools of rainwater glow in weird shades of red and orange.

The debate over strip mining has been gathering since the late nineteen-fifties, when larger and larger earth-moving machinery made its growth economically feasible and gave it a cost advantage over under-

ground mining. With a passion that coal men tend to see as mysticism, conservationists say that stripping destroys the very roots of men's souls—the land. The mining industry sees it with similarly strong conviction as the best way to tap a vital national resource which, as one strip mining executive put it recently, "God put there for man's use—it's a sin to waste it."

According to one Government geologist here, the six states and others in the West—Oklahoma, Texas and even a patch of Washington State—"are on the brink of, not years, but generations of strip mining for coal that will make the excavation for the Panama Canal look like a furrow in my backyard vegetable garden."

The first wave has begun. In 1970, for the first time in the 100-year history of coal mining in America, a Western mine—the Navajo strip mine of the Utah Construction and Mining Company near Farmington, N. M.—became the largest single producer in the country. Its output from Indian coal lands was more than six million tons for the Four Corners Electric Power Complex, an environmentally controversial steam-electric station serving New Mexico, Arizona, Nevada and Southern California.

Near Centralia, Wash., 30 miles south of Olympia and just beyond the foothills of Mount Rainier, a 5,000-acre, 135-million ton deposit of coal that was only nibbled at by tunneling from 1870 into the nineteen-fifties for pre-deisel locomotive fuel for the Northern Pacific and Union Pacific Railroads, is being turned into one of the biggest strip mines in the country. The planned rate of production is five million tons a year for a 700,000 kilowatt generating station of the Pacific Power and Light Company and the Washington Waterpower Company.

Pacific Power and Light also owns rights to an estimated 1.6 billion tons of strip mine reserves in Wyoming and Montana. The company expects to rank among the top five coal producers in the country by 1977 with production of 23 million tons a year. Its president has said that the company will go slow on expensive investment in nuclear power stations because "we've got coal running out our ears."

Even Texas lignite—lignite is the lowest rank of coal in energy per ton and it has never generated more than an asterisk in Government coal production statistics—is having a sudden boom.

Three electric utilites—Texas Power and Light, Dallas Power and Light and Texas Electrical Service, Inc.—announced two months ago that they would begin a 35-year strip mine operation on 17,500 acres of lignite beds in Freestone County, near Fairfield, to fuel the new Big Brown steam-electric station east of Waco. Other lignite-fired plants are scheduled for Rusk and Titus Counties.

Western coal is low in sulphur—a boon to electric utilities caught between soaring power demand and new air pollution regulations that forbid the burning of sulphur-contaminated fuel. Accordingly, also for the first time last year, some low-sulphur western coal was hauled by rail as far east as Chicago.

But according to Government coal men, an immense strip mine explosion west of the Mississippi River that, by comparison, will make this excavation for electric power stations look like a mere desert gulch, is coming in the nineteen eighties for a giant new coal consuming industry, gasification.

Official forecasts here say that 20 years from now perhaps 300 million tons of coal a year—half of last year's total United States production—will be processed at huge, refinery-like plants, surrounded by massive strip mines in the Western coal fields. The product will be quadrillions of cubic feet of pipeline quality, pollution-free gas. The Government and the mining and gas industries are now committed to this basic change.

Vast Coal Beds in West

Coal gasification will replace the country's dwindling supply of natural gas from wells, now

Sulphur Content of Strippable Coal Reserves

Millions of tons by sulphur content

	*Grade	#Low	Medium	High	Total
Wyoming	B	13,377	65	529	13,971
Montana	B,C	6,133	764	0	6,897
New Mexico	B	2,474	0	0	2,474
North Dakota	C	1,678	397	0	2,075
West Virginia	A	1,138	669	311	2,118
Texas	C	625	684	0	1,309
Kentucky (East)	A	532	189	60	781
Colorado	A	476	24	0	500
Arizona	B	387	0	0	387
South Dakota	C	160	0	0	160
Virginia	A	154	99	6	258
Washington	B	135	0	0	135
Alabama	A	33	74	27	134
Arkansas	A,C	28	118	28	174
California	B	25	0	0	25
Oklahoma	A	10	44	57	111
Utah	A	6	136	8	150
Tennessee	A	5	43	26	74
Michigan	A	0	0	1	1
Maryland	A	0	8	13	21
Ohio	A	0	126	907	1,033
Iowa	A	0	0	180	180
Kansas	A	0	0	375	375
Pennsylvania	A	0	225	527	752
Kentucky (West)	A	0	0	977	977
Indiana	A	0	293	803	1,096
Missouri	A	0	0	1,160	1,160
Illinois	A	0	80	3,167	3,247
Total		31.787	4.036	9.161	44.986

*A—BITUMINOUS, B—SUB-BITUMINOUS, C—LIGNITE

The New York Times   Aug. 22, 1971

The westward movement of strip mining has resulted from low-sulphur reserves west of the Mississippi that promise less pollution in fuels to meet the energy crisis.

estimated to be only about a 15-year reserve. Consumed in power plant and industrial boilers in the East, the gas will reduce air pollution. And pumped through pipelines that might otherwise be empty, it will save the pipeline industry from collapse.

Millions, perhaps billions, of dollars are thus finally ripening in coal beds under Western sagebrush, where the mineral has lain for geologic time, 130 million years.

The speculative market in Western strip mine leases to dig it, and in permits to explore for more, has suddenly become a bonanza.

In the 12 months that ended in July, 1970, the increase in prospecting permits issued by the Interior Department's Bureau of Land Management for coal exploration on Federal land — national forests, grassland, desert and range—shot up by 50 per cent to the greatest number in history, covering 733,576 acres. That is the area of all of New York City and Long Island, with Westchester and Rockland counties thrown in.

Prospecting permits on Indian reservations, issued separately by the Bureau of Indian Affairs, went from none to exploration rights covering 500,-000 more acres. Such permits are convertible to firm mineral leases if coal is found.

Coal-Fired Turbines

Nearly one million acres of public and Indian coal land in the West are already leased. Leases by private owners, chiefly by the transcontinental, land-grant railroads, are unknown but may cover an equal area.

The forces behind the sudden migration of coal mining to the West are complex, and the reasons for them are probably as irresistible as money.

First, despite the wide acceptance during the nineteen-sixties of visionary forecasts for nuclear electric power, half the nation's electricity is still generated by coal-fired steam turbines.

Dr. Glenn T. Seaborg, the retiring chairman of the Atomic Energy Commission, recently conceded that the poor record of the A.E.C.'s vaunted nuclear-electric program means that coal will fuel an even greater portion of the enlarged generating capacity required for the next three decades.

Other important factors are mining costs and mining volume.

Strip mine production of coal in the country as a whole has advanced very rapidly in the last few years, from about one-third ot the annual tonnage in 1968 to 40 or 42 per cent last year. According to the United States Bureau of Mines, the cost

advantage over deep mined coal is on the order of three to one.

Productivity per worker runs as high as five to one in favor of strip mining, and is going higher under the Federal Coal Mine Health and Safety Act of 1969, which requires deep mines to take expensive steps to curb the high rate of death and injury underground.

Moreover, particularly for gasification, huge guaranteed volumes of cheap, strip-mined coal are essential.

77 Per Cent of Reserve

The Bureau of Mines has just cautiously disclosed in an unpublished compendium that beneath 13 states west of the Mississippi River there lies 77 per cent of the country's total of economically strippable coal reserves of 45 billion tons. The Western coal is in seams 12 times thicker, on the average, than in the East. And 25.5 billion tons of it is low-sulphur coal.

Wyoming and Montana, together, contain 21 billion tons of the entire Western reserve of low-sulphur coal. Wyoming's low-sulphur reserve, alone, is eight times West Virginia's and Kentucky's put together.

The Government has apparently pre-empted most of one of Colorado's major strip mine fields by building the Air Force Academy on top of it at Colorado Springs. But Colorado still contains nearly half a billion tons of the highest grade of low-sulphur strip mine coal.

And still undisturbed beneath the wheat and grasslands of western North Dakota wait 50 billion tons of lignite—the leanest rank of coal, but equivalent in total energy to all the better grades of coal left to be mined in the four largest producing states, West Virginia, Kentucky, Pennsylvania and Illinois.

The Bureau of Mines has recently disclosed that Pennsylvania and Illinois have no low-sulphur stripping coal left at all. The reserve in West Virginia is only about 1.2 billion tons, one twenty-fifth of the national reserve.

For a hundred years the traditional coal field regions of the United States have been there —in the Appalachian east and south and across southern Indiana and Illinois, tapering off into Missouri, Kansas and eastern Oklahoma.

Billions of tons of coal and billions of dollars of investment in immovable tools and tunnels remain in these traditional coal areas, and depletion of total coal reserves is not the most important factor in the move to the West.

But although the Eastern and Midwestern fields now supply 94 per cent of the 600 million ton-a-year coal production, they contain only 17 per cent of the remaining reserve of strippable low-sulphur coal.

Energy System Shifting

It is this arcane statistic, the

83 per cent of shallow, strippable, low-sulphur coal beneath the Western states, that is starting what the United States Geological Survey calls "a massive change" in the whole national fuel and energy system.

Until the air pollution crisis of the nineteen sixties and seventies the West's low-sulphur coal was as worthless as a coyote. Coal is the cheapest of fossil fuels and, accordingly, freight is a large part in its cost to consumers. Longhaul reserves were not cost-competitive.

But now that many urban pollution abatement laws forbid the burning of coal or oil containing more than 1 per cent sulphur by weight — and the Federal Environmental Protection Agency has said the limit may have to be pushed to 0.7 per cent—the ancient economic maxims of coal, a $3-billion a year indusrty, are caving in.

Already, in a break with transportation tradition, the historic flow of coal from Appalachian mines to Lake Erie port to docks at Superior, Wis., or Duluth, Minn., has begun to turn around.

For example, Burlington Northern, Inc., the merged railway system—and also one of the largest private owners of Western coal reserves through 19th century Federal land grants —has been loading low sulphur coal from the Peabody Coal Company's Big Sky strip mine at Colstrip in eastern Montana. The coal goes by train to the docks at Superior and is shipped lake steamer to Tasonit Harbor, Mich., a movement that would have been economically unthinkable a few years ago.

It is the prospect, however, of prodigious volumes of strip-mined coal to supply gasification plants that lies behind the frantic scramble by coal, petroleum and pipeline interests—and by land brokers and speculators who expect to profit at their expense — to assemble leases and rights to large tracts of Western coal for future stripping.

The scope of this Western stripping for gasification—large both on a plant-by-plant basis and also in the area to be affected by big new surface mines — is suggested by what the American Gas Association calls its "very confidential" study of potential gasification sites.

Apparently for fear of stimulating price gouging in mineral leases and arousing conservationist opposition, the association will not discuss the study beyond acknowledging its existence. Association officials will not even say which states have been identified as gasification sites, much less which counties.

But it is known that the association report pinpoints 176 prospective plant locations — each to require a $200-million to $300-million investment in strip mine and coal processing facilities—and industry officials say variously that "a large majority" or "nearly all" of them lie west of the Mississippi.

A Government geologist who has seen the association study says that 156 of the 176 sites— all but 20—are in "the Rocky Mountain West." Enough of them are to be developed by 1985, the study suggests, so that gasification by then will materialize as a $1-billion-a-year industry on the West's open spaces.

According to Interior Department reports, coal for future gasification is spurring recent transactions like these:

¶In response to a United States Bureau of Land Management invitation to bid on 6,560 acres of Federally owned coal land in Campbell County near Gillette, Wyo.—the bureau delicately described the 10-square-mile area as "susceptible to stripping"—the Cordero Mining Company won the coal leases with a record high price of $505 an acre. In recent years, some Federal coal leases have gone for under $1 an acre. Cordero is a subsidiary of the Sun Oil Company.

¶On the same day last December, the Mobil Oil Company bid $441 an acre for leases on 4,000 acres of bureau land adjoining the Cordero site. The United States Geological Survey had estimated its worth at $35 an acre.

Lease Prices Soar

¶Bureau lease prices have advanced so rapidly that a short time earlier a successful bid of $257.50 an acre by a land-buying affiliate of the Ashland Oil Company—$1.9-million for coal rights to 7,600 acres, or 13 square miles, of Carbon County near Hanna, Wyo.—was being called a "precedent-shattering high price." The $257.50 precedent lasted two weeks, when Cordero doubled it.

But particularly on Indian reservations, there have also been what one official of the Bureau of Indian Affairs here calls "some damn lucky breaks" for Eastern coal companies bidding for leases of tribal coal reserves.

Last September, Westmoreland Resources, Inc., a year-old Western strip mining partnership of the Philadelphia-based Westmoreland Coal Company, Penn Virginia, Inc., the Kewanee Oil Company, the Morrison-Knudsen Company, and the Kemmerer Coal Company of Wyoming, had to bid an average of only $7.87 an acre for 32,300 acres of coal rights held by the Crow Indian reservation in the Sarpy Creek area of Treasure and Big Horn Counties, Mont.

Within months, the syndicate had sold options to buy 300 million of its 900 million tons of Montana coal reserves to the Colorado Interstate Gas Company, the pipeline division of the Colorado Interstate Corporation. The company is a major pipeline company and may be one of the first to erect a coal gasification plant, presumably near Hardin, Mont.

Other Vast Reserves

Other vast coal reserves in the West are owned by the railroads. Government land grants to the railroads, which were originally meant to encourage and finance the construction of track to the West but which have remained dormant and unsalable for 100 years, are suddenly valuable.

The Union Pacific, for example, has become a profitable lessor of its 10-billion-ton to 12-billion-ton reserve of coal on land given the company by the Federal Government under the railroad land grants of the last century.

But by far the greatest acreage of coal leaseholds is being acquired on speculation for later sale to the coal gasification industry.

An unpulished "working paper" prepared at the Interior Department shows that the 10 largest holders of Federal coal leases control 49 per cent of the 773,000 acres of public domain turned over to mining interests or land speculators as of July 1, 1970, and that very little of their acreage is being mined. Some of the inactive leases have been held at little cost since the nineteen-twenties but most are about five years old.

The 10 largest lease holders, in order of the acreage of their coal rights, are listed as the Peabody Coal Company; the Atlantic Richfield Company; the Garland Coal and Mining Company; the Pacific Power & Light Company; the Consolidation Coal Company; the Resources Company; the Kemmerrer Coal Company; the Utah Construc-

tion and Mining Company; Richard D. Bass, a Dallas geologist and land investor, and the Kerr McGee Corporation.

Drastic Change Seen

The Interior study says that, of all the Federal coal acreage under lease, those 10 lease holders control 97 per cent of the leases in Montana and North Dakota, 91 per cent in New Mexico and Oklahoma, 79 per cent in Utah, 75 per cent in Colorado and 77 per cent in Wyoming. Peabody and Atlantic Richfield together hold one-third of all the federally leased coal land in Montana and North Dakota.

Federal coal leases, many at bargain rates, are not the only incentives that the Government has provided for the development of Western coal.

On Aug. 4, the Interior Department signed an agreement with the gas industry that will add $80-million in Federal funds to $40-million from gas and pipeline companies for a four-year acceleration of existing work on small-scale but working pilot coal gasification plants. Some $176-million more in Federal money has been set aside for the next step — construction of a full-scale demonstration plant.

Meanwhile, the coal industry is working hard to picture the environmental prospect for the West as benign, if not uplifting.

Carl E. Bagge, a former member of the Federal Power Commission who now heads the National Coal Association, an influential Washington-based industry group, has been making an unusual number of trips into the West to preview the "new prosperity" in Western coal and to inveigh in speeches against "reckless," "radical," "emotional" conservationist attacks on strip mining.

Mr. Bagge has been pointing out in his Western travels that the strip mining industry genuinely means to do better there than in the ravaged coal fields of the East, and that the tempo of Western nature is slower — there is less timber, less rainfall, less visual discontinuity in stripping buttes and badlands than Appalachian hickery forests or Indiana cornfields.

One coal industry suggestion, put forward earlier this year at a session of the Rocky Mountain Mineral Law Institute, was that tourists might have some interest in visiting the scarred and barren "badlands" created by strip mining.

August 22, 1971

A.E.C. WILL REVIEW REACTOR PERMITS
By RICHARD D. LYONS Special to the New York Times

WASHINGTON, Sept. 3—The Atomic Energy Commission bowed to environmental pressures today and ordered reviews of construction permits and operating licenses that had been given to nuclear power plants designed to produce more than 100 million kilowatts of electricity.

The A.E.C. now must review the permits and licenses it granted to 96 nuclear power reactors throughout the country, including Consolidated Edison's Indian Point 2 plant on the Hudson River, after consideration of all environmental factors, not just radiological ones.

About 100 million kilowatts of electricity, more than a quarter of the nation's current generating capacity, was to have been produced by 91 of these plants that are either being built or have received construction permits.

The five remaining plants, with a capacity of over three million kilowatts are already in operation. Today's action threatens to close them down.

The five are: The Northeast Utilities' Milestone Plant No. 1 at Waterford, Conn.; Commonwealth Edison's Dresden No. 3 at Morris, Ill.; the Northern States Power Company's Monticello plant at Monticello, Minn.; the Carolina Power and Light Company's H. B. Robinson Unit No. 2 at Hartsville, S. C., and Wisconsin Michigan Power Company's Point Beach Unit 1 at Two Creeks, Wis.

The A.E.C.'s action in effect ordered that the thermal effects of atomic power plants on the environment must be considered before the generating stations could either be built or go into operation.

Thermal pollution — the re-

lease from power plants of large amounts of heat, especially into surrounding waterways—can and has upset the environments by causing the deaths of, for example, fish and aquatic life.

For almost two years the A.E.C. had contended that only radiation hazards must be considered when a nuclear power station applied for construction permits and operating licenses.

The practical result of today's action could lead to the installation of such costly equipment as cooling towers in those plants already in operation, and the redesign of those under construction or being planned to include such facilities to offset thermal pollution.

Today's action by the A.E.C. has the further effect of delaying for at least a year all the permit and licensing paperwork necessary for the operation of the plants, whose power output had been counted upon to meet shortages of electrical energy.

Court Decision Instrumental

The A.E.C.'s action in setting up new regulations aimed at settling the question of environmental hazards that such plants pose stemmed from a Federal Court of Appeals decision on July 23 that sharply criticized the environmental policies of the agency.

In a decision involving a suit over the controversial Calvert Cliffs plant under construction

on Chesapeake Bay in Maryland, Federal Judge J. Skelly Wright found that the A.E.C. had made "a mockery" of the National Environmental Policy Act of 1970.

One of Judge Wright's main points was that the A.E.C.'s rules, which conflicted with the wording of the act, prohibited an outside party from raising nonradiological issues up to 14 months after the act was passed.

Today's new regulations are aimed at complying with the Court of Appeals decision, which the A.E.C. chose not to appeal.

The regulations do not apply to those plants in operation on Jan. 1, 1970, the date that the National Environmental Policy Act went into effect. The regulations apply only to those plants given construction permits and operating licenses after that date.

Harold L. Price, director of hte A.E.C.'s division of regulation, declined during a news conference here to estimate the possible cost to power companies that the regulations could produce.

Mr. Price said it would also be impossible to determine if any of those five plants that went into operation after Jan. 1, 1970, or seven others nearing completion, would have to be closed. Among the seven is the new Indian Point facility.

The new regulations in ef-

fect place the burden of proof that a nuclear plant does not threaten the environment on the company that wants to build it.

A key section of the regulations reads as follows:

"Within 40 days each permit holder or licensee affected by the court decision will be required to submit information to the commission showing why his permit or license should not be suspended in whole or in part pending completion of the ongoing NEPA [National Environmental Policy Act] review.

"The commission's decision regarding whether or not to suspend permits or licenses will be published, and licensees and affected members of the public will have the opportunity to request a hearing on this determination.

"Considerations involved in such a determination will include whether there is a significant adverse impact on the environment and its nature and extent; whether continued operation or construction during the period of the NEPA review would foreclose subsequent adoption of alternatives in facility design or operation; and the effect of delay in facility construction or operation on the public interest, such as the need for power, alternative sources of power, and costs of delay to the licensee and the consumer."

September 4, 1971

Justices Bar States' Rule On Nuclear Plant Safety

By RICHARD D. LYONS

Special to The New York Times

WASHINGTON, April 3—Efforts by local governments to exert control over nuclear safety issues received a setback today when the Supreme Court affirmed a lower court ruling leaving nuclear standards to Federal jurisdiction.

By a 7-to-2 decision, the Court held in effect that the Atomic Energy Commission alone has the authority to regulate the discharge of radioactive debris from nuclear power plants.

In the case before the Court, Minnesota had asserted its right to set more rigid limits

on the radioactive discharges of the plant built by the Northern States Power Company at Monticello, 30 miles north of Minneapolis.

The company had requested and received an A.E.C. operating license that would have permitted a "stack release" of 41,400 curies a day in radioactive emissions. But the Minnesota State Pollution Control Agency had set a limit of 860 curies a day, only 2 per cent of the A.E.C. limits.

Northern States sued on the ground that it would be either impossible or prohibitively ex-

pensive to meet the limits that had been set by the state. Northern States won the first round in Federal District Court in St. Paul and the second in the United States Court of Appeals for the Eighth Circuit.

The Supreme Court's action today, in which Justices William O. Douglas and Potter Stewart dissented, affirmed the decision of the appeals court, which had held that "the Federal Government has exclusive authority" to set controls on nuclear power plants.

More than a dozen states had sided with Minnesota in the dispute. Other states not connected with the litigation in Minnesota have in recent years taken an increasing interest in local regulation of nuclear power plants.

Minnesota argued that it had "the power and duty to regulate and to prevent pollution of its

lands, waters and the air above it." The argument was that both the Federal and state governments had the right to set safety standards, even if it meant that the state standards would be stricter.

Northern States replied that Minnesota had improperly usurped the authority of the A.E.C. by setting any standards.

The lower court ruling, now upheld by the Supreme Court, was that Congress, in passing the Atomic Energy Act, had pre-empted state control over the nuclear power field.

Attorneys for the Department of Justice have contended that local control would lead to a proliferation of different regulations.

New York, New Jersey and Connecticut are not affected by the ruling. Only power plants are covered by the decision, not other hazards.

April 4, 1972

Administration Pushes Emergency Licensing of Nuclear Plants

By EDWARD COWAN
Special to The New York Times

WASHINGTON, April 9—The Nixon Administration, fearful that summer power shortages may irritate voters, is pressing Congress for legislation giving it emergency authority to license newly built nuclear power plants that have not fully satisfied the requirements of the National Environmental Policy Act of 1969.

Congressional committees are expected to take up the legislation next week.

With such authority to start up the new plants, it is said, extra power, notably in the Midwest and the Southeast, would be available if a heat wave caused unusually heavy use of air-conditioners.

The New York area, too, faces the possibility of reduced-voltage "brownouts" or even "load shedding," the shutdown of power flowing to big industrial users, but there is no new nuclear power station in the area that could be turned on this summer.

Doubt Expressed

Environmentalists ask whether power shortages are probable or the danger has been deliberately exaggerated to facilitate the weakening of environmental safeguards. Whatever the danger of shortages, and it depends in part on how hot it gets this summer, the environmentalists would prefer temporary brownouts to

what they regard as long-term subversion of their legislative cornerstone, the 1969 policy act. They fear thermal pollution, the heating of lakes and rivers.

The Administration, represented by the Atomic Energy Commission, is urging Congress to authorize the A.E.C. to issue operating licenses to nuclear power plants even if the commission has not issued detailed environmental impact statements for the plants as required by the environmental policy act.

The Administration bill and a similar but more restrictive bill introduced by Representative John D. Dingell, Democrat of Michigan, would permit emergency licensing only if the commission foresaw no "significant, adverse impact on the quality of the environment" or if it found other public-interest factors, such as a need for power, to be overriding.

Limit in Dingell Bill

Mr. Dingell's bill would make the licenses expire when the authority to issue them lapses, probably in the autumn of 1973. For that reason, he describes his measure—an amendment to the policy act, which he co-authored—as "a neat surgical incision that would be self-sealing."

The issue has created a rift between environmentalist groups and their friends in Congress, including Mr. Dingell. He and other like-mind-

ed Congressmen are unhappy about weakening the policy act, even temporarily but they argue that the cause of environmentalism might suffer graver damage if power shortages occur and the environmentalists are blamed for them.

Moreover, many Congressmen, particularly House members, share the Administration's anxiety about the effects of power shortages on Election Day results.

"A guy's got to be responsive just to protect himself," a House aide commented. "Suppose people are trapped in an elevator when the power fails—and the guy voted against the bill and he has to stand for re-election in November."

The environmentalists, despite the warnings of a sympathetic Congressmen that in this case discretion would be the better part of idealism, are expected to fight the legislation.

"I've gotten a lot of pressure to take a hard line on this," said a Washington staff man of a well-known group. "It's too early for our people to compromise."

Another such spokesman, Sam Love coordinator of Environmental Action, said the environmental movement was already "taking public relations beatings" from the better-financed utility industry.

Power shortages, if they occur, Mr. Love said, will be the result of inadequate planning by the utilities. The way

to avoid them, Mr. Love added, is for the power companies to encourage the public to cut back on the use of air-conditioners on the hottest days of the year.

The House Merchant Marine and Fisheries Committee is expected to approve Mr. Dingell's bill next week. Floor passage might come as early as April 17. Early hearings and favorable action are expected in the Senate Committee on Interior and Insular Affairs.

The Joint Committee on Atomic Energy is expected to consider the Administration's bill and a similar one by Representative Chet Holifield, Democrat of California, on Tuesday. Under these bills an amendment to the Atomic Energy Act would authorize the commission to issue the licenses and also to avoid time-consuming hearings.

The commission's regulations now provide for interim licensing, but a Federal court ruled last December that the agency was subject to the full constraints of the environmental policy act. Some environmentalists contend that the commission is still in a position to deal administratively with emergencies, but the agency says it wants unequivocal authority from Congress, lest it be tied up at length in the courts by new challenges.

April 10, 1972

Energy Crisis Ahead

The repeated warnings that an energy crisis looms ahead for the United States are becoming ever harder to ignore. Interior Secretary Morton's report for 1971 adds new evidence. Partly because of the recession, United States power consumption last year grew by only a relatively modest 2.3 per cent, yet even that moderate gain required an increase of more than 25 per cent in oil imports.

If 1971 had been a boom year, energy demand might have leaped 5 or 6 per cent, and the long-term average annual growth of this demand is generally expected to be 4 per cent. Meanwhile both the American Petroleum Institute and the American Gas Association have recently reported that proved United States reserves of both fuels declined last year. Current oil reserves in the lower 48 states are now at the lowest point in twenty years, while natural gas reserves are at the lowest level since 1957.

The oil and gas industry claims that higher prices would intensify prospecting and presumably lead to the discovery of additional deposits. It also wants increased freedom to do offshore drilling and is pressing for an Alaska pipeline to bring North Slope oil to the lower 48 states. These interests also warn of increasing dependence on imported fuel, with all the political dangers that involves. Atomic energy enthusiasts, of course, call for more nuclear power plants and the removal of present obstacles to these installations. Fearing politically damaging power shortages this summer, the Nixon Administration is pressing Congress for emergency authority to license newly built nuclear power plants that have not fully satisfied the requirements of the National Environmental Policy Act.

In contrast, consumer advocates demand continued availability of cheap energy, while the environmentalists have demonstrated skill in retarding the introduction of nuclear power plants, fighting offshore drilling and forcing the nation to question the wisdom of strip-mining.

What has been demonstrated by this debate and the developments flowing from it is that the United States cannot continue to enjoy limitless cheap power without exhausting its domestic reserves of oil and natural gas and without doing serious additional damage to its increasingly polluted environment.

In this dilemma, there needs to be much more serious discussion of alternatives for retarding the growth of energy demand, and even of reducing it. In general, power and fuel prices do not reflect the costs of environmental damage. A substantial additional tax on all fuel and power could discourage frivolous energy consumption and provide funds for environmental reconstruction.

Alternatively, the possibility has to be faced that eventually fuel and power may have to be rationed, perhaps by setting an upper limit per person on family electricity consumption. Or commercial and public buildings might be rationed on air conditioning. None of these or similar steps to reduce energy demand is attractive, nor is a Presidential election year the ideal time to expect politicians to discuss such a touchy issue candidly.

But sooner or later the problem will have to be faced, and how well it is handled will depend on the extent of earlier debate. The first limit on growth, it is now evident, will be the energy limit; and if it is a meteorologically hot summer this year New Yorkers and others will discover how close those limits are right now.

April 10, 1972

Washington:
June 6, 1972
THE PRESIDENT
Power. The President signed a bill giving the Atomic Energy Commission authority to issue temporary operating licenses for nuclear power plants in areas threatened by power shortages. The interim licensing will mainly affect operation of new plants in Eastern and midwestern areas threatened by potential brownouts in times of peak power demand this summer, next winter and the summer after. An A.E.C. spokesman said it would take a little time to determine the plants that will receive the interim licenses.

June 7, 1972

SCIENTISTS OPPOSE BREEDER REACTOR

31 Experts Urge Rejection of Funds for Project

By EDWARD COWAN
Special to The New York Times

WASHINGTON, April 25 — Thirty-one scientists and other professional persons urged Congress today to deny the Nixon Administration's request for funds to start building a $500-million demonstration model of a nuclear breeder reactor to generate electricity.

"Too many serious questions exist about the safety and environmental impact of such a project to make a commitment at this point to the commercial development of this technology," the scientists said in a statement.

The statement was drafted and issued by Environmental Action, an organization that said it was part of a coalition of similar groups "formed to halt funding of the planned prototype breeder reactor."

The environmentalists have pitted themselves against a program that President Nixon described in his energy message of last June 4 as "our best hope today for meeting the nation's growing demand for economical clean energy."

Plutonium From Uranium

A breeder reactor is so called because in the process of producing electric power from the fission of uranium atoms it would also convert the uranium into plutonium, itself a nuclear-power fuel.

This process, the President said, "could extend the life of our natural uranium fuel supply from decades to centuries, with far less impact on the environment than the power plants which are operating today."

Among the 31 persons who opposed the breeder reactor program were Dr. Linus Pauling, who has won Nobel Prizes for chemistry and for peace; Dr. Harold C. Urey, a Nobel laureate in chemistry; Barry Commoner and Dr. Paul Ehr-lich, biologists who have been prominent in the environmentalist movement; Dr. John Holdren, a physicist at the California Institute of Technology; Dr. William Nicholson, a physician at Mount Sinai Hospital in Manhattan, and Dr. Glenn Paulson, an environmental scientist at the City College of New York.

Three Safety Factors

The signers raised questions about the safety of three aspects of breeder reactors: the plants themselves, the handling of plutonium, and the disposal of plutonium waste products. They also suggested that, with large quantities of plutonium in use, some of it might be clandestinely diverted to the manufacture of nuclear weapons.

The Atomic Energy Commission plans to locate a breeder-reactor demonstration plant on the Tennessee Valley Authority power grid, probably near Rogersville, Tenn. The estimated cost is $400-million to $500-million. Critics say that the cost has been underestimated.

The demonstration plant is supposed to be finished by 1980.

Disputing the idea that breeder-reactor power development must go forward now to avert an energy crisis in the nineteen-eighties, the scientists and spokesmen for the environmental groups said the money should be channeled into reactor safety research and into development of other energy sources, notably solar energy and coal.

Coal-Use Urged

"We should concentrate on coal," said George L. Weil, a Washington-based nuclear energy consultant who spoke for Environmental Action. Noting that the United States has enormous coal reserves, Mr. Weil said: "It can be cleaned up, hopefully. My own preference is to have a little coal pollution rather than radioactive pollution."

Asked to comment on the scientists' statements, a spokesman for the A.E.C. said, "These breeders can be designed to operate safely."

The commission's view is that mishaps are inevitable but that they will not produce the dire consequences foreseen by the critics. Officials comment that the public has never been harmed by difficulties at any nuclear facility.

April 26, 1972

Geothermal Energy Held A Vast World Reservoir

By JOHN NOBLE WILFORD
Special to The New York Times

UNITED NATIONS, N.Y., Jan. 10—Energy experts from around the world met here this week and generally agreed that one of the most promising new sources of relatively nonpolluting power is the natural heat of the earth's core—geothermal energy.

Within 50 years, according to one optimistic estimate, geothermal energy may become a resource even more significant than petroleum. At least 80 nations are thought to have geological conditions indicating a substantial reservoir of such energy.

But the experts also noted a number of obstacles to the full development of geothermal energy. these include the continuing reluctance of governments and industry to take geothermal energy seriously, the lack of systematic exploration of its potential, and the failure of most nations to exchange information on the subject.

The three-day seminar on geothermal energy, which ended today, was sponsored by the United Nations Department of Economic and Social Affairs with the assistance of the Center for Energy Information, a nonprofit foundation involved in supporting the development of nonconventional energy sources.

Observers From Industry

Among the 250 participants were a number of officials of the American petroleum and utility industries. Their interest has been whetted by the Department of the Interior's plan to lease 58 million acres of public lands in the West for geothermal exploration.

At present, the only exploited geothermal field in the United States is at The Geysers, near San Francisco.

The source of most geothermal energy is the molten rock, or magma, in the earth's interior. When underground water comes into contact with the magma, hot water and steam are produced. Where this occurs in large quantities and within a few miles of the surface, the steam and hot water can be tapped and used to turn the turbines that generate electricity.

The most promising areas for exploration lie near earthquake faults, volcanic regions and hot springs and geysers.

At the United Nations meeting, the energy experts attempted to lay to rest what they said were three misconceptions about geothermal energy—that it was rare in nature, was generally uneconomical to exploit and was primarily a source of electric power.

Not 'a Freak of Nature'

Although fewer than 15 countries are attempting to tap such energy, and so far only on a small scale, Dr. Joseph Barnea, director of resources and transport at the United

Nations, said new estimates indicate it is not "a freak of nature."

Soviet experts, Dr. Barnea said, have estimated that the geothermal potential in their country "is probably equal to the combined U.S.S.R. resources of petroleum, coal and lignite."

Through United Nations technical assistance programs, Kenya and Ethiopia are tapping the geothermal energy stored in the African rift valley. Similar United Nations efforts are being made in Turkey, Chile, El Salvador and Nicaragua. Italy, Japan, Iceland New Zealand and Mexico already have geothermal power plants.

A report last year by the National Science Foundation and the University of Alaska estimated that 132 million kilowatts of geothermal electricity could be generated in the United States by 1985 and 395 million kilowatts by the year 2,000. The latter figure would represent an output greater than the total electricity generating capacity of the United States today.

"It should be borne in mind," Dr. Barnea said, "that we have geothermal energy in practically every geological environment, whereas petroleum is restricted to the sedimentary areas of the world. In 50 years, geothermal energy will be recognized as an energy resource of even greater significance than petroleum."

Lower Costs Foreseen

As for the economics of the energy, the experts heard a report prepared by the Public Service Commission of the State of New York in which it was estimated that by 1975 the capital costs of nuclear power plants would rise to $500 per kilowatt capacity, compared with $200 to $300 for coal or oil plants and $100 to $150 for the geothermal plants.

The experts conceded, however, that they had no basis for estimating the costs of exploring geothermal fields.

Dr. Barnea emphasized that geothermal energy should not be thought of solely as a source of electricity. Some of the lower-termperature hot waters may not be suitable for power generation, he said, but they could be used in desalination, mineral extraction and house heating.

As a matter of fact, interjected Dr. Robert W. Rex of the Pacific Energy Corporation, it might even be possible to heat the United Nations headquarters by drilling some 20,000 feet into the Manhattan bedrock.

Heat from the decaying radioactive elements (potassium, thorium and uranium) in the Manhattan schist, Dr. Rex said, probably gives off enough heat to act as an underground boiler for the heating system of the United Nations and other New York buildings.

"It would be a dramatic way of showing the possibilities of geothermal energy," Dr. Rex added.

January 11, 1973

PRESIDENT OFFERS POLICY TO AVERT AN ENERGY CRISIS

He Ends Oil Import Quotas — Message to Congress Asks Industry Subsidy

WARNING ON SHORTAGES

Nixon Urges Legislation to End Curbs on Wellhead Natural Gas Prices

By EDWARD COWAN
Special to The New York Times

WASHINGTON, April 18— President Nixon outlined to Congress today an energy policy that he said was designed to minimize shortages of fuels and power while the United States strives for greater development of its domestic energy resources, especially coal and offshore oil and natural gas.

In a long message devoid of major surprises, Mr. Nixon warned Americans that "in the years immediately ahead, we must face up to the possibility of occasional energy shortages and some increase in energy prices."

As expected, he announced that he was ending as of May 1 the controversial, 14-year-old mandatory quotas on imports of oil. Instead, he said, the Government is starting a system of license fees that will eventually apply to all imports of oil and gasoline.

Call to Check Trends

Asserting that "if present trends continue unchecked, we could face a genuine energy crisis," the President urged Congress to do the following:

¶End Federal regulation of wellhead prices of natural gas as an incentive to exploration. Wells newly discovered or newly dedicated to interstate markets would be free of Federal Power Commission jurisdiction immediately. Wells already producing for interstate delivery would become free when their present contracts expire.

Mr. Nixon implied strongly that he thought the Supreme Court majority was wrong in the famous 1954 Phillips case when it held that the Natural Gas Act of 1938 gave the commission jurisdiction over wellhead prices paid by interstate pipelines as well as the prices the pipelines charge to their customers.

"Ill-conceived regulation" has followed, the President said, whereby prices have been kept low for "America's premium fuel" causing industry and utilities to use it instead of coal or oil.

¶Give the oil industry an additional tax subsidy in the form of a tax credit for exploration outlays. Mr. Nixon described this as an extension to oil and gas drilling of the existing 7 per cent credit, or reduction in taxes owed, for investment in business equipment.

Treasury Secretary George P. Shultz said the credit would be 7 per cent for "dry holes" and 12 per cent for "wet holes." He estimated the intial revenue loss at $60-million a year but conceded that he did not know how high it could go if exploration increased as intended. By silence, the President and the Secretary indicated they contemplated no change in the existing depletion allowance of 22 per cent.

¶Authorize the Interior Department to license deep-water tanker terminals offshore. "We can expect considerably less pollution if we use fewer but larger tankers and deep-water facilities," Mr. Nixon said.

All three legislative proposals were considered certain to run into stiff opposition in the Democratic - controlled Congress.

Reaction from energy industries to the message was generally favorable, but qualified. The National Coal Association regretted the absence of "a massive commitment to coal research." The Independent Petroleum Association of America, long a supporter of import quotas, said their absence could hamper domestic resources development.

Enviromentalists took a critical view. Ann Roosevelt of Friends of the Earth's Washington office said the message favored "economics over ecology."

Mr. Nixon spoke affirmatively about environmental values several times, but his emphasis was on getting more energy flowing, especially from domestic sources.

Urging the states to encourage the use of coal, America's most abundant fuel, the President suggested that they take their time about putting into effect secondary air-pollution standards — and he assured them that they would not be hurried by Washington under the Clean Air Act.

Secretary Shultz, who arrived two hours late for a White House briefing with 100 journalists and offered no explanation, said on the environment question. "We have to face up to some of these tradeoffs and take them one by one." The Secretary was late because he was briefing members of Congress, it was learned.

Mr. Shultz noted that secondary standards, by definition do not involve "anyone's health and safety."

The Secretary was decidedly cool when asked about the importation of liquefied natural gas from the Soviet Union. "We know the gas is there," he said, but importing to this country is "a long-term proposition that being studied."

As expected, Mr. Nixon announced that the Interior Department would increase the sale of offshore leases for exploration. He said the annual acreage leased would triple by 1979.

In a highly speculative statement, he said, "by 1985 this accelerated leasing rate could increase annual energy production by an estimated 1.5 billion barrels of oil," or 16 per cent of this country's anticipated oil needs, and five trillion cubic feet of natural gas, or 20 per cent of needs.

To reassure those worried about the environmental risk of offshore drilling, Mr. Nixon said that "new techniques, new regulations and standards and new surveillance capabilities enable us to reduce and control environmental dangers substantially."

Asserting that "we as a nation must develop a national energy conservation ethic," the President argued that "energy conservation is a national necessity, but I believe that it can be undertaken most effectively on a voluntary basis."

Mr. Nixon suggested that "all workers and consumers can help by continually saving energy in their day-to-day activities — by turning out lights, tuning up automobiles, reducing the use of air conditioning and heating and use energy efficiently."

Industry should make more efficient products, he said and the Government will develop "a voluntary system of energy efficiency labels for major home appliances."

The President also announced that he was establishing an Office of Energy Conservation in the Department of the Interior to "educate consumers on ways to get the greatest return on their energy dollar."

Mr. Nixon reaffirmed his commitment to development of many more nuclear power plants—he spoke of their producing half the country's electric energy by the year 2000— and said that he would propose ways to shorten the time-consuming licensing procedures that have delayed many plants.

On energy research and development, he reviewed the 20 per cent increase in outlays that he announced in his January budget. He spoke enthusiastically of the potential of oil shale reserves and the harnessing of geothermal energy, but reserved judgment on both until he gets more information on environmental effects.

Mr. Nixon reiterated his commitment to early construction of an Alaska oil pipeline and attempted to squelch interest in a Canadian route, saying it would take longer to build.

The principal executive action disclosed by the message was the termination of the mandatory oil import quotas and the introduction of a graduated system of license fees on imports in excess of previously authorized 1973 volumes.

In 1959, when he was Vice President and President Eisenhower imposed the quotas, Mr. Nixon recalled, the United States had excess oil production capacity and world prices were below domestic prices.

Now, he went on, "we must import ever larger amounts to meet our needs." Quotas no longer benefit the country, Mr. Nixon said, actually hamper the meeting of energy needs and have caused "general dissatisfaction."

As outlined by William E. Simon, Deputy Secretary of the Treasury and chairman of the Oil Policy Committee, the new oil import program is designed to let the country import what it needs to avoid shortages and at the same time encourage domestic exploration and expansion of refinery capacity.

Secretary Shultz conceded, however, that imports might not avert gasoline shortages this summer. "We do face some important potential problems there," he said.

The new "license-fee quota" program consists of the following elements:

¶The tariffs of 10½ cents on a 42-gallon barrel of crude (a quarter-cent a gallon) and 1¼ cents a gallon on gasoline are terminated. The President has authority to take such action under the national-security provision of the Trade Expansion Act.

¶Imports previously authorized for 1973 may be brought in without payment of duty or fee. However, these volumes will be gradually reduced to zero over seven years.

¶Imports in excess of 1973 quota amounts, and in excess of the reduced amounts in future years, will be subject to license fees, a euphemism for protective tariffs. The fees will start at 10½ cents a barrel for crude oil and 52 cents a barrel for gasoline. They will rise by semiannual steps to levels at of Nov. 1, 1975, of 21 cents for crude, 63 cents for gasoline and 63 cents for all other finished products.

April 19, 1973

The Energy Crisis and How One Urban Family Lives

By WAYNE KING
Special to The New York Times

PHILADELPHIA, April 26 —Without really noticing, Bob Alotta has come a long way from the brash days when he was a chubby kid growing up in South Philadelphia 20-odd years ago.

The life-style in those days was a little rough-and-tumble —Bob Alotta suffered a tender nose in a tiff with a lad named Alfred Arnold Coccazza, who later became known as Mario Lanza—and by today's standards, Mr. Alotta feels, most of the people would be considered working class and just on the safe side of poor.

"But nobody seemed to notice," he recalls. "Most people seemed to live pretty much the same way."

Mr. Alotta's life today is significantly different. His salary as an employe of the Philadelphia Housing Authority is just under $20,000 a year, enough to allow him to buy a new home, a spacious, 1840-vintage, four-story brownstone in downtown Philadelphia, and to indulge his hobby of collecting tomes of American history.

And although he and his wife and two children have never really thought about it, Mr. Alotta's upward mobility has also made possible a great leap forward in an area most Americans have traditionally taken more or less for granted—the consumption of energy.

Some Exotic Necessities

Energy demands, which doubled between 1950 and

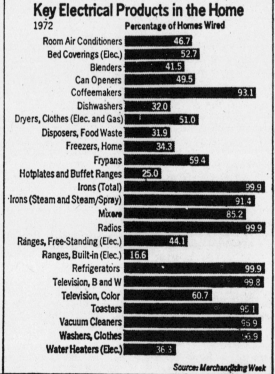

Key Electrical Products in the Home

1972

Product	Percentage of Homes Wired
Room Air Conditioners	46.7
Bed Coverings (Elec.)	52.7
Blenders	41.5
Can Openers	49.5
Coffeemakers	93.1
Dishwashers	32.0
Dryers, Clothes (Elec. and Gas)	51.0
Disposers, Food Waste	31.9
Freezers, Home	34.3
Frypans	59.4
Hotplates and Buffet Ranges	25.0
Irons (Total)	99.9
Irons (Steam and Steam/Spray)	91.4
Mixers	85.2
Radios	99.9
Ranges, Free-Standing (Elec.)	44.1
Ranges, Built-in (Elec.)	16.6
Refrigerators	99.9
Television, B and W	99.8
Television, Color	60.7
Toasters	95.1
Vacuum Cleaners	95.9
Washers, Clothes	95.9
Water Heaters (Elec.)	36.3

Source: Merchandising Week

The New York Times/April 27, 1973

1970 and are expected to double again by 1985, are straining the nation's fuel supplies and causing persistent shortages, greater dependence on imports and increasing concern for the environmental costs of a high-energy society. These are the elements of today's energy crisis.

Personal requirements are estimated to account for more than a third of all the energy the nation expends each year, more if one would count such things as public transportation that might be said to serve individual needs.

The growth in personal energy use is most pronounced among people like the Alottas, who, although not conspicuous consumers, can buy most of the things they feel they need.

And those needs, insofar as energy is concerned, are far more exotic than those of their parents—and significantly higher for the Alottas themselves today than they were just a few years ago.

Of 24 items listed as basic electrical appliances by the appliance industry, the Alottas have only 11. Even so, the list is impressive.

"It's surprising when you actually start listing the things," says Mr. Alotta. Of items on the basic list, the Alottas have an electric blender, coffee maker, freezer, frying pan, two irons, eight radios, a refrigerator, two black-and-white television sets, two toasters, and three vacuum cleaners.

In addition, the Alottas have two small record players and a stereo set, two electric toothbrushes, two electric clocks, two tape recorders, an electric sewing machine, an electric "water pick," an old electric train and a woodburning set.

A weekend handyman, Mr. Alotta has two electric saws, a sander and a drill, but he gets by with a hand-powered lawnmower. He often works at home and has an electric typewriter.

Both he and his 11-year-old son, Peter, use electric hair dryers—Peter's is a hot comb, his a blower type— that they use from once a day to once a week.

Many Items Passed Up

On the other hand, they do not have air-conditioners and they have passed up items like sunlamps, electric knives, can openers, waffle irons, food waste disposers, hotplates and buffet ranges, rotisseries, slide and movie projectors, popcorn

The New York Times/Bill Wingell

Bob and Alice Alotta using two of their electric appliances at their home in downtown Philadelphia

poppers, electric garage doors, outdoor lawn lights, humidifiers and what not.

Nonetheless, the conveniences the Alottas enjoy through the use of electric power contrast sharply with those of his parents.

When he was growing up in South Philadelphia, Mr. Alotta recalls, he lived in a six-room house that had not so much as an electric fan, and the idea of a vacuum cleaner was identified with affluence.

"We had one radio in the house, a gas stove in one room and a refrigerator and a toaster," he remembers.

But Mr. Alotta's recollections of 2½ decades ago, although sharp in their contrast with his life-style today, do not fully tell the story of the rapid acceleration of

electric living, as the power companies like to call it.

Ten years ago, 1 per cent of the families living in homes wired for electricity had room air-conditioners. Today the figure is 47 per cent. The number of families with electric blankets has doubled to more than 50 per cent. Nearly half the nation's families have electric can openers compared with 11 per cent 10 years ago.

In 1964, only 5 per cent of Americans had color television sets; now the figure is 51 per cent.

Today, 99.9 per cent of families in wired homes have electric irons, radios, refrigerators and black and white television sets, and 97 per cent have clothes washers and vacuum cleaners.

Moreover, the number of wired homes has grown from 53.7 million in 1962 to 67.3 million today—more Americans using more power at an increasing rate.

Electric power, of course, is far from the whole story. The Allotas cook and heat with gas and their projected gas consumption is expected to double in their new home.

The Alottas have one family car and Mr. Alotta has a second for business use. If they follow the past national pattern, they will each year both drive farther and use more gas per mile—a trend in part related to the growth in popularity of such automotive niceties as air-conditioning, power-operated radio antennas and stereo tape decks—not to mention such

things as automatic door openers being marketed by at least one auto accessory manufacturer.

All told, gas and electric consumption in the United States has increased about 25 per cent from 1962 to 1972, with electric consumption almost doubling, according to figures provided by the American Gas Institute and the Edison Electric Institute.

Pondering the energy crisis, to which his home is an infinitely small contributor, Mr. Alotta says he feels there is no practical way to turn back home energy use.

"If things really get bad," he says, "maybe I'll just move into the ice house"—a spacious relic left from the days of his home's original construction 130 years ago.

April 27, 1973

A SAVING DRIVE URGED BY NIXON IN ENERGY PLAN

A CRISIS FEARED

By JOHN HERBERS
Special to The New York Times

SAN CLEMENTE, Calif., June 29—President Nixon announced today his plan to meet the nation's energy needs. It included a voluntary conservation drive in which the Federal Government would set the example, expansion of research to find new energy sources and Government reorganization to give higher priority to energy matters.

He named Gov. John A. Love, Republican of Colorado, to direct from the White House a new office that will be responsible for forming and coordinating energy policies throughout the executive branch.

"America faces a serious energy problem," Mr. Nixon said in a statement. "While we have only 6 per cent of the world's population, we consume one-third of the world's energy output. The supply of domestic energy resources available to us is not keeping pace with our ever-growing demand, and

unless we act swiftly and effectively, we could face a genuine energy crisis in the foreseeable future."

Bars Mandatory Rationing

Governor Love, who said he would resign his Colorado office within a week, met with the President this morning and then held a news conference in which he rejected the idea of mandatory rationing of fuels and Government controls that would limit the horsepower of automobiles.

The Nixon policy is to increase the supply and induce voluntary restraints on the use of fuel.

"I am launching a conservation drive to reduce anticipated personal consumption of energy resources across the nation by 5 per cent over the next 12 months," Mr. Nixon said. "The

Federal Government will take the lead in this effort, by reducing its anticipated consumption by 7 per cent during this same period."

In this regard, he sent a directive to all departments and agencies ordering them to reduce the level of air-conditioning in office buildings, to relax dress standards, use more efficient and smaller automobiles, reduce business trips and cut down on unnecessary lighting.

Mr. Nixon said he had directed the Department of Transportation to seek a reduction in the speed of commercial airline flight and "where possible, the frequency." Governor Love said the airlines had initiated such restrictions, but there was no indication as to the extend of this.

The President called on private citizens to help by driving more slowly, using more car

277

pools and public transportation, and raising the thermostats of air conditioners by four degrees. He urged the nation's Governors to reduce highway speed limits, and he ordered Cabinet officials to meet with representatives of industry to urge them to find ways to cut back on energy consumption.

Spokesmen for the White House said in response to questions that no major advertising campaigns were planned to speed voluntary conservation and they knew of no plans to cut down on Presidential travel to set an example.

The goal of a 5 per cent reduction, Mr. Nixon said, can be achieved "by making very small alterations in our present living habits, for steps such as those we are taking at the Federal level can be taken with equal effectiveness by private individuals. We need not sacrifice any activities vital to our economy or to our well-being as a people."

To increase the energy supply, Mr. Nixon said he was calling for a $10-billion research effort over five years. But he was vague as to how that amount of money would be raised. He said he would receive recommendations on this from his advisers by Dec. 1 for preparation of the budget for the fiscal year 1975.

Meanwhile he added $100-million to the $770-million energy research budget for the fiscal year that starts July 1, directing that the new money be set aside for energy research, mostly for finding ways to produce clean 'liquid fuels from coal. He did not say where the money would come from.

On government reorganization, Mr. Nixon said that the duties of the Special Energy Committee and the National Energy Office that he set up two months ago to plan energy strategy would be combined under an expanded Energy Policy Office, to be headed by Governor Love, in the White House. Mr. Love will be both director of the office and an assistant to the President.

"He will spend full time on

United Press International

President Nixon with Gov. John A. Love, who was named to head new Energy Policy Office. With them at San Clemente, Calif., is Melvin R. Laird, left, Presidential aide.

this assignment and report directly to me," Mr. Nixon said. "My special consultant on energy matters, Charles DiBona, will continue in his present advisory capacity, working within the new office."

Governor Love said that in coming to the Federal Government at $42,500 a year, he would divest his oil interests, which he described as "so small they are almost embarrassing." An interest in a gas well provides him $1,000 a year. He has 300 shares of oil company stock and other investments that produce less than $100 a year,

Governor Love said.

In the news conference, he indicated he had not yet studied the complex questions concerning his new office.

Mr. Nixon said he would continue to press, with some refinements, creation of a new department of energy and natural resources to take over many responsibilities of the Department of Interior and some functions of the Department of Agriculture and other agencies. This proposal has floundered in Congress for two years.

The President further proposed creation of an energy re-

search and development administration that would carry out the proposed new research.

"The present functions of the Atomic Energy Commission, except those pertaining to licensing and related regulator responsibilities, would be transferred to [the new administration] as would most of the energy research and development programs of the Department of Interior."

The Atomic Energy Commission would continue under a new name, the Nuclear Energy Commission.

June 30, 1973

CRISIS

ARABS CUT OIL EXPORTS 5% A MONTH

U.S. CHIEF TARGET

By RICHARD EDER
Special to The New York Times

BEIRUT, Lebanon, Oct. 17— The Arab oil-producing nations proclaimed tonight a monthly

cut in exports of oil, with the burden to fall on the United States and other nations considered to be unfriendly to the Arab cause.

The long-awaited formal decision to use oil as a weapon

in the Middle East conflict was announced at the end of an eight-hour meeting in Kuwait of ministers from 11 countries.

The monthly export reduction was set at 5 per cent off each previous month's sale,

starting with the level of sales in September. The measure was at once more modest, more flexible and vaguer than had generally been predicted.

A Significant Shift
"It was about as mild a step

as they could have taken," said one oil expert who had talked to the participants. At the same time, to have finally come to the use of oil as a weapon, as had been threatened for years, marks a significant evolution in Middle Eastern affairs.

The cuts would continue, month by month, until Israel evacuated the territories occupied in the 1967 war, and made provision to respect Arab rights. This deliberately imprecise formulation alludes to the claims of the Palestinian refugees.

France May be Exempt

There was no specific mention of any country on the "unfriendly" list other than the United States. This was one of many flexible aspects of the decision. It allows the Arab states to grade customers in order of their support of the Arab cause. The participants promised to insure that the 5 per cent monthly export cut would not reduce sales to "friendly" countries, but again they did not say which countries these were.

Observers at the meeting assumed that France, for example, would not be subject to reductions. West Germany, presumably, might be. Japan, whose position was described by one participant as that of "odious neutrality," might experience some difficulties. It was hard to say what treatment would be given to Britain, which has also tried to be neutral.

The 5 per cent cutback would be computed against the previous month's exports.

The cut is less than it would be if this 5 per cent were computed from some single point. Thus, after six months the actual reduction would be 23 per cent instead of 30, and at the end of a year, 43 per cent instead of 60.

The 11 countries involved in the decision, not all of which are oil-producing countries, were Abu Dhabi, Algeria, Bahrain, Dubai, Egypt, Kuwait, Iraq, Libya, Qatar, Saudi Arabia and Syria.

Egypt and Saudi Arabia, opposing more militant proposals, are reported to have insisted on avoiding measures that would put relations with the United States beyond "the point of no return," a phrase used by the Egyptian President, Anwar el-Sadat, in his speech yesterday.

Reduction Is Modest

Tonight's decision appears to take account of this view. The dimensions of the cut were considerably more modest than the kind of all-out action called for by countries such as Syria and Iraq.

The United States uses some 17 million barrels of crude oil and refined products each day, and some 6.4 million barrels of this are imported. From the Arab countries the United States takes a total of crude and heating oil estimated variously at 1.5 million to 1.9 million barrels a day.

This week the United States released figures purporting to show that Americans would not be seriously affected even by major cuts in Arab oil production.

William E. Simon, chairman of President Nixon's oil policy committee, said that the United States could decrease its consumption of oil by as much as three million barrels a day if it made the necessary effort.

October 18, 1973

PRESIDENT ASKS CONGRESS FOR ENERGY-CRISIS ACTION;

FOR 50-M.P.H. LIMIT

Year-Round Daylight Time and Reduced Heating Urged

By EDWARD COWAN
Special to The New York Times

WASHINGTON, Nov. 7—President Nixon asked Congress tonight to give him a variety of far-reaching powers to deal with the deepening shortage of oil.

Endorsing proposals made earlier by his energy policy director, Mr. Nixon asked in a television address that Congress pass an emergency energy act before it recesses in December.

The act would empower him to relax environmental standards, reduce automobile speed limits, regulate transportation schedules, impose daylight saving time the year-round and provide contingency plans for rationing gasoline.

The President appealed to the American people to help cope with the oil shortage by driving cars less and urged that the states reduce speed limits to 50 miles an hour.

6-Degree Cut Sought

He also asked Americans to lower temperatures in the home by 6 degrees, to a daytime average of 68 degrees.

And in the Administration's first acknowledgement that the fuel shortage could cause loss of sales, production and income, Mr. Nixon asked offices, factories and stores to achieve the equivalent of a 10-degree reduction by lowering the thermostat or curtailing working hours.

Mr. Nixon reiterated his plea for quicker construction and licensing of nuclear power stations and again called for creation of an energy research and development agency. And he asked Congress to enact, before its recess, several energy bills the Administration has sought for months.

Mr. Nixon made no new major policy proposals tonight. In the main, he asked Congress to enact proposals outlined in the last fortnight by John A. Love, the director of the Energy Policy Office.

The President said that the aim of his program was to begin to move the nation away from any dependence on Middle East oil by 1980. The current shortage has been worsened by reductions in production by Arab states and their total embargo on shipments to the United States.

Although his principal purpose was to stress the gravity of the shortage and the need for action by Congress and state governments, Mr. Nixon sought to reassure the country about what he called the prospect of "the most acute shortages of energy since World War II."

"The fuel crisis need not mean genuine suffering for any American," he said, "but it will require some sacrifice by all Americans."

Briefings for Officials

Mr. Nixon, whose popularity is at its lowest ebb because of the Watergate controversy, sought to rally politicians from both major parties to support his energy program. He invited members of Congress, Governors, Mayors and county executives to White House briefings today in advance of the address.

Returning to the theme of self-sufficiency that he articulated in his April energy message, Mr. Nixon said that the country must strive to meet its needs without relying on foreign countries.

Mr. Nixon spoke only in passing of the Arab oil embargo, which he said would cause a supply gap of 10 to 17 per cent in oil supplies this winter.

Oil accounts for about half the total energy consumed in this country. Of the 17 million barrels of crude oil and refinery products that have been burned every day, about two million have come, directly and indirectly, from Arab states.

Mr. Nixon spoke deliberately of "Middle Eastern" producers in what officials had said was a conscious attempt to avoid anything that might ring in Arab ears as a tone of rancor or retaliation. The President's strategy, the officials said, was to show the Arab producers that the United States cannot be crippled by the embargo and that it will adapt successfully.

"We have an energy crisis," Mr. Nixon said, "but there is no crisis of the American spirit." It was the first time he had called the fuel shortage a "crisis."

Appeal Is Repeated

The President renewed an appeal to Congress to authorize the five-year, $10-billion energy research and development program he proposed on June 29, and to create as a separate agency an energy research and development administration.

To dramatize the proposal, Mr. Nixon likened the push to develop new forms of energy to the World War II Manhattan Project to develop an atomic bomb and to the Apollo effort that put men on the moon.

Returning to his theme of energy self-sufficiency, he suggested that the research effort be called "Project Independence."

November 8, 1973

President Signs Measure To Allow Alaska Pipeline

By EDWARD COWAN
Special to The New York Times

WASHINGTON, Nov. 16 — President Nixon signed the Alaska pipeline bill today and hailed it as a first step toward making the United States wholly self-sufficient for its energy supplies by 1980. An official of the Department of the Interior expressed confidence that the law authorizing the 789-mile pipeline from Prudhoe Bay on Alaska's North Slope to the warm water port of Valdez would survive any challenge in court.

Two of the environmental organizations that had earlier blocked the pipeline through the courts issued a strongly disapproving statement, but they stopped short of saying they would continue their legal fight.

The organizations left the door open for further court action, however, saying:

"The environmental issues that have been awaiting judicial determination should be left to the courts, in the absence of any adequate consideration of those issues in Congress."

The statement was issued by The Wilderness Society and Friends of the Earth. They and other plaintiffs had successfully challenged the pipeline earlier on the ground that it would have violated right-of-way limitations in the Minerals Leasing Act of 1920.

After the United States Court of Appeals for the District of Columbia so held last February, without reaching environmental issues, the Administration asked Congress for special legislation. The bill that Mr. Nixon signed today revised the right-of-way limitation, but more importantly, it authorized the Secretary of the Interior to grant exceptions.

The measure also declares it to be the will of Congress that the pipeline be built "promptly without further administrative or judicial delay or impediment." Seventeen members of Congress were present for the signature ceremony in the White House.

The measure seeks to confine judicial review to the question of constitutionality, and it orders that any such challenge be tried on an expedited basis with any appeal to go directly to the Supreme Court.

Jared G. Carter, a Deputy Under Secretary of Interior, said, "We rate our chances of success, if a lawsuit is brought, as excellent."

Mr. Nixon said he was signing the bill even though it contained "a couple of clinkers"— provisions that had nothing to do with the pipeline and that he disliked. They included expansion of the authority of the Federal Trade Commission to take companies to court and repeal of the authority of the Office of Management and Budget to oversee the collection of data by regulatory agencies.

The President said that he hoped the Democratic Congress would undo those parts of the measure, but there was little prospect that it would. Democrats attached those provisions to the pipeline bill because they believed that was a way to get Mr. Nixon to sign them into law.

In a prepared statement that he did not read during the Oval Office ceremony, Mr. Nixon saluted the $4.5-billion venture as "the largest single endeavor ever undertaken by private enterprise."

The President said the line would be completed by 1977, but officials of the Alyeska Pipeline Service Company, the consortium of seven oil companies that will build the line, said that the 1977 date was only a hope, not a certainty.

The oil companies and their interests in Alyeska are: Atlantic Richfield, 28.08 per cent; Standard Oil of Ohio, in which British Petroleum has a controlling interest, 28.08 per cent; Exxon, 25.52 per cent; Mobil, 8.68 per cent; Phillips, 3.32 per cent; Union Oil, 3.32 per cent, and Amerada Hess, 3 per cent.

The company estimates that it must receive some 1,100 permits from Federal agencies in the course of construction. For example, Interior Secretary Rogers C.B. Morton said, the Environmental Protection Agency must license the emissions from the 12 pumping stations that will make the crude oil flow through the line at a rate of seven miles an hour.

Mr. Morton said at a news briefing after the White House ceremony that he would sign no permit until Alyeska came to terms about payment of $12-million that Mr. Morton said was owed to the Government for extraordinary expenses incurred in the preparation of the project's environmental impact statement.

The pipe's interior diameter is 48 inches, giving it a transmission capacity of two million barrels a day. Originally, Alyeska had contemplated that such output would not be reached for years after the completion of construction, but more recently, there has been talk of compressing that schedule.

The consortium's Washington ofice said that the flow might reach 1.2 million barrels a day within a year after completion of the pipeline but that "the owner companies have not made a decision to do anything other than the 600,000-barrel initial rate."

As it is now developed, the Prudhoe Bay oilfield can produce only 1.6 million barrels a day.

The Prudhoe Bay discovery, announced in February, 1968, has been conservatively rated at 9.6 billion barrels. A White House "fact sheet" said: "Additional development is expected to increase this figure to 15 billion." If pumped out at the rate of two million barrels a day, 15 billion barrels would last about 20 years.

To the west of Prudhoe Bay lies the Naval Petroleum Reserve No. 4, largely unexplored but believed to hold tens of billions of barrels of oil. Mr. Morton, when asked, said extensive exploration of the reserve would be "in the national interest."

Mr. Nixon offered to Senator Henry M. Jackson, Democrat of Washington, the first of the four pens he used to sign the bill. Mr. Jackson, who had sponsored the legislation and worked hard for its passage, demurred, saying the pen should go to Mrs. William Pecora, who was present, the widow of an Under Secretary of the Interior who was closely associated with the pipeline project.

The other pens went to Alaska's Congressmen—Senators Mike Gravel, a Democrat, and Ted Stevens, a Republican, and Representative Donald E. Young, a Republican.

November 17, 1973

Douglas Links Lobbies To the Energy Crisis

BUFFALO, Nov. 28 (UPI) —Supreme Court Justice William O. Douglas said last night that the current energy crisis had been caused by powerful corporate lobbies.

In a speech at the State University at Buffalo, the 74-year-old jurist said that Federal bureaucracies responsible for dealing with energy problems were more responsive to corporate interests than to the public interest.

He added that the nation's tax system was "designed to protect those out to destroy our natural resources."

"We, the people, through tax concessions, are financing the destruction of the environment," he said.

Justice Douglas said that 25 oil companies in the United States owned most of the coal, gas and uranium.

"We have a fuel monopoly but no monopoly on solar energy and hydrogen fusion," he said. "That is why they are not being promoted."

He said that one solar energy plant could be built to supply enough electrical power to serve the entire country.

November 29, 1973

Nixon Seeks 15% Cut in Gasoline Output And Reduced Deliveries of Heating Oil

Sunday Sales and Holiday Lights to Be Forbidden

By EDWARD COWAN

Special to The New York Times

WASHINGTON, Nov. 25 — President Nixon told the American people tonight that he would take a variety of actions to reduce consumption of energy, including a cutback in home heating oil deliveries and a 15 per cent reduction in gasoline production that is sure to cause shortages.

As expected, Mr. Nixon said that to cope with the widening shortage of crude oil and refinery products caused by the Arab oil embargo, he would prohibit sales of gasoline on Sunday and would lower highway speed limits throughout the nation to 50 miles an hour for cars and 55 for trucks and buses.

Heating oil deliveries will be cut by 15 per cent to homes, the White House said, 25 per cent to stores and other commercial customers and 10 per cent to industrial users.

The President also banned all outdoor Christmas lights, even those used to decorate houses.

For 'Full Cooperation'

Summarizing a package of energy conservation actions in a 14-minute address broadcast on television and radio, Mr. Nixon said he was calling for "the full cooperation of all the American people in sacrificing a little so that no one must endure real hardship."

He said that the measures he was announcing would reduce the shortage of oil from 17 per cent to 7 per cent, and that "additional actions will be necessary."

He referred to the Arab oil embargo only as "cutbacks in oil from the Mideast." He said nothing about any lifting of the embargo or the Government's efforts to get Arab-Israeli peace talks started in hopes that it would lead to at least a partial relaxation of the embargo.

Mr. Nixon also spoke of reduced deliveries of jet fuel and gasoline.

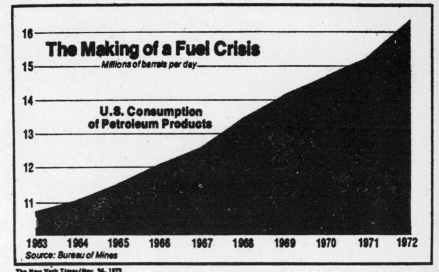

The Making of a Fuel Crisis
Millions of barrels per day
16
15
14
13
12
11

U.S. Consumption of Petroleum Products

1963 1964 1965 1966 1967 1968 1969 1970 1971 1972
Source: Bureau of Mines

The New York Times/Nov. 26, 1973

Reductions in jet fuel supplies on Dec. 1 and Jan. 7 are expected to result in still more cancellations of plane flights in addition to the cancellations announced earlier this month.

John A. Love, director of the Energy Policy Office, said at a White House briefing that the country had to cut its consumption of gasoline by 25 to 30 per cent and that the Administration hoped to do it without formal rationing.

It appeared certain that the Administration's request to refineries to decrease gasoline production by 15 per cent would lead to shortages and informal rationing of the kind experienced for a few weeks last spring—shorter hours at filling stations, pumps that run dry, retailer limits of 10 gallons a car and perhaps early-morning lines of cars.

Lines of patrons waiting to buy gasoline on Saturday appeared to be a certain prospect. So did increased use of buses and trains for weekend trips.

However, interstate trucks that use diesel fuel will be able to get it on Sunday. These pumps will not be closed. It is possible that a few service stations on interstate highways will, after some experimentation, be allowed to continue to sell gasoline on Sundays to service trucks and to meet genuine emergencies.

As expected, Mr. Nixon also said that he would prohibit promotion, display and ornamental lighting by commercial establishments. Stores and other places of business will be allowed to keep signs lighted during open hours. But all-night neon flashers and shop window display lights will be turned off.

The gasoline shortage and the Sunday sales ban, which is expected to close nearly all 220,000 filling stations, were expected to have especially hard economic effects on turnpike restaurants, motels, ski slopes, weekend resorts and other businesses that depend on weekend driving.

Heating Oil Price Rise

Officials indicated that increases in heating oil prices might be allowed to give the refiners an added incentive to produce less gasoline, normally their most profitable product, and more heating oil. The question has been referred to the Cost of Living Council, which administers price controls. Some sources said the increase would not exceed two cents a gallon.

Mr. Love announced that the Government was making deep cuts in fuel for general aviation—50 per cent for pleasure and instructional flying, 40 per cent for business flying, including corporate jets, and 20 per cent for air taxi and industrial aviation, such as crop dusting or ferrying of supplies.

President Nixon left open the

John A. Love briefing newsmen at the White House. He heads newly formed Energy Emergency Action Group.

United Press International

date when the Sunday sales ban would go into effect because he is waiting for Congressional passage of the National Energy Emergency Act, which would specifically authorize such actions. However, he asked filling stations to close voluntarily next weekend, from 9 P.M. Saturday to 12:01 A.M. Monday, the proposed hours of the ban.

8-to-10-Day Wait

Mr. Love and his associates in the Energy Policy Office would have liked the Sunday sales shutdown to start on Dec. 2, and they counseled that it could be imposed under the Economic Stabilization Act. However, Mr. Nixon evidently took the course recommended by his political advisers and decided to wait for a fresh legislative mandate, in effect making sure that Congress explicitly shared in the responsibility for a measure that may cause much inconvenience.

The Senate approved the National Energy Emergency Act last Monday and the House may pass it this week. Allowing time for a House-Senate conference to resolve differences, officials do not expect the bill, which would also authorize the ban on ornamental

lighting of homes, to reach Mr. Nixon's desk for eight to ten days.

Mr. Nixon did not discuss gasoline rationing, a topic that has been much debated in Washington and around the country.

On Nov. 17 he said the American people would resent peacetime rationing "very, very much," and that he personally disliked it because of the power it gave to the officials who administer it.

The President gave no hint of what other measures he may announce. Rationing is still a possibility but it appears that the Administration will first try other steps, such as restricting gasoline stations' weekday hours, putting schools on a four-day week, and asking people to give up all but essential driving.

Having recently denounced Congress for not doing enough in the energy sphere, Mr. Nixon sounded a more approving and conciliatory note tonight.

"The Congress also has been moving forward on the energy front," he said early in the address.

He cited the passage of the Alaska pipeline bill, which he signed on Nov. 16, and the

mandatory fuel allocation bill, which he said he would sign on Tuesday. It would require allocation of all crude oil and oil products.

To help administer these programs, the Government is summoning to temporary active Federal service 250 oil and natural-gas executives who belong to the National Defense Executive Reserve. Their companies can go on paying their salaries, and most are expected to, or the reservists can collect a Federal salary or they can offer to work for a nominal dollar a year.

Mr. Nixon appealed for "sustained and serious" action and cooperation "by millions of men and women" to cope with the fuel shortage, and he pledged "to participate personally and on a regular basis in the work of my energy advisers."

The President said that intercity buses and heavy-duty trucks "operate more efficiently at higher speeds" and therefore would be allowed to roll at 55 miles an hour. Mr. Love said fuel efficiency, not safety or the economic cost of delayed bus arrivals and truck deliveries, was the controlling reason for the exception.

25% Cut in Jet Fuel

The cumulative reduction in airline jet fuel will be 25 per cent, Mr. Nixon said, requiring "a careful reduction in schedules combined with an increase in passenger loads."

He specifically included in the ban on ornamental lighting outdoor gas lights, such as have been installed on many front lawns in the Greater New York area in recent years. Some householders have said that they put in the lights to discourage burglars and muggers.

"The energy consumed by ornamental gas lights alone in this country," Mr. Nixon said, "is equivalent to 35,000 barrels per day of oil. This is enough fuel to heat 175,000 homes daily."

Mr. Nixon reiterated the home-heating goal he set forth in his Nov. 7 energy address—a national daytime average of 68 degrees, six degrees below what has been normal. He gave this warning: "Those who fail to adopt such a cutback risk running out of fuel before the winter is over."

Some retailers have advised their customers that deliveries this winter will be based on a lower thermostat setting.

November 26, 1973

Oil Allocation Act Signed As Nixon Ends Opposition

Regulations for Crude Oil and Refinery Products to Be Set Within 30 Days— More Gasoline Cutbacks Possible

By EDWARD COWAN
Special to The New York Times

WASHINGTON, Nov. 27— President Nixon signed into law today the Emergency Petroleum Allocation Act of 1973, a bill his Administration had opposed for many months.

The bill requires the President to establish within 30 days supply-management, or allocation, programs for crude oil and all refinery products to make sure that the fuel shortage does not fall with unfair severity on any region or on "independent" refiners and distributors not affiliated with major oil companies.

In fact, the Government already allocates heating oil, diesel and jet fuel under the Economic Stabilization Act and

has announced plans to allocate gasoline. The new law would extend allocation to residual fuel oil, the heavy fuel burned by utilities and industry, and to crude-oil producers.

Officials re-affirmed today the prospect of additional gasoline cutbacks. They pointed out that the White House cited such an outlook Sunday when President Nixon announced that gasoline supplies in the first quarter of 1974 would fall 15 per cent below normal consumption. They noted that John A. Love has said that the ultimate cutbacks could approach 25 to 30 per cent.

November 28, 1973

E.P.A. LOSES POWER TO LIMIT RADIATION

In a Policy Reversal, Nixon Shifts Authority on Plant Standards to the A.E.C.

By RICHARD D. LYONS
Special to The New York Times

WASHINGTON, Dec. 11 — In a reversal of policy, President Nixon has shifted the authority to set radiation standards for individual nuclear power plants from the Environmental Protection Agency to the Atomic Energy Commission.

The environmental agency, which was assigned the power three years ago, was on the verge of setting stricter safety standards for individual plants while the A.E.C. has generally taken a more relaxed attitude toward radiation protection.

In September, the environmental agency was to have issued new radiation standards that would have drastically reduced the permissible limit of radiation-emitting materials that could be released from a nuclear power plant. The Administration has, in effect, canceled these standards by assigning the responsibility elsewhere.

Several critics of the atomic power industry said the effect of the move would be to allow more freedom in the construction and operation of nuclear plants, thus helping to open more generating stations to ease the energy crisis.

But these critics also foresee that looser standards may expose the general public to more radiation from the proliferating number of atomic power stations, and that this may eventually expose more people to the development of cancer and will eventually lead to genetic defects.

Dr. Ralph E. Lapp, a writer who is an expert on nuclear power, said he did not believe that the change in authority would be harmful because "the amount of radioactive materials emitted from a plant are very low."

But Dr. Henry W. Kendall, a nuclear physicist at Massachusetts Institute of Technology who is an official of the Union of Concerned Scientists, said that the switch in authority puts the country in the position "of having the goat guard the cabbages."

The President's intent is contained in a memorandum dated Dec. 7 from Roy L. Ash, director of the Office of Management and Budget in the White House, to Russell E. Train, administrator of the E.P.A. and Dr. Dixy Lee Ray, chairman of the A.E.C. A copy of the memorandum was obtained today by The New York Times.

The Ash memorandum states, in part, that a decision was needed "so that the nuclear power industry and the general public will know where the responsibility lies for developing, promulgating and enforcing radiation protection standards for various types of facilities in the nuclear power industry."

"We must, in the national interest, avoid confusion in this area," the memo continued, "particularly since nuclear power is expected to supply a growing share of the nation's energy requirements; and it must be clear that we are assuring continued full protection of the public health and the environment from radiation hazards.

"E.P.A. has construed too broadly its responsibilities to set 'generally applicable environmental standards for the protection of the general environment from radioactive material.'"

"On behalf of the President," the memo went on, "this memorandum is to advise you that the decision is that A.E.C. should proceed with its plans for issuing uranium fuel cycle standards, taking into account the comments received from all sources, including E.P.A.; that E.P.A. should discontinue its preparations for issuing, now or in the future, any standards for types of facilities; and that E.P.A. should continue, under its current authority, to have responsibility for setting standards for the total amount of radiation in the general environment from all facilities combined in the uranium fuel cycle, i.e., an ambient standard which would have to reflect A.E.C.'s findings as to the practicability of emission controls."

Charles L. Elkins, an official in the environment agency's office of hazardous materials control, said the agency "was going to set standards for the entire uranium fuel cycle," that is mining, fabrication of fuel elements, operation of atomic power plant reactors and reprocessing of the spent fuel.

Mr. Elkins said the agency's original intent had been to set safety standards for each step, but that under the President's directive the agency may only set one standard for the sum total of all the radiation emitted by all the steps.

"We wanted standards on the lines of what people are actually exposed to from radiation, while the A.E.C., through its licensing procedure, attempts to control this by telling a power company how to build a plant to minimize radiation loss," he added in explaining the difference between the two approaches to regulation.

Dr. Kendall said, "The hope we had was that the E.P.A. would set and enforce standards because the A.E.C.'s track record has been extremely poor in the areas of reactor safety and radioactive waste disposal."

He noted that A.E.C. would have responsibilities for not only developing and promoting atomic power, but also regulating it as well.

"This is a serious backward step that well dishearten many of the people in the controversy," he added.

The radiation safety question arose four years ago when two scientists at the A.E.C.'s Lawrence Radiation Laboratory at Livermore, Calif., called for a 10-fold reduction in the amount of radiation considered safe for the public.

These scientists, Dr. John W. Gofman and Dr. Arthur R. Tamplin, asserted that if current limits were maintained about 32,000 additional deaths from cancer could be expected annually.

December 12, 1973

U.S. Approves Con Ed Request for Burning of Coal on Staten Island, and City May Go Along

By RICHARD SEVERO

The Federal Environmental Protection Agency gave the Consolidated Edison Company its approval yesterday for coal-burning at the company's Arthur Kill plant on Staten Island.

The city, which has opposed the use of coal despite approval by the State Department of Environmental Conservation, appeared ready to go along with the idea soon. Without city approval, the utility would be barred from burning coal.

For health and environmental reasons, the Federal agency limited the coal-burning to Arthur Kill and did not approve Con Edison's request to be allowed to burn coal at its Ravenswood plant in Queens.

It said approval for that would not come until it was clear that not enough oil would be available to keep Ravenswood going.

On Nov. 25, Consolidated Edison said it was going ahead with the conversion of both plants to coal despite the city's opposition and, at that time, lack of Federal approval.

In its announcement yesterday the Federal agency relaxed some air-pollution standards in New Jersey to permit four plants to convert to coal, but it denied applications in seven other instances because of health and environmental concerns.

The denials were praised by New York environmental officials as evidence of Federal sensitivity to the city's air-pollution problems.

"There doubtless will be some increase in mortality" because of the use of coal, said the Federal agency's Administrator, Russell E. Train, at a White House briefing in Washington. He said he thought there would be an increase in respiratory ailments because of the Federal decision, but he emphasized that the decision would not be a total defeat for clean-air advocates and that air quality would return to no lower than 1972 levels. No coal has been burned by Con Edison since 1970.

The Federal announcement was released both in Washington by Mr. Train and the Federal Energy Administrator, William Simon, and in New York by Deputy Administrator Eric B. Outwater.

It came 16 days after New York State's Environmental

283

Conservation Commissioner, Henry Diamond, agreed to let Con Edison burn coal at both Ravenswood and Arthur Kill. City officials refused at that point to budge on their position that coal should not be used, at least not until the utility company had convinced them it was absolutely necessary.

Yesterday Herbert Elish, the city's Environmental Protection Agency Administrator, said the city was prepared to "act quickly" if Con Edison presented the proper evidence.

A spokesman for the company said yesterday it was asking the city for a reconsideration, but added: "Unfortunately, six weeks have elapsed since we first sought approval to burn coal. The market for coal is tight and we are not certain we will be able to obtain sufficient supplies in a timely fashion."

Mr. Train underscored what he regarded as the urgency of the situation when he said; "I would encourage these power plants to convert to coal as quickly as possible to help meet essential needs."

Mr. Outwater said he doubted Con Edison would be able to burn coal before early January; and emphasized that the resulting air pollution would be closely monitored "all winter long, so we can seek new alternatives if the situation threatens to bring serious health effects."

He added:: "We do know that the health of the people in New York and New Jersey would not be improved if the heat goes off and the lights go out."

If the city goes along with the Federal action, Con Ed would be able to burn dirtier fuel oil with a sulphur content of up to 3 per cent until next March 31. The limit now is 0.3 per cent. The coal burned in New York would have a sulphur content of not more than one pound per million British thermal units, with an ash content of 15 per cent or less.

In New Jersey, the Public Service Electric & Gas Company plant in Bergen would be able to convert to coal entirely and three units of its plant in Burlington could use coal. Atlantic City Electric Company's Beesleys Point plant could convert to coal, as could three units of its facility at Deepwaters. But units in Newark, Woodbridge, Kearny and Jersey City could not burn coal for now. In any event, all permits for New Jersey stand to expire March 15.

December 14, 1973

U.S. Energy Crisis Stirs Self-Interest of Regions

By JAMES P. STERBA
Special to The New York Times

NEW ORLEANS, Dec. 19 — The energy crisis has spawned a new regional self-interest in the major oil and natural gas producing states where, for the first time since World War II, fuel for local industry, homes and farms is running short and getting expensive.

State officials, sometimes skirting legality, are scrambling to find ways to keep more locally produced oil and gas for in-state customers. Southern politicians, sensing political ramifications as temperatures drop, are railing against environment-conscious Easterners whom they accuse of using relatively clean and cheap Southern energy while refusing to burn coal or to drill for their own oil and gas off the Atlantic coast.

And local residents in the South, traditionally unsympathetic to Northerners, are spreading the sentiment with a bumper sticker that reads, "Let the Bastards Freeze In The Dark."

The feeling is strongest in Texas, Louisiana and Oklahoma, which together produce three-fourths of the nation's domestic crude oil and natural gas. But ironically, they are having difficulty supplying energy to customers within these states.

This is because the bulk of the oil and gas that comes out of thousands of wells that dot these states is committed, under Federal control, to out-of-state customers, primarily in the Northeast.

Oklahoma farmers with oil wells on their own land, for example, are short of diesel fuel for their tractors and propane gas for their home heaters. Utility companies, which have always produced electricity with the locally produced natural gas, now face expensive conversions to coal —coal imported from Wyoming.

Texas has already turned off the lights on its Capitol dome in Austin, and home builders cannot get new natural gas hook-ups in some areas. Louisiana, which, like Texas, lured industry in with promises of endless supplies of cheap natural gas, cannot meet the gas demands for utilities or industry, including crucial sugar refineries.

And in all these states, oil and gas drillers themselves are having trouble getting enough diesel fuel to run drilling rigs used to find more energy.

Deregulation Urged

Gov. Edwin W. Edwards of Louisiana has emerged as the most vocal spokesman for deregulating oil and gas prices and providing for local needs first. He puts it this way:

"We cannot tolerate a situation where a few people who are not really aware of what the oil and gas industry is all about are able to inhibit the development of additional reserves in their own backyards and on their own shorelines.

"Meanwhile they expect the traditional producing states to continue to do so, depleting our reserves and furnishing gas and oil to them at regulated prices while we have to buy their unregulated automobiles, suits, refrigerators and other products which they make on the other end of the pipeline."

Clint Pray, energy adviser to Governor Edwards, says, "We're not trying to be Arab sheiks down here, we just expect other states to do their share."

Direct attempts to withhold energy from out-of-state customers have been ruled illegal in the past because they hampered interstate commerce, which is prohibited by the Constitution. However, producing states are trying other ways to preserve their oil and gas supplies, and some officials like Governor Edwards have threatened to slow down production to put pressure on other states to develop their own resources.

"This is an attitude that concerns me deeply," John A. Love, then President Nixon's energy adviser, said two weeks ago. "It is a form of regionalism and we cannot allow this to become effective."

Setting Production Rates

The most powerful tool producing states have is in setting production rates for oil and gas wells. State boards determine what the maximum efficient rate is for each well or field so that it will yield the most oil or gas in the long run. Pumping oil or gas too fast from a well may cut its total production greatly.

Because determining the maximum efficient rate involves judgments, states have been accused in the past of manipulating these rates to force price rises. Now, they could lower maximum efficient rates, thus cutting over-all production as a way of applying pressure on other states and the Federal Government.

However, the Federal Government is moving to take this power away from state governments. Under the President's new emergency energy bill, which cleared a House-Senate conference committee yesterday, the Federal Government could require production above the maximum efficient rate in some cases.

"A Federal-state confrontation is developing along these lines," said Jim C. Langdon, chairman of the Texas Railroad Commission, which regulates that state's oil and gas industry. "Our fields are already producing at the maximum efficient rate. Excess production would permanently damage the reservoirs and cause the loss of literally millions of barrels of oil."

Impact of Cutbacks

Sharp production cutbacks imposed by state governments would hurt everyone in the short run. However, producing states reason that they are in the same position as Arab producers, in that oil and gas left in the ground will be even more valuable later on.

State officials reasoned that in the long run oil and gas reserves are more valuable left in the ground. By turning down production on their wells, the wells will last longer and yield more, they feel, thus providing for state needs in the decades ahead.

When a cheap Louisiana natural gas runs out, Pennsylvania, for example, can begin mining its coal again. But Louisiana has no coal, so it has an interest in keeping gas available for its industry for as long as possible.

Production cutbacks would

284

be a drastic step—one that the Federal Government would challenge almost certainly. But by simply threatening cutbacks, producing states hope to prod other states into developing their own energy resources.

In the meantime, state governments are trying various other means to retain fuel for local customers. The Louisiana Legislature recently passed bills allowing that state to collect royalties on oil and gas production in kind rather than in money.

In-State Customers First

New Mexico and Louisiana have both passed bills requiring new oil and gas discoveries on state land to be offered to in-state customers first.

Ironically, pricing policies of the Federal Power Commission are doing the most to keep natural gas within producing states. The F.P.C. sets prices at the wellhead for all natural gas committed to interstate pipelines, and these prices are

generally low. However, natural gas produced and consumed within the same state is not under commission control and can be sold at free market prices.

Over the last year, as demand rose, intrastate prices have climbed far above prices set by the commission for interstate sales. For example, in Louisiana, newly discovered natural gas bound for out-of-state customers sells for 26 cents per 1,000 cubic feet. The F.P.C. set that price. At the same time, however, natural gas staying in Louisiana was bringing $1.01 per 1,000 cubic feet. The free market set that price.

The result of these price differences is that virtually all newly discovered natural gas in Louisiana, Texas and Oklahoma is being sold within the states.

'Our Own Interests'

"It would be in our own selfish interests for the Federal

Government to keep on regulating natural gas prices because nobody is going to sell it out of state when he can get more for it in Texas," said Gov. Dolph Briscoe of Texas in an interview.

However, as intrastate price rises are passed on to customers, they may have to pay more soon for natural gas coming out of their own backyard than is being paid by gas customers thousands of miles away.

As intrastate prices rise and shortages persist, elected officials in producing states envision adverse consequences at voting booths come next election. Gov. David Hall of Oklahoma said people in his state now tend to blame the Federal Government for fuel shortages.

But that could change, because under the Federal allocation program, state governments are allowed to allocate 10 per cent of the fuel supply.

"What's going to happen

when everyone starts calling up the State House demanding part of that 10 per cent," said one Oklahoma state aide. "Most of them are going to have to be turned down, and they're not going to be very friendly."

For the moment, Southern politicians are urging and threatening more than they are acting. They want other states to share the environmental risks inherent in offshore drilling, refinery construction, coal-burning and nuclear energy production.

They argue that everyone in the nation will suffer equally unless all states do their share to develop the energy reserves they have, even if it means more pollution.

Governor Edwards of Louisiana offers this summation, "It'll do little good to tell Millie down at the funeral home that at least her husband froze to death in clean air."

December 20, 1973

OIL PRICE DOUBLED BY BIG PRODUCERS ON PERSIAN GULF

6 States That Supply 43% of Western World Needs to Shift Rates on Jan. 1

OTHERS LIKELY TO ACT

Cost of U.S. Imports Could Rise to $8-Billion From Present $2.5-Billion

By BERNARD WEINRAUB
Special to The New York Times

TEHERAN, Iran, Dec. 23—A sharp increase in the price of Persian Gulf crude oil effective Jan. 1 was announced here today.

The measures, which double the price of a barrel of oil, were disclosed after a two-day meeting of ministers from the six gulf states that provide about 43 per cent of the non-Communist world's petroleum.

Associated Press
The Shah of Iran at his news conference yesterday

Posted prices, today's announcement said, will reach about $11.65 a barrel. Just two months ago the ministers unilaterally raised the price for a barrel of crude oil by 70 per cent to $5.11. Less than three years ago the posted price for a barrel of oil on the Persian Gulf was $1.80.

Expert 'Appalled'

One Western oil expert said tonight: "I'm shocked and appalled. We did not expect it to go this high. The United States will tough it out but it's really going to hit Europe. There's going to be severe economic consequences."

Another predicted that the cost of United States oil imports may climb as high as $8-billion next year because of the price increase, from the present level of about $2.5-billion a year. Most oil exporting nations are expected to follow the lead of the Gulf nations.

The "posted" price of a barrel of oil is, in the words of experts, a "legal fiction" that has endured for years. No oil is actually paid for at the quoted level.

Essentially, the posted price is the figure commonly used by oil-producing nations to calculate tax and royalty payments due from the Western companies who do the actual production work.

The posted level may reach anywhere from 40 to 60 per cent above the Government take.

Government Take

The Government take under the new price system, starting Jan. 1, will be $7 for one barrel of oil. At present, a barrel costs about $3.09. These represent, in a sense, the "real" value of the oil.

Officials here expected posted prices to reach about $9, but some Western oil experts tonight were really jolted at the $11.65 figure. "This price is a hell of a rough one," a knowledgeable source said.

Europe depends on the Middle East for 73 per cent of its oil imports. Japan depends upon the Arab nations and Iran for 88 per cent of her petroleum needs. This year, about 15 per cent of the oil consumed in the United States came from the Middle East, and the percentage has been rising in recent years.

One expert said tonight that in 1972 the Persian Gulf nations earned about $16-billion from oil. Next year, when the new prices begin, earnings of the gulf nations could reach as high as $60-billion.

The gulf states are Iran, Iraq, Kuwait, Saudi Arabia, Qatar and Abu Dhabi. The group, meeting privately to set oil prices, are members of the powerful Organization of Petroleum 'Exporting Countries. Other members of O.P.E.C., founded in 1950 as a bargaining agent for oil producers, are Algeria, Indonesia, Ecuador, Libya, Nigeria, Venezuela and Gabon. The organization's members provide 85 per cent of the world's oil exports.

At a news conference today in the Niavaran Palace, north of Teheran, Shah Mohammed Riza Pahlevi, observed: "The industrial world will have to realize that the era of their terrific progress and even more terrific income and wealth based on cheap oil is finished."

The 54-year-old Iranian leader added in a soft voice: "They will have to find new sources of energy, tighten their belts. If you want to live as well as now, you'll have to work for it.

"Even all the chidren of well-to-do parents who have plenty to eat, have cars, and are running around as terrorists throwing bombs here and there —they will have to work, too."

December 24, 1973

End-Game
By Anthony Lewis

BOSTON, Dec. 26—Other ages have had comets and new religions and revolutions. The change of the year is always likely to remind us that change is the law of life. But the sense of change, and the apprehension of it, cannot often have run so deep as they do this year. Many of us feel what an Englishman said in a letter to the editor the other day: "Our world is not ever going to be the same again."

One fundamental that is shifting is the relationship between the industrialized nations of the world and the others. We have been told again and again that the gap between rich and poor countries is dangerous. Now something is actually happening to redress the balance: the growing shortages and rising prices of basic resources.

Oil is the headline example, but it is not the only one. Underdeveloped countries that sell bauxite and copper and other basics are moving toward controlled sales and higher prices, following the Arab example. And the industrialized world is quite simply dependent on those resources.

Even with its advantageous mineral deposits, the United States must increasingly look to foreign suppliers. An official study forecasts that by 1985 we shall depend on imports for more than half our iron, lead, tungsten, aluminum, chromium, manganese, nickel, tin and zinc. We spent only $5 billion on metal imports in 1970; the bill is projected at $18 billion in 1985, $44 billion by 2000.

The average citizen does not need to know such figures to sense the looming reality of profound change— economic, political, social. That prospect, unsettling at best, is the harder for us to bear because we do not get honest or even competent answers from our political leaders.

The politicians in almost every developed country are relying on hope, public relations and luck to get them through the next weeks and months— anything except dealing with the long-term reality. In a time demanding extraordinary vision and patience, they seem more frivolous and short-sighted than ever.

Consider the Nixon Administration's tactics on oil. The message it keeps sending us is that the Arab embargo is easing, things are looking up, we'll get by this winter without real pain and then everything will be all right.

There is no serious student of the problem who believes that, and indeed the Administration knows better. On Dec. 13 William E. Simon, the energy administrator, told a group of Governors: "When the embargo ends, the crisis is not over. We will have a shortfall of five million barrels a day by 1978."

Experts in the oil industry and outside think world oil production will hit its peak within the next 25 years, then start to decline. In the last generation oil consumption has been doubling every decade. Put those two facts together, and you have a recipe for disaster. Prof. Arthur M. Squires of the City University of New York puts it:

"The dislocation of the world's economy and politics that will follow the collision between rising demand and a topping-out of oil production capability are mind-boggling. There is simply no chance at all for us if we do not learn quickly how to make do with less. The technological fixes can only ease the pain somewhat."

The politicians have been grossly deceptive in suggesting that there can be quick technological fixes. To go on with our pattern of expanding energy use would require some new source that could be developed even more rapidly than oil has been—and that is just not on.

There is a lot of talk, for example, about getting oil from shale. A suggested American target is one million tons of such oil a day by 1985, or 10 per cent of our projected oil demand. That would require mining two million tons of rock a day, which is about the total U.S. mining capacity today.

Where shall we get the water for mining or other imagined ventures on such a scale? What would we do with the waste? What about the effect on our desperately needed crop land?

The point is not that we should give up research on new sources of energy or other resources—only that we be realistic about the costs, the likely time-spans involved and the connection with everything else on one earth. The day of what has been called the cowboy economy is over—the easy exploitation of physical frontiers and backward peoples. Now, everything connects; everything has a price.

When the Arab embargo produced its short-run oil crisis, those who had been thinking about the problems of energy were actually encouraged. They hoped that the shock, the sudden confrontation with geophysical and economic reality, would make the political leaders of the great industrial societies begin to reckon with the future.

That is why the reaction of the politicians is so depressing. No one expects them to embrace bad news. But there is ample room for challenge and optimism in the world of the future, a world perforce living within its limits, if we begin to adjust to it. Instead, they still talk as if we could go on endlessly with the exploding material world of highways and suburbs and things. That way lies trauma.

An economist, E. F. Schumacher, said something recently about economists that applies even more acutely to politicians. Mostly, he said, they spend their time "optimizing the arrangement of the deck-chairs on the Titanic."

December 27, 1973

A.E.C. Promulgates Stiffer Safety Rules On Power Reactors

By EDWARD COWAN
Special to The New York Times

WASHINGTON, Dec. 28 — The Atomic Energy Commission tightened today the safety criteria that are meant to prevent the growing number of nuclear power reactors from overheating, melting their shields and emitting radiation that could injure and kill people in nearby communities.

Although the commission rejected proposals of the industry for less stringent criteria, its order in a controversial, two-year-old proceeding was immediately assailed as "cosmetic" by the Union of Concerned Scientists.

A senior official, speaking for the commission, said at a news briefing that the commission had revised the criteria without knowing what the economic impact on the manufacturers and utility companies would be.

"We have asked for this information to be submitted" after the ruling is studied, the official said. He acknowledged that some utilities might have to "derate"—reduce power output

—but that the commission would consider exemptions on a case-by-case basis. The cutbacks could come next summer, when power consumption reaches a seasonal peak in many regions.

"We believe that our decision affords the required reasonable assurance of protection for the public health and safety with a substantial margin," the five-member commission said in a unanimous opinion.

"The commission's decision," said Dan F. Ford and Henry W. Kendall of the Union of Concerned Scientists, an intervenor in the safety proceeding, "represents a continuation of the A.E.C.'s cover-up of critical safety problems." Ralph Nader, the consumer advocate who has allied himself with the critics, joined in the statement.

The issue of reactor safety is a troubling one because it involves public health, few citizens can understand its technical substance and a major expansion of nuclear power stations is under way. To meet the country's ravenous appetite for energy, the industry is likely to build 1,000 power stations by the turn of the century, the A.E.C. has estimated.

Operating Criteria

The commission's review of emergency, or back-up, cooling systems tightened the operating criteria that these systems must meet. For example, the commission lowered to 2,200 de-

grees Fahrenheit from 2,300 degrees the maximum temperature that the cladding, or reactor-core jacket, may attain if the primary cooling system fails.

Water is the primary coolant in 38 of the 40 operating nuclear power stations. It is also the emergency coolant. Gas is the other primary coolant.

The commission added a new safety criterion that limits the oxidation of the cladding, that is, its vaporization under high heat, to 17 per cent of its thickness.

It is also required that in calculating what might happen in a particular reactor in the event of loss of primary coolant the utilities and reactor manufacturers must take into account additional technical factors, such as swelling of the cladding.

The commission did not estimate in its 134-page opinion the probability of a loss-of-coolant accident that would activate the emergency core cooling system. The Union of Concerned Scientists, based in Cambridge, Mass., and other intervenors argued that such a calculation should be part of the proceeding.

Probability of Accidents

They were overruled, essentially on the ground that the probability is very low and that the interim criteria, published on June 29, 1971, were highly conservative. The commission also refused to include design characteristics in the proceed-

ing, which was confined to operating criteria.

However, an estimate was cited by a commissioner, William A. Anders, in a concurring opinion. That estimate was that one such event would occur in 10 million "reactor-years," or one million years of power output by 10 reactors.

The commission gave the industry six months to run new, complex calculations on computers and make power stations conform to the new criteria. The A.E.C. staff had proposed four months.

Commissioner Anders took a cautious view of "rapid implementation" within six months.

"The advantage of a further reduction in an already 'negligible' risk must be weighed critically" against adverse effects, such as reduced output, possible power outages and the environmental costs of any switch to coal, he wrote.

Mr. Ford said the intervenors would have preferred "substantial derating," a suspension of approval of construction permits for new stations, design review and "a greatly expanded program of research."

The critics said in their statement that the A.E.C. had "once again decided to side with the nuclear industry and to disregard the profound reservations of its own experts." They said that "the nuclear industry could hardly survive imposition of stringent safety standards."

December 29, 1973

Siting Unpopular Plants

Decisions on Land and Water Use Evaded When Communities Object

By GLADWIN HILL

The current friction between coastal states and Federal officials over offshore oil development is one more symptom of a far bigger governmental and constitutional problem that seems likely to crystallize soon.

News Analysis That is the "fair share" issue: the question of where many indispensible, if objectionable, national activities and fixtures should be located, which areas should bear the burdens, and above all how these matters should be decided.

The question involves not only offshore oil operations but also deep-water ports, onshore refineries, strip-coal mining,

power plants, heavy industry and even population itself.

At the recent White House conference on offshore oil leasing, a New Jersey official said in effect, "We don't want that sort of thing lousing up our resort business."

Last spring the college town of Durham, N. H., rejected an oil refinery proposed by Aristotle Onassis. The state backed up the community. And Mr. Onassis is still shopping for a site.

If Everyone Does

The decision was quite understandable from Durham's standpoint. But, governmental specialists are asking, suppose communities all over the country took the same position.

Where would the nation put its refineries?

Delaware in 1971 barred heavy industry from a belt along its 100-mile shoreline. Environmentalists cheered. But the same question arose: If all states did that, where would shoreline-oriented industry be able to locate?

The prospect of large-scale strip mining of coal in a half-dozen Western states has evoked a strong current of sentiment in some areas: "Let them do it somewhere else—we don't want it."

Scores of communities across the country, plagued with problems of growth, in the last few years have erected the array of barriers to newcomers — zoning restrictions, building-permit moratoria, prohibitive development fees. But, some courts have said, in invalidating such limitations, if all communities did that, where could future population settle?

While some development is of arguable necessity, there is no question that the nation is going to need more factories, power plants, power lines, air-

ports, coal, refineries, residential areas and other facilities that in the last few years have acquired pronounced unpopularity.

Small Coastline

Yet Delaware, for instance, when taxed with refusing to bear its "fair share" of shoreline industry, replied to critics: "We only have a small coastline, and a lot of it already is devoted to industry. If there's some elements of national responsibility here, what about states that have thousands of miles of coastline?"

The nation appears to be approaching a point where somebody has to decide what should go where, in terms of feasibility, resource conservation and, above all, fairness in distributing national burdens.

But who is "somebody"?

City councils are not qualified or objective enough to make decisions affecting other communities. Courts, state or Federal, are hardly in a position to prescribe national distribution patterns for things like power plants and population.

287

State governments are equally unpromising arbiters. Their inexorable inclination will be to do what New Hampshire did and buck the problem to some other state.

Thus the "fair share" question, being essentially an interstate matter, inevitably ends up at the Federal level. But to date, neither Congress nor national administrations have indicated much eagerness to tackle the problem.

Congress did, in 1972, enact the Federal Coastal Zone Management Act, calling for a modicum of Federal-state cooperation in shoreline develop-

ment. But it does not attempt to deal with location of facilities in terms of interstate equity or the over-all national interest.

The Nixon and Ford Administrations have pushed for, and Congress has been wrestling with, legislation covering deepwater ports. But the prospect is that the location of these will be as contentious as offshore drilling.

As far back as 1970 some Congressional leaders foresaw the fair share problem arising, and made a gingerly stab at it in a Federal land-use bill laying down some broad criteria

for siting of key developments.

But the measure, watered down to little more than a grant program, was scuttled this year as an unwarranted intrusion on state prerogatives.

As things stand, where development should be located is being determined either unilaterally by cities and states or by a dozen Federal agencies in thousands of uncoordinated decisions.

Officials of the Federal Power Commission and Atomic Energy Commission rule on the siting of power plants. The Federal Aviation Administration approves airport locations. The

Department of Transportation routes highways. The Department of Housing and Urban Development influences the location of both residential and commercial expansion. The Environmental Protection Agency, with its air and water pollution constraints, says in effect where certain things cannot go.

But few if any of these decisions are geared to any concept of "fair sharing" among states and localities — demands for which show every likelihood of growing more strident and acrimonious until the problem is systematically addressed.

December 2, 1974

Year-Round Daylight Saving Begins

HAROLD M. SCHMECK Jr.

The United States has begun a nearly nationwide experiment to see whether pushing back nightfall by an hour will help depress America's winter thirst for fuel.

Daylight Saving Time Began at 2 A.M. Today

Clocks Should Have Been Set Ahead One Hour

Government officials hope the return to daylight saving time, as of 2 A.M. today, will reduce the nation's energy consumption by a small but significant amount. Some experts also hope for a slight decline in automobile fatalities and violent crime by the shifting of one hour of day-

The New York Times/Tyrone Dukes

As daylight saving time was about to go into effect, a trainload of coal arrived at the Consolidated Edison Company generating station in Arthur Kill, S. I. The coal first had to be taken to a thawing shed, rear, where infrared lamps melt the snow so that the coal can be dumped in small pieces. Con Edison said that daylight saving time would save the system about 40 million kilowatt hours annually.

light from morning to evening. But no one can say definitively what the effects will be.

The strategem worked in World War II and in Britain in the late nineteen-sixties. Neither case offers an exact parallel to the situation of the United States in 1974.

The Federal Energy Office estimates that the energy individual Americans expended this weekend in pushing clock hands ahead one hour will be offset by a national saving of 100,000 to 150,000 barrels of oil a day. In the first quarter of the year, a spokesman for the office said, the anticipated oil use, without daylight saving, would be about 18½ million barrels a day.

The saving is expected to come mostly from electricity used for lighting. The assumption is that Americans use more light for their early evening activities than they do in the predawn period. There may be some saving in fuel for heating, too, but experts appear to be less certain of that.

The law providing for the experiment in year-round daylight saving time continues through April, 1975. In the interval, studies are to be done to determine whether or not the change does save energy.

Some parts of the country are exempted at the start. Hawaii, Puerto Rico, the Virgin Islands and most of Indiana are automatically exempted. An announcement Friday from the Uniform Time Office of the Department of Transportation said that Arizona would also receive an exemption through April, 1975, while the northeast corner of Oregon and most of Idaho, except for its northern panhandle, would receive temporary exemptions.

The purpose of the temporary exemptions is to give the state legislatures a chance to decide whether they want to be excluded for the entire year and a quarter. The Governor of Kentucky will be authorized to redraw the time zone lines within his state to put on Central time everything except 12 counties near Cincinnati and Huntington, W. Va. The 12 will be on Eastern time. With these changes, the entire state will go on daylight saving.

Requested by Nixon

The return to daylight time as an energy-saving move was requested by President Nixon in November. Bills to write the change into law were pushed through Congress largely on the initiative of Senators Warren G. Magnuson of Washington and Adlai E. Stevenson 3d of Illinois, both Democrats, and Representatives Craig Hosmer, Republican, and John E. Moss, Democrat, both of California. The Congressional push for daylight saving however, started before the President's message.

In opening hearings before the Senate Commerce Committee in November, Mr. Magnuson said that the nation's continuation on standard time might be a classic example of unnecessary energy consumption, considering that the alternative of daylight saving existed.

In testimony before the committee, John H. Gibbons, director of the Department of the Interior's Office of Energy Conservation, said that the nation had no workable options that would save a great deal of energy in the short term.

He cited the time-change idea as one of the most promising ways of saving a modest amount of energy immediately.

Mr. Gibbons said that the move to daylight time could save fuel for generating electricity Americans used daily. This, he explained, could be expected because of a spreading out and consequent lowering of the daily peak load.

He said that electrical generation was more efficient at lower load levels and that the morning peak load was generally lower than that of the evening.

Estimates of electricity savings his office has collected from available data, Mr. Gibbons testified, ranged from "negligible" to about 40 per cent of total normal demand. He said that most estimates ranged from a few tenths of a per cent to 1 per cent.

Con Edison Viewpoint

A saving of three-tenths of 1 per cent in national electricity generation, he said, would represent 0.07 per cent of total national energy use and would be equivalent to the energy used by one large modern steam power plant. Such a plant producing 1,000 megawatts of electrical energy would consume nearly 10,000 tons of coal a day, he said.

Luis H. Roddis Jr., of the Consolidated Edison Company of New York, Inc., told the Senate committee that the change would save his system about 40 million kilowatt hours annually. Nationwide, he said, the saving should be three billion to five billion kilowatt hours a year, roughly equivalent to the amount his company produces for New York City and Westchester County in an average month.

William R. Harris, of the Rand Corporation, of Santa Monica, Calif., estimated the probable national saving at from two-tenths to four-tenths of 1 per cent of national fuel needs in the October-April period.

Hopes for a reduction in crime and in traffic fatalities have been repeatedly expressed.

The rationale, so far as traffic is concerned, is that driving in the dark is more hazardous than in daylight and that the evening rush hour is intrinsically more hazardous than the morning rush because of the factors of fatigue and alcohol.

Problem for Schools

Crime is considered more prevalent in the evening darkness than in the morning and, presumably, there would be fewer victims available in the early evening dark period if most people got home before twilight was over.

Even proponents of daylight saving time concede that it would be an added inconvenience and perhaps sometimes an added hazard to schoolchildren in the predawn darkness unless schools change their hours. Some are believed to be planning to do this.

The main objection to 12-month daylight saving, however, comes from farmers who must do their work according to the schedule of light and darkness, not by the clock.

One farmer, who has been widely quoted and paraphrased, remarked, "You can legislate daylight savings time until you are blue in the face, but the dew is still going to dry off the fields on standard time."

January 6, 1974

A Reordering of Values

As lines at gas stations get longer and prices everywhere go higher, it becomes clear that this country is facing more than problems of energy and inflation. These are symptoms of something more serious and deep-seated: an uneasy reckoning with the American way of life. The word crisis is overworked, but what is involved is a forced reordering of values and life-style, something Americans are not facing at all while they talk of lowering thermostats.

The American way of life, for a vast majority, is comfortable and profligate, based on overconsumption and overindulgence and a belief in endless growth. The puritan virtues have been largely replaced by a sliding morality of upward mobility, eroded by easy credit and easier living. Americans on the whole have embraced the unfulfilled expectations and the self-indulgent individualism of suburbia and a consumer society, at a price that is just beginning to be realized.

This change has not been brought about purely by chance nor even by choice; it is national preference reinforced by national policy. The dissolution of the city and the pursuit of the suburban dream have been encouraged by Federal housing and highway programs. Behind the impulse for greener pastures is the legislation that hastens abandonment of the cities.

Exacerbating the energy crisis are those patterns of growth that use power with irresponsible abandon, backed by public action. Tax laws virtually mandate deterioration of the cities and destruction of open space. Transportation policies ignore mass needs in favor of individual indulgence.

America has lost more than cheap energy. It has lost the capacity to see itself as a society, or as a system of values, or to recognize its traditional ideals. The real issue is that this erosion of standards, particularly in the area of land use and urban growth, have been encouraged and institutionalized by law. The real crisis is failure of government, and of the people government represents. And the real answer may well be in some wrenching changes in national policy and values at the Federal level, not merely in ration cards.

January 13, 1974

Tom Zetterstrom

Energy Through Wind Power

By R. Buckminster Fuller

The inventions that impound solar energy through sun-reflecting or lensing devices are fascinating to the imagination but relatively insignificant in potential. In fact, they work only a few hours daily when the sun is at a favorable angle.

Even if half of Arizona were turned into a direct-sunlight-energy-converting mechanism, the production would be negligible in comparison with wind-power sources.

Wind power is in a class by itself as the greatest terrestrial medium for harvesting, harnessing and conserving solar energy. The water and air waves circulating around our planet are unsurpassed energy accumulators whose captured energy may be used to generate electrical, pneumatic and hydraulic power systems.

Windmills produce power from the sun-generated differentials of heat, which are the source of all wind, with far greater efficiency than do attempts to focus and store direct solar radiation. But the most comprehensive consideration regarding wind power is not technological. Rather it is an appreciation that wind power is by far the most efficient way to recapture solar power.

In the first place, three-quarters of the earth is covered with water, and the remaining quarter is land area consisting largely of desert, ice and mountains. Only about 10 per cent of our planet's area has terrain suitable to cultivation, in which vegetation can impound the sun's radiation by photosynthesis.

Among the solar-energy impounders in vegetation, none can match corn's performance. Corn converts and stores as recoverable energy 25 per cent of the received ultraviolet radiation, whereas wheat and rice average only 18 to 20 per cent. From these stores of solar energy humans can produce commercial alcohol, or they can leave the energy to the production of fossil fuels in the earth's crust, which requires millennia.

But one-half of the vegetation-producing area of the earth's surface is always in the shadow, or night, side, which reduces to 5 per cent the working area of the earth's surface on which vegetation impounds the sun's energy. Though theoretically 5 per cent of the area can impound energy at any one time, only an average of one per cent of the sun's energy is actually being converted because of local weather conditions and infrared and other energy-radiation interferences.

The area of the surface of a sphere is exactly four times the area of the sphere's great-circle disk, as produced by a plane cutting through the center of the sphere. The surface of a hemisphere is, then, twice the area of the sphere's great-circle plane. When we look at the "full" moon, we are looking at a surface twice the area of the seemingly flat, circular disk in the sky.

All of the earth's energy comes from the stars, but primarily from the star sun, as radiation or as inter-astro-gravitational pull. Twenty-four hours a day the sun is drenching the outside of the hemisphere of the cloud-islanded atmosphere's 100-million-square-mile surface area, which is twice that of the disk of the earth's profile.

This gives us one billion cubic miles on the sunny side and one billion cubic miles on the shadow side. The atmospheric mass is kinetically accelerated in the hemisphere constantly saturated by the sun, while simultaneously the atmospheric kinetics in the night hemisphere are decelerated.

All around the earth, yesterday's sun impoundments perturbate the atmosphere by thermal columns rising from the oceans and lands. The shadow side consists of one billion cubic miles of contracting atmosphere, while the one billion cubic miles on the sunny side is sum-totally expanding. This rotation of the earth brings about a myriad of high-low atmospheric differentials and world-around semi-vacuumized drafts, which produce the terrestrial turbulence we speak of as the weather.

The combined two billion cubic miles of continual atmospheric kinetics converts the solar energy into wind power. Wind power is sun power at its greatest, by better than 99 to 1.

All biological life on planet earth is regenerated by star energy, and overwhelmingly by the sun's radiation. The sun radiates omnidirectionally 92 million miles away from earth, with only two-billionths of its total radiation impinging upon earth. The radiation arrives at a rate of two calories of energy per each square centimeter of earth's sun-side hemispherical surface per each minute of time. About half of that is reflected back omnidirectionally to the universe. The other half, i.e., one calorie per minute per square centimeter, is impounded by our planet's biosphere in ways making them available to human use.

No matter how dubious one may be of such logical realizations of our potentials, the fact remains that our net receipt and impoundment of cosmic energy amounts to 168 quintillion horsepower a minute, which can also be stated as 125 quintillion kilowatts a minute, which, with 525,600 minutes a year amounts to 66 septillion kilowatts a year. This is 66 x 10^{24} kilowatts, eleven - billionfold the world's present 5 x 10^6 kilowatt production of electric-energy power.

If all humanity enjoyed 1973's "highest" living standards—that of the United States—each human on earth would consume 200,000 (2 x 10^5) calories a day.

Assuming five billion (5 x 10^9) humans by A.D. 2000, each consuming 2 x 10^5 calories daily, we will need 1 x 10^{15} calories a day, while our actual daily terrestrial income of cosmic energy is 72 x 10^{20} calories.

Our planet's usable daily energy income is therefore 72 x 10^5, or seven millionfold our daily requirements of A.D. 2000.

—R. Buckminster Fuller

R. Buckminster Fuller is an inventor, philosopher and occasional poet.

January 17, 1974

Decades of Inaction Brought Energy Gap

Lack of Coherent Policy Left Vacuum the Oil Industry Was Eager to Fill

By LINDA CHARLTON

Barely three months ago, President Nixon told the nation that its energy problems had become an "energy crisis." Since then, thermostats and speed limits have dropped obediently, but the initial shock has not faded—only soured in perplexity, disbelief and cynicism.

For the last month and a half, Americans have been lining up at gasoline stations, worrying whether next week — or next summer, or next year — will be worse.

During the last two weeks oil companies have been reporting soaring profits, and more than a half-dozen Congressional committees have been looking into the methods by which they were made.

Ministers to Meet

And tomorrow the foreign ministers of 13 major oil-consuming nations and organizations meet in Washington in search of common solutions to the problem.

Many Americans do not believe that a "crisis" exists. Many others are convinced of conspiracy in the back rooms of government, Arab sheiks, and huge international oil companies. Almost everyone, skeptical or simply puzzled, wants to know how the United States got here from there.

From a recent past in which Americans were urged to drive bigger cars farther, live and work in artificial climates and gauge their national success in terms of the expenditure of

apparently boundless energy, they have moved into a bleak present, a time when kilowatts and mileage must be measured with care.

An investigation by The New York Times found a complex but traceable pattern—not so much of conspiracy as of national complacency, and, above all, Government inaction going back decades. Not so much bad policy—although hindsight exposes that, too—as no policy at all. What was left was a vacuum that the oil industry was only too eager to fill, especially in the absence of significant countervailing forces.

Secret 1950 Decision

There were decisions made and decisions avoided. A secret, top-level Government decision in 1950 made it more profitable to drill for oil overseas. A 1956 national commitment to a concrete web of highways made more people even more dependent on the larger, less economical cars happily provided by Detroit.

A blinkered consciousness led environmentalists to ignore the possible consequences of their successes. A crucial Supreme Court ruling made natural gas too attractive to the wrong users. A system of oil import quotas instituted in the Eisenhower Administration for "national security" endured through 14 years of political bargaining, through a "little arrangement" in the Kennedy

years and a "just darned complicated" decision in the Nixon Administration.

"I wish it were as simple as a conspiracy," said Walter J. Levy, a noted independent oil economist, in a recent interview.

"Just damn foolishness," is the diagnosis of John F. O'Leary, a former deputy assistant Secretary of the Interior for Mineral Resources.

Stewart Udall, Interior Secretary in the Kennedy and Johnson Administrations, admits: "We didn't have a national energy policy."

And Frank N. Ikard of the American Petroleum Institute says, "The Federal Government absolutely refused to recognize or come to grips with this problem until 90 days ago.

"Government energy policy has been formulated in Dallas and Houston and rubber-stamped here in Washington," said S. David Freeman, a former White House energy adviser who now heads the Ford Foundation's energy project.

"The ad hoc, diffuse and often conflicting approaches to individual energy issues that have characterized the past will not be adequate for the future," M. A. Wright of Exxon told the Senate Interior Committee in 1971.

"We got to where we are by mistaken public policy," said

> *This article is based on reporting by Ben A. Franklin, Grace Lichtenstein, William D. Smith and Miss Charlton.*

Prof. M. A. Adelman of the Massachusetts Institute of Technology, a leading academic oil economist.

There have been many warnings that a shortage was coming. They have been largely ignored.

There was the 1952 report by the President's Materials Policy Commission, better known as the Paley Report, warning of the "extraordinarily rapid rate at which we are utilizing our materials and energy resources."

There was the 1966 annual report of the Atlantic Richfield

Company which, while underestimating the growth in demand for oil, warned flatly that "the nation faces the prospect of a domestic energy gap."

There was a National Intelligence Estimate sent to the White House in the spring of 1973 predicting that renewed conflict in the Middle East would surely mean an oil cutoff.

"There was just no way to get people interested in energy when prices were low," said Mr. Freeman, the former head of the Energy Policy Staff of the Office of Science and Technology at the White House. "If the lights are on and the bills are low, nobody cares."

Now, everyone cares.

Four Decades of Growth

Like a number of others, Mr. Freeman looks back for root causes to the New Deal days of the nineteen-thirtis, when, in his words, there were "great social reasons for priming the pump" — for encouraging the use of power by making it available cheaply.

As David Lilienthal, first chairman of the Tennessee Valley Authority, exulted, no function of T.V.A. was more vital than that of electric-rate yardstick, to demonstrate "that drastic reductions in electric rates [would] result in hitherto undreamed-of demands for more and more electricity..."

The New Deal premise, based on faith in unlimited growth and unlimited resources, has worked. Energy consumption in the United States has soared beyond the fondest hopes of the New Deal's visionaries. It has more than doubled in the last 20 years, so that by 1972 the United States, with 6 per cent of the world's population was using one-third of the total world production of energy.

Since 1947, annual consumption of petroleum products has gone from 1.9 billion barrels to 5.6 billion; the use of electricity went from 1,774 kilowatt

291

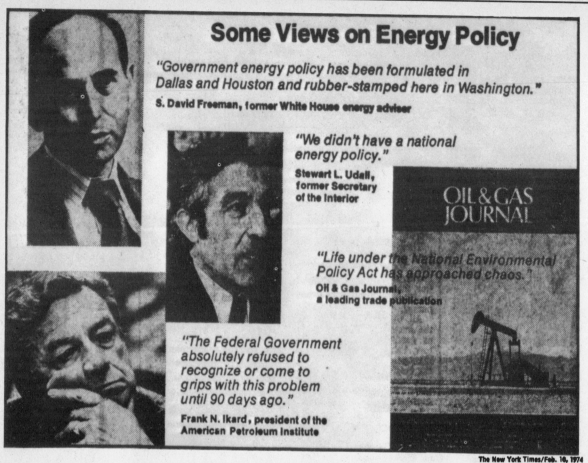

Some Views on Energy Policy

"Government energy policy has been formulated in Dallas and Houston and rubber-stamped here in Washington."

S. David Freeman, former White House energy adviser

"We didn't have a national energy policy."

Stewart L. Udall, former Secretary of the Interior

"Life under the National Environmental Policy Act has approached chaos."

Oil & Gas Journal, a leading trade publication

"The Federal Government absolutely refused to recognize or come to grips with this problem until 90 days ago."

Frank N. Ikard, president of the American Petroleum Institute

The New York Times/Feb. 10, 1974

hours per person in 1947 to 7,800 per person in 1971, when of course, the population was larger. If the rest of the world consumed energy at the same rate, it has been estimated that the world's total energy resources would be depleted by the year 2010.

As important as the amount of energy in looking for the roots of the present situation is the type of energy consumed. Coal, the country's most abundant source of energy, comprising 88 per cent of total reserves, became more difficult to extract and was being superseded even before the new environmental awareness of the nineteen-sixties dealt a major blow to its usefulness.

Oil and Gas Essential

The United States now is a country that runs on petroleum and natural gas. Together they supply about three-quarters of its energy needs — partly because of a designed way of life predicated on the notion of supplies so ample as to be, in effect, limitless.

At the end of World War II, Americans took to the skies and the highways. The railroads, overworked and undermaintained during the war, were exhausted, inefficient and unattractive. They could not— and, many say, did not really try to—compete with the new glamor of airplanes for long-distance travel; worn-out rolling stock had no allure for vet-

erans who had shining new automobiles to take them to their split-level houses in that post-war phenomenon, the suburbs.

And suburbs, where most Americans live now, were themselves designed on the premise of the individual automobile that has become a national gospel. Since the end of the war, ridership of trains has declined by 83 per cent, although trains are 12 times more efficient than cars in terms of fuel consumption.

During these same postwar years natural gas, a by-product of oil drilling that for many years was simply "flared" or burned off, emerged as a major factor.

Clean-burning and cheap, and heavily promoted by the industry with the "Gas Heats Best" slogan, it has been the fastest-growing energy source, in terms of consumption, over the last 20 years. Today it provides about one-third of the country's total energy consumption.

But by 1972, natural gas was scarce; in 21 states, it was so scarce that no new customers were accepted. Since 1968 more natural gas has been sold in the United States each year than has been found in terms of new reserves, or potential supply. Gas production, despite its accelerating popularity, has declined.

One reason for this—and in the view of many, one of the

milestones on the road to the present situation—was a 1954 Supreme Court decision known as the Phillips Petroleum ruling. This, in effect, ratified and expanded the powers of the Federal Power Commission to regulate the price of natural gas at the wellhead—that is, to keep it down.

Cites Phillips Decision

Mr. Ikard of the American Petroleum Institute, asked for his views of the factors contributing to the present situation, cited the Phillips decision immediately, saying that it set "an unrealistic price for natural gas, encouraged the [uneconomic] use of gas. It was not the kind of regulation that would do anything but have the disastrous result of discouraging the development of new supplies."

Joseph C. Swidler, the retiring chairman of the New York State Public Service Commission and chairman of the Federal Power Commision from 1961 through 1965, put the matter differently. In settlement conferences with gas companies over pricing, he said, the companies "would never commit themselves to increasing the supply. Their spirit was, 'When are you guys going to recognize we've got the trump here and we're not going to drill until we get our price?' "

Lee White, chairman of the Federal Power Commission from 1966 to 1969, agreed that

natural gas had been widely misused and over-used in industry and by utilities. "It's a national scandal, when a utility uses gas as a boiler fuel," he said.

In the same year as the Phillips case there was another, little noticed government decision. According to Mr. O'Leary, who is a fuel economist and also served as director of the Bureau of Mines in the Interior Department, the oil industry—not then, as now, deeply invested in the coal industry—persuaded the Eisenhower Administration to abandon its investment in research aimed at advancing the gasoline-from-coal technology developed in Germany during the war.

Mr. O'Leary believes that this, and the failure to continue other fuel-conversion research, was "the most serious error in energy policy made during the postwar years."

It was two years later, in March, 1956, that one of the many ignored prophets spoke his piece. M. King Hubbert, then a petroleum geologist with Shell, now with the United States Geological Survey, told a meeting of petroleum engineers in Texas that, on the basis of past consumption and of his estimate of reserves, the peak of United States oil production would be reached by 1971 at the latest. It was, in fact, reached in 1970.

This prediction flouted the

popular widsom of the time, which was, he said, "in effect, that we didn't have to worry about oil in our lifetime." Such was the disbelief and consternation, he said, that Shell deleted his prediction when it published his paper.

In frantic efforts to "avoid this unfortunately ominous date," he said, new graph curves were devised projecting reserves at far higher levels that "postponed this to the end of the century." And it was this higher figure that was incorporated in a report to the President by the National Academy of Sciences made in January 1963.

Mr. Udall lamented: "It's a kind of commentary that I didn't know about him [Hubbert] in all my years in Interior, never spoke to him."

In the same year as Mr. Hubbert's prophecy, the United States committed itself to its biggest peacetime public works project, building the 42,000-mile interstate highway network. At the time, it had widespread support—from both political parties, from economic planners, from industry and labor. Now, many see it as a turning-point, a national commitment to the automobile that was an inducement to build them larger and more powerful, and a powerful disincentive to expansion of mass transit.

By 1970, indeed, transportation accounted for one-quarter of total United States energy consumption, and the automobile used 55 per cent of that amount.

In 1959, the Eisenhower Administration adopted one of the most controversial policies in the long history of oil company-Government relations. For reasons stated as being based largely on the desire to protect the "national security" against an overdependence on foreign oil, oil imports were to be restricted on a quota basis.

The Environmental Movement

There are some, especially in the auto industry, who lay a significant part of the blame for the energy crisis on the fledgling environmental movement. Irving J. Rubin, legislative planning manager of the Ford Motor Company, calls the Clean Air Act of 1965—the first national-level victory for the movement — a "significant" factor.

What Mr. Rubin and others in Detroit point to are the increasingly tough emission controls for new cars mandated by the act and its 1970 amendments. What they do not point out are the made-in-Detroit factors that have cut fuel effi-

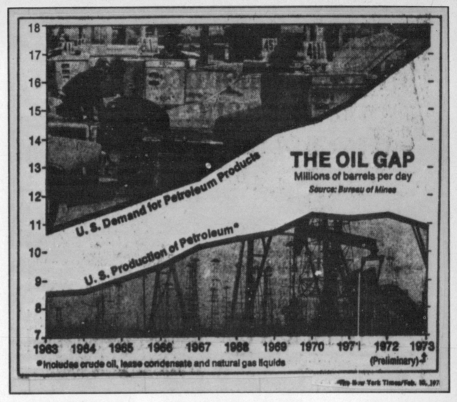

THE OIL GAP
Millions of barrels per day
Source: Bureau of Mines

U. S. Demand for Petroleum Products

U. S. Production of Petroleum*

18 17 16 15 14 13 12 11 10 9 8 7

1963 1964 1965 1966 1967 1968 1969 1970 1971 1972 1973 (Preliminary)

*Includes crude oil, lease condensate and natural gas liquids

The New York Times/Feb. 15, 197

ciency even more significantly, such as air-conditioning, which can cut mileage by as much as 20 per cent, and the increasing weight of automobiles. According to a study by the Environmental Protection Administration, "the most popular standard-size passenger cars have gained about 800 pounds from 1962 to 1973."

Most environmentalists concede freely that they were oblivious to the energy consequences of the conservation measures they saw as so desperately needed. Until recently, admitted Joe B. Browder, executive vice president of the Environmental Policy Center, "all of us were approaching the problem from the point of view of what was gumming-up the environment, not energy flows."

Certainly, newly aroused public awareness played a role in the increasing difficulty experienced by oil companies in finding sites for their refineries, and by utilities for their power, such as Con Edison's proposed Storm King hydroelectric plant, and in the Government decision—after the 1969 Santa Barbara oil spill—to curtail offshore drilling for oil. It has also curbed nuclear power development.

When the first commercial nuclear power plant became operative in 1957, a major role for nuclear power was foreseen—and still is. But now, 17 years later, nuclear power provides only the smallest fraction of the country's electric power, less than 1 per cent of the total. Its development has been

slowed by disputes about public and private ownership and by growing concern about safety and pollution, and—not least—by difficulties in developing the technology, which proved to be more complex than anticipated, and more costly.

The Clean Air Act also meant that, between 1965 and 1972, about 400 utilities switched their boilers from coal to oil. Now, the Government is encouraging as many as are able to re-convert, back to coal.

The most celebrated confrontation of energy development versus environmental protection was the three-year battle against construction of the trans-Alaska oil pipeline. The industry asserts that if the pipeline had started on schedule, Alaska today would be supplying as much oil as the Arabs have cut off.

The pipeline is expected to carry about two million barrels a day when it is "on stream." Total present United States daily consumption is about 18 million barrels.

The oil industry holds the environmentalists responsible for a considerable portion of the country's present problems.

"Life under the National Environmental Policy Act [a 1969 measure that required the environmental impact of any action to be taken into account in all Government decision-making and which was a legal basis for the opposition] has approached chaos," the Oil & Gas Journal editorialized in 1972. The Journal is generally regarded as a reliable reflection of the industry viewpoint.

Oil, Arabs and Taxes

Both domestically and internationally, the mixture of oil and politics has long been a blend of public interest and venality, of patriotism and profit. It was Winston Churchill who, in 1914, first definitely identified the control of oil with national interest when he urged the British Government to buy a controlling interest in the Anglo-Persian Oil Company.

The American oil industry, with a little help from the State Department, first gained a foothold in the Middle East in 1928. It won a concession in Saudi Arabia in 1933 — now known as the Arabian American Oil Company (Aramco) and the richest oil operation in the world.

World War II interrupted the pursuit of petroleum riches overseas but only heightened the importance of control over petroleum resources. Harold Ickes, Secretary of the Interior under President Roosevelt, strongly suggested that the United States should enter the oil business directly as Britain had done — a suggestion strongly discouraged by the oil industry.

For the major multinational oil companies, and the oil-consuming countries, the nineteen-fifties and most of the nineteen-sixties were years in which there were few challenges to their dominance from the oil producing countries. What few there were, such as the attempted nationalization of the Iranian oil industry in 1953, were quickly beaten back.

During the postwar era—indeed for 25 years—the price of oil held steady for the most part, at less than $2 a barrel. Oil company profits, however, did not—they rose. The tax benefits enjoyed by the industry contributed significantly to this prosperity, although they did provide a focus for what grumbling there was about the industry.

The best-known of the industry's benefits was the 27.5 per cent depletion allowance that went into effect in the nineteen-twenties and was to remain at that level until the late nineteen-sixties, when it was cut to 22 per cent.

Depletion allowances permit producers of oil and more than 100 other minerals to take a specified percentage tax deduction against the income they receive from each producing property — oilwells, copper mines, and the like — on the theory that the producers' assets are being depleted when the mineral is removed from the ground.

Far less well-known, however, and especially profitable to the industry, was the foreign tax credit decision.

This allowed the companies to describe part of the royalty they paid to the Government of the oil producing countries as a tax, and to then credit his amount against their United States income taxes on a dollar-for-dollar basis.

Until very recently, it had been generally believed that this procedure became effective with a 1952 private ruling by the Internal Revenue Service. At recent Congressional hearings, however, it was disclosed that the decision was made secretly two years earlier and at a far higher level—by the State and Treasury Departments in 1950.

The effects were dramatic. In 1950, Aramco paid Federal taxes of $50-million; in 1951, $6-million. In 1950, Aramco paid $66-million to Saudi Arabia; in 1951, this jumped—by precisely $44-million—to $110-million.

Tax policies such as these, in the view of many, have constituted strong disincentives to domestic exploration, drilling and refining. "For a long time, our domestic corporations, the multinationals, have been operating abroad not because they wanted to but because how could they resist," commented Lee White the former F.P.C. chairman.

The per-barrel price of crude oil was, until the late nineteen-sixties, determined unilaterally by the oil companies, not by the producing countries. A producing-state group, the Or-

The Profits Boom
1973 Profits of the 10 Largest U. S. Oil Companies

Company	Profit	% increase over 1972
Exxon	$2.4-bil.	59.5
Mobil	843-mil.	46.8
Texaco	1.3-bil.	45.1
Gulf	*570-mil.	60.0
Standard of Calif.	844-mil.	54.2
Standard Oil (Ind.)	511-mil.	36.4
Shell	333-mil.	27.7
Continental	243-mil.	42.6
Atlantic Richfield	270-mil.	38.4
Phillips	230-mil.	55.3

*First nine months   Source: Company reports

The New York Times/Feb. 10, 1974

ganization of Petroleum Exporting Countries, was formed in 1960, in an effort to gain some control over prices and push them up, but for most of the decade it posed no threat to the industry's autonomy. There was a brief, abortive attempt at an oil embargo just after the 1967 Arab-Israeli war, whose failure only encouraged the view that O.P.E.C. was powerless.

In September, 1969, the power began to shift. Col. Muammar al Qaddafi, a young, devout Moslem who was also devoutly anti-Communist and anti-West, seized power in Libya. A year later in January, 1971, Col. Qaddafi began demanding higher prices for his oil. Rebuffed with what he considered an insultingly low offer, he resorted to what are now known as the "Arab salami" tactics, not against the powerful major companies but against the smaller independents.

The first company to yield, because it was the weakest, was Occidental Petroleum, which gave the Libyan Government a 30-cent-a-barrel increase and a higher tax rate. The other independents toppled quickly, and finally the majors fell.

What the colonel had won, the King of Saudi Arabia, the sheiks of the oil emirates in the Persian Gulf and the Shah of Iran had to have. "The companies simply held no cards," according to Walter J. Levy, the oil economist. From then on, price demands by the producing countries have leapfrogged upward.

Attempts at a United Front

There have been attempts by the multinationals to form a united front, attempts supported by the United States Government. In 1970, they were given — reportedly over the

unavailing protests of some in the Justice Department's Antitrust Division—an unpublished "business letter of review" by the department. This amounted to a guarantee of immunity from civil antitrust prosecution for banding together to negotiate jointly with Libya.

Not until June, 1971, did the energy matter seem important enough in itself for any President to devote even one "message," or major policy-making speech, to the subject. At that time, the United States was importing about 28 per cent of its oil, more than one-third of it from the politically volatile Middle East. And consumption continued to soar unchecked.

As imports rose, there was impetus to reconsider the existing import quota program, which restricted oil imports to a set percentage of domest production but also, some said, inhibited the expansion or construction of refineries in this country because of a resulting uncertainty about the supply of crude oil.

There had been questions about the program's efficacy before. In 1962, there was a sub-Cabinet level review of the program, and resulting suggestions for change. The system was changed, but changed precisely to the oil industry's order.

"If there was anything high on the agenda [at that time] it was the Trade Expansion Act," said one man who was deeply involved in the Administration end of things. "In order to get the votes, President Kennedy had to concede a good deal—a bloc of votes hinged on this. Senator Long [Russell B. Long, a Democrat from the oil-producing state of Louisiana] and I worked out this little arrangement."

The "little arrangement" revised the formula for the quota and, in effect, cut imports, which was a major gain for the domestic producers.

Senator Long is one of a cadre of powerful men in Congress, their ranks now thinned somewhat, who represented oil-producing states and, inevitably, oil interests. This platoon of power also included the late Senator Robert Kerr of Oklahoma (and the Kerr-McGee Oil Company), House Speaker Sam Rayburn of Texas, and Senate Majority Leader — later President — Lyndon B. Johnson.

The oil interests are still as Roland Homet Jr., chief counsel for the 1969 review of import quotas, describes them, "a very effective lobby, fueled by campaign financing," despite the loss of some of their Capitol Hill friends. Some 70 of the 125 members of the National Petroleum Council, an advisory committee almost all of whom are oil-company executives, contributed a total of $1.2-million to President Nixon's 1972 re-election campaign.

Pressure to Go Along

No single incident better illustrates the industry's clout than the fate of a Nixon Cabinet task force report on the import quota system. The group, headed by then-Labor Secretary George Shultz, was set up in March, 1969, and released a report in February, 1970, whose majority opinion was that the quota system should be eliminated and replaced with a tariff system. President Nixon, in August, rejected the recommendation. The quota system was ultimately done away with by Mr. Nixon in April, 1973.

Mr. Ikard of the oil industry association conceded that he himself still felt it would have been "a fatal error" to have adopted the Shultz committee's recommendation.

An advance copy of the Shultz report, according to Mr. Homet, was leaked to Standard Oil of New Jersey, now Exxon, to give this largest of all oil companies time to prepare counter-arguments. A vice president of the company admitted to Mr. Homet having seen the report — whose statistical basis was figures supplied, as usual, by the oil companies themselves.

Philip Areeda of Harvard Law School, who was executive director of the task force, recalled: "One high official [of an oil company] told me he regretted having given us the optimistic-pessimistic data as distinct from the pessimistic-pessimistic data. In other words, he had drawersful of data. I don't mean to suggest that the oil companies are run by crooks. The nature and meaning of figures is, within any enterprise, subject to some dispute and difference of opinion. . . ."

But the report did recognize that "the Government is profoundly ignorant" in matters of reserves and other statistical material. One of its major recommendations was that "steps be taken to gather data" independently of the oil companies.

Just why the report was rejected—first delayed "to await the outcome of discussions" with other "affected nations" and finally, in August, turned down definitely—was and still is unclear. General George Lincoln, then the director of the Office of Emergency Preparedness, said it was "just darned complicated." The opposition of many within the Administration, the explosive situation in Libya at that time and domestic politics were all thought to be factors in the decision.

Among the political reasons was George Bush, now chairman of the Republican National Committee, who was running for the Senate in Texas that fall. It was felt that his chances would be hurt if the Administration went along with the recommendation — and against the oil interests.

There is widespread agreement that, in Mr. Areeda's words, "Had the report been adopted in 1970, there would have been less question of an assured source of supply. Therefore, refineries would have been built," and some part of the present crisis eased. There is less universal caution that it might have further increased this country's dependence on Arab oil.

The 'Crisis' Begins

Between the rejection of the Shultz report and last October's oil embargo, which forced the designation of "crisis" on a situation that many Americans had been only dimly aware of as even problematical, there were two Presidential energy messages, a plethora of Congressional hearings and investigations, and legislative proposals that rarely progressed beyond the proposal stage.

The first Presidential message — in 1971 — was an effort to balance accelerated growth of demand for energy with "the new emphasis on environmental protection." There were somewhat vague urgings to conserve energy, combined with new emphasis on "clean" fuel technology, in particular the development of a newer and more efficient source of nuclear power, the breeder reactor. In the second message, in April, 1973, Mr. Nixon announced the demise of the import quota program.

The first nip of what was to become the energy crisis was felt in the winter of 1972, when a severe heating-oil shortage hit parts of the nation.

'In Ehrlichman's Desk'

It was, according to Martin Lobel, former energy adviser to Senator William Proxmire,

a shortage "contrived" by the majors through cutbacks of inventory to force independents out of business in anticipation of what they saw as inevitable Arab nationalization. The Arab producing states could not get direct access to United States markets if there were no independents to handle their sales.

The oil industry, disputing all charges of "contrivance," says that last winter's experience was just the first widely visible sign of the difficulty the country was headed for.

Despite a widespread perception of White House inaction, John Ehrlichman, formerly President Nixon's chief domestic adviser and the man in charge of energy problems at the White House, said recently that energy was a "front-burner project" in December, 1972.

But Elmer F. Bennett, a Washington lawyer with several years' experience in a high post in the Office of Emergency Preparedness, recalled a proposal that, it was believed, would provide strong incentives to expand refineries. It was "discussed and hashed out," he said, and was finally sent to the White House in the summer of 1972. It is "probably still in Erlichman's desk," he said. No action was ever taken.

Mr. Ehrlichman himself conceded, in an Associated Press interview, that the energy problem "lay pretty flat" from March, 1972, when Watergate became a White House crisis, until Gov. John Love of Color-

ado was appointed energy czar at the end of June. And Mr. Love, he added, "didn't have the levers [of power] available to him."

During that same hectic Watergate spring, a National Intelligence Estimate, a form of report that pulls together the thinking of all parts of the intelligence community from the State Department to the Central Intelligence Agency and the Pentagon, was sent to the White House. It warned, according to a Government official, that if there were another Arab-Israeli conflict, "there would very likely be a serious interruption in the flow of oil from the Middle East."

The chances of war in the next few months, he said, were estimated as "a little less than even," although some, particularly in the State Department, were less sanguine. And the report predicted not an embargo but "cutting off the flow of oil physically," by destroying pipelines, wells and refineries.

The apparent lack of response to this warning, he said, and to other "ample warnings" that "we were getting overcommitted to imports which were in jeopardy," was due in part to "a very simplistic view" of the Arabs that tended to discount their threats as bluff and bluster. Another factor, he added, was "the preoccupation of the Administration with other things, which just prevented them from focussing on this basic issue. . ."

February 10, 1974

Ecologists and Energy: The Unheeded Warnings

To the Editor:

Your Feb. 10 energy report by Linda Charlton stated: "Most environmentalists concede freely that they were oblivious to the energy consequences of the conservation measures they saw as so desperately needed." Begging to differ, I recall environmentalist warnings about rising consumption and waste of energy while energy producers were promoting inefficient all-electric homes and gasoline sales for speed and power on the highways. Specifically:

¶Environmental groups appealed at least five years ago for vastly expanded research by government and industry into clean and reliable energy sources.

¶Environmentalist testimony on the Alaska Pipeline suggested other ways

of bringing Alaskan oil to market, means of safeguarding alternative supplies, new petroleum import and storage policies and better energy sources. Above all, environmentalists said that orderly petroleum conservation (vs. today's shock treatment) could save more oil than would be provided by a pipeline and highway across our last great wilderness.

¶Environmental groups have gone into considerable detail on the availability, fuel value and sulphur content of deep-mined vs. strip-mined coal. They have compared true costs of stripped coal with real surface restoration to true costs of deep coal with safety for miners. They have examined U.S. coal export policy, trade-offs among alternative means of using coal energy and other relevant supply issues.

¶Promotional advertising and rate structures of energy utilities have been under environmental attack for years

as generators of waste and of inequitable energy allocation.

¶Population growth, increasing per capita consumption and careless technology have been decried by environmental leaders since Fairfield Osborn wrote "Our Plundered Planet" in 1948. The environmental issue has been coupled constantly with concern for natural resource supplies: The growth-is-good society would soon outstrip the resource supplies obtainable without intolerable damage to life systems.

Far from being oblivious to the energy consequences of conservation measures, most environmentalists have been deeply disturbed by prospects of shortage for which our economic and social system was not prepared.

SYDNEY HOWE
Executive Director
Center for Growth Alternatives
Washington, Feb. 13, 1974

February 26, 1974

Nuclear salvation or nuclear folly?

By Ralph E. Lapp

Nuclear power is a reality—some 40 plants are turning out badly needed electricity today—but the argument about the safety of this new power source has continued to smolder, and as Senator John O. Pastore, chairman of the Joint Committee on Atomic Energy, put it recently: "The public is absolutely confused."

Lately, two figures have dominated the debate. The Atomic Energy Commission, the Federal agency so hated by critics of nuclear power, a year ago gained a new and colorful chairman, Dr. Dixy Lee Ray. The agency's sagging credibility improved considerably with the appointment of Dr. Ray, an environmentally conscious marine biologist.

At the same time, Ralph Nader, the consumer advocate, has begun to exploit the nuclear safety

Ralph E. Lapp is an energy consultant and a member of the Sierra Club's energy policy committee. His most recent book, "The Logarithmic Century," is about growth and energy use.

issue. Although he finds nuclear power too hazardous from the moment uranium is dug from the ground until its radioactive ashes are earth-interred, he emphasizes: "The underlying point is that no society should rest its energy future in a fragile nuclear-fission basket when the risks of accident and sabotage are at a point of catastrophic consequence unparalleled in the history of mankind."

Nader's attacks have not always been temperate. "The problem with Dixy Lee Ray" he told a press conference in San Francisco last November, "is that she is suffering from professional insanity. She is locked into a bureaucratic momentum that has so distorted her capacity for reason that she is leading the Atomic Energy Commission into this drive for technological suicide, through nuclear fission." Madam Chairman Ray (she dislikes the word "chairwoman") responded to that tirade with softer language, saying that Nader was wrong and had based his case on "innuendo and inaccuracies."

So the stage is set for a protracted debate on the safety of nuclear power at the very time when new plants will be "coming on stream," as utility executives put it, at the rate of one a month. By 1980, if the nuclear program proceeds at full speed, one-fifth of all U.S. electricity will be generated from uranium. Some parts of the country are already quite dependent on nuclear power. In Illinois, for example, where Commonwealth Edison dominates the energy picture, the company's executive vice president, James J. O'Connor, estimates: "Our seven operating nuclear power plants will provide more than a third of our electric power in 1974. Uranium fuel will substitute for 12 million tons of coal, or 41 million barrels of residual oil, this year."

Yet Nader argues: "If the public knew what the facts were and if they had to choose between nuclear reactors and candles, they would choose candles."

Just what are the facts? And whose interpretation of them is more valid, Nader's or the A.E.C.'s?

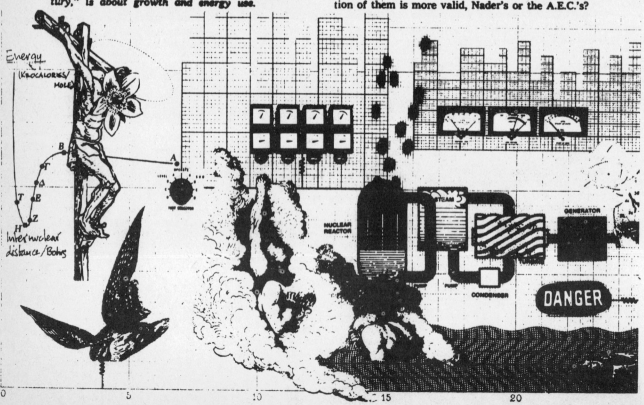

This article is an attempt to dig into the substance of the issue, and to place the problem in perspective.

It is essential to describe the anatomy of a nuclear power plant if we are to understand how accidents might happen and how designers have provided mechanisms to prevent them and to lessen their consequences. In conventional electric plants, fired by coal, oil or natural gas, the fossil fuel is burned in the firebox and the flaming heat serves to generate steam in a boiler. This steam—the historic propellant of the Industrial Age—roars into a huge turbogenerator at high pressure, sometimes reaching 5,000 pounds per square inch, and at a searing temperature of up to 1,050 degrees. The steam spins the huge turbine, whose shaft is coupled to a mammoth generator which produces electricity. The new steam-electric plants generate more than 1 million kilowatts of power. Their fossil-fuel appetite is prodigious; a coal-burner requires more than 400 tons each hour.

A nuclear plant "burns" a flameless fuel, uranium; heat is released as atoms split in a controlled chain reaction. Since there is no combustion, there is no exhaust gas and, of course, no sulphurous pollution. The nuclear firebox consists of a compact core smaller than a living room. It contains a year's supply of nuclear fuel—about 100 tons or so of uranium-oxide pellets of thimble size. About 10 million of these tiny pellets are neatly arranged in 12-foot tubes or fuel rods sheathed to prevent leakage of radioactivity.

The nuclear fuel used in modern power plants is a very dilute form of the weapon-grade material, and there is no danger that a reactor will explode like a bomb. However, it's still potent stuff—a single half-ounce pellet releases the same energy as 160 gallons of oil. The energy is released during a chain reaction—a self-sustaining sequence of atom splittings in which particles released when one atom splits collide with other atoms, causing them to split. To start a chain reaction, all you need to do is to put together in a core enough fuel and enough of a light substance such as water. When a uranium atom splits, it releases two or three nuclear particles called neutrons; these particles are speedy when first born, but oddly enough, they are more effective at splitting other uranium atoms when they are slowed down by bumping into atoms of other elements. Water is added to a reactor core to slow down the neutrons and thus promote the chain reaction.

Certain elements, like boron, tend to swallow up neutrons, and their presence in the core opposes a chain reaction; they may be said to act as poisons. Reactor designers are careful to make sure that poison-loaded rods are positioned in the reactor core before it is filled with water. With programed withdrawal of these control rods, the chain reaction develops, the quiescent reactor "goes critical"; further movement of the rods increases the power output of the machine. At the control panel of a reactor there are two red buttons marked SCRAM, which an operator can push in the event of a need to shut down the machine in a hurry. Actually, the reactor system is carefully monitored by instruments and any abnormal behavior triggers automatic scrams. To scram the reactor, control rods are thrust back into the core, terminating the chain reaction.

Water in the core serves two purposes—it acts to promote the chain reaction and also to remove heat from the uranium fuel. The whole core structure fits inside a huge pressure vessel made of steel—a great pot with a domed head that is bolted down and opened up only once a year for replacement of fuel.

The six-inch-walled vessel is fitted with huge pipes that serve one of two purposes, depending on the reactor's design. In one case, they convey water heated by the reactor core, but pressurized to prevent its turning to steam, to a heat exchanger, where it transfers its heat to another vessel of unpressurized water that boils and provides steam to power a turbine. In the other, water in the core is permitted to boil, and

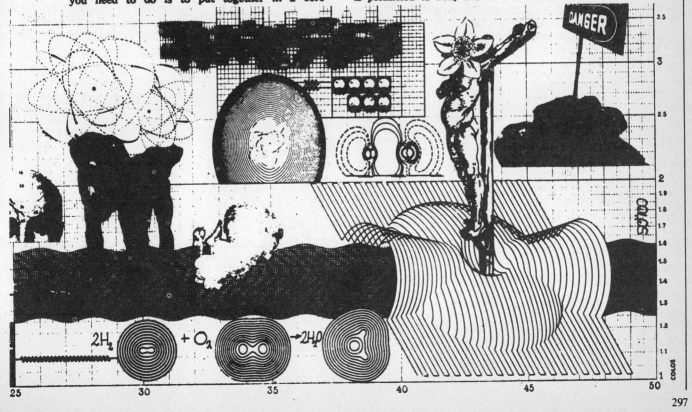

$2H_2 + O_2 \rightarrow 2H_2O$

the pipes convey steam directly to a turbine.

Inevitably, the water surrounding the core becomes somewhat radioactive. This occurs because as uranium breaks down during a chain reaction, other radioactive elements are created. The sheathing of the fuel rods is intended to prevent these radioactive elements from passing into the water, but a small amount always escapes through tiny flaws in the sheathing. Fears about the safety of nuclear power begin with the fact that these radioactive elements are continually being generated at the heart of the reactor. Under the most catastrophic circumstances, they could be released to the environment in large amounts.

Dr. Henry W. Kendall, a high-energy physicist at the Massachusetts Institute of Technology, who is probably Nader's most astute adviser, expresses the radiation hazard of a nuclear power plant (reactor) as follows:

"The radioactive accumulation in a large power reactor is equivalent to the fallout from thousands of Hiroshima-size nuclear weapons. . . . Consider, for example, that 20 per cent of a reactor's radioactive material is gaseous in normal circumstances and, if released to the environment in one way or another, could be swept along by the winds for many tens of miles to expose people outside the reactor site boundaries to what could be lethal amounts of radioactivity. The lethal distance may approach 100 miles."

And yet this view obviously is not shared by the Atomic Energy Commission. The commission approved the siting of three powerful nuclear reactors just 26 miles north of New York City at Indian Point, south of Peekskill. If Dr. Kendall's estimate of the situation is anywhere near accurate, such a siting would have to be viewed as an act of Federal and corporate recklessness. So what is the actual risk, and why is there such a difference of opinion about it?

The reactor accident most commonly visualized by nuclear engineers is known as the loss-of-coolant accident,

or LOCA; it would begin with a break in one of the heavy pipes that carry water and steam to and from the nuclear core. Such a break might occur because of faulty construction and maintenance, sabotage, or natural disaster. (The A.E.C. requires utilities to design reactor installations strong enough to withstand the assaults of earthquakes, hurricanes, tornadoes and even giant sea waves, or tsunamis.)

The first consequence of such a break would be a reservoir of water, condensing of a mixture of water and steam that would carry with it any radioactivity that had leaked through the protective fuel-rod sheath. Reactors are designed with containment systems to prevent this primary expulsion of radioactivity from reaching the atmosphere. In one design, the escaping water/steam is conducted through large-diameter pipes to a huge metal doughnut-shaped container, where the steam exits into a reservoir of water, condensing and reducing the pressure and temperature within the containment. In another design, the kind built at Indian Point for Con Edison, the primary containment takes the form of a tremendous silo made of reinforced concrete, which is designed to be big enough and strong enough to absorb the pressures and temperatures created by a severe blowdown.

But handling the violent release of high-pressure, high-temperature steam in the event of a pipe break is not the only problem, for a reactor core deprived of its coolant might melt. At this point we must explain a strange quality of the nuclear core. Unlike an automobile engine, which stops generating heat when the pistons cease moving and no gasoline is burned, a nuclear reactor core cannot be turned off completely—it keeps giving off heat when it is fully scrammed and there is no vestige of a chain reaction. This afterheat is produced by the radioactive disintegration of the split atoms which accumulate within the fuel rods. Reactor engineers have to handle it whenever a plant is shut down, whether routinely or accidentally.

When the chain reaction is

terminated, a reactor still generates about 7 per cent of the power it had been producing just prior to shutdown. Normally it is dissipated as water in the reactor cools off the core. Now, if water is blown out of the core by an accident, the afterheat of fuel pellets will build up within the fuel rods and, in the absence of further cooling, the rod sheaths may deteriorate. If this happens, then continued fuel melting could cause the pellets in the core to form a molten mass which could eat its way through the bottom of the thick steel vessel. Such a melt-through could spill hundreds of thousands of pounds of superhot radioactive debris into the primary containment vessel. Since the vessel would be no match for the viscous "dropped core," it might melt itself, permitting a glowing mass to penetrate into the earth, perhaps reaching a spherical diameter of 100 feet. Where would it stop? It would depend upon the nature of the substratum underlying the site, but experts have coined the phrase "the China syndrome" to describe the fate of the dropped core. It would take many years to cool off.

Experienced reactor experts admit that this sequence of events could happen, but contend that it's dependent on a series of highly improbable events. For example, unless the molten material concentrates first at the bottom of the reactor vessel and then at one point in the concrete base, there would be no melt-through. Further, as we shall see shortly, engineers contend that safeguards will prevent fuel from melting in the first place.

A melt-through is a nightmare for reactor designers, but it is not the China syndrome that worries the experts as much as the breakout of radioactive gases and particulates that would occur within the first hour or two. A great puff of radioactivity released through the fractured containment would be at the mercy of the winds and the weather. This is the severe sort of accident that forms the basis of Nader's concern that nuclear power is too risky an energy source for

man.

This kind of hazard was on the minds of physicists even before the first chain reaction was achieved on Dec. 2, 1942. Moreover, it was of great concern to utilities in the mid-nineteen-fifties as they contemplated building nuclear power plants to generate electricity. Utilities prefer to site power plants close to metropolitan centers, yet such siting might involve great liability in the event of an accident. Accordingly, the Atomic Energy Commission set out to provide some estimate of the hazard, and in March, 1957, it published a 105-page report called WASH-740 and titled "Theoretical Possibilities and Consequences of Major Accidents in Large Nuclear Power Plants." Its findings included the following two paragraphs:

"For the three types of assumed accidents, the theoretical estimates indicated that personal damage might range from a lower limit of none injured or killed to an upper limit, in the worst case, of almost 3,400 killed and about 43,000 injured.

"Under adverse combination of the conditions considered, it was estimated that people could be killed at distances up to 15 miles, and injured at distances of about 45 miles. Land contamination could extend for greater distances."

Here it should be noted that the power plants assumed in the WASH-740 analysis were seven times less powerful than the modern machines being ordered by utilities. In addition, today's reactors are being sited closer to large populations than assumed in the 1957 report. Taking such differences into account, it would seem that Dr. Kendall's estimate of a 100-mile lethal distance is not off the mark.

"That 1957 report doesn't apply today," asserts Dr. Dixy Lee Ray. "For one thing, reactors today feature a whole panoply of safeguards built into the basic design so as to prevent accidents and to mitigate their consequences should one occur. But even more importantly, the WASH-740 study assumed conditions that were so extreme that they would be virtually impossible to accomplish even if there was some unimaginable reason to try to do so. Such an accident would require instantaneous meltdown of a reactor core with no safeguards operable and a breach of the containment system."

In order to understand the basic reason for Dr. Ray's confidence we need to look at a safeguard which is the primary mechanism for preventing fuel melting in a reactor core. It is known as the emergency core-cooling system (ECCS), and it is a set of mechanisms for flooding a reactor core with water in the event of an accident, such as a big pipe break. For example, one emergency cooling system consists of pumps and water supply to inject water at high pressure back into the core; another operates at low pressure to flood the core. These systems are powered by independent sources of electricity run by sets of diesel generators that start up automatically if a situation requires ECCS action.

Nuclear engineers maintain that these emergency systems will work to reflood the water-starved core in the event of an accident, while critics argue that such assurance is based on models and computer codes and not on actual testing under real accident conditions. The A.E.C. as developer of power reactor designs has an elaborate nuclear safety research program that has already cost over half a billion dollars. One safety program is called LOFT, for loss of fluid test; it is a nuclear reactor built at the A.E.C. Idaho reactor test site, and it's designed to be wrecked deliberately. Sometime next year the LOFT reactor will be subjected to a planned pipe break or LOCA, and the whole sequence of events following the accident will be examined in a systematic manner.

Critics of LOFT take a skeptical view, saying that it is "too little, too late." The wreckable reactor, they say, is small compared to today's giants, which are 70 times more powerful. Therefore, any results from the LOFT experiment will have to be scaled up and may not apply to the actual power units in operation. Antinuclear spokesmen also call attention to the

fact that some 60 nuclear plants will be in operation by the time LOFT experiments are analyzed. This, they assert, means that the A.E.C is conducting a *post factum* safety research program.

It was a small scale experiment in the LOFT program that served to precipitate the argument over emergency core-cooling. This experiment took place in Idaho late in 1970 and yielded initial results that were interpreted by some as meaning that a water-deprived core would not be reflooded; thus doubt was cast upon the models and computer codes that the industry was using for predicting the performance of their emergency safeguards. Dr. Kendall, spearheading the efforts of a tiny group known as the Union of Concerned Scientists, challenged the A.E.C. criteria for emergency core-cooling, and in January, 1972, the A.E.C. began public hearings that dragged on through the summer of 1973. Some 22,200 transcript pages of testimony were taken down—some of it could as well be Sanskrit even to many physicists. Here, for example, is one of many assumptions spelled out in the testimony of an expert: "The time to DNB is computed using the W-3 critical heat flux correlation in both the CRAFT and THETA 1-B codes, except that the B&W-2 correlation is used in THETA 1-B for the non-vent valve plants." That's enough to cross a lawyer's eyes—and lawyers abounded in the protracted A.E.C. hearing on emergency core-cooling.

On Dec. 28, the A.E.C. concluded its review of the emergency core-cooling issue and put forth a 29-page acceptance criteria backed up by a 140-page exposition of technical factors involved in core-cooling. Basically, the commission judges nuclear power plants to be safe, but will require utilities to re-examine the margin of safety in their operation; in the event it is judged inadequate, the power level of the plant involved would be reduced. Although the A.E.C.'s decision on nuclear safety embraced objections of Dr. Kendall and constituted a minor victory for him, he was joined by Ralph Nader in an instant response

that the decision "represents a continuation of the A.E.C. cover-up of critical safety problems."

When Dr. James Schlesinger, now Secretary of Defense, took over command of the A.E.C. in the summer of 1971, he recognized that the agency's failure to revise its 1957 report on theoretical accidents in nuclear plants allowed critics to scale up the hazard estimate of that 1957 analysis to correspond to large modern reactors, intensifying fears of a nuclear accident. He set up a task force of technical experts to assess the probability of a serious accident in a large modern reactor and to define its consequences. Dr. Norman Rasmussen, an M.I.T.-trained nuclear expert, was picked to head up this group, and it is understood that he will avoid the "worst case" approach of the 1957 report; instead, speculations will be based on "best engineering judgment."

Dr. Rasmussen sums up his philosophy of accident analysis this way:

"Uncertainties are treated by developing realistic probability values of all possible outcomes rather than choosing the worst possible values. This approach will lead to a prediction of the most likely consequence and the probability of smaller or larger consequences. This should provide a more complete and accurate view of nuclear accident risks than previous studies that computed only 'worst case' values."

To illustrate what is called the probabilistic approach, let's assume, as do many conservative experts, that, in a given year, the chance of a major pipe break in a reactor leading to a loss of water from the core is 1 in 1,000. The Rasmussen task force then goes on to analyze the other links in the accident chain and, in particular, the probability that the emergency core-cooling will fail to function. Nuclear experts in industry claim that there is only one chance in a thousand of its failing.

The combination of these two probabilities — that the core will dry out under accident conditions and that safeguard systems will fail to reflood the core — is the product of the two individual probabilities, or 1 in 1 mil-

lion. Projecting ahead to 1980, when over 100 reactors are scheduled to be in operation, this should mean that the chance of *any* nuclear accident would be 1 in 10,000 for a single year. And from the viewpoint of a single community near a reactor this chance would be only 1 in 1 million.

Presumably, this is the kind of conclusion that the Rasmussen analysis will reach when the final report is published this summer. The report is going to be backed up by many technical appendices documenting the task force findings, but it may take a nuclear Rosetta stone to translate the prose into something understandable to the man in the street.

For their part, critics charge that the nuclear industry's record does not instill high confidence that a nuclear accident of serious magnitude can be avoided, whatever the probabilities calculated by experts. They point to the A.E.C.'s own extensive tabulation of "abnormal occurrences" in reactor construction and operation as evidence that quality control in this new industry is inadequate. For example, Con Ed's Indian Point 2 plant has been plagued with incidents, the latest of which was a pipe break in the steam system that affected the reactor building, but not the core area, and may have caused damage to the steel surface of the huge containment structure.

The A.E.C. admits that there have been numerous abnormalities in plant construction and operation, but maintains that their detection is in fact proof that quality control is being exercised. Far from covering up such incidents, the A.E.C. maintains, the record is made public in various publications, such as an annual compilation titled "Safety-Related Occurrences in Nuclear Facilities." Each utility is required by the A.E.C. to submit timely reports on all unusual occurrences at nuclear plants, and the commission's regulatory staff studies these reports and issues analyses of them.

Quite apart from the difficulty of comprehending the complex technical niceties of this debate, the layman must also face the bewildering question: How safe is safe enough?

No group of experts can pass judgment on the acceptability of a public risk. The A.E.C.'s accident-consequences report will only put the risks in better perspective.

Modern life confronts people with a multitude of risks; some of these are obvious—like the risk of accident in an automobile collision or an airplane crash—others are subtle, remote and not immediate in ill effects.

For example, the risk of being bitten by a rattlesnake in Times Square is low, but not zero. The chance of being hit by a car is fairly high. People accept the risk of death involved in traveling on a common carrier, such as an airline—that risk is about one in a million, for each flight.

But calculations of the odds are not always the last word. Suppose you go out to Shea Stadium to watch a ball game. How safe is that in light of the chance that an airplane might crash into the stadium? The probability of that happening has been calculated as very low, yet strange things can and do happen: One night not long ago an airplane was circling to land at LaGuardia Airport. The weather was soupy and the pilot mistook the lights of Shea Stadium for those of the airport. He nosed his 727 down to land and only at the last minute did he realize his error. Fortunately he was able to zoom the 727 upwards and the event was recorded as a "near miss."

We tend to react to this problem of risk by making choices based on the magnitude of risk as we perceive it and the benefits to be gained from accepting the risk. The public apparently judges the convenience of commercial air travel to be worth the risk that results in 200 fatalities per year; the convenience of driving an automobile is considered worth much higher levels of risk. Sometimes the public judgments are not especially rational. About 49 million Americans continue to smoke cigarettes despite the clear warning of risk to their health printed on each package.

If we assume, as we have to, that public demand for

electric power will continue to increase, then we must consider the question of nuclear safety in this risk-benefit context.

Furthermore, we have to realize that we do not have a great many options. Environmentalists argue that the problems of current power-producing technology might be avoided by developing alternatives such as solar power, geothermal energy or fusion energy. But a utility burning 9,000 tons of coal per day would need a collection area of 18 million square yards of level ground to absorb the necessary solar heat —that's almost 6 square miles! If one assumes that the heat can be trapped and converted to electric power at 30 per cent efficiency, this would mean paving over 20 square miles with solar catchers. That's not an option for Con Ed even if the technology for

solar conversion to electric power were available at reasonable cost. Geothermal energy is available in limited amounts at relatively few geographic sites, unless one wishes to exploit "hot rock" beneath the earth's crust—a still unproved technology. As for fusion power, the hoped-for generation of energy by fusing atoms of hydrogen in a controlled reactor, $1-billion in research and development has not yet realized the first practical laboratory demonstration of this scientific principle. And even if scientists succeed in the laboratory it will take another two decades to engineer massive energy conversion equipment so that it constitutes a significant contribution to energy production.

This means that for our future electric-power needs, we will have to rely either on nuclear power or else on fossil fuels—and as it looks now, the main fossil fuel will be coal. Right now utilities de-

pend on fossil fuels for four-fifths of their energy; coal makes up more than half of this total, natural gas about a quarter and oil about a fifth. Gas and oil are in short supply and can be expected to remain so.

Recognizing these facts permits us to restate the problem of "nuclear safety" in more sophisticated terms: In the years to come, how will the balance of risks and benefits of nuclear power generation compare to the risks and benefits of heavily increased reliance on coal?

The benefit of nuclear power is that it is based on a new fuel that does not need to be mined in disruptive quantities, and whose flameless "fire" does not pollute the air. The risks involve the chance of catastrophic reactor accidents as discussed above, plus the additional problem of waste disposal. Many fear that the radioactive ashes left over

when nuclear fuel is spent cannot be disposed of safely and will build up the level of radioactivity in the environment to intolerable levels. (Dr. Ray of the A.E.C. disputes this, however, contending that technology exists to store the wastes safely.)

The benefit of coal is that it is abundant in America and that it does not pose the risk of catastrophic accident or residue that emanates dangerous radiation. But it does pose other risks. Most obviously, it contributes in a big way to air pollution. Except in certain locations, coal is a deeply buried resource, and the mining of it is one of the most dangerous occupations in the country. In this century, more than 100,000 miners have lost their lives digging coal out of the ground. Millions more have been injured or afflicted with "black-lung" disease. Where coal beds come close to the earth's surface, it may be extracted by strip-mining, but that practice so grossly

lacerates the earth's surface that it has caused widespread popular reaction. Yet it is only through strip-mining, particularly in the rich Fort Union Formation of the Upper Missouri basin, that the coal industry can increase production enough to meet the growing demands of utilities.

So is nuclear power too risky? I believe that nuclear power plants can be operated safely if their designs are carefully checked out, if high quality control is exercised in their construction, and if their operation is subject to vigilant regulation at all times. I also believe that a further margin of safety could be gained by siting plants so that in the improbable event of an accident, the radiation risk to the population nearby is minimized. Standardization of plant design can help in assuring that the licensing time can be shortened and quality control in construction can be increased. And the plants can be clustered at "nuclear parks" where greater nuclear security can be achieved.

When opponents of nuclear power argue that it should be stopped dead in its tracks unless, as Nader testified before the Joint Committee on Atomic Energy Jan. 29, reactors "are safe beyond any question of doubt and superior to other energy alternatives," they jeopardize the whole environmental movement, because their short-term alternatives—usually solar power and geothermal energy—are not real options for an energy-hungry society. And when Nader suggests that coal replace uranium, he seems to have no concept that the amount of coal required for such replacement, at least two billion tons annually by the end of the century, involves immense environmental assault as well as exorbitant social costs for the nation.

Given the alternatives, given their availability and their risks, I see nuclear power not only as an acceptable risk for the United States, but also as the only practicable energy source in sight adequate to sustain our way of life and to promote our economy. ∎

> Beyond the technical complexities of reactor operation and design, the layman must face an even more bewildering problem: How safe is safe enough?

Nader on nuclear alternatives

To the Editor:

Ralph Lapp believes that nuclear power plants can be safe if they are designed, constructed and operated safely "at all times" ("Nuclear salvation or nuclear folly," Feb. 10). This is transforming the hypothetical into belief. Unfortunately, the evidence doesn't hold. There have been too many A.E.C.-documented design defects, construction blunders and operating "near misses" (to quote a phrase from an A.E.C. document) affecting present nuclear plants. Serious as the sporadic spills, emissions and leaks of radioactive materials around the country are, the unresolved problem of safely transporting deadly radioactive wastes and then storing them securely for tens of thousands of years is utterly awesome.

Dr. Lapp can be forgiven for glossing over the risks of sabotage and theft of weapons-grade material as nuclear technology diffuses throughout the land. For such risks lend themselves even less to the meaningless exercises in probability by the A.E.C. The agency needs to be reminded of the earlier estimates of space technology accidents before the accidents happened. More people need to learn about the almost catastrophic Fermi plant accident near Detroit in 1966. The Fermi accident was not even supposed to have been able to happen, according to the A.E.C.'s calculations.

When discussing alternatives to nuclear power, it is necessary to put the latter into quantitative focus. The A.E.C.'s most optimistic figures put nuclear power as the source of 25 per cent of the nation's energy needs by the year 2000. This projection assumes no "big accidents" as described by Lapp. For one "big accident," with its devastations of human life, property and land, and genetic inheritance, would produce a stop on nuclear power and plunge the nation into a radioactive, as well as an energy, crisis if the economy was then heavily reliant on these plants.

But given Lapp's assumption that all will go well, what could take the place of this amount of energy? Briefly and incompletely: use of waste heat, burning of trash (as is done in Paris), reduction of massive energy waste by industry and commerce (as studied by the Office of Energy Conservation), more efficient vehicles, and energy-conserving buildings and housing. Continued use of petroleum products from the recovery of vast and conventional domestic reserves is the central 30-year energy source. Further, the National Science Foundation is much more impressed by the share of energy needs which could be met by geothermal resources over the next quarter century than is Lapp. Solar energy for building heating and cooling can become a significant factor during the next generation with additional breakthroughs for efficient 'conversion to electricity likely, if given modest research and development funds.

As for the nation's enormous coal reserves, if much more coal has to be used to the end of the century as a last resort against nuclear power, why does Dr. Lapp have such little faith in the application of much safer underground (where 95 per cent of the coal lies) mining techniques? Improving the safety of coal mining isn't a fraction of the difficulty of making nuclear technology safe forever, which he believes is possible.

RALPH NADER
Washington

Ralph E. Lapp replies:

It pains me to respond to Ralph Nader's letter because I have such a high regard for Nader as a social critic and because I feel that both of us are dedicated to assuring the public interest in matters of nuclear safety. This is a field with which I have been concerned for many years and I welcome Nader's recent manifestation of interest in nuclear issues.

However, it is imperative that critics of nuclear power do their homework and attack technical issues on a basis that is fair, objective and responsible. The nation's energy crisis is not a transitory affair and we must learn to live within our energy means while maintaining a viable economy and environment. Energy is too important to be treated in a casual manner. Ours is a time for responsible criticism, sensitive to the need to protect the public health and safety while taking into account the needs of a dynamic economy. It is a time for judicious appraisal of energy supply choices, prospects and consequences.

I am sorry to say that Mr. Nader's letter, together with material that he has either authored or signed in the past year, leaves me with no alternative but to conclude that he is perilously weak in his understanding of energy technology. Allow me to pinpoint a single instance in his letter to illustrate how he has failed to do his homework. He claims that the Fermi accident "was not even supposed to have been able to happen, according to the A.E.C.'s calculations." Apparently, he never took the trouble to research this proposition. If he had—and it would not have been a difficult chore—he would find a complete safety evaluation that contradicts his contention. Hazards Analysis of the Enrico Fermi Reactor, Nov. 24, 1962, Public Docket No. 50-16, available in Washington, D. C., details the A.E.C. estimates of fuel melting that are entirely consistent with the accident that actually happened. It should be noted that the accident consequences fell within the prescribed limits of the A.E.C. safety evaluation and that there was no radioactive release of any significance from this accident.

Other points raised by Nader in his letter could be equally well confuted, were I to have space for adequate argument.

Human error

To the Editor:

The most powerful argument against nuclear power I've ever read was Ralph Lapp's defense of it. The fact that meltdowns and lethal clouds are possible only after a "highly improbable sequence of events" is small comfort when I consider that the men and women who design, build, and operate those plants are people exactly like me; that is to say, people who make three or four mistakes every day of their lives.

MALCOLM B. WELLS
Cherry Hill, N. J.

Lucidity

To the Editor:

I want to put in a word of congratulations to Ralph Lapp for converting the numbing complications of the subject into lucid form and for cutting through some very dense emotional fog in "Nuclear salvation or nuclear folly?"

JOHN MCPHEE
Princeton, N. J.

Disutility

To the Editor:

Ralph Lapp's sophisticated, accurate, and responsible analysis does not fully take into account, in computing the risk-benefit of nuclear reactors, the size of the disutility. To say that reactors have a 1 out of 10,000 chance to blow, each year (or 1 out of 1,000,000 per community), which makes them about as safe as flying, does not take into account the number of persons to be killed in a nuclear disaster.

As the size of the disutility is often overlooked in the discussion of risk probabilities, the point deserves illustration. Most persons who would accept a $10 bet at odds of 99 to 1 in their favor, would hesitate if the bet was $1,000 at the same odds, and refuse a $100,000 bet at identical odds.

Why? Only because the disutility changed.

Similarly, in the case of nuclear reactors. If the loss is one person's life, 1 out of 1,000,000 might be quite tolerable. The same odds to lay waste a city seem quite unacceptable. Perhaps for an individual the difference is

theoretical; he will be just as dead, but for public policy the difference between losing some lives on a flight and—a city, are monumental.

AMITAI ETZIONI
Director, Center for Policy
Research, Inc.;
Professor of Sociology,
Columbia University
New York City

Ralph E. Lapp replies:

Dr. Etzioni, a valued critic, makes the point that there is a significant difference between nuclear risks and those involved in an aircraft crash. He is justified in this distinction, based on the A.E.C.'s published data on accident consequences. However, last January Dr. Dixy Lee Ray previewed the results of the Rasmussen study and indicated that the two risks may be comparable. Those of us who are outside the A.E.C. must await publication of the study to judge the fairness of this estimate.

Sierra Club opposition
To the Editor:

Ralph Lapp is of course entitled to his opinions about the need for and safety of nuclear power. However, the fact

that he is a member of the Sierra Club's energy policy committee does not mean that he speaks for the club. On this issue he does not. In January the Board of Directors adopted the following addition to the Sierra Club's energy policy:

"The Sierra Club opposes the licensing, construction and operation of new nuclear reactors utilizing the fission process, pending: (a) development of adequate national and global policies to curb energy overuse and unnecessary economic growth; (b) resolution of the significant safety problems inherent in reactor operation, disposal of spent fuels, and possible diversion of nuclear materials capable of use in weapons manufacture; and (c) the establishment of adequate regulatory machinery to guarantee adherence to the foregoing conditions."

ELISA K. CAMPBELL
Chairperson,
Mt. Holyoke Sierra Club
Mt. Holyoke, Mass.

Columbia reactor
To the Editor:

Our group has been fighting against the activation of a small Triga II nuclear reactor on the campus of Columbia University for six years. If we can raise the money, we will take our case to the U. S. Supreme Court. While Ralph Lapp's article was quite informative in describing the catastrophic effects of major accidents, it does not mention the effects of day-to-day, normal operation. It has been determined that during normal operation of even a small reactor, radioactive gases such as argon-41 would be released through the stack of the reactor into the atmosphere. These gases would be breathed into the lungs and enter the blood stream. These radioactive gases would be harmful to anyone unlucky enough to live near Columbia's reactor, but would be especially harmful to human fetuses, resulting in stillbirths, leukemia, genetic defects and

death in the first year of life. These probabilities have been examined by Dr. Ernest J. Sternglass of the University of Pittsburgh, based on studies of populations near small reactors operating in three locations. What happens to people living near big reactors during normal operation?

Mr. Lapp's article ends with a statement about perfect design, perfect construction and perfect operation. We know that combinations of materials designed, built and operated by humans are subject to rather complete failures. Let us remember the "unsinkable" Titanic, the gas tanks that exploded on Staten Island a year ago and the Apollo spaceship that burned up in 1969, to mention only three examples of human failure.

ROBERT W. HEDGES
Ad Hoc Committee Against
Columbia's Reactor
New York City ■

March 24, 1974

MOST ARAB LANDS END BAN ON OIL SHIPMENTS FOR U.S.;

By JUAN de ONIS
Special to The New York Times

VIENNA, March 18—Most of the Arab oil countries announced officially today that they would lift the embargo against shipments to the United States that they had imposed during the Middle East war of last October.

The formal decision, which had been made in principle last week, was accompanied by predictions from some Arab oil ministers that shipments to the United States would be restored to nearly the level that had existed before the war.

This appeared to be the prospect despite the refusal of Libya and Syria to go along with the seven-nation majority in the conference of nine Arab oil countries here.

New Meeting Is Set

In addition, Algeria said she was lifting the embargo provisionally until June 1. Arab oil ministers are scheduled to meet again on that date in

Cairo to review the oil situation.

With the official announcement of the lifting of the embargo, Saudi Arabia pledged an immediate production increase of a million barrels a day, with that oil going to the United States market.

American imports of oil are currently running at about five million barrels a day, or two million below estimated requirements, a shortage that is caused by the Arab embargo. Thus, a million barrels a day from Saudi Arabia would erase half that deficit.

Production Rise Expected

Ahmed Hillal, Egypt's oil minister, said he expected that the Arab countries that ended the embargo would raise their production so as to supply the United States with 90 per cent of the oil they were shipping to it before the October war. There were predictions also that increased supplies would lead to lower prices.

The Arab oil ministers announced, however, that their countries would continue the embargo against the Netherlands and Denmark. At the same time, they said West Germany and Italy had been placed on the list of "friendly" countries, which are eligible for full oil supplies to meet their needs.

"I think the United States will get enough oil for its requirement," said Ahmed Zaki al-Yamani, Saudi Arabia's Minister of Petroleum Affairs, at a news conference where he joined Belaid Abdesalam, Algeria's Minister of Energy and Industry, the chairman of the conference, in announcing the decision.

This action, undertaken in recognition of American peace efforts in the Middle East, confirmed the decision that the seven oil countries had made in principle in Tripoli, Libya, last week, with Libya and Syria dissenting, as they did today.

A statement explaining the

political reasons for ending the embargo that had been imposed because of American support for Israel said that American policy, "as evidenced lately by the recent political events, assumed a new dimension" toward the Arab-Israeli conflict.

"Such a dimension, if maintained," the statement said, "will lead America to assume a position which is more compatible with the principle of what is right and just toward the Arab-occupied territories and the legitimate rights of the Palestinian people."

The announcement that Saudi Arabia would immediately raise production by a million barrels a day above her present output of more than seven million barrels a day, was made by Sheik Yamani, who led the fight to end the embargo.

In reply to a question of how soon Arab oil would be reaching the United States, Sheik Yamani said, "The time it takes for tankers to go from export terminals to United States ports."

This is about two months from the Persian Gulf and a month from Algeria, which is an important supplier of crude oil to refineries in the Caribbean that ship products to the United States market.

March 19, 1974

Shale-Oil Search Intensified in a Reaction to Arab Action

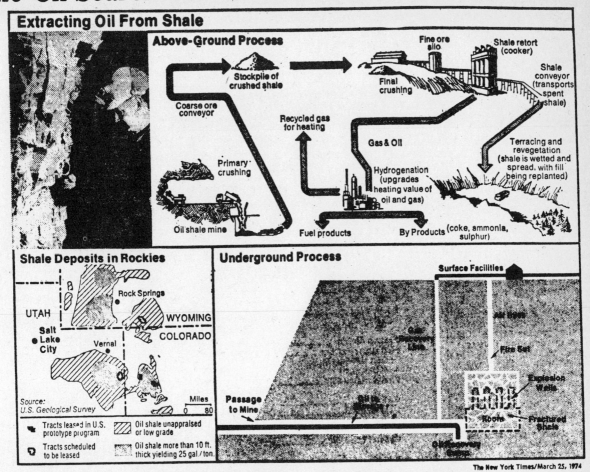

Extracting Oil From Shale

Above-Ground Process

Stockpile of crushed shale

Coarse ore conveyor

Fine ore silo

Final crushing

Shale retort (cooker)

Shale conveyor (transports spent shale)

Recycled gas for heating

Primary crushing

Gas & Oil

Hydrogenation (upgrades heating value of oil and gas)

Terracing and revegetation (shale is wetted and spread, with fill being replanted)

Oil shale mine

Fuel products

By Products (coke, ammonia, sulphur)

Shale Deposits in Rockies

UTAH

Rock Springs

WYOMING

COLORADO

Salt Lake City

Vernal

Source: U.S. Geological Survey

Miles 0 80

■ Tracts leased in U.S. prototype program

▨ Oil shale unappraised or low grade

◯ Tracts scheduled to be leased

▦ Oil shale more than 10 ft. thick yielding 25 gal./ton.

Underground Process

Surface Facilities

Air Vent

Fire Set

Explosion Wells

Passage to Mine

Room

Fractured Shale

Oil Recovery Sump

The New York Times/March 25, 1974

Developing shale-oil industry is based on deposits in the Rockies. Fuel is extracted by superheating shale. The above-ground method involves bringing the shale to the surface for processing, while the underground method leaves the fragmented shale in place. A room is carved out at end of horizontal passage and shafts are drilled upward from room for planting explosives. After explosions fragment the laminated rock, gas is introduced from surface through air vent and ignited. Fuel from superheated shale flows into a sump and is pumped out.

By VICTOR K. McELHENY

Inside the waterless, rocky wall of a Western Colorado valley, engineers are bringing the long-discussed idea of extracting oil from shale underground — without removing the shale from the mine — closer to reality.

On a 4,000-acre private site near DeBeque, in a corner of the rich Piceance formation, the Garrett Research and Development Company, a wholly oowned subsidiary of the Occidental Petroleum Company, has been experimenting since 1972 with a patented underground process inspired by earlier work of the United States Bureau of Mines.

The typical above-ground shale-mining process involves transporting the shale to processing facilities, where the oil is cooked out and the shale residue disposed of. The underground process involves fragmenting the shale inside the mine by explosives, heating it inside the mine and then letting the oil flow out, with the residue remaining in place.

Garrett Process Studied

The Garrett underground process, if it succeeds, would be better on several counts. It would be cheaper, it would be more productive, it would cause less pollution of the environment, and it would not use so much valuable water.

But the big question now is: Will it succeed? Those who are proceeding with the established system would gladly adopt the Garrett system, and pay the fee for it, if indeed it proved to be better, but at this moment they are skeptical.

Thus far, the experimental Garrett installation has been producing 25 to 30 barrels a day from the first of three 32-foot-wide chimneys of explosively crushed rock within the valley wall.

This oil comes from formations bearing about half a barrel of oil a ton or less. Garrett experts expect a forth chimney to be produc-

ing 500 to 1,000 barrels a day by the end of the year and intend to build steadily with more chimneys to 30,000 barrels a day within the next few years.

If the United States could exploit all of the nearly 2,000 billion barrels of oil locked in Western shale formations — several times the total world reserves of petroleum —the shale could supply its current annual consumption of petroleum for a century and a half.

But enormous obstacles lie in the way—not only technical and environmental problems but also fierce competition with other energy industries for scarce men, money and equipment. Experts agree that development of the shale industry will be slow.

Indeed, even with strong Government support, oil-industry sources estimate that the production of oil from shale is unlikely to exceed one million barrels a day in the next decade unless underground processing expands

unexpectedly.

With 1985 energy needs projected as the equivalent of 10 million barrels a day of oil more than the current level (not counting contribution from nuclear power), shale would account for only a fraction—though a substantial one—of the extra need.

As Dr. Charles Prien of the Denver Research Institute said in an interview: "There's going to be a continuing energy shortage. I don't think we should kid ourselves. But shale sure is worth doing. Every million barrels is a tremendous contribution."

By 1985, according to current plans, the industry will consist of the six Government-owned areas in Colorado, Utah and Wyoming that are being leased this year and privately held tracts in western Colorado owned by the Union Oil Company, the Colony group of oil companies and Occidental Petroleum. The investment in these areas is expected to reach several billion dollars.

303

A Fossil Remnant of Lakebeds

Both the underground and the above-ground methods of extracting a crude petroleum substitute from shale involve the elementary physical process of changing a solid into a liquid by heating it.

The shale is a fossil remnant of ancient lakebeds where the hydrocarbon remains of livin gthings were laced like the chocolate in chocolate whirl ice cream between layrs of grit that, over millions of years, became pressed into marlstone.

The hydrocarbon called kerogen, a solid black substance is a complex polymer. The kerogen molecule must be broken down to make a substitute for petroleum. With this process there is usually an excess of nitrogen, ueless for burning, which mut be removed by processing.

Whether heated underground or on the urface in cookiers called retorts, the shale must be brought to about 900 degrees Fahrenheit to break bonds between on carbon atom and another and so vaporize and break down the kerogen.

If the rock gets much hotter than 900 degrees, much heat energy would be wasted on such unwanted reactions as converting calcium carbonate in the marlstone into calcium oxide. If the rock stays cooler than 900 degrees, according to Dr. Mark Atwood of the Oil Shale Corporation's laboratories in Rocky Flats, Colo., "the shale just sits there and looks at you."

Once the kerogen is vaporized, its byproducts are cooled, and a substitute petroleum condenses. There are other byproducts, such as a kind of coke and also gases that can often be used for burning to maintain the shale cooking. In one shale process in Brazil, the gas is even used as a source for sulphur, which is scarce there.

Rock Must Be Crushed

There are many ways to do the job. But common to all of them is a need to crush the rock so that air, which is necessary for burning, can get at it.

In above-ground processes, the crushing requirements vary enormously. Some processes, such as one invented by the Bureau of Mines, can use huge hunks of rock; others can use only tiny pieces of uniform size. Some processes can tolerate a lot of fine material, such as the one invented by the Oil Shale Corporation, and others, such as the one invented by the Union Oil Company, cannot.

After crushing, the shale is fed either in batches or in a stream into vertical, stationary cookers, or rotating, horizontal ones.

The shale in the cooker is heated sometimes by an external source, such as an oven for heating aluminum-rich ceramic pellets that then join the crushed shale in a rotating horizontal retort, as in the Oil Shale Corporation's TOSCO-2 process, which has been tried by the Colony group of oil companies on a 1,000-ton-per-day scale at Parachute Creek in Colorado.

In other heating processes, such as the one tried at the same scale by the Union Oil Company on Parachute Creek in the nineteen-fifties, heat is provided by combustion within the cooker.

Major problems with these processes are to feed the shale smoothly into the retort, heat the shale evenly, keep things from getting too hot, conserve energy wherever possible, and produce a "spent," shale waste that can be disposed of with minimum alteration to the environment.

Waste Stays in Mountain

With underground processing, the waste says inside the mountain. With above-ground retorting, the waste must be wet down, terraced in some nearby valley, and then fertilized and planted to trees and grasses.

A technological ancestor of most ways of processing shale, above ground and below, is the so-called Nevada-Texas-Utah retort developed by the Bureau of Mines. The bureau currently runs a $2-million-a-year program on shale, concentrating now on underground processing.

Just north of Laramie, Wyo., the bureau still operates a 10-ton and a 150-ton retort to simulate what might happen underground, according to Dr. Gerald U. Dineen, who directs the bureau's research center there.

In addition, the bureau experiments near Rock Springs, Wyo., with fracturing rock underground, either with injections of water down through wells or with chemical explosives.

The bureau's shale researchers also cooperate with Lawrence Livermore Laboratory in Livermore, Calif., on studies—so far only on paper—of how deep-lying shale formations might be fractured with strings of nuclear explosives resembling those detonated in Colorado to stimulate production of natural gas from formations that grip the gas too tightly.

The underground work that the Garrett-Occidental operation is doing near DeBeque extends the Laramie studies to far bigger batches of rock, within a mountainside.

The process is made effectively continuous because each batch is so large that it is trapped for weeks on end, finishing just ahead or just behind the batches on either side.

Announcements of success with the Garrett process, starting last September, have been greeted with caution in the in-dustry. Many are skeptical because of what they consider splashy, high-pressure publicity used by Dr. Armand Hammer, head of Occidental Petroleum.

But there are technical reservations as well. The immediate reactions of a number of engineers, few of whom have actually visited the Garrett site and most of whom work for competing companies, focus on three things: the small scale of production achieved so far, potential contamination of ground water and possible limitations on the height (and thus productive capacity) of underground chimneys.

On the latter point, the engineers fear that in rich shale formations, melting kerogen could contaminate air passages in the crushed rock and prevent passage of cumbustion gases, in effect shutting off the chimney.

Proponents of the Garrett process maintain that the waterless formation at DeBeque, with its relatively "lean" admixture of kerogen, may be free of both the chimney height and water contamination problems. They note that there are numerous sites like the one at DeBeque.

And as for those who point to the small scale of production—Dr. Prien of the Denver Research Institute, for example, calls the Garrett process "embryonic" and says it has "to go through a few more years" —Dr. Donald Garrett, inventor of the process, waves them aside as "doubting Thomases."

The process, he says, has worked well so far, shows no evidence of limitations, and can be scaled up rapidly on private land, hindered neither by the lack of a Government lease nor of capital from the parent company.

Dr. Garrett said in a recent interview that a complete run of the 70-foot-tall first chimney had been "an absolutely resounding success."

The second chimney, designed to double the production rate of the first, is on the point of being set afire, he said, and the third, 120 feet tall, will not be far behind.

Work on the fourth chimney, some 120 feet across and at least 250 feet high, has begun, Dr. Garrett said. He called this "the development room" and referred to it as "the beginning of commercial operation."

$50-Million Pact Reported

Knowledgeable insiders said that Garrett Research and its parent, Occidental, had been negotiating a long-term, $50-million contract with a potential customer for the oil in order to support commercial use of the Garrett process. Until now, much of the support for the work has come from outside the company, including an undisclosed amount from Southwestern Public Service, a utility based in Amarillo, Tex.

The patented Garrett process differs from other underground shale processing concepts in that it concedes that a mine shaft must be dug into the formation from the side and that some shale rock must be mined out, at least temporarily, to ready things for cooking the shale within the mountain.

At the end of the shaft, a square room and an oil collecting sump are excavated. Then wells are drilled into the chimney of shale and explosives are placed in the wells and also next to any pillars supporting the roof of the room. The explosives are set off, fracturing the shale into fairly uniform pieces and collapsing it into the mined-out room.

The collapse of the fractured shale opens air passages so that burning gases may reach the shale to vaporize and break up the kerogen in the rock, and then allow the vapors to flow down to cooler rocks below and condense into oil for collection.

The shale is set on fire through an airshaft drilled from above the formation.

70 Per Cent Recovery Rate

Dr. Garrett said that although the shale formation at DeBeque was "groungey," or very low in kerogen content, the slow process, which takes several weeks to work its way through a chimney, ended by recovering 70 per cent of the hydrocarbons in a given chimney. He said this was a much higher rate of recovery than was possible with conventional "room and pillar" underground mining for shale to be processed on the surface.

On the 4,000-acre site, Dr. Garrett said, "we have a shot at about a billion barrels." He added: "If all has gone well, this property can support an eventual output of about 100,-000 barrels a day." It is possible that such a rate may be reached before conventional surface plants do so, he indicated.

John Savage, a consulting chemical engineer who lives in Rifle, Colo., and who is independent of the major shale interests, said of the apparent early success of the Garrett process: "It has sort of changed matters. You go in, use an area, leave it and go on instead of being stuck with an immobile $500-million plant on the surface."

Mr. Savage estimated that a square chimney, 400 feet tall and 206 feet on each side—that is, an acre in area—would produce 250,000 barrels in six weeks to two months.

This would imply a rate between 4,000 and 6,000 barrels daily, nearly as large as the anticipated flow from a commercial - scale above - ground plant of the sort planned by the Colony group with its so-called TOSCO-2 process.

Five to seven such chimneys producing simultaneously

would reach the Occidental goal of 30,000 barrels of daily output. Mr. Savage said: "You should have a series of rooms in every stage, and finish one a week."

Meanwhile, the burgeoning shale effort is emerging as a significant bargaining counter in the world energy economy. Prof. Hendrik Houtthaker of Harvard University told a Senate hearing last summer, for example, that a shale effort was worth supporting if only to help hold down the world price of petroleum.

In the short run, sharp increases in the price of petroleum by the Arab producing countries have made shale oil appear economic. In the long run, with the inevitable exhaustion of easily available petroleum, a shale industry would probably come to dominate the petroleum industry some day regardless of world oil prices.

What the Arabs seem to have done is speed up the process.

March 25, 1974

A-PLANT HEARINGS TERMED CHARADE

By EDWARD COWAN
Special to The New York Times

WASHINGTON, Mach 25—A study of citizen-group participation in Atomic Energy Commission hearings on nuclear power plants concludes that the "whole process as it now stands is nothing more than a charade."

The outcome, a victory for the utility sponsoring the nuclear generating station, "is for all intents and purposes predetermined," the study finds.

The authors are Steven Ebbin, a 41-year-old political scientist who is director of the Environmental Policy Study Group at George Washington University here, and Raphael Kasper, a 31-year-old nuclear engineer on the staff of the National Academy of Sciences. Their study was financed by the National Science Foundation.

The authors have presented their findings and recommendations in a 289-page book, "Citizen Groups and the Nuclear Power Controversy," published this month by The MIT Press at $6.95.

Their appraisal of the motives, methods and effectiveness of citizen groups that challenge nuclear power projects is regarded as the most comprehensive such inquiry made. In light of the electric power industry's plans to build scores of nuclear stations in the next 25 years, the book is seen as timely.

Lessons for Communities

Moreover, there may be lessons in it for coastal communities faced with the prospects of oil refineries, and inland communities designated as sites for plants to gassify or liquefy coal. The energy industry has said that it must build a number of installations in both categories in the next decade to keep abreast of the country's growing appetite for energy.

The Ebbin-Kasper book argues that "there is a common set of interests" shared by Atomic Energy Commission hearing and licensing boards, the agency's staff, the utilities and the reactor manufacturers, and that "effectively makes them allies."

The authors assert that the agency's atomic safety and licensing boards have "a predisposition" to issue construction permits and approve operating licenses.

However, the book commends the commission as having "distinguished itself among Federal regulatory agencies in its efforts to provide opportunities for citizen participation in its hearing process."

Nevertheless, the authors say, the agency's hearing process, essentially an adversary proceeding in which lawyers spar with technical witnesses and with one another, is a "failure" in virtually every respect.

Fears Not Allayed

It fails, they say, to give citizen groups "substantive due process," to "allay community fears," to resolve environmental issues or to test the merits of the opposition to a nuclear plant.

The hearings, they say, "are legal proceedings designed by lawyers for lawyers and are inppropriate to "arriving at scientific or technological truth."

Most citizen groups, the study finds, are loosely organized, inadequately financed by voluntary contributions and lacking a community mandate.

The utility industry and the reactor manufacturers, the study finds, are far superior in "scientific, technological and economic resources" and consequently are more likely to make a convincing case.

Moreover, the authors say, "citizens do not become involved in the nuclear power plant licensing process until quite late," because the applicants must first secure a string of preliminary approvals from the commission.

Officials believe that earlier citizen involvement would be possible under the commission's recent proposal for early designation by states of sites for future nuclear plants.

To correct what they find is an imbalance in favor of the industry, Drs. Ebbin and Kasper recommend that:

¶The public be informed of plans for nuclear facilities at least five years in advance of the start of construction.

¶"Legitimate citizen-group interventions" be financed by the utility, which would be assessed an application fee.

¶Public hearings "be made accessible to a broader public by requiring evening or weekend sessions to the extent possible."

¶Participants be required to summarize their cases "in language understandable to educated laymen."

In place of the adversary type of hearing, the authors would substitute appraisal by "independent assessment centers." These centers, to be financed by Congress and the states, would be "independent of any mission-oriented agency."

March 26, 1974

Energy Crisis: In One Town, They Think It's Over

By JOHN KIFNER
Special to The New York Times

PITTSFIELD, Mass.—Keith LaCasse is among the entrepreneurs who did not make a killing on the energy crisis.

Last fall, as fears that oil and gasoline would run out swept the country, he invested in two chain saws. Now he has a yard lined with neatly split firewood.

As suddenly and mysteriously as it appeared, the Great Energy Crisis seems, to people here, to have disappeared. What has remained is a marked increase in their bills.

So, as the first traces of green try to push through the remains of a late snowstorm in the Berkshire Hills, there is a sense, among many, of amusement and a certain cynicism.

There was a run on chain saws in the local hardware stores early in the winter, and on weekends their snarl could be heard in the wooded hills outside this city. Here, as in much of New England, firewood normally sells for around $40 a cord. By midwinter, the price had climbed to $65 or $70. In Boston, some were paying $90 or more.

Now, Mr. LaCasse allowed with a rueful smile as he contemplated the long stack of birch and maple he had cut and chopped, "if you were talking $35 I might think about it."

Ms. LaCasse, a 23-year-old student at Berkshire Community College, said that so many people had gone into the firewood business that it brought the price down. But, he added with a shrug, there was another factor: "With everybody being able to get all the oil they wanted . . ."

According to Leonard Lipton of the Lipton Oil Company, there was a "tremendous abundance" of fuel oil in the area. Mr. Lipton, whose company is one of the area's major suppliers, said that some dealers were turning back oil because their inventories were full.

But, he added, each month the cost of the oil he buys from British Petroleum has gone up.

Mr. Lipton attributes the lack of a shortage mostly to people's efforts to conserve fuel but also to the unusually mild winter—his charts show that it was 302 degree-days warmer last winter than the previous one, which was itself mild. A degree-day, a basic unit in the heating industry, is a day in which the average temperature falls one degree below 65 degrees Fahrenheit.

Last month, the gas stations in Pittsfield were either open a few hours early in the morning, or were shut completely. Along the roads outside the city and in the surrounding small towns, many gas stations—particularly independents such as Gas Land or Tulsa—were boarded over.

The gas stations are back in business. They are open

305

for regular hours, and there are even a few "24 hour" signs. Many have signs saying "no limit." The price, which fluctuates between stations, is running around 50 cents a gallon.

The Shoppers Speak

At Ron's Shell, near one of the main gates of the General Electric plant, cars lined up in the darkness last month and a crush began from 6 A.M. to 9 A.M. each day.

Now, although the station allotment has been cut back another 15 per cent this month, there is sometimes difficulty in selling the quota. The fear that drove people to turn their cars into "rolling storage" by continually topping off their tanks seems to have gone.

At the Adams Supermarket, in one of the shopping centers on Balton Road east of town the other day, shoppers voiced considerable skepticism about the alleged crisis.

"Half of it's been made up

in order to increase prices," said Mrs. Mary Zarn, who was grocery shopping with her husband, Philip. "When gas goes to 60 cents a gallon, there will be plenty."

She said that her heating oil bills had gone from 17.4 cents a gallon to 31 cents a gallon this year.

"I think it's been dreamed up," said Mrs. Shirley Resster. "When they get the prices up to where they want them, then we'll stop having a

crisis."

Emil Bonanate, who is retired and whose fuel bills have gone up each month, leaned on the handle of his shopping cart and said: "I don't think it's real—there's too much monkey business going on It stinks."

"What energy crisis?" asked another woman. "I told my husband I wasn't going to turn our thermostat down. So everybody believed them and froze. Ha!"

March 24, 1974

Two Test Projects to Seek Power From Ocean's Heat

By WALTER SULLIVAN

Groundwork is being laid for realistic testing of the hypothesis that substantial amounts of energy could be derived, at low cost and with no pollution, from temperature differences within the oceans.

Two conceptual designs for oceanic power plants of this type are in preparation on an academic level, and the National Science Foundation, which is financing these studies, is offering $1.8-million for further development, chiefly by industry.

It has been calculated that the heat being carried by the Gulf Stream through the Florida Straits between Miami and the Bahamas could be harnessed to produce all the electricity now used by the United States.

The proposed plants would use warm surface water to vaporize a "working fluid," such as propane or ammonia, that vaporizes at a temperature as low as that of tropical surface water. The vapor would drive power plant turbines and then be condensed back into a fluid by frigid water brought up from great depth.

The warm water and cold water would flow through the system in great volume, whereas a much smaller amount of working fluid would be constantly recycled through the turbines.

Of the conceptual designs the one being prepared by Dr. William E. Heronemus and his colleagues at the University of Massachusetts in Amherst is the most detailed. The plant would be moored some 25 miles off Miami, drawing up cold

water through a conduit attached to its tether line.

Warm water would be swept through the system by the natural flow of the Gulf Stream. The working fluid, at least initially, would be propane and the generated power would be transmitted to the Miami electrical system by submarine cable.

The other design was described at the New York Academy of Sciences, 2 East 63d Street, recently by Dr. Clarence Zener of Carnegie-Mellon University in Pittsburgh. In the 1930's Dr. Zener laid the theoretical basis for the zener diode, a basic component of modern electronics.

His group at Carnegie-Mellon has been using computer models of various schemes to assess their costs, net energy production, pumping demands and materials requirements. As Dr. Zener pointed out in an interview before his talk, a variety of subtle effects must be taken into account, such as the slow conduction of heat through smoothly flowing liquids.

Economic Question

The key problem, he added, is to determine whether the units (evaporators) that transfer heat from warm water to the working fluid and those (condensers) that transfer heat from that fluid to the cooling water can be built economically. Because the temperature difference between the two water streams is not great, the flow must be massive to produce useful energy.

Surface water throughout the tropics remains at about 77 degrees Fahrenheit and the deep water is at about 40 degrees. The necessary flow would be comparable to that through a hydroelectric plant producing the same amount of electricity.

One way to reduce the cost of the huge evaporators and condensers needed for such a plant would be to submerge each such unit at a depth where the water pressure equals the pressure within that unit, making it possible to use light construction. For a system using propane as the working fluid, according to the calculations of J. Hilbert Anderson, this would require the evaporator, drawing water from near the surface, to be deeper than the condenser, drawing water from the depths. The latter would be 154 feet down, whereas the evaporator would be 278 feet below the surface.

At $165 a Kilowatt

In 1966 Mr. Anderson and his father, who formed Sea Solar Power, Inc., in York, Pa., estimated that such a plant would cost $165 per kilowatt of generating power. Dr. Zener believes that, despite inflation, this is still valid in view of the savings in more recent designs.

By contrast, he says, a fossil fuel plant costs about $340 per kilowatt, but its delivery of power is far simpler than that for a plant out at sea.

Dr. Zener is cool to the idea of placing the initial plants in the Florida Straits. While it is said that such plants would not seriously affect the heat load of the Gulf Stream, the latter plays so critical a role in ameliorating European climate that any tampering with it would raise political problems.

The swiftness of the current there is also a challenge. While it would obviate the need to pump warm water into the system—a saving of perhaps 5 per cent in energy demands—it would place a heavy strain on the moorings. Dr. Heronemus himself estimates the tension as high as 22 million pounds, despite streamlining of his submerged units.

To achieve such streamlining the condensers and evaporators

would lie horizontally, which to Dr. Zener is a disadvantage. His own scheme provides for vertical tubes in the evaporator so that the bubbles of newly formed vapor can rise unimpeded. His plant would be placed in less swift-moving waters where warm seas flow into the Caribbean between the island chain of the Antilles.

From such remote sites the energy would have to be "packaged" for transport. The favored scheme is to use the generated electricity to separate water into its components —oxygen and hydrogen. If this were done at depth in the sea, the gases would already be compressed and could be shipped, in that state, via tanker.

Hydrogen is regarded as a potential fuel of great efficiency and there are a variety of demands for oxygen.

Two deadlines have been set by the National Science Foundation for more advanced proposals. The first, on May 7, concerns schemes for testing design concepts, subsystems and components. Initial tests would be ashore. Then "proof of concept" experiments would be carried out at sea and perhaps, initially, on an island or coastal site close to deep water. For the initial studies $300,000 is available.

The second deadline, July 9, relates to more specific problems such as the design of pumps, problems of corrosion and fouling by marine life, construction methods, anchoring, environmental effects, energy delivery (for example, in the form of hydrogen, compressed air or batteries) and by-products such as fresh water or shellfish. For this category $1.5 million is available.

According to Dr. Lloyd O. Herwig, director of advanced solar energy research and technology at the N.S.F., a number of large industrial concerns are showing an interest in the project and considerable support is expected from the Navy.

April 22, 1974

New Energy Agency To Allot Gasoline Approved by Nixon

WASHINGTON, May 7 (UPI)—President Nixon signed today legislation creating the Federal Energy Administration to replace the Federal Energy Office he established by Executive order last year to deal with the energy shortage.

The law signed by Nixon gives the agency a two-year existence. It would carry out any emergency energy rationing program and be responsible for allocating gasoline to the states.

The new law requires energy-producing companies to furnish information on their supplies and the way they arrive at pricing formulas.

Mr. Nixon said at a signing ceremony that the agency's first priority would be to work with other Government agencies to prepare a comprehensive plan to make the nation self-sufficient in energy by 1980. Warning of a need for "continued conservation," he said that, with the end of the immediate oil shortage, "many Americans believe that good conservation habits can be forgotten."

Treasury Secretary William E. Simon headed the original energy office when it was created last December. His former deputy, John C. Sawhill, succeeded him recently when Mr. Simon was named Treasury Secretary recently.

Despite the end of most price controls a week ago, the energy office retained price jurisdiction over energy products.

The American Automobile Association, meantime, reported that its latest survey of gasoline availability showed that supplies continued to improve slightly throughout the nation, while a predicted rise in prices had not yet taken place.

The association said that a check of 5,908 gasoline stations in all states but Alaska showed the average prices for regular and premium gasoline were unchanged this week for the fourth consecutive week, at 54 and 58 cents a gallon, respectively.

May 8, 1974

Nuclear Fusion Reported In Lab With Aid of Laser

By HAROLD M. SCHMECK Jr.
Special to The New York Times

WASHINGTON, May 13—A significant step toward the long-range goal of nuclear fusion as a source of almost limitless energy was announced today by a private research company.

The reported step was the achievement of nuclear fusion on a very small laboratory scale with the aid of intense beams of laser light.

The company involved, KMS Industries, Inc., of Ann Arbor, Mich., described the accomplishment as a "definite step." In answer to a query, a spokesman for the Atomic Energy Commission called it "a small, but significant initial step toward the achievement of laser fusion power."

Many physicists regard nuclear fusion as the long-range answer to world energy problems because, potentially, the fuel for fusion is almost as plentiful as the hydrogen in the water of the world's oceans.

Fission and Fusion

Unlike fission, in which the atomic nucleus is split to release energy, the basic fusion reaction requires that two nucleii of heavy hydrogen—called deuterium — fuse to make an atom of helium. Fission is the process that powered the original atom bomb. Fusion is the energy source for the much more powerful thermonuclear devices — hydrogen bombs.

For about a quarter of a century, scientists have been grappling with the huge problem of harnessing the hydrogen bomb reaction in the laboratory. The problem is to generate heat of over 100 million degrees to start such a reaction and, then, to maintain the reaction long enough and strongly enough to produce a surplus of energy. No one has achieved this yet and many more years of work may be required.

The accomplishment announced today was the release of high-energy atomic particles called neutrons in laboratory fusion experiments in which beams of intense laser light heated and compressed tiny pellets of deuterium.

A Fleeting Reaction

The announcement from the research concern's subsidiary, KMS Fusion, Inc., said their scientists had "obtained high energy neutrons unambiguously from a process of laser fusion, involving shock compression and heating of a deuterium pellet" The announcement, in other words, said that a laser-driven fusion reaction had been achieved, although only fleetingly and on a very small laboratory scale.

Experts say this has not been achieved before in the United States and perhaps not anywhere in the world. Soviet scientists have been working intensely in the same field and have reported release of neutrons in laser-fusion experiments. Among United States physicists there has been some uncertainty concerning the Soviet accomplishment.

In a letter to Dr. Dixy Lee Ray, chairman of the Atomic Energy Commission, Dr. Keeve M. Siegel, chairman of KMS Industries, said their first successful experiment was on May 1 and that it was repeated once on May 3 and twice on May 9.

In answer to queries to the A.E.C. today, Dr. Richard Schriever, chief of the laser fusion branch of the Office of Laser and Isotope Separation Technology, issued this statement:

"We have received a letter from KMS describing their recent laser fusion experiments. We have not received the detailed scientific data which would be required for us to make a thorough evaluation of their results. However, it appears that KMS has made a small, but significant, initial step toward the achievement of laser fusion power. We hope in the near future to receive more detailed information which will permit us to make a thorough evaluation of their work."

Broadly speaking, there are two main streams of fusion research directed toward harnessing the thermonuclear reaction for power. One of these uses the intense and highly organized beams of light from lasers as a source of heat, and tiny droplets of frozen deuterium about as thick as a human hair as the target.

The other realm of fusion research involves attempts to produce fusion in an intensely heated gas, called a plasma, confined by magnetic forces in a finite volume.

Both the laser fusion research and the efforts in the magnetic containment field appear to be making progress at about the same pace.

The Michigan company said in its announcement that it hoped for commercial application of its research by the early nineteen-eighties.

The concern's fusion research is being done under an A.E.C. contract in which no Government funds are provided. The company's chief scientific advisor in fusion is Dr. Robert Hofstadter of Sanford University, Nobel Prize winner in physics.

May 14, 1974

307

POWER REACTORS FACE SAFETY TEST

By DAVID BURNHAM
Special to The New York Times

WASHINGTON, Sept. 21 — The Atomic Energy Commission has ordered 21 of the 50 nuclear reactors producing commercial electric power in the United States to close down within the next 60 days to determine whether cracks are developing in the pipes of their cooling systems.

Meanwhile, a leading nuclear safety expert announced he was quitting his job with the commission "in order to be free to tell the American people about the potentially dangerous conditions in the nation's nuclear power plants."

Suggests Complete Halt

Carl J. Hocevar, author of one of the commission's basic methods of analyzing nuclear power plant safety, said in his letter of resignation to the commission chairman, Dixy Lee Ray, "In spite of the soothing reassurances that the A.E.C. gives to the uninformed, misled public, unresolved questions about nuclear power safety are so grave that the United States should consider a complete halt to nuclear

power plant construction while we see if these serious questions can, somehow, be resolved."

Mr. Hocevar, one of several safety research experts who have recently resigned from the commission's Idaho Safety Research Center, said in a statement that he planned to work with critics of nuclear reactors, such as the Union of Concerned Scientists in Massachusetts and Ralph Nader, the consumer advocate, to inform the public that the commission was using what he termed "wholly unacceptable" methods to judge the dangers of reactors.

The order to shut down power plants was relayed by telephone last Thursday to power companies in 15 states, including New York, New Jersey and Connecticut, after cracks were discovered in the pipes of three boiling water reactors within the last 10 days.

The commission ordered the inspection of the 21 reactors because the failures found in the quarter-inch-thick stainless steel walls of the pipes in the three plants raised the possibility of such problems in all the reactors.

Hazard Held Doubtful

Although the failures were not believed to pose a serious safety hazard, they could mean long shutdowns for the plants while the cooling systems were repaired.

Twenty of the reactors were

manufactured by the General Electric Company and one by the Allis-Chalmers Corporation.

A commission official said that he recalled three other mandatory shutdowns and inspections of similar magnitude for various kinds of technical problems with boiling water reactors.

About 6 to 7 per cent of the country's electric power is generated by atomic facilities.

The first indication of a potentially widespread problem in the reactor cooling systems came a week ago yesterday when an alarm sounded in the Dresden No. 2 reactor of the Commonwealth Edison power plant near Morris, Ill. The alarm indicated a leak of radioactive water that is used to cool the nuclear core of the reactor; in this case, 5 gallons a minute, according to the commission.

As a result of this leak, Commonwealth Edison examined similar piping in its Quad Cities No. 2 unit near Cordova, Ill., which at the time was shut down for routine maintenance. Using special equipment, the inspectors found cracks that had not yet penetrated the pipe wall.

Would Shut Down

Then last Tuesday, a crack that appeared to have been leaking was found in a cooling pipe of the Millstone No. 1 unit of the Northeast Nuclear Energy Company in Connecticut.

A spokesman for the commission said that if a serious leak suddenly developed in an operating sector the radioactive cooling water would be caught in special catch basins and the

reactor would be automatically shut down.

According to the commission, the 21 reactors that must be inspected produce about 14,000 of the 34,000 megawatts of electricity developed by atomic reactors in the United States. A megawatt is a million watts and refers to the amount of power that a facility can produce at any one time.

The commission estimated that each inspection would require a week. A spokesman for General Electric said that one day would be enough. Spokesmen for both the commission and the company also noted that the inspection order would require a special shutdown for only about half the 21 reactors because 10 of them either are now shut down for routine maintenance or are scheduled to be shut down during a 60-day inspection period.

Following is a list of the affected facilities:

Duane Arnold, Iowa Electric Light and Power.
Big Rock Point, Consumers Power of Michigan.
Browns Ferry No. 1 and 2, Tennessee Valley Authority, Alabama.
Cooper Station, Nebraska Public Power District.
Dresden No. 1, 2 and 3 and Quad Cities 1 and 2, Commonwealth Edison Company, Illinois.
Humboldt Bay, Pacific Gas and Electric, California.
La Cross, Dairyland Power Cooperative, Wisconsin.
Millstone No. 1, Northeast Nuclear Energy Company, Connecticut.
Monticello, Northern States Power Company, Minnesota.
Nine Mile Point Unit No. 1, Niagara Mohawk Power Corporation, New York.
Oyster Creek, Jersey Central Power and Light.
Peach Bottom No. 2 and 3, Philadelphia Electric Company.
Pilgrim No. 1, Boston Edison Company.
Vermonte Yankee, Vermont Yankee Nuclear Power Corporation.
Hatch No. 1, Georgia Power Company.

September 22, 1974

50% CUT PROPOSED IN ENERGY GROWTH

A Ford Foundation Report Sees No Economic Harm

By EDWARD COWAN
Special to The New York Times

WASHINGTON, Oct. 17—A major energy study sponsored by the Ford Foundation recommends a Government-led national commitment to conserve energy and argues that by such an undertaking the United States can put off for 10 years "massive new commitments" to offshore drilling, oil imports, nuclear power or development of coal and shale in arid areas of the West.

The final report of the foundation's Energy Policy Project,

released today, contends that the energy consumption growth rate can be cut by more than 50 per cent without hurting the national economy.

"Energy growth and economic growth can be uncoupled," said S. David Freeman, the project director, at a news conference.

Among the report's key recommendations for conservation are the following:

Higher energy prices, resulting from elimination of tax breaks for producers, such as the percentage depletion allowance, from enactment of pollution taxes and from charging consumers for "the costs of stockpiling oil" to protect the country against another embargo by foreign oil producers.

¶Federal assistance to low-income families, perhaps including "energy stamps" similar to food stamps or special payments or fuel allocations "to low-income persons who

demonstrate potential hardship as a result of shortages or price increases for energy."

¶Enactment by Congress of minimum gasoline economy standards for cars, with a goal of raising average fuel economy to 20 miles a gallon by 1985.

¶Federal loans to help householders and small businesses install insulation or energy-saving equipment.

¶Redesign of electricity rates "to eliminate promotional discounts and to reflect peak load costs."

Eventually, new energy sources will have to be tapped, the report says. But it asserts that a major conservative effort, including redirection of investment funds from energy supply to energy efficiency, would give the nation time to learn more about environment, health and economic risks of new supplies and make the wisest choice.

Views of Staff

The report, entitled "A Time

to Choose," is 343 pages long. In addition, 62 pages are devoted to the comments, critical and laudatory, of an advisory board. The report itself reflects solely the views of the project staff.

The volume is published by Ballinger at $3.95 in paperback and $10.95 in hard cover.

The report, which with associated research cost the Ford Foundation $4-million, is considered a major contribution to the emerging national debate about future energy needs.

Arguing that there is not much more time to debate and that it is now time to choose, the report says that the best thing to do at the outset is cut deeply into the country's appetite for energy, which has been exploding at a rate of 4.5 per cent a year for eight years.

"Slower energy growth," the report says, "can work without undermining our standard of living and can also exert a powerful positive influence on environmental and other problems closely interwined with energy," such as foreign relations.

The report was expected to add to pressures on the Ford Administration to adopt conservation actions stronger than the President's appeal last week for voluntary restraint.

The book published today went beyond a preliminary report issued in June that outlined three options but expressed no preferences. McGeorge Bundy, president of the Ford Foundation, said in a foreword to the book that he and the project's advisory board had changed their initial position and had decided to permit recommendations "precisely because this is a time to choose."

The report expressed skepticism about public interest in energy issues. "It remains to be seen," the introductory chapter said, "whether citizens, in the absence of shortages, will sustain their interest in energy."

Traditionally, with energy cheap and abundant in the United States, "most citizens were content to let industry make the major decisions," the report continued. It then added:

"But the fact is that the private interests of energy companies and the broader public interest do not always coincide. What is good for business is not always good for the rest of the country."

The study contends that a so-called technical-fix scenario, or package of conservation actions, would slice the energy growth rate to 2 per cent a year, by 1985, and that negligible growth thereafter could be achieved.

Electricity growth would be somewhat higher than 2 per cent, the report foresees, because "the nation is gradually moving towards a predominantly electric energy economy."

Nevertheless, the authors find, electricity growth would be about only half the customary rate of 7 per cent. Consequently, "power plants now on order for completion by 1980 could satisfy the demand for electricity until 1985," making possible "a pause of several years in new power plant starts."

Attack by Utilities

These findings were immediately attacked by the Edison Electric Institute, the New York-based association of privately owend electric utilities. W. Donham Crawford, president of the institute, said that such a pause would be "hazardous."

Mr. Crawford's four-page statement that "the report is replete" with the views of Mr. Freeman and is not "the truly fresh and independent study of our energy situation that could have been provided."

The report was also criticized by the American Petroleum Institute. Its president, Frank N. Ikard, said in a statement that following the course recommended by the report "would be like putting all our eggs in one basket — a basket that might have a hole in its bottom."

William P. Tavoulareas, president of the Mobil Oil Corporation and the only oil company executive on the Ford Foundation's Energy Policy Project, issued a statement yesterday taking issue with the Ford Foundation report.

Mr. Tavoulareas disputed the validity of the report's premises and conclusions and criticized the manner in which it was prepared, and, characterized the project's report "as a formula for perpetual economic stagnation."

Understanding Problems

Summarizing an extended discussion of safety problems associated with nuclear power plants, the report said:

"We do not advocate an absolute ban on new nuclear plants because the problems posed by using fossil fuels instead are also serious. But a conservation-oriented growth policy will provide breathing room so that we can gain a better understanding of nuclear power problems" before making "major new expansions."

Other findings included these:

¶"The United States energy-supplying industry does not constitute a monopoly by economic standards, although there are indications of diminished competition in some areas." The report recommended changes in offshore leasing policy to stop big companies from bidding jointly, and stronger antitrust enforcement.

¶The Interior Department has made available to private interests "vast amounts of the resource base" with "a grossly inadequate return to the public treasury."

¶Regulation of electric power rates by state regulatory commissions "is inadequate to cope with utilities that largely operate in regional power grids." Regional commissions "could assure that utility expansion plans were integrated into regional grids so as to meet regional needs with maximum efficiency."

¶Because energy makes up a large and growing part of poor people's budgets, "the social equity implications of high energy prices should be resolved by a national commitment to income redistribution measures, such as a guaranteed minimum income or negative income tax."

October 18, 1974

Clean Coal

Technology is now available to use this country's billions of tons of high-sulphur coal without harm to the environment. This announcement by the Environmental Protection Agency refutes the almost hysterical advertising campaign of the American Electric Power System, which would have the public believe that the only hope for adequate energy lies in strip-mining the low-sulphur coal of the West.

The technology known as the "'scrubber" is essentially a system for passing smokestack gas through sprays that remove most of its sulphur oxides and particulate matter before it is allowed to pass harmlessly out of the system. Use of the scrubber has long been decried by the utilities industry as unproved, unreliable and prohibitively expensive. But it is going to be hard to sustain that line of argument now that one of the industry's major units—the Philadelphia Electric Company—has decided to install scrubbers in three of its generating plants. Other utilities have already adopted the system—the number of scrubbers in use has gone up from 44 to 93 in the last eleven months—and the Philadelphia company has expressed confidence in them to the extent of investing $68 million.

That is not the kind of money that utility companies spend on tentative schemes for the purpose of pleasing environmentalists. It is a rebuttal, rather, to those operators who are pressing Congress for changes in the Clean Air Act that would allow them to improve slightly the air quality in their immediate neighborhood, at the expense of other neighborhoods, by dispersing their sulphur emissions over a wider area through the use of tall chimneys. Notable among these operators—and agitators—is the American Electric Power System, whose newspaper advertisements have been a running assault on scrubber technology. Misleadingly, these advertisements cite the report of a single state agency as unchallengeable proof that the scrubber is unacceptable, disregarding all evidence to the contrary.

Should the scrubber prove as effective as the E.P.A. and the Philadelphia Electric Company expect it to be, there should be a quick de-emphasizing of the need for Western coal. Most of the Appalachian coal, which without some cleaning technology might be unusable under the Clean Air Act, is taken from conventional deep mines, obviating the need to strip and ravage the countryside. It can be brought to the Eastern utilities, moreover, without the cost of long-haul transportation.

Often in the past, segments of industry have fought off improved technologies because of their immediate costs, only to find in the end that the improvements have been a boon to them as well as the country. The scrubber seems destined to be placed in this category.

October 24, 1974

A.E.C. Files Show Effort To Conceal Safety Perils

By DAVID BURNHAM
Special to The New York Times

WASHINGTON, Nov. 9 — Atomic Energy Commission documents show that for at least the last 10 years the commission has repeatedly sought to suppress studies by its own scientists that found nuclear reactors were more dangerous than officially acknowledged or that raised questions about reactor safety devices.

One key study, which the commission kept from the public for more than seven years, found that a major reactor accident —should one occur — could have effects equivalent to a "good-sized weapon," killing up to 45,000 persons, and that "the possible size of such a disaster might be equal to that of the state of Pennsylvania."

In addition, the documents show that the commission ignored recommendations from its own scientists for further research on key safety questions. And they show that on at least two important matters the commission consulted with the industry it was supposed to be regulating before deciding not to publish a study critical of its safety procedures.

Memos Back to 1964

Details of the commission's efforts to avoid publishing reports on the potential reactor hazards have emerged from an examination by The New York Times of hundreds of memos and letters written by commission and industry officials since 1964. Additional material was found in the record of an obscure commission hearing in 1972.

Some of the documents were originally leaked by A.E.C. officials to the Union of Concerned Scientists, a Boston-based research group that has questioned many commission policies. Others became available as a result of suits and threats of suits under the freedom of information law by such critics of the commission as David Dinsmore Comey of the Chicago-based group, Business and Professional People for the Public Interest.

In response to an inquiry about the commission's information policies, L. Manning Muntzine, director of regulation, said that "there is no agency as dedicated to opening up as the A.E.C." He conceded that there had been "bad examples" of secrecy in the past, but he said the beginning "three years ago we created a revolutionary openness — we may not be perfect, but we're a lot better."

Increasing concern about the inherent conflicts in the A.E.C.'s twin roles of regulating atomic power and promoting its use played a role this year in the Congressional decision to split the commission into two agencies—one to sponsor energy research and one to monitor the nuclear industry.

Questions Are Posed

But the documents, some of them written by staff members still in the Government's atomic energy bureaucracy, raise a number of continuing questions. Among them are these:

¶Just how safe are the millions of persons who live close to the aproximately 50 reactors now operating in the United States?

¶In its effort to deal with the sharp rise in world oil prices and the pollution probems of caol, should the United States Government continue to push for the construction of about 900 more reactors in the next 25 years?

¶Why did the Government agency responsible for protecting the public from the hazards of reactors try to suppress studies dealing with the potetntial dangers of these reactors?

The extent of th ealleged failure of the A.E.C. to do required safety research was commented on five years ago by D. H. Imhoff, head of development engineering for General Electric's nuclear division, in a letter to a top commission official.

"We believe future safety

"We believe future safety program plans should give more attention to possible future needs and that some funding should be available to resolve important safety issues before—rather than after-the-fact," Mr. Imhoff wrote.

In other words, Mr. Imhoff, then a top official in one of the two major reactor manufacturers in the world, was complaining that the commission should do safety research before rather than after building reactors.

Over and over again, the internal memos of the A.E.C. officials indicate that they were apparently more concerned about the possible public relations impact of safety studies than the actual safety of reactors.

In September, 1971, for example, Steven H. Hanauer, a top commission official, wrote to colleagues that a paper by A.E.C. experts questioning the commission's method of estimating the effectiveness of reactor safety systems had been "temporarily forestalled" but that further action dealing with the paper was required.

"The present goal should be a paper that can be published without hurting the A.E.C. and without inciting a cause célèbre for squelching a paper because of technical dissent," Dr. Hanauer wrote.

In January, 1972, the commission was forced by critics to hold a public hearing on the standards it had adopted for nuclear power plant cooling systems. These systems are supposed to prevent a massive release of radioactive material should the reactor's nuclear core overheat. One of the witnesses during the protracted hearings was Milton Shaw, the head of the agency's reactor development and technology division.

Mr. Shaw was asked if it was not a fact that his division had been "censoring" the monthly reports of the commission's safety laboratory in Idaho.

"Censoring?" Mr. Shaw replied. "If you want to use that terminology in the sense I think you are using it, yes."

On the next day of the hearing, J. Curtis Haire, the general manager of the Idaho laboratory's safety program, was asked why the Washington officials were "censoring, in your judgment, free and open discussions of Aerojet's views on nuclear safety."

"Well, I believe that R.D.T. is trying to avoid the problems or burden, if you will, of having to spend a lot of time answering public inquires that are addressed to Congress and referred to them."

"On nuclear safety?"

"On general questions of nuclear safety," Mr. Haire replied. Within a few months of his public testimony, Mr. Haire was relieved of his duties in the A.E.C.'s safety research program—as a result of his candor, many in the commission believe.

Even more recently, on April 17, 1973, a group of A.E.C. staff members met with representatives from six major power companies to discuss a policy paper the commission was considering on the proper location of reactors in relation to population centers.

"The consensus of the meeting," a report by the A.E.C. said, "was that the principal impact of the policy would be the potentially adverse reaction to any action which indicated that the safety of reactors was in question."

Study Not Published

Despite the urging of some senior A.E.C. officials, the commission apparently agreed with the concerns of the utility officials and the so-called reactor siting study was not published.

One year ago, an internal A.E.C. task force on the reactor licensing process completed a critical study of the commission's effort to provide safe reactors.

"The large number of reactor incidents, coupled with the fact that many of them had real safety significance, were generic in nature, and were not identified during the normal design fabrication, erection and preoperational testing phases, raises a serious question regarding the current review and inspection practices both on the part of the nuclear industry and the A.E.C.," the task force report concluded.

A copy of this report, completed in October of 1973, was given last January to the Union of Concerned Scientists, which in turn made the document available to the press. Following publication of the document, the A.E.C. put out an official version that modified or deleted many of the key conclusions of the original.

A finding that safety problems were "besieging reactors under construction and in operation" was entirely removed.

An extensively documented case in which the commission suppressed one of its own scientific studies concerns a $120,000 research project undertaken by the A.E.C.'s Brookhaven National Laboratory in 1964, updating a previous study done by the same group on the estimated damages of a major reactor accident.

The findings of the 1964 update, which Government officials came to refer to as the Wash-740 revision, were grim. In one memo written on Nov. 13, 1964, an A.E.C. official, Stanley A. Szawlewicz said: "The results of the hypothetical Brookhaven National Laboratory accident are more severe than those equivalent to a good-sized weapon and the correlation can readily be made by experts if the Brookhaven National Laboratory results are published. . . ."

Area For a 'Big Accident'

Several months later, the advisory committee reviewing the Wash-740 revision received a Jan. 6, 1965, memo from an A.E.C. official that said that "Mr. Smith has prepared isotope curves for given releases and meteorological conditions that show the areas involved: For a big accident the area would be the size of the State of Pennsylvania."

Mr. Sazwlewicz, who is still an atomic energy official, was aware of the possible impact the Brookhaven study might have on reactor construction. "The impact of publishing the revised Wash-740 report should be weighed before publication," he wrote to U. M. Staebler, another commission official, on Nov. 27, 1964.

A week later, on Dec. 4, Howard G. Hembree, now retired from the A.E.C., wrote a memo about Mr. Szawlewicz's view to those working on a rewrite of the Wash-740 revision.

"One concern that Szawlewicz expressed was that the reactor chosen by Brookhaven could generate an accident whose consequences could be projected downward to planned reactions, such as Nine-Mile Point and Oyster Creek, and that such projections could affect their building and site locations."

The Nine Mile Point reactor, which is situated 36 miles south of Oswego, N.Y., began generating commercial power in 1969. Oyster Creek, nine miles north of Toms River, N.J., also went commercial in 1969.

Just before Christmas 1964, Mr. Szawlewicz wrote another memo to Mr. Staebler saying that the review committee had agreed to submit copies of the draft report to the Atomic Industrial Forum after its next meeting. The forum is the major industrial lobbying organization of companies manufacturing reactors or otherwise involved in nuclear matters.

In the same memo, Mr. Szawlewicz said, "The results of the study must be revealed to the commission and the Joint Atomic Energy Committee without subterfuge, although the method of presentation to the public has not been resolved at this time."

Recently, in response to questions, Mr. Szawlewicz said he did not feel the commission had attempted to suppress the Wash-740 revision. "We just held up the report because we wanted to get more data," he said.

On March 17, 1965, C. K. Beck, the former assistant director of regulation, wrote a summary memo to the full commission, then headed by Dr. Glenn Seaborg.

Mr. Beck told the commission that it was an "inescapable calculation" that, given the hypothetical reactor accidents considered in the original Brookhaven National Laboratory study and the subsequent growth of reactors, "damages would possibly 10 times as large as those calculated in the previous study."

Problem Facing Commission

"The problem facing the commission, therefore, at this time, is the choice among the few alternate methods which might be selected for presenting the results of this newest Brookhaven study in 'proper perspective," Mr. Beck continued.

The official then told the commission that a special committee of the Atomic Industrial Forum—the industry's lobbying group—had twice met with the commission's review staff and that they "strongly urge" that "the revised Brookhaven report not be published in an yform at the present time" but that the study be extended for "another year or two."

The forum, Mr. Beck continued, recommended that the commission "at the present time simply report in a very brief letter to the Joint Atomic Energy Committee that if major accidents are assumed to occur without regard to the improbability of such events, very large damages, of course, would be calculated to happen. . . ."

The official added that a draft of the letter "along the lines discussed between the forum and the steering committee members has been prepared for discussion" of the commission.

On June 13, 1965, Dr. Seaborg sent such a letter to the joint committee and no public announcement was made about the Brookhaven findings. Eight years later, June 25, 1973, the commission responded to a threat of a freedom of informa-

tion suit by Mr. Comey, the nuclear critic in Chicago, and released selected parts of what it called "the final draft" of its report on nuclear reactor safety.

Despite all the statements to the contrary in the A.E.C. files, the commission press release said the Wash-740 revision done by Brookhaven "was never completed."

The press release summarized the Brookhaven study as finding that the possible damages of a reactor accident "would not be less and under some circumstances would be substantially more than the consequences reported in the early study."

On the third page of the press release, the commission said that in one extreme case examined by Brookhaven using "grossly unrealistic assumptions" it had been found that "45,000 fatalities could result from such an accident."

There is a sharp contrast between the conclusions of the original Wash-740 revision—made public about seven years later—and even the press release and many of the public statements by top commission officials.

On July 21, 1971, for example, Mr. Seaborg told a Washington audience that though there will be some failures, "I believe that just as has been the case in the past, these problems will only cause a temporary shutdown of the plant for the necessary repairs and corrective action and will not harm the public."

During a recent telephone interview, Dr. Seaborg denied that his 1971 statement conflicted with the Wash-740 revision but "in retrospect, I wish we had published it sooner."

The long-time head of the commission, now a professor at the University of California at Berkeley, explained that "we didn't want to publish it because we thought it would be misunderstood by the public.

"Even when the laboratories operated by the commission developed important reports raising questions about safety,

the commission staff in Washington sometimes ignored it.

On April 2, 1971, for example, the A.E.C.'s Idaho laboratory submitted a complex analysis of the computer methods then being used to estimate what would happen to a reactor if it lost its coolant.

"The analysis of a loss of coolant accident in a nuclear reactor is an extremely complex problem," a summary of the April report said. "The complete and correct analysis is beyond the scope of currently used techniques and in some areas beyond present scientific knowledge. Because of the complexity of the problem, simplifications are often made in the analyses and defended on the grounds that the simplifications make the predicted results 'conservative.' However, it is difficult to ascertain what is 'conservative' if the correct and complete answer is not available."

In A.E.C. jargon, a "conservative" judgement is one that leans toward overwhelming safety.

During the hearings on emergency core cooling standards almoste a year later, Mr. Hanauer, the A.E.C. official who had been concerned to avoid a cause célébre, was asked whether he had read the report in question.

"I leafed through it. I did not read it," Dr. Hanauer replied.

"And you dismissed it as not helpful?"

"It did not seem to help me any," he said.

According to A.E.C. internal documents, the report from the Idaho laboratory was intended to provide the technical support for an important statement on safety policy that the commission wanted to issue.

Instead, the report tended to undermine the proposed policy. The commission resolved the matter by publishing the policy statement without the technical backup that had been prepared.

November 10, 1974

A.E.C. HEAD DENIES SUPPRESSING DATA

Cites 'Openness' of Policy on Nuclear Plant Safety

By DAVID BURNHAM
Special to The New York Times

WASHINGTON, Nov. 15 — Dixy Lee Ray, the outgoing chairman of the Atomic Energy

Commission, said today it is "demonstrably not true" that that the A.E.C. suppresses information on the safety of nuclear plants.

Dr Ray added that "while there may be some validity for such accusations in the past, the situation has changed today."

Dr. Ray made the statements defending the A.E.C. in a 700-word statement issued by the commission in apparent response to an article in Sunday's

editions of The New York Times.

The article said commission documents showed that for at least the past ten years the agency has repeatedly sought to suppress studies by its own scientists which found that reactors are more dangerous than officially acknowldged or raised questions about reactor safety devices.

Dr. Ray said in her statement that "there has been an unprecedented effort, especially during the last two years, to

provide the public lwith ful documentation on all questions of nucear power plant safety."

Her statement concerning the recent improvement of the atomic agency was similar to the remarks quoted in The Times article by L. Manning Muntzing, the commission's director of regulation.

Conceding there had been what he called "bad examples" of secrecy in the past, Mr. Muntzing said that beginning about "three years ago we created a revolutionary openness," adding, "We may not be perfect, but we're a lot better."

Dr. Ray's statement today cited what she called "a few examples of the openness that characterized the A.E.C. today."

The chairman's first example was the release this year of some 25,000 pages of documents developed during the deliberations of the Advisory Committee on Reactor Safety

A second example cited by Dr. Ray was the release in 1973 of what she called "an uncompleted 1965 study" that sought to update a 1957 report on the hypothetical consequences of a major accident at a large nuclear power plant.

"Anyone who looks objec-

tively at the record would find it difficult indeed to conclude that the A.E.C. has been anything other than responsive to public concerns about the development of nuclear power," Dr. Ray concluded. "Public disclosure is a policy that we not only preach; we practice it every day."

Daniel Ford, an official with the Union of Concerned Scientists, a Boston-based group critical of many A.E.C. policies, did not agree that the examples cited by Dr. Ray proved the openness of the commission.

Mr. Ford said in an interview that the commission began releasing the 25,000 pages of doc-

uments from the Advisory Committee on Reactor Safety only after being threatened with a suit under the Freedom of Information Act. He further noted that many pages of the documents released so far had been heavily edited, with the opinions of members of the committee being deleted.

Concerning the second example cited by Dr. Ray, Mr. Ford said a large number of internal commission memorandums had shown that the 1965 study was deliberately not published by the commission for seven years after its completion.

He noted that this study also

was published under the threat of a Freedom of Information Ad suit and asserted that the press release that accompanied the study when released was "highly misleading."

The A.E.C. is being split into two agencies, the Energy Research and Development Administration and the Nuclear Regulatory Commission President Ford's nominations to head the new agencies are pending before the Senate.

Dr. Ray has been named an Assistant Secretary of State for Oceans, and International Environmental and Scientific Affairs.

November 16, 1974

As the West Sinks Slowly Into the Sun...

Steve Salameri

By Barry Commoner

When engineers want to understand the strength of a new material they apply stress to it to the breaking point and analyze how it responds. The energy crisis, a kind of "engineering test'" of our economic system, has revealed a number of deep-seated faults.

Although energy is useless until it produces goods or services, and although nearly all the energy that we use is derived from limited, nonrenewable sources that will eventually run

312

out (all of which pollute the environment), we have perversely reduced the efficiency with which fuels are converted into goods and services.

In the last thirty years, in agriculture, industry and transportation, those productive processes that use energy least efficiently have been growing most rapidly, driving their energy-efficient competitors off the market.

In agriculture, the older, energy-sparing methods of maintaining fertility by crop-rotation and manuring have been replaced by the intensive use of nitrogen fertilizers synthesized from natural gas.

In the same way, synthetic fibers, plastics and detergents, made from petroleum, have captured most of the markets once held by wood, cotton, wool and soap—all made from energy-sparing and renewable sources.

In transportation, railroads—by far the most energy-efficient means of moving people and freight—are crumbling, their traffic increasingly taken over by passenger cars, trucks, and airplanes that use far more fuel per passenger-mile or ton-mile.

What has gone wrong? Why has the postwar transformation of agriculture, industry and transportation set us on the suicidal course of consuming, ever more wastefully, nonrenewable sources of energy and destroying the very environment in which we must live?

The basic reason is one that every businessman well understands. It pays. Soap companies significantly increased their profits per pound of cleaner sold when they switched from soap to detergents; truck lines are more profitable than railroads; synthetic plastics and fabrics are more profitable than leather, cotton, wool or wood; nitrogen fertilizer is the corn farmer's most profit-yielding input, and as Henry Ford has said, "Minicars make mini-profits."

All this is the natural outcome of the terms that govern the entry of new enterprises into the economic system. Regardless of the initial motivation for a new productive enterprise—the entry of synthetics into the fabric market, of trucks into the freight mar-

ket—it will succeed only if it can yield a greater return on investment than the older competitor.

At times, this advantage may be expressed as a lower price for the new goods, an advantage that is likely to drive the competing ones off the market. At other times, the advantage may be translated into higher profits, enabling the new enterprise to expand faster than the older one, with the same end result.

However, the deepest fault that is revealed by the impact of the energy crisis on the economic system is not that we are running out of energy but out of capital. As oil wells have gone deeper, petroleum refineries have become more complex, and power plants have given up the reliability of coal- or oil-fired burners for the elaborate, shaky technology of the nuclear reactor, the capital cost of producing a unit of energy has sharply increased.

According to the National Petroleum Council, the total United States output is projected to increase from about 57,000 trillion B.T.U.'s in 1971 to about 92,000 trillion B.T.U.'s in 1985 —an increase of about 60 per cent. This would require annual capital expenditures for energy production to rise from about $26.5 billion to $158 billion over that period.

This trend coupled with the growing inefficiency in energy use means that energy production will consume an increasing amount of the total capital available in the United States for investment in new enterprises—in the factories that use energy and produce energy-using goods, not to speak of homes, schools and hospitals.

One projection based on present maximum estimates of energy-demand indicates that energy production could consume as much as 80 per cent of all available capital in 1985. This is, of course, an absurdly unrealistic situation in which the energy industry would, in effect, be devouring its own customers.

Thus, the compounded effects of a trend toward enterprises that ineffi-

ciently convert capital into energy production threaten to overrun the system's capacity to produce its most essential factor—capital.

This may well explain why, according to a recent New York Stock Exchange report, we are likely to be $650-billion short in needed capital in the next decade.

In a sense there is nothing new here, only the recognition that, in the United States economic system, decisions about what to produce and how to produce it are governed most powerfully by the expectation of enhanced profit.

What is new and profoundly unsettling is that the thousands of separate entrepreneurial decisions that have been made in the United States regarding new productive enterprises in the last thirty years have, with such alarming uniformity, favored those that are less efficient in their use of energy and capital and more damaging to the environment than their alternatives.

This is a serious challenge to the fundamental precept of private enterprise—that decisions made on the basis of the producer's economic self-interest are also the best way to meet social needs.

That is why the environmental crisis, the energy crisis, and the multitude of social issues to which they are linked suggest—certainly as an urgently-to-be-discussed hypothesis—that the operative fault, and therefore the locus of the remedy, lies in the design of our profit-oriented economic system.

Barry Commoner is director of the Center for the Biology of Natural Systems, Washington University, St. Louis, and chairman of the board of directors of the Scientists' Institute for Public Information.

November 20, 1974

Of Fuel Waste and Profits

To the Editor:

Barry Commoner makes a persuasive argument for the proposition that "we have perversely reduced the efficiency with which fuels are converted into goods and services" by replacing crop rotation with nitrogen fertilizers, natural materials with synthetic ones and railroads with cars, trucks and airplanes (Op-Ed, Nov. 20). He is far less persuasive, however, when he suggests that "the operative fault, and therefore the locus of the remedy," lies in a profit-oriented economy.

Suppose that, for the last fifty years, our economy had been managed by a benevolent agency totally unconcerned

with profits, interested only in public satisfaction and happiness. How would such an agency have acted? Surely it would have promoted the use of nitrogen fertilizers, because they make possible a greatly increased production of food. It would have encouraged the manufacture of synthetic materials, because of their highly desirable and adaptable use characteristics. It would have provided cars to large numbers of people because they are so convenient, and it would have replaced trains with airplanes because they are so much faster. But those are exactly the results achieved by the profit-oriented system, and of course that's no coincidence. Profits have been made from inefficient uses of fuels precisely

because the consuming public has preferred those uses to their alternatives.

Consumers might have made different choices, though, if more knowledge had been available about the consequences of their behavior and if that knowledge had been effectively conveyed to them. The fault lies with science and education, and that is where the remedies must be sought.

I don't mean to imply that there is anything sacrosanct about profits, but neither is there anything to be gained by striking at the wrong targets.
ROBERT A. FELDMESSER
Lambertville, N. J., Nov. 21, 1974

December 11, 1974

PRESIDENT VETOES STRIP MINING BILL, OIL TANKER PLAN

By JOHN HERBERS
Special to The New York Times

VAIL, Colo., Dec. 30—President Ford refused today to sign two highly controversial bills. One would have put stringent new restrictions on strip mining of coal and the other would have required that 20 per cent of the oil imported into the United States be carried on United States tankers.

On the surface mining control and reclamation bill, Mr. Ford said he was withholding his approval because˜ the measure would hurt domestic coal production "when the nation can ill afford significant losses from this critical energy source."

He said he could not approve the measure that would give preference to American ships because it would hurt the United States economy and its foreign relations and would create serious inflationary pressures by increasing the price of oil. He said it would also serve as a precedent for other countries to increase protection of their industries.

Congress Can't Override

Mr. Ford rejected both meas-ures by means of pocket vetoes. Under the Constitution, when Congress is not in session lack of Presidential signature on a bill facing a signing deadline means the measure is dead and there is no way that Congress can override his action. The pocket veto is usually a milder form of disapproval than the outright veto.

The White House had announced earlier that Mr. Ford would reject the strip mining bill but it was not known until today what he would do about the ship preference measure.

The controversy on that bill had been increased by the fact that some members of Congress who favored the United States maritime industry wanted to tie it to the foreign trade bill, which contains a provision sponsored by Senator Henry M. Jackson, Democrat of Washington, intended to ease the emigration of Soviet Jews.

Mr. Ford has strongly favored the trade bill, which is now on his desk for his signature. A deal was made, Administration officials confirmed, in which the President would not oppose the ship preference bill in Congress in order to get the trade bill.

No Deal on Veto

However, no compromise was made, both Congressional and Administration officials said, that the President would not veto the bill. It was promoted at the urging of the United States maritime industry, which already operates under heavy Federal subsidies and protection.

In a memorandum of disapproval issued by the White House, Mr. Ford said the measure would have had "the most serious consequences."

"It would have an adverse impact on the United States economy and on our foreign relations," he wrote. "It would create serious inflationary pressures by increasing the cost of oil and raising the prices of all products and services which depend on oil. It would further stimulate inflation in the ship construction industry."

"In addition," he continued, "the bill would serve as a precedent for other countries to increase protection of their industries, resulting in a serious deterioration in beneficial international competition and trade. This is directly contrary to the objectives of the trade bill which the Congress has just passed. In addition, it would violate a large number of our treaties of friendship, commerce and navigation."

When Mr. Ford was Republican minority leader in the House he was a strong supporter of the maritime industry. In his memorandum today, Mr. Ford said he wanted to "reiterate my commitment to maintaining a strong United States merchant marine. I believe we can and will do this under our existing statutes and programs."

Mr. Ford rejected the strip mining bill over the objections of both his Interior Secretary and chief energy adviser, Rogers C. B. Morton, and his director of the Environmental Protection Agency, Russell E. Train. Mr. Morton had predicted that unless the President signed this bill an even more restrictive measure would be passed by Congress next year.

In his memorandum of disapproval, Mr. Ford said, "By 1977, the first year after the act would take full effect, the Federal Energy Administration has estimated that coal production losses would range from a minimum of 48 million tons to a maximum of 141 million tons."

These figures have been disputed by a number of members of Congress and officials in the Department of the Interior, who said that any production loss, while difficult to estimate at all, would be considerably less.

The bill would have taxed the surface coal industry for reclaiming mined land and would have established new regulations and environmental studies.

"We are engaged in a major review of national energy polices," Mr. Ford wrote. "Unnecessary restrictions on coal production would limit our nation's freedom to adopt the best energy options."

The bill, he said, would increase United States dependence on expensive foreign oil.

Since he became President on Aug. 9, Mr. Ford has vetoed 20 bills, and three of his vetoes have been overridden by Congress.

December 31, 1974

Energy Plan to Raise Consumer Prices

By EDWARD COWAN
Special to The New York Times

WASHINGTON, Jan. 15—President Ford submitted to Congress today a many-sided energy independence plan that would mean higher energy prices for all consumers.

Officials estimated that decontrol of all crude oil prices and new taxes proposed by Mr. Ford would push up the cost of petroleum products by an average of 10 cents a gallon. Gasoline might go up more, heating oil less. Electricity rates would rise by an average of 15 per cent, the officials said.

Mr. Ford has said that he would take the first price-raising action Feb. 1 by putting a fee of $1 a barrel on imported crude oil, with the fee rising to $2 March 1 and $3 April 1.

By that date Mr. Ford wants Congress to put a tax of $2 a barrel on domestic crude oil and a comparable tax on natural gas.

Additionally, in his State of the Union Message to Congress today, the President said he would ask Congress "to authorize and require tariffs, import quotas or price floors to protect our energy prices at levels which will achieve energy independence."

The other main points of the President's energy plan—some of which were made known by him and other officials in the last two days—are as follows:

¶Legislation to authorize production from the naval petroleum reserves at Elk Hills, Calif., and in Alaska.

¶Administrative action by the Interior Department to compel holders of coal leases on public lands who have not been mining them to do so, and to write production requirements into new coal leases. There has been much criticism of the department because existing leases let companies hold coal back for higher prices. Some officials, however, fear that a retroactive condition on existing leases could be challenged successfully in the courts.

¶An amendment to the Clean Air Act to block any tightening of auto emission standards for the model years 1977-81. Frank G. Zarb, the Federal energy administrator, said that the Big Three auto makers had made "a deal," in Mr. Zarb's words, with the Administration that in exchange for such legislation they would increase auto gasoline efficiency by 40 per cent.

¶Other amendments to the Clean Air Act to delay compliance with state air-quality standards more stringent than the primary Federal standards, which are designed to protect public health. Mr. Zarb stressed that he had reached agreement on this with Russell E. Train, administrator of the Environmental Protection Agency. "I'm sure he will tell you," Mr. Zarb said, that the Administration's proposals "do not sacrifice environmental goals."

¶Expanded leasing of prospective oil and gas acreage on the ocean floor. The White House estimated that offshore fields could add 1.5 million barrels a day of crude oil, plus natural gas, to domestic energy supplies by 1985.

¶Enactment of a surface mining act like the one Mr. Ford vetoed a few weeks ago, but with removal of environmental standards he regarded as too stringent.

314

¶Legislation to assist utilities financially, by a high investment tax credit, tax deductions for preferred stock dividends. rate increases for construction in progress and other changes in state utility regulatory policies and practices. The states are considered certain to object.

The President held out to the Federal lawmakers the prospect that enactment of the plan would move the United States toward being an important exporter of energy and energy technology by the end of this century, a development that could assure the United States of continued status as an economic and political superpower.

President Ford devoted a large part of his message to Congress to energy issues — measures aimed at diminishing United States energy consumption and oil imports and measures seeking to foster development of indigenous energy resources and technologies.

In the latter category is the proposal on tariffs; import quotas and price floors. This proposal to protect domestic energy producers from cut-price foreign competition, should it develop, is intended to encourage private investment in new oilfields, shale oil, coal and other ventures.

The proposal is believed certain to meet stiff resistance in Congress, as are other elements of the Ford program.

Many Democrats oppose Mr. Ford's proposals to end controls on crude oil prices, put taxes of $2 a barrel on all crude oil and 37 cents a thousand cubic feet on natural gas prices, and end Federal regulation of wellhead prices of new natural gas supplies.

In an initial Democratic move to oppose the President, Senator Edward M. Kennedy and Representative Thomas P. O'Neill, the House majority leader, introduced a joint resolution to prevent Mr. Ford from acting on his announced intention to impose by executive action "fees" of $3 a barrel on imported crude oil and $1.20 a barrel on imported petroleum products.

A Curb on Ford's Power

The resolution of the two Massachusetts Democrats, if passed by both houses and signed by the President, would effectively suspend the President's authority under the national security provision of the 1962 Trade Expansion Act to impose import fees on petroleum.

Senator Henry M. Jackson, Democrat of Washington, a potential Presidential aspirant, is expected to argue at a news conference tomorrow that rather than discourage consumption by higher taxes and higher prices, the Government should curtail imports by a quota and use allocation or rationing to manage the resulting shortage.

Mr. Zarb, the Federal energy administrator, who has played a key role in shaping the energy package, said the Administration preferred the tax-and-price approach because it would better contribute to restoring to the United States "material influence in petroleum price markets."

In a similar vein, the President depicted the United States as having been a balancing force in world energy markets in the nineteen-sixties, when "this country had a surplus capacity of crude oil."

"Our excess capacity neutralized any effort at establishing an effective cartel, and thus the rest of the world was assured of adequate supplies of oil at a reasonable price," he said.

A high-ranking official, who asked not to be identified, said that Mr. Ford's reference to making the United States capable of supplying "a significant share of the energy needs of the free world by the end of this century" was part of what the official called "a grand design."

"The only way to get price stability is to have an alternative to Persian Gulf oil," the official said. "The alternative to that oil is United States technology. It is a grand design." The official added that Mr. Ford had been alluding to possible exploration of both energy, and new energy technology, such as floating nuclear power plants.

This "grand design" bore the imprint of Secretary of State Kissinger, who talked of exporting nuclear technology at the 12-nation energy conference here a year ago.

Under the heading of "Energy Conservation Actions" the Ford plan, as elaborated in a White House "fact sheet," contemplated improvement in auto efficiency, legislation to set national building standards, a tax credit for installation of home insulation, cash grants to pay for insulation for the poor, legislation to require that appliances bear energy efficiency labels and administrative action to get appliance manufacturers to improve efficiency.

Under the heading of "Emergency Preparedness," Mr. Ford said that he would propose legislation to authorize a strategic reserve of 1.3 billion barrels of oil, including 300 million earmarked for the military; standby rationing and allocation authority, and authority to compel increases in domestic oil production and the establishment of petroleum product inventories at levels set by the Government.

Mr. Zarb said that the Administration wanted to let the law authorizing price control and allocation authority, the Emergency Petroleum Allocation Act of November 1973, expire as scheduled on Aug. 31, 1975. The proposed legislation would confer new rationing and allocation authority. The Administration wants no price control authority.

Mr. Zarb also disclosed that he and William J. Casey, president of the Export-Import Bank, a Government agency, had agreed to curtail export financing for drilling rigs so that more rigs might be kept in this country to hasten exploration. Mr. Zarb said that he and Mr. Casey would review each application individually and decide "case by case."

January 16, 1975

Experts Skeptical on Ford's Energy Plan

By VICTOR K. McELHENY

President Ford's program to free the United States by 1985 from the threat of an oil embargo has aroused serious doubts among leading energy experts as to its costs, relevance and achievability.

The Administration plan, envisioning the construction of scores of oil refineries, synthetic coal plants and nuclear power plants, relies on such devices as import tariffs and tax incentives both to reduce energy consumption and develop new energy sources.

In the month and a half since it was first outlined, the Ford program has been bitterly criticized by many members of Congress and Governors as placing too heavy an economic burden on low-income groups.

The technical critics take a different stance, however. They hold that the program puts too much emphasis on purely economic incentives, then fails to "plow back" such revenues as it generates into energy investment.

They say that current discussions overemphasize conservation and fail to emphasize the urgency of getting new energy supplies quickly, particularly because of new, pessimistic estimates of oil and gas supplies.

They say that the nation's need for new energy supplies in the next 10 years is being grossly understated and that its ability to achieve conservation without severe, short-term dislocations is being grossly overstated.

And they say that White House planners are using excessively optimistic estimates of the nation's reserves of oil and gas and that, because of a still-ignored natural gas crisis, the nation needs immediate, crash efforts to gasify coal and to increase direct industrial use of low-sulphur coal.

In interviews and public meetings since the President announced his program, these scientists and engineers have criticized it as a "business-as-usual" effort to manage a problem that threatens both short-term and long-term economic decline.

The specialists suspect that the Ford program, despite its economic orientation, does not provide for enough money, material or manpower to achieve the huge increases in production of World War II cited by Mr. Ford when he said, "They did it then. We can do it now."

The huge construction effort called for by President Ford would require investment of more than $400-billion. It would involve between 140 million and 160 million tons of steel, fabricated into hundreds of large steam turbine generators for electricity and draglines for the surface mining of coal, thousands of coal-cutting machines for underground mining, dozens of tankers for Alaskan oil and gas, between 8,000 and 10,000 locomotives and up to 260,000 hopper cars for coal trains, and thousands of miles of well-linings and refinery tubing and pipelines for oil and gas.

The size and speed of the effort has excited much skepticism. As John O'Leary of the Mitre Corporation, a former director of the Bureau of Mines, said in New York in January, "I think there's a very real chance that we won't take the appropriate action even in a crisis environment. The appropriate action is very painful and entails quite long lead times."

NATURAL GAS

The President proposed that prices on new supplies be decontrolled, that an excise tax be imposed, and that new exploration be encouraged.

315

The experts expressed particular urgency about natural gas. The supply is dwindling, particularly for interstate shipments, and in their view attention to the immediate crisis is being diverted by Mr. Ford's focus on a date 10 years in the future.

The decline in the nation's over-all gas supplies is estimated at about 5 per cent a year, with interstate shipments going down at twice that rate. Because of this, according to Dr. Philip H. Abelson, editor of Science magazine, the East and West Coasts and Middle West all face sharply increasing difficulty in allocating gas between homes and industry.

According to an estimate by the Columbia Gas Company, printed in the Nov. 4, 1974, issue of Oil and Gas Journal, if the current practice of giving top priority for gas supplies to home heating persists, the nation could virtually run out of all industrial gas by 1980.

At the annual meeting of the American Association for the Advancement of Science last month in New York City, Dr. Henry Linden, director of the Institute of Gas Technology in Chicago, said the decline in gas supplies, amounting to nearly 1 trillion cubic feet annually out of a total of 23 trillion, is adding 500,000 barrels of oil a day to the nation's oil demand.

To moderate the effects of the gas crisis, observers like Drs. Abelson and Linden urge a massive, immediate drive to use existing technology to make synthetic gas from strip-mined Western coal, along with rapid conversion of many industrial furnaces and boilers to coal.

OIL

The President proposed tariff increases to achieve reductions in imports of 1 million barrels of oil daily by the end of this year and a further million by the end of 1977; civilian and military stockpiles to reduce vulnerability to an embargo; decontrol of domestic crude oil prices; a "windfall" profits tax; the construction of 30 new refineries, and the drilling of thousands of offshore and Alaskan wells.

The energy experts, while regarding President Ford's program as more realistic than former President Nixon's call for an end to all oil imports by 1980, doubt that even the scaled - down aims can be achieved through the operations of the market place, the main reliance of the Administration's program.

They think that the United States reserves of oil and gas are not only much lower than the official estimates of the Geological Survey but that such reserves as do exist will take longer than many realize—five to eight years—to bring into production.

Recently, the National Academy of Sciences endorsed such a pessimistic view of the resources, a view close to that frequently expressed by M. King Hubbert, an employe of the Geological Survey.

By Mr. Hubbert's projections, the total United States resource is about 170 billion barrels, compared to the Geological Survey projection of 600 billion. Of the 170 billion, 100 have already been found. Of the remaining 70, about 65 might be used by 1985 if Federal projections are correct.

At the University of Washington, Dr. Abelson said, "Almost all the undiscovered recoverable oil is located in regions where exploitation will be controversial, difficult and slow."

Dr. Abelson was skeptical that higher prices would bring out more oil. During 1974, he said, "in spite of a doubling of prices, production has dropped about 5 per cent. Perhaps in time the higher prices will bring out more oil, but for the moment the credibility of those who have argued for higher prices is suspect."

Of efforts to increase the proportion of oil and gas recovered from a well by pumping water or detergent into a subterranean petroleum formation, Dr. Abelson said, "Thus far, there has been no magical solution, only slow progress. The winning of additional oil through such methods will be slow and costly."

As for Government estimates that $11 a barrel represented a long-term upper limit on oil prices, both Drs. Abelson and Linden were skeptical. In a session at last month's science association meetings Dr. Linden said the oil-exporting nations were selling their petroleum at "substantially below replacement cost," that is, what synthetic oil from coal or shale or tar sands would cost. "In fact, the O.P.E.C. nations have been very kind to us," Dr. Linden said.

At the University of Washington, Dr. Abelson said that experience with spot prices during the embargo of 1973-74 showed that "a price of $20 or more a barrel as an upper limit would be more realistic" than the range between $7 and $11 used in Government estimates.

COAL

The President proposed opening 20 new synthetic fuel plants, most of them for turning coal into synthetic pipeline gas, achieving a total of 150 major coal-fired electric power plants, and some flexibility on standards for strip-mining and cleaning sulphur from smokestack gases.

The Ford proposal would require doubling United States coal production in less than 10 years. The key to reaching this

target is agreement on environmental rules, the experts assert. They doubt that the nation's political leaders are ready to agree on them. Without the rules, it is said, there can be no investment in the new energy supplies.

Dr. Albert Osborn, a former director of the Bureau of Mines, told the science association meeting that the coal industry was eager for passage of any strip-mining bill, regardless of its exact requirements for land reclamation, so that it could open up vast new Western resources. "The coal industry cannot move ahead unless it knows what the rules are," Dr. Osborn said. He now works at the Carnegie Institution of Washington.

The limestone "scrubbers," which the Environmental Protection Agency is requiring for coal-fired plants, are "an outrage to the system," Dr. Osborn said. "This is not the way to go."

The sulphurous sludge byproduct of the scrubbers, which are to be installed on some 86 million kilowatts of coal-fired power plants expected by 1980, would cover 50,000 acres to a depth of 20 feet, Dr. Osborn said.

Eric Reichl of the Conoco Coal Development Company said the limestone stack-gas scrubbers were "technically out in left field. We need something far more sophisticated to make it practical."

Both Mr. Reichl and Dr. Osborn urged development of a so-called citrate process of stack-gas cleaning that would lead to pure sulphur rather than sulphate sludge as the byproduct. But Mr. Reichl warned, "We don't want to confuse or fool ourselves. The cost of doing this is just monumental."

To encourage development of synthetic fuels from coal, Mr. Reichl said, the Government might need to build and own the factories for a time, then sell them off to private industry, as was done with synthetic rubber and aviation fuel factories after World War II.

Mr. Reichl strongly urged that the United States not wait for a "second generation" of coal gasification plants in hopes of "something better around the corner" that would cut costs substantially. The cost reductions from new processes now being developed in the United States would amount to 10 to 15 per cent, Mr. Reichl predicted.

The Federal Energy Administration's "Project Independence Blueprint" took a pessimistic view of the role that synthetic gas from coal might play before 1985. Because of this, the document urged heavy stress on supplying electricity to so-called "heat pumps" in homes to ease demand for home heating oil or natural gas—both now expensive and scarce.

Dr. Linden attacked this analysis, countering with one of his own. An Institute of Gas Technology study last year indicated that home-heat from electricity in 1980 would cost $11.52 per million British Thermal Units, compared with only $5.85 from gasified coal.

NUCLEAR POWER

The President proposed that a total of 200 nuclear power plants be built by 1985 (against 50 today), that investment tax credits be extended, and that guidelines be imposed on state utility regulations affecting profitability and hence investor interest in new power plants.

To meet President Ford's goal, new nuclear power plants would have to be commissioned about once every three weeks for the next 10 years. Several commentators have said in recent weeks that 100, or one every five weeks, would be a more realistic target. They cited the same delays and cancellations—because of acute financial difficulties—that President Ford mentioned in his address Jan. 15.

Milton Levenson, in charge of nuclear industry problems at the Electric Power Research Institute in California, said in a telephone interview, "It's one thing to say 200 plants, and another to build them."

Mr. Levenson added, "The utility industry is in no position to build 100 plants at $600-million each. The Government has to make available the capital through some device."

Many commentators on the nuclear industry have expressed doubt that there will be time, talent or money enough to build up the nuclear industry. Mining activity will have to more than triple in a decade. New factories for making and reprocessing nuclear fuel will have to be built. The pace of development of the "breeder," a type of nuclear plant vital to the long-term nuclear economy, is in doubt.

CONSERVATION

The President proposed a general cutback in the nation's energy growth rate and a shifting away from oil and gas. He would defer tightened automobile pollution standards for five years in favor of a 40 per cent improvement in new-car fuel efficiency by 1980. And to encourage electric heating of homes, he would provide a tax credit for installing insulation, nationally mandated heating standards for new buildings, and help for low-income families in buying insulation.

In a news conference in New York late in January, Dr Chauncey Starr, director of the Electric Power Research Institute, estimated that 10 to 15 per cent was all the "cream" that could be skimmed off the economy by energy conservation. He said, "Conservation buys, at best, a few years. It does not do more than delay the onset of the problem."

Dr. Starr was one of many energy specialists concerned that energy-conservation steps might not only seriously erode the real income of Americans in the next 10 years, but actually increase the amount of energy needed per unit of economic

output instead of decreasing it.

In this view, there could be a long-term recession as the nation sought a rapid reduction in its "classical" 4 per cent annual growth rate in energy-use.

In testimony last Dec. 10 before the President's Energy Resources Council, Dr. Linden said of energy-saving measures, "Conservation is much too benign a label for this public poli-

cy option in the near term. Rather, it should be presented to the public as what it likely will be for the foreseeable future: a policy of lower real income, reduced standard of living, reduced mobility, and possibly, higher unemployment."

True conservation, Dr. Linden said, would mean that less energy would be needed than today to produce a given amount of economic output. In today's

economy, with some 78 per cent of the energy coming from oil and gas, the United States needs 90,000 British Thermal Units, somewhat less than the energy in a gallon of gasoline, for each constant 1958 dollar of gross national product. That dollar would now be worth about $1.50.

March 1, 1975

Scientists Say Nation Badly Needs Atom Plants

Special to The New York Times

WASHINGTON, Jan. 16 — Thirty-four prominent scientists, 11 of them nuclear physicists, released today a joint statement saying there was no reasonable alternative to the increased use of nuclear power to satisfy the United States' energy needs.

"The U.S. choice is not coal or uranium; we need both," said the statement made public at a news conference here at which three of the scientists spoke.

"Today's energy crisis is not a matter of just a few years, but of decades," said the state-

ment read by Dr. Hans A. Bethe, its chief author. "It is the new and predominant fact of life in industrialized societies."

Dr. Bethe, a Nobel Prize winner in physics, is a professor of theoretical physics at Cornell University.

Dr. Frederick Seitz, the president of Rockefeller University, and Dr. Richard Wilson, a professor of physics at Harvard University, also spoke.

The scientists' statement said that they believed that the United States was in the most serious situation it has faced since World War II. They said

that they "deplore the fact that the public is given unrealistic assurances that there are easy solutions."

Defend Nuclear Power

The three scientists agreed, at a news conference that research on such alternative energy sources as solar, geothermal and wind power, must be done, but they expressed doubts that any of these could make a great impact on the energy situation during the remainder of this century.

Much of the discussion during a question and answer period was devoted to questions

of environmental impact and safety. Dr. Bethe and his colleagues said that all forms of energy production involved problems in these areas. The three scientists defended nuclear power as being no worse, and in some respects better, than alternative energy sources in those respects.

Spokesmen for Ralph Nader's Union of Concerned Scientists took issue with the statements of the three speakers.

"We believe that a wide range of public safety problems must be resolved before nuclear power plant construction proceeds in the U. S." said a statement released by the union.

January 17, 1975

Defect in a Reactor Leads U.S. to Order 23-Plant Shutdown

By DAVID BURNHAM
Special to The New York Times

WASHINGTON, Jan. 29—The discovery of cracks in the emergency cooling system of an atomic reactor in Illinois has forced the Government to order utilities operating half of the nation's reactors to shut down within the next 20 days and search for similar possible faults in their power plants.

The order marked the second time in four months that most of the same utilities had been ordered to inspect for possible cracks.

The 23 reactors involved in today's action generate 14,283 megawatts of electricity, somewhat less than half of the 35,000 megawatts produced by the 52 reactors now licensed to operate in the United States. The total power created by all reactors equals about 7.5 per cent of the nation's electrical demands.

The order to inspect the 10-inch-diameter pipes with half-inch walls was given to the utilities by the Nuclear Regu-

latory Commission, the recently renamed part of the now abolished Atomic Energy Commission responsible for licensing the commercial use of atomic power.

The regulatory commission took the action after five small cracks had been found in a stainless steel pipe of the Dresden Unit-2 reactor operated by the Commonwealth Edison Company near Morris, Ill.

The operators of the 23 plants will have 20 days in which to shut down and inspect the pipes in question as well as other specific pipes in the primary and backup cooling systems.

Two of the reactors undergoing inspections are in New York, one in Connecticut and one in New Jersey. A spokesman for the General Electric company, which manufactured all but one of the boiling water reactors, said that the company would cooperate fully with the Government and the utilities.

The commission estimated that each plant would be shut down for about two weeks to make the required inspections.

The pipe that developed the five small cracks in the Dresden plant was part of the emergency core cooling system. This system is designed to flood the

reactor core with an emergency supply of water should the primary cooling system fail.

Last September, the commission ordered the utilities operating 23 reactors to inspect a separate group of pipes when a four-inch diameter pipe—also at the Dresden 2 reactor—was found to have developed a leak.

Other Cracks Found

Since the discovery of the 5-gallon-per minute leak of radiated water in the Dresden 2 plant, subsequent inspections have discovered cracks or preliminary indications of cracks in the four-inch pipes of a total of seven other boiling water reactors.

The experts are examining a number of possible explanations for the cracks including the geometry of the by-pass pipes, the materials used in the by-pass pipes, the welding procedures and the chemistry of the water passing through the pipes.

The N.R.C. said that the latest leak had not resulted in the release of any radioactivity to the environment.

David Comey, a nuclear critic with the Chicago-based nonprofit group, Business and Professional People for the Public Interest, said he felt that the "new cracking problem may be the most ominous event yet in the atomic reactor program."

"If more than a third of the nation's reactors have to be shut down, it indicates the re-

liability problem may be increasing," Mr. Comey said.

Following is a list of the reactors ordered to make the inspections, the states where they are located and the utilities which operate them.

Big Rock Point unit 1, Michigan, Consumers Power Company.
Browns Ferry units 1 and 2, Alabama, TVA.
Brunswick, unit 2, North Carolina, Carolina Power & Light Company.
Cooper, Nebraska, Nebraska Public Power District.
Dresden units 1, 2 and 3, Illinois, Commonwealth Edison Company.
Duane Arnold, Iowa, Iowa Electric Light & Power Company.
Fitzpatrick, New York, Niagara Mohawk Power Corporation.
Hatch Unit 1, Georgia, Georgia Power Company.
Lacrosse, Wisconsin, Wisconsin Dairyland Power Corporation.
Humbolt Bay, California, Pacific Gas & Electric Company.
Millstone unit 1, Connecticut Northeast Nuclear Energy Company.
Monticello, Minnesota, Northern States Power Company.
Nine Mile Point Unit 1, New York, Niagara Mohawk Power Corporation.
Peach Bottom Units 2 and 3, Pennsylvania, Philadelphia Electric Company.
Pilgrim, Unit 1, Massachusetts, Boston Edison Company.
Oyster Creek, New Jersey, Jersey Central Power & Light.
Quad Cities, Units 1 and 2, Illinois, Commonwealth Edison.
Vermont Yankee, Vermont, Vermont Yankee Nuclear Power Company.

January 30, 1975

EXPERTS QUESTION POLLUTANT LEVELS

By JANE E. BRODY

A panel of scientists seriously questioned yesterday the adequacy of both the scientific and the economic bases for regulatory decisions on allowable levels of environmental pollutants.

The question of where to set such levels has become especially pertinent in recent months as the nation's energy crisis deepened, leading to calls for the lifting of air pollution control measures and posing the prospect of an increasing dependence on coal and nuclear power, both of which produce pollutants known to be hazardous to human health.

The question the scientists addressed was, "Are there thresholds in the effects of pollutants on health?"—that is, are there levels below which a given pollutant is not damaging or toxic.

The answer, according to the scientists who spoke to the annual meeting of the American Association for the Advancement of Science, at the Americana Hotel here, is that current knowledge indicates that there is no known safe level of any harmful substance.

With radioactivity, for example, said Dr. George A. Sacher, a biologist at Argonne National Laboratory in Illinois, "the smallest amount of radiation produced is still capable of damaging cells."

Thus, in establishing an allowable limit of a pollutant—whether radiation from a nuclear power plant or sulfur dioxide from coal—the public must assume some risk, the panel members said.

Combination of Pollutants

The issue is further complicated by the fact that people are simultaneously exposed to many potentially harmful substances, and the separate effects of these pollutants may be additive or the pollutants may interact in ways that multiply their risks.

"We tend to talk about levels of pollutants as if we live in a pure and pristine state." remarked Dr. David Rall, director of the National Institute of Environmental Health Sciences in North Carolina. "We're already living with a lot of carcinogens [cancer-causing substances] in the environment. If you add more, you can assume their effects will be additive to already existing ones."

Dr. Rall said in an interview that science was only beginning to find practical ways to measure the combined actions of several substances.

Another problem in setting allowable pollutant levels is extrapolating from the results of experiments on laboratory animals to probable effects on human beings.

"Reactions to various agents are determined by the inherited properties of an animal," noted Dr. Jerzy Neyman, director of statistics at the University of California, Berkeley. "Some reactions in mice are unknown in humans, and I would be most surprised if the reverse was not also true."

Dr. Neyman suggested that "the best hope for securing reliable information" lay in establishing an agency to coordinate a large, interdisciplinary study that would consider many pollutants at the same time in the settings of many different communities.

At the moment, he said, decisions about pollutants are being made, not on the basis of reliable information, but "in response to tumult and shouting."

Dr. Rall said, however, that the scientific data for such decisions were infinitely more sound than the economic data used to estimate how much it would cost to achieve a given level of pollution control.

An accurate economic data base is especially critical, Dr. Rall said, because the economic costs or benefits of a regulation are immediate, but the health costs are often paid decades later.

January 31, 1975

STUDY FORECASTS MAJOR SHORTAGES

By HAROLD M. SCHMECK Jr.

Special to The New York Times

WASHINGTON, Feb. 11—The world faces a future plagued by shortages in those resources vital to modern industrial civilization, according to a major study made public today by the National Academy of Sciences.

"Man faces the prospect of a series of shocks of varying severity as shortages occur in one material after another, with the first real shortages perhaps only a matter of a few years away," said the report, titled "Mineral Resources and the Environment."

The report concluded that it was "essentially impossible" for United States oil production to rise enough in the next decade to make the nation independent of foreign supplies.

There will probably never again be large annual increases in United States production of gasoline and natural gas, the report said. Conventional onshore production will inevitably decline and the development of other sources, such as oil on the continental shelf, will be slow and difficult, it said.

In fact, Government estimates of total American oil and gas resources have been over-optimistic, Dr. Brian J. Skinner, chairman of the academy's study panel, said at a news conference today. He is a professor of geology and geophysics at Yale.

The new report estimated total United States petroleum resources, discovered and undiscovered, at roughly 150 billion barrels; most of it undiscovered but presumed to exist in Alaska and on the continental shelf. Some Government estimates have been more than twice that high.

The report predicted that most of the world's oil supplies would be used up within 50 years and that the "enormous" reserves in the Middle East would be gone in 30 years at the present and prospective rates of use.

'What Kind of Culture?'

"The Arab countries are entitled to ask themselves, and us, what kind of economy and culture they will have achieved by the time this transient bounty runs out," the report said.

The study that led to the 348-page analysis of resources and their use was by a committee of experts assembled by the National Research Council, operating arm of the academy, which is the nation's most prestigious organization of scientists. The study took two years. Dr. Skinner said there would be further reports on various aspects of the materials-energy situation.

The central conclusion of the report, which was released today, he said, was that the United States must emphasize conservation of energy and other resources to a degree hitherto unparalleled. Indeed, he said it should become almost a religion. In answer to questions at the news conference, he said there was little evidence that the Government was pursuing the conservation goal as vigorously as would be necessary.

Energy and Resources

"Because of the limits to natural resources as well as to means for alleviating these limits," the report said, "it is recommended that the Federal Government proclaim and deliberately pursue a national policy of conservation of material, energy and environmental resources, informing the public and the private sectors fully about the needs and techniques for reducing energy consumption, the development of substitute materials, increasing the durability and maintainability of products, and reclamation and recycling."

The committee recommended a national policy of stockpiling critical materials to sustain both civilian and military needs in case of emergencies. It also said better ways of making long-range demand forecasts should be developed.

'Threatened List'

Although the report did not predict which material resources would be in short supply first, it did list several on a "threatened list." Among those it cited asbestos, helium and mercury because they have vital industrial uses and special properties that are not duplicated by any other substances. Tin was on the list because of a "potential general worldwide shortage."

Although the study noted that the United States had abundant copper reserves it predicted that the nation would not be able to depend entirely on domestic supplies much longer. It recommended developing means of obtaining copper from metallic nodules that occur naturally on the ocean bottom.

The report also said it was no longer safe to assume that technology could solve every problem of material shortage.

The committee said the report should not be viewed as a counsel of despair. It said the United States had abundant resources that should insure it a continued strong position in the world.

But the report said efforts to increase supplies should be made concurrently with policy aimed at decreasing demand, that progress in substitution and recycling should be stimulated along with, not independently of, encouragement of the conservation ethic.

The committee said United States and world dependence on coal would increase in the years ahead and that the United States had huge reserves of this valuable resource. The report continued, however, that the mining and burning of huge amounts of coal could have serious environmental and health consequences that must be taken into account.

February 12, 1975

Physicists Voice Long-Term Worry Over Safety of Reactors

By WALTER SULLIVAN
Special to The New York Times

WASHINGTON, April 28—A team of physicists, convened last year by the American Physical Society to assess the safety of American nuclear reactors, has found no reason for "substantial short-term concern," but is critical in terms of long-range prospects.

The study focused on the water-cooled reactors that are the standard energy source in American atomic power plants. Fifty-four are in operation, as well as one gas-cooled reactor. A total of 236 water-cooled reactors have been built or are under construction or projected.

The physicists noted that so far the safety record of such reactors "has been excellent, in that there has been no major release of radioactivity," and said they had uncovered no reasons for substantial short-range concern regarding risk of accidents. But they did express concern for long-term operation of an increasingly large number of such power plants when the likelihood of seemingly "improbable" accidents becomes greater.

The tone of the physicists' report, particularly in dealing with the short-term outlook, contrasted with the views of critics of nuclear power plants such as the Union of Concerned Scientists and environmental organizations such as the Sierra Club.

Attention has been drawn by some critics to mishaps that have occurred, including limited releases or spillages of radioactive material and the recent discovery of cracked piping in several plants.

That the hairline cracks were found was cited by the panel of physicists as evidence of vigilance in the reactor safety effort.

One major source of concern to the panel is the absence of any realistic, full-scale test of what would happen in case of the most serious accident—a core melt-down. If all the cooling systems failed, the reactor core and its fuel rods could heat sufficiently to melt.

The reactors are enclosed in pressure vessels designed to contain the radioactive gases released if melting occurred. If heat or internal pressure became sufficient to rupture the pressure vessel, winds could carry this lethal debris hundreds of miles.

The physicists, chosen to be independent of any connections with the reactor safety program, accepted estimates of the former Atomic Energy Commission that the likelihood of such a casualty was very small. The calculated probability that it would happen to any one reactor in any one year ranged from one in 50,000 to one in 50 million.

The study found, however, that estimates of cancer deaths resulting from such an event should be 50 times higher than those of the A.E.C. It also regarded the A.E.C. estimates of genetic injuries as much too low.

The physicists' higher figure for cancer deaths derived from consideration of two factors not considered by the A.E.C.

One was the long-term effect on populations downwind of radioactive material deposited on the ground. The other was the effect of radioactive debris on organs such as lungs and thyroid gland. In the extreme case considered, the physicists estimated that from 20,000 to 300,000 residents would have thyroid glands damaged from exposure to radioactive iodine.

Although this is less frequently fatal than other such exposures, it would help raise the total cancer deaths to 10,000 to 20,000 in the 10,000 to 20,000 square miles downwind from the plant, the physicists estimated. These estimates assume a population density of 300 a square mile

It was pointed out here today by Dr. Wolfgang K. H. Panofsky, director of the Stanford Linear Accelerator in California, that since the exposed population was very large, these additional cancer deaths would add only 0.1 per cent to the 20 per cent of the exposed population who would be expected to die of cancer normally, according to national statistics.

Since many of the deaths would occur long after the accident, it would be hard to pin down the blame. It would also be difficult to persuade people to leave their homes in the contaminated area if the added risk of their dying of cancer was less than 1 per cent.

Dr. Panofsky presided over a session of the Physical Society's spring meeting here at which results of the study were presented by some of its dozen participants. Dr. Panofsky was chairman of a trio of internationally known physicists who reviewed the study's findings.

The two others were Drs. Hans Bethe of Cornell University and Dr. Victor F. Weisskopf of the Massachusetts Institute of Technology. Like Dr. Panofsky, both are former presidents of the society. The published report was also made public yesterday

At a news conference before the session at the Shoreham Hotel, Dr. Harold W. Lewis of the University of California at Santa Barbara, chairman of the year-long study, noted that it was focused entirely on the safety of reactors cooled by ordinary water, such as those now in use in this country.

It did not assess other reactor types or the relative merits of other energy sources.

"We studied risks; we didn't study benefits," he said.

Publication of the report followed an assessment last week by the Environmental Protection Agency suggesting that development of breeder reactors be delayed from 4 to 12 years. Such reactors would "breed" new atomic fuel

Their chain reactions would not only generate power but also would release neutrons with sufficient energy to penetrate the nucleus of Uranium 238, which is useless as fuel. This would convert it into plutonium 239, which can power a reactor — or a bomb. Breeder reactors are already functioning in Europe and the Soviet Union. Opponents of the program fear that, if many are built, enough plutonium will be produced to tempt seizure of the material by terrorists.

The E. P. A. report said that new estimates of national power needs through the year 2020 indicated that the requirements would not be so great as assumed in earlier planning. It proposed, therefore, that the breeder program could be delayed while alternatives were assessed.

Today, Elmer B. Staats, Controller General of the United States, submitted to Congress a report on the primary breeder development effort—that of a liquid metal fast breeder reactor. He called it "our nation's highest priority energy program." While his report is a background paper that makes no recommendations, it finds projections by the Energy Research and Development Administration "optimistic and possibly unrealistic."

The administration, which has acquired many of the functions of the defunct A.E.C., has projected operation of the first breeder reactor in 1987. It has also projected that 186 commercial-size breeders will be operating by the year 2000, and 1,178 by the year 2020.

The cost to the Government of breeder development so far has been close to $2-billion. By 2020 it has been projected to $10.7-billion. However, inquiries within the atomic industry indicated, according to Mr. Staats, "that few utilities would be willing to commit large amounts of capital until they were fairly certain that Lmfbr's [breeder reactors] would be technically and economically viable."

Major Steps Proposed

Members of the Physical Society's study group cited a long delay in carrying out a test designed to provide data needed to assess what would happen if the cooling system of present reactors failed. Much concern has been expressed in recent years about the adequacy of the emergency core cooling systems provided in case of such failures.

The performances of various elements of the reactor system in an emergency have been simulated in computer models. However, the panel found, there has been no such modeling of the system as an interrelated whole.

In view of the large number of water-cooled reactors now operating and being planned, the panel said, "we believe it is important that the reactor safety research program quickly take major steps to bring about a convincing resolution of the uncertainties in [emergency core cooling] performance."

Likewise it was found that operating personnel had not been trained adequately to respond quickly and correctly to unexpected emergencies.

As an example of a "weak point" found in the safety program it was noted that diesel generators, like those at air traffic control centers, were provided at reactor sites in case of a power failure. However, inquiries of the Federal Aviation Administration showed that such generators fail to start up 3 per cent of the time. Two are needed for high reliability.

Among the recommendations of the study were the following:

¶Greater automation of reactor operation to reduce to a minimum the danger of human error.

¶Consideration of controlled venting to prevent explosion of the containment vessel after an accident.

¶Preparations to lessen the effects of a major accident, such as providing the population with iodine pills to reduce uptake of radioactive iodine.

¶Defense measures against sabotage, including provisions for shutdown before invaders could reach their goal.

¶"Careful assessment" of benefits and costs of placing atomic plants underground or in remote "nuclear park" settings.

The study also recommended a major expansion of research on many aspects of reactor safety, including quality control of components and biological effects of an accident.

April 29, 1975

A Wyoming Coal Pipeline Starts New Energy Clash

By GRACE LICHTENSTEIN
Special to The New York Times

CASPER, Wyo., May 22 — A major skirmish in the energy crisis war has been joined here and in Washington over a proposed coal slurry pipeline, a new technological system for transporting coal from the mineral-rich fields of Wyoming to the power-hungry electric plants of the Southeast.

It is a battle that pits an odd combination of forces against each other. Railroads, railroad unions and Democrats are on one side. An energy transport company, environmentalists and Republicans are on the other. It involves five states as well as Stanley K. Hathaway, the former Wyoming Governor who has been nominated to be the new Secretary of the Interior.

Its outcome could determine whether coal slurry pipelines become the principal carriers of Western coal in future decades. Four additional ones are on the drawing boards.

At its root lies the natural resource Wyoming considers more precious than all its mineral deposits: water.

"After all," said Tom Howard, publisher of The Casper Star-Tribune, as he pored over a map of the state's strip mines, "out here people kill over water."

Coal slurry pipelines are a relatively recent innovation. They are underground tubes similar to oil pipelines that pump coal pulverized to the consistency of sugar mixed with water from mines to power plant sites. There, the water is filtered out and the coal used to fire the generating plants. The single coal slurry pipeline now in operation in the United States runs 238 miles from the Black Mesa mine in Arizona to southern Nevada.

The pipeline being debated would be the longest, most expensive ever constructed. It would carry 25 million tons of coal a year from near Gillette in northeast Wyoming to White Bluffs, Ark., 1,036 miles away.

Cheapest System

The builder, Energy Transportation Systems, Inc., with offices in San Francisco and Casper, argues that the pipeline would be the cheapest and least environmentally destructive system available to bring Wyoming's vast deposits of strip-mined coal to the Southeast's big population centers.

Utility company figures contend that over a 30-year period, the average delivered cost would be $7.50 a ton of coal via the pipeline, compared with $28.50 a ton for railroad transport.

But before building, the company must tap huge amounts of water from the Madison formation, an underground reservoir that lies beneath the near-barren plains of Wyoming, Montana and South and North Dakota. In addition, the company must get approval to bury the pipeline under 49 railroad tracks owned by nine railroads. In both cases, it has run into stiff opposition.

In 1974, under the leadership of Governor Hathaway's administration, the Wyoming Legislature granted the company the right to dig 40 deep wells in its territory. Five test wells have been drilled, but some Democrats who opposed the plan now want to repeal those permits.

"The idea of taking our precious water table, which we really know very little about, and sending it to Arkansas is a very bad concept," said William G. Rector, a Democratic State Senator from Cheyenne.

He and other opponents say the water permits were pushed through the Legislature despite conflicting reports on how much good quality water would be withdrawn from the underground reservoir.

Some of that water is now used to serve several towns in the region, including Gillette and Edgemont, S. D. Legislators, including Wyoming's only House Representative, Teno Roncalio, fear there is not enough hard information on the effect of the pipeline wells on future municipal water needs.

According to Mr. Roncalio, a Democrat, one study suggests that the wells of Edgemont might dry up as a result. South Dakota is considering a Federal suit to protect its share of the water.

The railroads and their unions are even more adamant. They have refused to grant all but nine permits for the pipeline to cross their rights-of-way. They are afraid the pipeline will ruin their profitable coal freight business.

Harry D. Evans, special representative for the Burlington Northern Railroad in Casper, predicted that the pipeline would cost the railroad $225-million in gross freight revenue a year.

He said there was a "strong possibility" that the railroads would have to abandon their plans for a new 116-mile line from Gillette to Douglas, Wyo., a line that would provide more than 200 additional jobs.

"The future of the railroad is the future of the union," said Andrew J. Mulhall of the United Transportation Union, which also opposes the pipeline.

The pipeline company counters these objections with a barrage of statistics and ecological arguments. According to Frank Odasz, the Energy Transportation representative in Casper, the pipeline is a "water conservation project."

An Alternative

The eventual alternative to pumping coal to Arkansas or elsewhere, he said, would be mine-mouth power plants that would consume seven times as much water.

In addition, he said, company and government geological reports show that the Madison water is replenished continually by seepage from surface streams and snow, more than enough to make up for the water taken out by pipeline wells.

Furthermore, he argued that the pipeline would need such a small number of workers that it would not further disrupt the sparsely populated Wyoming coal regions the way ugly boomtowns created by massive, polluting power projects have done.

However, Energy Transportation must still deal with the railroads. A bill before the House Interior Committee would overrule their opposition by allowing the Interior Secretary to grant the pipeline company the right of eminent domain, the same kind of right the railroads originally got to build their lines.

Hearings on the bill are scheduled to continue next month. Earlier this spring, Representative Roncalio, a member of the committee, spoke out against the bill until further studies were taken.

Export of Water

"Wyoming cannot afford to export its water in the form of a carrier agent," he said.

Meanwhile, Western environmentalists find themselves somewhere in the middle. While they like the idea of a "clean" invisible pipeline as an alternative to more smokestacks, trailer park boomtowns and diesel-burning trains, they, too, worry about the water.

Their stance has angered Keith Henning, the state director of the American Federation of Labor and Congress of Industrial Organizations, whose attitude reflects that of Westerners concerned about a reasonable trade-off between industrial development, its inevitable pollution, and work.

"I don't want to see this state raped but I do want quality growth," Mr. Henning said. "With millions of people out of work in this country, we need that growth. We need jobs so we can keep our young people from leaving Wyoming.

"The environmentalist would like to build a wall around this state so the wealthy can sit around in their log cabins and look out at the trees."

May 25, 1975

Studies Seek Ways to Store Energy

By WALTER SULLIVAN

In Germany preparations are being made to fill a cavernous salt mine with compressed air. In California the construction of flywheels weighing many tons that can spin at 2,000 revolutions a minute is being considered.

Elsewhere exotic electric batteries and superconducting magnets are under development. In all cases the purpose is to find better ways to store energy for use during periods of peak electric power demand.

Such demands vary radically, not only between hours of the day but between days of the week and seasons of the year. The power drain by air conditioning can double the load on a hot day leading to brownouts or a total power failure.

Present methods for meeting energy demands, such as the use of gas turbines or the lighting off of additional furnaces, are wasteful of fuel and tend to pollute. In the current fiscal year, the Energy Research and

Development Administration is spending some $6-million on the search for new energy storage methods.

Storage Systems Stored

Among the projects being supported by the energy development agency in this regard is one being conducted in Newark, N. J., by the Public Service Electric and Gas Company. Using the methods of systems analysis it is comparing the merits of various storage systems in terms of efficiency, cost and environmental impact relative to one another and to storage systems in current use.

At the same time, it is collecting from American utilities data on variations in power demand on day-night, seasonal and annual time scales. The patterns of these variations will determine the nature of surplus power available in off-peak periods to recharge a storage system.

An energy storage system familiar to many New Yorkers because of a long-standing controversy is pumped storage of water. The proposal of the Consolidated Edison Company to build such a system near Storm King Mountain on the Hudson River has been delayed more than a decade by the protests of those who feel the most scenic stretch of that river would be blemished.

The scheme was to use surplus power during off-peak periods, such as late at night, to pump water from the river into a lofty reservoir. At times of peak demand, water would flow back down, turning the turbines of a hydroelectric plant.

Landscape Preserved

It was noted this week by specialists from the energy development office that such a system could be operated underground with no effect on the landscape. The storage reservoir would be at surface level and, during times of heavy power demand, its water would flow into a deep underground reservoir, such as an abandoned mine. The power plant would have to be relatively deep in the mine.

In pumped water systems,

for every three kilowatt hours of energy stored only about two kilowatt hours can be recovered.

A major slice of the budget of the energy resources office in this area—roughly $3-million —is allocated to research on batteries. These include such advanced concepts as high temperature sodium-sulfur and lithium-sulfur batteries.

If these batteries prove feasible, according to energy specialists, they probably could propel an automobile 200 miles at 60 miles an hour without recharging. Research on lithium-sulfur batteries is being done at Argonne National Laboratory in Illinois—one of the former Atomic Energy Commission centers taken over by the Energy Research and Development Administration.

Hydrogen Production

Another major area of research, accounting for about $1-million in agency support, is in hydrogen production, storage, transmission and utilization. During periods of low energy demand hydrogen would be extracted from coal, water or some other source.

Numerous private concerns are working on such batteries, whose operating temperatures range from 550 to 750 degrees Fahrenheit.

Large-scale batteries could be installed at strategic points in the power distribution system, avoiding the costly construction of additional power lines. In this respect they would have an advantage over pumped storage systems that would tend to be at remote sites requiring new power lines.

Another major area of research, accounting for some $1-million in support from the energy development agency, is in hydrogen production, storage, transmission and utilization. During periods when surplus power is available, it could be used to extract hydrogen from water or from a fuel such as coal.

The hydrogen could then be held in reserve for employment in one of several ways during peak demand periods. It could, for example, be used to "stretch" natural gas or could be burned directly.

Also under investigation are various ways to store heat—a technology also applicable to solar energy, which is available only intermittently.

Air compression could be used in combination with gas turbines. The operation of such turbines depends on a supply of compressed air normally provided by compressors that burn up a large part of the power produced by the plant. In off-peak hours, the compressors could be used to compress air for storage in a reservoir for use during peak hours. The compressors could then stand idle at such times, greatly increasing net power production. The reservoir would be an abandoned mine, a natural cavern or an aquifer—a geologic formation with a capacity for water storage.

Aquifers often have a capacity for storing gas or air, as well as water. The feasibility of storing compressed air underground has already been demonstrated by similar storage of natural gas.

According to the energy development office, a salt mine near Darmstadt, Germany, is to be used in a demonstration project as a compressed air receptacle.

For flywheels to store enough energy to be useful for peak utility demand periods they will have to be massive and rotate extremely fast. Consequently, the technical demands are severe. They must, for example, be built of a fiber composite that will not disintegrate under the resulting stresses.

A Major Problem

A major problem is that where the fibers are aligned, the material has great strength along their axis but not at right angles to that axis. In one design, by Dr. Richard Post of the Lawrence Livermore Laboratory, the fibers are aligned circumferentially and the flywheel is divided into concentric rings that are elastically connected.

Their purpose is to minimize centrifugal stresses on any one ring. A prototype three feet in diameter is being built by William A. Brobecek and Asso-

ciates in Berkeley, Calif.

The flywheel will be spun inside a vacuum chamber at rates as high as 25,000 revolutions a minute. No attempt is being made at this stage to perfect a design for the bearings or for retrieval of stored energy.

The experiment is being carried out under a contract with the Electric Power Research Institute in nearby Palo Alto.

The Institute is supported by the utilities and, with the energy development agency is funding the project of Public Service Electric and Gas in Newark.

In another scheme, by David Rabenhorst of Johns Hopkins University in Baltimore, the fibers are aligned radially like the bristles of a circular brush.

50-Ton Flywheels

According to Richard G. Stone of the Lawrence Livermore Laboratory, which is conducting research on such materials, flywheels weighing 20 to 50 tons will have to spin at from 2,000 to 3,000 revolutions a minute to be useful as energy reservoirs. They should be from 30 to 50 feet in diameter if possible, he added a few days ago. The laboratory is operated by the University of California at Livermore on behalf of the energy development office.

A method of energy storage not expected to bear fruit for some time would make use of superconducting magnets. In such magnets, cooled to within a few degrees of absolute zero —the total absence of heat— electricity can circulate with no loss of power.

Research on such a method is being conducted at another center of the energy development agency, the Los Alamos Scientific Laboratory, operated by the University of California in New Mexico.

Energy storage methods applicable to electric utility systems might also solve the problem of storing energy in systems whose power source is intermittant, such as wind and solar energy.

June 8, 1975

Solar Energy Is Called the Solution to Power Pollution

By GLADWIN HILL
Special to The New York Times

LAS VEGAS, Nev., Sept. 20 —New calculations of the pollutional residues of power generation indicate that eventually the world will have to turn almost entirely to solar energy, an international environmental conference was told this week.

Dr. Harold I. Zeliger, a New York chemist, said that an admittedly over simplified and partial computation of global pollution in terms of heat calories suggested that pollution had at least tripled in the last 35 years.

"Our challenge is to at least triple the efficiencies of our

methods of power production just to keep from further polluting our environment," he said.

However, he added, the principal systems of power generation—including far-in-the-future nuclear fusion—all involved energy wastes of up to 91 per cent, which represented pollution in one form or another.

Since these inefficiencies are largely inherent in the power-production processes and cannot be substantially reduced, he continued, the only ultimate alternative is pollution-free solar energy, obtained from such sources as the sun's heat, the winds and ocean currents.

Dr. Zeliger's findings were set forth in a paper presented to the International Conference on Environmental Sensing and Assessment, sponsored by the World Health Organization and the Environmental Protection Agency.

The five-day conference, attended by some 800 scientists, engineers and public officials from 40 nations, explored many aspects of global ecological monitoring, one of the principal agenda agreed on by the 1972 United Nations environmental conference at Stockholm.

Dr. Zeliger is a chemical and environmental consultant in Spring Valley. Marsha Funk, then a mathematics major, collaborated in the study, which was carried out while Dr. Zeliger was teaching at Sarah Lawrence College in Bronxville.

Dr. Zeliger said in an interview that it was virtually impossible to determine the exact amount of pollution but that a useful approach seemed to be in terms of wasted calories of heat, chiefly in combustion processes.

"In determining the extent of environmental pollution from the inefficiencies of different energy transformations, one is confronted with a serious dilemma," he said. "How, for example, are the effects of nuclear fallout to be compared with hydrogen oxide exhaust from an automobile engine for environmental impact?

"Such comparisons may be made by defining pollution in units of energy released. For combustion of carbonaceous fuel, the heat of reaction does not account for the total energy impact on the environment.

"Resulting carbon dioxide and water vapor cause a greenhouse [translucent lidding] effect in the atmosphere, and sulphur dioxide and oxides of nitrogen are further oxidized and hydrated in the air into corrosive acids."

Using standard data on combustion processes, the research team calculated that in typical oil combustion, 89 per cent of the fuel's intrinsic energy was dissipated in waste; with coal, 82 per cent; with nuclear fission, 71 per cent; and with nuclear fusion, 91 per cent.

Altogether, on the basis of fuel consumption projections of the National Academy of Sciences, the researchers said, man's current annual energy release, expressed in calories, was about 45,000,000,000,000,000,-000—"only a small fraction of the solar energy reaching and leaving the earth's surface, but too much for the globe to handle, as evidenced by polluted streams and fouled air, etcetera."

This release, the report continued, was three times that of the National Academy of Sciences estimate for 1940, arbitrarily taken in the study as "the last non-polluting year."

"As the world's energy needs grow," the study said, "we shall have to increase energy conversion efficiencies. Even if this were possible, the growing demand is such that the requirement of 100 per cent efficiency could be reached and surpassed in a very few years. It seems obvious that the only long-term solution to the world's energy-pollution problem is the use of solar energy."

Dr. Zeliger emphasized that his caloric measurement did not imply an eventually insupportable release of heat itself, as some scientists have projected, but represented a compound of many forms of pollution.

"Although nuclear fission yields the smallest polluting energy," the study concluded, "it, like the other processes, is unable to supply man's power needs without polluting his environment to the point where it will be unable to sustain life. Only solar energy seems to provide any real hope for the future."

September 21, 1975

Hope Fades for Cheap and Abundant Atomic Power

By DAVID BURNHAM
Special to The New York Times

WASHINGTON, Nov. 15—The long-held dream that nuclear power would give the United States and the world an endless stream of low-cost electric power has faded, according to a growing number of economists, technical experts and utility officials.

In the years immediately after World War II, people believed that the miracle of the atom could produce automobiles gliding through smogless cities. As recently as 1969, a leading nuclear scientist was predicting that the cheap energy of nuclear power might very well set man free. Just two years ago, President Nixon held out nuclear power as one of the key weapons in the American battle for energy independence by 1980.

But now, that nuclear dream is clouded by problems, some great and some small, such as the soaring increase in the cost of building reactors to an expected $1,135 per kilowatt in 1985, from $300 per kilowatt in 1972; a growing concern about the problems and costs of protecting reactors and their waste products from sabotage; the rising price of uranium, and a possible requirement of new and expensive safety devices for the nation's reactors as a result of a fire last spring in a reactor at Brown's Ferry, Alabama.

Though national defense considerations and environmental restrictions may still make the atom more attractive than fossil fuels such as coal and oil, many experts have become convinced that substantial subsidies will be required if the United States is even to come close to the Ford Administration's stated goal of building 620 reactors in the next 25 years.

Providing such a subsidy, in fact, is a prime objective of the Administration's proposed $100-billion Energy Independence Authority and several other possible aid plans under consideration.

"I agree there was a dream, and five years ago, when we were generating power at $100 a kilowatt, the dream seemed justified," said Dr. Ivan M. Weinberg, an independent consultant who is one of the nation's most distinguished nuclear scientists, in an interview.

"Right now," Dr. Weinberg said, "it looks like the dream has ended, but I caution you all the returns aren't in. At this moment, though, it is probable that nuclear energy is going to be a great deal more expensive than enthusiasts such as myself first thought."

In an article in The New York Times in 1969, Dr. Weinberg said that "recent technical developments suggest that H. G. Wells's vision of a 'world set free' by very cheap energy must be taken seriously."

Dr. Carl Walske, president of the Atomic Industrial Forum, a pronuclear lobbying group, acknowledged that there had been widespread stories that "electricity would be as free as water" but did not agree that there had been a widespread public dream of endless cheap power.

Unrealistic Claims

But Dr. Walske said he did believe that overly optimistic expectations about the potential of nuclear power were created during the 1950's by the unrealistic claims of the various competing reactor manufacturers that their product "was better and cheaper than that of the others."

"This industry right now has incredibly serious problems," said Irvin C. Bupp, a professor at Harvard University's Graduate School of Business, the co-author of a recent study analyzing the relative costs of generating power by nuclear and coal-fired plants.

"Publicly available information on the costs of nuclear power versus other alternatives tends to strongly overstate the case for nuclear power and understate the case for the alternatives," a report to the Energy Research and Development Administration concluded recently.

"We noted a distinct tendency in the nuclear energy literature to underestimate nuclear power costs, more often than not by simply omitting some costs, or neglecting the potential effects on costs of practical or operational experience such as significantly lower capacity factors than theoretical projections would suggest," said the report, by Richard J. Barber Associates, a Washington consulting firm.

"All things considered, it appears that purely on economical grounds and ignoring shortage problems resulting from state regulation of electricity rates, the future of the United States nuclear reactor industry is less bright than most recent Government forecasts indicate," an article in the forthcoming issue of the Bell Journal of Economics and Management Science concludes.

The article was written by Paul L. Jaskow, an associate professor of economics at the Massachusetts Institute of Technology, and Martin L. Baughman, associate director of energy modeling at the University of Texas at Austin.

"Right now, if you can build your plant near a railroad that provides a cheap supply of coal, coal looks much better than nuclear," said William Kriegsman, a former member of the Atomic Energy Commission who is now with the consulting firm of Arthur D. Little.

The use of nuclear power to generate electricity in New England, for example, may still be economic because of the remoteness of the Northeastern states from coal and oil. But when the generator is to be built near the mouth of a coal mine, the equation is said to go against nuclear reactors.

The apparent fading of the dream of cheap power, the difficulties faced by some utilities in raising capital and the slowdown in the traditional growth pattern of the use of electricity in the United States have prompted a number of utilities to postpone or cancel plans to build new reactors.

Even officials with many years of experience in nuclear matters appear to have lost some of their enthusiasm. John D. Selby, the new president of the Michigan Consumer Power Corporation, who was formerly the deputy general manager for nuclear fuel and reactors at General Electric, said in an interview:

"I don't feel that consumer power should commit [build] another nuclear plant at this time. Nuclear power is a viable form of power. But I feel coal is equally viable."

The Ford Administration is still committed to nuclear power as a key part of its drive to make the United States independent of foreign sources of energy, particularly the oil produced by the Arab nations in the Middle East.

"The figures I have show that nuclear reactors today are more than competitive," said Merrill J. Whitman, an energy systems expert in the Energy Research and Development Administration.

"The problem," he added, "is the tremendous front-end costs, raising the money to build the reactors."

Nuclear proponents have long argued that the advantage of nuclear power over fossil-fired plants is that, although it costs more to build a reactor, the cost of producing electricity over the life of the system is lower because uranium is cheaper than coal or oil.

This basic thesis is now being challenged on several fronts.

One challenge was stated in a paper on the economics of nuclear power written by Professor Bupp of the Harvard Business School and Jean-Claude Derian, Marie-Paul Donsimoni and Robert Treitel of the Center for Policy Alternatives at M.I.T.

Their statistical study of reactors and coal-fired plants found that while the cost of constructing both kinds of generators is increasing, the cost of nuclear reactors is increasing faster than the cost of coal plants.

Construction costs for coal plants, Professor Bupp and his colleagues said, increased at an average of $13 per kilowatt per year between 1969 and 1975,

while the cost of nuclear plants increased at $31 per kilowatt per year.

They noted that the more rapidly rising capital costs of nuclear plants would mean increasing production costs for amortization of the investment.

The industry likes to blame the Federal Government for the high nuclear construction costs, contending that complex licensing and safety requirements have resulted in lengthening the construction time for a reactor to 10 years. The Nuclear Regulatory Commission blames the industry.

Whoever is responsible, the long construction period means the utilities have to absorb the high cost of borrowing the huge amounts of required capital for long periods before the reactors begin earning their keep.

A second challenge has emerged in the sharply increasing price of uranium, which several weeks ago prompted the Westinghouse Electric Corporation to tell 20 utility customers that it would no longer provide uranium after 1978. Uranium costs have quadrupled from their mid-1973 level of about $7 a pound, and many observers believe the cost may reach $50 a pound for deliveries in the 1980's.

Yet another serious unresolved problem confronting the nuclear industry is what to do with spent fuel—in effect 'the nuclear ashes—created as the reactors heat the water to produce the steam that turns the turbines and generates electricity.

For many years, the Government and industry had proposed chemically treating this spent fuel in a complex process

that considerably reduces the bulk of the waste while at the same time extracting plutonium. The plutonium then would be used to fuel the reactor.

Complex Process

The process of extracting the plutonium on a large-scale commercial basis, however, now appears more complex than was foreseen. One of the problems is that small amounts of plutonium could be turned into home-made nuclear bombs by a small group of terrorists. This potential has prompted the Nuclear Regulatory Commission to delay approving "plutonium recycle" until it determines how many guards and fences and other expensive safeguards will be required.

The delay in licensing has prompted the owners of a major reprocessing plant now nearing completion in Barnwell, S.C., to ask the Energy Research and Development Administration to buy them out. This request—in effect for another subsidy for the nuclear industry—is under active consideration by the agency.

The delay has also prompted the Edison Institute, the New York-based trade association of the nation's privately owned utilities, to undertake a study dealing in part with the actual benefits that would result from reprocessing and re-using plutonium.

One of the chairmen for the Edison Institute study is Bernard H. Cherry, manager of fuels for the General Utilities Corporation of New Jersey.

In Favor of Nuclear

While confirming a report by National Public Radio that a preliminary draft of the study had concluded that plutonium recycle was "marginally economic," Mr. Cherry said that concerning his company's operations in New Jersey he still felt that nuclear power had more advantages than coal.

"The decision is tougher than a few years ago, but it still seems to flop over to nuclear if you examine all the costs for the entire life of a plant," he said.

Mr. Cherry said one reason he was "still inclined toward nuclear" was the excellent operating experience his company had had with its reactors. The Three Mile Island facility, for example, near Harrisburg, Pa., generated 79.4 percent of the electricity it was designed to produce during the first six months of 1975, he said.

This was far superior to the over-all record of the 53 reactors now operating in the United States, which on the average produced only 58.3 percent of their designed capacity during the first half of 1975.

A nuclear reactor vessel on route to installation. Increased costs, arising from unforeseen problems, have created doubt as to nuclear devices being answer to energy problem.

November 16, 1975

FORD SIGNS BILL ON ENERGY THAT ENDS POLICY IMPASSE AND CUTS CRUDE OIL PRICES

IMPORT FEE STOPS

Gasoline and Fuel Oil Costs Expected to Level Off at First

By PHILIP SHABECOFF
Special to The New York Times

WASHINGTON, Dec. 22— President Ford ended a year-long stalemate with Congress over energy policy today by signing into law a bill that will roll back crude oil prices and help stabilize gasoline and fuel oil prices for consumers, at least temporarily.

The President also announced that he was removing, effective today, the $2-a-barrel import fee on crude oil that he had imposed to discourage imports.

In announcing his decision, the President said that "the single most important energy objective for the United States today is to resolve our internal differences and put ourselves on the road toward energy independence."

"It is in that spirit," Mr. Ford then said, "that I have decided to sign the Energy Policy and Conservation Act."

Repudiates Earlier Policy

The legislation he signed today repudiates the original energy policy he announced last year. That policy called for sharp increases in fuel prices to encourage production and discourage consumption.

The President conceded that "this legislation is by no means perfect." He said, "It does not provide all the essential measures that the nation needs to achieve energy independence as quickly as I would like."

Democrats in Congress said that the enactment of the legislation represented a victory for them and a setback for the President.

Cut in Crude Oil Price

The new law will force a reduction in the average price of crude oil produced in the United States from the current $8.75 a barrel to $7.66 a barrel. The change will go into effect in February.

However, Frank G. Zarb, the Federal Energy Administrator, briefing reporters after the President's statement at the White House, said that consumers would see little if any decrease in the price they pay at the gas pump or in their fuel bills.

Mr. Zarb said that rising costs would eat away much of the rollback as far as consumers were concerned and that the real price reduction would come to a cent a gallon at most. There had been estimates that the bill would save consumers 2.5 to 3 cents a gallon on gasoline and fuel oil initially.

However, if the President had vetoed the bill, an action urged by many of his advisers, price controls on oil might have expired and fuel costs to consumers would have shot up rapidly.

The new energy law will allow oil prices to rise by a maximum of 10 percent a year at the discretion of the President. However, after February 1977 Congress could halt any increase in fuel prices above the inflation rate for the economy as a whole.

The law would expire at the end of 40 months. At that time, oil prices would be decontrolled or Congress could vote a new price controls bill.

Other Major Provisions

The new law contains the following other major provisions:

¶Automobile manufacturers must comply with mandatory gasoline mileage standards. The gradually rising standards would require an average automobile mileage of 27.5 miles a gallon for 1985-model cars.

¶Appliance manufacturers must meet efficiency standards and label their products to show fuel consumption.

¶The bill establishes "stra-tegic petroleum reserves," including storage of at least 150 million barrels of petroleum within three years and up to 400 million barrels within seven years.

¶It would authorize the President to develop contingency plans for energy emergencies, including possible fuel rationing.

¶Loan guarantees of up to $750 million would be authorized for coal operators investing in new underground mines.

The legislation also gives the General Accounting Office, the Congressional investigative agency, the power to audit the records of oil companies.

Ford Explains Signing

The President said that he had signed the bill because, after balancing its "inadequacies and merits," he found that "this legislation is constructive and puts into place the first elements of a comprehensive national energy policy."

Administration officials said today that the bill was a compromise. Mr. Zarb pointed out that it contained four of 13 provisions asked for by the President and that Congress was close to enacting six more of those provisions.

Mr. Zarb also suggested that the need for sharp price increases to discourage consumption had eased because oil consumption had increased less rapidly than forecast and was down by about $1 billion a day from projections this year. The savings, he said, were a result of conservation and warmer-than-average weather.

Mr. Ford faced pressures from both directions in deciding whether to sign or veto the energy bill. He was lobbied intensively by oil and automobile companies who strongly urged a complete removal of price controls, mandatory standards and other restraints. Many of the Southern conservatives in the Republican Party, whose support he needs to win the nomination next year, had urged a veto.

A number of his own aides reportedly wanted him to allow the price mechanism of the free market to determine fuel prices, a position that Mr. Ford is sympathetic to ideologically.

But if he vetoed the bill and allowed fuel costs to jump, the results could have helped touch off a new wave of inflation. In the Northeastern states, including New Hampshire, where he faces an important primary election in February, passage of the energy bill was a major issue.

Oil Group Protests

The Independent Petroleum Association of America, which represents the nation's small producers, issued a statement calling Mr. Ford's decision "unfortunate for consumers and the economy."

"It is disappointing that the President has compromised his firmly stated concept of relying on the free market as the best allocator, conserver and solicitor of energy supplies and has gone along with this unwise decision by Congress," the statement said.

The Exxon Company issued a statement saying that the energy bill signed by President Ford "may hinder efforts to increase domestic oil and gas production and lead to increased dependence on foreign oil imports."

A fact sheet put out by the White House today estimated that oil imports would rise by about 150,000 barrels a day now because of initially lower prices. However, it said that imports would probably be about 200,000 barrels a day less after three years because of price increases allowed by the bill. There are 42 gallons in a barrel of oil.

Under the bill, "old oil," drilled from wells in operation before 1972, would continue to be controlled at about $5.25 a barrel. Prices for "new oil" will be initially controlled at $11.28 a barrel compared with the prevailing $12.50 to $13.

Although he signed the bill continuing price controls on oil, the President, through his spokesman, said that he was opposed to the philosophy that the Federal Government should make economic decisions for Americans.

Answering questions about a report by the Joint Economic Committee of Congress that warned of serious consequences to the economy if President Ford insisted on a Federal budget limited to $395 billion next year, Ron Nessen, the White House press secretary, said that the committee's warning was based on an "increasingly discredited economic philosophy."

He said that President Ford hoped to make a start next year in implementing "a really clear-cut historic difference in economic philosophy."

The basis of Mr. Ford's philosophy, as explained by Mr. Nessen, who said he had discussed it with the President, is that deficit spending by the Federal Government cannot bring prosperity to the nation.

He said that President Ford hoped to reverse economic practices of the last 40 years.

December 23, 1975

The Age of Nuclear Energy: A Prolonged Adolescence

By ROBERT GILLETTE

Nuclear power by any measure is an awesome technology. Not even the men and women who design, build, regulate and operate power reactors are immune to the emotions these half-billion-dollar machines can engender.

Most of them seem to react with a sense of pride in the mastery of nature's darker forces and of confidence in the reliability of these great congeries of concrete, steel and radioactive fuel. Yet for some, awe gives way to a deeply felt fear that, where a modern 1,000-million-watt atomic plant is concerned, simple and perhaps inevitable human error may lead to catastrophe. Such would seem to be the case for Dale Bridenbaugh, Richard Hubbard and Gregory Minor, the engineers who recently quit their jobs at General Electric's nuclear division—and for Robert D. Pollard, one of the Federal Nuclear Regulatory Commission's 48 project managers who followed suit a few days later.

These four engineers were not the first, and they are probably not the last, to renounce promising careers to tell the public that nuclear power is unsafe, or at any rate not safe enough. David E. Lilienthal, the first chairman of the Atomic Energy Commission, was an early skeptic: "What public policy demands or justifies going ahead with a program in which there are still unresolved risks to human health and safety?" Mr. Lilienthal asked in 1963.

In the late 1960's, John Gofman and Arthur Tamplin, two scientists at the Atomic Energy Commission's Lawrence Radiation Laboratory at Livermore, Calif., became celebrated if embittered critics of nuclear power. In 1972 a small band of A.E.C. safety researchers appeared in public hearings to voice concern about the adequacy of emergency cooling equipment then being used in large new power reactors. Still others in the industry and in government have covertly supplied newspaper reporters and groups of critics, such as the Union of Concerned Scientists, with technical advice and embarrassing documents.

What is the public to make of these acts of technological apostasy?

A cynic might say that jeopardizing one's job security may testify to a man's sincerity but not necessarily to his prescience. Yet there is no disputing that 30 years into the nuclear age this once-promising technology remains burdened with a host of unresolved problems. Mr. Pollard and the three General Electric engineers invoked the leading ones in explaining their public resignations: What to do with nuclear waste that remains lethal for centuries? How to mitigate the risks of theft and sabotage in a plutonium economy? And, most important, how to settle the scores of engineering questions that still cloud the day-to-day operation of the nation's 57 licensed nuclear power reactors?

Although some progress has been made in calculating the real risks involved in these problems, their urgency is still a matter of subjective engineering judgment: The best that can be said is that honest engineers disagree wildly. But it is worth asking why, at this late stage, the debate is still going on.

There are several answers. For one, the current debate began belatedly. The basic directions of American nuclear power development were set in the late 1940's and early 1950's, long before battles over pesticides and other forms of pollution had sensitized the public (to say nothing of government) to the adverse affects of big technology. Nor was public participation in the making of arcane technological policy as much in fashion then as it is now. For example, Glenn Seaborg, Atomic Energy Commission chairman from 1961 to 1971, has noted that in the 1950's, when it came time to decide which of several reactor technologies was to be emphasized by civilian power programs, the choice was left up to industry. Industry (mainly General Electric and Westinghouse) chose to scale up water-cooled reactors originally developed for nuclear submarines, not because this technology was most appropriate, but because it was the most familiar.

Reactor safety was always of interest to the old Atomic Energy Commission. But in 1966, safety became the subject of an important internal argument. Ironically, this argument began with Consolidated Edison's application for a permit to build the Indian Point No. 2 reactor, the plant that is a focus of Mr. Pollard's concern.

Located only about 24 miles from New York City, Indian Point Plant No. 2 was among the first of a large new generation of power reactors whose ability to withstand a major "loss of coolant" accident without a catastrophic melting of its radioactive core was only dimly understood. Nevertheless, the A. E. C. granted Con Edison its permit, then set about assessing the safety engineering questions it raised.

From there it was all downhill. The A. E. C.'s safety research program, afflicted with declining budget and growing tensions between headquarters and the national laboratories, stagnated. By the early 1970's, the Atomic Energy Commission had accumulated a list of 139 unsettled safety questions, 44 of which it designated as "very urgent."

The Atomic Energy Commission, of course, was disbanded in late 1974 and the new Nuclear Regulatory Commission assumed its responsibility for safety research. The new commission's research budget has risen to $100 million this year, industry is spending $60 million more on commercial reactor safety, and some critics acknowledge that the Federal program's management is much improved. Even so, a study by the respected American Physical Society last year disclosed deficiencies as well as a continuing paucity of hard experimental evidence to back industry's assertions that reactor safety systems are adequate. This sparsity of data, and a consequent heavy reliance on computer simulation of reactor accidents, is a major worry of Mr. Pollard and the General Electric engineers.

New Designs, New Problems

Not all the unsolved problems that worry them are part of a decade-old backlog, however. Like the aircraft industry, nuclear power is an evolving technology; new designs raise new problems or reveal previously unsuspected flaws in older plants. Until last year, for instance, hardly anyone would have guessed that a workman with a candle could start an electrical fire that would knock out two of the nation's largest reactors and narrowly miss causing a disastrous meltdown. Yet that is precisely what happened to the Tennessee Valley Authority's Browns Ferry plant near Decatur, Ala. Mr. Minor says that his faith in the redundancy of nuclear safety systems evaporated with that fire. Says Mr. Pollard, "In so many near accidents we've had, what caught us up was something no one had thought of."

In effect, they are saying that reactor designs have evolved faster than have the abilities of scientists and engineers to understand their frailties.

Other engineers counter that designs are "conservative" enough to tolerate human error and to compensate for uncertainties in a reactor's performance. To that, Mr. Pollard and other critics contend that not enough experimental evidence exists to know whether design assumptions are really conservative or not.

Even so, Mr. Minor, for one, is not unsympathetic with Nuclear Regulatory Commission, a new agency eager to prove its mettle, and one which the nuclear industry has criticized for precipitous action in closing down nuclear plants deemed unsafe.

"They're under tremendous pressure to make difficult decisions in a complex environment," Mr. Minor says. "They are under pressure from utilities and reactor vendors to keep plants on line, from the President to speed things up and from environmental groups to slow things down. I don't envy them."

Robert Gillette, a reporter for Science magazine, is currently a Nieman Fellow at Harvard.

February 15, 1976

BACTERIA VIEWED AS POWER SOURCE

Sunlight Is Converted Into Chemical Energy Without the Help of Chlorophyll

By WALTER SULLIVAN

The progress made in understanding how some bacteria convert sunlight into chemical energy without help from chlorophyll has been sufficent to encourage hope that the process may eventually be used as a power source.

This conversion process, in a limited sense a form of photosynthesis, occurs in the membranes of bacteria that live in extremely salty water, such as that of the Dead Sea or the bright red salt pans south of Oakland on San Francisco Bay.

When stimulated by sunlight these membranes remove protons from the bacteria and eject them into the surrounding fluid. Since the protons, which are the nuclei of hydrogen atoms, are positively charged, their removal gives the organism a negative electric charge.

This apparently enables the bacteria to synthesize ATP (adenosine triphosphate), the so-called "energy currency" of life. Once synthesized, ATP delivers energy to the parts of the cell needing it.

A Clue to Desalting

The ability of the membrane, under the influence of sunlight, to remove protons from the cell is seen as a potential way of removing salt from water.

Yesterday, Dr. Walther Stoeckenius, professor of cell biology at the University of California Medical School in San Francisco, who has been working on the process for more than a decade, described recent findings at a news conference there.

They were also outlined in a news release by the Ames Research Center of the National Aeronautics and Space Ad-ministration in nearby Mountain View, where Dr. Stoeckenius has been working with a group of scientists.

The proton-emitting process is performed by a bacterial form of rhodopsin, the pigment of the retina also known as "visual purple" that figures in night vision. It occurs in the Halobacterium halobium bacterium.

According to the NASA report, molecules of the pigment can transfer 250 protons a second across the bacterial membrane. The elucidation of the role of the pigment, known as bacteriorhodopsin has been done step-by-step over the last several years.

Sunlight's Energy Used

Photosynthesis, in the most widely used sense, is the synthesis of carbohydrates by plants, using the energy of sunlight. Carbon dioxide is combined with hydrogen that, typically, is derived from water, and oxygen is released. Thus plants "inhale" carbon dioxide and "exhale" oxygen.

Actually, the process involves many stages and great complexity. One step apparently synthesizes ATP (through the addition of a phosphate to adenosine diphosphate, making it energy-rich adenosine triphosphate). It is this that is stimulated by the pigment under study, but Dr. Stoeckenius proposed yesterday that other elements of photosynthesis could also be achieved with the aid of the pigment.

His co-workers have included Drs. Richard Lozier, Roberto Bogomolni and Janos Lanyi at Ames and Dr. Efraim Racker at Cornell University in Ithaca, N. Y.

The work was called "important" yesterday by Dr. Melvin Calvin of the University of California at Berkeley, who won a Nobel Prize for deciphering the chemistry of photosynthesis.

The membrane pigment, Dr Calvin said, seems to "organize itself" in an ordered configuration that makes possible the proton transfer. To him, this suggested the long-term possibility of creating membranes tailored to derive energy from sunlight—preferably, he said, by transferring electrons rather than protons.

March 3, 1976

New Fusion Reactor Likely to Break Even in Fuel Use

By WALTER SULLIVAN

Unexpectedly rapid progress in recent months in a critical area of fusion research, known as neutral beam injection, has virtually guaranteed "break-even" power production for a new fusion reactor to be built at Princeton.

This is the view of specialists at the Princeton Plasma Physics Laboratory, site of the $228-million project. When completed in 1981, it will be the most ambitious effort in the American program to harness the hydrogen atom that began at Princeton just a quarter of a century ago.

Fusion —the release of nuclear energy by fusing small atoms such as hydrogen into larger ones such as helium — could provide virtually unlimited energy.

The original line of attack was magnetically "bottling" a hot gas, or plasma, composed of the heavier forms of hydrogen, squeezing and heating the mixture until fusion occurred.

A more recent approach has been to use converging laser beams to crush pellets of fusion fuel to the required density and temperature. The main American effort, however, is on magnetic confinement of plasma with special emphasis on the Soviet-originated tokamak design.

'Bottled' Plasma

Neutral beam injection makes it possible to fire a high energy

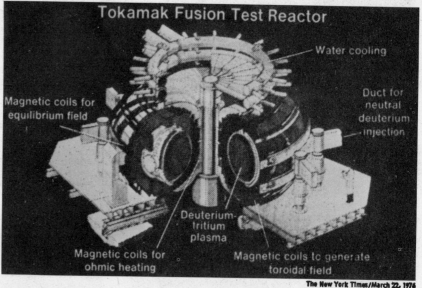

Tokamak Fusion Test Reactor

Water cooling
Duct for neutral deuterium injection
Magnetic coils for equilibrium field
Deuterium-tritium plasma
Magnetic coils for ohmic heating
Magnetic coils to generate toroidal field

The New York Times/March 22, 1976

beam into the magnetically "bottled" plasma to heat and enrich the plasma fuel. Recent advances in this area have made it seem that the new Princeton reactor will be able to "break even" —that is, release as much fusion energy as is needed to operate it.

Nevertheless Dr. Melvin B. Gottlieb, director of the laboratory, which is operated by Princeton University for the Energy Research and Development Administration, cautioned in an interview this week that a key question remains: "How big?"

It already appears that a working fusion reactor of the tokamak variety will have to be large and costly. Despite the optimism and enthusiasm evident in the laboratories where fusion research is under way, Dr. Gottlieb noted continuing uncertainty as to whether or not fusion would become an economic energy source.

Last week he, his colleagues and others in fusion research told of recent advances here and abroad. They reported, for example, that the Russians, as the next step in their ambitious program, were planning a hybrid fusion-fission reactor, the Tokamak-20.

Fission is the atom-splitting process that powers the nuclear plants of today. The atoms split are typically those of uranium 235, although plutonium 239 can also be used.

Hybrid Reactors

In the hybrid reactor, the intense flow of high-energy neutrons produced by fusion would convert uranium 238 (which is unsuitable as reactor fuel) into plutonium.

A similar "breeding" of new fuel is caused by neutrons from fission reactions in breeder reactors, a few of which are now operating. The Tokamak-20, to be completed by 1982, would produce 3,300 pounds of plutonium yearly.

In Japan, the JT-60 will be even larger than the projected TFTR, or Tokamak Fusion Test Reactor, at Princeton. The purpose is to test the behavior of hydrogen plasma in so large a machine. No fusion reactions are expected.

Other big machines, planned or built, include the Joint European Tokamak, or JET, as well as devices in Italy, Germany and elsewhere.

The TFTR will fuse the two heavy forms of hydrogen: deuterium and tritium. The deuterium nucleus contains a neutron in addition to the single proton characteristic of all hydrogen atoms. Tritium contains yet another neutron. When they fuse, a left-over neutron is released at high energy.

In the reactor, tritium plasma will be confined inside a magnetic "bottle" and a beam of deuterium will be fired into the bottle, heating the deuterium-tritium mixture to the fusion point. The deuterium beam can pass through the magnetic wall because it is electrically neutral.

Yet the beam cannot be accelerated when neutral. Hence the deuterium is first stripped of electrons, leaving its atoms positively charged. They are accelerated, then fired through an electron-rich gas to become neutral again, and aimed into the magnetic "bottle."

It has been the rapid development of this technique, notably at Fontenay-aux-Roses in France, the Oak Ridge National Laboratory in Tennessee and the Lawrence Livermore Laboratory in California, that has

led to new optimism in this area of fusion research.

This week Dr. Frederic Coensgen, in charge of the project at Livermore, said pulses a hundredth of a second long are firing a billion billion deuterium nuclei into the 2X-IIB machine there. Beams seven times that strong will be needed and are under development at Livermore's sister laboratory at the University of California in Berkeley.

The 2X-IIB is a mirror machine —a simpler design than the tokamak — which has recently re-emerged as a contender, thanks to the development of neutral beams. Livermore now hopes that $100 million will be provided in the Federal budget for the fiscal year 1978 to build the MX —a large machine of this type.

Fusion Reactors

In the present concept of a fusion reactor, neutrons ejected by the fusion of deuterium and tritium would be trapped in a surrounding blanket of lithium 6, heating the latter and, at the same time, producing new tritium. The latter is relatively short-lived and so, in contrast to deuterium, is not found in naturally occurring hydrogen. The heat would produce steam to drive a power plant.

A limiting factor, if this approach to fusion becomes practical, may be the cost of lithium. The latter is aboundant, but the extent of easily accessible sources, according to an assessment in the March 12 issue of the magazine Science, is uncertain.

As pointed out this week by Dr. Edward A. Frieman, associate director of the Princeton laboratory, some fusion processes do not require tritium, such as the combining of helium 3 with deuterium. This, he said, is the "dream"

process that produces charged particles.

The latter could be used for direct generation of electricity, free from the inefficiency and heat-disposal problems of steam. The requirements to achieve such fusion are much more severe than for the deuterium-tritium process but if the latter is achieved, Dr. Frieman said, the other may ultimately prove feasible.

A few days ago, Ebasco Services Inc., was chosen as industrial subcontractor for the TFTR project with the Grumman Aerospace Corporation as subcontractor to Ebasco. There had been six bids for the contract.

The device will be roughly twice the size of the Princeton Large Torus, which began operation in December. The latter represents an intermediate step in exploring what happens as the scale of tokamak systems is greatly enlarged.

The next step after TFTR would be the EPR, or Engineering Power Reactor, to be built in the mid-1980's for $1 billion to test such elements as the lithium blanket and heat extraction systems. The first demonstration power plant would come near the end of the century.

Explaining the relationship of the Princeton effort to work at other American centers, Dr. Frieman said: "We are tramping down the main road." In building devices so large and costly, radical innovations uncertain of success are unacceptable. Therefore other laboratories are being more experimental.

He cited the $28 million Doublet-III machine being built by the General Atomic Company in San Diego, which will test a novel plasma chamber, wasp-waisted in cross section. Operation by early 1978 is projected. The Los Alamos

Scientific Laboratory in New Mexico is working on the Scyllac machine in which plasma is subjected to a sudden magnetic pinch.

At Princeton, the PDX, or Poloidal Divertor Experiment, is to be built in the next year or two at a cost of $17 million to test a magnetic method for removing plasma particles that otherwise would hit the chamber walls. At present such collisions knock carbon and oxygen atoms out of the walls to pollute the plasma.

Lead Time in Work

The PDX plasma cloud will be doughnut-shaped, as in all tokamak devices, but relatively thick —some three feet in diameter. It is typical of the long lead times in such work, that PDX results will not come in time for use in the TFTR, although they will be applicable to later machines.

The TFTR will operate in pulses one to five seconds long. Fusion power plants with pulses hundreds of seconds long are hoped for, with a possibility, as well, of continuous operation.

It is possible that the reactor will incorporate elements of the "Bitter coil" that has enabled the Alcator of the Massachusetts Institute of Technology to produce extremely powerful magnetic fields. The Alcator is relatively small, but larger models —Alcator A and B —are projected.

Federal energy officials, testifying before the Joint Congressional Committee on Atomic Energy this week, said recent advances come close to assuring that useful power will be extracted from fusion reactors by the year 2000.

March 22, 1976

Big Power Plan Dropped After Long Fight in Utah

By GLADWIN HILL
Special to The New York Times

LOS ANGELES, April 14—A consortium of electric companies unexpectedly announced today the cancellation, after years of controversy, of the $3.5 billion Kaiparowits power plant project in a scenic area of southern Utah.

The move, attributed to economic, regulatory and environmental reasons, represented a major victory for conservationists and a significant setback in the national energy development program.

The three-million-kilowatt generating plant—large enough to supply a community of three million persons — was planned to provide electricity for Arizona and Southern California, and would have been the largest coal-fired plant in the country.

Consortium officials told Secretary of the Interior Thomas S. Kleppe in Washington late yesterday of the decision to "withdraw," after several days of conversations. The project would have been on Federal land and involved several In-

terior Department agencies.

The partners in the project were the Southern California Edison Company (40 percent ownership), the San Diego Gas and Electric Company (23.4 percent) and the Arizona Public Service Company (18 percent), with 18.6 percent uncommitted. The uncommitted portion is allocated among the partners proportionately.

San Diego Gas and Electric said that it concurred with Southern California Edison in the decision, but Arizona Public Service said it was "disappointed." It said that the move "is a blow to the energy needs of the Southwest, and precludes the use at this time of a most valuable resource."

The plant, at Four-Mile Bench 30 miles north of the Glen Canyon on the Colorado River, would have burned more than 1,000 tons of coal an hour and would have emitted about 300 tons a day of atmosphere contaminants into an area that has eight national parks and three national recreation areas within 200 miles. The Glen Canyon National Recreation Area adjoins the plant site, and other preserves in the area are Bryce, Zion, Capital Reef and Canyonlands National Parks.

The plant's emissions would have exceeded permissible limits if the parklands were to be classified as atmospheric "nondegradation" areas under the Clean Air Act of 1970. They have been temporarily classified as "Class II" areas, which permits some new air pollution in the vicinity, and under this classification the emissions would have been within Federal and state limits, according to an Interior Department final environmental impact analysis published a few weeks ago.

The National Park Service, the Bureau of Land Management and the Environmental Protection Agency all had submitted formal judgments that the emissions would adversely affect the region.

Conservationists Surprised

The cancellation came as a surprise to conservationists, who had been expecting Secretary Kleppe to give approval for the plant in a decision scheduled later this month.

Michael McCloskey, executive director of the Sierra Club, said in San Francisco:

"Kaiparowits was a project at the wrong time and in the wrong place. It deserved to be dropped.

"But the deep-mined coal of Utah may still have a place in our energy future if technical breakthroughs can be made in the next 10 or 15 years in perfecting coal gasification and other techniques to overcome air pollution problems."

The cancellation was an economic blow to Utah. In addition to prospective construction expenditures, the project envisioned the creation of a new permanent community of 15,000; an increase in payrolls in Kane County of more than $100 million a year, and eventual, additional tax revenues of $28 million a year.

Comment by Governor

However, Gov. Calvin L. Rampton, one of the leading supporters of the project, had only this brief comment today: "I certainly understand the decision in view of escalating costs and the unreasonably long delay."

The project had been in the planning stage and in negotiation since 1963. Strenuously opposed by such organizations as the Sierra Club and the Environmental Defense Fund, the project was rejected by Secretary of the Interior Rogers C.B. Morton in 1973, but revived under pressure from Utah officials and the power companies.

The consortium had invested some $10 million in preliminary planning and design. Last Dec. 30 the group announced it was suspending the project, originally scheduled to start operation in 1981, for a year because of "delays in regulatory approvals."

Since then, one official said today, "the difficulties had simply piled up until they seemed too formidable." Consortium officials had been conferring almost daily for the last two weeks.

In a terse announcement today, Southern California Edison's executive vice president, William R. Gould, said that the company "has removed the

The New York Times/Gary Settle

A view of the area in southern Utah near the Bryce Canyon and Zion National Parks, where the Kaiparowits power plant project was to have been built.

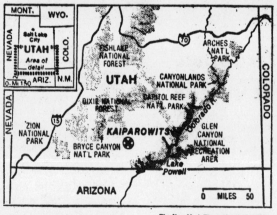

The New York Times/April 15, 1976

proposed Kaiparowits project in southern Utah from its financial and resource planning schedule."

The action effectively precluded the minority partners' continuing with the project.

Mr. Gould ascribed the decision to "a series of uncertainties, particularly relating to ultimate costs."

The uncertainties, he continued, "involve delays in regulatory approvals, one environmental lawsuit and the antici-

pation of other lawsuits, and anticipated legislative and regulatory opposition at the Federal and California state levels which would result in additional delays."

Speaking for the consortium, Mr. Gould added that "the participants will maintain their interests in coal and other rights relating to the project, and will determine at a future time how the project might be continued either in the present form or reconstructed in another form." The extent of the "rights" was not specified.

Congressmen Ask Delay

Recently there had been a succession of adverse developments for the project.

Among them was a staff report of the California Public Utilities Commission, which might have to license the importation of power into California, that urged that the commission investigate the many pros and cons of the project and require certification of "convenience and necessity," covering the need for the power.

And last Saturday 31 members of Congress from various parts of the country formally requested Secretary Kleppe to defer approval of the project pending more study of its controversial aspects.

Congress currently is deliberating amendments to the Clean Air Act dealing, among other things, with the classification of Federal preserves in regard to pollution limitations. Conservationists have challenged both the prospective environmental impacts of Kaiparowits and California's need for the additional energy.

The state's energy needs for the next 30 years are disputed, with some experts saying as many as 30 new nuclear power plants will be needed — there are now three — and others

contending that with efficient conservation measures, the state could get along with a few more conventional power plants.

The cancellation was the latest of dozens throughout the country in the last couple of years, due variously to a drop-off in the growth of electricity consumption — which had been doubling every decade—increasing construction costs, difficulties in raising capital and environmental objections.

Illustrating the cost escalation problem, Southern California Edison officials noted that, as conceived in 1963, the Kaiparowits plant would have been more than 50 percent larger — five million kilowatts —and would have cost only $700 million.

Currently the cost of the plant itself was put at $2.7 billion, with $800 million more for related facilities such as

the nearby coal mine, the new-community development, and "environmental protection equipment," estimated at $600 million.

Southern California Edison officials said any near-term needs for the plant's canceled output probably would be met by "combined cycle" generating plants, which burn a type of kerosene and can be built relatively quickly.

For the longer term, the company has been counting heavily on the current tripling of its nuclear generating capacity on the coast at San Onofre, and on the construction of additional nuclear plants.

The question of more nuclear plants in the state is the subject of intense controversy, which will come to a climax in an initiative measure on the June 8 primary ballot. The measure would impose strin-

gent restrictions on any nuclear development.

Southern California Edison has been in the forefront among utility and allied interests opposing the initiative's proposals, and the elimination of Kaiparowits presumably will lend some strength to the company's argument for nuclear development.

The construction of coal-fired plants in California is generally considered to have been virtually precluded by air pollution problems and stringent restrictions on development along the coast, where there is plenty of the cooling water that power plants must have.

These factors, along with the dearth of water in inland parts of the state, have reduced potential power plant sites to a minimum, which is why the utility companies went to Utah.

April 15, 1976

Controversy Over Liquefied Gas Pits Energy Need Against Danger

By GLADWIN HILL

Every few weeks an unusual ritual takes place in Boston Harbor.

As Coast Guard craft scuttle about like sheep dogs, herding other harbor traffic aside, a tank ship from Algeria moves slowly up the channel to a pier at Everett. Other vessels are kept clear of her course for two miles ahead and a mile astern, as if the ship were laden with explosives.

Her cargo is a substance in everyday use in many factories and most of the households of the country, and one that might be imported in increasingly large volumes to help ease the energy crisis: natural gas. In transit, it is in a form far less explosive than in the kitchen: condensed to one six-hundredths of its normal volume and refrigerated to a liquid 260 degrees below zero. In this state it cannot be ignited even with a match.

However, a large-scale spill of liquefied natural gas, or LNG, might lead, if not to an explosion, to an immense fireball possibly several miles in extent, with commensurate calamity to people and objects in the vicinity.

This fact is the crux of a gathering national debate paralleling in a striking degree the running controversy over the hazards of atomic power generation.

If many big utility companies have their way, the cautious docking procedure currently found only in Boston will be regularly emulated at many United States ports, including New York, to avert what utility executives foresee as some very uncomfortable and disruptive shortages in the nation's already constricted energy supply.

Hundreds of thousands of jobs in industry, and gas service for cooking and heat-

ing homes, are said to hinge on increasing gas imports. Hundreds of millions of dollars have already been invested in LNG facilities to handle the expected increase.

Long Delays Possible

But if the feelings of many environmental specialists, governmental officials, legislators and apprehensive citizens prevail, the commencement of large-scale LNG importation may be delayed indefinitely or confined mainly to terminals remote from habitation rather than the metropolitan installations now contemplated.

In Washington, the Federal Power Commission has before it a half-dozen applications for approval of big, long-term projects for importation of LNG from as far away as Indonesia.

The F.P.C. also has before it a recent petition from the Attorneys General of New York, New Jersey, Pennsylvania and Delaware asking the agency to promulgate uniform national standards for the siting and safe operation of LNG marine terminal facilities. The state officials suggested that these terminals should be confined to areas of low population density, with suitable buffer zones maintained around them.

When the F.P.C. may decide these questions is uncertain. The agency, noted for the glacial pace of its proceedings, has been grappling with the LNG question for some six years.

It has tentatively approved terminal facilities at Cove Point, Md., 42 miles south of Washington, and at Savannah, Ga. But other proposals involving Providence, R.I., Staten Island, Logan and West Deptford Townships in New Jersey, Lake Charles, La., and three places in California are still up in the air.

Along with safety, the F.P.C. has to

consider the effects large-scale importation may have on the gas and other energy markets, and on national dependence on foreign energy sources. The White House suggested to the commission that, to avoid overdependence, imports be limited to 10 percent of annual consumption, which has been running about 20 trillion feet. Beyond that, state, county and municipal agencies must approve LNG operations.

Danger of Tank Rupture

Gas now provides nearly one-third of the nation's energy and one-half of the energy for industry. Up to now nearly all of it has come from domestic sources. Domestic reserves and production have been declining, and prices to users—including 41 million of the nation's households—have been rising.

With an early go-ahead on importations, gas industry executives say, LNG from a half-dozen foreign countries could, by 1980, meet as much as 15 percent of the nation's energy needs—more than three times the amount now obtained from hydroelectric power.

The big question is safety.

Contrary to a popular misconception, LNG does not have to be kept under heavy pressure: The customary amount is only one and a half times atmospheric pressure. It just has to be kept cold—and contained. LNG tanks are heavily insulated double-walled containers like thermos bottles.

The chief hazard expected in LNG handling is an accident, such as a ship collision, an airplane crash or an earthquake, that would rupture a tank, causing the freezing liquid to run out.

At the outer margin of the resulting spill, the liquid would warm up into gas and combine with air; the mixture containing 5 to 15 percent gas is the only ratio in which the gas is burnable. If not accidentally ignited, the spilled gas could evaporate harmlessly. If ignited, the heat could turn progressively more of the liquid into gas, starting a chain reaction that could extend for miles.

Staten Island Blast

The principal situation in which this might occur would be on the ocean. On shore, storage tanks are surrounded by dikes to contain any spill. Even if spilled,

329

Boston Harbor is empty of traffic except for tank ship Descartes bringing in shipment of liquefied natural gas

The New York Times/Arthur Grace

the gas inherently is not explosive unless ignited in a confined space.

Apprehensions of such an occurrence stem from two major accidents in the last generation—which engineers say could not recur with current technology.

At Cleveland in 1944 an early LNG tank ruptured and spilled millions of gallons that flowed into the sewer system and caught fire, killing 128 people and injuring 300. This was attributed to use of a weak alloy in the tank and inadequate diking.

In 1973 a big LNG tank in the Bloomfield section of Staten Island blew up, killing 40 workmen on a repair crew. The tank supposedly had been empty for a year. The explosion was attributed to residual fumes trapped in an unsatisfactory experimental plastic tank lining.

The big question mark in projecting LNG hazards is like the one hanging over possible nuclear power plant accidents: It is virtually impossible to deliberately stage a full-scale test accident to see what would happen.

Instead, studies have consisted, as with nuclear estimates, largely of computer calculations of the probabilities of accidents—even the possibility of an airplane hitting an LNG tank ship.

The answers that have emerged resemble those in the nuclear debate: millions-to-one odds against major accidents, with the mathematical premises disputed among experts.

One study done for the F.P.C. conjectured that an LNG tanker spill fire in New York Harbor might cause 807,000 casualties.

But a study done for the sponsors of a proposed Los Angeles terminal concluded that a person living within five-eighths of a mile would have only one three-hundredths the chance of being killed by an LNG fire as by an ordinary fire.

An independent study done by the Rand Corporation, the research organization in Santa Monica, Calif., concluded: "The prudent course of action would be to locate all facilities for handling LNG at remote sites until better estimates of risk can be made. Otherwise experience may be accumulated at enormous cost."

Dr. Edward Teller, the hydrogen bomb scientist, told a recent state legislative hearing in California that technical

knowledge of possible accidents with LNG stood now about where parallel knowledge about nuclear reactors was 25 years ago. He recommended going ahead cautiously with LNG shipping, but with a greatly accelerated safety research program.

The uncertainties about LNG safety involve some odd contradictions.

The current importation of LNG at Boston, by the Distrigas Corporation, is a result of a bureaucratic tangle. The F.P.C. originally approved the importation, doubting its own jurisdiction to prevent it. Later the agency reversed itself and now has scheduled hearings on whether the Boston imports should be allowed to continue.

94 Storage Facilities

Meanwhile, however, unknown to many citizens, some 94 LNG storage stations have been operating over the last decade without incident across the country, including New York City and New Jersey.

Forty-six of these facilities are plants that liquefy ordinary pipeline gas and store it against heavy cold-weather demand. Forty-eight are satellite facilities that simply store LNG made elsewhere and delivered to them, generally by truck.

Trucks are filled from storage tanks through hoses, with the loader's main concern being to avoid getting his hands frost-bitten handling the coupling.

A spill from an LNG truck could present a fire hazard, but considerably less, engineers say, than from a truckload of gasoline. Nevertheless, some communities, including New York City, restrict LNG trucking in central city areas.

A typical liquefying facility is the large plant of the San Diego Gas and Electric Company at Chula Vista, Calif. Its LNG customers, serviced by truck, range from trailer parks without conventional gas connections to the San Diego Zoo where a satellite facility fuels sightseeing vehicles, which vaporize the liquid from engine heat as it is needed.

Situation in New York Area

In New York, the Brooklyn Union Gas Company has been operating two LNG storage facilities at Greenpoint, and the Consolidated Edison Compay has one in Astoria, Queens. A $100 million marine LNG terminal, not yet licensed to operate, is nearing completion in the Rossville section of Staten Island.

A number of gas companies have been proceeding with large-scale importing plans as if eventual public sanction was a foregone conclusion. Lately this assumption has appeared more problematical.

The New York State Legislature this year assigned jurisdiction over LNG transportation and storage to the Department of Environmental Conservation. The department this month began the first specified step in implementing the law by studying the capabilities of the New York City Fire Department to cope with LNG accidents. Public hearings on the whole LNG question are to be held by January.

The New Jersey Legislature is also considering laws to fill various gaps in the regulation of LNG. The state claims jurisdiction over public utilities' LNG storage tanks, but not over those owned by pipeline companies. On utility companies' tanks, the state's Department of Labor and Industry has applied engineering criteria of the National Fire Protection Association and a requirement of buffer areas up to 600 feet between tanks and the nearest property line.

One of the biggest pending proposals, by the Pacific Lighting Company, to bring gas by ship from both Indonesia and Alaska to southern California, currently is involved in approval proceedings at Federal, state, county and municipal levels, with the safety question looming ever larger.

The company has put $60 million into planning terminals in Los Angeles Harbor, at Oxnard, 40 miles up the coast, and at Point Conception, 70 miles farther up. But the cities of Los Angeles and Oxnard are concerned about the safety aspects, and state legislators are drafting a bill that would restrict terminals to remote areas.

Regional gas shortages are a perpetual possibility in many parts of the country because pipeline delivery contracts from the principal production areas of Texas, Louisiana and New Mexico generally have loopholes permitting diversion of gas to more profitable intrastate markets without Federal price ceilings.

The Federal ceiling on interstate gas

is 52 cents for 1,000 cubic feet at the wellhead. By the time it has moved across country the wholesale price may go to more than 90 cents, and the consumer will pay around $1.20. A typical household will use about 300 cubic feet a day..

But the same gas, sold within the state where it originates to big users like petrochemical companies, will bring $1.75 to $2. Hence the lessening interstate supply. New England, Middle Atlantic and Middle Western states all have experienced pinches. Southern California was especially hard hit because of its heavy dependence on gas to lessen air pollution problems. A reduced supply from Texas has already forced Pacific Lighting to curtail service to some big southern California industrial customers.

Many big gas companies foresee regional shortages as being most simply alleviated, at prices competitive with domestic levels, by foreign gas, which can be imported only by LNG tank ships. A big LNG tanker can carry the condensed equivalent of 2.8 billion cubic feet of gas—enough to supply nine million households for one day.

Warnings on Curtailments

Joseph Rensch, Pacific Lighting's president, has testified at Federal and state hearings that without additional gas sources such as the contemplated LNG imports, by 1980 at the latest the company will have to curtail service to 38,000 smaller industrial and commercial customers not equipped to convert to any other fuel. He said a shutdown of these

enterprises would put 390,000 employees out of work and jeopardize the jobs of some 300,000 more related workers—altogether, 14 percent of the area's employment.

Even in the unlikely event that importation of gas from foreign countries was excluded for national policy reasons, LNG issues would persist in connection with Alaska.

The big gas deposits in the Prudhoe Bay oilfields may come south by pipeline. But there are extensive deposits being developed in many other parts of Alaska where transport by ship is the most feasible method. A liquefying plant of Phillips Petroleum on the Kenai Peninsula near Anchorage has been shipping LNG to Japan for several years.

October 7, 1976

Lag in Oil Conservation Troubles Aides of West's Energy Agency

By CLYDE H. FARNSWORTH
Special to The New York Times

PARIS, Nov. 1—The men charged with energy security in the West are growing increasingly concerned over the skimpy results of oil conservation programs and what they feel is a dangerous complacency in the public mind about the problems of future supplies and prices.

Three years after prices were quadrupled, at a moment now when the Organization of Petroleum Exporting Countries is considering still another increase, the warnings of troubles ahead are becoming louder and more insistent.

"As the 1973-74 energy crisis recedes from public memory," said Dr. Ulf Lantzke, executive director of the International Energy Agency in Paris, "there is a tendency to believe that the energy crisis is behind us. Nothing could be further from the truth."

The thinking in the 192-nation organization, established on an American initiative after the 1973-74 Arab oil embargo, is that the consumer nations are less than a decade away from the point at which their demand will meet the maximum levels the producing nations are willing to supply.

The implications are even greater price pressure and new interruptions of supply. Given the long lead times of energy research and development and energy investment, the West cannot, in the agency view, rely on the rapid availability of energy alternatives on a commercial basis.

This is not yet an official forecast of the agency. But the sobering drift of analysis by the economists and other experts in the Western energy coordinating body shows, in Dr. Lantzke's words, that "the world energy situation must remain a matter of serious concern to all of us."

Oil remains today the dominant energy source. Last year it supplied half of the total primary energy requirements of the industrial democracies. More than two-thirds of their oil requirements were met by imports.

The energy organization has given pretty close to flunking grades to efforts by the United States, consumer of a third of the world's oil supplies, to curb demand.

The conservation report card, handed out recently as the American election campaign was getting into high gear, barely caused a ripple.

Much of the agency's current depression over Western energy prospects stems from the attitudes in the biggest consumer country, where fuel prices remain well below world levels and where road transport holds the dubious record of the world's highest ratio of consumption per passenger mile.

"Unless the industrialized democracies implement together a coherent international strategy," said Dr. Lantzke, a 49-year-old West German legal official, "we will face another and, in the long term, perhaps more serious energy crisis."

The Paris-based agency sees substantial scope for increased energy saving through measures that would: price energy at world market levels; introduce effective automobile and other transportation efficiency standards; monitor energy conservation in industry more closely, and apply insulation standards and energy-saving building codes more effectively.

Cuts in Energy Use

Ratio of Total Primary Energy to Gross National Product*

	1973	1975	Index (1973-100)		1973	1975	Index (1973-100)
Belgium	1.52	1.36	89.5	Italy	1.28	1.23	99.0
Norway	1.49	1.36	91.3	Japan	1.32	1.27	96.2
Denmark	1.15	1.05	91.3	New Zealand	1.39	1.35	97.1
Britain	1.67	1.54	92.2	Canada	1.92	1.95	99.0
Sweden	1.36	1.26	92.6	Spain	1.48	1.48	100.0
Austria	1.39	1.30	93.5	United States	1.54	1.54	100.0
W.Germany	1.27	1.20	94.5	Switzerland	1.00	1.03	103.4

* The ratio is the number of energy units needed to produce one unit of G.N.P. Thus it took Belgium 1.52 units of energy to produce one unit of G.N.P. in 1973, but only 1.36 energy units per G.N.P. unit in 1975. If 1973 is taken as 100 on an index, Belgium would rate 89.5 in 1975, a cut in consumption of 11.5.

Source: International Monetary Fund

The New York Times/Jan M. Rosen

Filling up at a gas station in Austria, one of the countries that has been able to cut its consumption of energy.

November 2, 1976

331

Nuclear Power: No Green Light at the Polls

By GLADWIN HILL

Spokesmen for the atomic power industry are suggesting that the rejection of nuclear regulatory proposals by voters in six states represents a "green light" for nuclear development.

A number of realities suggest, however, that this interpretation is on the euphoric side, and that realization of President Ford's 1974 plea for "200 more nuclear power plants by 1985" still faces an array of political, technical and economic obstacles.

News Analysis

Although the nuclear industry spent millions spreading the idea that Nov. 2 ballot propositions in Ohio, Montana, Colorado, Washington, Oregon and Arizona amounted to a vote for or against atomic power per se, this was not the fact.

There is no great argument about public sentiment on that point. Repeated surveys have shown that a strong majority of citizens—71 percent in a Gallup Poll last July—approves of nuclear power in principle.

Concern Over Safety

However, the same surveys have shown that a comparable majority is unsatisfied with safety precautions in current nuclear development. In the July poll, only 34 percent thought present safety regulations "safe enough."

The focal national issue is under what regulatory constraints atomic power should be allowed to continue and grow. Some 50 plants are now operating.

The Nov. 2 ballot propositions, and one similarly rejected in California on June 8, attempted to address this question. They set forth a rather complex proposal conditioning future nuclear development upon approval by state legislatures only after demonstration that various safety criteria had been met.

It was this particular proposal, assuming citizens understood what they were voting for, that was in question, not either an unconditional halt or an unconditional go-ahead for nuclear power.

Situation in California

This distinction was amply illustrated in California. Voters there rejected the ballot proposal only after the State Legislature enacted three quite restrictive laws on nuclear development that tended to make the ballot proposition superfluous.

Perhaps the most significant thing about the seven citizen-initiated proposals—something calculated to give the nuclear industry pause—is that they occurred at all. Atomic power is supposed to be under strict Federal regulation. The emergence of the ballot proposals was clearly a broad citizen expression of "no confidence" in the Federal regulation, or in the industry's implementation of Federal requirements.

Moreover, unlike most voting, which is like a winner-take-all marbles game, the fact that a majority of voters in seven states rejected a certain regulatory proposal manifestly does nothing to alter the beliefs of those on the losing side that better regulation is needed.

They, along with like-minded millions in the other 44 states, will continue to protest, with time on their side. Activists in Maine, Michigan and other states are already at work on ballot-initiative petitions for future elections, with the moral support of a sizable segment of the scientific community, including, according to the latest reports, a number of concerned experts within the Nuclear Regulatory Commission itself.

'A Funny Situation'

"It's a funny situation, where we're losing all the battles but winning the war," one conservation official remarked. "Even when these proposals go down to defeat, we've educated more millions of people about the problems we see and get more people on our side."

Another reality confronting the nuclear industry is that the same election that saw the ballot proposals defeated also brought victory to Jimmy Carter. He supported the initiatives in principle: He said he would have voted for the one in Oregon, which was similar to the others.

He favors subordination of nuclear power to other energy sources; has promised to formulate a coherent energy policy in place of the present welter of "options" in which nuclear power has predominated, and has vowed to realign the tangle of Federal energy agencies whose combined efforts have given nuclear development impetus.

In recent weeks it has become apparent that misgivings about nuclear development are not just maunderings of a few dissidents in this country but are an international concern. Waves of objection have swept through France, Germany and Sweden, and a British Royal Commission turned in a decidedly adverse report on nuclear power. Ironically, all the public debate about atomic power is in a sense academic because the industry is up against far more pressing economic constraints.

Construction costs have risen rapidly since the nuclear push started in the 1950's. Industry spokesmen continue to put forward figures indicating that nuclear power is cheaper than other sources. But there is endless debate about the data, and there are even executives in the power industry who doubt the economic feasibility of nuclear power.

One glaring gap in the industry's equations is the still-unknown cost of recycling radioactive fuel elements and disposing of radioactive wastes. At the same time the industry has been saying nuclear power is economical, it has been pressing for Federal subsidies for these operations on the ground that they are too expensive for industry to bear. After years of work and many false starts, national arrangements for these two essential functions have not yet materialized.

Setback in Development

In the face of such problems, President Ford's prospectus of 200 new plants by 1985—20 new plants a year—is far from being fulfilled. Last year actually brought a retrogression in development: 11 new projects were announced, while 13 were canceled.

The ballot initiatives, if they had been approved, would have put obstacles in the way of a number of nuclear plants under construction or planned.

But the failure of the initiatives does not appear to have dispelled any of the basic problems confronting the industry or to have opened the way for any great leap forward in nuclear development toward the goal of its yielding 25 percent of the nation's electric power in 1985.

All in all, rather than a green light, the signal confronting atomic power looks more like, at best, the amber light of caution.

November 9, 1976

Study Finds Energy Projects Off Jersey and Delaware Pose Serious Ecological and Technological Problems

By BAYARD WEBSTER

The proposed development of huge energy systems off the coasts of New Jersey and Delaware to produce oil, gas and electricity will pose serious ecological, technological and political problems that state and Federal agencies are not equipped to handle, a Congressional study indicates.

Meanwhile in New Jersey, the Public Service Electric and Gas Company announced that it had received a license to operate the first unit of its Salem generating station at Hancocks Bridge. There will eventually be four units in what will become one of the largest nuclear installations in the world.

The study, released to Congress yesterday by the Office of Technology Assessment, surveys three energy projects that would be situated from three to 150 miles off the coasts of the two states.

The projects encompass floating nuclear power plants, the exploration for and development of oil and gas resources, and the construction of deepwater ports to handle imports of crude oil from abroad.

The liabilities in these proposed ocean-sited systems, the report states, includes a current lack of knowledge of the consequences of a nuclear accident in the marine environment, the increased possibility of major oil spills from supertankers using a deepwater port, and the absence of regulations requiring the use of the newest and safest equipment in drilling for oil and gas in the mid-Atlantic ocean floor.

In addition, the two-year study noted a continued lack of significant Federal support for energy conservation and alternative energy sources such as solar energy and fusion will mean that proposed offshore programs will have to be enlarged, creating the potential for even greater problems.

"No significant damage to the environment or changes in patterns of life in either New Jersey or Delaware is anticipated during operation of the three systems at presently projected levels," the

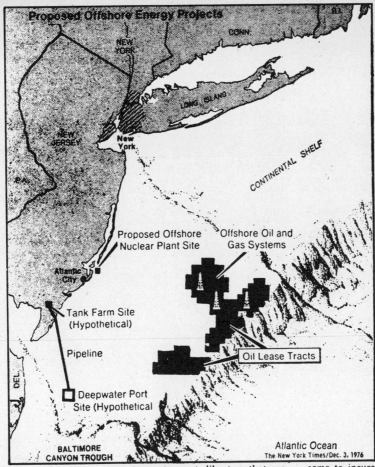

Proposed Offshore Energy Projects

NEW YORK
CONN.
R.I.

LONG ISLAND

New York

NEW JERSEY

CONTINENTAL SHELF

Proposed Offshore Nuclear Plant Site

Offshore Oil and Gas Systems

Atlantic City

Tank Farm Site (Hypothetical)

Pipeline

Oil Lease Tracts

DEL.

Deepwater Port Site (Hypothetical

BALTIMORE CANYON TROUGH

Atlantic Ocean
The New York Times/Dec. 3, 1976

report states.

But the report quickly adds that expansion of ocean projects in the absence of new energy sources could create serious conflicts among users and impose excessive ecological burdens on ocean and coastal environments.

The energy system closest to being activated is the oil and gas drilling and transport project in the Baltimore Canyon Trough on the outer continental shelf where some 500,000 acres of lease tracts have already been sold for more than $1 billion. Drilling by oil companies is expected to start within a few months. Two more sales of lease tracts amounting to another 500,000 acres are planned for the near future.

Estimate on Oil Extraction

Studies by the United States Geological Survey have estimated that between 1.8 billion and 4.6 billion barrels of oil could be obtained from the already-leased Baltimore Canyon tracts over a period of 14 years. This would be about as much as 7 percent of the total United States production in 1973.

The next most imminent system involves the plan by the Public Service Electric and Gas Company of New Jersey to build two 1,150-megawatt floating nuclear power plants behind a breakwater 2.8 miles off Little Egg Inlet 12 miles northeast of Atlantic City. Completion of these plants is tentatively projected for 1985.

The third project, advanced some years ago when there was a threat that foreign oil supplies would be cut off, is not expected to be acted on before the next decade. It is a deepwater port that might be situated 32 miles southeast of the southern end of the New Jersey coast with an undersea pipeline connecting it to an oil tank farm in Cape May County.

The deepwater port, like two that are already on the drawing board for the Gulf of Mexico, would consist of a huge floating steel buoy connected to the buried pipeline. Supertankers would tie up to the buoy, hook up with its floating rubber hoses and pump oil into the pipeline.

Danger of Huge Oil Spill

In its assessment of the deepwater port system, the study noted that substituting supertankers at the offshore deepwater port for smaller tankers using shallower coastal port facilities would tend to reduce the overall amount of oil spilled. But a huge oil spill with major impact would become a possibility with the supertankers, the report said, adding, "Yet tanker regulations are less strict than port regulations."

In addition to pointing out that the Nuclear Regulatory Commission had not looked at the detrimental possibilities of large-scale production of floating nuclear power plants, the report said:

"No detailed procedure or design standards have been developed for transporting fuel to a floating plant or for carrying irradiated fuel and other radioactive wastes to shore."

Risks of Oil Spills Assessed

In its appraisal of offshore oil and gas development, the Congressional agency indicated that "there probably would be at least one major oil spill during development of the Baltimore Canyon fields." The risk assessment, based on 10 years of offshore oil operation in the Gulf of Mexico, is that the odds of an oil slick, even from a major spill at a platform 50 miles offshore, reaching the Delaware or new Jersey shoreline would be one in 10.

Political and legal problems add to the thicket of obstacles that must be overcome to insure safe and efficient operation of any of the three systems, the study indicated, in the case of offshore drilling, it was found, Federal management of offshore tract leasing is fragmented, existing liability laws are inadequate, and the limited role of state governments in the decision-making process may cause bad planning and delays.

There are also potential conflicts between offshore oil and gas activities and shipping and fishing interests, but no coordinated study has been made to identify all the conflicts and to find solutions for them, the study noted.

Alternatives Examined

The report, a two-volume document totaling more than 1,000 pages, examined available alternatives for the three proposed systems and concluded "No new technology is likely to be developed sufficiently to provide more than a small percentage of the total energy required in the two states before the year 2000."

But the study added that, among other possibilities, conservation, "more efficient use of waste heat from power plants, increased insulation and improved automobile efficiency, could hold down the growth in energy demand."

The Office of Technology Assessment, an advisory arm of Congress, has undertaken a variety of assessments including food, materials resources and health. The current report, entitled "Coastal Effects of Offshore Energy Systems," was made at the request of Senator Ernest F. Hollings, Democrat of South Carolina. It can be obtained through the Superintendent of Documents, U.S. Government Printing Office, Washington, D.C. 20402, for $4.45.

December 3, 1976

Technology

Sweden—An Energy-Use Model for U.S.

By VICTOR K. McELHENY

The people of Sweden, who live about as well materially as average Americans consume a lot less energy in pursuit of the good life.

Attention to the Swedish success in using energy at only 60 percent of the American rate has intensified this year among conservationists and other energy analysts in the United States.

These experts know that the easy energy savings have been achieved since the 1973 multiplication of world oil prices. Yet the nation's domestic oil and gas production continue to erode, increases in coal mining are slow and environmental, financial and technical problems hobble the construction of new coal or nuclear electricity plants.

While environmentalists continue to urge drastic change in American life styles, others, including executives of electric utilities, are beginning to search for new ways to save energy with a minimum of change in the American home or workplace.

Earlier attempts to compare American energy use with Western Europe were dismissed as too sketchy, or neglecting differences in population densities and industrial patterns.

• • •

But the attention of energy analysts has turned to Sweden, with a standard of living, industrial "mix" and population distribution closely comparable to the United States.

In the last two years, three comparative studies have been made in this country, by Andres Doernberg of Brookhaven National Laboratory, S. I. Kaplan of Oak Ridge National Laboratory in Tennessee, and by Lee Schipper and Dr. Allan L. Lichtenberg of the Energy and Resources Group of the University of California at Berkeley.

The California study, most detailed and recent of the three, is receiving widest attention. It was issued first in April as a report of the Lawrence Berkeley Laboratory and then was published this month in the weekly, Science.

This week, in an interview, Dr. Lichtenberg elaborated on the report, whose comparisons chiefly involved data from 1971, before the oil crisis intensified pressures for conservation. He said the report took about a year's work, including a special trip to Sweden by his Swedish-speaking collaborator, Mr. Schipper.

Reviewing a mass of published information from the two countries, the California researchers found that the Swedes achieved their energy savings in many sectors of the economy, most dramatically in transportation.

The average weight of automobiles in Sweden was only 2,400 pounds, 60 percent of the American figure of 3,700 pounds. Apparently because of reliance on better mass transit, there were only 0.3 cars per person in Sweden, compared with 0.45 in the United States.

In Sweden, where "second cars are replaced by mass transit and a significant number of families have no car at all," people used their long-lived cars 14 years versus less than 10 in the United States, for only 55 per cent of their trips shorter than six miles, compared with 90 percent in the United States.

• • •

Dr. Lichtenberg said, "It appears that the best and highest use of a car is recreational." This is what Swedes tend to use their cars for, relying on mass transit for shopping and commuting.

Stiff taxes on gasoline and the actual weight of cars put on pressure for low-maintenance, high-efficiency, low-weight cars. Sales taxes were about $500 for a 2,500-pound car, and rose to $900 for the American average weight. Annual user charges began at $32 for the 2,400-pound car and increased $8.40 for each additional 220 pounds.

Overall, the California researchers found, the American transportation sector used about 24,000 out of a total budget of 100,000 kilowatt hours of heat per person. In Sweden, transportation required 7,800 kilowatt hours of a total of 60,000, and the proportion of all energy used for transportation, 13 per cent, was just over half that of the United States.

In 1971, the United States "burned" 17 kilowatt hours of heat for each 1972 dollar of economic output, and Sweden consumed 10. The United States figure had been virtually level since the 1950's, after a rapid shift from reliance on coal to inexpensive domestic oil and natural gas.

Sweden, having shifted from coal to expensive imported oil, and lacking natural gas, experienced a 25 percent increase in energy use per unit of output in the 1960's.

In the interview, Dr. Lichtenberg said the contrasting energy policies of the United States and Sweden seemed to have been specified by contrasting energy-price history. For the United States, it had always been cheap, for Sweden always expensive.

Examples of the pressure came from Swedish homes, which take an average of 9,200 "degree-days" of heating each winter to maintain a temperature of 68 degrees fahrenheit, in contrast to an average of 5,500 such days in the United States. The Swedish figure is comparable to North Dakota's.

The average heat loss through the walls of a Swedish home, either apartments or single-family dwellings, was half the United States figure. And so, overall household heat-energy use in Sweden was 10 percent below that of the United States.

Of the Swedish home-heating energy in 1971, 19 percent came from hot water piped in from neighboring heating or electric power plants. The energy-saving from two-fold use of hot water in Swedish electricity plants amounted to 2 percent of that nation's United States level.

Although Swedish industry concentrates on such energy-intensive products as steel, paper and cement, virtually every sector of Swedish industry used less energy per pound of output than its American counterpart. Overall Swedish industrial energy use per inhabitant was 17 percent below the total energy budget.

With such data in hand, American energy policymakers may have a harder time in future dismissing the effectiveness of taxes as incentives to conservation.

United Press International

Sweden has been able to hold down its energy bill by making use of small cars and efficient public transportation. This scene is in Stockholm.

December 8, 1976.

Solar Energy's Future: Optimism Is Restrained

By VICTOR K. McELHENY

Optimism among experts about the future role of solar energy is still restrained, despite cost-cutting advances in technology and sharply increased Federal Government support for work toward harnessing sunshine. Attacked as "pessimistic" by such critics as the Congressional Office of Technology Assessment, the United States Energy Research and Development Administration's plan for solar work, published in June 1975, predicts that various methods of tapping the sun's rays will contribute 6 percent of the nation's energy supply in the year 2000.

According to the Energy Administration, the solar energy would come in many forms. Rooftop collectors for heating water or air would provide about 6 percent of the heating—and cooling—energy requirements of homes and commercial buildings. An even larger supply of energy might come from the burning of forest and other wastes or fuels manufactured from them.

Solar electricity is expected to come from such sources as windmills, silicon solar cells (costing $15.50 per watt compared to $200 five years ago), temperature-gradients in the sea or arrays of mirrors focusing sunlight toward the top of a tower where air or helium gas would be heated to run a power plant.

Spending Appropriations Included

To study such possibilities, the agency's budget of more than $6 billion for the fiscal year that began Oct. 1 includes Congressional appropriations for $179 million in spending and authority to write contracts totaling $290 million. Both figures were far above the Ford Administration's request, transmitted last January, and were equal to ERDA's requests to the Office of Management and Budget.

Just two years ago, the solar budget of the then newly created agency was $42 million in contract-writing authority and $15 million in spending.

Nonetheless, the tone of scientists in the field is subdued. A panel of 30 faculty members at the Massachusetts Institute of Technology said in a report earlier this year: "We are enthusiastic about the potential of solar energy, particularly if it combines with widespread energy conservation. But we caution that the major impact of solar energy will probably not occur soon."

For this reason, the M.I.T. group said that ERDA's solar program "places undue emphasis on demonstration of already-known technologies. A larger effort should be devoted to long-range research aimed at greater economic savings."

A Combination Is Urged

Among the better-established solar technologies is the use of rooftop collectors to heat water for heating homes, schools and commercial buildings. Experts now are urging that such systems be combined with so-called "heat pumps" for greater efficiency and a lower need for electricity as a backup.

On Wednesday, ERDA released a study by the Mitre Corporation indicating that solar heating for individual new homes is now judged economically competitive with baseboard, or "resistance" electric heating in a dozen urban areas, led by New York City. The study said solar energy could be competitive in many more areas if rooftop units could be financed at relatively attractive mortgage rates and if energy costs rose at 10 per cent annually.

The Mitre study, the most exhaustive of its kind by the Government, also raised the points that make many observers predict only a modest share of the market for solar heating units.

The estimates for a new solar home-heating unit ranged between $4,000 and $12,000, and the minimum period in which such an investment would be recovered—in New York City—through savings on electricity was estimated at 10 years. This is longer than the seven-year median period of home ownership.

Such costs and periods for recovering the investment through savings, according to the agency's 1976 research plan document, are major obstacles to rapid adoption of solar heating by owners or home-builders.

The energy agency said: "The problem is intensified by the current absence of consensus standards on construction performance, modification in current construction practices, and lack of information on system reliability and maintenance requirements."

Despite this, Arthur D. Little Inc., a Cambridge, Mass., consulting firm heavily involved in solar technology, predicts that the market of $40 million to $60 million for solar heating in 1976 will multiply to between $800 million and $1.5 billion by 1985.

A second factor limiting the optimism of solar experts is the slow pace at which the United States and other advanced nations tear down old homes and buildings and replace them with new dwelling units. The rate is about 1 per cent a year, far below the rate of nearly 10 percent in the motor vehicle industry.

A less-noticed restraining factor derives from the close alliance between solar technologies and efforts at overall energy conservation. Because the sun shines intermittently, captured solar energy must be stored, in insulated tanks of hot water or chemicals, heated rocks, caverns filled with compressed air or electric storage batteries.

But such storage devices are also intended for general savings in building all types of energy-generating equipment. Electric utilities are beginning to seek such savings through a process they call load-leveling, which includes inducements to customers to use appliances in off-peak hours. Such savings could help stimulate the solar market, and then limit it.

December 31, 1976

GAS CRISIS SPREADS

OHIO SCHOOLS CLOSE

Businesses in Pennsylvania Shutting—Hundreds of Thousands to Be Idle

By STEVEN RATTNER

The two-week-old natural-gas crisis bit more deeply for millions of Americans yesterday as the effects of the bitter, persistent cold were forcing shutdowns of stores and factories in Pennsylvania and causing school closings there and in Ohio, new layoffs for hundreds of thousands of people in a dozen states and warnings of possible further hardships.

A day after 2.6 million children and their teachers were turned out of school for at least three days, Gov. Milton J. Shapp of Pennsylvania appeared on television last night to ask all nonessential businesses to close until noon Monday. Factories, stores, restaurants and entertainment centers were all called on to shut down.

Compliance Reports Mixed

While Governor Shapp's call had no legal force and early reports of compliance were mixed, in Pittsburgh the Equitable Gas Company said it would stop service to "nonessential" businesses from Saturday to Monday. In Philadelphia, officials ordered 2,500 companies employing 130,000 people to cease all "nonessential" usage of natural gas.

In Ohio, Gov. James A. Rhodes declared an "energy crisis," and most schools prepared to close as a result of appeals by four of the five gas utilities in the state. With an estimated 75,000 people already jobless because of the cold, the Ohio Association of Manufacturers estimated that the elimination of service to all industries — planned for today—would bring the layoffs to 250,000.

In both states, the authorities were acting as part of a desperate effort to avoid more dramatic disruptions if temperatures plunge this weekend, as predicted. A new storm, sweeping in from the Northwest, was expected to produce minus-30 degree readings in North Dakota and Minnesota and snow and subzero temperatures throughout the Great Lakes area.

For the first time since the cold wave swooped down on Jan. 17, energy officials are raising the specter of widespread fuel cutoffs to homes.

"It's more a possibility than it has been in my experience," said John Borrows, director of utilities at the Ohio State Public Utilities Commission. "The situation is very tight."

Nor were Pennsylvania and Ohio alone. In the New York metropolitan area, the first widespread industrial and commercial curtailments were imposed. In Virginia, as many as 30,000 workers may be laid off and hundreds of schools closed as the State Corporation advanced a curtailment slated for Monday and ordered gas withheld from all nonessential customers in an eastern area, including Richmond and Norfolk.

The cold has brought added hardships, not all of them energy related. Parts of the Maryland and Virginia shore have been declared a disaster area because frozen rivers have put boatmen and fishermen out of work. Prices for fresh vegetables are reported to be soaring in many places as a result of crippling frosts in Florida.

The first freezing of the Ohio River in 30 years continued to stall dozens of barges loaded with fuel oil, road salt and other products in short supply. A few communities in Pennsylvania have reported running out of heating oil and officials fear the shortages could mushroom across the upper Middle West.

Div Inventories Drop

The prospect of a more widespread fuel oil shortage remained as the American Petroleum Institute reported yesterday that inventories continued to drop last week and the Petroleum Industry Research Foundation warned of serious problems from continuing cold weather and gas shortages.

In Ohio, a grim Governor Rhodes appeared before reporters a day after he called on residents to pray and said, "What we are talking about tonight is the survival of Ohio."

He upgraded his Sunday declaration of an "energy emergency," which asks for only voluntary cooperation, to an "energy crisis," which gives him broad powers without legislative approval. He said that at present he would only lift the ban on burning high sulfur coal. However,

few industries are able to switch from gas to coal.

Ohio School Superintendent, Martin Essex, estimated that more than 400 of the state's 627 school districts will close, at least through Monday. Leaders of the General Assembly said they would issue a statement on the problem of the lost school time.

All gas service to industry, except for plant protection, was ended at midnight last night and with reserves low and prospects of added supplies at best uncertain, no one was offering predictions on when the crisis would ease.

The Ohio Public Utilities Commission ordered utilities to begin media campaign to tell homeowners and apartment dwellers what to do if the flow of natural gas stops.

In Pennsylvania, Governor Shapp told his constituents, "If we do not take severe measures during the next few days we may find ourselves in a truly catastrophic situation."

Part of the gas would be diverted to residences in the hope of preventing a sharp drop in distribution pressure, if demand soars over the weekend as predicted. A fall in pressure could eliminate service to thousands of homes, and restoring service might require individual house calls, a process that could take days.

"We hope it's enough to keep the home fires burning," said Thomas Clift, a gas supply technician at the State Public Utilities Commission.

Keeping Protection Levels

Some gas would also be used to keep industry, which has already lost virtually all of its service, from falling below minimum protection levels. With 50,000 people already out of work in Pennsylvania and many thousands of additional layoffs predicted, the loss of this supply would mean weeks rather than days of joblessness because of the time required to restart kilns and furnaces.

In Indiana, environmental restrictions on burning high-sulfur coal were also lifted as gas shortages idled 22,000.

During the bitter cold last week, loss of heat to homes, except in a few isolated instances, was avoided as thousands of factories closed to reduce natural gas usage. It had been hoped that service would be gradually restored to companies without alternative fuel, to reduce the economic burden.

But now, as stockpiles of gas grow lower daily, gas officials believe that goal has become unrealistic.

Economists have not yet estimated what the combined impact of cold weather, high energy bills and shortages is likely to be on the economy, but early indications, ranging from steel production to reports from retail companies, indicate that the effect will be deep and

lingering.

Just yesterday, Ward's Automotive Reports, the industry's statistical agency, said that the effects of cold weather supply shortages and natural gas curtailments had already cost the production of 30,000 cars this month. The General Motors Corporation said that a Georgia assembly plant with 4,000 workers remained closed because of gas shortages and that two plants near Buffalo, employing 8,000, had been crippled by a blizzard.

Overshadowed by the widening gas shortage, but equally worrisome to energy experts, are declining supplies of home heating oil, particularly in the Northeast and Upper Midwest.

Yesterday, the Petroleum Industry Research Foundation calculated that if the drawdown in supplies of the last five weeks continued through the rest of the winter, shortages could develop.

The foundation attributed the problem to a combination of cold weather, delivery problems and a shift by customary users of natural gas to fuel oil.

"None of these problems is of major proportions as of now," it said. "But continuing subnormal weather could shortly make them so."

Inventories of home heating oil now stand at 162 million barrels and have been declining in the last month by an average of 7.5 million barrels per week. Analysts believe that problems will occur at the 100 million-barrel level.

On the price side, the Federal Energy Administration took action yesterday to slow the steady rise in fuel prices—3-1 cents per gallon between last June and December, according to its own estimates.

It plans to require domestic refiners to subsidize the cost of imported heating oil for February, March and possibly April. Agency officials contended that the move might shave almost 4 cents a gallon from heating oil prices, although they acknowledged that the final effort would depend on the severity of the winter and on the level of foreign imports required.

Fear of Controls

The agency argued that without the subsidy, East Coast distributors might refrain from importing as much oil as their customers need for fear that the high-priced fuel would lead to reimposition of price controls. Prices are now within a penny a gallon of the level at which the F.E.A. has promised to take action.

The freeze has also crippled freight and passenger railroad operations in much of the East and Midwest. Some of the worst disruptions are occurring at Baltimore and at Newport News, Va., where coal has frozen in the railroad hopper cars and cannot be loaded onto ships heading abroad.

January 28, 1977

President Signs Natural Gas Bill, Marking First Legislative Victory

By EDWARD COWAN
Special to The New York Times

WASHINGTON, Feb. 2—President Carter signed into law today the bill he had requested to let the Government shift natural gas from surplus areas to shortage areas to keep houses warm and essential services operating.

Mr. Carter's first legislative victory came with final approval of the bill by Congress earlier today, his 14th day as President. The vote in the House was 336 to 82. The Senate approved the bill by unrecorded voice vote.

In addition to allocation authority, the bill suspends until July 31 Federal price ceilings for extra gas volumes purchased by interstate pipelines to replenish reserves depleted by this winter's bitterly cold weather in all regions east of the Rocky Mountains.

The cold weather in the New York metropolitan area, meanwhile, has sharply increased demand for home heating oil. According to major dealers, prices have risen about 12 percent since last winter and show no signs of abating.

In his television address tonight, President Carter congratulated Congress "for its quick action" on the emergency legislation, but he added, "The real problem—our failure to plan for the future or to take energy conservation seriously —started long before this winter and will take much longer to solve." Mr. Carter said that by April 20 he would submit an overall energy program to Congress that would emphasize conservation of fuel and "development of coal safeguards."

The President said: "Oil and natural gas companies must be honest with the people about their reserves and profits. We will find out the difference between real shortages and artificial ones. We will ask private companies to sacrifice, just as private citizens must do."

With factories shutting down and with gas supplies to homes threatened, Mr. Carter began planning the legislation and his Jan. 21 appeal for 65-degree temperatures in homes even before he was sworn in on Jan. 20.

The Senate this morning and the House later today approved a "compromise" bill hammered out quietly yesterday evening in a joint legislative conference dominated by the Senate and its new majority leader, Robert C. Byrd of West Virginia. Meeting informally with members of the House Commerce Committee before a crackling log fire in the office of the Secretary of the Senate, the legislators reached agreement on what was essentially the Senate bill even before Mr. Byrd put before the Senate a motion to appoint conferees—a motion he evidently planned to withhold unless he had an agreement with the House in hand.

The Senators had a powerful ally— President Carter. Mr. Byrd argued that, if the House didn't withdraw a price-ceiling amendment it had written into the White House bill, there would be no act of Congress for Mr. Carter to report to the American people in his televised talk tonight. Democrats in Congress savored the prospect of the new President showing the country fast action by a Government whose executive and legislative branches are again both controlled by the Democratic Party.

The House amendment, sponsored by two Texas Democrats, had sought to protect consumers in gas-producing states from experiencing steep runups in energy costs if interstate pipelines bid up prices under the emergency powers in a desperate effort to replenish depleted winter reserves. To that same end, the conferees wrote into the joint bill language intended to prevent the triggering of escalation clauses in intrastate gas-purchase contracts.

The bill has two key provisions. One gives the President authority until April 30 to order shifts of gas among interstate pipelines for the sake of keeping houses warm, averting damage to industrial equipment that would be damaged by a sudden loss of heat and maintaining fuel flows to small stores, hospitals and other essential public services. The second provision would let interstate pipelines take delivery through July 31 of extra gas for which they pay more than the Federal

Power Commission ceiling price of $1.44 a thousand cubic feet.

In addition, the bill requires intrastate carriers—for example, Lovacca, a pipeline wholly within Texas—to transport gas that is under allocation orders from one interstate pipeline to another. This authority was regarded as necessary to bring eastward gas from West Texas that normally flows to California, which has been having a relatively mild winter and can spare some reserves.

West Coast gas and some gas from the producing states—chiefly Texas, Louisiana, Oklahoma, Kansas and Arkansas—will flow mainly to the Southeast and to industrial states of the Northeast, especially Ohio, Pennsylvania, New Jersey and New York. Northern Illinois and the Detroit area of Michigan, which have had relatively good supplies, are expected to give up some gas.

The Federal Power Commission last night ordered the Columbia Gas Transmission Corporation to sell to the Southern Natural Gas Company one-fourth of 250 million cubic feet of Canadian gas that Columbia imports daily on an emergency basis.

In another development, the Senate Interior Committee approved the nomination of John F. O'Leary to be Federal Energy Administrator. No obstacle to confirmation by the Senate itself was foreseen.

James R. Schlesinger, the White House energy chief, repeatedly told Congress and the public that the allocation bill would help to protect the most essential uses of fuel but would not cure the deep shortage that has shut factories and processing plants and caused layoffs of hundreds of thousands of persons. Some of these shutdowns may last for weeks, depending on how cold it is.

The bill was criticized by members from hard-hit states, notably Ohio, because it did not authorize the President to shift gas for the sake of reopening some plants, such as glass-making shops, that must use the steady, high heat of natural gas. For the most part, however, Congress accepted the Administration's view that a shortage could not be cured simply by moving meager supplies around and that the minimal additional volumes of gas that might be drawn from intrastate markets at premium prices would not be enough to give much help to industry.

The emergency pricing authority was made to last until July 31 so that the pipelines and local distribution companies would have some time after the heating season to buy extra gas to replenish their reserves for next winter. This provision led some liberals, notably Senator James Abourezk, Democrat of South Dakota, to denounce the measure as "a deregulation bill."

Members of Congress from gas-producing states repeatedly reminded their colleagues—and the courts, should a case arise—that the bill could not be construed as extending Federal price controls to gas consumed in its state of origin (so-called intrastate gas) which sells for $2 or $2.25 a thousand cubic feet compared with the maximum Federal price of $1.44 for gas that moves across state lines.

The bill's Democratic managers, Senator Adlai Stevenson of Illinois and Representative John D. Dingell of Michigan, repeatedly assured Congress that the Government was not trying to extend its reach to intrastate gas.

The vexed jurisdictional and pricing issues will be the subject of a furious battle later this year when Congress takes up permanent amendments to the Natural Gas Act. President Carter has said he favors ending Federal regulation of producer prices for new gas but not for gas under contract already.

Resolving doubts about the pricing provisions of the emergency bill required satisfying producing and consuming states in ways contrary to their usual purposes. Senator Stevenson and other Northerners feared that a low price ceiling would inhibit intrastate sales of extra gas to the North. Members from producing states feared that the energy costs of their constituents would soar. They also feared the possibility of price rollbacks after delivery by Presidential order.

A letter from Mr. Schlesinger to Senator Stevenson promised that as a general rule the President would permit sales at or just above prevailing intrastate levels, with higher prices subject to review. "It is our intention to provide price certainty," Mr. Schlesinger wrote, "so that once a specific transaction has been authorized by the President there would be no risk of a price rollback for that transaction."

In addition, the House-Senate conference added to the bill language intended to strengthen an original provision to prevent emergency prices from triggering renegotiation clauses in intrastate contracts. Under these clauses, prices escalate with the market. The original provision insulated such contracts from emergency purchase prices under the Presidential authority. The additional language said that the renegotiation clauses would also not be triggered by prices paid during the emergency for sales under the Federal Power Commission's standard 60-day emergency purchase authority.

Comments From Industry

Spokesmen for the natural gas industry said they were generally pleased by the passage of the emergency legislation but warned that there might be little gas available to share because of tight supplies across the country.

"Columbia believes it will not significantly ease the gas supply crisis for the balance of the winter," said a spokesman for the Columbia Gas System. "It may prevent severe outages of service if unusually cold days occur during February and March. By authorizing 180-day emergency purchases, it should help Columbia to rebuild its storage during the 1977 summer."

Other industries called for deregulation of the gas business. "We emphasize that the emergency legislation adds nothing to overall natural gas availability," said a spokesman for the Union Carbide Corporation. "Only through permanent deregulation of the price of new natural gas—combined with citizen and industry conservation actions—can we expect to withstand the rigors of future severe winters."

February 3, 1977

Gas Shortage a Fundamental, Long-Term Economic Threat to U.S., Experts Say

By JAMES P. STERBA

Special to The New York Times

HOUSTON, Feb. 21—America's natural gas shortage is fundamental, according to many experts in the industry and in government, and the resulting economic disruptions and personal hardship cannot be prevented in the next several years.

But, these experts say, there is no choice but to end the nation's addiction to natural gas as a cheap power source. The pain can at best be eased and spread more thinly by mandatory conservation and allocation measures. But it cannot be stopped. Its severity depends largely on the weather.

While this winter's crisis has eased and the severity of next winter's is anybody's guess, members of Congress and the Administration are trying to figure out whom to blame for the shortages and why they exist. Two House subcommittees have scheduled hearings this week on whether natural gas producers are curtailing production in an attempt to get higher prices.

In addition Interior Secretary Cecil Audrus told his department last week to find out why production has ben cut back in offshore natural gas fields.

The long-term outlook for natural gas supplies is also bleak, although there is sharp debate over how bleak it is. Many experts agree, however, that even a successful massive search for new gas fields cannot reverse the current decline in production and reserves. Higher prices to stimulate that search will, at best, ease long-term shortages, these experts believe.

There is unanimous agreement among producers that if President Carter keeps his campaign promise to lift Federal price controls on new gas discoveries, free-market pricing mechanisms will lead to both increased discoveries and decreased consumption. Many of them doubt, however, that they will be able to find as much natural gas each year as the nation is now using—roughly 20-trillion cubic feet a year.

At Best, Minimizing the Decline

"At this point, the best we can hope to do is minimize the decline," said F. E. Ellis, vice president for North American production at the Continental Oil Company.

L. D. Wooddy Jr., natural gas manager for Exxon U.S.A., the nation's largest gas producer, said he personally agrees:

"You can be real wrong in this business. But based just on statistics and the fact that every time you find a gas field there is one less to find, it seems hard for me to believe that we could ever get back up to a finding rate of 20-trillion cubic feet per year."

In a single generation, the United States States has gone from natural gas feast to famine. What was burned off as an oil producer's nuisance 30 years ago, cannot be bought at any price by some consumers during the current frigid winter. Because it was—and still is—the cheapest, cleanest and most efficient fuel available, and because it was promoted as such, gas consumption soared after World War II to become the country's single biggest source of energy, accounting for a third of the nation's energy needs.

Because Federal price controls kept it cheap, exploration patterns have in the past been distorted in favor of the search for oil. One barrel of oil contains roughly the same amount of energy as 6,000 cubic feet of natural gas. At current Federal price ceilings, the same amount of energy can be sold for $12 as oil but only $4.64 as gas. In the past, the price difference has been even greater.

Gas production peaked in 1973 at 22,600 billion cubic feet for the year and has been declining since. Last year, it was 19,700 billion cubic feet, according to the United States Geological Survey. The basic reason was that some older gas fields were running dry while not enough new fields were being discovered and produced to make up for the lost volumes.

Complications in Shipping

Unlike oil, natural gas cannot simply be pumped into tankers in the Middle East and shipped to the United States to make up for shortages. Gas requires expensive liquefication plants and special ships, which so far bring only tiny amounts at prices ranging from $4 to

Dwindling Reserves of Natural Gas

Proven reserves in trillions of cubic feet

291 279 266 250 237 228

1970 1971 1972 1973 1974 1975*

*Current estimates, based on U.S. Geological Survey figures, show reserves of 215 trillion cubic feet.

Source: American Gas Association

The New York Times/Feb. 22, 1977

$5 per 1,000-cubic feet—double the highest prices domestically.

So-called "proven" domestic gas reserves—new discoveries that can be produced economically—have also declined steadily for the last nine years. Since 1968, the nation has consumed more gas than has been discovered. In that year, the United States produced and consumed about 19,000 billion cubic feet but found only about 13,000 billion cubic feet to replace it. In 1975, the nation consumed about 20,100 billion cubic feet but found only 6,300 billion cubic feet. Thus, reserves have steadily shrunk to less than a 10-year supply at current consumption rates. In other words, if no more gas were found, the amount in the country that has been discovered but not yet produced would last less than 10 more years.

By the end of last year, the United States had produced and consumed roughly 520,620 billion cubic feet has been consumed since John Glenn first orbited the earth in 1962.

Current estimates of "proven reserves" —which are not proven at all, but are only educated guesses —range around 215,000 billion cubic feet. Thus, in all, producers have discovered 735,000 billion cubic feet of gas.

Guessing About Supplies

How much gas is left to be discovered? Guesses of these potential reserves are far less educated than those of proven reserves. They are the subject of heated debates and constant revisions. Between 1950 and 1965, for example, there were at least 45 supposedly authoritative studies that concluded potential reserves ranged from zero to 2,959,000 billion cubic feet. This range has been used to support every argument ranging from the contention that we are virtually out of gas to the position that we have a 150-year supply waiting to be discovered.

The last Geological Survey estimate, made two years ago, was this: Given available technology and current economics, there is a 95 percent probability that 322,000 billion cubic feet can be located

and produced: there is a 5 percent probability that 655,000 billion cubic feet can be located and produced. The statistical mean, or best guess, is that 484,000 billion cubic feet can be found and eventually produced. That is roughly a 25-year supply at current consumption rates. The more the nation conserves, the more years it will last.

Significantly, if the low figure is accurate, the United States has already discovered three-fourths of all the produceable natural gas that it will ever find. And that amount was the cheapest and easiest to find. What is left to discover is generally in smaller fields, deeper fields, fields further offshore, or fields in hostile environments such as the Alaskan Arctic. And that means it will be more expensive to find and produce. Inflation makes it even more expensive.

On top of the mountain of geological guesswork are two major factors that constantly cause gas producers to revise their estimates of proven and potential reserves: technology and economics. Since reserves are defined as amounts produceable given the available technology and current prices, each technological breakthrough and each price rise alters the estimates.

Uneconomical Production

For example, geologists estimate that there are some 300,000 billion cubic feet of gas—a 15-year supply—trapped in tightly packed rock layers under the Colorado Rockies. But they can't figure out how to get it out economically. Two expensive and controversial underground nuclear explosions were tried. They failed. The gas would not seep out of the rocks fast enough to produce in economical volumes.

Similarly, drillers find lots of other oil and natural gas that cannot be produced economically. About one out of every nine exploratory holes drilled turns up some hydrocarbons, either oil or gas or both. But only about one in 50 finds enough to be produced commercially. The rest are classified as "dry holes," even though they are not necessarily dry at all. Roger D. Stanwood, vice president of Transcontinental Gas Pipe Line Corporation explained it this way two years ago:

"Many new gas wells that were classified as dry holes a few years ago, when gas may have sold for 15 cents per thousand cubic feet or less, may not be economically 'dry' today in areas where intrastate prices have reached $1.50 and higher per thousand cubic feet. The 'dry' in dry holes is not necessarily an indication that a well has found no hydrocarbons. It simply means that any hydrocarbons present would cost too much to produce to make a profit at the prevailing price."

Economic Realities

Since producers are in business to make a profit, they are naturally unwilling to spend more money developing a gas field than they could make selling the gas they could produce from it. A simple example:

Suppose a producer found a small field containing a billion cubic feet of gas, and that it would cost a million dollars to drill a producing well to it. At last year's federally controlled price ceiling of 52 cents per 1,000 cubic feet, the most money he could hope to get selling the gas would be $520,000. He would lose $480,000. At this year's Federal price of $1.44, he could sell it for $1,440,000. At current unregulated prices in the intrastate market of about $2 per thousand cubic feet, he could double his money.

Unfortunately, producers point out, there are myriad other factors to be evaluated in deciding whether the discovery is commercial. In any case, there is serious doubt whether the volumes of gas contained in small fields are great enough to reverse overall production declines, although they could ease the decline as price rises make them economical to produce.

An estimated 30,000 billion cubic feet of gas has been discovered on Alaska's North Slope. But the gas cannot be produced immediately for two major reasons. First, there is no pipeline to carry it to consumers. Second, it is situated on top of oil that is scheduled to begin flowing this summer. The gas is needed underground as pressure to force the oil out of the wells. To relieve that pressure by producing the gas first would decrease the amount of oil that would come out. Thus, say the oil companies, to maximize the oil production, the gas must be produced at some later date. The same principle applies to many discoveries in which oil and gas are contained in the same underground rock layers.

Question of Transportation

Once a gas field is found, a pipeline must be built to deliver it to consumers. Underwater pipelines now cost $1 million a mile. If a gas field is too far away from an existing pipeline and has too small a production capacity to pay for the cost of the connecting pipeline, it simply will not be built and the gas cannot be produced. In Federal offshore waters, that decision is made by the Federal Power Commission because pipeline costs affect the price consumers pay for regulated gas utilities.

The numbers change from day to day, but at one point last year, according to industry figures, there were 137,594 producing gas wells in the United States. They produced an average of 54 billion cubic feet per day.

Producers must decide what the maximum efficient rate of production is for each well. They contend they set these rates primarily to insure that they can eventually get out the maximum amount of gas—roughly 80 percent of what's there. An old rule of thumb was that gas could be produced at one-fourth or one-third of the volume that would naturally flow out of the well. It is a matter of judgment.

It is also a matter of serious controversy, with some investigators charging that companies are producing some wells at rates far below maximum in order to worsen shortages, force price rises and increase profits. It is impossible to tell for sure. On Federal leases, the United States Geological Survey approves

production rates annually based on data supplied by the producer. The rates change over the life of the well. The data can be interpreted to support virtually any argument, as several investigations have concluded.

Danger in Producing Too Fast

The idea is to produce at the maximum rate without damaging the reservoir. If produced too fast, for example, water might flow into the hole and block off the gas flow.

Producers acknowledge that they withhold certain volumes of gas for their own uses and periodically shut in producing wells, but not to create shortages.

Exxon, for example, says it produces about 10 percent of the nation's gas supply. It has roughly 2,800 gas wells and 12,000 oil wells that also produce some natural gas. These wells produce an average of 5.5 billion cubic feet per day, of which about 5 billion cubic feet is sold under more than 2,000 different contracts (most for 20 years) at prices ranging from 18 cents to $2 per 1,000 cubic feet.

Thus, Exxon keeps 500 million cubic feet each day for itself. Roughly half of that is reinjected into oil wells to create pressure to force more oil out of the ground. The rest is used for a variety of purposes such as fueling compressors, pumps and other equipment.

Of Exxon's total wells, about 40 to 50 are shut down at any given time for maintenance work, the company says. Exxon says it performs these "workovers" on about 800 wells a year.

Considering the Variables

Whether a gas discovery is commercial to produce depends on where it is, the amount discovered and the price for which it can be sold.

Suppose an estimated 30 billion cubic feet was discovered in Federal waters 200 feet deep in the Gulf of Mexico. Mr. Wooddy, of Exxon, estimates that it would cost roughly $40 million to pay for a production platform at that depth and to drill the development wells. Under the Federal price ceiling of $1.44 per 1,000 cubic feet, the gas could be sold for $43.2 million. If there were no other expenses such as operating and maintenance costs or interest on loans, Exxon could earn $3.2 million over the life of the well—say 10 years. A savings account would pay much more.

That same 30 billion cubic feet discovery on shore, however, would be commercial because development costs would probably be less than $20 million. The gas could be sold to Texas customers for the unregulated intrastate price of about

$2 per 1,000 cubic feet, for a total of $60 million. Exxon's earnings: $40 million.

"Now that would be a dandy," said Mr. Wooddy.

Although natural gas is often found together with oil, as in the case of Alaska's North Slope, about 70 percent of the nation's proven gas reserves and 80 percent of gas production comes from fields that contain little or no oil, according to Mr. Stanwood, of Transco.

For many years, the price differences between oil and gas favored the expiration for oil-prone fields instead of gas-prone fields. Since oil could be sold at a higher price than the equivalent amount of energy in gas, companies could get their investments back quicker and make more profits sooner.

"In the past, the roughly threefold higher field price of crude oil over gas made it the more valuable commodity and has greatly influenced exploration and production research and development toward oil," Mr. Stanwood said.

Diverging Arguments

The extent to which this influenced Federal offshore lease-bidding patterns and subsequent gas discoveries is a matter of debate within the industry. Most companies insist now that they are trying to find as much of both oil and gas as they can. However, since producers have limited exploration budgets, the incentives favor bidding on an oil prospect instead of a gas prospect of equal value.

But again, the geologists are only guessing with the imprecise scientific data available to them.

"That's the reason why you see these funny bids when they have these lease sales," said Mr. Wooddy.. "Somebody will bid a bunch of money on one prospect and no one else will bid a thing. That's because their geologists look at it differently." In other words, one company's geologists believe a field is oil-prone while the others think it is gas-prone.

"Right now, we'd be more interested in oil than we would gas," said Mr. Wooddy, "although we're trying to find as much of both as we can."

Geologists never know for certain what's down there until a hole is actually drilled. That's why they equate oil and gas exploration to gambling. The deeper the prospects, the farther from shore, the more money they will have to gamble in drilling the hole, and the more money they will have to spend to produce whatever they find. That, inevitably, means higher prices.

February 22, 1977

PRIVATE REPORT SAYS GAS PRODUCERS HELD RESERVES OFF MARKET

By DAVID BIRD

A private report commissioned by leading gas utilities has concluded that gas producers failed to bring to market proven reserves of gas in the Gulf of Mexico that would have been equal to much of this winter's cutbacks to consumers.

The report was in the form of a two-page "executive summary" for members of the Associated Gas Distributors, a utili-

ty trade organization. The summary was to be deleted when the full report was forwarded to Government officials.

Inadvertently, however, the summary was included in material sent to the Task Force on Natural Gas of the New York State Assembly that opened hearings yesterday on why there was such a severe natural gas shortage this winter.

The summary report said of 105 offshore Louisiana leases that had been held by producers for at least five years, only a third, or 35, had been brought to production although all had been classified as producible by the United States Geological Survey.

Reserve Potential Estimated

"We place the reserve potential of the leases under study at some eight trillion cubic feet," the private summary said, "with a potential deliverability of a [billion] cubic feet per day (a substantial

fraction of the 1976-77 winter level of interstate pipeline curtailments), which seemingly could be made available in relatively short order, it would appear, were emphasis placed on promptly bringing these known reserves to market."

The summary went on: "Pipeline connections do not appear to be a problem, since a vast majority of the leases are within a few miles of existing offshore pipelines."

The release of the previously confidential summary yesterday by Assemblyman Melvin Zimmer, the chairman of the Task Force, added to the growing controversy over whether the gas companies were deliberately withholding supplies in the hope of selling at much higher prices in the future.

At his news conference last Wednesday, President Carter said that detailed studies would be undertaken by the Government

339

to determine whether the producing companies were giving accurate data on supplies.

Reaction of President

"I think it's obvious to all of us that there are some instances where natural gas is withheld from the market," the President said, but he said it was "understandable" that producers withheld natural gas to wait for higher prices.

As to why there was an attempt to withhold the "executive summary" stating that producers had failed to market gas that was available and needed, an official of Associated Gas Distributors said he was concerned that "people would take it out of context."

The official, Charles Neumeyer, who is chairman of the executive committee of the Associated Gas Distributors as well as senior vice president of the Brooklyn Union Gas Company, said there was no way that the gas in the reserves could have been put on the line immediately to relieve the situation this winter.

Mr. Neumeyer said that this was the coldest winter in 100 years and that no one could have been ready for it.

Pressure Held Lacking

Mr. Neumeyer, interviewed outside the Task Force hearing at the New York State Chamber of Commerce and Industry, 65 Liberty Street, said the utilities had been trying to prod the Interior Department into taking a tougher stand on getting the lease holders to produce the gas, which is on Federally-owned lands, instead of holding it in reserve for higher prices.

"Interior's regulation is not too tight," Mr. Neumeyer said, "so there's not enough pressure on people to develop within a set time."

Mr. Neumeyer expressed concern, however, that the summary might be interpreted as meaning that there was really no cause for concern about shortages of gas.

The utilities that sell to consumers have

joined the producers in urging for development of more fields to assure future supplies and Mr. Neumeyer indicated that he did not want to blunt that joint cause with statements that might be interpreted as devisive.

Asked if the shortage this winter could have been eased if those fields in the Gulf of Mexico had been developed before this year, Mr. Neumeyer replied that there still would have been a shortage.

If that extra gas had been available previously, Mr. Neumeyer said, his company would have sought out additional customers to buy it and those added customers, too, would have been affected by the current shortage.

"It would be prohibitively expensive to make sure you had enough supply on hand to take care of a winter that occurs only once in a hundred years," Mr. Neumeyer said.

February 26, 1977

The Battle for Western Coal

By STEVEN RATTNER

GILLETTE, Wyo.—For a year the giant earthmovers that are carving out five mammoth coal mines from endless acres of scrubland in the heart of the coal country here in Campbell County were silenced as a suit by environmentalists worked its way through the courts. Now the suit has been settled, the mines are nearing completion and, although initial production seems assured, no one is quite sure what will happen next.

The bloom is off Western coal. Three years after a post-oil-embargo energy frenzy sent coal prices skyrocketing, development work on nearly half the nation's coal reserves—more than 200 billion tons—here in the Powder River basin and in neighboring states seems to have slowed below the most pessimistic predictions of 1974.

A variety of factors are being blamed for the slowdown. Environmental lawsuits have impeded work not only on the five mines here, but on a rail line to bring the coal out and on a variety of other coal-related projects. With oil still plentiful, demand for coal remains sluggish as electric utilities resist conversions. Transportation continues to be uncertain and expensive—railroads grow more congested and a proposed coal pipeline project had been shelved. More exotic projects, such as coal gasification, appear farther and farther out of reach.

"Utilities don't believe they have to convert to coal or they are putting it off for as long as possible," said William S. Cook, executive vice president of the Union Pacific Corporation, which has major Western coal reserves. "And with a new Administration, they're adopting a wait-and-see attitude." Mr. Cook said that new contracts for coal have slowed to a virtual standstill.

Nor, despite campaign statements, has President Carter yet produced any help for Western coal. A new, tough stripmining bill should clear Congress

this year, with help from the White House, perhaps shifting the economics somewhat in favor of deep-miner Eastern sources.

The Department of the Interior planned to resume coal leasing in 1977 —the moratorium has been an industry peeve—but with postponements ordered for a number of oil leases in recent weeks, the resumption appears far from assured. Last week's budget amendments, while aiding conservation and hindering nuclear power, apparently did nothing for coal.

To be sure, many of these factors have contributed to a general slowdown in coal development across the country. One indication of the fortunes of coal might be the spot market. In September 1973, the Federal Power Commission reported, steam coal for electric utilities sold for $10.67 a ton. Prices peaked in November 1974 at $31.95 a ton and, by last August, had fallen sharply to $21.35 a ton.

Coal production in all regions of the United States climbed only slightly last year, to 665 million tons from 648 million tons, according to the National Coal Association. Low-sulfur Western coal, once billed as the solution to America's energy problems but thus far unable to produce the growth expected of it, amounted to only about 90 million tons last year. Earlier projections of 300 million tons of production by 1985 have now been cut to about 200 million.

For a while it looked as if Western coal might not leave the starting gate. In January 1975, just as the first concerted development was beginning, the Sierra Club won an injunction halting development of four giant strip mines ringing the city of Gillette (and indirectly causing a postponement of a fifth) on the grounds that the Federal Government should be forced to prepare environmental impact reports on the entire coal-rich region, not just on the individual mines.

A year and a half later, that injunction was lifted by the United States

Supreme Court and construcion resumed. At the Atlantic Richfield Company's 6,500-acre Black Thunder Mine, giant parts for 170-ton trucks and steam shovels—wildly out of scale like a scene from Gulliver's Travels—are jumbled around, awaiting assembly. (The trucks will never be driven on a public road; once assembled, they will spend their careers shuttling coal from the ever-moving mine pit to the loading station.)

The five mines are due to begin operations within the next year, but that's hardly the end of the problems. For one thing, when the injunction was lifted by the Supreme Court, all five mines leaped into activity, taxing local services and vexing company executives.

"Local suppliers just have so much capability and once it's exceeded, they have difficulties," said Charles B. Smith, manager of the Black Thunder Mine. "Skilled craftsmen, electrical subcontractors, that type of supplier is where it's difficult to find enough skilled people."

Beyond the local problems, a big impediment to the development of Western coal, experts agree, is transportation. Railroads, expected to carry most of the coal, at least over the short-term, may not be able to handle the burden. The Burlington Northern, possibly with the Chicago & North Western, wants to build a 116-mile line through the Powder River Basin here where the new giant mines are, but the line's prospects are uncertain because of a legal challenge by the Sierra Club to the Interstate Commerce Commission's approval of the project.

●

Then there are the experts who question whether, even without regulatory impediments, the railroads are up to the task of hauling the coal. A fairly typical assessment of the problem came in late December when the Electric Power Research Institute released a study suggesting that, by 1985, bottlenecks on both railroads and inland

340

waterways—which are handling a significant amount of coal barge traffic—may develop.

For example, "Some of the utilities from time to time indicated a general concern about the Burlington Northern," said R. Gale Daniel, vice president, of Arco's Thunder Basin Coal Company. "Even now, the B.N. will need a lot of improvements to stay up with the business coming out of the area."

The railroads dispute this scenario. "Can we move all the coal that has to be moved?" Norman Lorentzsen, president of Burlington Northern Inc., asked rhetorically. "The answer is yes."

Another possible coal-hauler, a slurry pipeline, has been proposed to run from here to Arkansas, but even its supporters now acknowledge that its future appears dim. The railroads have refused to give it permission to cross their tracks and have blocked attempts by slurry proponents to win the right of eminent domain in Congress for the lines.

Backers of the project contend that the system, which moves a mixture of coal and water with the consistency of a thick syrup, would mean transportation costs as much as 75 percent below railroad tariffs. Environmentalists maintain that vast amounts of precious water would be exported from the dry Northern plains via the line. The railroads say that the slurry line would "skim the cream" from them.

"We still see success over a period of time," said Ward Woods, a partner in Lehman Brothers, the investment banking house, and a director of Energy Transportation Systems Inc., which would build the line. "We're not sure when that time will be."

Looking further ahead, analysts feel that the use of Western coal can be made most efficient by not moving the coal itself, but instead gasifying it or turning it into electricity via mammoth mine-mouth power plants. But both of these options face increasing problems.

The timetable for commercial gasification is being steadily pushed back as costs escalate sharply. Congress last year killed a potentially major Federal effort to aid development of gasification and other synthetics.

●

The prospects for on-site electrical plant ("coal by wire") are equally uncertain. Last year, the most prominent proposal, for a giant complex on the Kaiparowits Plateau in Utah, was dropped due to pressure from environmentalists. Battles of similar intensity over other projects are expected. The environmental movement appears increasingly to accept the inevitability of strip mining and instead is concentrating on stopping the electric plants and the development they would bring.

"In general, you bring industry to the energy," said Carolyn Johnson, chairman of the Mining Workshop of the Colorado Open Space Council. "Certainly some don't intend to move,

but that's the way it worked. Maybe not for the first 10 years but in the 15th year, you do it."

In the sparsely populated West, where several states were actually losing population before the energy boom began in the 1960's, the frenetic development that brings thousands of job-seekers, mushrooming growth, local economic strains and social problems ranging from alcoholism to prostitution has become an acute issue. Gillette, the center of the coal-mining operations for the area, has been wrenched by the surge in activity and also by on-again, off-again development caused by the various legal actions.

In 1970, Gillette had 7,194 people. Today it has an estimated 14,000, 38 percent of whom live in mobile homes. The unemployment rate is 2.1 percent and the workmen in the coal mines or tending the incessant railroad traffic in and out of the downtown yard contrast vividly with the cowboy-like ranchers and their shotgun-adorned pick-up trucks.

"I like Wyoming—better job opportunities and fewer people," said Al Boger, a 28-year-old welder as he worked on the base of a giant shovel at the Thunder Basin Mine. Since he started here last August, Mr. Boger has earned $9.35 an hour; his last job in a saw mill payed $3.67. "This is the best company I've found so far. I've bought a house and I'm going to stay here."

February 27, 1977

Technology
Underground Coal-Gasification Progress

By VICTOR K. McELHENY

Setting controlled fires in a 30-foot-thick coal seam more than 250 feet beneath Wyoming's high, dry plains, United States Government researchers have produced up to 11.5 million cubic feet daily of a gas mixture that would be suitable for running a local electric power plant.

In tests at Hanna, Wyo., last year, the heating value of the gas reached the highest level achieved so far in many years of underground gasification tests in the Soviet Union, Britain and the United States, 175 British thermal units per cubic feet.

In a six-month test scheduled to begin in September, researchers of the Laramie Energy Research Center hope to maintain the heating value and double the average daily output of gas.

They are preparing for a year-long run in 1979 designed to produce 90 million cubic feet daily.

The 30-year-old Laramie center is part of the Energy Research and Development Administration. Its encouraging results in producing coal gas from pairs of so-called linked vertical wells has raised hopes of exploiting vast deposits of deep-lying or thin-seam coal that would be otherwise uneconomic to mine.

The deposits that can be strip-mined on the surface or extracted by subsur-

face mining are estimated to be only 10 percent of the nation's total reserves of 4,000 billion tons of coal within 6,000 feet of the surface.

The exploitation of these economic deposits is being pushed. But there are environmental, health and social costs that tend to inhibit expansion of production from both types of mines beyond the current level of 660 million tons a year.

The Carter Administration intends to press for the substitution of coal for oil and gas both for generating electrical power and for industrial processes and to favor coal over uranium as a source of expanded electrical capacity.

If underground gasification proves economically attractive and environmentally acceptable in tests at Hanna and other locations in Wyoming and West Virginia, it could provide an additional route to harnessing coal for electric power.

According to energy agency officials in Washington and Laramie, the key problems to be overcome are the sagging of the surface above the gasified coal seam, and the numerous toxic chemicals produced during gasification. Unless contained in or near the seam, these chemicals could contaminate ground water.

The officials point to other possibilities rising from underground gasifica-

tion. Pure oxygen and steam, they say, could be injected into the wells that penetrate the coal seam in order to produce a gas with an intermediate heating value, 400 or more B.T.U.'s per cubic foot.

Gas produced this way could be economically exported to greater distances than the 175-B.T.U. gas. It could be used as raw material for feed stock, for chemical plants as well as for electric power—or it could be upgraded in a surface plant to a synthetic pipeline gas with a heating value of 1,000 B.T.U.'s per cubic foot.

The land on which the Hanna experiments have been conducted since March 1973 has been made available by the Rocky Mountain Energy Company, a subsidiary of the Union Pacific Railroad.

For the fiscal year that began last Oct. 1, the Hanna project was budgeted at $2.5 million out of the total Federal underground coal gasification budget of $8.2 million.

A budget totaling $11 million has been requested for the next fiscal year, including an estimated $3.3 million for the Hanna work. Before last October, spending at Hanna totaled $4 million.

The energy agency's Washington administrators for underground gasification are Larry Burman, an assistant director of the oil, gas and shale division, and Dr. Paul Wieber, a branch chief.

The director of the Laramie center is Dr. Andrew W. Decora and the manager of the Hanna project is Dr. C. F. Brandenburg.

● ● ●

In an interview, Dr. Wieber said underground coal gasification may have some unexpected environmental advantages beyond that of leaving the coal in the ground.

The sub-bituminous coal at Hanna, with a heating value around 8,000 B.T.U.'s per pound instead of the 11,000 B.T.U.'s typical of bituminous coals in the East, is saturated with water, and the gas at Hanna also contains much water.

Dr. Wieber thought that this water, containing many burnable solid particles, could be usefully recycled. It could be used in a steam-oxygen process to produce gas of a higher heating value than is possible using air, which is only 20 percent oxygen.

As for the danger of polluting ground water, Dr. Wieber said that laboratory

work by the Lawrence Livermore Laboratory indicated that coal remaining in the underground seam may trap many of the toxic byproducts of gasification.

The Lawrence laboratory is investigating underground gasification at a site at Hoe Creek, near Gillette, Wyo. It has tried using explosives to fracture underground coal to create passages through which air can penetrate the seam and promote gasification.

• • •

This technique is different from the one that has been established at Hanna. There, air passages between the feet of two wells are created by what is called reverse combustion.

In the reverse combustion process, a fire is started at the foot of the first

well, and air is forced down the second under high pressure. The fire burns through the coal toward the second well, the source of oxygen, and the coal is turned into a porous mass. Now gasification can begin.

With the passage opened, even larger volumes of air are now forced down the second well, driving a "combustion front" back toward the first well along with the gas.

Although this gas is about 50 percent inert nitrogen from the air, and between 10 and 15 percent unburnable carbon dioxide, the rest is combustible. The burnable portion includes 12 to 18 percent carbon monoxide and 15 to 20 percent hydrogen.

March 30, 1977

CARTER CALLS FOR FIGHT ON ENERGY WASTE

Widespread Impact Foreseen on Standards of Living and Key Industries

By CHARLES MOHR
Special to The New York Times

WASHINGTON, April 20—Saying "the time has come to draw the line" on unfettered use of energy, President Carter proposed tonight a national energy policy designed to increase the cost of fuels, penalize waste and bring important changes in industrial habits and in some of the ways Americans live.

But in a nationally televised speech delivered to a joint session of Congress in the chamber of the House of Representatives, Mr. Carter argued that the plan "can lead to an even better life for the people of America" rather than debasing life styles and living standards.

The energy program that he proposed is complex, controversial and faces major political obstacles. It involves a vast scale of political and social engineering that would take billions of dollars from citizens' pockets to discourage overconsumption of fuels and energy. But it would return much of that money in the form of tax credits, tax rebates and in incentives to those who do the most to conserve.

'It's Our Job'

The President, attempting to enlist the support of Congressmen who for economic, ideological or regional reasons may find much to dislike and who may fear the wrath of their constituents, said the task of fashioning the new policy has been "a thankless job, but it's our job."

In an apparent attempt to fend off or to limit the probably inevitable efforts

to modify or soften the plan's provisions, Mr. Carter said that the program depended "for its fairness on all its major component parts."

House Speaker Thomas P. O'Neill Jr. predicted that the proposals would bring "the toughest fight this Congress has ever had," but White House sources believe that the Speaker and other Democratic leaders can pass most, if not all, of it.

Still, the proposal will be perhaps the greatest test of Mr. Carter's capacity for personal leadership, of his ability to capture and direct public opinion and of his untried skills in parliamentary maneuver.

'Greatest Domestic Challenge'

Calling the energy problem the "greatest domestic challenge that our nation will face in our lifetime," Mr. Carter predicted to his audience that he would get little applause. Indeed, although he was interrupted seven times by applause, the reception he received was relatively restrained.

Several of the bursts of applause came as Mr. Carter, dressed in a dark-blue suit and reading his speech from off-camera prompting devices, inveighed against windfall oil company profits and said that there was now too little competition in industry.

Mr. Carter both defended and minimized the probable impact of what he called "one of the most controversial and most misunderstood" parts of the plan. This calls for a standby tax increase on gasoline that could gradually raise taxes by 5-cent increments to a total of 50 additional cents by 1987 if the country does not meet Mr. Carter's goal of a 10 percent reduction in gasoline consumption by that date.

The President said "this gasoline tax will never have to be imposed" if Americans meet the challenge to conserve, in large part through shifting to more efficient automobiles. "I know and you know it can be done," he told Congress.

Politically, the program is planned to retain to the maximum extent possible the support of private citizens, who are offered a considerable number of financial awards to balance the burdens. The

program is likely to arouse much more vehement anger and opposition in the oil industry, among public utilities, the automobile industry and other economic groups whose expected profits and whose production and consumption patterns could be changed.

Mr. Carter argued that, with proper planning, continued economic growth, expanded employment and "a higher quality of life" could result from the program. Most economists, however, disagreed with the White House contentions that the inflationary and other possibly adverse effects of the plan would be minimal.

Many people may find the plan much less draconian or harsh than they had feared. Mr. Carter, for instance, did not seem to be threatening to take Americans out of their automobiles, and he even predicted an increase in miles driven each year. His measures would, however, punish with increasing severity those who drive 'gas-guzzling' cars, until by 1985 someone buying a new car that gets only 10 miles a gallon would have to pay a $2,500 excise tax.

Some of the major elements of Mr. Carter's plan, in addition to the standby gasoline tax, are as follows:

¶The graduated excise tax on more inefficient automobiles and corollary graduated rebate for those who purchase automobiles that exceed Federal efficiency standards. These are 18 miles a gallon this year and will rise to 27.5 miles a gallon by 1985. A senior energy adviser to the President conceded at a news briefing this afternoon that this would probably cause the "aggregate take of the auto industry to decrease," even though he argued that total sales of cars might continue to increase.

¶A new tax at the wellhead on domestic oil production that will, over a three-year period, raise the price of oil, now controlled at two lower price tiers, to the prevailing world price, which is about $14 a barrel. One effect will be to raise the price of oil, and of gasoline made from it, an estimated 7 cents a gallon.

¶A plan to return the money collected from this oil tax to citizens with a "per capita" rebate that could reach $75 a year for most Americans but that would gradually diminish as "old" oilwells were exhausted. The money collected by the standby gasoline tax, if it is put into effect, would also be rebated "progressively," a White House statement said. The White House was unwilling to specify precisely what sort of plan it had in mind. If the full 50-cent tax ever came into play, by 1987 it would generate $60 billion. Most of this would be returned to citizens, but not necessarily in the amounts paid by individuals, whose automotive habits and incomes vary widely.

¶Measures to encourage better insulation and other energy-conserving improvements in homes and buildings. If tax and other incentives do not spur enough voluntary action, mandatory improvements in residential efficiency will be considered, a White House "fact sheet" on the program said.

¶Mandatory and "stringent" efficiency standards for household appliances by 1980.

¶Reform in public utility rate structures to end "promotional" rates that give big industrial users better prices than small residential users.

¶A special 10 percent tax credit for industries that undertake energy conservation measures, including so-called "cogeneration" of electricity and usable steam and heat.

¶The setting of "new" natural gas prices at about $1.75 per thousand cubic feet and subjecting, for the first time, gas sold within a state to the same price as interstate gas. Mr. Carter contended that this would reduce the advantage of producing states, like those in the Southwest, that have more—though more expensive—supplies of gas than areas like the Northeast. He said, "We must not permit energy shortages to divide, or balkanize, our country."

¶Taxes and laws that will force industry to shift largely from burning oil and gas to coal.

¶ Tax incentives to encourage two and one-half million homes to install solar power units for heating and cooling.

Mr. Carter said in the speech, his first to a joint session of Congress, that he preferred voluntary conservation measures with a "minimum of coercion." But he added that the problem was so large and time so short that some mandatory "penalties and restrictions to reduce waste are essential."

A scene on Capital Hill after Mr. Carter's speech seemed to give his message ironic emphasis. Platoons of traffic policemen, waving flashlights that cast red beams, directed traffic as top congressional leaders, cabinet members, dip-

lomats and others were driven away by chauffeurs in gas-guzzling, energy-inefficient, black limousines.

In a speech Monday night the President had used a doomsday tone about the gravity of the energy crisis—mainly because public opinion poll data convinced him that his first task was to convince a skeptical majority of people that the problem posed by ever-escalating demand and finite supplies of oil really existed.

The tone tonight was less somber and more keyed to positive aspects of his program, although he again warned that to continue to fail to act "would subject our people to an impending catastrophe."

One political problem that worries White House aides is the possibility of strange new coalitions, what one official called "unholy alliances," forming against Mr. Carter's plan. An example cited was a possible coalition of the United Automobile Workers union with conservative Southwestern politicians in opposition. Mr. Carter promised tonight that auto workers would "not bear an unfair share of the burden."

Mr. Carter attempted to deflect one likely counterproposal from the Republican Party and from the oil industry when he said that "immediate and total decontrol of domestic oil and natural gas prices would be disastrous for our economy and also for working American families."

In a move that will probably please those distrustful of the oil industry and cause gloom in Houston's Petroleum Club, Mr. Carter recommended that energy companies be required to furnish more complete data on their operations. While he carefully avoided embracing liberal suggestions to "break up" the oil companies, he said tonight that "strict enforcement of the antitrust laws based on this data may prevent the need for divestiture" of integrated operations from oil well to gas pump.

Along with the President the chief architect of the overall proposal was James R. Schlesinger, the White House energy adviser.

A senior energy adviser, who insisted on anonymity, told reporters today that "very ample" incentives were left for continued exploration and production of oil and that only windfall profits that would accrue from the sharp increase in price on oil from old wells were being taxed away.

The senior adviser was asked why, when the United States had delayed so

long in trying to curb the energy crisis, he seemed reasonably optimistic about adoption of Mr. Carter's sweeping plan.

It often seems, he said, that "nations and men will not do the rational thing until they have exhausted every possible alternative," adding, "We may have reached that point."

A warning that some national habits, including a reduction in the number of commuters driving alone to work, would need to change was underlined by the Administration fact sheet issued earlier in the day.

It said that, even if passed by Congress in full and "implemented effectively," the President's plan would save only about 4.6 million barrels of oil a day by 1985 and thus leave projected imports of foreign oil at a level of seven million barrels a day.

To reach the President's stated goal of reducing such imports to a level of six million barrels daily in 1985 "would require voluntary conservation efforts by the American public," the sheet said. The targets for gradually decreasing consumption of gasoline—targets that will touch off the imposition of standby gasoline taxes if not met—also "assume some additional reduction in consumption through such items as observing speed limits and more carpooling," the fact sheet said.

'Achievable' Goal

It also said that the goal for reduced gasoline consumption was "achievable," despite an expected increase in total miles driven by year by 1987. However, it then went on to say the sum of the programs requested by the President would "not achieve the President's goal of reducing gasoline consumption by 10 percent," and it added that if "voluntary efforts" did not make up for the expected shortfall, "then mandatory measures, such as a commuter tax, will be considered."

The program encourages homeowners to undertake such conservation improvements in their dwellings as insulation, both by promising tax credits for the expenditure and by making financing for loans more readily accessible. Public utilities would be required to offer their customers a "conservation service" program to be repaid through additions to the customer's monthly utility bills.

April 21, 1977

SOVIET ALSO SEEKING TO CONSERVE ENERGY

But World's Largest Oil Producer Has Trouble Convincing Users to Take Appeals Seriously

By CHRISTOPHER S. WREN
Special to The New York Times

MOSCOW, April 25—The call to conserve energy has been heard in the Soviet Union, too, these days. But it often

sounds like a voice crying in the wilderness, for however urgent the tone, relatively few Russians seem ready to take it seriously.

This is mainly because the Soviet Union abounds in natural resources, including fossil fuels, but also because Russians have far fewer private cars, private homes and household appliances that would consume energy.

The Soviet Union has already outstripped the United States as the world's largest oil producer, with wells now yielding an average of 10.5 million barrels a day. One-fourth of the output is exported.

More than half the world's proved energy reserves are said to be in the Soviet Union. Yet Soviet economic planners are stressing conservation as they look to

the future. A recent Soviet study of more than 500 enterprises concluded that more "rational use" of energy by Soviet industry could save at least 5.6 million tons of fuel a year.

The current five-year plan calls for oil and natural gas, which account for nearly two-thirds of the nation's energy consumption, to be used less as primary fuels and more as raw materials for technological use or export. The demand will be met partly by greater production of coal and also by atomic energy, with nuclear reactors accounting for a fifth of new electric capacity between now and 1980.

Imbalance Between East and West

Several reasons appear behind a drive for frugality in the midst of apparent plenty. Most of the fossil fuel reserves and hydroelectric power are concentrated in Siberia and other Asian regions while cities and industries are clustered in the western, European part of the Soviet Union.

Sergei Yatrov, the director of the Institute of Complex Fuel and Energy Problems in Moscow, said in an interview last year that while over two-thirds of the economy's production capacity was in the west, fully 80 percent of energy resources were east of the Ural Mountains.

The transport of the fossil fuels is expensive and their extraction is increasingly difficult, particularly in the subarctic environment of Siberia, which now yields about 40 percent of the nation's oil and 15 percent of the gas.

There is another inducement for prudence. Oil and gas have become Moscow's most lucrative earners of hard currency, which is needed to buy Western technology. This has put a squeeze not only on domestic allocations but also on shipments to the Soviet-bloc allies. Although

Moscow still provides most of the oil needed by its allies, it has advised them to look for additional supplies elsewhere.

Conservation Drives Mounted

To admonish domestic consumers to be more thrifty with energy, the Soviet press has mounted a succession of conservation campaigns. But the average Russian, exposed to reports of economic successes in oil and gas production, has responded casually in the absence of an energy pinch.

Even in winter, central heating is warm enough so that many Moscow apartment dwellers leave open a small window pane to let in the frosty air. Fuel itself is state-subsidized. Natural gas for cooking costs a little more than 21 cents per person monthly regardless of how much is used.

Gasoline prices for the relatively few car owners are equally low. Regular octane costs 38 cents a gallon and premium octane 48 cents. The lines at gasoline

stations result more from a shortage of stations than of gasoline.

The demand made on fuel resources by the average citizen is substantially less than in the West. The entire country is believed to have only a few million private cars, about as many as New Jersey, and these are mostly smaller models that can average better than 23 miles to the gallon. With little private housing, the Government, as the nation's landlord, can decide where to trim heating needs.

Criticism carried in the official press indicates that far more energy is wasted by state-run industry than by consumers. A Smolensk factory was discovered doling out gasoline to users without keeping proper records. In Baku, enterprises have shown energy savings on paper by exaggerating their theoretical energy use rates, a trick that was uncovered among some Soviet Government ministries last year.

April 26, 1977

WORLD OIL SHORTAGE IS CALLED INEVITABLE

Study Says Shift to Other Fuels Is Vital and Urges Conservation

By EDWARD COWAN
Special to The New York Times

WASHINGTON, May 16—Energy company executives and analysts from 15 countries warned today that non-Communist countries must make enormous investments in coal, nuclear power and energy conservation to offset an inevitable world oil shortage and a risk of war.

The warning emerged from a two-and-a-half-year study sponsored by the Massachusetts Institute of Technology that was conceived and directed by Professor Caroll L. Wilson.

"The free world," Mr. Wilson said in summarizing the findings, "must drastically curtail the growth of energy use and move massively out of oil into other fuels with wartime urgency. Otherwise, we face foreseeable catastrophe."

The study was titled "Energy: Global Prospects 1985-2000."

Carter's Theme Recalled

The theme of the 291-page book, published by McGraw-Hill in soft cover at $6.95, is similar to that of President Carter's energy program — that there is a grave danger of an oil shortage in the 1980's, soaring oil prices and economic depression for industrial and developing countries. But the tone employed by Mr. Wilson and his associates, whom he styled the Workshop on Alternative Energy Strategies, was grimmer than Mr. Carter's.

Energy could "become a focus for confrontation and conflict," the report warned. "Even with prompt action the margin between success and failure in the 1985-2000 period is slim."

Mr. Wilson explained at a news conference in Washington that the 15 years from 1985 were chosen as the temporal focus of the study because there was little that could be done to alter energy supplies before 1985.

"Large investments and long lead times are required to fill the prospective shortage of oil, the fuel that now furnishes most of the world's energy," the report declared. It added: "The task for the world will be to manage a transition from dependence on oil to greater reliance on other fossil fuels, nuclear energy and, later, renewable energy systems," such as sunshine and atomic fusion.

The principal participants in the study were chosen for the most part from big corporations and universities and governments. Asked why he had not included among the 35 principals some figures who might have been expected to dissent from the general direction of the report, Mr. Wilson said he had formed "a group who could work together intensively" and share "a global perspective."

The interdependence of nations in cop-

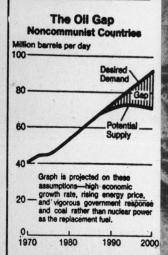

The Oil Gap
Noncommunist Countries

Million barrels per day

Desired Demand

Gap

Potential Supply

Graph is projected on these assumptions—high economic growth rate, rising energy price, and vigorous government response and coal rather than nuclear power as the replacement fuel.

100
80
60
40
20
0
1970   1980   1990   2000

A Comparison of Oil and Coal Reserves
Billion Barrels Oil Equivalent

OIL

Proven Reserves

2,000
Ultimately Recoverable

COAL

Economically Recoverable

Potentially Recoverable
12,000

A tanker filled with liquefied natural gas, the Hili, unloading more than 700,000 barrels on Saturday in Sodeaura, Japan. An energy study sponsored by the Massachusetts Institute of Technology warns that the non-Communist countries must develop alternatives to offset a world oil shortage.

Associated Press

ing with energy problems was one of the report's themes. For example, it proposed that the United States triple or quadruple its coal production by the turn of the century to permit large exports to Europe and Japan that would diminish their needs for oil.

The report skirted controversial nuclear questions, including the safety of nuclear power plants and President Carter's recent decision not to develop a plutonium fuel cycle in this country and not to build a sodium-cooled fast-breeder reactor, the Clinch River project, at Oak Ridge, Tenn.

Instead, the study pointed to the contribution that nuclear power could make to filling the oil gap, depending on policy decisions by governments and communities. "Shortfalls in nuclear power would most likely increase the prospective shortage between desired fuels and probable supply," the study declared.

One of the American authors, Thornton Bradshaw, president of the Atlantic Richfield Company, a major oil company, described coal and nuclear power as "bridging fuels, the only fuels that can get us through this transition period". from oil to renewable resources in the 21st century.

Hopes for an Early Review

Asked at the news conference if he thought that President Carter's plutonium and fast-breeder policy was compatible with this "bridging" function, Mr. Bradshaw replied, "no." He said that he hoped for an early review of the decision.

Mr. Wilson said there were differences on the nuclear issues among the authors and that they had agreed to a noncommittal formulation that separated development of conventional light-water reactors from later decisions about plutonium recycling from spent uranium fuel rods and about the breeder, which converts uranium into plutonium.

"We can get an awful lot of energy out of them before we cross over into reprocessing and fast breeders," Mr. Wilson said when asked for a personal view. That is also the Carter Administration's position. Opponents of the breeder fear that it will erode if other countries reject Mr. Carter's appeal for renunciation of plutonium as a fuel.

"No means of rendering plutonium useless for [nuclear] weapons is known," the study declared.

The Allied Chemical Corporation, a partner in an unfinished $250 million Barnwell, S.C. plant to process plutonium, was one of the principal financial backers of the project. Others included Atlantic Richfield, the General Motors Corporation, and a number of foundations, including the Ford, Rockefeller, Mellon and Sloan foundations.

Support for Study Acknowledged

The study acknowledged "financial or professional support" from companies, research institutes and government agencies in foreign countries. It noted that three of the 15 countries represented by the authors were oil exporters—Iran, and Venezuela, which belong to the 13-country oil cartel, and Mexico. The book represented the "unanimous conclusions" of the 35 authors, a news release stated.

In assessing the outlook for oil, the book argued, as did a Central Intelligence Agency report released by the White House last month, that much would hinge on the willingness of Saudi Arabia to expand production.

But it is only a question of time before an oil shortage occurs, the study argued. Even if Saudi Arabia doubles its present production rate of 10 million barrels a day, "the shortage shows up" by 1989, the study said.

Oil demand could exceed supply by 1983 if present production ceilings are maintained, Mr. Wilson said in a foreword, but the shortage might not occur until 1995.

The study said that, "although the resource base of other fossil fuels such as oil sands, heavy oil and oil shale is very large, they are likely to supply only small amounts of energy before the year 2000."

Mr. Wilson said synthetic gas and oil from coal were unlikely to come into "substantial production" without much higher prices than at present or substantial Federal financing.

The study found it unlikely that sunshine, wind power and tidal action—so-called renewable or inexhaustible resources—would "contribute significant quantities of additional energy during this century at the global level, although they could be of importance in particular areas. They are likely to become increasingly important in the 21st century."

May 17, 1977

President Signs Strip-Mining Bill, But Cites Defects

By BEN A. FRANKLIN
Special to The New York Times

WASHINGTON, Aug. 3 — President Carter climaxed a 10-year struggle over the strip mining of coal by signing into law today the first uniform Federal controls on strip mining, until now a prerogative of coal-state governments often dominated by mining interests.

But at a brief ceremony this morning in the White House Rose Garden, Mr. Carter told about 300 guests who have fought the strippers that "in many ways this has been a disappointing effort."

The new law will not be generally effective at all strip mines until January 1979. And about 80 percent of the mines, qualifying as "small mines," won an 18-month extension for compliance with most reclamation standards.

Replanting Required

In the end, however, the law will require strip-mine operators to restore the approximate original contour of disturbed land, to backfill and grade "highwalls," or sheer rock faces left by excavation, to replant trees and grass and to prevent siltation and pollution of streams.

In addition, no mining state can maintain a coal price advantage over another by stinting on the cost of reclamation.

Meanwhile, the House of Representatives turned back an effort to deregulate the price of natural gas and accepted a proposal by President Carter on ceilings for gas sold outside the state of production combined with regulation of the instate market

The strip-mine law also commands a start, to be financed by a new Federal tonnage tax effective Oct. 1, on the massive job of reclaiming the wrecked acreage left by earlier, uncontrolled stripping.

Millions of coal tax dollars will be distributed by the Government to contractors to restore the abandoned, desolate land. Other millions will flow back through the states to ease the social impact of the "boom town" phenomenon that huge new strip mines have brought to the Far West.

The long-range mined land reclamation program is to be financed by a sliding-scale Federal tax of 35 cents a ton on strip-mined coal and 15 cents a ton on deep-mined coal, a plan designed to encourage deep mining. But the work will be slow going. The Interior Department estimates the cost of reclamation at $1,100 to $5,000 an acre, depending on the difficulty of the terrain.

Particularly because of the exemptions for "small mines," those that produce 100,000 tons a year along the Appalachian highlands, the bill passed by Congress this year is significantly weaker than versions approved by large majorities every year since 1972 by one or both houses of Congress, but vetoed by President Ford in 1974 and 1975.

A 'Watered Down' Bill

The President's other main objection to the bill is that it allows the mining industry to cut off the tops of Appalachian mountains to reach entire seams of coal, which are about 80 feet below the summit. Environmentalists had wanted to limit the mining companies to cutting a ridge or "bench" around the top of the mountain, from which they would be able to extract only part of the coal.

"I would have preferred a stricter strip mine bill," the President said, amid applause as he sat at a table with the thick Surface Mining Control and Reclamation Act before him.

But he said that this year's "watered down" bill would enhance the legitimate and much-needed production of coal and also assuage the fears that the beautiful areas where coal is produced were being destroyed."

Such fears were not assuaged, however, among some of the most militant citizens' groups that have battled for years for a national strip mine law against intense lobbying by the coal industry.

A group of them, called the Applachian Coalition, had labeled the 80,000-word measure the President signed today as "a blatant travesty" and had urged yet another veto. But in his Presidential campaign, Mr. Carter had pledged to sign such a bill.

Today, the President noted the long, behind-the-scenes lobbying struggle over the strip mine issue and reached across the table to kiss Louise C. Dunlap, head of the Environmental Policy Center and a proponent since 1971 of stringent controls.

August 4, 1977

345

Poll Finds Doubt On Energy Crisis

By ANTHONY J. PARISI

Too few people believe the nation's energy problems are serious enough—and even fewer understand them well enough—to provide broad support for the Administration's energy program.

This is the overriding conclusion that can be drawn from a New York Times/CBS News poll on energy taken one month ago and released last night.

The results come at a critical point in President Carter's effort to get Congressional approval for his national energy plan. The House of Representatives has already passed almost the entire package of proposals Mr. Carter unveiled in April, with the exception of a rebate for small cars and a standby tax in gasoline. The Senate—the last and biggest hurdle—takes up the package this month.

A tough fight is expected, and the poll seems to mirror this mixed Congressional reaction.

At first glance, the results appear to show that the public largely favors most of Mr. Carter's key proposals. But a closer look suggests that the respondents were actually superficially choosing energy options that looked to them like quick, easy and painless solutions. When pressed, they often contradicted themselves with answers that appeared to reflect a more believable opinion.

Moreover, the key variable of the survey seems to be education: The better informed the respondents, the more likely they were to consider the energy problem serious and the more willing they said they were to adapt as needed.

Most people, however, were skeptical and far fewer among the less educated groups seemed willing to adapt.

Of 1,463 people interviewed by telephone from coast to coast, 38 percent said they thought the energy problem was real. Thirty-three percent believed the situation was as bad as Mr. Carter depicted it.

In a comparison with six other serious problems facing the nation — crime, education, health care, unemployment, the high cost of living and defense—respondents ranked energy second in every instance.

Moreover, they were surprisingly ignorant of some basic energy facts. Despite all the publicity over the last four years on rising oil imports, which currently account for almost half the country's total needs, one third of those polled thought the United States produced all the oil it requires. Another 19 percent couldn't say. Only 48 percent knew that the United States must import oil.

When directly asked what percentage of the nation's oil was now imported, less than a fourth came within 10 percentage points of the correct answer.

Likely to Exceed Supplies

And although several well publicized reports have recently predicted that global demand for oil is likely to exceed available supplies during the 1980's, only 12 percent of those polled thought the world would run short of oil in less than 15 years.

Only 26 percent of the respondents lacking high school diplomas believed there was a shortage against 58 percent of the college graduates. Only 26 percent of those without high school diplomas thought the situation was as bad as the President said, compared with 50 percent of those with college degrees.

Asked whether they would buy a small, medium, or big car today if they were replacing their present model, only 26 percent of those without high school diplomas opted for a compact. By contrast, college graduates settled for a small car 43 percent of the time.

As in any such poll, the better-educated respondents also tended to have higher family incomes. Thus, in many cases similar trend lines could be drawn for affluence as for education. One seemingly obvious conclusion might be that the affluent—those with greater stakes in the system—would more readily cooperate with energy solutions than deprived Americans.

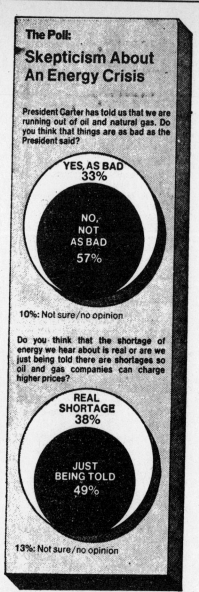

The Poll: **Skepticism About An Energy Crisis**

President Carter has told us that we are running out of oil and natural gas. Do you think that things are as bad as the President said?

YES, AS BAD 33%
NO, NOT AS BAD 57%

10%: Not sure/no opinion

Do you think that the shortage of energy we hear about is real or are we just being told there are shortages so oil and gas companies can charge higher prices?

REAL SHORTAGE 38%
JUST BEING TOLD 49%

13%: Not sure/no opinion

The New York Times/Sept. 1, 1977

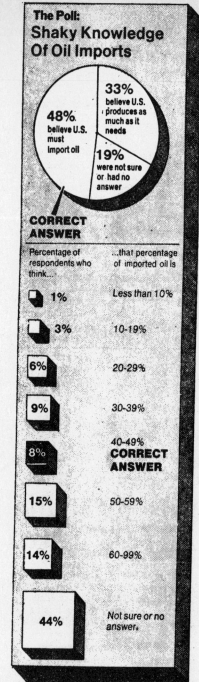

The Poll: **Shaky Knowledge Of Oil Imports**

48% believe U.S. must import oil
33% believe U.S. produces as much as it needs
19% were not sure or had no answer

CORRECT ANSWER

Percentage of respondents who think...	...that percentage of imported oil is
1%	Less than 10%
3%	10-19%
6%	20-29%
9%	30-39%
8%	40-49% **CORRECT ANSWER**
15%	50-59%
14%	60-99%
44%	Not sure or no answer

The New York Times/Sept. 1, 1977

The survey strongly suggests, however, that the public's willingness to shift priorities for the sake of energy depends not so much on what they can afford to do, as on how convinced they are of the need to do so.

A Recurring Theme

This understand-more, cooperate-more theme crops up again and again:

¶Asked if they approved of allowing natural gas companies to increase prices to finance more exploration, respondents sounded an overwhelming no (59 percent against, 34 percent for and 7 percent unsure). But those who thought the shortage was real split on the issue, with 46 percent approving and 47 percent disapproving.

¶Asked whether energy or the high cost of living was the greater problem,

58 percent of all respondents cited living costs and only 38 percent chose energy. But those who thought the problem real picked energy over living costs more than twice as frequently as the skeptics.

¶Asked if they thought it would be fair or unfair to increase gasoline taxes until most people drove less, 66 percent of the respondents replied unfair, only 26 percent fair. But among those who believed things were as bad as President Carter said, 39 percent considered such gasoline taxes fair.

¶Asked if they thought gasoline rationing fair or unfair, 49 percent replied unfair and 42 percent fair. But of those who believed the situation was as bad as Mr. Carter depicted it, a clear majority, 57 percent, thought rationing fair.

More Consistent Bias

And in each of these comparisons the bias by education was stronger, or at least more consistent, than the bias by income.

In question after question on some key Carter proposals such as using more coal, building more nuclear power plants, providing rebates for small cars and granting tax credits for insulation, re-spondents voiced clear support (though one big exception was the gasoline tax). But a scrutiny of the numbers suggests that this professed support was illusory.

Proposals that posed some tradeoffs, such as rebates for small cars or more lenient air pollution laws, won approval by a much lower margin. And painful proposals such as higher prices and rationing were roundly rejected.

Furthermore, respondents often contradicted themselves. To one question, 55 percent said they would approve of more strip mining to produce coal for industry; only 26 percent were opposed. To another question, 48 percent said they would approve of relaxing air pollution laws to permit more coal burning; 41 percent would not.

Environment by 5 to 3

Yet when asked which they thought was more important, producing energy or protecting the environment, respondents chose the environment by a margin of 5 to 3.

Pollsters generally place more credence on questions that confront respondents with alternatives. People, they say, have a tendency to approve of things when their only option is to disapprove. Faced with a less-than-ideal alternative, how-ever, their assurance often slips.

The poll also showed that Mr. Carter's support is slipping. A large portion of the respondents, 47 percent, still said they approved of the way the President was handling the energy problem in general. But supporters were more numerous when this question was asked in a Times/CBS poll taken in April after Mr. Carter's TV message on energy. At that time, 58 percent approved.

Interestingly, opposition to his handling of the problem has dropped, too. In April, 27 percent disapproved of his actions. This time, 24 percent did.

The President's lost supporters have slipped into the "unsure" group. In April, only 15 percent of the people were unsure. Last month 29 percent were.

Nonetheless, the public seems to remain optimistic, judged by their responses to this outlook question: "Do you think Jimmy Carter will or will not be able to handle the energy problems effectively?"

In April, 48 percent predicted he would, 38 percent thought he would not, 14 percent said they did not know. In the new poll, 49 percent said he would, 29 percent said he would not and 22 percent were unsure.

September 1, 1977

LIMITS TO GROWTH?

Pollution's Tie to Prosperity Disputed

By GLADWIN HILL
Special to The New York Times

CHICAGO, June 9—Environmentalists disputed today economists who depict both booming population and pollution as integral parts of American prosperity.

Excess population is lowering living standards rather than raising them, and the net impact of much of the nation's production involves unnecessary degradation of the environment, they said.

Their forum was the First National Congress on Optimum Population and Environment, a conference of hundreds of representatives of organizations concerned about global population growth and related ecological problems.

The more militant environmentalists have been advocating stabilizing or reducing the nation's population of 200 million and turning to a "no-growth economy." There has been little study of how a non-expanding economy might work and some economists have expressed alarm at the idea.

"If per capita income is used as a measured, we have probably surpassed the optimum population size," said Philip E. Sorensen, an economist at the State University of New York at Brockport. "Other things staying equal, per capita income will grow at least one percentage point more per year given a zero rate of population growth rather than a 2 per cent rate population growth."

"Some sectors of the economy stand to reap substantial benefits from a population boom, most particularly real estate and the resource-based industries," he went on. "But a redistribution of the economic benefits of zero population growth would more than compensate for the losses of these sectors."

Larry Barnett, a sociologist at the California State College at Los Angeles, said, "The continued growth of population in the United States threatens to impose a crushing burden on the nation's taxpayers."

With about 85 per cent of California's state budget based on population size, he said, from 1953 through 1967 "straight government expenditures increased almost twice as fast as population (with inflation taken into account), and the per person expenditure increased 34 per cent faster than per person personal income."

Dr. Paul Ehrlich, a Stanford University biologist, said, "The American advertising industry is essentially communistic — much of its work is creating demand for products that are identical by imbuing them with fictitious differences — many of which nobody needs, anyway."

Robert E. Jenkins, of Harvard's Museum of Comparative Zoology, declared, "The public is being constantly bombarded with irresponsible messages which openly advise them to exploit and consume in certain ways, and subliminally appeal to instinctive competitive, sexual or acquisitive urges. We need to replace, if possible, and to counter in any case, this sort of ongoing inducement to deleterious behavior with a class of messages which urge healthy action at the conscious level."

Former Secretary of the Interior Stewart L. Udall, now an environmental consultant in Washington, said that an obvious counterattack was recently launched against the campaign for what he termed a "quality-of-life economy."

Admonitions by Udall

Commenting that he sought to "open up a dialogue between the nation's economists and its environmentalists, he voiced some admonitions to the economists.

"Don't turn us off by oversimplifying our arguments," he said. "We are not, for example, proposing a return to a 'Waldon Pond.' But we believe something is seriously wrong with a system that creates a fat life of empty affluence but can't even learn to recycle its wastes or build livable cities.

"Don't dismiss us as an angry wave of machine smashers. We are simply convinced the time has come to civilize and restrain technology and make it serve men.

"Don't assume that we propose zero population growth as an economic panacea. As we believe another generation of more-of-the-same economic growth is a prescription for catastrophy, we are convinced that one sure way to reduce long-term demands on the life-support system of the planet, and make all of our social problems more manageable, is to level off the population increase in this country."

"The current economic system is based on an ideology of maximum production and maximum consumption," Mr. Udall went on. "It may have been appropriate in 1932 or 1945, but it could lead us down a path to disaster in the remaining years of this century.

"I would ask our friends in the world of economics to weigh our arguments, and, over the long haul of life on this planet, it is the ecologists, and not the bookeepers of business, who are the ultimate accountants."

347

Dr. Garrett Hardin, a University of California biologist, remarked, "The population of American people increases at 1 per cent a year. While our population doubles every 70 years, our energy requirements double every 10, increasing the per capita pollution from electric power plants."

"The population of American automobiles increases at 5 per cent per year," he continued. "Surveying cities of various sizes, we discover that the number of crimes increases faster than the size of the city —and this in spite of the fact that big cities also spend proportionately more on crime control."

Dr. Hardin, while stating that it was too soon to impose legal limits on procreation, said that "in the long run" voluntary birth control would probably not work because "the unjustifiably optimistic tend to have more children than do the realists."

The conference, at the Pick-Congress Hotel, will continue through Thursday.

June 10, 1970

The Canary Has Fallen Silent

By HERMAN E. DALY

In 1857 John Stuart Mill argued that a "stationary state" (by which he meant constant population size and a constant stock of *physical* wealth) was both necessary and desirable. During the century since Mill wrote most economists have argued that the stationary state was either unnecessary, undesirable, or both.

There are many reasons for believing that the times have caught up to Mill, and that today the big question for the social and physical sciences is to invent institutions and technologies which will allow us to reach and maintain a stationary-state economy.

Since our world is finite nothing physical can grow forever—neither the population of human beings, nor the "population" of physical commodities. However, growth in nonphysical things, well-being, satisfaction, leisure time, can continue. In the stationary state technological progress must be aimed at nonphysical growth, or at improving the quality of life.

One may theoretically agree that eventually the stationary state is necessary, but yet argue that it is only relevant to the remote future. How do we know that the time is now? Coal miners used to take a canary into the mines with them. When the dust and gases had so fouled the subterranean environment that the canary had trouble staying alive, then the miners knew it was time to stop. With entire species of birds and animals unable to survive in our environment today (and with human life expectancy falling in some areas) it would seem that mankind is very much in the position of the miner whose canary has just died.

Depletion and pollution of the environment are inevitable by-products of production and consumption. Matter and energy can be neither created nor destroyed. Production must deplete the environment. Consumption must pollute it.

Populations of organisms and commodities have much in common: both are born (produced) and both die (are consumed); both draw primary inputs from the environment for production and maintenance, and both return to the environment the waste products resulting from maintenance and finally death. The flows of depletion and pollution are necessary to maintain the stocks of wealth and people, but they interfere with ecological processes

and limit the carrying capacity of the environment.

The limiting factor which determines carrying capacity is not space, but the least abundant necessary material, the most fragile link in the chain of ecological interdependence. Nor is this limit necessarily approached gradually and continuously. Critical thresholds and "trigger relations" are common in ecology.

Dick Saunders/Scope

Constant stocks should therefore be maintained with the lowest possible rates of input and output (which, of course, are equal when the stock is constant and may be jointly referred to as the rate of throughput). The lower the rate of throughput the lower the rates of depletion and pollution, and the higher the carrying capacity of the environment.

Low rates of throughput imply increased life expectancy. The slower the water flows through a tank the more time an average drop spends in the tank. The lower the birth and

death rates of a constant population, the longer an average person lives. The lower the production and consumption rates of a constant stock of commodities, the longer an average commodity lasts. If it is good for people to live longer and for commodities to last longer, then it is good to minimize the rate of throughput by which the stock is maintained.

Minimizing the rate of throughput which maintains a stock is equivalent to maximizing the life expectancy or durability of the stock. This can be done in two ways: increasing the durability of the individual commodities (instead of planning for obsolescence and self-destruction), and designing commodities and distribution channels so that the "corpse" of commodities can be recycled and "reincarnated" in a new commodity (the opposite of throw-away technology).

The rate of throughput, which is the same as the annual physical flow of production, which in turn is the same as real gross national product, is thus seen to be the cost of maintaining the stock of wealth. Since production is a cost it would logically be minimized in the stationary state, not illogically maximized as today. The same largely holds true for the growing economy. As Kenneth Boulding has argued for many years, it is the services of the stock of wealth which satisfy human wants. Production is a deplorable activity made necessary by the fact that wealth wears out or is used up, and must be replaced. It is not the increase of production, but the increase of capital stock which makes us rich. The social and economic implications of minimizing rather than maximizing production are decidedly radical. Ecological conservatism breeds economic radicalism.

In the past growth has been a solvent for sticky income problems. As long as everyone gets absolutely more the fight over relative shares will be less intense. But in the stationary state relative and absolute shares move together and the focus will shift to distribution of the stock of wealth, which is a far more radical issue than the distribution of income.

The main justification for inequality in income and wealth, namely that it facilitates saving and provides incentives, both of which are necessary for growth, would not be relevant in the stationary state. Also full-employment policies, which allow us to maintain the income-through-jobs principle of income distribution, require investment to stimulate aggregate demand. Investment means growth. The result of zero growth would be mass unemployment. Since production is minimized, the demand for factors, including labor, will be less. In fact the tendency will be to use ever less labor. Some form of supplementary non-wage income must be the rule, not the exception. Again this is all quite logi-cal—labor is a cost, why not minimize it, even if it requires a different system for distributing income and wealth?

By keeping the "pie" constant we make fewer demands on our environmental resources. But in sharing the constant "pie" we make much greater demands on our moral resources.

Economists have generally assumed that moral resources are the scarcest of all and should never be relied on very heavily. But in today's world of doomsday machines, cybernetics, mass media brainwashing, genetic control, etc., there is no alternative. Indeed, the physically stationary economy must be a morally growing economy.

Herman E. Daly is Associate Professor of Economics at Louisiana State University.   October 14, 1970

A Blueprint for Survival

This document was drawn up by a group of British scientists and philosophers professionally involved in studying global environmental problems. Their full report, which is excerpted here, first appeared in last month's issue of The Ecologist. The avowed aim is to "herald the dawn of a new age" in which "Man will learn to live with the rest of Nature rather than against it."

LONDON—The principal defect of the industrial way of life with its ethos of expansion is that it is not sustainable. Its termination within the lifetime of someone born today is inevitable—unless it continues to be sustained for a while longer by an entrenched minority at the cost of imposing great suffering on the rest of mankind. We can be certain, however, that sooner or later it will end (only the precise time and circumstances are in doubt), and that it will do so in one of two ways; either against our will, in a succession of famines, epidemics, social crises and wars; or because we want it to—because we wish to create a society which will not impose hardship and cruelty upon our children—in a succession of thoughtful, humane and measured changes.

Unfortunately, we behave as if we knew nothing of the environment and had no conception of its predictability, treating it instead with scant and brutal regard as if it were an idiosyncratic and extremely stupid slave. We seem never to have reflected on the fact that a tropical rain forest supports innumerable insect species and yet is never devastated by them; that its rampant luxuriance is not contingent on our overflying it once a month and bombarding it with insecticides, herbicides, fungicides, and what-have-you. And yet we tremble over our wheatfields and cabbage patches with a desperate battery of synthetic chemicals, in an absurd attempt to impede the operation of the immutable "law" we have just mentioned—that all ecosystems tend towards stability, therefore diversity and complexity, therefore a growing number of different plant and animal species until a climax or optimal condition is achieved.

There are half a million man-made chemicals in use today, yet we cannot predict the behavior or properties of the greater part of them (either singly or in combination) once they are released into the environment. We know, however, that the combined effects of pollution and habitat destruction menace the survival of no less than 280 mammal, 350 bird, and 20,000 plant species. To those who regret these losses but greet them with the comment that the survival of *Homo sapiens* is surely more important than that of an eagle or a primrose, we repeat that *Homo sapiens* himself depends on the continued resilience of those ecological networks of which eagles and primroses are integral parts.

Industrial man in the world today is like a bull in a china shop, with the single difference that a bull with half the information about the properties of china as we have about those of ecosystems would probably try and adapt its behavior to its environment rather than the reverse. By contrast, *Homo sapiens industrialis* is determined that the china shop should adapt to him, and has therefore set himself the goal of reducing it to rubble in the shortest possible time.

The United Nations Food and Agriculture Organization's program to feed the world depends on a program of intensification, at the heart of which are the new high-yield varieties of wheat and rice. These are highly responsive to inorganic fertilizers and quick-maturing, so that up to ten times present yields can be obtained from them. Unfortunately, they are highly vulnerable to disease, and therefore require increased protection by pesticides, and of course they demand massive inputs of fertilizers (up to 27 times present ones).

We must beware of those "experts" who appear to advocate the transformation of the ecosphere into nothing more than a food-factory for man. The concept of a world consisting solely of man and a few favored food plants is so ludicrously impracticable as to be seriously contemplated only by those who find solace in their own willful ignorance of the real world of biological diversity.

Continued exponential growth of consumption of materials and energy is impossible. Present reserves of all but a few metals will be exhausted within 50 years, if consumption rates continue to grow as they are. Obviously there will be new discoveries and advances in mining technology, but these are likely to provide us with only a limited stay of execution.

At the same time, we are sowing the seeds of massive unemployment by increasing the ratio of capital to labor so that the provision of each job becomes ever more expensive. In a world of fast diminishing resources, we shall quickly come to the point when very great numbers of people will be thrown out of work, when the material compensations of urban life are either no longer available or prohibitively expensive, and consequently when whole sections of society will find good cause to express their considerable discontent in ways likely to be anything but pleasant for their fellows.

There will be those who regard accounts of the consequences of trying to accommodate present growth rates as fanciful. But the imaginative leap from the available scientific information to such predictions is negligible, compared with that required for those alternative predictions, laughably considered "optimistic," of a world of 10,000 to 15,000 million people, all with the same material standard of living as the United States, on a concrete replica of this planet, the only moving parts being their machines and possibly themselves. Faced with inevitable change, we have to make decisions, and we must make these decisions *soberly* in the light of the best information, and not as if we were caricatures of the archetypal mad scientist.

Possibly because government sees the world in fragments and not as a totality, it is difficult to detect in its actions or words any coherent general policy, although major political parties appear to be mesmerised by two dominating notions: that economic expansion is essential for survival and is the best possible index of progress and well-being; and that unless solutions can be devised that do not threaten this notion, then the problems should not be regarded as existing. Unfortunately, government has an increasingly powerful incentive for continued expansion in the tendency for economic growth to create the need for more economic growth. This it does in six ways:

Firstly, the introduction of technological devices, i.e. the growth of the technosphere, can only occur to the detriment of the ecosphere, which means that it leads to the destruction of natural controls which must then be replaced by further technological ones. It is in this way that pesticides and artificial fertilizers create the need for more pesticides and artificial fertilizers.

Secondly, industrial growth, particularly in its earlier phases, promotes population growth. Jobs must constantly be created for the additional people—not just any job, but those that are judged acceptable in terms of current values.

Thirdly, no government can hope to survive widespread and protracted unemployment, and without changing the basis of our industrial society, the only way government can prevent it is by stimulating economic growth.

Fourthly, business enterprises, whether state-owned or privately owned, tend to become self-perpetuating, which means that they require surpluses for further investment. This favors continued growth.

Fifthly, the success of a government and its ability to obtain support is to a large extent assessed in terms of its ability to increase the "standard of living" as measured by *per capita* gross national product.

Finally, confidence in the economy, which is basically a function of its ability to grow, must be maintained to ensure a healthy state of the stock market. Were confidence to fall, stock values would crash, drastically reducing the availability of capital for investment and hence further growth, which would lead to further unemployment. This would result in a further fall in stock-market values and hence give rise to a positive-feedback chain-reaction, which under the existing order might well lead to social collapse.

For all these reasons, we can expect our government (whether Conservative or Labor) to encourage further increases in G.N.P. regardless of the consequences, which in any case tame "experts" can be found to play down. It will curb growth only when public opinion demands such a move, in which case it will be politically expedient, and when a method is found for doing so without creating unemployment or excessive pressure on capital.

The emphasis must be on integration. If we develop relatively clean technologies but do not end economic growths then sooner or later we will find ourselves with as great a pollution problem as before but without the means of tackling it.

Our task is to create a society which is sustainable and which will give the fullest possible satisfaction to its members. Such a society by definition would depend not on expansion but on stability. This does not mean to say that it would be stagnant —indeed it could well afford more variety than does the state of uniformity at present being imposed by the pursuit of technological efficiency. We believe that the stable society, as well as removing the sword of Damocles which hangs over the heads of future generations, is much more likely than the present one to bring the peace and fulfillment which hitherto have been regarded, sadly, as utopian.

February 5, 1972

An Answer to 'Blueprint'

The following is excerpted from two editorials that appeared last month in Nature magazine under the titles "The Case Against Hysteria" and "Catastrophe or Change?"

LONDON—Britain is being assaulted by the environmentalists. The new magazine *The Ecologist* published what it called "A Blueprint for Survival" which reflects and sometimes amplifies a good many of the half-baked anxieties about what is called the environmental crisis. On this occasion, the doctrine that dog should not eat dog notwithstanding, the magazine deserves to be taken to task if only for having recruited a "statement of support" from 33 distinguished people, many of them scientists, at least half of whom should have known better.

The abiding fault in these discussions is their naiveté, and nowhere is this more true than in speculations about the social consequences of the phenomena over which *The Ecologist* wrings its hands. Starting with the assertion that the developed nations have already collared the raw materials with which developing nations might seek to improve their standards of living, the journal goes on to say that "we are altering people's aspirations without providing the means for them to be satisfied. In the rush to industrialize, we break up communities, so that the controls which formerly regulated behavior are destroyed. Urban drift is one result of this process, with a consequent rise in antisocial practices, crime, delinquency and so on. . . ."

The truth is, of course, that this is merely speculation. All the attempts which there have been in the past few years to discover correlations between such factors as population density and prosperity per head of population with the tendency to violence, either civil or international, have been fruitless. Who will say that the crowded Netherlands are more violent than the uncrowded United States? And who will say that the forces which have in the past 2,000 years helped to make civilized communities more humane can now be dismissed from the calculation simply because a new generation of seers sees catastrophe in the tea leaves?

The error in supposing that they constitute a proof of imminent calamity is the assumption that administrative and social mechanisms which exist already or which are in the course of being developed will do nothing to fend them off, but this is to ignore the beneficent tendencies already apparent —the rapid decline of fertility in the past decade in Southeast Asia and the Caribbean and the working of the classical economic laws of scarcity, originally described by the great Victorians, to strike a balance between exploitation and conservation and the way in which governments in North America and Western Europe have succeeded in improving the quality of urban air and water by laying out money on pollution control. In short, those who prophesy disaster a century or more from now and ask for apocalyptic remedies overlook the way in which important social changes have historically been effected by the accumulation of more modest humane innovations.

February 5, 1972

To Grow And to Die

By ANTHONY LEWIS

LONDON, Jan. 28 — Our diverse worlds — developed, underdeveloped, East, West—have at least one article of faith in common: economic growth. For individuals, for economic enterprises and for nations, growth is happiness, the specific for ills and the foundation of hope. Next year our family will be richer, our company bigger, our country more productive.

Now the ecologists have begun to tell us that growth is self-defeating, that the planet cannot long sustain it, that it will lead inevitably to social and biological collapse. That was the central thesis of the recent "Blueprint for Survival" published in Britain, and it is a theme increasingly found in analytical studies of the earthly future.

The proposition is so shocking that the natural reaction is to wish it away. Some economists, the apostles of growth, do just that. There was an especially acute example of wishfulness in a Newsweek column by Henry C. Wallich, Yale professor and former U.S. economic adviser, condemning the opposition to growth as dangerous heresy.

"It is an alarming commentary on the intellectual instability of our times," Professor Wallich said, "that today mileage can be made with the proposal to stop America dead in her tracks. Don't we know which way is forward?"

As long as there is growth, he said, "everybody will be happier." By "allowing everybody to have more" and refusing to "limit resources available for consumption," we shall also have "more resources" to clean up the environment.

If Professor Wallich's opinion is representative of the American intellectual community, it is an alarming comment on our awareness of the most important facts of life today.. For he is evidently in a state of ecological illiteracy.

There are no such things as endless growth and unlimited resources for everyone and everything. We live in a finite world, and we are approaching the limits. Discussion of growth as an environmental factor has to begin with some understanding of such considerations.

The crucial fact is that growth tends to be exponential. That is, it multiplies. Instead of adding a given amount every so often, say 1,000 tons or dollars a year, the factors double at fixed intervals. That tends to be true of population, of industrial production, of pollution and of demand on natural resources—some of the main strains of planetary life.

The rate of increase determines the doubling time. If something grows 7 per cent a year it will double in ten years. Right now world population is growing 2.1 per cent a year; at that rate it doubles in 33 years. And with each doubling, the base is of course larger for the next increase. The world had about 3.5 billion people in it in 1970. At the present rate of increase, it will have seven billion in 2003.

Exponential growth is a tricky affair. It gives us the illusion for a long time that things are going slowly; then suddenly it speeds up. Suppose the demand for some raw material is two tons this year and doubles every year. Over the next fifteen years it will rise to only 32,768 tons, but just five years later it will be 1,048,576 tons.

That phenomenon is what makes it so hard for people to understand how rapidly we may be approaching the limits of growth. For as population and per capita consumption both grow, the curves of demand suddenly zoom upward.

Consider the case of aluminum as a sample of resource demand and supply. The known reserves of aluminum are enough to supply the current demand for 100 years. But the use is increasing exponentially, and at the rate of increase the supply will be enough for only 31 years. Moreover, the multiplying demand is a much larger factor, mathematically, than any likely discovery of new sources of supply. If reserves were multiplied by five, the same growth of demand would still exhaust them in 55 years.

The example of aluminum is not especially chosen to disturb, for there are others that even more dramatically indicate the way exponential growth can run up to projected limits. One is simply arable land. At the present rate of world population growth, the supply of land necessary for food production will run out by the year 2000. If agricultural productivity were doubled, the limit would be pushed back thirty years.

Those estimates are taken from drafts of what is likely to be one of the most important documents of our age. It is a report made for the Club of Rome, an eminent international group of industrialists, economists, scientists and others. Entitled "The Limits of Growth," it was done by scientists using world system models developed at the Massachusetts Institute of Technology. It will be published in March by Potomac Associates of Washington.

The report's authors would never insist on any particular figure. They know that they are dealing with variables, and they have indeed leaned way over backward to make optimistic assumptions in their projections.

But *every model they build assuming continuation of the present world philosophy of growth ends in collapse.* To ignore that tendency, to pretend that growth can go on forever, is like arguing that the earth is flat. Only the consequences are more serious.

January 29, 1972

A World Without Growth?

By HENRY C. WALLICH

NEW HAVEN, Conn. — Anthony Lewis, in two recent issues of The New York Times, warns us of the deadly consequences of growth. Running out of resources, running into total pollution, running to the point of total exhaustion and collapse — those are the ultimate rewards of growth. We must stop growth, not just of population, but of production and income.

The group of ecologists who generated this well meaning scare are members of an old club. Its founder, the Rev. Thomas Malthus, issued dire warnings of inevitable starvation in 1798. This having proved a poor bet, the emphasis today shifts to a dearth of all natural resources and mounting pollution.

It does not take an ecologist to explain that if the world's population doubles every so many years, after a while there will be Standing Room Only, at least on the surface of this planet. Likewise, it is fairly obvious that if we deplete existing resources without discovering new sources, developing methods of recycling, and inventing substitutes, we shall some day

run out. But perhaps an economist can be helpful in clarifying why these problems are not top priority today.

In the first place, the economy will simply substitute things that are plentiful for things that become scarce. If we run out of aluminum, the price of aluminum will go up. That will encourage manufacturers to use something else, and will stimulate research and development to produce substitutes.

Some scientists believe that matter and energy are fundamentally interchangeable in many forms, but as a layman, I would not bet on any near-term miracles. The simple processes of economics will keep us going. If they don't, the ecologists' advice to slow down will not be worth much— it would only postpone the day of disaster without avoiding it.

In the course of centuries, more basic adjustments will probably be needed. Population may stop growing, production may stop growing. The chances are that the world will adapt to the changing environment gradually. Lack of space will cause families to shrink, if families then still exist. Great per capita income will reduce interest in producing and consuming more. We do not need to rely on "misery," as the Rev. Malthus thought, to bring about the adjustment.

The real question is at what time this transition will have to be faced. New York restaurants carry signs to the effect that occupancy by more than some maximum number is unlawful. If half a dozen persons were to gather in an otherwise empty restaurant with such a sign and discuss heatedly the urgency of keeping newcomers out, they would be in something like the position of Americans debating the zero growth notion. To stop growing now, generations before the real problems of growth arise if ever, would be to commit suicide for fear of remote death.

The ecologists do not seem to be aware of what it would mean to freeze total income anywhere near today's level. Do they mean that the present income distribution is to be preserved, with the poor frozen into their inadequacies? Would that go for the underdeveloped countries too? Or do they have in mind an equalization of incomes? It will take pretty drastic cuts in upper income bracket standards to bring them down to the average American family income of about $10,000, to say nothing of a cut to average world income. We can and perhaps should approach this condition over generations. Trying to do it quickly would create completely needless problems.

The ecologists also do not seem to be aware of what their prescriptions, contrary to their wishes, might do to the environment. If growth came to a halt, it is obvious that every last penny of public and private income would be drawn upon to provide minimal consumer satisfactions. There would be very little left for the cleaning-up job that needs to be done. Growth is the main source from which that job must be financed.

I would like to end with a quote from my Newsweek column on which Mr. Lewis commented. "A world without growth, that is, without change, is as hard for us to imagine as a world of everlasting growth and change. Somewhere in the dim future, if humanity does not blow itself up, there may lie a world in which physical change will be minimal . . . hopefully a much more humane and less materialistic world. We shall not live to see it."

Prof. Henry C. Wallich, of Yale's department of economics, is a columnist for Newsweek. February 12, 1972

Mankind Warned of Perils in Growth

By ROBERT REINHOLD
Special to The New York Times

CAMBRIDGE, Mass., Feb. 26—A major computer study of world trends has concluded, as many have feared, that mankind probably faces an uncontrollable and disastrous collapse of its society within 100 years unless it moves speedily to establish a "global equilibrium" in which growth of population and industrial output are halted.

Such is the urgency of the situation, the study's sponsors say, that the slowing of growth constitutes the "primary task facing humanity" and will demand international cooperation "on a scale and scope without precedent." They concede such a task will require "a Copernican revolution of the mind."

The study, which is being sharply challenged by other experts, was an attempt to peer into the future by building a mathematical model of the world system, examining the highly complex interrelations among population, food supply, natural resources, pollution and industrial production.

The conclusions are rekindling an intellectual debate over a question that is at least as old as the early economists, Thomas Malthus and John Stuart Mill:

Will human population ultimately grow so large that the earth's finite resources will be totally consumed and, if so, how near is the day of doom?

The study was conducted at the Massachusetts Institute of Technology under the auspices of the Club of Rome. In the findings, to be published next month by the Potomac Associates under the title "The Limits to Growth," the M.I.T. group argues that the limits are very near—unless the "will" is generated to begin a "controlled, orderly transition from growth to global equilibrium."

The study would seem to bolster some of the intuitive warnings of environmentalists. In Britain last month, for example, a group of 33 leading scientists issued a "blueprint for survival," calling on the nation to halve its population and heavily tax the use of raw materials and power.

But others, particularly economists, are skeptical.

"It's just utter nonsense," remarked one leading economist, who asked that he not be identified. He added that he felt there was little evidence that the M.I.T. computer model represented reality or that it was based on scientific data that could be tested.

Another economist, Simon S. Kuznets of Harvard, a Nobel Prize-winning authority on the economic growth of nations, said he had not examined the M.I.T. work first hand, but he expressed doubt about the wisdom of stopping growth.

"It's a simplistic kind of conclusion—you have problems, and you solve them by stopping all sources of change," he said.

Others, like Henry C. Wallich of Yale, say a no-growth economy is hard to imagine, much less achieve, and might serve to lock poor cultures into their poverty.

Malthus Again and Again

"I get some solace from the fact that these scares have happened many times before — this is Malthus again," he said.

Malthus, the 19th-century British economist, theorized somewhat prematurely that population growing at exponential rates that could be graphically represented as a rising curve would soon outstrip available food supply. He did not foresee the Industrial Revolution.

Don't Have Alternative

Prof. Dennis L. Meadows, a management specialist who directed the M.I.T. study—which is the first phase of the Club of Rome's "Project on the Predicament of Mankind" — conceded that the model was "imperfect," but said that it was based on much "real world" data and was better than any previous similar attempt.

The report contends that the world "cannot wait for perfect models and total understanding." To this Dr. Meadows added in an interview: "Our view is that we don't have any alternative — it's not as though we can choose to keep growing or not. We are certainly going to stop growing. The question is, do we do it in a way that is most consistent with our goals or do we just let nature take its course."

Letting nature take its course, the M.I.T. group says, will probably mean a precipitous drop in population before the year 2100, presumably through disease and starvation. The computer indicates that the following would happen:

¶ With growing population, industrial capacity rises, along with its demand for oil, metals and other resources.

¶ As wells and mines are exhausted, prices go up, leaving less money for reinvestment in future growth.

¶ Finally, when investment falls below depreciation of manufacturing facilities, the industrial base collapses, along with services and agriculture.

¶ Later population plunges from lack of food and medical services.

All this grows out of an adaptation of a sophisticated method of coming to grips with complexity called "systems analysis." In it, a complex system is broken into components and the relationships between them reduced to mathematical equations to give an approximation, or model, of reality.

Then a computer is used to manipulate the elements to simulate how the system will change with time. It can show how a given policy change might affect all other factors. If human behaviour is considered a system, then birth and death rates, food output, industrial production, pollution and use of natural resources are all part of a great interlocking web in which a change in any one factor will have some impact on the others.

Interrelations Studied

For example, industrial output influences food production, which in turn affects human mortality. This ultimately controls population level, which returns to affect industrial output, completing what is known as an "automatic feedback loop."

Drawing on the work of Prof. Jay W. Forrester of M.I.T., who has pioneered in computer simulation, the M.I.T. team built dozens of loops that they believe describe the interractions in the world system.

They then attempted to assign equations to each of the 100 or so "causal links" between the variables in the loops, taking into account such things as psychological factors in fertility and the biological effects of pollutants.

Critics say this is perhaps the weakest part of the study because the equations are based in large part on opinion rather than proved fact, unavailable in most cases. Dr. Meadows counters that the numbers are good because the model fits the actual trends from 1900 to 1970.

The model was used to test the impact of various alternative future policies designed to ward off the world collapse envisioned if no action is taken.

For example, it is often argued that continuing technological advances, such as nuclear power, will keep pushing back the limits of economic and population growth.

To test this argument, the M.I.T. team assumed that resources were doubled and that recycling reduced demand for them to one-fourth. The computer run found little benefit in this since pollution became overwhelming and caused collapse.

Assumptions Tested

Adding pollution control to the assumptions was no better; food production dropped. Even assuming "unlimited" resources, pollution control, better agricultural productivity and effective birth control, the world system eventually grinds to a halt with rise in pollution, falling food output and falling population.

"Our attempts to use even the most optimistic estimates of the benefits of technology," the report said, "did not in any case postpone the collapse beyond the year 2100."

Skeptics argue that there is no way to imagine what kind of spectacular new technologies

are over the horizon.

"If we were building and making cars the way we did 30 years ago we would have run out of steel before now I imagine, but you get substitution of materials," said Robert

M. Solow, an M.I.T. economist not connected with the Club of Rome project. "It is true we'll run out of oil eventually, but it's premature to say therefore we will run out of energy," he added.

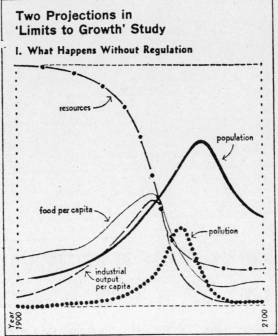

Two Projections in 'Limits to Growth' Study

I. What Happens Without Regulation

The New York Times/Feb. 27, 1972

M.I.T. group used a computer global model to project trends in five key growth factors to year 2100. This computer "run" shows rapidly diminishing resources eventually slowing growth, assuming no major change in physical, economic, or social relationships. The time lags in decline of population and pollution are attributed to natural delays in the system. Rise in population is finally halted by an increase in death rate.

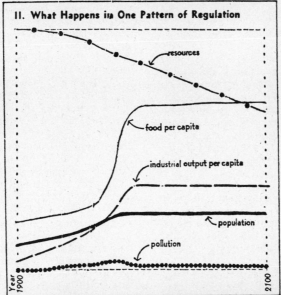

II. What Happens in One Pattern of Regulation

The New York Times/Feb. 27, 1972

This computer "run" projects relatively stable future on assumption that "technology policies" are combined with other growth-regulating mechanisms. Study says policies would include resources recycling, pollution control devices, increased lifetimes of all forms of capital, methods to restore eroded and infertile soil.

At any rate, the M.I.T. group went on to test the impact of other approaches, such as stabilizing population and industrial capacity.

Zero population growth alone did very little, since industrial output continued to grow, it was found. If both population and industrial growth are stabilized by 1985, then world stability is achieved for a time, but sooner or later resource shortages develop, the study said.

System Suggested

Ultimately, by testing different variations, the team came up with a system that they believe capable of satisfying the basic material requirements of mankind yet sustainable without sudden collapse. They said such a world would require the following:

¶Stabilization of population and industrial capacity.

¶Sharp reduction in pollution and in resource consumption per unit of industrial output.

¶Introduction of efficient technological methods — recycling of resources, pollution control, restoration of eroded land and prolonged use of capital.

¶Shift in emphasis away from factory-produced goods toward food and nonmaterial services, such as education and health.

The report is vague about how all this is to be achieved in a world in which leaders often disagree even over the shape of a conference table.

Even so, critics are not sanguine about what kind of a world it would be. Dr. Meadows agrees it would not be a Utopia, but nevertheless does not foresee stagnation.

"A society released from struggling with the many problems caused by growth may have more energy and ingenuity available for solving other problems," he says, citing such pursuits as education, arts, music and religion.

Many economists doubt that a no-growth world is possible. Given human motivations and diversity, they say, there will always be instability.

"The only way to make it stable is to assume that people will become very routine-minded, with no independent thought and very little freedom, each generation doing exactly what the last did," says Dr. Wallich. "I can't say I'm enamored with that vision."

"Can you expect billions of Asians and Africans to live forever at roughly their standard of living while we go on forever at ours?" asked Dr. Solow.

Dr. Wallich terms no-growth "an upper income baby," adding "they've got enough money, and now they want a world fit for them to travel in and look at the poor."

The M.I.T. team agrees there is no assurance that "humanity's moral resources would be sufficient to solve the problem of income distribution." But, they contend "there is even less assurance that such social problems will be solved in present state of growth, which is strain-

ing both the moral and physical resources of the world's people."

The report ends hopefully, stating that man has what is physically needed to create a lasting society.

"The two missing ingredients are a realistic long-term goal that can guide mankind to the equalibrium society and the human will to achieve that goal,"

it observes.

Collaborating with Dr. Meadows in writing "The Limits to Growth," were his wife, Donella, a biophysicist; Jorgen Randers, a physicist, and William W. Behrens 3d, an engineer. They were part of a 17-member international team working with more than $200,-000 in grants from the Volkswagen Foundation in Germany.

The major conclusions of the study have been circulating among experts for a few months. The full details are to appear in next month's publication and in future technical documents. This Thursday, a symposium on the study will be held at the Smithsonian Institution in Washington.

February 27, 1972

Continuous growth or no growth?
What the ecologists can teach the economists

By Bertram G. Murray Jr.

The intensifying debate between the "pessimists" and the "optimists" reached a peak recently with the publication of "The Limits to Growth," a report prepared by an M.I.T. team headed by systems analyst Dennis L. Meadows for the Club of Rome's "Project on the Predicament of Mankind." The authors, using an admittedly simplified world model, fed data into a computer and concluded "with some confidence that, under the assumption of no major change in the present system, population and industrial growth will certainly stop within the next century, at the latest." In taking exception to this conclusion, those who are optimistic about future economic and population growth point out what Meadows's team had already admitted: the lack of hard data, the difficulty of modeling the effects of technological innovation and the impossibility of modeling changes in human value systems. In addition, some progrowth adherents emphasize the potentially disastrous economic consequences of a no-growth policy; for instance, Peter Passell and Leonard Ross, writing in this magazine last March 5, conclude: "Quite simply, growth is the only way in which America will ever reduce poverty."

There can be no doubt that Americans are being offered a choice—no, they are being told they must make a choice—between a continuous-growth economic system and a no-growth economic sys-

tem. What's more, Americans will make a choice, even unwittingly, because acceptance of current economic policies is one alternative. But how are Americans to decide who is right, the optimists or the pessimists? Is the argument entirely esoteric or academic? What evidence is there to support one side or the other?

First, Americans must come to understand the nature of forecasting the future. All forecasts are derived from models of the real world. Simplified assumptions are always made, whether one is predicting the consequences of economic growth or the need to build nuclear power plants now to satisfy the energy demand in 1992. Scientific models are evaluated on how well they predict and describe. But models are evaluated by human beings, and therefore models are often accepted or rejected on emotional grounds, whether they describe economic, biological or physical relationships. Even Albert Einstein could ignore the later developments of Quantum Theory because their statistical nature conflicted with his belief that "God does not play dice with the world."

In the social sciences, models are also evaluated according to the goals and values of the system. Capitalism or socialism, democracy or fascism are neither good nor bad except as they conform to the goals and values of the societies that practice them or to those of neighboring societies.

Both ecologists and economists have developed models that describe the cause and effect relationships within their respective systems. It seems incredible that ecologists and economists have not shared their ideas before now because both study the same phenomena, albeit in different populations. Ecologists study competition between individuals and between populations for resources,

Bertram G. Murray Jr. is assistant professor of zoology in the Department of Science of Rutgers University.

Overpopulation

the growth of populations and the movement of materials (*e.g.*, water and minerals) in ecological systems (ecosystems). Economists study competition between producers for markets, the growth of production and the movement of goods and resources within economic systems. In each of these areas, ecologists and economists have models which have entirely different consequences. A comparison of these models may enable us to understand better the choice we are going to make between continuous-growth and no-growth economic systems.

Growth

Biological growth of all kinds has a characteristic pattern with respect to time. Regardless of the nature of the population—whether birds, or bees, or protozoans, or the cells of human beings—its numbers grow slowly at first, then increase rapidly before slowing and leveling off at some equilibrium level at which the cells of the tissue or organism, or the animals in a population, are dying at the same rate that new ones are being formed or born. This is the so-called steady state.

The economic model of American businessmen, economists and politicians demands continuously increasing growth, as reflected by the goal of increasing the Gross National Product. Undeniably, economic growth has brought Americans the highest standard of living anywhere in the world. By contrast, a no-growth, or steady-state, economic system has such consequences as declining material wealth and increasing unemployment as population increases. Given such a choice most reasonable persons would select the continued growth of the American economy. But can a 4 per cent annual increase in the G.N.P., which requires doubling the production of goods and services in 17 years and quadrupling them in 34, be sustained?

Such continuous growth curves are not unknown in biological and physical systems. When cells continue multiplying in animal tissues, we call them cancer cells. Their growth is not indefinite. Indeed, they eventually kill the host organism. Populations of animals that are increasing have a similar fate,

a decisive population crash. A classic case is that of the deer population inhabiting the Kaibab Plateau on the northern rim of the Grand Canyon. In 1907, game managers began removing the deer's natural predators, the mountain lions, wolves and coyotes. The deer population increased rapidly from 4,000 to an estimated 100,000 in 1924. This growing population seriously depleted the resources of its environment, and it finally crashed. Sixty thousand deer died of starvation and disease in the winters of 1925 and 1926. The population continued declining, finally leveling off at around 10,000. (The implication of this and similar cases with respect to the world's rapidly increasing human population should be clear, but this is not the point at issue here.)

In physical systems, an example of such exponential growth is the chain reaction of fissioning uranium-235 nuclei. A single neutron splits a uranium nucleus, giving off two or three more neutrons (average, 2.5), which in turn split two or more uranium nuclei, giving off increasingly large numbers of neutrons that split increasingly large numbers of uranium nuclei, resulting in a nuclear explosion that generates vast amounts of energy for a short period of time.

Thus, in biological and physical systems, continuous growth can lead to disasters—death from cancer, a sharp increase in death rate leading to population decline, inefficient use of energy in nuclear explosions. Conditions are evidently optimal for increasing growth for a time. If a cancer cell, a deer or a neutron could think and talk, it might say, "My, things couldn't be better, for we seem to be thriving." But we human beings *can* think, and we know by observation that these conditions are short-lived. In nature most populations are in equilibrium. By some means or another, an increase in numbers is followed by a decrease. After accumulating uranium-235 atoms in a bomb or power plant, we humans are careful in regulating the production of neutrons.

In biological and physical systems, then, the consequences of increasing growth are precisely those forecast in "The Limits to Growth" for human population and industrial growth. Although

Stopping pollution

this forecast could be considered "theoretical," there is ample observational evidence from analogous systems to increase the probability of its correctness. Because of numerous modifying factors, the accuracy of the forecast with respect to the timing of the collapse is more difficult to evaluate. Thus, we can be fairly certain that collapse will occur but less certain as to when.

Movement of materials

A second area of interest to ecologists is the biogeochemical cycles. These describe the movement within ecosystems of minerals, water, oxygen, carbon dioxide and other nutrients essential for life. For example, carbon dioxide in the air is incorporated into organic molecules (carbohydrates, etc.) by photosynthesis in plants. The plants are eaten by animals, which in turn may be eaten by other animals. Carbon dioxide is returned to the air through the chemical breakdown of organic molecules through metabolism and decomposition. Once back in the air, the carbon dioxide can be incorporated again into new organic molecules. In other words, carbon dioxide follows a cycle between the atmosphere and living organisms. The other nutrients also re-cycle through an ecosystem but often in more complex ways. An ecosystem such as a pond, a field or a forest is maintained because of this recycling of essential nutrients, which results because one species' waste is some other species' food. But recycling is not 100 per cent efficient. In the course of time there is a net

change in the chemical make-up of the ecosystem. This results in a continuously changing environment, a process called succession. The ecological status quo cannot be maintained without perfect cycling.

In a simplified man-made ecosystem, we can better observe the consequences of interfering with nutrient recycling. The minerals removed from the soil by a corn crop, for example, find their way to market either directly as corn or indirectly via corn-fed pigs rather than back into the soil. Several consecutive crops seriously deplete the fertility of the soil, at least for corn plants. Smart farmers rotate their crops, each crop replacing those minerals the previous crop removed. One crop's waste (what it puts in the soil) is another's nutrient.

Man's complex technological society demands from its environment not only food but also large amounts of raw materials for building houses,

factories, cars, television sets and so on. The recycling of these materials is virtually zero. Iron, for example, is mined where it is concentrated in the earth, processed into steel and formed into cars, which after a few years' use are allowed to rust away in some field. The iron in this dispersed state is no longer mineable. A technological ecosystem that does not recycle materials is no more likely to be sustained indefinitely than a cornfield.

Recycling is increasing in the United States (as with soda bottles and newspapers) but it is not yet a way of life. Just look at our overflowing dumps.

Competition

Ecologists and economists have strikingly different views on the effects of competition in ecosystems and economic systems, respectively. A cornerstone of ecological theory is the competitive exclusion

Starving deer

Recycling

principle. Simply, this principle states that competing species cannot coexist indefinitely. If two species are utilizing a resource that is in short supply, one of them will be eliminated as a competitor, either by being forced out of the ecosystem or by being forced to use some other resource. As a result, in communities of animals, ecologists normally find that each species differs from the others in its utilization of the resources in the environment.

The competitive exclusion principle is consistent not only with observations in natural situations but also with laboratory experiments. In the nineteen-thirties, G. F. Gause, the Russian ecologist, demonstrated the "struggle for existence" between species of yeast cells and between species of protozoans. Later Thomas Park and his colleagues at the University of Chicago undertook an elegant series of experiments with flour beetles. In each case only one species could survive. Again and again, the evidence seems to indicate that competition reduces the number of competitors.

The economists' model of competition is notably different. Competition is supposed to maintain diversity and stability in economic systems; it is assumed that with numerous producers competing for the market, no single producer could control the industry and therefore fix prices and limit the entrance of new producers into the business. In competing for markets, producers would increase efficiency and reduce prices or increase quality at the same price. Either way, the consumer benefits. Or so say the optimistic economists, businessmen and politicians.

To the contrary, the evidence suggests that competition in economic systems has the same effect as competition in ecosystems. It reduces the number of competitors. The more efficient or larger producers force the less efficient or smaller out of business or buy them out, resulting in monopoly. This was recognized, practically if not theoretically, in the 19th century, and Congress passed the Sherman Antitrust Act of 1890, which

unfortunately does not eliminate industry dominance by one or a few corporations. The number of competitors continues to become smaller, prices and profits increase, and the huge corporations and conglomerates are more difficult if not impossible to manage efficiently. The railroads and Lockheed are examples.

The future

Americans must face several unpleasant facts squarely. First, they must understand that the American economic model—one that values increasing growth, waste (nonrecycling of essential materials) and competition—is inherently unstable. This model violates, either actually or theoretically, ecological principles that have been established by observation: Competition results in elimination of competitors; continually growing populations eventually collapse; and ecosystems whose essential materials are not recycled cannot be sustained. There is no reason to believe that economic systems based on principles that bring col-

lapse to ecological systems are immune to a similar fate. Indeed, some economists implicitly agree. Leonard Silk reported in The Times last spring that at a Bryn Mawr College symposium on "Limits to Growth," economists suggested that the businessmen might manage to hang on for several decades longer rather than write capitalism off already as a lost cause. There is no question that collapse is inevitable. The question seems to be how much longer businessmen can continue making profits at the expense of future generations. For nonbusinessmen a reasonable course of action would be to take steps now that will prevent or at least ameliorate the effects of the crash.

A second unpleasant fact that we Americans are going to have to face is the necessity of reducing our current standard of living. The current disparity between the conditions of living in the developing countries and those in the developed countries, especially in the United States, will not be tolerated. Inasmuch as the American standard of living depends upon imports of raw materials from the developing countries, the decision

357

Dice game?

to reduce our level of affluence may not be ours to make but will be forced upon us as the developing countries increase their demands on the earth's resources for their own consumption. There is great danger here. Without international regulation of economic systems, the countries will be competing with one another for resources, a condition that often leads to war. With today's weapons such competition, unlike biological competion, does not guarantee a survivor.

Compounding this situation is the necessity to permit increasing economic growth and a rising standard of living in the developing countries while the developed countries are scaling down their own standard of living. The United States may still remain Number 1, but Numbers 2 and 10 and 100 will not be so far behind as they are now.

Like it or not Americans must now decide between two alternatives: (1) the status quo —an economic system that offers short-term profits to a few and disaster within generations, if not decades (for some it has already arrived), and (2) an economic system that provides for the long-

term survival of humankind at a decent standard of living. Whatever is decided in the immediate future in the United States will have long-term global effects.

It probably should be mentioned that neither socialism nor communism is an acceptable economic system for providing for man's long-term economic survival. It turns out that all of man's economic systems are progrowth and wasteful, and therefore unstable. The alternative economic system has yet to be formulated. In broadest outline, this economic system will have to be consistent with ecological theory. First, it will have to be, over-all, a no-growth system. Some growth will occur in individual enterprises as they develop new technologies, but other enterprises will have to be curtailed. Such a no-growth economic system will have to be tied in with a no-growth population policy. Second, the new system must demand the recycling of as much as possible of those materials we now call waste. For instance, old autos must be made into new autos. And, when raw materials cannot be extracted from worn-out finished products, these products must be converted

into other products. And human and animal wastes must be converted into useful fertilizer. And so on. Third, the effects of competition must be recognized and controlled. Competing businesses must be ecologically responsible in their consumption of raw materials, production of pollution and reuse of materials.

Such an economic system would have to be highly regulated, and because both modern economies and environmental problems cross international boundaries, a worldwide, environmentally responsible economic system would have to be managed by an international team of planners, most reasonably organized by the United Nations.

Government legislation affecting business practices is not a new idea. Capitalism has been evolving for more than a century. At one time, the entrepreneur's sole responsibility was to make a profit. Without restriction, the quest for profit led to extensive abuses of labor, which became sufficiently widespread and evident by the late eighteen-hundreds to move governments to pass laws that forced socially responsible management policies on busi-

ness. The conflict between management and labor continues but no one now suggests that child labor, 14-hour days, and 10-cents-a-day wages should be reinstituted. The quest for profit also led to extensive abuses of our environment, which are only now becoming sufficiently evident to lead to the notion that government should pass laws that force environmentally responsible policies on business. The time to begin planning and instituting changes is now, if we are going to avert the disaster inevitable under our present economic policies.

A pessimistic outlook

I am pessimistic about the future—not because the problems we face are technically or economically insurmountable but because they seem humanly insurmountable.

The generation in political power grew up during the period of man's greatest "progress." More people are living better now than ever before. This increase in living standards undeniably resulted from economic growth and competition, sufficient reason for understanding why those persons who lived through that stage in man's history should be reluctant to give up their faith in economic growth and competition.

At the same time technological innovation, in agriculture, medicine and industry, for example, has bordered on the miraculous, culminating in putting a man on the moon. These successes have engendered a faith that in technology lies the solutions to today's problems. But the technology of the past made use of nature's laws rather than denied them and it must do so in the future as well. Technologists will not be able to use physical laws to get around ecological laws. Still, the myth of technological innovation remains.

A second hurdle in the way of rational decision making is the misunderstanding that Americans have regarding the relationship of economic systems to governing systems. There is a widespread notion that capitalism and democracy are synonymous. But economics deals with the production and flow of goods, whereas government deals with the means by which people establish policies to live by. A government-regulated economy is not inconsistent with a democratic form of government. If the American people evaluate the advantages and disadvantages of a no-growth economy against those of a continuous-growth economy and decide that environmentally responsible legislation is necessary for our long-term survival, such legislation could be determined democratically by the people through their representatives. Indeed, the issue is not whether or not govern-

ment should interfere with economic policy but what that economic policy should be. Should the government encourage economic growth, as it has in the past, or should it favor ecologically sound economic policies?

Finally, increasing the public's understanding of the alternatives being considered is proceeding too slowly. Many people refuse to listen to any suggestion that conflicts with the long-cherished American ideals of production, profit, competition and growth. Perhaps the outstanding example of the conservationists' failure to educate even those with a direct economic interest is the whaling industry. Despite the ex-

termination of local populations of whales from the eastern Atlantic, the northern Atlantic, and then the northern Pacific oceans during the preceding three centuries, the whaling fleets of 17 nations continued the intensifying hunt into the southern oceans following the end of World War II. One after another species approached extinction, and one after another whaling fleet ceased operations until now only Japan and Russia maintain active fleets. If it proved impossible to convince the whaling companies that conservation of whales through regulating the kill was in their own long-term interest, what chance is there to convince the city dweller

of the need to conserve wilderness, or the suburban and rural resident of the need to improve living conditions in the cities, or industralists, businessmen and people in general of the need to reduce production and consumption?

The future is not entirely black. At least the problems of overpopulation, overconsumption and pollution are being discussed by a diverse group of persons. But whether the ultimate causes of our problems ever become understood, whether the solutions become clear and whether those solutions are ever carried out remain to be seen. ▪

December 10, 1972

Growth vs. Environment

Anxiety on Dangers Undermining Faith In Economic Expansion and Competition

By LEONARD SILK

Concern about the impact of population and industrial growth on the environment and limited resources of "Spaceship Earth" is causing some economists to re-examine their first principles.

Probably the oldest battle cry of the economics profession is "TINSTAAFL"—"There is no such thing as a free lunch." This means you can't get something for nothing; natural resources, capital and labor must be used to produce anything of value. Economists cite the principle as proof of their hard-boiled realism. But Prof. Nicholas Georgescu-Roegen of Vanderbilt University, a Distinguished Fellow of the American Economic Association, says TINSTAAFL is not nearly harsh enough.

Economic Analysis

He contends that "the cost of any biological or economic enterprise is always greater than the product." In other words, "You never get as much out of a lunch as you put into it."

The explanation of this more pessimistic principle is that the economic process normally consists of taking free energy—

such as coal or oil—and ultimately converting it into forms of "bound" energy — such as pollution or wastes that are no longer readily available to produce heat or mechanical power. In brief, economic production is a process of downgrading the resources needed to support life.

This is the process known to physicists as the Law of Entropy — and Professor Georgescu-Roegen has described it in his book, "The Entropy Law and the Economic Problem," which was published last year by the Harvard University Press.

He contends that, precisely because man has always felt, however unsophisticatedly, that his life depends on scarce, irretrievable resources he has nourished the hope that he would eventually discover a self-perpetuating force, a perpetual-motion machine, an inexhaustible fuel. In the past, that hope always failed.

The discovery of atomic energy has again raised hopes that man has finally found a self-perpetuating power that could unlock bound energy without limit.

But nuclear energy has thus far been anything but cost-

less. Professor Georgescu-Roegen says: "The shortage of electricity which plagues New York and is gradually extending to other cities should suffice to sober us up. Both the nuclear theorists and the operators of atomic plants vouch that it all boils down to a problem of cost."

As he sees it, economists and the society as a whole are now compelled by the growing environmental damage resulting from production to bring their thinking into line with physical principles, especially the Second Law of Thermodynamics, which states that entropy—the amount of bound energy—of a closed system continuously increases.

Economic activity turns order into disorder. The so-called "utilities" that economic production create blow away with the wind, or turn into junk—Chevys and Fords rusting in dumps.

Recycling can slow the exhaustion process, but, says Professor Georgescu-Roegen, "there is no free recycling just as there is no wasteless industry." It takes energy to recycle.

Population and economic growth have, in the last two centuries, enormously accelerated the exhaustion of terrestrial resources. The spread of industrialization throughout the world is increasing the rate at which free energy is being used up.

This, says the Vanderbilt economist, "is the main problem for the fate of the human species."

He holds that, even with a very parsimonious use of ter-

restrial resources, the industrial phase of man's evolution will end many hundreds of millions of years before the sun ceases to support life on earth.

In any case, he insists, the higher the rate of economic development, the faster the rate of depletion of terrestrial resources and the shorter the expected life of the human species.

Those economists, scientists, engineers and businessmen who refuse to accept this economic pessimism that makes the Rev. T. R. Malthus sound like Polyanna put their trust in technology to save man from the exhaustion of usable resources and from suicide by pollution.

Some would capture needed resources from outer space — first the moon. Some put their hope in controlled nuclear fusion. Others believe that, sooner or later, if technology is to rescue the human race, it must shift man's reliance from terrestrial to solar energy.

Professor Georgescu-Roegen himself thinks it is "quasicertain" that, in the struggle for survival, man will discover the means of transforming solar radiation into motor power directly.

"Certainly," he says, "such a discovery will represent the greatest possible breakthrough for man's entropic problem, for it will bring under his command also the more abundant source of life support."

Before anyone relaxes too much, however, it is well to remember that the big breakthrough that would provide enough direct use of solar energy to support the world's growing billions has not yet been made, and that the supply of low-entropic materials is shrinking fast, as industry advances.

Seen as Lesser Evil

And, around the world, pollution appears to be growing even faster. Some of the strongest resistance to environmental controls is coming from poor countries whose food supplies are dependent, or soon may be dependent, on heavy use of fertilizers, pesticides and herbicides. They see pollution as a lesser evil than hunger and the uncontrollable spread of diseases.

As last summer's United Nations Conference on the Human Environment in Stockholm demonstrated, the world is still a long way from being prepared to accept international controls on pollution. Among the obstacles to global controls is the fear of many industries and national governments that controls will put them at a competitive disadvantage in world markets or impose excessive costs upon them.

Nevertheless, growing anxiety about environmental hazards is beginning to undermine blind faith in the dogmas of economic growth and unrestricted competition on which the optimism of classical economics was founded.

December 20, 1972

NO-GROWTH IDEA REJECTED BY FORD

He Calls for Compromise on Use of Resources

SPOKANE, Wash., Aug. 15 (AP)—President Ford said today that "zero growth" environmental policies flew in the face of human nature and must be rejected in favor of reasonable compromises.

The President said that the energy crisis of last winter demonstrated that the nation must mine and use more coal, drill for more oil on the ocean's continental shelf, develop oil shale resources and speed construction of nuclear power plants.

"There are some well-meaning people who see the environmental issue as an 'either, or' proposition," Mr. Ford said in a statement read at the Expo '74 world fair by Interior Secretary Rogers C. B. Morton.

Mr. Ford said that, although environmentalists argued for zero growth as the only course for world resource salvation, their goal was impossible because "man isn't built to vegetate or stagnate—we like to progress, we have ideas, we have hopes and dreams of a better world, and that better world includes jobs and a better environment."

The zero growth argument says man cannot obtain a clean environment and continue to expand at the same time, Mr. Ford said, adding that it contends man is exhausting all resources, including the environment.

"But it fails to take into consideration the one inexhaustible resource: man's creative ability," the President said. "As far as we can tell, the frontier of creative scientific knowledge is limitless."

Mr. Ford said that, because of scientific potential, it was possible for the nation to have both environmental protection and economic expansion.

August 16, 1974

Growing With Energy

A strong policy to conserve energy, implemented through higher prices, quotas, taxes or rationing, does not necessarily mean fewer jobs and deteriorating living standards. It is certainly true that major cutbacks in energy consumption may bring immediate hardship and unemployment. But it does not follow that energy industry spokesmen are on solid ground in reiterating their self-serving claims that a healthy economy requires ever more energy supplies to keep it growing.

What started as a paradox is now increasingly accepted in logic and fact by energy experts: As the national economy grows, energy consumption grows more slowly. Energy needs can come close to zero-growth—as more efficient plants and processes become the norm in industry, and as higher living standards call forth more output from service industries than the energy-intensive basic industries.

Support for this once-radical thesis has come in an important new study by the Conference Board, a prestigious business research organization aimed at top management. "Energy use and economic growth are certainly not independent of one another, but the link between them is more elastic than is commonly assumed," the

board's report concludes, "provided time is allowed for the necessary adjustments in production and consumption that will permit less energy to be consumed per unit of output."

That proviso, and the necessity of compensatory measures during the period of adjustment to new and lower levels of energy demand, is what makes today's policymaking so complex.

The Federal Energy Administration's report on Project Independence found that for industry a 10 per cent rise in energy prices brings a 2.1 per cent decrease in demand over the first year or so. In succeeding years, however, demand will decline 7 per cent without any further price disincentives, according to the report.

"The forces of technological change and response to higher prices are quite likely to produce a much lower rate of growth of energy use in the next ten years than has prevailed in the preceding two decades," the Conference Board study declared. That suggests an annual growth rate as low as 1.5 per cent, less than half the long-prevailing rate. It also suggests that the industrial problems for today's policymakers involve focus on the short-term adjustment process, not on the fear that energy conservation will necessarily stunt long-term economic growth.

February 25, 1975

Where We Stand

A Weekly Column of Comment on Public Education

by Albert Shanker President, United Federation of Teachers

[Mr. Shanker's guest columnist today, a nationally-known leader in the civil rights movement, is President of the A. Philip Randolph Institute.]

Those "Liberal" Enemies of Growth

By Bayard Rustin

It has become commonplace for certain liberals to dismiss the concept of economic growth as a regressive, even destructive force in society. The affluence generated by an expanding economy, we are told, has led to the despoilment of the environment, brought deprivation to hundreds of millions in underdeveloped nations, and in general made America and the rest of the world an unpleasant place to live.

One must be charitable when discussing growth's critics. For they are, by and large, individuals who support in the abstract the objectives of social progress and racial equality.

But somehow they have missed the fundamental—and obvious—contradiction in the notion that the condition of those at the bottom can be raised when society itself is standing still. No doubt they have failed to consider the implications of their creed: that while a no-growth economy may protect the fields and streams (which in itself is a dubious claim) it will most certainly result in untold misery for thousands of ordinary people, many of whom are the black poor of America and the poverty-stricken masses of Asia, Africa and Latin America.

Those who doubt that this is so need only examine the most recent unemployment statistics published by the federal government. In a year of economic stagnation — with growth significantly curtailed — unemployment among blacks is at its worst level since before the civil rights era. For the future, it is quite likely that the continuation of economic retrenchment over the next several decades would triple or quadruple the unemployment lines, with the most shattering effect on minorities.

The goals of growth's critics are commendable. But the elimination of air and water pollution and the safeguarding of the wilderness can and is being accomplished through stronger government regulation.

As for the problems of the underdeveloped world, they will not be resolved by the weakening of the American economy. On the contrary, realization of a no-growth economy would exacerbate the suffering of the peoples of India, the sub-Saharan countries, and other areas in desperate need. The fact is that America is not going to play a more humanitarian role in the world by diminishing its capacity; it can be much more convincingly argued that America's productive capacity should be increased, while government policy is transformed so that it will assist and encourage the development of the poorest nations.

A few years ago the term "elitist" was often attached to those liberals who had seemingly lost touch with the values and aspirations of working people. Clearly growth's critics are guilty of defining social problems from an elitist point of view. As intellectuals and middle class professionals, they have as yet to feel the consequences of recession, and no doubt believe themselves immune to the brutalization of an inequitable economic structure.

But the attack on growth is far more than just elitist: it is conservative, even reactionary to its very heart, and should be recognized as such. Within the context of a rigid, stagnant economy, social equality could only be achieved by an across-the-board *lowering* of living standards; what is much more likely is that class divisions would harden, with the poor and working people locked forever in their unequal states.

It will be said, of course, that society can be remade by "reordering priorities," taking money from defense spending and allocating it to human needs.

We heard this argument before, in 1965 to be precise, when A. Philip Randolph proposed the Freedom Budget for All Americans—a ten year program of massive social and economic change. At that time a number of liberals refused to endorse the Freedom Budget because, they said, all that was needed to solve racial inequality was the end of the Vietnam War.

The war drew to an end, and no "peace dividend" ever materialized. It never materialized because it never existed, except as a diversion from much more complex and difficult challenges.

Those who argue today that society can advance without economic growth are the same people who preached the doctrine of the peace dividend ten years ago. And they are just as wrong now as they were then. If America is to take up anew the challenge posed by the civil rights movement, it must be prepared to address the basic inequities that are woven into the fabric of our economic system. The goal, however, is not to cripple our economy, but rather to transform it along humanitarian lines. In this task we will be struggling against those with a vested interest in the status quo—the rich and privileged. And, should they persist, we shall also find ourselves struggling against the enemies of economic growth — no matter how liberal or radical they pretend to be.

July 20, 1975

No Growth or No Waste?

A column that recently appeared in the pages of this newspaper as a paid advertisement deserves attention both for the importance of the subject and the prestige of its author, Bayard Rustin. Unfortunately, it deserves rather less admiration.

Dr. Rustin, who is president of the A. Philip Randolph Institute and a social analyst of note, inveighs in the column against "certain liberals" who are opposed to economic growth; who fail to see that a no-growth economy that may protect the environment "will most certainly result in untold misery for thousands of ordinary people," many of them black and poor. What he himself fails to see—or at least to note—is that most environmentalists do not argue for no growth; what they want is no waste — especially an end to a waste of energy that is now so great that its elimination could by itself close the country's fuel gap by 1985.

They want, further, an environmental program calling for a change in national priorities which, far from spelling "misery for thousands," would provide jobs for hundreds of thousands. Dr. Russell W. Peterson, chairman of the President's Council on Environmental Quality, recently estimated the costs of effective environmental protection for the next decade at $195-*billion*—to be spent on machinery and equipment, materials and metals, wages to labor, and industrial plants.

The money pumped into these activities obviously would not stop with environmental improvements. It would go into the general economy via the pockets of those who produced these goods and services. An investment of such magnitude, making jobs and raising the general living standard, would be an enormous impetus toward economic recovery, to the advantage of the poor about whom Dr. Rustin is rightly and genuinely concerned.

The best of such an approach is that the upsurge would not come at the expense of a livable world—as it would from a program, say, of building more needless highways and dams, more gadgetry with built-in obsolescence and more cars of the sort that burn energy far beyond what is really required to transport their purchasers. To slow the growth of energy-wasting products is neither to slow the economy itself nor to freeze the disparity between rich and poor. Quite the reverse.

These are the concerns of environmentalists—as they are, no doubt, of Dr. Rustin. It is regrettable that he did not address himself to them instead of to strawmen—those "élitist" enemies of economic growth "no matter how liberal or radical they pretend to be." There is more to worry about from those who would waste our declining resources—energy included—to the point where the country is forced to an economy of scarcity. Such an economy, by the very costs it would impose, would truly favor the "élite."

July 24, 1975

Less Energy, Better Life

By René Dubos

Bob Adelman

Current discussions about energy are focused on problems of its cost and supplies, almost ignoring its influence on the quality of life. We have slipped into the habit of regarding human problems from a technological point of view and seem to consider it an obvious truth that the more energy we can afford to use, the better off we are.

Yet, the evidence on that score is far from clear. In any kind of society, the healthiest, happiest and most creative persons are likely to be found among those who consume least. And even granted that high levels of energy consumption have accelerated the growth of technological civilization in the past, there are reasons to believe that we have now reached the point of diminishing returns. In many situations indeed, the more energy we use, the more problems we create.

In the United States, the average consumption of energy per person is today approximately double what it was thirty years ago and double also what it is now in Europe. Does anyone really believe that this difference is reflected in more happiness, less suffering, greater longevity among present-day Americans, or in a more rapid progress of American civilization toward more desirable goals? A recent study based on measurements of various social indicators in 55 countries failed to reveal any beneficial effect of increased energy use on the quality of life; if there was a correlation, it was that the greater the energy consumption, the larger the percentages of divorces and suicides!

An abundant supply of energy is, of course, essential for the production of more and more industrial goods, but this is not all that there is to happiness and civilization. If one judges on the basis of civic virtue, sophistication of thought, quality of writing, charm of landscapes, architectural styles and perhaps even of average comfort, I see little evidence that our civilization has been made more appealing by the recent phenomenal increases in the use of energy.

I shall go even further and claim that in the highly industrialized parts of the world a decrease in energy use could have a multiplicity of beneficial effects in the long run. These would include improvements in physical and mental health, sounder agricultural practices based on ecological principles, architectural styles more interesting because they are better adapted to local conditions, policies of rural and urban planning that would favor a revival of community spirit—and of course a less disturbed global ecology. Since I cannot present here the evidence from which I make these predictions I shall limit myself to a dogmatic statement of their theoretical basis.

We have made it a universal practice to inject industrial energy into human and natural systems as a substitute for the adaptive responses that these systems would make if allowed to function naturally. As a consequence of the overuse of energy there are fewer and fewer occasions for the body and the mind to make creative responses to environmental challenges; architecture and planning have become duller because there is less need for ingenuity in coping with climatic and topographic constraints; the biological forces that used to contribute to soil fertility have no chance to operate. In others words, overuse of energy tends to interfere with the adaptive and creative mechanisms of response that are inherent in human nature and in external nature.

A large percentage of the energy we use today is not for creative activities but for reducing and eliminating, wherever possible, the efforts required to deal with environmental challenges. This practice makes for an easier life, but it impoverishes our experience.

We live only to the extent that we face up to the world with all our faculties and as directly as possible. "Energy is Eternal Delight," William Blake wrote in "The Marriage of Heaven and Hell," but he had the wisdom to add that "Energy is from the Body." In principle, energy from external sources can enrich our contacts with the world, but in practice we use it in such a manner that it weakens our contacts with reality.

The energy crisis will be a blessing if it compels us to develop ways of life that encourage fuller expression of the adaptive and creative potentialities that are present in us and in nature. Let me add my voice to those who proclaim: "There is no wealth but life. Let it flower."

René Dubos, scientist and author, is Distinguished Professor at Polytechnic Institute of New York and professor emeritus at The Rockefeller University.

January 7, 1975

'Trend Is Not Destiny'

By René Dubos

In 1575—400 years ago!—the French scholar Louis Le Roy published a learned book in which he voiced despair over the upheavals caused by the social and technological innovations of his time, what we now call the Renaissance. "All is pell-mell, confounded, nothing goes as it should." We, also, feel that our times are out of joint; we even have reason to believe that our descendants will be worse off than we are.

The earth will soon be overcrowded and its resources exhausted. Pollution will ruin the environment, upset the climate, damage human health. The gap in living standards between the rich and the poor will widen and lead the angry, hungry people of the world to acts of desperation including the use of nuclear weapons as blackmail. Such are the inevitable consequences of population and technological growth *if* present trends continue. But what a big *if* this is!

The future is never an extrapolation of the past. Animals probably have no chance to escape from the tyranny of biological evolution, but human beings are blessed with the freedom of social evolution. For us, trend is not destiny. The escape from existing trends is now facilitated by the fact that societies anticipate future dangers and take preventive steps against expected upheavals.

In the past, disasters caught humankind by surprise; now future situations are discussed long before the event, especially if they are likely to be dangerous. One of the fashionable intellectual games of our time consists in imagining the symptoms of "future shock" that people will experience when their ways of life are transformed either by man-made changes or by natural catastrophes. But the very fact that these symptoms have been publicized in advance makes it unlikely that they will occur as described. A few examples will illustrate the range of potentially dangerous situations which modern societies have anticipated and for which they are developing preventive measures.

During the winter of 1917-1918, the Spanish flu spread explosively throughout the world, killing more than 20 million people. The epidemic was unexpected; it was like an apocalyptic visitation of mysterious origin and nothing could be done to stop it. Other outbreaks of influenza have occurred since that time, but any new strains of it, the Hong Kong flu for example, can now be detected early in its spread and steps can be taken to tame its virulence.

During the 1950's, environmental degradation and population growth reached critical levels in many parts of the world. These problems are still with us, but progress is being made toward their control wherever the public realizes the dangers of present trends.

For example, Algeria and continental China are carrying out reforestation programs based on the experience of the Dust Bowl in the United States; population growth has begun to slow down in several industrialized countries.

Urban agglomerations are in a state of crisis, but efforts are being made everywhere to reform urban life. Old cities are rediscovering the value of their ancient buildings and traditions; large new cities are being created, especially in Europe, each with its own economic and cultural identity; even New York City may eventually establish a sounder budgetary basis on which to build its future.

These examples are typical of our times in that they do not correspond to final solutions of problems but rather symbolize a kind of social ferment generated by public concern for the future.

Despite the widespread belief that the world has become too complex for comprehension by the human brain, modern societies have often responded effectively to critical situations.

The decrease in birth rates, the shelving of the supersonic transport, the partial banning of pesticides, the rethinking of technologies for the production and use of energy are but a few examples illustrating a sudden reversal of trends caused not by political upsets or scientific breakthroughs, but by public awareness of consequences.

Even more striking are the situations in which social attitudes concerning future difficulties undergo rapid changes before the problems have come to pass—witness the heated controversies about the ethics of behavior control and of genetic engineering even though there is as yet no proof that effective methods can be developed to manipulate behavior and genes on a population scale.

One of the characteristics of our times is thus the rapidity with which steps can be taken to change the orientation of certain trends and even to reverse them. Such changes usually emerge from grassroot movements rather than from official directives; they are less a result of conventional education than of the widespread awareness of problems generated by the news media.

There is the danger, admittedly, that such awareness is not always sufficient for rapid enough feedback to prevent critical processes from overshooting and causing catastrophes. But hope against the danger of overshooting can be found in the fact that most biological and social systems are extremely resilient. It is this resilience which leads me to reject the myth of inevitability and reaffirm that, wherever human beings are concerned, trend is not destiny.

November 10, 1975

On Growth

By René Dubos

Much of contemporary gloom concerning the future originates from the belief that there are "limits to growth" —an expression which has penetrated deeply into the public subconscious from the catchy title of a much-publicized book. One of the themes of this book is that the resources of the Earth are limited and that shortages will soon reach critical levels.

Although the phrase "limits to growth" appears self-explanatory, it is in fact deceptive because it hides assumptions and has static connotations that are incompatible with human behavior.

It implies that growth means producing more and more of what industrial societies have been producing at an obscene rate, and that it will therefore require more of the same kind of resources that were used in the past.

History shows, however, that social evolution continuously drives human activities into new channels and that each age creates the resources it needs.

Resources are not as "natural" as usually assumed. They are derived from raw materials that acquire value only after they have been separated from the earth to serve human purposes.

Gold and copper became resources very early bcause these metals can be extracted and manipulated by simple techniques. Iron did not become a resource until much later because it requires more complex technologies. Aluminum became a resource only after sophisticated methods had been developed to derive it from bauxite at the turn of the century. And so it goes for other metals.

Agricultural lands, also, had to be created out of the wilderness by human ingenuity and labor. In North America this involved clearing the forests that used to cover a large part of the continent, using the plow "that broke the plains," draining marshes and irrigating semidesert areas. Much of what is called nature was for ages some aspect of wilderness that has been transformed by human efforts.

One kind of growth is simply the exploitation of the materials stored in the earth; another more interesting kind results from the transformation of raw materials into resources through a continuous evolutionary process.

To a large extent, in other words, growth means the evolution of the man-made. The creativeness of social evolution is strikingly evident in the change of attitudes regarding sources of energy.

For millennia, all work was done by human and animal muscles. During the Middle Ages, mechanization began with the use of water mills and windmills. The Industrial Revolution operated its machines first with wood, then with fossil fuels such as coal, petroleum products, and recently uranium.

At present, studies are going on all over the world to determine what sources of energy are best suited for each individual purpose and what are the safe limits in the production and use of energy.

The awareness that the supplies of fossil fuels are limited is now directing thought to renewable sources—for example, nuclear fission and the sun—and to the vital contribution made by the wilderness to the energy balance of the earth.

In any given year, the total amount of energy derived from the sun by the photosynthetic activities of wild vegetation greatly exceeds the total amount used to support human life and to drive technologies. The problem of energy supplies thus leads back to concern about the preservation of nature.

The meaning of the word "growth" has evolved also with regard to human existence. Quantitative growth, for its own sake, is no longer socially acceptable because it threatens the quality of life and of the environment.

In the countries of Western civilization, many members of the upper and middle classes are beginning to recognize the merits of a less-consuming society. Just as this bellwether group led the movement toward smaller families, so it may eventually transmit new social values to the rest of the population.

In matters of growth, the new mentality is more important than advances in science and in technology. The fact that a good environment is now considered one of the "inalienable rights" will probably influence the design of future technologies as much as scientific discoveries have in the past.

Even though the phrase "quality of life" does not define a social philosophy, it symbolizes an attitude that can be contrasted with the following statement from the guidebook prepared for the 1933 Chicago World's Fair: "Science finds, industry applies, man *conforms*."

Today, 40 years after the 1933 World's Fair, no one would dare state that humankind must conform to technological imperatives. The goal is rather to make technology conform to human needs and aspirations.

This involves a kind of qualitative growth for which there are no discernible limits, because social evolution is more inventive than biological evolution and more creative of resources really valuable for human existence.

November 11, 1975

Suggested Reading

Allen, Shirley W. *Conserving Natural Resources.* New York: McGraw-Hill, 1955.

Baker, Gladys L., Wayne D. Rasumssen, Vivian Wiser, and Jane M. Porter. *Century of Service: The First 100 Years of the United States Department of Agriculture.* Washington: USGPO, 1963.

Barnett, Harold J. and Chandler Morse. *Scarcity and Growth.* Baltimore: Johns Hopkins Press, 1963.

Bennett, Hugh Hammond. *Elements of Soil Conservation.* New York: McGraw-Hill, 1955.

Bruchey, Stuart. *The Roots of American Economic Growth.* New York: Harper and Row, 1965.

Bruchey, Stuart. *Growth of the Modern American Economy.* New York: Harper and Row, 1975.

Bruchey, Stuart, advisory ed. *Use and Abuse of America's Natural Resources.* A reprint series of 41 books. New York: Arno Press, 1972.

Butcher, Devereux. *Exploring Our National Parks.* Boston: Houghton-Mifflin, 1956.

Callison, Charles H. *America's Natural Resources.* New York: Ronald Press, 1963.

Carson, Rachel. *Silent Spring.* Boston: Houghton-Mifflin, 1962.

Clawson, Marion. *Land and Water Recreation: Opportunities, Problems and Policies.* Chicago: Rand McNally, 1963.

Clepper, Henry, ed. *The Origins of American Conservation.* New York: Ronald Press, 1966.

Clepper, Henry and Arthur B. Meyer. *American Forestry: Six Decades of Growth.* Wash., D.C.: Society of American Foresters, 1960.

Colman, E. A. *Vegetation and Watershed Management.* New York: Ronald Press, 1953.

Coyle, David Cushman. *Conservation: An American Story of Conflict and Accomplishment.* New Brunswick, N.J.: Rutgers U. Press, 1957.

Dana, Samuel Trask. *Forest and Range Policy: Its Development in the United States.* New York: McGraw-Hill, 1956.

de Bell, Garrett, ed. *The Environmental Handbook.* New York: Ballantine, 1970.

Disch, Robert, ed. *The Ecological Conscience: Values for Survival.* Englewood Cliffs, N.J.: Prentice-Hall, 1970.

Esposito, John C. *Vanishing Air: Ralph Nader's Study Group Report on Air Pollution.* New York: Grossman, 1970.

Fabricant, Neil and Robert M. Hallman. *Toward a Rational Power Policy: Energy, Politics, Pollution.* New York: Braziller, 1971.

Frank, Bernard. *Our National Forests.* Norman, Oklahoma: U. of Oklahoma Press, 1955.

Gabrielson, Ira N. *Wildlife Conservation.* N.Y.: Macmillan, 1959.

Gates, Paul W. *History of Public Land Law Development.* N.Y.: Arno Press, 1978.

Goldman, Marshall I., ed. *The Spoils of Progress: Environmental Pollution in the Soviet Union.* Cambridge, Mass.: M. I. T. Press, 1972.

Greeley, William B. *Forests and Men.* Garden City, N.Y.: Doubleday, 1951.

Harbaugh, William H. *Power and Responsibility; the Life and Times of Theodore Roosevelt.* N.Y.: Farrar, Straus & Cudahy, 1961.

Hays, Samuel P. *Conservation and the Gospel of Efficiency: The Progressive Conservation Movement, 1890-1912.* Cambridge, Mass.: Harvard U. Press, 1959.

Hornaday, William T. *Thirty Years War for Wildlife.* N.Y.: Arno Press, 1971.

Hornaday, William T. *Our Vanishing Wildlife: its Extermination and Preservation.* N.Y.: Arno Press, 1971.

Hosmer, Charles B., Jr. *Presence of the Past; History of the Preservation Movement in the United States Before Williamsburg.* N.Y.: G. P. Putnam's Sons, 1965.

Humphrey, Robert R. *Range Ecology.* N.Y.: Ronald Press, 1962.

Ise. John. *The United States Forest Policy.* New Haven: Yale, U. Press, 1920.

Ise, John. *The United States Oil Policy.* N.Y.: Arno Press, 1972.

James, Harlean. *Romance of the National Parks.* N.Y.: Arno Press, 1972.

Jarrett, Henry. *Environmental Quality in a Growing Economy.* Baltimore: Johns Hopkins Press, 1966.

Kinney, J. P. *The Development of Forest Law in America.* N.Y.: Arno Press, 1972.

Klein, Louis. *Aspects of River Pollution.* N.Y.: Academic Press, 1957.

Landsberg, Hans H., L. Fishman and J. L. Fisher. *Resources in America's Future.* Baltimore: Johns Hopkins Press, 1963.

Lord, Russell. *The Care of the Earth: A History of Husbandry.* N.Y.: Thomas Nelson & Sons, 1962.

Lowenthal, David. *George Perkins Marsh, Versatile Vermonter.* N.Y.: Columbia U. Press, 1958.

Marsh, George Perkins. *The Earth as Modified by Human Action.* N.Y.: Arno Press, 1971.

McGee, W. J., et. al., eds. *Proceedings of a Conference of Governors in the White House, Washington., D.C. May 13-15, 1908.* N.Y.: Arno Press, 1972.

Mead, Elwood. *Irrigation Institutions.* N.Y.: Arno Press, 1972.

Meadows, Donella H., Dennis L. Meadows, Jorgen Randers, and William W. Behrens. III, *The Limits of Growth.* N.Y.: Universe Books, 1972.

Moreell, Ben. *Our Nation's Water Resources—Policies and Politics.* N.Y.: Arno Press, 1972.

Morgan, Dale L. *Jedediah Smith and the Opening of the West.* Indianapolis: Bobbs-Merrill, 1953.

Murphy, Blakely M., ed. *Conservation of Oil and Gas: A Legal History.* Arno Press, 1972.

Newell, Frederick H. *Water Resources: Present and Future Uses.* N.Y.: Arno Press, 1972.

Nimmo, Joseph Jr. *Report in Regard to the Range and Ranch Cattle Business of the United States, May 16, 1885.* N.Y.: Arno Press, 1972.

Nixon, Edgar B. ed. *Franklin D. Roosevelt & Conservation, 1911-1945.* N.Y.: Arno Press, 1972.

Passell, Peter and Leonard Ross. *The Retreat from Riches: Affluence and Its Enemies.* N.Y.: Viking Press, 1973.

Peffer, H. Louise. *The Closing of the Public Domain.* N.Y.: Arno Press, 1972

Pinchot, Gifford. *Breaking New Ground.* N.Y.: Harcourt, Brace, 1947.

Puter, S.A.D. and Horace Stevens. *Looters of the Publ Domain.* N.Y.: Arno Press, 1972.

Record, Samuel J. and Robert W. Hess. *Timbers of th New World.* N.Y.: Arno Press, 1972.

Reid, George K. *Ecology of Inland Waters and Estuaries.* N.Y.: Reinhold Pub. Corp., 1961.

Richardson, Elmo R . *The Politics of Conservation Crusades and Controversies, 1897-1913.* Berkeley & Los Angeles: Univ. of Calif. Press, 1962.

Rosenberg, Nathan. *Technology and American Economic Growth.* N.Y.: Harper and Row, 1973.

Rounsefell, George A. and W. Harry Everhart. *Fishery Science—Its Methods and Application.* N.Y.: John Wiley & Sons, 1953.

Scott, G. R . *Studies of the Pollution of the Tennessee River System.* N.Y.: Arno Press, 1971.

Shankland, Robert. *Steve Mather of the National Parks.* N.Y.: Alfred A. Knopf, 1951.

Stegner, Wallace. *Beyond the Hundredth Meridian: John Wesley Powell and the Second Opening of the West.* Boston: Houghton Mifflin, 1954.

Stoddart, Laurence A. and Arthur D. Smith. *Range Management.* N.Y.: McGraw-Hill, 1955.

Swain, Donald C. *Federal Conservation Policy, 1921-1933.* Berkeley: U. of California Press, 1963.

Toole, Ross. *The Rape of the Great Plains: Northwest America, Cattle and Coal.* Boston: :Little, Brown, 1976.

Udall, Stewart L. *The Quiet Crisis.* N.Y.: Avon Books, 1964.

Van Sickle, Dirck. *The Ecological Citizen: Pollution Survival and Activist's Handbook.* N.Y.: Harper and Row, 1971.

Ward, Barbara and Rene Dubos. *Only One Earth.* N.Y.: W.W. Norton, 1972.

Webb, Walter P. *The Great Plains.* N.Y.: Grosset & Dunlap, 1931.

Whitaker, J. Russell and Edward A. Ackerman. *American Resources: Their Management and Conservation.* N.Y.: Arno Press, 1971.

Wilson, Carroll L. and William H. Matthews, eds. *Man's Impact on the Global Environment: Assessment and Recommendations for Action.* Cambridge, Mass.: M. I. T. Press, 1970.

Wolman, Abel. *Water Resources.* Wash., D.C.: National Academy of Sciences-National Research Council, 1963.

Zwick, David and Marcy Benstock. *Water Wasteland: Ralph Nader's Study Group Report on Water Pollution.* N.Y.: Grossman, 1971.

Index

Muskie, Edmund: and air pollution, 107, 129-30; and auto pollution, 95-96; and environment, 226; and phosphates in water, 174; and pollution control, 121; and water pollution, 143, 148, 155

museums, 29

Nader, Ralph: and environment, 222; and mercury, 161; and nuclear power plants, 287, 296, 300-301

National Association of Manufacturers, 139

National Conservation Association, 10, 11

national parks: addition to, 57; recreation budget, 71; visitors to, 28-29, 57-58; watersheds, 71; wilderness preservation, 30, 71-72; wildlife, 71

National Park Service, 29, 58

National Resources Board, 35

natural gas: distribution system, 246; Gerald Ford policy, 315-16; Jimmy Carter bill, 336-37; liquefied, 262-63, 329-31; need for, 116-17; rules, 242; scarcity, 292, 335-39; supply, 250, 263-64, 266; Supreme Court ruling about, 247-48; waste, 237-38; withholding of, 339-40

natural resources: adequacy of, 46-47; and business cooperation, 17; conservation of, 6-7, 8, 9; federal control of, 12; underdeveloped, 14-15; see also environment; individual natural resources

nature, 48

navigation, 42

Nelson, Gaylord, 226

nene goose, 49

Neuberger, Richard L., 44

Neumeyer, Charles, 340

New England, 40

New Jersey, 40, 117

New York City, New York: auto pollution, 128; garbage-dumping, 133, 151-52; smog emergency, 101-3; water, 40, 134

New York, upstate, 40

nitrilotriacetic acid, 169

nitrogen, 21

nitrogen oxides, 117, 124

Nixon, Richard: and air pollution, 116; and Alaska pipeline, 280; clean water bill, 176, 183-84; on DDT, 203; and Earth Day, 226; and energy crisis, 275-79, 281-82, 291, 294-95, 307; and Great Lakes Pact, 174; and industrial pollution, 66-67; and national forests, 70; and nuclear power, 273-74; and oil spills, 155-56; and pollution, 59; and public lands, 61-62; and radiation authority, 283

noise: Concorde controversy, 215-16; effects of, 212-15; regulation in New Jersey, 212

North Atlantic Ocean, 112

North Cascades National Park, 57

nuclear power plants: breeder reactor, 268, 274; complexity of, 325; controversy, 254-55, 265, 296-302, 317, 332; cost, 322-23, 359; defective reactors, 317; emergency licensing, 273-74; fusion reactor, 326-27; hazards, 266, 268, 298-302, 310-12, 333; in Idaho, 254; laser light, 307; in Long Island City, 255; safety, 254, 287, 308, 319; Supreme Court ruling on, 251, 272; threat to coal, 253-54; see also atomic power

ocean, 84, 112, 154-55, 202

oceanography, 74

Ohio Valley, 134, 135

oil: Alaskan pipeline, 280; allocations, 282; Arabian, 267, 278-79, 285-86, 293-95, 302; consumption, 18, 242, 244; development, 287-88; environmental movement, 293; Gerald Ford policy, 316, 324; increased prices, 285-86; liquefied natural gas, 262-63; low sulphur fuels, 105-6; replacing coal, 246; shale, 303-5; shortage, 235-39, 245, 250, 275-76, 280-82, 291-95, 331; U.S. tankers, 314; world shortage, 344-45

oil sludge: and birds, 131-32; danger of, 333; ocean contamination, 85, 154-55; and Richard Nixon, 155-56; Santa Barbara beaches, 147-48, 149; tanker accident, 145-46, 188-89

Olds, Leland, 43

Osborn, Albert, 316

Osborn, Fairfield, 39

Ottinger, Richard, 151-52

Outwater, Eric B., 283-84

overpopulation, 347-48, 349, 351-55, 360

oxygen, 114

ozone, 114, 120

Pacific Northwest water supply, 40

packaging, 193

Paley Report, 291

Paley, William S., 46

Panama Canal, 8

Panofsky, Wolfgang, 319

Paradise, Kentucky, 255-61

Pardee, George S., 9, 11

Parker, James A., 43

parks. See national parks

Parthenon, 93

Patterson, Clair C., 98

Pearl, Milton A., 61-62

Pennsylvanian coal, 15

pest control, natural, 204, 209-11

pesticides: alternative to, 204, 209-11; and cancer, 208; curb on, 204; export of, 230; kepone, 185; Memphis, Tennessee sewers, 138; residue, 201; selective use of, 204; study of, 59; use of in U.S., 203; and water pollution, 144; with mercury banned, 209; see also DDT; dieldrin; endrin; Silent Spring (Carson)

Petrified Forests, 29

petroleum products, 250

Philadelphia, Pennsylvania, 40

Phillips Petroleum Company, 247-48

phosphates, 142, 169, 174

photochemical oxidants, 117, 125

Pinchot, Gifford: Albert Fall and, 24; and conservation, 11-12; and national forests, 33; and Richard Ballinger, 13; and water power, 9, 16

pipelines, 253, 280, 320

plutonium, 274

Pointer, Sam C., 118

Poland, Joseph, 79